Advanced Lectures in Mathematics
Volume 38

Complex Geometry from Riemann to Kähler–Einstein and Calabi–Yau

Edited by Lizhen Ji

International Press
www.intlpress.com

高等教育出版社
HIGHER EDUCATION PRESS

Advanced Lectures in Mathematics, Volume 38
Complex Geometry from Riemann to Kähler–Einstein and Calabi–Yau

Volume Editor:
Lizhen Ji (University of Michigan, Ann Arbor)

This edition published by International Press in November 2018.

ISBN: 978-1-57146-352-4

Printed in the United States of America.

21 20 19 18 9 8 7 6 5 4 3 2 1

ADVANCED LECTURES IN MATHEMATICS

Preface

Complex geometry has been extensively studied and developed since the 19th century. It is now one of the central subjects in mathematics. History has played a more important role in mathematics than in other disciplines. The reason is that whatever was proved in mathematics will remain true forever, and hence there is a direct continuity of the past, the present and the future of mathematics. On the other hand, in the huge literature of mathematics, only a few works stand out and exert a big influence on the development of mathematics. A book consisting of seminal papers in complex geometry together with some commentaries will be a valuable source for both students and experts, who want to learn and appreciate a global and historical picture of the subject.

Since Bernhard Riemann is one of the originators of complex geometry and differential geometry, this book includes a selection of the thesis of Riemann on the foundation of complex analysis and his masterpiece on the foundation of geometry. It also includes papers of modern masters: Atiyah, Bott, Chern, Calabi, Chow, Donaldson, Hirzebruch, Kähler, Kodaira, Siu, Uhlenbeck, and Yau. To motivate the selection of these papers, a brief history of complex geometry based on extensive quotes from many experts on both mathematics and research in mathematics is included at the beginning of this book. At the beginning of each selected paper, we have included a brief explanation for the selection.

To see the development of contemporary mathematics through the eyes of one of the most active practitioners and leaders, the book contains a set of commentaries written by Prof. Shing-Tung Yau from his personal perspective on the broad subject of geometric analysis, in particular complex geometry, and its many applications, especially to mathematics physics.

Since only a limited number of papers are reprinted in this book, we have also compiled a list of important papers in geometric analysis in the 20th century. This list is the fruit of many intense and extensive email correspondences with Prof. Shing-Tung Yau and should reflect the perspective of Prof. Yau on the broad subject of geometric analysis. It appears as the last chapter and, we hope, should serve as a guide to geometric analysis and some related subjects.

Lizhen Ji
August 2016

Contents

Part I
Summary

Complex Geometry from Riemann to Kähler-Einstein and Calabi-Yau
ALM 38, pp. 3–75

A Historical Glimpse of Some Topics in Complex Geometry

Lizhen Ji

Contents

1 Introduction

In this book we have collected some important papers in the broad subject of complex geometry which will shed light on complex geometry and its interaction with algebraic geometry and algebraic topology. We have also put some emphasis on topics related to Kähler-Einstein metrics for reasons to be explained below. By arranging them in the historic order together with some commentaries, we hope that this will give a brief historical tour of complex analysis and complex geometry.

There are several motivations for editing this book:

(1) Complex analysis and complex geometry have been intensively studied by many of the best mathematicians in the world since the 19th century. The history of complex analysis has also been discussed by many historians and some mathematicians. See for example, the books [BoG], [Gra1], [Gra2], [Bot], and a survey paper [Wel]. Several general books on history of mathematics such as [Boy], [Kat], and [Kli] also mention or discuss complex analysis. But all these discussions basically stop at the beginning of the 20th century. To the current generation of mathematicians, some might be curious about an overall picture of what has happened after that. Therefore, it seems interesting to take a look at some highlights of complex geometry in the 20th century. Instead of a comprehensive picture of the whole subject of complex geometry, we will concentrate on those parts which are connected with algebraic geometry, global differential geometry, differential equations, algebraic topology and mathematical physics.

(2) Understanding properly the history of mathematics is complicated, but it can help us understand better mathematical ideas on a more personal level and appreciate more significance of individual results. The history can also help people see a bigger picture and hence look into the future.

(3) We can argue that given a manifold or a variety, it is important to understand its geometry, and the geometry has much to do with metrics on it. One reason is that the classical (or ancient) geometry depends on the Euclidean metric. Another reason is that Riemannian metrics were introduced by Riemann in 1854 before the notion of differential manifolds was formally formulated and hence might be considered more fundamental. The basic question of what are "optimal Riemannian metrics" on smooth manifolds is natural and intrinsically important. It turns out that Kähler-Einstein metrics are the right choice for complex manifolds for various reasons. Their existence is difficult and subtle. On the other hand, once they exist, they have many unexpected or expected applications.

(4) It seems valuable to understand better the above questions and find their answers by reading the original papers written by the originators and main contributors of the topics. This is probably one of the most effective ways to learn complex geometry and is one concrete adaption of Abel's famous saying: "By studying the masters, not their pupils."

We will address the points raised in the above summary in the following sections and hence give a historical outline of some major results in complex geometry in the 20th century.

2 What are optimal Riemannian metrics

In this section, we want to provide several answers to the question: which Riemannian metric is optimal.

The reason for asking this is that given a smooth manifold M, to study the geometry of M, we need to put metrics on M. On the other hand, there are many Riemannian metrics on M. This leads to the basic question: *What is the best Riemannian metric on M? How is the existence of special metrics related to algebraic and topological properties of the manifold M?*

In dimension 2, there is only one notion of curvature: the Gauss curvature. One direct condition is to require that the curvature is constant. It turns out that every surface has metrics of constant curvature, and the sign of the curvature is determined by the topology of the surface. With the exception of the sphere S^2, most surfaces have many non-isometric Riemannian metrics of constant curvature.

On the other hand, an exact analogy is impossible in higher dimensions. Requiring constant sectional curvature is too strong. For example, in dimension 3, there are eight homogeneous geometrics, and three dimensional manifolds of constant sectional curvature account for only three of them. Other topological operations such as connected sum will also produce more examples of three-dimensional manifolds which do not support Riemannian metrics of constant sectional curvature.

Replacing the sectional curvature by scalar curvature and imposing the condition of constant scalar curvature seems to be too weak, since the positive solution to Yamabe conjecture implies that every compact smooth manifold admits Riemannian metrics of constant scalar curvature and such metrics of constant scalar cannot be used effectively to study the manifold.

It turns out that the condition that the Ricci curvature is constant is the right one for several reasons:

(1) Both the metric tensor and the Ricci curvature tensor depend on the same number of parameters. Therefore, the condition that the Ricci curvature is constant imposes the right constraint in terms of parameter count.

(2) For any compact smooth manifold M, when $\dim M = 2$, moduli spaces of Riemannian metrics of constant curvature on M are finite dimensional.

On the other hand, when $\dim M \geq 3$, moduli spaces of Riemannian metrics of constant scalar curvature are nonempty and infinitely dimensional. In such dimensions, the moduli space of Riemannian metrics on M of a fixed sectional curvature is either empty or can be described simply. If M admits a metric of sectional curvature 0, then M is isometric to a flat torus or a finite quotient of it, and the moduli space of flat metrics on tori of volume 1 can be described by the locally symmetric space $SL(n, \mathbb{Z}) \backslash SL(n, \mathbb{R}) / SO(n)$. Moduli spaces of other flat manifolds can be described using flat tori with symmetry. If M admits a metric of sectional curvature 1 or -1 (only one is possible), then the moduli space of such metrics consists of only one point, due to the Mostow strong rigidity theorem.

For some important classes of compact smooth manifolds of dimension greater than 3, Riemannian metrics of constant Ricci curvature exist and they form finite dimensional spaces.

(3) For manifolds of dimension 2, the existence of such metrics of constant curvature allows one to study and understand many questions and properties about the manifolds. It is also helpful in understanding their moduli spaces. For manifolds of suitable high dimensions, existence of metrics of constant Ricci curvature also leads to topological restriction on the manifolds, and hence can be used to understand better the topology of the manifolds.

(4) For manifolds of dimension 2, the existence of such constant metrics amounts to the uniformization theorem for Riemann surfaces, which is one of the most important theorems in the history of mathematics. Though an analogous uniformization theorem cannot be expected to hold in high dimensions, the existence of Einstein metrics and Kähler-Einstein metrics are probably the closest result one could expect for Kähler manifolds.

(5) Finally, the main justification for metrics of constant Ricci curvature comes from Einstein's field equation in general relativity.

Because of the last reason, Riemannian metrics of constant Ricci curvature are called *Einstein metrics.* For Kähler metrics on complex manifolds, the corresponding condition on the Ricci curvature gives rise to *Kähler-Einstein metrics.* They are heroes of this book. See [Bes, pp. 1-5] for more detail about these special metrics and explanation for considering these Einstein and Kähler-Einstein metrics.

In some sense, Kähler-Einstein metrics are even more special and important than Einstein metrics for two reasons:

(1) It is not clear if the existence of Einstein metrics imposes topological restriction on manifolds of dimension equal to or above 5. (In dimension 4, existence of Einstein metric does impose topological restriction, for example, $\chi(M) \geq \frac{1}{2}|p_1(M)|$.)

    On the other hand, the existence of Kähler-Einstein metrics does impose some topological restriction on the manifolds.

(2) It is not easy to construct non-obvious Einstein and Kähler-Einstein metrics. But thanks to the celebrated solution by Yau to the Calabi conjecture, there are many Kähler-Einstein metrics.

Riemannian manifolds of constant sectional curvature are locally symmetric spaces and hence locally homogeneous Riemannian manifolds. Homogeneous manifolds also provide Einstein manifolds. Up to 1980, there was no general series of compact Einstein manifolds other than the homogeneous ones. It was even more difficult to construct Ricci flat compact manifolds.

Yau's solution of the Calabi conjecture provides a large class of non-homogeneous Ricci flat compact Kähler-Einstein manifolds through methods in algebraic geometry and complex geometry. They are called *Calabi-Yau manifolds.*[1]

For example, any smooth complex hypersurface in $\mathbb{C}P^{n+1}$ of degree $n+2$ admits a flat Kähler-Einstein metric. But understanding such metrics is still a challenge. For example, it will never be the induced Kähler metric from the Study-Fubini on $\mathbb{C}P^{n+1}$.

These Ricci flat Kähler-Einstein metrics obtained through Yau's solution to the Calabi conjecture cannot be obtained in other ways yet, since Riemannian homogeneous manifolds are never Ricci flat unless they are already flat.

It is helpful to quote from Yau [LW, p. 1]:

> It is therefore difficult to construct such manifolds by exploiting the symmetry group. In fact, for a long time, geometers believed that no compact Ricci-flat manifolds existed beyond these flat space-forms. In particular, it was a general belief that the conjecture of Calabi on Kähler manifolds was too good to be true, as it implied that the only obstruction for the existence of a Ricci-flat Kähler metric comes from the cohomological condition of the vanishing of the first Chern class. This conjecture of Calabi was in fact proved by the author in [90] in 1976.
>
> On a compact Kähler manifold, the space of Kähler metrics with zero Ricci curvature is then parameterized by those complex structures with zero first Chern class and their Kähler classes. (It is not known whether every compact Ricci-flat metric on a Kähler manifold is Kähler or not.) This moduli space has played an important role in the effort to compactify string theory.
>
> Until now, nobody has been able to find an explicit formula for any nontrivial Ricci-flat metric on a compact Kähler manifold. It is possible that the recent conjecture of Strominger-Yau-Zaslow [78] may solve this question.

Yau's work has a big impact on topology of algebraic varieties such as inequalities on Chern numbers, and rigidity of $\mathbb{C}P^n$. Probably the most important application of Calabi-Yau manifolds so far is in string theory, especially the mirror symmetry.

To get a feeling of the immediate impact of Yau's solution to the Calabi conjecture, it is interesting to quote from the book [Bes, p. 5] about its motivation:

> In September 1979, a symposium on Einstein manifolds was held at Espalion, France. It was there we realized that a book on the subject could be worthwhile. The subject seemed ripe enough: certainly most basic questions were still open, but good progress had been made, due in particular to the solution of Calabi's conjecture (by T. Aubin for

[1] The name *Calabi-Yau manifold* first appeared in [CHSW]. It also appeared in [Wit2]. These physicists found these manifolds constructed by Yau eight years ago to be important in string theory after Yau had already succeeded in finding many applications of Kähler-Einstein metrics in algebraic geometry and differential geometry.

the negative sign and S. T. Yau for both negative and zero signs) and to N. Koiso's results on the moduli space.

Once Kähler-Einstein metrics are constructed, one natural problem is to understand their uniqueness property and to classify them. The latter is essentially the moduli problem of Kähler-Einstein metrics. It seems that work on moduli spaces of Kähler-Einstein metrics in high dimensions lags very much behind the moduli space of one dimensional Kähler-Einstein metrics, i.e., the moduli space $\mathcal{M}_g$ of compact Riemann surfaces of genus g and the Teichmüller space $\mathcal{T}_g$. (Note that each Riemann surface has a canonical Kähler-Einstein metric up to scaling.)

3 The origin of complex geometry

As is well-known, complex numbers make many things easier and more elegant. For example, polynomial equations are always solvable in complex numbers. On the other hand, complex numbers can only be properly understood in terms of geometry, and Gauss made important contribution to this. Similarly, many geometric problems can be solved through complex number.

Riemann was probably the first person in history who combined geometry and complex analysis systematically and created the theory of geometric function theory. For example, Riemann's thesis in 1851 is titled "Foundations for a general theory of functions of a complex variable". In this thesis, he introduced the Cauchy-Riemann equation, the concept of Riemann surfaces, used the Dirichlet principle to derive the Riemann mapping theorem.

On the other hand, complex analysis and some of the related complex geometry were studied before Riemann by Jacobi and Abel. To understand better elliptic integrals, in particular periodic properties of their inverses, and to study elliptic functions, Jacobi and Abel introduced complex methods and initiated the theory of elliptic functions and hence of elliptic curves.

In a systematic history of the complex analysis, Bottazzini and Gray wrote on [BoG, p. 2]:

> We have tried both to describe the rise of complex analysis and to explain it. That is why a major and novel aspect of our work is the emphasis it places on elliptic functions as one of the principal impetuses for a theory of complex functions. The arrival by 1830 of elliptic functions in the hands of Abel and Jacobi caused tremendous excitement, and our history starts with that event and its immediate consequences because elliptic functions were such a spur to the development of complex function theory. They were in an essential way complex functions, and yet their foundation in the inversion of an integral containing a two-valued integrand was soon felt to be both inadequate and mysterious. Gauss felt this acutely and took considerable steps to create a properly complex theory of doubly periodic functions that did not start from an integral; Jacobi turned to the theory of theta functions.

> But they were not the only spur: Cauchy was drawn to the subject as an analyst, someone who had taken profound steps towards a rigorisation of the calculus and knew very well the way complex issues had become entangled with problems in evaluating integrals. Cauchy, however, was not so strongly drawn to the theory of elliptic functions, and his insights into the integration of complex functions took over 20 years to suggest a rich, organised theory to him. It was left to others, Liouville and Hermite, to find a way to bring enough analysis to bear to re-found elliptic functions as doubly periodic complex functions....
>
> We have dealt differently with the subject of topology, because it was not a pre-existing discipline to be enriched by the new ideas. Rather, it was a new subject, created in part to meet the increasing need for rigour in complex function theory itself. From Riemann's intuitive vague and yet profound idea of a "Riemann" surface to the work of Harnack on connectedness and path-connectedness, complex function theory generated a number of basic concepts in both geometric and point set topology.

One of the reasons for Riemann to introduce the notion of Riemann surfaces and related topological notions was to handle multi-valuedness of solutions of algebraic equations, elliptic integrals studied by Jacobi and Abel, or more general abelian integrals.

Bottazzini and Gray also wrote on [BoG, pp. 7-8]:

> The piecemeal discovery of complex function theory as a theory of analytic functions defined on the complex plane was disrupted so completely by Riemann that it was a long time before the implications of his ideas were fully understood, even though—contrary to the impression sometimes given—they were promptly and energetically studied....
>
> Riemann's visionary presentation left much for his successors to do. Some of his bolder claims collapsed under careful criticism. For a generation his use of Dirichlet's principle, which he had tried to prove, was held in great suspicion....
>
> By the 1850s complex function theory was increasingly well established in France because of the work of Cauchy and in Germany through the work of Riemann. The next major figure with a vision for the subject was Karl Weierstrass, whose lifelong ambition was to create a theory of abelian functions, as the functions obtained by inverting an arbitrary algebraic integral were called in honour of Abel.

Once Riemann surfaces were studied more systematically by Klein and others and finally formally defined by Weyl in 1913 in the classical book [Wey], the complex geometry was established.

On the other hand, it should be made clear that the meaning of complex geometry, or even geometry, is not precisely defined.

According to the preface to a book [DEG] titled *Different Faces of Geometry* by Donaldson, Eliashberg, and Gromov,

> "Geometry" is famously hard to pin down, and means many different things to different mathematicians: it is probably best interpreted as a way of thought rather than a collection of specific subject areas—there is, perhaps, no branch of mathematics which cannot be considered a part of geometry, when approached in the right spirit.

In spite of the above concern, it seems safe to say that one important part of the complex geometry is to understand Riemannian (or Hermitian) metrics on complex manifolds and their interactions with algebraic geometry and topology.

Among all metrics on complex manifolds, Kähler-Einstein metrics are special and important. Their existence is guaranteed by the Calabi conjecture, which was solved by Yau in 1977. These Calabi-Yau metrics are the central point reaching out to many parts of the mathematical world.

The above consideration gives us the title of this book:

Complex Geometry from Riemann to Kähler-Einstein and Calabi-Yau.

Most papers selected in this book fit into the above general framework. But we have also selected them to emphasize the historical development of the subject. For this purpose, we also give a brief historical summary of some major results in complex geometry since the beginning of the 20th century. It is definitely a big challenge to the editor of this book. To avoid unnecessary mistakes (or saying some nonsense) and to follow the basic principle of learning from masters, we have tried, whenever possible, to quote from writings of experts or authorities on topics under discussion.

4 Why study history of mathematics and read classical papers

Maybe some explanation is needed for emphasizing the historical development. It is perhaps best to start with a quote from the great mathematician Weil, who was also a great and serious historian of mathematics. In his plenary talk at ICM in 1978 titled *History of mathematics: why and how*, Weil wrote [Wei, pp. 229-236]:

> It is time to come back to Leibniz and consider the value of mathematical history, both intrinsically and from our own selfish viewpoint as mathematicians. Deviating only slightly from Leibniz, we may say that its first use for us is to put or to keep before our eyes "illustrious examples" of first-rate mathematical work.
>
> Does that make historians necessary? Perhaps not. Eisenstein fell in love with mathematics at an early age by reading Euler and Lagrange; no historian told him to do so or helped him to read them. But in his days mathematics was progressing at a less hectic pace than now. No doubt a young man can now seek models and inspiration in the work of his contemporaries; but this will soon prove to be a severe limitation. On the other hand, if he wishes to go much further back, he may find himself in need of some guidance; it is the function of the historian, or

> at any rate of the mathematician with a sense for history, to provide it.
>
> The historian can help in still another way. We all know by experience how much is to be gained through personal acquaintance when we wish to study contemporary work; our meetings and congresses have hardly any other purpose. The life of the great mathematicians of the past may often have been dull and unexciting, or may seem so to the layman; to us their biographies are of no small value in bringing alive the men and their environment as well as their writings....
>
> Leibniz recommended the study of "illustrious examples", not just for the sake of esthetic enjoyment, but chiefly so that "the art of discovery be promoted"....
>
> We know only too well, for instance, that one should not invariably assume a mathematician to be fully aware of the work of his predecessors, even when he includes it among his references; which one of us has read all the books he has listed in the bibliographies of his own writings? ...
>
> In other words, a quality of intellectual sympathy is required, embracing past epochs as well as our own. Even quite distinguished mathematicians may lack it altogether; each one of us could perhaps name a few who resolutely refuse to be acquainted with any work other than their own. It is also necessary not to yield to the temptation (a natural one to the mathematician) of concentrating upon the greatest among past mathematicians and neglecting work of only subsidiary value....
>
> The mathematician does his reading mostly in order to be stimulated to original (or, I may add, sometimes not so original) thoughts; there is no unfairness, I think, in saying that his purpose is more directly utilitarian than the historian's. Nevertheless, the essential business of both is to deal with mathematical ideas, those of the past, those of the present, and, when they can, those of the future. Both can find invaluable training and enlightenment in each other's work. Thus my original question "Why mathematical history?" finally reduces itself to the question "Why mathematics?", which fortunately I do not feel called upon to answer.

In an interview [Mi1, p. 406], Milnor also commented on the history of mathematics:

> I certainly enjoy trying to track down just when and how the ideas that I work with originated. This is, of course, a very special kind of history, which may concentrate on obscure ideas which turned out to be important, while ignoring ideas which seemed much more important at the time. History to most scientists is the history of the ideas that worked. One tends to be rather bored by ideas that didn't work. A more complete history would describe how ideas develop and would be

> interested in the false leads also. In this sense, the history I would write is very biased, trying to find out where the important ideas we have today came from, who first discovered them. I find that an interesting subject.

In an interview [Se1, p. 12], Serre made a short yet insightful comment on the history of mathematics:

> I can see that it takes more time to write a paper on the history of mathematics than in mathematics itself. Still, history is very interesting; it puts things in the proper perspective.

In the preface to the book which gave the formal definition of Riemann surfaces and of manifolds [Wey, p. IX], Weyl wrote: "the younger generation is only too inclined to forget the connection of the new with the old!"

In the foreword to the Chinese edition of Atiyah's Collected Works, Chern wrote:[2]

> No matter how refined or improved a new account is, the original papers on a subject are usually more direct and to the point. When I was young, I was benefited by the advice to read Henri Poincaré, David Hilbert, Felix Klein, Adolf Hurwitz, etc. I did better with Wihelm Blaschke, Elie Cartan and Heinz Hopf. This has also been in the Chinese tradition, when we were told to read Confucius, Han Yu in prose, and Tu Fu in poetry.

In an interview [At1, pp. 9], Atiyah commented on how he learned mathematics:

> But I had a good education, and I had plenty of opportunity to meet good mathematicians; I had a good background in that sense....
>
> I used to enjoy just going into the library to read, being surrounded by all of those books. It was an impressive atmosphere; it captured my imagination.

Meeting and communicating with great people is important. If we cannot meet them, reading their original papers is a good replacement or approximation. In any case, being surrounded by collections of good papers is always enjoyable, and reading them is even more beneficial. Yau has often told us the Chinese saying, "It is always beneficial to open any volume." This might be one justification for editing this book.

---

[2] Atiyah explained the original motivation to publish his Collected Works in its preface (see for example [At4]), "The initiative for these particular volumes came in fact from a different direction. A few years ago Professor Chern, who is now in active retirement trying to help China rebuild its mathematics, suggested that collections of mathematical papers made available in China would be most helpful to the younger Chinese mathematicians. Following on from this proposal the Oxford University Press agreed to publish my collected works and to make suitable arrangements to ensure their availability in China through the World Publishing Corporation in Beijing."

5 Joy and pain in mathematics

Many mathematicians study mathematics because they enjoy it. It is fun to quote from several distinguished mathematicians about this.

Sarnak said in [Co, p. 150], "I feel fortunate to be able to earn a living doing something that I enjoy." In the same book [Co, p. 152], Faltings gave a more concrete explanation:

> The work I do is very rewarding because I find myself back in the product I produce. It's very satisfying if you can make your own project and finish it, having accomplished something that others couldn't. Your name is then attached to the accomplishment, which is more satisfying than most people experience in their work. I think that I am very privileged.

In an interview, Milnor [Mi1, p. 408] said the best part of being a mathematician:

> It is trying to understand things, trying to explain them to yourself and to others, to interchange ideas and watch how other people develop new ideas. There is so much going on that no one person can understand all of it, but you can admire other people's work even if you don't follow it in detail. I find it an exciting world to be in.

But mathematical life is probably not always so pure and pleasant. In the same interview, Milnor [Mi1, p. 408] said about the competition in mathematics:

> Competition can be very unpleasant if there are several people fighting for the same goal, especially if they don't like each other. If the pressure is too great and if the reward for being the successful one is too large, it distorts the situation. I think, in general, most mathematicians have a fair attitude. If two different groups produce more or less the same results at more or less the same time, one gives credit to everyone. I think it's unfortunate to put too much emphasis on priority. On the other hand, if one person gets an idea and other people claim credit for it, that becomes very unpleasant.

Deligne also commented on priority issue in [Co, p. 156]:

> We did not care much about priority because Grothendieck had the ideas we were working on and priority would have meant nothing. I later met other areas of mathematics where people were worried about doing something first and were hiding what they were doing from one another. I didn't like it. There are all kinds of mathematicians, even competitive ones.

About these issues on writing, both S. S. Chern and C. N. Yang quoted in the prefaces of their selected papers the following famous sentence by the great Chinese poet Du Fu:

文章千古事，得失寸心知.

A direct, almost word-for-word, translation is:

An article is a matter for a thousand generations,
the one inch heart knows its gain and loss.

It is interesting to take a look at their own slightly different and better translations, which probably reflect their own perspectives on their works.

Chern wrote in [Che]: *Literary writings matter for a thousand generations. Only the inner heart cherishes their success or failure.*

And Yang wrote in [Yan]: *A piece of literature is meant for the millennium. But its ups and downs are known already in the author's heart.*

Manin also wrote something quite provocative and different in the preface to his selected papers [Man] about writings and their legacy:

> The Publishers suggested that I supply some comments. For the most part, I have chosen to give references for further reading and to sketch the subsequent development of the ideas presented in my works rather than reminisce about their conception. The reason is that most mathematical texts have a transient value, if any. Their survival is secured only if their content becomes a part of oral tradition, even a commonplace. The author may lament that his ideas were misrepresented, misattributed, and misnamed, but who could care less.

The above quotes indicate a very natural and well-known fact: different people often have different interpretations and opinions about the same issue.

6 What is good mathematics according to several famous people

Since this book consists of some important papers in complex geometry, it seems reasonable to take a look at what some distinguished people have said about good mathematics. In some sense, we have kept some criteria mentioned below when we selected papers to be included in this book.

In an interview [At1, pp. 10-12], Atiyah commented on what is important in mathematics, and gave several criteria.

(1) *The importance of connection and unity.*

> My interest is in the fact that things in different subjects interact...
>
> I strongly disagree with the view that mathematics is simply a collection of separate subjects.... Mathematics is much more of an organic development. It has a long history of connections with the past and connections with other subjects.
>
> Hardcore mathematics is, in some sense, the same as it has always been. It is concerned with problems that have arisen from the actual physical world and other problems inside mathematics having to do

with numbers and basic calculations, solving equations. This has always been the main part of mathematics. Any development that sheds light on these topics is an important part of mathematics....

I believe that it is very important that mathematics should be thought of as a unity. And the way I work reflects that; which comes first is difficult to say. I find the interactions between the different parts of mathematics interesting. The richness of the subject comes from this complexity, not from the pure strand and isolated specialization...

I believe in mathematics as sort of an entity; a theorem is just a staging post. I know lots of nice nuggets, nice facts, nice things, but I don't attach much importance to them individually....

A theory is interesting because it solves lots of special problems and puts them in proper context; it enables you to understand them all. Quite often a theory evolves because somebody solved some very hard problem first and then you try to understand what is going on—you build a superstructure around it. Soft theory which has no hard problems in it is of no use.

(2) *The importance of interaction and impact.*

The importance of a part of mathematics is something one can judge roughly by the amount of interaction it has with other parts of the subject. It is a kind of self-consistent definition of importance....

In a given area of mathematics the things that are important are quite often not the most technically difficult parts—the hardest things to prove. They are quite often the more elementary things because those are the parts that have the widest interaction with other fields and other areas—they have the widest impact....

The specialists in a field, of course, get enamoured of the hard technical problems. But from the point of view of mathematicians as a whole, while they admire them, they don't use them.

(3) *The importance of understanding.*

If I'm interested in some topic then I just try to understand it... I don't pay very much attention to the importance of proofs. I think it is more important to understand something....

If I give talks, I try always to convey the essential ingredients of a topic. When it comes to writing papers or books, however, then it is much more difficult. I don't tend to write books. In papers I try to do as much as I can in writing an account and an introduction which gives the ideas....

Most books nowadays tend to be too formal most of the time, they give too much in the way of formal proofs, and not nearly enough in the way of motivation and ideas. Of course it is difficult to do that—to give motivation and ideas....

> I think it is very unfortunate that most books tend to be written in this overly abstract way and don't try to communicate understanding....

In an article discussing the work of Atiyah [Do4], Donaldson wrote:

> Nearly twenty years have passed since this period I have been recalling, and of course the landscape in this part of mathematics has evolved, although many of the topics mentioned above are still active areas of research and I will not attempt to summarize subsequent developments .... Looking back, I think that one of the distinctive things that we learnt from Atiyah was his broad view of mathematics. Technical specialisation, as an algebraic geometer, differential geometer, topologist or whatever, was not particularly encouraged; the great thing was to explore the interaction of these different areas.

Arnold also emphasized unity of mathematics in [Ar1]:

> The existence of mysterious relations between all different domains is the most striking and delightful feature of mathematics (having no rational explanation).
>
> The experience of the past centuries shows that the development of mathematics was due not to technical progress (consuming most of the efforts of mathematicians at any given moment), but rather to discoveries of unexpected interrelations between different domains (which were made possible by these efforts).

In the section *Advice to a young mathematician* in the book [Gow], Connes wrote:

> Where things get really interesting is when unexpected bridges emerge between parts of the mathematical world that were remote from each other in the mental picture that had been developed by previous generations of mathematicians.

In the Foreword to the book [Hub, p. xiii], Thurston wrote:

> My view is that in mathematics, an internal test is the best gauge for the significance of a subject. If it is rich and interconnected and if it grabs your interest, then it is very likely to become significant to you, even though in many cases you can't foresee how.

7 The importance of being connected with the sources

It is well-known that mathematics was motivated by the real world applications right at the beginning of civilization. Since then its relation with sciences, especially with physics and astronomy, has been indispensable for the development of mathematics.

In the interview [At1], Atiyah also emphasized the importance of applications and closeness to sources:

> You will find views like these expressed by von Neumann and Hermann Weyl, some thirty years ago. They worried about the way mathematics might be going; if it goes too far away from its sources then it might become sterile. I think that is fundamentally correct....
>
> The more I've learned about physics, the more convinced I am that physics provides, in a sense, the deepest applications of mathematics. The mathematical problems that have been solved or techniques that have arisen out of physics in the past have been the lifeblood of mathematics. And it's still true. The problems that physicists tackle are extremely interesting, difficult, challenging problems from a mathematical point of view. I think more mathematicians ought to be involved in and try to learn about some parts of physics; they should try to bring new mathematical techniques into conjunction with physical problems.

The views of von Neumann mentioned by Atiyah were given in an article titled *The mathematician* [Neu]. Specifically, John von Neumann wrote:

> The most vitally characteristic fact about mathematics is, in my opinion, its quite peculiar relationship to the natural sciences, or, more generally, to any science which interprets experience on a higher than purely descriptive level.
>
> Most people, mathematicians and others, will agree that mathematics is not an empirical science, or at least that it is practiced in a manner which differs in several decisive respects from the techniques of the empirical sciences. And, yet, its development is very closely linked with the natural sciences. One of its main branches, geometry, actually started as a natural, empirical science. Some of the best inspirations of modern mathematics (I believe, the best ones) clearly originated in the natural sciences. The methods of mathematics pervade and dominate the "theoretical" divisions of the natural sciences. In modern empirical sciences it has become more and more a major criterion of success whether they have become accessible to the mathematical method or to the near-mathematical methods of physics.

Bernhard Riemann also emphasized such connections between mathematics and sciences. One can see them in all his works, especially in his famous Habilitation Dissertation titled *The hypotheses on which geometry is based* in 1854. Riemann wrote [Rie, p. 257, p. 270]:

> I have accordingly set myself as a first task the construction of the concept of a *multi-dimensional quantity* from the general concept of *magnitude*. It emerges from this study that a multi-dimensional object is capable of being measured in different ways and that space is only

> a particular example of the case of a three-dimensional quantity. It necessarily follows from this that the theorems of geometry cannot be deduced from the general notion of magnitude alone, but only from those properties which distinguished space from other conceivable entities, and these properties can only be found *experimentally*....
>
> The answer to these questions can be found only by starting from the existing theories, whose foundations were laid by Newton and which were derived from a body of phenomena. These theories and concepts must then be gradually modified whenever facts are encountered which cannot be explained on the existing basis...
>
> .... This takes us into the realm of another science—physics.

Yau commented on this passage of Riemann in [Yau1, p. 2]:

> Riemann thinks a deep understanding of geometry should be based on concepts of physics. And this is indeed the case as we experienced in the past century, especially in the past 50 years development of geometry.

As we shall see later, Calabi-Yau manifolds were motivated by Einstein's field equation in general relativity and have also played a crucial role in the string theory.

8 From Kepler's laws of planetary motion to elliptic integrals

Following the emphasis on going back to the sources mentioned in the previous section, we consider the origin of one important source of Riemann surfaces: elliptic integrals.

As it is known, elliptic integrals led to elliptic functions and hence elliptic curves (or Riemann surfaces of genus 1). They were crucial to the development of complex geometry.

Almost every student of calculus has heard of elliptic integrals, since they arise from the problem of the arc length of an ellipse. This is one example of integrals for which one cannot find an explicit formula in terms of elementary functions.

From the mathematical point of view, they are integrals of functions of the form $R(x, \sqrt{P(x)})$ which cannot be computed explicitly, where R is a rational function of two variables, $P(x)$ is a polynomial of degree 3 or 4 with no repeated roots.

Probably many students may not wonder why it might be useful for real applications to consider such integrals or the arc length of an ellipse.

It turns out that there was some important motivation from astronomy. As is known, Kepler improved the heliocentric theory of Copernicus using elliptical orbits rather than circular orbits and proposed three Kepler's laws of planetary motion:

(1) The orbit of a planet is an ellipse with the sun at one of the two foci.

(2) A line segment joining a planet and the sun sweeps out equal areas during equal intervals of time.

(3) The square of the orbital period of a planet is proportional to the cube of the semi-major axis of its orbit.

It is also known that Newton developed his laws of motion, his law of universal gravitation, and calculus, and as an application, he derived these Kepler's laws rigorously.

One natural problem is to compute quantities associated with ellipses. According to [BoG, p. 18]:

> The contribution of Abel and Jacobi was to create a theory of functions at once traditional and novel; traditional in its algebraic aspect; novel in that it was, in ways that required to be understood, truly a theory of functions of a complex variable.
>
> As for the topic of elliptic integrals, it had a long-established place in contemporary astronomy. Since Kepler's second law asserts that an elliptical orbit is parameterised by a satellite sweeping out equal areas in equal times, mathematicians were led straight away into questions involving the rectification of an ellipse and so to elliptic integrals.

9 From elliptic integrals to abelian integrals and Riemann surfaces

Plane cubic curves give rise to elliptic curves. The Weierstrass form shows that every elliptic curve can be realized as a plane cubic curve.

Integrals of $R(x, \sqrt{P(x)})$ with the degree of P greater than 5 lead to Abel integrals. By definition, if x, y satisfy an algebraic equation $F(x, y) = 0$, integrals of functions $R(x, y)$ are called *abelian integrals.* As pointed above, if $F(x, y) = 0$ is an irreducible equation of degrees higher than 5, then the associated abelian integral is not an elliptic integral.

To give a general theory of abelian integrals and to solve the Jacobi inversion problem, Riemann introduced many important concepts and methods in his masterpiece in 1857 titled *The theory of abelian functions* [Rie, pp. 79-134]:

(1) Introduced the moduli space $\mathcal{M}_g$ of compact Riemann surfaces of genus g and counted the number of complex parameters needed to describe points in $\mathcal{M}_g$, which is equal to $3g - 3$ when $g \geq 2$.

(2) Defined the Betti number of Riemann surfaces and started algebraic topology of surfaces.

(3) Riemann-Hurwitz formula for coverings of Riemann surfaces.

(4) Riemann's inequality which was completed to the Riemann-Roch theorem by his student Roch.

(5) Riemann's bilinear relation for periods of Riemann surfaces.

(6) Riemann theta functions.

(7) Solution to the Jacobian inversion problem for all compact Riemann surfaces.

(8) Associated a compact Riemann surface to each algebraic curve, and proved that every compact Riemann surface comes from an algebraic curve.

This paper of Riemann was immediately and highly regarded by many people. In [BoG, p. 286], Bottazzini and Gray wrote:

> The depth of originality in Riemann's remarkable paper (1857c) can be measured by the fact that on reading it no less an authority than Weierstrass withdrew a paper of his own on the same subject, preferring to wait until he had assimilated what Riemann had to say. In it Riemann tackled a problem which had been outstanding since the time of Jacobi, and with which Weierstrass had made his name by resolving a special case of: the study of the integrals of arbitrary algebraic functions.

About Riemann, Bottazzini and Gray wrote on [BoG, p. 259]:

> Riemann is one of the three creators of the theory of complex functions, but he wrote relatively little. Unusually for a major mathematician, his Werke, even augmented as they are by notes from lecture courses and papers left unpublished at his death, fill only a single volume. Each of his papers on complex function theory extended the subject greatly and then applied it to solve a substantial problem. As a result his successors found themselves with much to do, yet it can be argued that the profundity of his best ideas hindered their immediate reception, and it was to be a generation before those ideas were finally taken up.
>
> Riemann's importance is not illuminated by knowing his biography, which has an independent interest, well conveyed in Laugwitz's account (Laugwitz 1999). It is the intellectual connections Riemann made and the depth to which he saw that marks his great contribution, and we shall discuss these below.

10 Riemann mapping theorem

In his thesis, Riemann proposed the Riemann mapping theorem and outlined a proof using his general results on functions on Riemann surfaces, in particular the Dirichlet principle.

It is interesting to quote from Riemann [Rie, pp. 36-37] to see exactly what he claimed:

> In order to clarify our general results, it is useful to give an example of their application.... Two given simply connected plane surfaces can

> always be related in such a way that each point of one surface corresponds to a point of the other, varying continuously with that point, with the corresponding smallest parts similar. One interior point, and one boundary point, can be assigned arbitrary corresponding points; however, this determines the correspondence for all points.
>
> When two surfaces T and R both correspond to a third surface S in such a way that similarity holds between corresponding smallest parts, this yields a correspondence between T and R that clearly has the same property. The problem of producing a correspondence between two arbitrary surfaces with similarity between the smallest corresponding parts now reduces to mapping an arbitrary surface onto a specific surface, with similarity in the smallest parts. To prove our result, draw a circle K in the plane B with radius 1 and center $w = 0$.

According to [BoG, p. 275],

> Given this emphasis on defining complex functions via harmonic functions and harmonic functions via continuous functions defined on the boundary of a region, it was natural for Riemann to consider the question of whether two simply connected regions whose boundaries correspond one-to-one can sustain significantly different families of complex functions....
>
> It [Riemann mapping theorem] establishes not only that there are complex functions defined on any simply connected region, and indicates how they can be defined, but also that any two such regions are equivalent for the purposes of Riemannian complex function theory....

The proof of the Riemann mapping theorem has a complicated history. See [Gra2] for a detailed discussion. For our discussion in this book, it leads to the uniformization theorem for Riemann surfaces and also motivates the general uniformization problem for complex manifolds.

11 Uniformization of Riemann surfaces

The uniformization theorem for Riemann surfaces is one of the most important theorems in mathematics and can be stated in several equivalent ways:

(1) Every simply connected Riemann surface is biholomorphic to, i.e., conformally equivalent, to, one of the three domains: the Riemann sphere $\mathbb{C} \cup \{\infty\}$, the complex plane $\mathbb{C}$, or the open unit disk in $\mathbb{C}$ (or the upper half-plane $\mathbf{H}^2$).

(2) Given any connected Riemann surface Σ, its universal covering space $\tilde{\Sigma}$ is biholomorphic to one of the three domains: the Riemann sphere $\mathbb{C} \cup \{\infty\}$, the complex plane $\mathbb{C}$, or the open unit disk in $\mathbb{C}$ (or the upper half-plane $\mathbf{H}^2$).

(3) Any connected Riemann surface admits a Riemannian metric which is conformal to the complex structure and has constant curvature, which can be normalized to be $1, 0, -1$.

There are many different proofs of this fundamental theorem now. For Riemann surfaces which are obtained from compact Riemann surfaces by removing finitely many points, for example, plane algebraic curves over $\mathbb{C}$, the uniformization theorem were announced in 1882 and 1883 independently by both Klein and Poincaré. This result was the high point of the strong competition between them, which led to the totally collapse of Klein and the end of his research career. See a description of this competition by one of the participants, see Klein's history book [Kle].[3]

Klein and Poincaré introduced some forms of Teichmüller spaces of Riemann surfaces and applied the method of continuity to prove. Due to lack of structure or knowledge of the Teichmüller spaces, their proofs were not complete. For a detailed discussion of their approach, see the book [Sai] and also a summary [Beg].

For a detailed historical treatment on the uniformization theorem and its relation with the monodromy groups of differential equations and other related topics, see the book [Gra2]. We will discuss the uniformization of complex manifolds of higher dimensions later.

12 Moduli space and Teichmüller space of Riemann surfaces

Classification is a basic problem in mathematics and hence moduli spaces appear naturally in many different contexts. Moduli spaces have been studied for over one and a half centuries and will continue to keep mathematicians busy for many generations to come.

Though the moduli space $\mathcal{M}_g$ of compact Riemann surfaces of genus g was introduced by Riemann in 1857, it was not obvious what kind of space it was. Putting suitable structures on it and understanding their properties is called Riemann's moduli problem and was solved many years later. Many people have tried to work on this problem and their efforts have contributed substantially to mathematics, especially algebraic geometry and complex geometry.

Teichmüller made several important contributions:

(1) Introduced markings on Riemann surfaces to kill nontrivial automorphisms of Riemann surfaces to construct Teichmüller space $\mathcal{T}_g$, and then undid the markings by dividing the Teichmüller space by the mapping class group.

(2) Introduced the notion of a fine moduli space to put a natural complex structure on $\mathcal{T}_g$ and hence on $\mathcal{M}_g$.

[3] In an interview [Ar2], Arnold said that: "One-half of the mathematics I know comes from the book of F. Klein *Lectures on the Development of Mathematics in the 19th Century*." This is the book cited as [Kle] here.

Both points are very important in the contemporary algebraic geometry, or the algebraic geometry after Grothendieck. To see the importance of the idea of marking Riemann surfaces, or adding additional data to kill nontrivial automorphism groups, consider a letter from Grothendieck to Serre. On November 5, 1959, Grothendieck wrote in [GS, p. 94]:

> The "Grand existence theorem" is progressing little by little; many technical difficulties remain, but I am more and more convinced that there is an absolutely marvelous technique at the end of all this. I have already come to the practical conclusion that every time that my criteria show that no moduli variety (or rather moduli scheme) for the classification of (global or infinitesimal) variations of certain structures (complete non-singular varieties, vector bundles etc.) can exist, despite good hypotheses of flatness, properness, and if necessary non-singularity, the only reason is the existence of automorphisms of the structure which prevent the descent from working... The remedy in moduli theory seems to me to be to eliminate bothersome automorphisms by introducing additional structures on the objects being studied: points or differential forms etc. on the varying varieties (a process which is already used for curves), trivializations at sufficiently many points of the vector bundles one wants to vary, etc.

One concrete application by Grothendieck is to construct Hilbert schemes, which are essential ingredients in the method to construct moduli spaces by geometric invariant theory in [Mu1]. In this construction, the notion of stable points (stable curves, stable varieties, and stable vector bundles over curves and algebraic varieties) is essential. This notion will also be essential to the discussion on Hermitian Yang-Mills connections later in this book.

Remark 12.1. *Such an approach is also the reason that for higher dimensional complex manifolds or algebraic varieties, moduli spaces only make sense for polarized complex manifolds and polarized varieties. Consequently, without being known to many people, Teichmüller's idea of adding additional data to varieties has far reaching consequences in algebraic geometry.*

Teichmüller introduced families of Riemann surfaces over complex manifolds and identified the Teichmüller space as the space which represents the functor of families of marked Riemann surfaces. This idea was put into the proper modern language of categories and functors by Grothendieck in his 10 lectures at the Cartan seminar in 1960—1961. For more detail, see [Ji].

Besides recovering the moduli space $\mathcal{M}_g$ as a quotient of $\mathcal{T}_g$ under the action of the mapping class group Mod_g of a compact oriented surface of genus g, the Teichmüller space $\mathcal{T}_g$ also provides a natural space for the group Mod_g, which has far reaching consequences in geometric topology and low dimensional topology. See [AJP] for a history of the Teichmüller theory.

In complex analysis and geometry, Teichmüller's work has also been very influential. In [Be], Bers wrote:

> This address is a progress report on recent work, partly not yet published, on the classical problem of moduli. Much of this work consists in clarifying and verifying assertions of Teichmüller whose bold ideas, though sometimes stated awkwardly and without complete proofs, influenced all recent investigators, as well as the work of Kodaira and Spencer on the higher dimensional case.

For example, the importance of Teichmüller's work on quasiconformal maps was described by Ahlfors in [Ah, pp. 72-73]:

> The theory of quasiconformal mappings is almost exactly fifty years old. They were introduced in 1928 by Herbert Grötzsch in order to formulate and prove a generalization of Picard's theorem....
>
> Grötzsch's papers remained practically unnoticed for a long time. In 1935 essentially the same class of mappings was introduced by M. A. Lavrentiev in the Soviet Union.... In any case, the theory of quasiconformal mappings, which at that time had also acquired its name, slowly gained recognition, originally as a useful and flexible tool, but inevitably also as an interesting piece of mathematics in its own right.
>
> Nevertheless, quasiconformal mappings might have remained a rather obscure and peripheral object of study if it had not been for Oswald Teichmüller, an exceptionally gifted and intense young mathematician and political fanatic, who suddenly made a fascinating and unexpected discovery. At that time, many special extremal problems in quasiconformal mapping had already been solved, but these were isolated results without a connecting general idea. In 1939 he presented to the Prussian Academy a now famous paper which marks the rebirth of quasiconformal mappings as a new discipline which completely overshadows the rather modest beginnings of the theory. With remarkable intuition he made a synthesis of what was known and proceeded to announce a bold outline of a new program which he presents, rather dramatically, as the result of a sudden revelation that occurred to him at night. His main discovery was that the extremal problem of quasiconformal mapping, when applied to Riemann surfaces, leads automatically to an intimate connection with the holomorphic quadratic differentials on the surface. With this connection the whole theory takes on a completely different complexion: A problem concerned with non-conformal mappings turns out to have a solution which is expressed in terms of holomorphic differentials, so that in reality the problem belongs to classical function theory. Even if some of the proofs were only heuristic, it was clear from the start that this paper would have a tremendous impact, although actually its influence was delayed due to the poor communications during the war. In the same paper Teichmüller lays the foundations for what later has become known as the theory of Teichmüller spaces.

In complex geometry, Teichmüller's work is also important. In [Re, p. 225], Remmert wrote:

> From the very beginning it was felt that Riemann's approach to complex analysis should also bear fruits in higher dimension. But only in 1932, at the International Congress in Zürich, did Carathéodory in [Carathéodory 1932] strongly advocate studying four dimensional abstract Riemann surfaces (as he called them) for their own sake. However, due to his rather cumbersome approach, there was no response by his contemporaries.
>
> Only after differentiable and real analytic manifolds had already been studied intensively, and with great success, was time ripe for complex manifolds. It seems difficult to locate the first paper where complex manifolds explicitly occur. In 1944 they appeared in Teichmüller's work on "Veränderliche Riemannsche Flächen", [Teichmüller 1944, p. 714]; here we find for the first time the German expression "komplexe analytische Mannigfaltigkeit". The English "complex manifold" occured in 1946 in Chern's work [1946, p. 103]; he recalled the definition (by an atlas) just in passing. And in 1947 we found "variété analytique complexe" in the title of Weil's paper [1947]. Overnight complex manifolds blossomed everywhere.

Teichmüller theory was also one of the motivations for the Kodaira-Spencer deformation theory for complex manifolds. In [KS, p. 331], Kodaira and Spencer wrote:

> During the hundred years which have elapsed since the publication of Riemann's memoir, the questions centering around the deformation of the complex structure of a Riemann surface have never lost their interest and, during the last twenty years, have received fresh stimuli from two sources at least, namely: from the theory of extremal quasi-conformal mappings which was initiated by O. Teichmüller (see references [38] listed at the end of this paper) and which has been developed by L. Ahlfors [1] and others; from the variational calculus for Riemann surfaces which began with a paper by M. Schiffer in 1938 and which has been developed and applied by several authors (see, e.g., [34]). In both of these developments the quadratic differentials of the surface play a fundamental role.

Riemann's moduli problem also motivated the Hurwitz space first studied by Clebsch and Hurwitz, which has played an important role in the enumerative geometry.

Moduli space $\mathcal{M}_g$ and Teichmüller space $\mathcal{T}_g$ are role models of moduli spaces in subjects such as algebraic geometry and differential geometry. For example, in the moduli spaces of marked K3 surfaces and unmarked K3 surfaces, and moduli spaces of Calabi-Yau manifolds.

13 Kähler manifolds

For a differential manifold, Riemannian metrics are essential to its geometry. For a complex manifold, Kähler metrics are the most natural metrics. One reason is that once endowed with a Kähler metric, a Kähler manifold has three mutually compatible structures: a complex structure, a Riemannian structure, and a symplectic structure. Interaction between these structures makes Kähler manifolds very special manifolds, and various techniques can be used to understand them. One important example is the Hodge theory for compact Kähler manifolds.

There is a large and natural supply of Kähler manifolds. For example, smooth projective algebraic varieties over $\mathbb{C}$ are Kähler manifolds with respect to the metric induced from the Study-Fubini metric of the complex projective space.

As mentioned before, Kähler-Einstein metrics are really special among Kähler manifolds, and this is one of the concepts which makes Kähler famous. Another famous concept which is partially named after him is that of K3 surfaces. Besides these contributions, Kähler works reached out to many different fields.

It is helpful to quote from a biography of Kähler, titled *Life of Erich Kähler*, in his collected works [Ka, p. 3]:

> His [Kähler's] contributions to mathematical physics, as well as algebraic and arithmetic geometry assure him a permanent place in the history of mathematics. Mathematics was for him all encomposing world-view. The unity of mathematics and its progress was his foremost concern. His passion to reach out beyond the traditional boundaries of mathematics can be understood from his biography. He had a happy childhood and adolescence, but his life was later influenced by unusually tragic events.

His paper on Kähler metrics included in this book is probably one of the most influential papers in complex geometry, since it started the subject of Kähler geometry. The above biography in his collected works also explained how it was written [Ka, p. 4]:

> Already in 1931, Blaschke had recommended Erich Kähler for a professorship in mathematics at the University of Rostock, but Kähler preferred to stay in Hamburg to participate in the inspiring atmosphere at the Mathematical Institute. One result of his stay in Hamburg was the paper.... In Hamburg Kähler also met Chern. This meeting led to a lifelong relationship of mutual respect and admiration.

In an article titled *A tribute to Herrn Erick Kähler* published in [Ka, p. 1], Chern wrote:

> I arrived at Hamburg in the summer of 1934. The university began in November and I attended, among other classes, Herrn Kähler's seminar on exterior differential systems.... I devoted much time to the subject and in 1935 wrote a thesis on the applications of exterior differential systems to web geometry.

> I received my *Doktor der Wissenschaften* in February 1936 and my thesis was published in the Hamburger Abhandlungen in the same year. Clearly it received much advice from Herrn Kähler, from whom I learned the subject of exterior differential calculus and what is now known as the Cartan-Kähler theory. We had frequent lunches together at the restaurant *Cuiro Haus* near the seminar. He told me many things, mathematical or otherwise. My gratitude to him cannot be overstated, At that time Herrn Kähler had just published his article *Über eine bemerkenswerte Hermitesche Metrik* in the Hamburger Abhandlungen which is the starting point of Kähler geometry. In this paper he showed a clear understanding of the complex structure and proved in particular that a Kähler metric (later name) has a potential relative to the operator $\partial\bar{\partial}$. This paper affects me in my conception on the Gauss-Bonnet formula and characteristic classes.

14 Hodge theory

One natural approach to understand the natural complex structure on the Teichmüller space $\mathcal{T}_g$ and hence on the moduli space $\mathcal{M}_g$ is to use the period map of Riemann surfaces. This was mentioned by Teichmüller, used later by Rauch, Ahlfors and Bers to define *rigorously* a complex structure on $\mathcal{T}_g$ for the first time.

The theory of harmonic analysis and period of Riemann surfaces was treated systematically and rigorously by Weyl in his classical book on Riemann surfaces [Wey].

One important consequence of the existence of a Kähler metric on a compact complex manifold is the existence of a Hodge structure on the cohomology groups of the manifold.

The Hodge theory of compact Kähler manifolds was initiated by Hodge, and presented in his book [Hod1]. Important contributions from Weyl and Kodaira complected the basic Hodge theory.

In a biography of Hodge [At2], Atiyah explained clearly the history of Hodge theory and Hodge's essential contribution:

> In one of his papers, Francesco Severi mentioned the importance of knowing whether a nonzero double integral of the first kind could have all its periods zero. By a stroke of luck, Hodge saw a paper by Solomon Lefschetz in the 1929 Annals of Mathematics in which purely topological methods were used to obtain the period relations and inequalities for integrals on a curve. It was clear to Hodge that Lefschetz's methods could be extended to surfaces to solve Severi's problem, and he found it incredible that this should have escaped Lefschetz's notice. It did not take Hodge long to work out the details and write "On Multiple Integrals Attached to an Algebraic Variety". This was the turning point in his career, and therefore it is perhaps appropriate to describe the essence of his proof and its relation to Lefschetz's paper....

In addition to the primitive state of topology at this time, complex manifolds (other than Riemann surfaces) were not conceived of in the modern sense. The simplicity of the proof indicated above owes much to Hodge's work in later years, which made complex manifolds familiar to the present generation of geometers. All this must have had something to do with Lefschetz's surprising failure to see what Hodge saw. In fact, however, this was no simple omission on Lefchetz's part, and he took a great deal of convincing on this point. At first he insisted publicly that Hodge was wrong, and he wrote to him demanding that the paper be withdrawn. In May 1931 Lefschetz and Hodge had a meeting in Max Newman's rooms at Cambridge. There was a lengthy discussion leading to a state of armed neutrality and an invitation to Hodge to spend the next academic year at Princeton. After Hodge had been at Princeton for a month. Lefschetz conceded defeat and, with typical generosity, publicly retracted his criticisms of Hodge's paper. Thereafter Lefschetz became one of Hodge's strongest supporters and fully made up for his initial skepticism....

Hodge in due course became Lefschetz's successor in algebraic geometry. Whereas the Princeton school inherited and developed Lefschetz's contributions in topology, his earlier fundamental work on the homology of algebraic varieties was somewhat neglected, probably because it was ahead of its time. Hodge's work was complementary or dual to that of Lefschetz, providing an algebraic description of the homology instead of a geometric one. The fact that Lefschetz's theory has now been restored to a central place in modern algebraic geometry is entirely due to the interest aroused by its interactions with Hodge's theory.

By the time Hodge came to Princeton, his mathematical ideas, arising from "On Multiple Integrals", had already progressed very significantly. In studying integrals in higher dimensions, he soon put his finger on the crucial point. Whereas for the Riemann surface of a curve the number of holomorphic 1-forms is half the first Betti number, there is no corresponding relation in higher dimensions for holomorphic p-forms with $p > 1$. Hodge discussed this point on a number of occasions with Peter Fraser until one day Fraser pointed out Georges de Rham's dissertation, which had just arrived in the Bristol University library. In later years Hodge described this as a stroke of good fortune; although it did not solve his problem, it helped him to see what was involved. In de Rham's theory, valid on any real differentiable manifold, the main result is that there always exists a closed p-form ω with prescribed periods, and that ω is unique modulo derived forms. On a Riemann surface there are natural choices given by the real and imaginary parts of the holomorphic differentials, and Hodge was looking for an appropriate generalization to higher dimensional algebraic varieties. He saw that the real and imaginary parts of a holomorphic 1-form on a Riemann

> surface are in some sense duals of one another, and he had a hunch that there should be an analogous duality in general. More precisely, for each p-form ω there should be an $(n-p)$- form $^*\omega$ (n being the dimension of the manifold); the preferred forms, later called harmonic would be those satisfying $d\omega = 0$ and $d(^*\omega) = 0$. The main theorem to be proved would be the existence of a unique harmonic form with prescribed periods. Once established at Princeton, Hodge tried to clarify and develop these vague ideas. He soon realized that the relationship of ω to $^*\omega$ was a kind of orthogonality and was able to make this precise in Euclidean space and, more generally, on a conformally flat manifold. He then attempted to prove the existence theorem by generalizing the classical Dirichlet methods....
>
> After his success with the signature theorem, Hodge worked steadily, polishing his theory and developing its applications to algebraic geometry. He also began to organize a connected account of all his work as an essay for the Adams Prize. He was awarded that prize in 1937, but the magnum opus took another three years to complete and finally appeared in book form (*The Theory and Applications of Harmonic Integrals*) in 1941. In the meantime he had published another approach to the existence theorem that had been suggested to him by Hans Kneser. It involved the use of the parametrix method of F. W. Levi and David Hilbert and was, as Hodge said, superior in all respects to his first attempt. Unfortunately this version, reproduced in his book, contained a serious error that was pointed out by Bohnenblust. The necessary modifications to provide a correct proof were made by Hermann Weyl at Princeton and independently by Kunihiko Kodaira in Japan. Hodge freely admitted that he did not have the technical analytical background necessary to deal adequately with his existence theorem. He was only too pleased when others, better-qualified analysts than himself, completed the task. This left him free to devote himself to the applications in algebraic geometry which was what really interested him. In retrospect it is clear that the technical difficulties in the existence theorem required not significant new ideas but a careful extension of classical methods. The real novelty, which was Hodge's major contribution, was in the conception of harmonic integrals and their relevance to algebraic geometry. This triumph of concept over technique is reminiscent of a similar episode in the work of Hodge's great predecessor Bernhard Riemann.

It is well-known that the most famous conjecture in algebraic geometry and complex geometry is the Hodge conjecture. In [De], Deligne wrote:

> If Z is a closed analytic subspace of X, of complex codimension p, Z is an integral cycle and, by Poincaré duality, defines a class $\mathrm{cl}(Z)$ in $H^{2p}(X, \mathbb{Z})$. The integration current on Z is a closed (p,p)-form with generalized function coefficients, representing the image of $\mathrm{cl}(Z)$ in $H^{2p}(X, \mathbb{C})$. The class $\mathrm{cl}(Z)$ in $H^{2p}(X, \mathbb{Z})$ is hence of type (p,p), in the

sense that its image in $H^{2p}(X,\mathbb{C})$ is. Rational (p,p)-classes are called Hodge classes. They form the group

$$H^{2p}(X,\mathbb{Q}) \cap H^{p,p}(X) = H^{2p}(X,\mathbb{Q}) \cap F^p \subset H^{2p}(X,\mathbb{C}).$$

In [6], Hodge posed the

Hodge Conjecture. *On a projective non-singular algebraic variety over* $\mathbb{C}$*, any Hodge class is a rational linear combination of classes* $\mathrm{cl}(Z)$ *of algebraic cycles.*

15 Algebraic varieties and topology

One of the main contributions of Lefschetz is to combine algebraic topology and algebraic geometry. For example, he is well known for the Lefschetz fixed point formula, the hard Lefschetz theorem in the Hodge theory, the Lefschetz hyperplane theorem, and the method of Lefschetz pencils in algebraic geometry.

To understand relations between them, their impact on algebraic geometry, and how Lefschetz achieved them, it is interesting to read a description by Hodge in [Hod2]. The quote from the previous section shows the great influence of Lefschetz on Hodge and his work. We hope that such a summary on the work of Lefschetz in algebraic geometry by Hodge is appropriate and insightful [Hod2, p. 3, p. 4, p.21]:

> If one attempts to give a systematic account of the work of a distinguished mathematician paper by paper, the final impression produced is apt to make this work appear as a museum piece, and this may be the opposite of what is really intended. If any such impression should be given by my remarks on Professor Lefschetz's work on algebraic geometry, it would be particularly unfortunate and misleading; for it is a fact that a number of the discoveries which he has made are of more vital interest to mathematicians at the present day than they have been since they aroused widespread excitement on their first appearance some thirty or thirty five years ago. I therefore think it would be wiser for me not to follow the historical method, but to select a number of the outstanding contributions which Lefschetz has made to algebraic geometry and to discuss these in relation to current mathematical developments; and I venture to suggest that this procedure will have the approval of Lefschetz himself, since, though no one is more generous in recognizing the merits of others, to him mathematics is more important than the mathematician....
>
> One of the most striking features of the whole range of Lefschetz's contributions to the topological and transcendental theories of algebraic varieties is the interplay between the various ideas, and it is clear that the next step in one line was inspired by some achievement in another direction. The essential fact is that Lefschetz clearly regarded the whole range which he covered as a single subject, and the keynote is its unity—a unity which he eventually carried into his work in pure

topology, where his fixed-point theorems are clearly derived from ideas familiar in algebraic geometry. But having made this point, it will be convenient to take up a number of aspects of his work separately.

One of the great lessons to be learned from a study of Lefschetz's work on algebraic varieties is that before proceeding to the investigation of transcendental properties it is necessary first to acquire a thorough understanding of the topological properties of a variety. Not only does this greatly simplify the technical problems encountered in developing the transcendental theory, but it is an absolute necessity if one is to appreciate the true significance of the difficulties to be overcome....

The importance of the topological structure of curves in the study of abelian integrals of algebraic functions of one variable was made clear by Riemann, and when Picard came to study the integrals attached to an algebraic surface he naturally made use of all the topological methods available.... his final analysis of the topology of an algebraic surface provided Lefschetz with the scheme for a direct investigation of the topology, not only of surfaces but of varieties of any dimension. Lefschetz's achievement was to obtain all of Picard's topological results by direct methods, and then to use them to simplify the transcendental theory. I wish to emphasize the directness of Lefschetz's methods, for it does seem to me to be most satisfactory to know the topological structure of a variety thoroughly before embarking on transcendental considerations. This is not intended as a criticism of elegant methods which have been employed to obtain results similar to those of Lefschetz by using harmonic integrals or the theory of stacks (though it must be remembered these methods only obtain the homology groups with complex (or real) coefficients), whereas the direct method enables us to use integer coefficients, but merely reflects my view that without a direct method of investigating the topology of an algebraic variety something important would be missing.

The method used by Lefschetz is a direct generalization of the classical process for studying the topology of an ordinary Riemann surface by reducing it to an open 2-cell by means of a system of cuts. Lefschetz similarly introduced cuts into a non-singular variety V of complex dimension d and reduced it to an open $2d$-cell.

An induction argument is used to reduce V to a cell, making use of the fact that on F there are systems of subvarieties of dimension $d-1$....

In considering the development of algebraic geometry since 1924 (the date of the publication of the Borel Tract) one is conscious of three main streams: first, the classical theory which we owe to the Italian school, in which the ideas of system of equivalence and canonical systems have emerged; secondly, the abstract algebraic geometry which we owe to Zariski and Weil; and thirdly, the transcendental-topological theory, which has, to a certain extent, been merged into the theory of complex manifolds, in which, however, the algebraic varieties keep on

> distinguishing themselves as the only manifolds on which some of the operations of the theory can be performed. No single person can claim to be the sole founder of the theory of complex manifolds, but when one considers how many of the properties originally derived by Lefschetz for algebraic varieties now hold their place in the theory of complex manifolds, and how he has influenced decisively so many who have contributed to this theory, one must accord him an honored place among the founders of a great branch of mathematics; and this without even taking account of his influence, through his work in pure topology, on the topologists who have helped to build the theory. It seems clear to me that Lefschetz by his work on the topology and transcendental theory of algebraic varieties has been a major influence in turning the minds of geometers in new and fruitful directions, and in so doing he has achieved what it is given to few to do.

16 Kodaira vanishing and Kodaira embedding theorems

Kodaira could be considered the real successor of Hodge in using analysis to understand algebraic varieties and complex manifolds. Among many of his deep results, two results stand out: Kodaira vanishing theorem and Kodaira embedding theorem.

To understand the work of Kodaira in a historical context, let us quote from Atiyah [At3]:

> Kodaira studied at Tokyo University, taking degrees in both mathematics and physics. From 1944 to 1951 he was an associate professor of physics at the University. His PhD thesis was published in the Annals of Mathematics [18], and it immediately attracted international attention. Essentially this filled a significant lacuna in the basic theorem of W. V. D. Hodge on harmonic integrals. Kodaira had worked on this for many years but, because of the war, his research was carried out in isolation from the international community and did not become known until much later.
>
> Hermann Weyl, who had been a keen supporter of Hodge's work, realised the importance of Kodaira's thesis, and arranged for him to come to the Institute for Advanced Study in Princeton in 1949. This was the start of Kodaira's 18-year residence in the United States, a fruitful period which saw the full blossoming of his research, much of it in collaboration with Donald Spencer....
>
> During his time at Princeton, Kodaira continued his involvement with harmonic forms, particularly in their application to algebraic geometry, the area which had also provided the motivation for Hodge's work. The 1950s saw a great flowering of complex algebraic geometry, in which the new methods of sheaf theory, originating in France in the hands of

> Leray, Cartan, and Serre, provided a whole new machinery with which to tackle global problems. Sheaf theory fitted with Hodge theory, so it was natural that Kodaira should have been well placed to exploit the new developments. This he did, in a rapid succession of papers written in collaboration with Donald Spencer. These papers altered the face of algebraic geometry, and provided the framework in which Hirzebruch and others of the younger generation were able to make spectacular progress. Large numbers of problems left unsolved or incomplete by the Italian geometers of the classical school were now disposed of in convincing fashion. In particular, a full understanding of the arithmetic genus in higher dimensions emerged, and the Hirzebruch–Riemann–Roch theorem was the culmination of a long story (and the beginning of a new one).
>
> But perhaps his most striking individual achievement was in the general characterisation of projective algebraic varieties in [38]. A few years earlier, Hodge had drawn attention to what he modestly called "Kähler manifolds of restricted type", ones where the periods of the Kähler form are integral. Hodge had shown that these possessed all the important properties of algebraic varieties that could be deduced from the theory of harmonic forms. For a short while, André Weil had appropriately christened these "Hodge manifolds". Unfortunately for Hodge, this terminology dropped out of use as soon as Kodaira proved in [38] that Hodge manifolds are all projective varieties. This striking theorem of Kodaira generalises the classical results, going back to Riemann and Siegel, characterising complex tori which are algebraic. The proof was typical of Kodaira's work, involving a masterly use of the new analytic methods together with a detailed understanding of the relevant geometry.

17 Algebraic geometry and complex geometry

One source of compact complex manifolds comes from smooth algebraic subvarieties of $\mathbb{C}P^n$. In dimension 1, Riemann showed that the converse is true, i.e., every compact Riemann surface is biholomorphic to a projective smooth curve over $\mathbb{C}$. One natural question is whether the converse is also true in higher dimensions. One application of the Hodge theory is that it is not true, for example, through the example of Hopf surfaces and more general Hopf complex manifolds.

The Kodaira embedding theorem mentioned in the previous section allows one to embed compact Hodge complex manifolds into a complex projective space. Then by appealing to Chow's theorem that *every analytic subspace of complex projective space $\mathbb{C}P^n$ that is closed in the ordinary topological sense is an algebraic subvariety*, one can conclude that compact Hodge complex manifolds come from complex projective varieties.

Perhaps it is helpful to note that another natural source of complex manifolds is to take quotients of bounded symmetric domains, i.e., Hermitian symmetric spaces of noncompact type, by discrete subgroups of Lie groups. In a paper by Kodaira in 1954 included in this book (see also the book [MK, p. 144]), Kodaira proved: *If the universal covering of compact complex manifold M is biholomorphic to a bounded domain in $\mathbb{C}^n$, then M is a complex projective variety.*

Besides the Chow theorem mentioned above, Chow also made deep contributions to other topics in mathematics. In an article in memory of Chow [Cho, p. 1118], Chern wrote:

> Wei-Liang was an original and versatile mathematician, although his major field was algebraic geometry. He made several fundamental contributions to mathematics:
>
> (1) A fundamental issue in algebraic geometry is intersection theory. The Chow ring has many advantages and is widely used.
>
> (2) The Chow associated forms give a description of the moduli space of the algebraic varieties in projective space. It gives a beautiful solution of an important problem.
>
> (3) His theorem that a compact analytic variety in a projective space is algebraic is justly famous. The theorem shows the close analogy between algebraic geometry and algebraic number theory.
>
> (4) Generalizing a result of Caratheodory on thermodynamics, he formulated a theorem on accessibility of differential systems. The theorem plays a fundamental role in control theory.
>
> (5) A lesser-known paper of his on homogeneous spaces gives a beautiful treatment of the geometry known as the projective geometry of matrices and treated by elaborate calculations. His discussions are valid in a more general context.
>
> Chow led a simple and secluded life, with complete devotion to mathematics and some other intellectual activities including philately. He was an authority on Chinese stamps and published a book on them.

More general results on relations between complex analytic geometry and algebraic geometry (GAGA) were treated systematically in a paper of Serre [Se2]. It is often refereed to as Serre's GAGA paper.

18 Riemann-Roch theorem and index theory

Riemann-Roch theorem is a key result in the theory of Riemann surfaces, since it allows one to produce nontrivial meromorphic functions on Riemann surfaces, which are essential to understanding Riemann surfaces, for example, embedding them into projective spaces.

Though its generalization to compact complex surfaces was obtained by Kodaira, the real breakthrough was the Hirzebruch-Riemann-Roch theorem for compact complex manifolds in all dimensions. This result of Hirzebruch was further generalized by Grothendieck to the Grothendiek-Riemann-Roch theorem and was also put into the general framework of the index theory developed by Atiyah, Singer, and Bott.

The history of the Hirzebruch-Riemann-Roch theorem and the Hirzebruch signature theorem is well explained by Hirzebruch in [Hir, p. 3]:

> I shall recall (§1, §2) how the general Riemann-Roch theorem for complex manifolds is related to the signature theorem for differential manifolds. The signature theorem could be proved by Thom's cobordisim theory ([25], [26]) and was then the basic tool for proving the Riemann-Roch theorem for algebraic manifolds [13]. This Riemann-Roch theorem (for complex manifolds) is nowadays a special case of the Atiyah-Singer theorem for elliptic operators ([5], [6], [7], [8]): namely, the index theorem is applied to the Cauchy-Riemann equation. Atiyah-Bott-Singer generalized the index theorem to the equivariant case ([2], [3], [5], [7], [8]). This is a fantastic generalization of the Lefschetz fixed point theorem. The equivariant index theorem can be specialized to the signature operator ([8], p. 582). This gives a fixed point theorem involving the signature and generalizing the old signature theorem.

In [At5, p. 2], Atiyah wrote:

> The Riemann-Roch theorem marks the turning point from the old style algebraic geometry to the new. Important in its own right, it also opened up new doors for the future. The first of these was the generalization of Grothendieck, involving the introduction of his K-groups and the construction of a functorial homomorphism
>
> $$f_! : K(X) \to K(Y)$$
>
> for a proper holomorphic map $f : X \to Y$ of algebraic varieties. Grothendieck's theorem gave a formula for the Chern character of $f_!$ and it reduced to the Hirzebruch-Riemann-Roch theorem when Y is a point.
>
> Emerging from the works of Grothendieck and Hirzebruch came first the development of topological K-theory, and then the index theorem for elliptic differential operators. In various guises these have gone on developing into the field of C^*-algebras. The Hirzebruch-Riemnn-Roch theorem is, through these descendants, still alive and kicking.

There are many variants of index theorems. This can be seen in many joint publications of Atiyah with others.

19 Uniformization of higher dimensional complex manifolds

Motivated by the uniformization theorem for Riemann surfaces, there have been many attempts to establish uniformization of manifolds of higher dimensions.

For real smooth manifolds, Thurston's geometrization program for three dimensional manifolds is a success story. By using methods of geometric analysis including some crucial ideas of P. Li and S. T. Yau, Perelman has proved Thurston's geometrization conjecture. See [Mi2], for example, for a lively history of low dimensional topology.

For complex manifolds, besides the question on existence of Kähler-Einstein metrics on complex manifolds, Yau made several conjectures on the general question of obtaining "normal forms" of spaces. For example, in the paper [Yau4, p. 265, p. 267], Yau wrote:

> In geometry, a lot of efforts have been devoted to provide a "normal form" to certain geometric objects. A famous example is the Weyl's embedding theorem which says that any abstractly defined closed surface with positive curvature can be isometrically embedded in $\mathbb{R}^3$ as a convex surface.
>
> In the process of finding a normal form, there are local problems and global problems. In this article, we shall concentrate only on the problem of uniformization of geometric structures on smooth manifolds. Classically we have the famous uniformization theorem that any conformal structure on a surface can be normalized in such a way that it is the quotient of S^2, $\mathbb{R}^2$ or D^2 by a group of conformal transformations. While the program of "uniformization" originated from function theoretic consideration, we shall take a geometric viewpoint. (It should, however, be stressed that the function theoretic method is actually an important tool in a major part of the program that we are going to discuss.)
>
> Our goal would be to generalize these two-dimensional theorems to higher-dimensional manifolds. In our proposal of uniformization, we should have special models. A typical model has the following form. Let G be a Lie group acting transitively on a manifold $\tilde{M}$. Let Ω be a domain in $\tilde{M}$ and Γ be a subgroup of G so that Ω/Γ is compact. (In many cases, compactness can be replaced by the concept of completeness.) Then Ω/Γ is our model space.
>
> The question of uniformization can be phrased as follows: Given a manifold M with certain structure (related to G) can we find a model space Ω/Γ so that M is isomorphic to Ω/Γ? Most of the known uniformization theorems are existent in nature; it is normally very difficult and interesting to give an explicit description of the uniformization. An interesting example is the following: Let M be an algebraic surface defined by an ideal of homogeneous polynomials in the complex projective space. Suppose that we know that M is uniformized by the ball in $\mathbb{C}^n$. Can we calculate the covering map mapping the ball onto M? This is clearly a difficult question and is largely unknown. In this article, we shall only discuss the existence problem.
>
> There are several levels of questions we can ask. The fundamental

question is to give an intrinsic topological characterization of manifolds which can be uniformized. An example is the famous theorem of Thurston that, in three dimensions, a compact manifold which is "atoroidal" and "sufficiently large" admits a hyperbolic structure and can be uniformized. For dimension greater than three, no theorem of this type exists. It is certainly a challenging problem to find an analogue of Thurston's theorem in higher dimension and in a wider category.

The question of uniformization becomes more tractible when the manifold has certain special structures. The most notable structures are complex structures, conformally flat structures, affine structures and projective structures.

From the point of view of Riemannian geometry, the most homogeneous and isotropic manifolds are manifolds with constant curvature. They are isometrically covered by the sphere, the Euclidean space or the hyperbolic space. From the point of view of complex geometry, the corresponding spaces will be complex projective space, the complex Euclidean space or the complex hyperbolic space....

When the manifold has a complex structure, the question of uniformization becomes more classical. Of course when the dimension of the manifold is greater than one, there are many types of questions we can ask. The fundamental question is to find a good sufficient condition for an almost complex manifold to be deformable to a complex manifold....

I. The compact case. The model spaces would be compact Hermitian symmetric spaces. It is natural to ask whether the complex structure over these spaces is "rigid" or not....

II. Higher dimensional generalization of the complex torus and the complex Euclidean space....

III. Algebraic manifolds of noncompact type. As a step toward understanding the topology of compact algebraic manifolds with infinite fundamental group, we propose to show that the universal cover of such a manifold is always birational to a subdomain of another compact algebraic manifold....

20 Calabi conjecture and Kähler metrics

The Calabi conjecture was first announced in 1954 as a theorem in the short note titled *The space of Kähler metrics*, Proc. Internat. Congress Math. Amsterdam 2, pp. 206–207. Later it was turned into a conjecture in 1957. His motivation was to find good metrics on complex manifolds, and Kähler-Einstein metrics are the optimal ones.

In [Ber, p. 21], Berger wrote:

> Eugenio Calabi is a geometer. What geometry is being hard to define today, I have quoted above three masters on the subject. Note that Weil is refereeing to Monge as well as Elie Cartan.
>
> Let me add that a number theorist is said to be the personal friend of every prime number. I can risk saying that Calabi is the personal friend of every geometric problem and of every geometrical concept.

Calabi is a very original mathematician, a rare combination of an analyst and a geometer. Besides his famous conjecture on Kähler metrics, he also made important contributions to several other topics in geometry. For example, Calabi pioneered the study of rigidity of locally symmetric space using the method of differential geometry. In an influential paper with Vesentini, they proved the rigidity of the complex structure of a compact quotient of a bounded symmetric domain. Calabi also proved deformation rigidity of compact hyperbolic spaceforms of dimension at least 3. He also made several pioneering contributions to the modern theory of affine differential geometry, and was the first one to use ideas of twistor theory in the study of minimal surfaces.

In connection with Kähler-Einstein metrics, Calabi introduced the important concept of extremal Kähler metrics. He also constructed some of the most interesting non-Kähler complex manifolds.

21 Applications of Yau's solution of Calabi conjecture in geometry and physics

As is known, the uniformization theorem for Riemann surfaces gives each compact Riemann surfaces of genus $g \geq 2$ a unique hyperbolic metric, and this metric provides an effective tool to understand Riemann surfaces and the related moduli space $\mathcal{M}_g$ and Teichmüller space $\mathcal{T}_g$. This was clearly visible in the earlier works of Klein, Poincaré, Fricke, Nielsen, Dehn etc., and also transparent in the more recent work of Thurston on surfaces and three dimensional manifolds.

The existence of Kähler-Einstein metrics on complex manifolds can naturally be considered as the *uniformization of complex manifolds of high dimensions.* This is an interesting and important result by itself. Furthermore, it also has many applications.

The paper of Yau in 1977 included in this book contains several striking applications. The following is a more updated and complete list:

(1) There exist compact simply-connected Kähler manifolds whose Ricci curvature is identically zero. Examples are given by a complex hypersurface in $\mathbb{C}P^n$ of degree $n+2$. Note that complex tori $\Lambda\backslash\mathbb{C}^n$ are flat Kähler manifolds but clearly not simply connected. The existence of the above simply connected manifolds is not obvious at all.

(2) There exist compact-simply connected Kähler manifolds whose Ricci curvature is equal to a negative constant.

(3) If M^n is a Kähler manifold with the first Chern class $c_1(M)$ negative or zero, then
$$(-1)^n c_1^{n-2}(M) c_2(M) \geq (-1)^n \frac{n}{2(n+1)} c_1^n(M).$$
This is the celebrated Yau inequality. When M is a Kähler surface with ample canonical bundle, it specializes to
$$3c_2(M) \geq c_1(M)^2.$$
This is usually called the Miyaoka-Yau inequality.

(4) Assume that M is a Kähler manifold with the first Chern class $c_1(M)$ negative or zero. Then the equality in the Yau inequality holds if and only if the universal covering space of M is biholomorphic to the unit ball in $\mathbb{C}^n$ if $c_1(M) < 0$, and the complex vector space $\mathbb{C}^n$ if $c_1(M) = 0$. This gives a natural characterization of such important Hermitian locally symmetric spaces. For algebraic surfaces of general type, the Yau inequality was proved independently by Miyaoka [Mi1]. But even in this special case, the characterization of the equality case was not given there, and has not been proved by methods from algebraic geometry up to now.

(5) If a complex surface M is homotopic to the complex projective plane $\mathbb{C}P^2$, then M is biholomorphic to $\mathbb{C}P^2$. This was conjectured by Severi. It also reminds one of the well-known Borel conjecture in geometric topology that two homotopic closed aspherical manifolds are homeomorphic. The proof of this result raised the problem of finding all fake projective planes. See the paper [PrY] for references and details.

(6) The complex projective plane $\mathbb{C}P^2$ is the only simply connected algebraic surface with positive definite index.

(7) If a Kähler manifold M is homeomorphic to the complex projective space $\mathbb{C}P^n$, then it is biholomorphic to $\mathbb{C}P^n$.

(8) Let N be a compact complex surface that is a quotient of the unit ball in $\mathbb{C}^2$. Then any complex surface M that is oriented homotopically equivalent to N is biholomorphsically isomorphic to it. This reminds one of the Mostow strong rigidity, which in fact was used in the proof. Based on this, Yau conjectured that compact Kähler manifolds with negative sectional curvature is rigid if its dimension is greater than or equal to two. Yau suggested to use harmonic maps to settle this, and Siu [Siu80] carried this out in a special case with a stronger curvature assumption.

(9) An easy consequence of Yau's solution to the Calabi conjecture gives the following structure theorem for Calabi-Yau manifolds: If M is a compact Kähler manifold with vanishing first Chern class, then there exists a finite covering of M which is the product of a complex torus with a simply connected Kähler manifold, whose first Chern class also vanishes.

(10) Proof of the Torelli theorem for $K3$-surfaces [Tod1] together with the fact that every $K3$-surface is Kähler [Siu83] (the Torelli theorem for algebraic $K3$-surfaces [PiS], Kähler $K3$-surfaces [BuR] was known).

(11) Proof of the surjectivity of the period map of $K3$-surfaces [Tod1], [Kul].

(12) The moduli space of compact Calabi-Yau manifolds M is smooth, i.e., every element in $H^1(M,T)$ is tangent to a uniquely determined complex path of deformation of complex structures on M, and hence the moduli space of the complex structures on M has canonical coordinates. (See also §28 below.)

(13) Calabi-Yau threefolds M give rise to important examples of models of universe with spacetime supersymmetry, and the non-linear σ-model based on these manifolds gives rises to a quantum cohomology group, i.e., a quantum deformation of the cohomology ring structure on $H^{1,1}(M)$. (More details are given later in §29.)

22 Yau's integration of analysis and geometry

Yau's solution to Calabi conjecture is one part of his contribution to geometric analysis. Throughout his work, he integrated geometric analysis with differential geometry, algebraic geometry and mathematical physics. He brought nonlinear differential equations to several subjects for the first time and solved many outstanding problems. The scope of his work is well summarized by the citations of several major prizes he has received.

When Yau received Crafoord Prize in 1994, a description of his work appeared in Notices of AMS. Let us quote from it [Yau3, pp. 795-796]:

> Differential geometry is a new branch of geometry which was developed in the last century. In the beginning, one worked mainly with curves, and, when formulating their properties, one used ordinary differential equations. Modern differential geometry studies manifolds, which include surfaces, and their global structure. Nonlinear partial differential equations emerge when one studies these global properties in terms of curvatures. Examples of such equations are minimal surfaces equations and Yang-Mills equations.
>
> As a result of Yau's work over the past twenty years, the role and understanding of basic partial differential equations in geometry has changed and expanded enormously within the field of mathematics. His work has had, and will continue to have, a great impact on areas of mathematics and physics, representation theory, and general relativity as well as differential geometry and partial differential equations.
>
> One of Yau's most important works is his solution for the the Calabi conjecture... These results represent a major step forward in the uniformization problem for complex manifolds...

> Traditionally, minimal surfaces were studied by geometers for their own intrinsic beauty. Yau was the first to pioneer the application of minimal surfaces to questions in geometry, mathematical physics, and topology.

The citation of the Crafoord Prize for Yau reads:

> The Royal Swedish Academy of Sciences awards the 1994 Crafoord Prize to differential geometry.
>
> The Prize is awarded to ... Shing-Tung Yau, Harvard University, Cambridge, MA, USA, for his development of non-linear techniques in differential geometry leading to the solution of several outstanding problems.
>
> Modern geometry is divided in two main directions, differential geometry which takes distances into account, and topology, which only cares about shape. Gradually it has become clear that there is in topology a clear division between geometry in five dimensions or more on the one hand, and geometry in lower dimensions on the other. In the fifties and sixties topology in five and higher dimensions was investigated and today could be said to be well understood. As for low dimensions, from one to four, they have all turned out to have their own peculiarities. One and two dimensions are, because of their perspicuity, relatively simple. Three-dimensional geometry, through considerably more complicated, behaves much like that of lower, particularly two, dimensions. Geometry of four dimensions seems to be a boundary case. It differs fundamentally from lower dimensions while the methods used for higher dimensions do not work.
>
> The study of five or more dimensions shows that in topology there is reason to take into account how strongly the form of geometrical objects are permitted to be changed. Continuity is the notion that is used when one allows the sharpest possible changes, differentiability the one used when only smooth changes are accepted. The understanding of this distinction in these dimensions is part of the achievements obtained in the fifties and sixties while this distinction may not be seen in one, two and three dimensions.
>
> Thanks to Shing-Tung Yau's work over the past twenty years, the role and understanding of the basic partial differential equations in geometry has changed and expanded enormously within the field of mathematics. His work has had an impact on areas of mathematics and physics as diverse as topology, algebraic geometry, representation theory, and general relativity as well as differential geometry and partial differential equations. Yau is a student of legendary Chinese mathematician Shiing-Shen Chern, for whom he studied at Berkeley. As a teacher he is very generous with his ideas and he has had many students and also collaborated with many mathematicians.

Yau received the Wolf prize in 2010, and the citation reads:

> THE 2010 WOLF FOUNDATION PRIZE IN MATHEMATICS
>
> The Prize Committee for Mathematics has unanimously decided that the 2010 Wolf Prize will be awarded to Shing-Tung Yau for his work in geometric analysis that has had a profound and dramatic impact on many areas of geometry and physics.
>
> Shing-Tung Yau has linked partial differential equations, geometry, and mathematical physics in a fundamentally new way, decisively shaping the field of geometric analysis. He has developed new analytical tools to solve several difficult nonlinear partial differential equations, particularly those of the Monge-Ampere type, critical to progress in Riemannian, Kähler and algebraic geometry and in algebraic topology, that radically transformed these fields. The Calabi-Yau manifolds, as these are known, a particular class of Kähler manifolds, have become a cornerstone of string theory aimed at understanding how the action of physical forces in a high-dimensional space might ultimately lead to our four-dimensional world of space and time. Prof. Yau's work on T-duality is an important ingredient for mirror symmetry, a fundamental problem at the interface of string theory and algebraic and symplectic geometry. While settling the positive mass and energy conjectures in general relativity, he also created powerful analytical tools, which have broad applications in the investigation of the global geometry of space-time.
>
> Prof. Yau's eigenvalue and heat kernel estimates on Riemannian manifolds, count among the most profound achievements of analysis on manifolds. He studied minimal surfaces, solving several classical problems, and then used his results, to create a novel approach to geometric topology. Prof. Yau has been exceptionally productive over several decades, with results radiating onto many areas of pure and applied mathematics and theoretical physics. In addition to his diverse and fundamental mathematical achievements, which have inspired generations of mathematicians, Prof. Yau has also had an enormous impact, worldwide, on mathematical research, through training an extraordinary number of graduate students and establishing several active mathematical research centers.

23 Yau's conjecture on Kähler-Einstein metrics on Fano manifolds

Unlike the 1-dimensional case, Kähler-Einstein metrics may not exist on compact complex manifolds whose first Chern class is positive, and the issue is quite subtle. There are obstructions to the existence, and nailing down the exact condition is not easy.

According to the paper [Yau1, p. 5] by Yau:

> When the first Chern class is positive, it is called a Fano manifold.... In this case, there is an obstruction for the existence of Kähler-Einstein metric due to Matsushima [60]: the Lie algebra of the automorphism group of the manifold must be reductive. On [28], Futaki introduced his beautiful invariant defined on this Lie algebra. The Futaki invariant soon became a fundamental tool to study Kähler-Einstein metric on Fano manifolds. On the other hand, It took a long while to find a necessary and sufficient condition for the existence of Kähler-Einstein metric on Fano manifolds....
>
> Right after I proved the Calabi conjecture on the existence of Kähler metric with prescribed volume form, I tried to work on the problem of the existence of Kähler-Einstein metric on Fano manifolds....
>
> In the meanwhile, in 1977, I realized that Bogomolov [8] used the concept of stability of bundles to prove Chern number inequalities for algebraic surfaces which were sharpened by Miyaoka [61] and myself [84] independently. I started to believe there has to be links between the concept of stability with the existence of Hermitian Yang-Mills connections on bundles....
>
> The Bogomolov inequality is optimal for general stable bundles. But it is not as sharp as the Miyaoka-Yau inequality when applied to the tangent bundle of the manifold. Hence I suspected that existence of Kähler-Einstein metric should be considered as a nonlinear version of the existence of Hermitian Yang-Mills connection, and the stability of bundle should be replaced by manifold stability. Therefore, only in 1980, I realized that the right condition for existence of Kähler-Einstein metric is the stability of the algebraic manifold.
>
> I made the conjecture that the existence of Kähler-Einstein metric is equivalent to stability of manifold. I told all my graduate students about this conjecture, especially to Gang Tian who showed interest in the problem of Kähler-Einstein metric. But it took a long time to convince him of the validity of my conjecture.

In [Do6], Donaldson wrote:

> One of the most influential questions raised by Yau in his problem lists concerns the existence of Kähler-Einstein metrics on Fano manifolds. In open problems in geometry, Problem 65, Yau conjectured that this should be equivalent to stability of the manifold in the algebro-geometric sense. Important progress was made by Tian [52]. On the personal level, this conjecture of Yau has been the guiding light for the bulk of the author's own work over the past 15 years and the conjecture was recently verified in joint work with Chen and Sun [15]. The result can be viewed as a version of the Kobayashi-Hitchin correspondence but one at a higher level of difficulty from those, related to Yang-Mills theory, considered above. In one direction, this difficulty is reflected in the fact that the relevant equations are fully nonlinear

(indeed, Yau has suggested that the Hermitian Yang-Mills equations can be seen as a partial linearisation of the Kähler-Einstein equations). In another direction, this difficulty is related to the fact that relevant infinite dimensional group K is the group of symplecto-morphisms and the complexification K^c does not exist. However one can still view the space of Kähler metrics (in a fixed cohomology class) as a symmetric space, formally equal to K^c/K. This leads to many interesting and challenging questions, for example involving the geodesics in K^c/K, as studied by Chen [13] and others.

The resolution of Yau's conjecture is really the beginning rather than the end of a story. One important problem is to understand the stability condition better, so that it can be verified in concrete cases. Another, as raised by Yau in open problems in geometry, Problem 88, is to understand the limits of Kähler-Einstein metrics and to relate these to algebraic geometry.

24 Hermitian Yang-Mills connections and Donaldson-Uhlenbeck-Yau correspondence

In studying differential manifolds M, the tangent vector bundles TM play an essential role. Many important invariants of manifolds are also defined through vector bundles over the manifolds. For example, K-groups of manifolds are defined directly using vector bundles over the manifolds. Therefore, it is natural that there is a lot of similarity between structures on manifolds and vector bundles.

As we discussed above, Riemannian metrics are basic structures on smooth manifolds, and Kähler-Einstein metrics are special metrics on complex manifolds for many applications. For vector bundles over manifolds, connections are basic structures on vector bundles. Besides their applications in physics, connections also allow one to pick out preferred sections. The analogue of Kähler-Einstein metrics for complex vector bundles over complex manifolds is Hermitian-Einstein connections, or Hermitian Yang-Mills connections.

As Uhlenbeck and Yau wrote in their paper [UY, p. S258]:

> The Hermitian Yang-Mills connection plays the same role for vector bundles as the Kähler-Einstein metric for complex manifolds. Historically, Calabi [6] based his conjectures on the same variational formalism used in Yang-Mills theory. He considered the L^2-norm of the curvature of the metric and showed that under the appropriate assumptions the critical metrics must be Kähler-Einstein. The same reasoning leads from Yang-Mills equations to Hermitian Yang-Mills connections. Likewise, the existence of an Hermitian Yang-Mills connection implies the Bogomolov inequality and strengthens the results by giving insight into the case of equality.

Since Yau's solution to Calabi conjecture motivated or inspired a lot of work and applications in the broad subject of geometric analysis, it is not surprising that it also has some definite impact on the question of existence of such connections on the vector bundles, especially given the similarity between manifolds and vector bundles.

It turns out that Hermitian Yang-Mills connections do not always exist on complex vector bundles over complex manifolds. There is a stability condition on the vector bundles.

As it is known in algebraic geometry [Mu1], in constructing moduli spaces of algebraic varieties and vector bundles over algebraic varieties, only those which are stable in a suitable sense can be considered in order to get Hausdorff moduli spaces.

It turns out that holomorphic vector bundles over compact Kähler manifolds admit Hermitian-Einstein connections, or Hermitian Yang-Mills connections, if and only if they are stable in a suitable sense. This result has been proved by several people, and decisive results are contained in papers of Donaldson [Do2], and Uhlenbeck-Yau [UY] included in this book. This is an important result for many reasons, both in mathematics and physics. It often goes under the name *Donaldson-Uhlenbeck-Yau correspondence* or *Donaldson-Uhlenbeck-Yau theorem.*

Let X be a compact complex manifold, and V a complex vector bundle over X. Assume that X is a compact Riemann surface. In [NS, p. 540], Narasimhan and Seshadri proved the following result:

> A holomorphic vector bundle of degree zero on X is stable if and only if it arises from an irreducible unitary representation of the fundamental group of X. As a consequence one sees that a holomorphic vector bundle on X arises from a unitary representation of the fundamental group of X if and only if each of its indecomposable components if of degree zero and stable.

By making use of the above result of Narasimhan and Seshadri, Atiyah and Bott established in [AB1, p. 11] and [AB2] a direct relation between Yang-Mills connections and stable vector bundles over Riemann surfaces. Using the reformulation of the result of Narasimhan and Seshadri in terms of Yang-Mills connections by Atiyah and Bott, Donaldson gave a new proof of the theorem of Narasimhan and Seshadri in [Do1].

When the complex manifold X is of dimension greater than or equal to 2, the result that the existence of an Hermitian-Einstein connection implies the stability was proved by Kobayashi in [Kob2] and Lübke in [Lu1].

The other direction that stable vector bundles over X admit Hermitian-Einstein connections is much harder. It was proved by Donaldson [Do2] when X is an algebraic surface.

As mentioned before, there is a close analogy between manifolds and vector bundles. Therefore, it is not surprising that Yau proceeded to attack this problem on the existence of Hermitian Yang-Mills connection right after he solved the Calabi conjecture on Kähler-Einstein metrics. In publications in early 1980s, Kobayashi and Hitchin also became interested in this problem based on the same motivation, though they had no specific ideas on how to approach the problem. But the real breakthrough was in the paper of Uhkenbeck and Yau [UY] which was finished in 1984 and printed in 1986. This has been the only proof that works for general Kähler manifolds.

Specifically, according to the paper [Yau1, p. 13] by Yau:

> In December of 1977, when I was preparing the talk for the ICM in Helsinki, I thought about the possible conditions for existence of Hermitian Yang-Mills connections. I concluded that it had to be related to the slope stability of the holomorphic bundle, as was motivated by the work of Bogomolov and Miyaoka on Chern number inequalities. I was informed much later that this possibility was also believed to be true by Hitchin and Kobayashi.

In the same paper [Yau1, p. 5], Yau wrote:

> Right after I proved the Calabi conjecture on the existence of Kähler metric with prescribed volume form, I tried to work on the problem of the existence of Kähler-Einstein metric on Fano manifolds....
>
> In the meanwhile, in 1977, I realized that Bogomolov [8] used the concept of stability of bundles to prove Chern number inequalities for algebraic surfaces which were sharpened by Miyaoka [61] and myself [84] independently. I started to believe there has to be links between the concept of stability with the existence of Hermitian Yang-Mills connections on bundles....
>
> The Bogomolov inequality is optimal for general stable bundles. But it is not as sharp as the Miyaoka-Yau inequality when applied to the tangent bundle of the manifold. Hence I suspected that existence of Kähler-Einstein metric should be considered as a nonlinear version of the existence of Hermitian Yang-Mills connection, and the stability of bundle should be replaced by manifold stability. Therefore, only in 1980, I realized that the right condition for existence of Kähler-Einstein metric is the stability of the algebraic manifold.

Furthermore, Yau wrote on [Yau1, p. 13]:

> As was acknowledged by Donaldson, the argument of Uhlenbeck-Yau is most natural and in fact, all the later development for Hermitian Yang-Mills connections for higher dimensional manifolds are based on the procedure of Uhlenbeck-Yau.
>
> Some later paper such as the one by Bando-Siu [5] used the Hermitian Yang-Mills flow to generalize our result, but the essential feature of Uhlenbeck-Yau procedure is still needed in an essential manner. It should also be pointed out that the continuity argument used by Uhlenbeck-Yau is just as convenient as the Hermitian Yang-Mills flow.
>
> A few years later, Carlos Simpson [69] generalized the Uhlenbeck-Yau argument to establish similar theorem when the Higgs field was introduced. Hermitian Yang-Mills connections were proposed by me to

Edward Witten in 1984 to study heterotic string, which had since became an important subject in mathematical physics. But from the very beginning, we knew the importance of Hermitian Yang-Mills connections, as it provides important Chern number inequalities, and also the conditions for the bundle to be projectively flat.

An email of Yau to many of us on November 21, 2015, contains some detail about the proof:

> The Uhlenbeck-Yau theorem is one of the most important and difficult theorems I have ever proved as the regularity was not easy at all! The applications are extremely important. It took many years for us to achieve the complete proof! Karen came to visit me many times and even locked herself up for a few days to think about the problem. I should say that the argument that we used is still the ONLY approach to solve this problem. This includes the later work on Higgs field by Simpson and Bando-Siu and many others. There are always ways to put the proof in a different way, using Hermitian Yang-Mills flow, for example. But the key argument is still the part that Uhlenbeck-Yau developed. (People tried to bury our part of argument within their proof, without mentioning that is the key part in the preface of their paper.) Up to now, nobody is able to avoid the argument devised by Uhlenbeck-Yau! The theorem of Higgs field by Simpson needs this approach in an essential manner and so are all works on non-Kähler geometry. It took Uhlenbeck and me many years of hard work to achieve this result.
>
> In 1984, I told Witten this work and suggested that it would be important for string theory. He did not believe it at the beginning, and wrote a very important paper in 1986 after he saw the importance of the work. And he called it DUY theorem.

In an email on March 10, 2016, Eduard Witten wrote:

> The Donaldson-Uhlenbeck-Yau theorem on existence of Hermitian Yang-Mills fields is very important in string theory because of its use in constructing supersymmetric compactifications of the heterotic string. The generic supersymmetric compactification of the heterotic string is constructed, at the field theory level, by using a suitable solution of the Hermitian Yang-Mills equation, on a Calabi-Yau manifold. This is then the starting point for a quantum analysis.

In an email on March 11, 2016, a leading expert on string theory Andrew Strominger wrote:

> The DUY theorem plays a fundamental role in making contact between string theory and physical reality. It is generically required in the construction of compactifications of heterotic string theory from ten to four dimensions on Calabi-Yau manifolds with chiral fermions.

In an email on March 16, 2016, another leading expert on string theory Hirosi Ooguri wrote:

> The Donaldson-Uhrenbeck-Yau theorem has played fundamental roles in many aspects of string theory and gauge theory in theoretical physics. Since the string theory is proposed as the ultimate unified theory of forces and matter in nature, we need to find a way to connect the theory to the Standard Model of Particle Physics. One of the most promising approach involves Calabi-Yau manifolds with Hermitian Yang-Mills connections. The Donaldson-Uhrenbeck-Yau theorem provides an essential tool for physicists to describe and analyze them in a powerful way and to demonstrate that the string theory indeed contains all the theoretical ingredients for the Standard Model of Particle Physics. The theorem has also played significant roles in our understanding of dynamics of D-branes in string theory, properties of the mysterious $(2,0)$ theory in six dimensions, aspects of F-theory and of supersymmetric theories in diverse dimensions. The theorem has had broad and deep impacts in theoretical physics, and we are grateful to Professor Yau for foreseeing the importance of vector bundle in complex manifolds and for providing this indispensable tool for us.

For a detailed discussion of the application of the Donaldson-Uhlenbeck-Yau theorem in physics, see the initial seminar paper [Wit1] by Witten. After this paper of Witten in 1985, Donaldson-Uhlenbeck-Yau theorem has been a prevailing name, and it has been cited in more than 400 papers according to a Google search in physics literature. In particular, it has been discussed in the influential book in string theory [GSW, §15.6.2].

More specifically, in an email on March 10, 2016, the distinguished physicist Michael R. Douglas wrote:

> The Donaldson-Uhlenbeck-Yau theorem is one of the cornerstones of the theory of compactification of superstrings. In brief, the compactification problem is to postulate a six-dimensional manifold M and find a solution of the equations of motion of superstring theory on M times four-dimensional Minkowski space-time. To a good approximation, these are the equations of supergravity, which includes general relativity coupled to Yang-Mills theory and a number of antisymmetric tensor gauge fields. While intractable in general, if we assume that the solution has supersymmetry, an assumption with many physical motivations, the problem is solvable. One can show that M must be Kähler, and that the Yang-Mills gauge field is a connection on a Hermitian line bundle, in this case Yau's proof of the Calabi conjecture, and the Donaldson-Uhlenbeck-Yau theorem, allow completely solving the general relativity and Yang-Mills parts of the problem, and the rest reduces to fairly simple PDE's.
>
> This picture of heterotic string compactification on a Calabi-Yau manifold was put forward in 1985 by Candelas, Horowitz, Strominger and

> Witten [CHSW]. In the 30 or so years since, there have been hundreds of physics papers which find compactifications, analyze their testable predictions, and make mathematical conjectures based on physical reasoning. One much-studied problem [Wit1] is to find stable rank 3, 4 or 5 bundles with specified Chern classes $c_1 = 0$, $c_2 \leq c_2(M)$ and $c_3 = 6$. This is the case which leads to the standard model gauge group with three generations of quarks and leptons. Another important physics question is to find the moduli space of Yang-Mills solutions — by the DUY theorem, equivalently the moduli space of stable holomorphic bundles — and certain cubic and higher order structure constants which become the physical "Yukawa couplings". An introduction to these questions can be found in [Dou], and some important recent developments are [Bra], [Ble].
>
> Finally, string theory leads to conjectural generalizations of the DUY theorem and the associated concepts. One can generalize the Yang-Mills equations to the case of gauge fields on Dirichlet branes, leading to the equations of open string field theory. Following a line of work by Kontsevich, Douglas and Bridgeland, the DUY correspondence is conjectured to generalize as follows: supersymmetric solutions to the open string equations are in one-to-one correspondence with a certain subset of Bridgeland stable objects in the derived category of holomorphic sheaves on M [ABCD]. Another generalization of the Hermitian Yang-Mills equations is obtained by considering the world-sheet of the heterotic string. This leads to the definition of $(0, 2)$ sigma model. This model and conjectures such as $(0, 2)$ mirror symmetry are reviewed in [Sh].

In an email on March 17, 2016, another distinguished physicist Gregory Moore explained:

> In the context of supersymmetric field theories the equations for (classical) vacua can be divided into "F-term" equations and "D-term" equations. The F-term equations are typically holomorphic equations on a symplectic and holomorphic variety. The D-term equations say that the moment map of a unitary isometry group vanishes. The latter are much more difficult to solve in describing the vacuum and, thanks to geometric invariant theory, one can often replace the F- and D-term equations by the quotient of stable solutions of F-term equations with respect to a complexification of the isometry group. This maneuver has led to a great deal of progress in understanding the vacua of both string theories and supersymmetric field theories. One way to understand the Donaldson-Uhlenbeck-Yau theorem is that it is an infinite-dimensional generalization of this maneuver. In particular, the replacement of solutions of the Hermitian Yang-Mills equations by the study of stable holomorphic vector bundles has led to a great deal of progress in understanding the compactification of heterotic string theory (and in heterotic M-theory), just as Yau's theorem replaced the

> difficult task of analyzing Einstein's equations on six-manifolds by the relatively easier task of analyzing three-folds with vanishing canonical class. As a measure on the power of these theorems one need only compare what we know about the compactifications of string theory on Calabi-Yau manifolds with what is known about the compactifications on G_2- and $Spin_7$-holonomy manifolds.

As the name suggests, Hermitian Yang-Mills connection came from the work of Yang and Mills in physics. In fact, relations between Yang-Mills connections and holomorphic vector bundles were first observed by Yang in [Yan2]. According to [UY, S257]:

> The techniques of partial differential equations were used by Taubes [25], [26] to construct solutions on arbitrary four-manifolds. These results and compactness theorems of the first author were incorporated into a general theory of Simon Donaldson to obtain beautiful and spectacular results on the differential structures on four manifolds; see [10], [11]. It is an observation dating essentially back to Yang's formalism of using $\mathbb{C}^2$ instead of $\mathbb{R}^4$ that Taubes solutions give rise to holomorphic vector bundles over Kähler surfaces; see [30].

Therefore, as explained above by M. Douglas, it is not surprising that the theorem of Uhlenbeck-Yau is also used in a substantial way in physics. For example, according to [LY, p. 143]:

> In their proposed compactification of superstrings [4], Candelas, Horowitz, Strominger and Witten took the matrix product of a maximal symmetric four dimensional spacetime M with a six dimensional Calabi-Yau vacua X as the ten dimensional spacetime; they identified the Yang-Mills connection with the $SU(3)$ connection of the Calabi-Yau metric and set the dilaton to be a constant. To make this theory compatible with the standard grand unified field theory, Witten [28] and Horava-Witten [20] proposed to use higher rank bundles for strong coupled heterotic string theory so that the gauge groups can be $SU(4)$ or $SU(5)$. Mathematically, this approach relies on Uhlenbeck-Yau's theorem on constructing Hermitian Yang-Mills connections on stable bundles [27]. Many authors, including Friedman, Morgan and Witten [18]; Donagi, Ovrat, Pantev and Reinbacher [12]; Andreas [1], Kachru [21] and others, have worked on this subject since then.

As mentioned above, both the result and proof of the Donaldson-Uhlenbeck-Yau theorem have important applications. In [Ph, p. 1299, p. 1301], D. H. Phong wrote:

> The question is to determine when Hermitian-Einstein metrics exists. This is answered completely by the theorem of Donaldson-Uhlenbeck-Yau [22, 23, 24] ... Almost as soon as the Donaldson-Uhlenbeck-Yau theorem becomes available, it found some immediate applications in

> compactifications of superstrings, just as was the case for the solution of the Calabi conjecture [3]. The method of Uhlenbeck-Yau itself has developed into an essential tool for further developments of the theory. These include extensions to the case of Gauduchon metrics $g_{\bar{k}j}$, to noncompact manifolds and interactions with Higgs fields [27], and especially to the very important question of describing the singularities of the (generalized) Hermitian-Einstein metric in terms of the Harder-Narashimhan-Seshadri filtration of the bundle E when E is not stable [28, 29, 30]. Very recently, there has been considerable progress on this last question.

In the history of mathematics, concepts, conjectures, and theorems have been introduced, made, observed, and proved by several people at about the same time from different perspectives. One explanation is that the time is ripe. This often is a sign that the thing under discussion is interesting and important. The great mathematician Kolmogorov edited a series of books on the history of the 19th century [Kol2]. It contains some interesting and insightful comments in the first volume of this series [Kol2, p. 36]:

> Here the historian of science is faced with the task of reconstructing the mysterious process of birth, growth and interaction of mathematical ideas on the basis of notes, papers, letters and recollections. What makes this task particularly difficult is that in the history of every great mathematical thought there is a period of implicit existence when, unbeknownst to contemporaries, it begins to turn up here and there in the masquerade garb of special cases and applications. And when it suddenly appears in all its beauty, it is not immediately possible—and at times altogether impossible—to determine who helped it make the decisive step of appearing on the stage.

A leading mathematics historian Jeremy Gray also made an interesting comment in a book on the geometric function theory titled *Linear differential equations and group theory from Riemann to Poincaré* [Gra2]. He wrote on [Gra2, p. xviii]:

> Mathematics is not just a body of results, each one attached to a technical argument, but an intricate system of theories and a historical process. History which proceed from problems to results and leave out the methods by which those results were achieved omit a crucial aspect of the process of this theoretical development.

25 Arithmetic and modularity of elliptic curves and Calabi-Yau manifolds

As we mentioned before, elliptic integrals and elliptic curves have played a central role in the development of complex geometry. It is well-known that the arithmetic of elliptic curves, especially structure of rational points of elliptic curves and the modularity of elliptic curves, has played a central role in number theory since the end of the 19th century.

Since Calabi-Yau manifolds and K3 surfaces are higher dimensional generalizations of elliptic curves, it is expected that the arithmetic of K3 surfaces and Calabi-Yau manifolds are interesting and should be taken seriously. For example, this point of view is emphasized in the book [Hus].

25.1 Arithmetic of elliptic curves

It was conjectured by Poincaré and proved by Mordell that the group of rational points of each elliptic curve is finitely generated. Though it has been intensively studied, many questions about this group are still open. For example, the famous Swinnerton-Dyer conjecture for elliptic curves is still open in spite of a lot of progress and partial results.

In [Wi], Wiles wrote:

> A polynomial relation $f(x, y) = 0$ in two variables defines a curve C_0. If the coefficients of the polynomial are rational numbers, then one can ask for solutions of the equation $f(x, y) = 0$ with $x, y \in \mathbb{Q}$, in other words for rational points on the curve. If we consider a non-singular projective model C of the curve, then topologically it is classified by its genus, and we call this the genus of C_0 also. Note that $C_0(\mathbb{Q})$ and $C(\mathbb{Q})$ are either both finite or both infinite. Mordell conjectured, and in 1983 Faltings proved, the following deep result.
>
> **Theorem** ([9]). *If the genus of C_0 is greater than or equal to 2, then $C_0(\mathbb{Q})$ is finite.*
>
> As yet the proof is not effective so that one does not possess an algorithm for finding the rational points. (There is an effective bound on the number of solutions but that does not help much with finding them.)
>
> The case of genus zero curves is much easier and was treated in detail by Hilbert and Hurwitz [12]. They explicitly reduce to the cases of linear and quadratic equations. The former case is easy and the latter is resolved by the criterion of Legendre. In particular, for a non-singular projective model C we find that $C(\mathbb{Q})$ is non-empty if and only if C has p-adic points for all primes p, and this in turn is determined by a finite number of congruences. If $C(\mathbb{Q})$ is non-empty, then C is parametrized by rational functions and there are infinitely many rational points.
>
> The most elusive case is that of genus 1. There may or may not be rational solutions and no method is known for determining which is the case for any given curve. Moreover when there are rational solutions there may or may not be infinitely many. If a non-singular projective model C has a rational point, then $C(Q)$ has a natural structure as an abelian group with this point as the identity element. In this case we call C an elliptic curve over $\mathbb{Q}$. (For a history of the development

of this idea see [19].) In 1922, Mordell [15] proved that this group is finitely generated, thus fulfilling an implicit assumption of Poincaré.

Theorem. *If C is an elliptic curve over $\mathbb{Q}$, then*

$$C(\mathbb{Q}) \cong \mathbb{Z}^r \oplus C(\mathbb{Q})^{tors}$$

for some integer $r \geq 0$, where $C(\mathbb{Q})^{tors}$ is a finite abelian group.

The integer r is called the rank of C. It is zero if and only if $C(\mathbb{Q})$ is finite. We can find an affine model for the curve in Weierstrass form

$$C : y^2 = x^3 + ax + b$$

with $a, b \in \mathbb{Z}$. We let Δ denote the discriminant of the cubic and set

$$N_p := \#\{\text{solutions of } y^2 \equiv x^3 + ax + b \mod p\},$$

$$a_p := p - N_p.$$

Then we can define the incomplete L-series of C (incomplete because we omit the Euler factors for primes $p|2\Delta$) by

$$L(C, s) := \prod_{p \nmid 2\Delta} (1 - a_p p^{-s} + p^{1-2s})^{-1}.$$

We view this as a function of the complex variable s and this Euler product is then known to converge for $\mathrm{Re}(s) > 3/2$. A conjecture going back to Hasse (see the commentary on 1952(d) in [26]) predicted that $L(C, s)$ should have a holomorphic continuation as a function of s to the whole complex plane. This has now been proved ([25], [24], [1]). We can now state the millenium prize problem:

Conjecture (Birch and Swinnerton-Dyer). *The Taylor expansion of $L(C, s)$ at $s = 1$ has the form*

$$L(C, s) = c(s-1)^r + \textit{higher order terms},$$

with $c \neq 0$ and $r = \mathrm{rank}(C(\mathbb{Q}))$.

In particular this conjecture asserts that $L(C, 1) = 0 \iff C(\mathbb{Q})$ is infinite.

Some other problems about rational points of elliptic curves are also open. For example, Mazur and Rubin wrote in [MR, p. 185]:

> Eighty years have passed since Mordell proved that the (Mordell-Weil) group of rational points on an elliptic curve F is finitely generated, yet so limited is our knowledge that we still have no algorithm guaranteed to compute the rank of this group.

Faltings' solution to the Mordell conjecture is especially interesting from the perspective of complex geometry, since the proof depends on both the hyperbolicity of the geometry of Riemann surfaces of negative Euler number and moduli spaces of algebraic curves.

The close relation between the arithmetic and complex geometric properties of algebraic curves was conjectured to hold for some algebraic varieties of higher dimensions in a suitably interpreted way by Lang. See his two survey papers [La1], [La2] and his book [La3], in particular Chapter 3. These results and conjectures shed a different light on the complex geometry. Specifically, Lang wrote on [La3, p. 176]:

> Complex differential geometry intervenes in diophantine problems through several factors. First, if one considers holomorphic families of varieties, the problem of determining whether there exist only finitely many sections can be studied from a complex geometric point of view. But it also turns out (conjecturally at the moment) that the property of being Mordellic for a projective variety can be characterized in terms of purely complex differential geometric invariants, or complex analytic invariants. For instance, I conjectured that a projective variety X defined over a subfield of $\mathbb{C}$ finitely generated over the rationals is Mordellic if and only if every holomorphic map of $\mathbb{C}$ into $X(\mathbb{C})$ is constant. It is known in many cases that certain projective varieties have this holomorphic property, but except for curves of genus ≥ 2 or subvarieties of abelian varieties which do not contain translations of abelian subvarieties of dimension > 0 (Faltings' theorems) or varieties derived from those by products or unramified coverings or quotients, it is not known that they are Mordellic. Thus one obtains complex analytic criteria for a variety to be Mordellic. Similarly, in Chapter IX, we shall get quantitative diophantine criteria by inequalities at one absolute value.
>
> By now, we see a pattern emerging, that *certain global diophantine properties* of a variety are controlled conjecturally, in first approximation and qualitatively, by *geometric conditions* (the Mordellicity of the complement of the special set), and also by conditions at one archimedean absolute value, which is a local condition.
>
> In addition, to deal with quantitative estimates, Vojta has taught us that not only was there a classical analogy between algebraic numbers and algebraic functions, but also there is an equally deep seated analogy between algebraic numbers and holomorphic functions, via Nevanlinna theory.

As it is well known, Wiles together with Taylor proved the last Fermat theorem by proving the modularity of some elliptic curves [Wi2], [TW], and the modularity of all elliptic curves defined over rational numbers was proved in [BCDT].

One natural generalization of elliptic curves consists of abelian varieties, and their arithmetic has also been extensively studied. See [La3] for example.

The arithmetic of general algebraic varieties is difficult. Given that Calabi-Yau manifolds, in particular K3 surfaces, are generalizations of elliptic curves, it is an important and interesting problem to understand the arithmetic of them. Arithmetic Hermitian locally symmetric spaces, or rather Shimura varieties, are natural generalizations of curves of higher genus (see [La1]), and their arithmetic has also been extensively studied. In studying the arithmetic of such varieties, the complex and algebraic geometric properties of the underlying spaces always play a crucial role.

Though K3 surfaces are relatively simple algebraic surfaces, their arithmetic is rich and waiting to be explored. Only for special K3 surfaces, more is known. In [Swi], Swinnerton-Dyer wrote under the section titled *K3 surfaces and Kummer surfaces*:

> K3 surfaces are the simplest kind of variety about whose number-theoretic properties very little is known; indeed they still present many unsolved problems even to the geometer. There are infinitely many families of K3 surfaces; the simplest of them, and the only one which will be considered in the present article, consists of all nonsingular quartic surfaces. An important special type of K3 surfaces consists of the Kummer surfaces, a phrase which can carry either of two related meanings:
>
> (1) The quotient of an abelian surface A by the automorphism -1; this has 16 singular points corresponding to the 16 2-division points of A.
>
> (2) The nonsingular surface obtained by blowing up the 16 singular points in the previous definition.
>
> One advantage of Kummer surfaces in comparison with general K3 surfaces is that for the former it is easy to determine $\mathrm{Pic}(\overline{V})$.

25.2 Modularity of Calabi-Yau manifolds

The arithmetic of Calabi-Yau manifolds, in particular the modularity of Calabi-Yau manifolds, has been studied but relatively little is understood. Maybe this is related to inadequate understanding of the geometry of Calabi-Yau manifolds. (Note that in studying arithmetic of abelian varieties, subabelian varieties play an important role. How about some special subvarieties of Calabi-Yau manifolds?)

According to Yau [Yau5, pp. 287-288]:

> In search of Calabi-Yau manifolds that distinguish themselves from the rest, their modularity become the focus of some researchers. This is interesting from the perspective of arithmetic geometry.
>
> For a d-dimensional projective Calabi-Yau manifolds X defined over $\mathbb{Q}$, it is said to be *modular* if the L-function of the Galois representation on the middle ℓ-adic étale cohomology group $H^d_{et}(X_{\bar{\mathbb{Q}}}, \mathbb{Q}_\ell)$ is equal to the product of L-functions of modular forms up to factors associated to bad primes. Part of Langlands philosophy is the conjecture that all motives,

in particular our X, are modular. When d is even, $H^d_{et}(X_{\bar{\mathbb{Q}}}, \mathbb{Q}_\ell)$ contains $d/2$-dimensional algebraic cycles, and the interesting part is the modularity of the sub-representation on the orthogonal complement of the images of algebraic cycles.

Calabi-Yau varieties of dimension 1 are elliptic curves. The modularity of elliptic curves over $\mathbb{Q}$ has been established by Wiles [284], and Taylor–Wiles [266]. They proved that the two-dimensional Galois representation associated to an elliptic curve over $\mathbb{Q}$ does come from a weight $2 = d + 1$ modular form.

Dimension 2 Calabi-Yau varieties are K3 surfaces. For a K3 surface X defined over $\mathbb{Q}$, $H^d_{et}(X_{\bar{\mathbb{Q}}}, \mathbb{Q}_\ell)$, which has dimension 22, factors into a direct sum $(NS(X) \oplus T(X)) \otimes \mathbb{Q}$ of the Néron-Severi group of algebraic cycles $NS(X)$ and the group of transcendental cycles $T(X)$.

There are partial result for small rank $T(X)$. The lattice $NS(X)$ has rank at most 20; when it is 20, X is called a *singular* K3 surface. (The term "attractive" K3 is sometimes also used in physics [223].) In this case, $T(X)$ has rank 2 and defines a two-dimensional Galois subrepresentation and the associated L-function $L(T(X), s)$. The modularity of $L(T(X), s)$ has been established by Livné [204] that $L(T(X), s)$ does come from a weight $3 = d + 1$ modular form of CM type. When $T(X)$ has rank 3, its modularity follows from the modularity of elliptic curves, because $T(X)$ is endowed with an orthogonal pairing, so that it is essentially the symmetric square of a $GL(2)$ representation.

For Calabi-Yau threefolds defined over $\mathbb{Q}$, much is known for the rigid case, where there is no complex structure deformation. More specifically, we say that a Calabi-Yau threefold X is rigid if the $H^3_{et}(X_{\bar{\mathbb{Q}}}, \mathbb{Q}_\ell)$ has dimension 2. In this case, there is a two-dimensional Galois representation associated to X; the modularity has been established, under some mild conditions, that $L(X, s)$ is determined by some weight $4 = d + 1$ modular forms.

It is worth noticing that currently more than 50 modular rigid Calabi-Yau threefolds over $\mathbb{Q}$ have been constructed, and expanding.

The modularity question for a non-rigid Calabi-Yau threefold X over $\mathbb{Q}$ poses more serious challenge as the dimension of $H^3_{et}(X_{\bar{\mathbb{Q}}}, \mathbb{Q}_\ell)$ gets larger.

For a book devoted the modularity of Calabi-Yau manifolds, see [Mey]. More recent surveys on arithmetic of K3 surfaces and Calabi-Yau manifolds are also given in [LSY] and [Sch].

26 Moduli spaces of complex manifolds, algebraic varieties and vector bundles

Moduli spaces are fundamental objects in mathematics, and they are concrete ways to classify objects under study. To determine a moduli space, it is more than just identifying it as a set. We need to put and understand natural structures, such as complex structures and algebraic structures on them.

Moduli spaces of elliptic curves were classical objects, and they are foundational to elliptic modular functions. Before the notion of moduli spaces were introduced by Riemann, the set of equivalence classes of Riemann surfaces of genus 1, or lattices in $\mathbb{C}$, were studied and understood through the reduction of quadratic forms by Legendre and Gauss, and they appeared in Gauss' famous book on arithmetic.

Moduli spaces were often studied by algebraic methods. But the work of Teichmüller on deformation of Riemann surfaces and the later deformation theory of Kodaira-Spencer brought complex geometry to bear on this important topic. In [Sp, p. 389], Spencer wrote:

> Deformation of complex structures was therefore much on my mind when I arrived in Princeton. Since quadratic differentials are connected to the moduli of Riemann surfaces, I wondered what mathematical objects were connected to moduli in higher complex dimensions. Finally, Frolicher and Nijenhuis showed, in an important paper, that the vanishing of the first cohomology group with values in the sheaf of germs of holomorphic vector fields implies the rigidity, or the nondeformability, of the complex structure on a compact manifold. At about the same time, J.-P. Serre proved a duality theorem which, applied to a Riemann surface, implies that the quadratic differentials are dual to this cohomology. This was the starting point of Kodairas and my joint work on the deformation of the structures on higher dimensional complex manifolds. Most of the work is described by Kodaira in his monograph *Complex Manifolds and Deformation of Complex Structures*, Springer Verlag, 1986.

The Kodaira-Spencer deformation theory was completed by the work of Kuranishi on versal deformation families.

Closely related to deformation of complex manifolds is the theory of variation of Hodge structures and period domains. One of the originators of this subject is Griffiths (see [Gri2] for a summary). It is interesting to note how important results on Riemann surfaces have been to his work. In the preface to his selected works [Gri1], Griffiths wrote:

> One of the projects that Don suggested turned out to be decisive for me in many respects — to study Herman Weyl's book *Die Idee der Riemannschen Fläche*. A revision edition of this 1912 classic had come out a few years earlier, and it is safe to say that mastering this volume was the most important and influential accomplishment of my early graduate studies.

In algebraic geometry, one effective approach to construct moduli spaces is the method of geometric invariant theory developed by Mumford in his book [Mu1]. It is perhaps interesting to note that in his ICM address [Mu2, p. 529] in 1963, Mumford treated both moduli spaces of algebraic curves and moduli spaces of vector bundles over an algebraic curve at the same time, though the latter were not discussed in the book [Mu1].

Moduli spaces of vector bundles were studied in the book [LeP], and closely related moduli spaces of sheaves are studied in the book [Huy2], according to which

> There are at least three good reasons to study moduli spaces of sheaves on surfaces. Firstly, they provide examples of higher dimensional algebraic varieties with a rich and interesting geometry. In fact, in some regions in the classification of higher dimensional varieties the only known examples are moduli spaces of sheaves on a surface. The study of moduli spaces therefore sheds light on some aspects of higher dimensional algebraic geometry. Secondly, moduli spaces are varieties naturally attached to any surface. The understanding of their properties gives answers to problems concerning the geometry of the surface, e.g. Chow group, linear systems, etc. From the mid-eighties till the mid-nineties most of the work on moduli spaces of sheaves on a surface was motivated by Donaldsons ground breaking results on the relation between certain intersection numbers on the moduli spaces and the differentiable structure of the four-manifold underlying the surface. Although the interest in this relation has subsided since the introduction of the extremely powerful Seiberg-Witten invariants in 1994, Donaldson's results linger as a third major motivation in the background; they throw a bridge from algebraic geometry to gauge theory and differential geometry.

As we emphasized above, the analogy between complex manifolds and complex vector bundles over complex manifolds has been crucial in studying existence of Kähler-Einstein metrics and Hermitian Yang-Mills connections, and interaction between them.

In constructing moduli spaces of curves and algebraic varieties, and vector bundles over them, the notions of *stability* are crucial. As discussed above in the section on Hermitian Yang-Mills connections and Kähler-Einstein metrics on Fano manifolds, the notions of stability are the right conditions for the existence of Hermitian Yang-Mills connections and Kähler-Einstein metrics.

At the beginning of this section, we have emphasized that Riemann surfaces (or algebraic curves) are role models of higher dimensional complex manifolds. But there is one important difference in considering moduli spaces. We have mentioned many times the moduli space $\mathcal{M}_g$ of compact Riemann surfaces of genus g, which is a complex orbifold and a quasi-projective variety. On the other hand, in higher dimensions, we can only consider moduli spaces of *polarized* complex manifolds. For example, the moduli space of all complex tori of a fixed dimension greater than 1 is a non-Hausdorff space, while the moduli space of abelian varieties of principal polarizations (or a fixed polarization type) are given by locally Hermitian symmetric spaces.

It is also perhaps important to point out that complex differential geometry has also been a crucial ingredient in some moduli spaces. Putting natural Kähler metric on moduli spaces can enrich structures of the moduli spaces and allow one to understand them better. The best example is the Weil-Petersson on Riemann's moduli space $\mathcal{M}_g$ of compact Riemann surfaces of genus $g \geq 2$.

27 Moduli space of K3 surfaces

Among all moduli spaces in algebraic geometry, the moduli space of K3 surfaces is special. One reason is that methods of both algebraic geometry and complex differential geometry are used effectively to understand this moduli space. It is also one example that the theory of period and period domains works out in a clean and satisfactory way, and hence the moduli space can be identified with an arithmetic locally symmetric space. Moduli spaces of K3 surfaces can also be used to study other moduli spaces.

In [DK, p. 44], Dolgachev and Kondo wrote:

> The development of the general theory of periods of integrals on algebraic varieties in the 1960s due to P. Griffiths raised a natural question on moduli-theoretical interpretation of other Hermitian symmetric spaces and their arithmetic quotients and the analogs of the Torelli theorem. For the spaces of classical types I–IV, this can be achieved by embedding any type of space into a Siegel half-plane, and introducing the moduli space of abelian varieties with some additional structure (like a level, complex multiplication or some tensor form on cohomology). All of this becomes a part of the fancy theory of Shimura varieties. However, a more explicit interpretation remained to be searched for. For the type IV Hermitian symmetric space of dimension ≤ 19 such an interpretation had been found in terms of moduli of complex algebraic surfaces of type K3. The fundamental result of I. Piatetsky-Shapiro and I. Shafarevich in the 1970s gives an analog of the Torelli theorem for polarized algebraic K3 surfaces. Although this gives only a realization of the 19-dimensional type IV space by a very special arithmetic group depending only on the degree of the polarization, a modified version of the polarization structure due to V. Nikulin allows one to extend this construction to type IV space of any dimension $d \leq 19$, with a variety of arithmetic groups realized as certain orthogonal groups of integral quadratic forms of signature $(2, d)$. Recently, some arithmetic quotients of type IV domain of dimension 20 have been realized as periods of holomorphic symplectic manifolds of dimension 4.
>
> .... The periods of hypergeometric functions allow one to realize some complex ball quotients as the moduli space of weighted semi-stable ordered point sets in projective line modulo projective equivalence. In some cases these moduli spaces are isomorphic to moduli spaces of other structures. For example, via the period map of algebraic curves,

> the hypergeometric complex ball quotients are mapped onto a subvariety of $\mathcal{A}_g$ parametrizing principally polarized abelian varieties with a certain cyclic group action. For example, the moduli space of 6 points with equal weights defines a curve of genus 4 admitting a cyclic triple cover of the projective line ramified at the 6 points. Its Jacobian variety is an abelian variety of dimension 4 with a cyclic group of order 3 acting by automorphism of certain type. The moduli space of such abelian varieties is a complex ball quotient. Another example is the moduli space of equally marked sets of 5 points which leads to the moduli space of marked del Pezzo surfaces of degree 4 [K3]. Other examples relating the arithmetic complex ball quotients arising in the Deligne-Mostow theory to moduli spaces of Del Pezzo surfaces were found in [MT], [HL]. It turns out that all of these examples are intimately related to the moduli space of K3 surfaces with special structure of its Picard group of algebraic cycles and an action of a cyclic group.... We conjecture that all Deligne–Mostow arithmetic complex ball quotients are moduli spaces of K3 surfaces.

Another importance of moduli spaces of K3 surfaces for us is that they provide a bridge between the classical and well-understood moduli space of elliptic curves and new and complicated moduli spaces of Calabi-Yau manifolds. In comparison with other algebraic varieties, their geometry and arithmetic are understood much better.

There are several systematic books on moduli spaces from algebraic geometry and complex geometry points of views. See for example the books [BHPV], [Huy1], [BBD], and [Bea].

It seems that one question about the moduli space of K3 surfaces has not been studied. It is known that the moduli space $\mathcal{A}_g$ of abelian varieties with principal polarizations is given by an Hermitian locally symmetric space, the Siegel modular variety $Sp(2g,\mathbb{Z})\backslash\mathfrak{h}_g$, where $\mathfrak{h}_g$ is the Siegel upper half-space. It is also known that moduli spaces of abelian varieties with other polarizations and additional structures are also given by arithmetic Hermitian locally symmetric spaces.

These are more understandable since abelian varieties are linear objects, and locally symmetric spaces are essentially linear.

One interesting question is to understand why the moduli space of K3-surfaces is given by an Hermitian locally symmetric space. For example, a locally symmetric space has many correspondences. In the case of the moduli space of abelian varieties, these correspondences can be understood in terms of isogenies of abelian varieties. One question is what is the corresponding symmetry for K3 surfaces.

28 Moduli spaces of Calabi-Yau manifolds

Calabi-Yau manifolds and their moduli spaces are currently being intensely studied. Besides in mathematics, many string theorists are studying and using them.

As the experiences with elliptic curves and K3 surfaces show, moduli spaces of Calabi-Yau manifolds will shed light on the arithmetic and geometry of each individual Calabi-Yau manifold.

According to Yau [Yau5, pp. 286-287]:

> The existence of the moduli of polarized Calabi-Yau manifolds was settled by the work of Viehweg [279]. The next question is the regularity of the moduli space. The first theorem was due to Bogomolov [30] who proved that the universal deformation space of a compact Kähler-Hamiltonian manifold is unobstructed. In one of his unpublished manuscript, he also claimed that the same is true for any projective manifold with trivial first Chern class. Todorov [274] and Tian [269] each confirmed this claim by proving that every Calabi-Yau manifold has unobstructed deformations. Both proofs used essentially the Calabi-Yau metric of the manifold to derive a differential geometric computational equality that allows them to solve the Kuranishi equation in analytic deformation theory...
>
> On the moduli space, there is a natural Kähler metric obtained from the variation of the Ricci-flat metric called the Weil-Petersson metric. The volume of the moduli space of polarized Calabi-Yau manifolds with respect to this metric was proved by Lu-Sun [209] and Todorov [275] to be finite....
>
> The study of the moduli space of complex structures on a Calabi-Yau threefold led Hitchin to study invariant functionals on differential forms [134]....

C.-L. Wang [Wang1] proved incompleteness of the Weil–Petersson metric on the moduli space of Calabi–Yau manifolds. For one parameter degenerations of Calabi–Yau manifolds, based on the minimal model program in birational geometry, he proved the correspondence between finite Weil–Petersson distance degenerations and Calabi–Yau varieties with at most canonical singularities [Wang2, p. 58]. This lead him to conjecture [Wang2, p. 68–69] that the metric completion of the Calabi–Yau moduli is closely related to Viehweg's quasi-projective moduli of polarized Calabi–Yau varieties with at most canonical singularities [Viehweg, p. 258].

29 Calabi-Yau manifolds and string theory

As we explained at the beginning, Kepler's laws of planetary motion motivated the study of elliptic integrals and elliptic curves and complex geometry, and Calabi-Yau manifolds are the natural higher dimensional generalization of elliptic curves. Besides being special and interesting complex manifolds, Calabi-Yau manifolds are basic building blocks in string theory and mirror symmetry. This makes the whole circle: come from physics and astronomy, and then go back to physics and the new model of the universe.

To learn more specific detail about this, let us read what Yau, one of the heroes of this book, wrote in [Yau5, pp. 288–289]:

> **4.1. Calabi-Yau manifolds in string theory.** Superstring theory is a unified theory for all the forces of nature including quantum gravity.

In superstring theory, the fundamental building block is an extended object, namely a string, whose vibrations would give rise to the particles encountered in nature. The constraints for the consistency of such a theory are extremely stringent. They require in particular that the theory takes place in a 10-dimensional space-time. To make contact with our 4-dimensional world, it is expected that the 10-dimensional space-time of string theory is locally the product $M^4 \times X$ of a 4-dimensional Minkowski space $M^{3,1}$ with a 6-dimensional space X. The 6-dimensional space X would be tiny, which would explain why it has not been detected so far at the existing experimental energy levels. Each choice of the internal space X leads to a different effective theory on the 4-dimensional Minkowski space $M^{3,1}$, which should be the theory describing our world.

It has long been argued that, in order to solve certain classic problems of unified gauge theories such as the gauge hierarchy problem, the 4-dimensional effective theory should admit an $N = 1$ supersymmetry. In a fundamental paper, Candelas-Horowitz-Strominger-Witten [58] analyzed what the constraint of that $N = 1$ supersymmetry would mean for the geometry of the internal space X. They found that, for the most basic product models with $N = 1$ supersymmetry, the space X must be a Calabi-Yau manifold of complex dimension 3. Shortly afterwards, Strominger [264] considered slightly more general models, allowing warped products. For these models, the $N = 1$ supersymmetry constraint results in a modification of the Ricci-flat equation of the earlier model.

4.2. Calabi-Yau manifolds and mirror symmetry. Around 1987—1988, physicists including Dixon [76], and Lerche, Vafa, and Warner [181] observed that in mapping an abstract $N = 2$ superconformal field theory to a possible geometrical realization as a Calabi-Yau sigma model, an ambiguity arose. A superconformal field theory has two natural rings (called (c, c) and (a, c) rings)) as does a Calabi-Yau sigma model (the Dolbeault cohomology naturally splits into even and odd dimensional forms). The question which came to light was which of the two possible pairings of the conformal field theory and geometrical rings is induced by the map between the superconformal theory and the Calabi-Yau sigma model. Lerche, Vafa, and Warner conjectured that maybe both pairings are realized because, they suggested, Calabi-Yau threefolds come in pairs in which the even and odd cohomologies are interchanged. (To be precise, the interchange is between the Dolbeault cohomology $H^{(p,q)}$ with $H^{(3p,q)}$ for Calabi-Yau threefolds.) At the time, the evidence in support of this conjecture was thin. Some suggested that a less radical solution to the observed ambiguity might be to keep the base Calabi-Yau manifold fixed and merely consider completing the geometrical model in two ways: by including its tangent bundle or its co-tangent bundle.

> Nevertheless, in 1989, Greene and Plesser [116], using the methods of conformal field theory as applied to Calabi-Yau sigma models realized as twisted products of $N = 2$ minimal models, were able to establish that certain pairs of Calabi-Yau manifolds come in pairs in which their Hodge diamonds are mirror reflections (through a diagonal) of one another. Moreover, Greene and Plesser were able to establish that these pairs of Calabi-Yau manifolds, even though topologically distinct, when used as the basis for Calabi-Yau sigma models, give the same physical string theory. They named such pairs of Calabi-Yau manifolds *mirror manifolds.* The existence of such pairs of Calabi-Yau manifolds with specified properties are known to mathematicians as the *mirror symmetry conjecture.*

In the paper [Yau5, p. 291], Yau continued to write about the impact of the mirror symmetry on mathematics:

> One area inspired by the mirror symmetry conjecture is the construction of various enumerative invariants of Calabi-Yau manifolds. They appeared in the counting formula of CDGP. This development has led to a proof of the mirror symmetry conjecture by independent works of Givental [110, 29, 232] and of Lian-Liu-Yau [195]. In 1994, Kontsevich [164] expanded and formulated his version of mirror symmetry as an equivalence between complex and symplectic geometry of Calabi-Yau manifolds in all dimension. The geometric approach to mirror symmetry was finally unveiled by Strominger, Yau and Zaslow [265] in their 1997 paper in which they proposed that mirror symmetry is a geometric version of the Fourier transformation along dual special Lagrangian tori fibrations on mirror Calabi-Yau manifolds. This SYZ proposal has guided many research works.

According to Yau [Yau1, p. 15]:

> More importantly the conformal field theory based on one Calabi-Yau manifold is dual to that of its mirror manifold.... The most remarkable consequence is that it solved an old problem in enumerative geometry.
>
> The reason is that the type IIB theory can be computed by deformation theory of Kodaria-Spencer while the type IIA has quantum corrections. The quantum corrections are provided by the rational curves on the Calabi-Yau manifold. Since type IIA theory of one manifold is isomorphic to the type IIB of the mirror manifold, we can compute the number of rational curves on the Calabi-Yau manifolds by the variation of Hodge structure for its mirror family.
>
> The initial theory was mostly based on physical intuition. But two groups of mathematicians proved such statements rigorously in 1996, by Lian-Liu-Yau [47] and Givental [30] independently. Despite that the proof is rigorous, the intuition from string theory played the most important role. The idea of super-symmetry has become one of the most

> fundamental philosophies underlying the current modern development of algebraic geometry.

He also explained mathematical approaches to mirror symmetry in [Yau1, p. 15]:

> As mentioned above, two different approaches were proposed. One is the famous Kontsevich's homological mirror symmetry conjecture [41] which says that the derived category of coherent sheaves of a Calabi-Yau manifold is equivalent to the Fukaya category of its mirror manifold. Fukaya pioneered the research to study the extensive complicated structure of the Florer theory of Lagrangian cycles through A_∞-algebra [26, 25], which is an important ingredient in the conjecture of Konsevich. Another approach was proposed by Strominger-Yau-Zaslow [71] that the Calabi-Yau manifold is fibered by special Lagrangian torus and the mirror manifold is obtained by replacing the torus by its dual torus. Much progress was made by Auroux, Seidel et al. in this direction. It is important that singularities are allowed in the fibration for both topological and more subtle reasons. The SYZ conjecture has much evidence to be true. Gross and Siebert made a lot of progress in the last few years using tropical methods.

It seems fitting to conclude this brief tour of complex geometry with the following summary and expectation by Yau [Yau1, pp. 16-17]:

> In conclusion, we should say that the beautiful subject initiated by Riemann in the nineteenth century on Riemann surfaces had deep influence on the development of complex geometry in the 20th century. While Hodge provided the fundamental structure relating complex analysis with topology via Hodge groups, Kodaira provided fundamental methods to construct holomorphic sections of bundles. With the works of Chern classes and Hirzebruch-Riemann-Roch formula, the works of Hodge and Kodaira have been developed to be most powerful tools in understanding Kähler geometry. The modern development has been emphasizing the use of nonlinear elliptic equations, relating the concept of Kähler-Einstein metrics and Hermitian Yang-Mills equations to various fundamental concepts of stability introduced to study moduli spaces.
>
> The most recent development on Calabi-Yau space due to cooperations between mathematicians and string theorists has been spectacular. Ideas of many fields in mathematics were used. We hope to see some more ideas of number theory in this beautiful subject. For many Calabi-Yau manifolds, the partition functions related to conformal field theory are related to modular forms.
>
> .... Manin has conjectured that Kähler-Einstein metric will play important role in arithmetic geometry. I believe that is the case. There is still much to learn about the relation between complex geometry, algebra and number theory.

Acknowledgements

In writing this long introductory chapter, constructive suggestions from S. T. Yau and D. H. Phong have been very helpful. I would like thank them for their advice and patience. I would also like to thank Bong Lian, Chin-Lung Wang, and Hao Xu for their help in tracking down several references.

References

[AJP] N. A'Campo, L. Ji, A. Papadopoulos, On the early history of moduli and Teichmüller spaces. Lipman Bers: A Life in Mathematics. Edited by Linda Keen, Irwin Kra, Rubi E. Rodriguez, 175–262.

[Ah] L. Ahlfors, Quasiconformal mappings, Teichmüller spaces, and Kleinian groups. Proceedings of the International Congress of Mathematicians (Helsinki, 1978), pp. 71–84, Acad. Sci. Fennica, Helsinki, 1980.

[Ar1] V. I. Arnold, Polymathematics: is mathematics a single science or a set of arts? Mathematics: frontiers and perspectives, 403–416, Amer. Math. Soc. Providence, RI, 2000.

[Ar2] S. H. Lui, An interview with Vladimir Arnol'd. Notices Amer. Math. Soc. 44 (1997), no. 4, 432–438.

[At1] R. Minio, An interview with Michael Atiyah. Math. Intelligencer 6 (1984), no. 1, 9–19.

[ABCD] P. Aspinwall, T. Bridgeland, A. Craw, M. Douglas, M. Gross, A. Kapustin, G. Moore, G. Segal, B. Szendrï, P. Wilson, Dirichlet branes and mirror symmetry. Clay Mathematics Monographs, 4. American Mathematical Society, Providence, RI; Clay Mathematics Institute, Cambridge, MA, 2009. x+681 pp.

[At2] M. Atiyah, Hodge, William Vallance Douglas. In Complete Dictionary of Scientific Biography, Vol. 17. Detroit: Charles Scribner's Sons, 2008. pp. 425–430.

[At3] M. Atiyah, Obituary: Kunihiko Kodaira. Bull. London Math. Soc. 31 (1999), no. 4, 489–493.

[At4] M. Atiyah, Collected works. Vol. 5. Gauge theories. Oxford Science Publications. The Clarendon Press, Oxford University Press, New York, 1988. xxiv+685 pp.

[At5] M. Atiyah, Friedrich Hirzebruch—an appreciation. Proceedings of the Hirzebruch 65 Conference on Algebraic Geometry (Ramat Gan, 1993), 1–5, Israel Math. Conf. Proc., 9, Bar-Ilan Univ., Ramat Gan, 1996.

[AB1] M. Atiyah, R. Bott, Yang-Mills and bundles over algebraic curves. Proc. Indian Acad. Sci. Math. Sci. 90 (1981), no. 1, 11–20.

[AB2] M. Atiyah, R. Bott, The Yang-Mills equations over Riemann surfaces. Philos. Trans. Roy. Soc. London Ser. A 308 (1983), no. 1505, 523–615.

[BHPV] W. Barth, K. Hulek, C. Peters, A. Van de Ven, Compact complex surfaces. Second edition. Ergebnisse der Mathematik und ihrer Grenzgebiete. 3. Folge. A Series of Modern Surveys in Mathematics, 4. Springer-Verlag, Berlin, 2004. xii+436 pp.

[Bea] A. Beauville, Complex algebraic surfaces. Translated from the 1978 French original by R. Barlow, with assistance from N. I. Shepherd-Barron and M. Reid. Second edition. London Mathematical Society Student Texts, 34. Cambridge University Press, Cambridge, 1996. x+132 pp.

[BBD] A. Beauville, J. P. Bourguignon, M. Demazure, Géométrie des surfaces $K3$: modules et périodes. Séminaires Palaiseau, Astérisque 126, Paris: Société Mathématique de France, 1985. pp. 1–193.

[Beg] F. Béguin, Henri Poincaré and the uniformization of Riemann surfaces. Henri Poincaré, 1912–2012, 193–230, Prog. Math. Phys., 67, Birkhäuser/Springer, Basel, 2015.

[Ber] M. Berger, Encounter with a geometer: Eugenio Calabi. Manifolds and geometry (Pisa, 1993), 20–60, Sympos. Math., XXXVI, Cambridge Univ. Press, Cambridge, 1996.

[Be] L. Bers, Spaces of Riemann surfaces. In Proceedings of the international congress of mathematics, 1958, 349–361. 1960. Cambridge Univ. Press, New York . Selected works, part I, pp. 355–367.

[Bes] A. Besse, Einstein manifolds. Ergebnisse der Mathematik und ihrer Grenzgebiete, 10. Springer-Verlag, Berlin, 1987. xii+510 pp.

[Ble] S. Blesneag, E. I. Buchbinder, P. Candelas, A. Lukas, Holomorphic Yukawa couplings in heterotic string theory. JHEP **1601**, 152 (2016) doi.10.1007/JHEP01(2016)152 [arXiv:1512.05322 [hep-th]].

[Bo] R. Bott, Raoul Bott: collected papers. Vol. 4. Mathematics related to physics. Edited by Robert D. MacPherson. Contemporary Mathematicians. Birkhäuser Boston, Inc., Boston, MA, 1995. xx+485 pp.

[Bot] U. Bottazzini, The higher calculus: a history of real and complex analysis from Euler to Weierstrass. Translated from the Italian by Warren Van Egmond. Springer-Verlag, New York, 1986. vi+332 pp.

[BoG] U. Bottazzini, J. Gray, Hidden harmony–geometric fantasies. The rise of complex function theory. Sources and Studies in the History of Mathematics and Physical Sciences. Springer, New York, 2013. xviii+848 pp.

[Boy] C. Boyer, A history of mathematics. Revised reprint of the second edition. With a foreword by Isaac Asimov. Revised and with a preface by Uta C. Merzbach. John Wiley & Sons, Inc., New York, 1991. xx+715 pp.

[Bra] V. Braun, P. Candelas, R. Davies, R. Donagi, The MSSM Spectrum from (0, 2)-Deformations of the Heterotic Standard Embedding. JHEP **1205**, 127 (2012) doi: 10.1007/JHEP05(2012)127 [arXiv:1112.1097 [hep-th]].

[BCDT] C. Breuil, B. Conrad, F. Diamond, R. Taylor, On the modularity of elliptic curves over Q: wild 3-adic exercises. J. Amer. Math. Soc. 14 (2001), no. 4, 843–939.

[CHSW] P. Candelas, G. Horowitz, A. Strominger, E. Witten, Vacuum configurations for superstrings. Nuclear Phys. B 258 (1985), no. 1, 46–74.

[Che] S. S. Chern, A mathematician and his mathematical work. Selected papers of S. S. Chern. Edited and with a foreword by S. Y. Cheng, P. Li and G. Tian. World Scientific Series in 20th Century Mathematics, 4. World Scientific Publishing Co., Inc., River Edge, NJ, 1996. xvi+707 pp.

[Cho] S. S. Chern, Wei-Liang Chow, 1911–1995. Notices Amer. Math. Soc. 43 (1996), no. 10, 1117–1124.

[Se1] C. T. Chong, Y. K. Leong, An interview with Jean-Pierre Serre. Math. Intelligencer 8 (1986), no. 4, 8–13.

[Co] M. Cook, Mathematicians: an outer view of the inner world. Portraits by Mariana Cook. With an introduction by Robert Clifford Gunning and an afterword by Brandon Fradd. Princeton University Press, Princeton, NJ, 2009. 200 pp.

[De] P. Deligne, The Hodge conjecture. The millennium prize problems, 45–53, Clay Math. Inst., Cambridge, MA, 2006.

[DK] I. Dolgachev, S. Kondo, Moduli of K3 surfaces and complex ball quotients. Arithmetic and geometry around hypergeometric functions, 43–100, Progr. Math., 260, Birkhäuser, Basel, 2007.

[Do1] S. Donaldson, A new proof of a theorem of Narasimhan and Seshadri. J. Differential Geom. 18 (1983), no. 2, 269–277.

[Do2] S. Donaldson, Anti self-dual Yang-Mills connections over complex algebraic surfaces and stable vector bundles. Proc. London Math. Soc. (3) 50 (1985), no. 1, 1–26.

[Do3] S. Donaldson, Infinite determinants, stable bundles and curvature. Duke Math. J. 54 (1987), no. 1, 231–247.

[Do4] S. Donaldson, Geometry in Oxford c. 1980–85. in a special issue Sir Michael Atiyah: a great mathematician of the twentieth century. Asian J. Math. 3 (1999), no. 1, xliii–xlvii.

[Do5] S. Donaldson, Remarks on gauge theory, complex geometry and 4-manifold topology. Fields Medallists' lectures, 384–403, World Sci. Ser. 20th Century Math., 5, World Sci. Publ., River Edge, NJ, 1997.

[Do6] S. Donaldson, Shing-Tung Yau's Surveys. Selected expository works of Shing-Tung Yau with commentary. Vol. II. pp. 1209–1220. Edited by Lizhen Ji, Peter Li, Kefeng Liu and Richard Schoen. Advanced Lectures in Mathematics (ALM), 29. International Press, Somerville, MA; Higher Education Press, Beijing, 2014. pp. i–xxxii and 705–1354.

[DEG] S. Donaldson, Y. Eliashberg, M. Gromov, Different faces of geometry. International Mathematical Series, 3. Kluwer Academic/Plenum Publishers, New York, 2004. xxii+404 pp.

[Dou] R. Douglas, Calabi-Yau metrics and string compactification. Nucl. Phys. B 898, 667 (2015) doi:10.1016/j.nuclphysb.2015.04.009 [arXiv:1503.02899 [hep-th]].

[Fr] R. Friedman, Algebraic surfaces and holomorphic vector bundles. Universitext. Springer-Verlag, New York, 1998. x+328 pp.

[Ga] P. Gauduchon, La topologie d'une surface hermitienne d'Einstein. C. R. Acad. Sci. Paris Sér. A-B 290 (1980), no. 11, A509–A512.

[Giv] A. Givental, Equivariant Gromov-Witten invariants. Internat. Math. Res. Notices (1996), no. 13, 613–663.

[Gow] W. Gowers, J. Barrow-Green, I. Leader, The Princeton companion to mathematics. Princeton University Press, Princeton, NJ, 2008. xxii+1034 pp.

[Gra1] J. Gray, The real and the complex: a history of analysis in the 19th century. Springer Undergraduate Mathematics Series. Springer, Cham, 2015. xvi+350 pp.

[Gra2] J. Gray, On the history of the Riemann mapping theorem. Rend. Circ. Mat. Palermo (2) Suppl., No. 34 (1994), 47–94.

[Gra2] J. Gray, Linear differential equations and group theory from Riemann to Poincaré. Reprint of the 2000 second edition. Modern Birkhäuser Classics. Birkhäuser Boston, Inc., Boston, MA, 2008. xx+338 pp.

[GSW] M. Green, J. Schwarz, E. Witten, Superstring theory. Vol. 2. Loop amplitudes, anomalies and phenomenology. 25th anniversary edition. Cambridge University Press, Cambridge, 2012. xii+596 pp.

[Gri1] P. Griffiths, Selected works of Phillip A. Griffiths with commentary. Part 1. Analytic geometry. Edited by Maurizio Cornalba, Mark L. Green and Wilfried Schmid. American Mathematical Society, Providence, RI; International Press, Somerville, MA, 2003. xxvi+654 pp.

[Gri2] P. Griffiths, Periods of integrals on algebraic manifolds: Summary of main results and discussion of open problems. Bull. Amer. Math. Soc. 76 (1970) 228–296.

[GS] A. Grothendieck, J.-P. Serre, Grothendieck-Serre correspondence. Translated by Catriona Maclean Pierre Colmez from French. American Mathematical Society; Bilingual edition. 2003. 600 pages.

[Hir] F. Hirzebruch, The signature theorem: reminiscences and recreation. Prospects in mathematics (Proc. Sympos., Princeton Univ., Princeton, N.J., 1970), pp. 3–31. Ann. of Math. Studies, No. 70, Princeton Univ. Press, Princeton, N.J. 1971.

[Hi] N. Hitchin, A note on vanishing theorems. Geometry and Analysis on Manifolds: In Memory of Professor Shoshichi Kobayashi. Progress in Mathematics, Vol. 308, edited by T. Ochiai, T. Mabuchi, Y. Maeda, J. Noguchi, A. Weinstein Birkhäuser, 2015, pp. 373–382. arXiv:1408.0885.

[Hi2] N. Hitchin, Geometry and physics: a personal view. The many facets of geometry, 1–10, Oxford Univ. Press, Oxford, 2010.

[Hod1] W. Hodge, The theory and applications of harmonic integrals. 2d ed. Cambridge, at the University Press, 1952. x+282 pp.

[Hod2] W. Hodge, Professor Lefschetz's contributions to algebraic geometry: an appreciation. A symposium in honor of S. Lefschetz, Algebraic geometry and topology, pp. 3–23. Princeton University Press, Princeton, N. J., 1957.

[Hub] J. Hubbard, Teichmüller theory and applications to geometry, topology, and dynamics. Vol. 1. Teichmüller theory. With contributions by Adrien Douady, William Dunbar, Roland Roeder, Sylvain Bonnot, David Brown, Allen Hatcher, Chris Hruska and Sudeb Mitra. With forewords by William Thurston and Clifford Earle. Matrix Editions, Ithaca, NY, 2006. xx+459 pp.

[Hus] D. Husemöller, Elliptic curves. Second edition. With appendices by Otto Forster, Ruth Lawrence and Stefan Theisen. Graduate Texts in Mathematics, 111. Springer-Verlag, New York, 2004. xxii+487 pp.

[Huy1] D. Huybrechts, Lectures on K3 surfaces. Cambridge University Press. 2016. 480 pages.

[Huy2] D. Huybrechts, M. Lehn, The geometry of moduli spaces of sheaves. Second edition. Cambridge Mathematical Library. Cambridge University Press, Cambridge, 2010. xviii+325 pp.

[IN] M. Itoh, H. Nakajima, Yang-Mills connections and Einstein-Hermitian metrics. Kähler metric and moduli spaces. pp. 395–457. Edited by T. Ochiai. Advanced Studies in Pure Mathematics, 18-II. Academic Press, Inc., Boston, MA; Kinokuniya Company Ltd., Tokyo, 1990. xii+457 pp.

[Ji] L. Ji, The story of Riemann's moduli space. The legacy of Bernhard Riemann after one hundred and fifty years, HEP-IP, 2016, pp. 307–358. It also appears in ICCM Not. 3 (2015), no. 2, 46–73.

[Ka] E. Kähler, Mathematische Werke/Mathematical works. Edited by Rolf Berndt and Oswald Riemenschneider. Walter de Gruyter & Co., Berlin, 2003. x+971 pp.

[Kat] V. Katz, A history of mathematics. Pearson; 3 edition, 2008, 992 pages.

[Kle] F. Klein, Development of mathematics in the 19th century. With a preface and appendices by Robert Hermann. Translated from the German by M. Ackerman. Lie Groups: History, Frontiers and Applications, IX. Math Sci Press, Brookline, Mass., 1979. ix+630 pp.

[Kli] M. Kline, Mathematical thought from ancient to modern times. Vol. 2. Second edition. The Clarendon Press, Oxford University Press, New York, 1990. pp. ixx, 391–812 and i–xxii.

[Kob1] S. Kobayashi, Differential geometry of complex vector bundles. Publications of the Mathematical Society of Japan, 15. Kano Memorial Lectures, 5. Princeton University Press, Princeton, NJ; Iwanami Shoten, Tokyo, 1987. xii+305 pp.

[Kob2] S. Kobayashi, Curvature and stability of vector bundles. Proc. Japan Acad. Ser. A Math. Sci. 58 (1982), no. 4, 158–162.

[Kob3] S. Kobayashi, First Chern class and holomorphic tensor fields. Nagoya Math. J. 77 (1980), 5–11.

[KS] K. Kodaira, D. Spencer, On deformations of complex analytic structures. I, II. Ann. of Math. (2) 67 (1958) 328–466.

[Kol2] A. N. Kolmogorov, A. P. Yushkevich, Mathematics of the 19th century. Mathematical logic. Algebra. Number theory. Probability theory. Second revised edition. Translated from the 1978 Russian original by A. Shenitzer, H. Grant and O. B. Sheinin. Translation edited by Shenitzer. Birkhäuser Verlag, Basel, 2001. xiv+308 pp.

[KT] T. Kotake, T. Achiai, Non-linear problems in geometry: Proceedings of the Sixth International Symposium. Division of Mathematics, the Taniguchi Foundation, 1984, Dept. of Mathematics, Tohoku University, Sendai 980, Japan.

[Kul] V. Kulikov, Surjectivity of the period mapping for $K3$ surfaces. Uspehi Mat. Nauk 32 (1977), no. 4(196), 257–258.

[La1] S. Lang, Higher dimensional diophantine problems. Bull. Amer. Math. Soc. 80 (1974), 779–787.

[La2] S. Lang, Hyperbolic and Diophantine analysis. Bull. Amer. Math. Soc. (N.S.) 14 (1986), no. 2, 159–205.

[La3] S. Lang, Number theory. III. Diophantine geometry. Encyclopaedia of Mathematical Sciences, 60. Springer-Verlag, Berlin, 1991. xiv+296 pp.

[LSY] R. Laza, M. Schütt, N. Yui, Arithmetic and geometry of $K3$ surfaces and Calabi-Yau threefolds. Proceedings of the workshop held at the Fields Institute and University of Toronto, Toronto, ON, August 1625, 2011. Fields Institute Communications, 67. Springer, New York; Fields Institute for Research in Mathematical Sciences, Toronto, ON, 2013. xxvi+602 pp.

[Lan] S. Lang, Some history of the Shimura-Taniyama conjecture. Notices Amer. Math. Soc. 42 (1995), no. 11, 1301–1307.

[LW] C. LeBrun, M. Wang. Essays on Einstein manifolds. Lectures on geometry and topology, sponsored by Lehigh University's Journal of Differential Geometry. Surveys in Differential Geometry, VI. International Press, Boston, MA, 1999. x+423 pp.

[LeP] J. Le Potier, Lectures on vector bundles. Translated by A. Maciocia. Cambridge Studies in Advanced Mathematics, 54. Cambridge University Press, Cambridge, 1997. viii+251 pp.

[LY] J. Li, S. T. Yau, The existence of supersymmetric string theory with torsion. J. Differential Geom. 70 (2005), no. 1, 143–181.

[LLY] B. H. Lian, K. Liu, S.-T. Yau, Mirror principle. I, Asian J. Math., 1 (1997), 729–763.

[Lu1] M. Lübke, Stability of Einstein-Hermitian vector bundles. Manuscripta Math. 42 (1983), no. 2-3, 245–257.

[Lu2] M. Lübke, Chernklassen von Hermite-Einstein-Vektorbündeln. Math. Ann. 260 (1982), no. 1, 133–141.

[LT] M. Lübke, A. Teleman, The Kobayashi-Hitchin correspondence. World Scientific Publishing Co., Inc., River Edge, NJ, 1995. x+254 pp.

[Man] Y. Manin, Selected papers of Yu. I. Manin. World Scientific Series in 20th Century Mathematics, 3. World Scientific Publishing Co., Inc., River Edge, NJ, 1996. xii+600 pp.

[Mar] C. Margerin, Fibrés stables et métriques d'Hermite-Einstein (d'après S. K. Donaldson-K. K. Uhlenbeck-S. T. Yau). (Séminaire Bourbaki, Vol. 1986/87. Asterisque No. 152-153 (1987), 5, 263–283 (1988).

[MR] B. Mazur, K. Rubin, Elliptic curves and class field theory. Proceedings of the International Congress of Mathematicians, Vol. II (Beijing, 2002), 185–195, Higher Ed. Press, Beijing, 2002.

[Mey] C. Meyer, Modular Calabi-Yau threefolds. Fields Institute Monographs, 22. American Mathematical Society, Providence, RI, 2005. x+195 pp.

[Mi1] M. Raussen, C. Skau, Interview with John Milnor. Notices Amer. Math. Soc. 59 (2012), no. 3, 400–408.

[Mi2] J. Milnor, Topology through the centuries: low dimensional manifolds. Bull. Amer. Math. Soc. (N.S.) 52 (2015), no. 4, 545–584.

[Mi1] Y. Miyaoka, On the Chern numbers of surfaces of general type. Invent. Math. 42 (1977), 225–237.

[MK] J. Morrow, K. Kodaira, Complex manifolds. Reprint of the 1971 edition with errata. AMS Chelsea Publishing, Providence, RI, 2006. x+194 pp.

[Mu1] D. Mumford, Geometric invariant theory. Ergebnisse der Mathematik und ihrer Grenzgebiete, Neue Folge, Band 34 Springer-Verlag, Berlin-New York 1965 vi+145 pp.

[Mu2] D. Mumford, Projective invariants of projective structures and applications. 1963 Proc. Internat. Congr. Mathematicians (Stockholm, 1962) pp. 526–530 Inst. Mittag-Leffler, Djursholm.

[NS] M. S. Narasimhan, C. S. Seshadri, Stable and unitary vector bundles on a compact Riemann surface. Ann. of Math. (2) 82 (1965), 540–567.

[Neu] J. von Neumann, The mathematician. Edited for the Committee on Social Thought by Robert B. Heywood. The Works of the Mind, pp. 180–196. The University of Chicago Press, Chicago, Ill., 1947.

[Ok] C. Okonek, Book Review: Differential geometry of complex vector bundles. Bull. Amer. Math. Soc. (N.S.) 19 (1988), no. 2, 528–530.

[Ph] D. H. Phong, Commentary on the legacy of Shing-Tung Yau in geometric partial differential equations. pp. 1291–1308, in Selected expository works of Shing-Tung Yau with commentary. Vol. II. Edited by L. Ji, P. Li, K. Liu and R. Schoen. Advanced Lectures in Mathematics (ALM), 29. International Press, Somerville, MA; Higher Education Press, Beijing, 2014. pp. i–xxxii and 705–1354.

[PiS] I. Piatetskii-Shapiro, I. Shafarevich, Torelli's theorem for algebraic surfaces of type $K3$, Izv. Akad. Nauk SSSR Ser. Mat. 35 (1971), 530–572.

[PrY] G. Prasad, S. K. Yeung, Fake projective planes. Invent. Math. 168 (2007), 321–370. Addendum, ibid 182(2010), 213–227. (See the corrected version posted on arXiv.)

[Re] R. Remmert, From Riemann surfaces to complex spaces. Matériaux pour l'histoire des mathématiques au XXe siècle (Nice, 1996), 203–241, Sémin. Congr., 3, Soc. Math. France, Paris, 1998.

[Rie] B. Riemann, Collected papers. Translated from the 1892 German edition by Roger Baker, Charles Christenson and Henry Orde. Kendrick Press, Heber City, UT, 2004. x+555 pp.

[Sai] H. P. de Saint-Gervais, Uniformisation des surfaces de Riemann. Retour sur un théorème centenaire. ENS Éditions, Lyon, 2010. 544 pp.

[Sch] M. Schütt, Arithmetic of K3 surfaces. Jahresber. Deutsch. Math.-Verein. 111 (2009), no. 1, 23–41.

[Se2] J. P. Serre, Géométrie algébrique et géométrie analytique. Université de Grenoble. Annales de l'Institut Fourier 6 (1956) 1–42.

[Sh] E. Sharpe, A few recent developments in 2d (2, 2) and (0, 2) theories. arXiv:1501.01628 [hep-th].

[Siu1] Y. T. Siu, Lectures on Hermitian-Einstein metrics for stable bundles and Kähler-Einstein metrics. DMV Seminar, 8. Birkhäuser Verlag, Basel, 1987.

[Siu2] Y. Siu, The complex-analyticity of harmonic maps and the strong rigidity of compact Kähler manifolds. Ann. of Math. (2) 112 (1980), no. 1, 73–111.

[Siu3] Y. T. Siu, Every K3 surface is Kähler. Invent. Math. 73 (1983), 139–150.

[Sp] D. Spencer, Kunihiko Kodaira (1915–1997). Notices Amer. Math. Soc. 45 (1998), no. 3, 388–389.

[Swi] P. Swinnerton-Dyer, Diophantine equations: progress and problems. Arithmetic of higher-dimensional algebraic varieties (Palo Alto, CA, 2002), 335, Progr. Math., 226, Birkhäuser Boston, Boston, MA, 2004.

[TW] R. Taylor, A. Wiles, Ring-theoretic properties of certain Hecke algebras. Ann. of Math. (2) 141 (1995), no. 3, 553–572.

[Tod] A. Todorov, Applications of the Kähler-Einstein-Calabi-Yau metric to moduli of K3 surfaces. Invent. Math. 61 (1980), 251–265.

[UY] K. Uhlenbeck, S. T. Yau, On the existence of Hermitian Yang-Mills connections in stable vector bundles. Frontiers of the mathematical sciences: 1985 (New York, 1985). Comm. Pure Appl. Math. 39 (1986), no. S, suppl., S257–S293.

[Viehweg] E. Viehweg, Quasi-Projective Moduli of Polarized Manifolds. Springer-Verlag, New York, 1995.

[Wang1] C.-L. Wang, On the incompleteness of the Weil-Petersson metric along degenerations of Calabi-Yau manifolds. Math. Res. Lett. 4 (1997), no. 1, 157–171.

[Wang2] C.-L. Wang, Quasi-Hodge metrics and canonical singularities. Math. Res. Lett. 10 (2003), 57–70.

[Wei] A. Weil, History of mathematics: why and how. Proceedings of the International Congress of Mathematicians (Helsinki, 1978), pp. 227–236, Acad. Sci. Fennica, Helsinki, 1980.

[Wel] R. Wells, The origins of complex geometry in the 19th century. arXiv: 1504.04405.

[Wey] H. Weyl, The concept of a Riemann surface. Translated from the third German edition by Gerald R. MacLane. Dover Books on Mathematics, 208 pages, 3rd ed. edition, 2009.

[Wi] A. Wiles, The Birch and Swinnerton-Dyer conjecture. The millennium prize problems, 31–41, Clay Math. Inst., Cambridge, MA, 2006.

[Wi2] A. Wiles, Modular elliptic curves and Fermat's last theorem. Ann. of Math. (2) 141 (1995), no. 3, 443–551.

[Wit1] E. Witten, New issues in manifolds of $SU(3)$ holonomy. Nuclear Phys. B 268 (1986), no. 1, 79–112.

[Wit2] E. Witten, Symmetry breaking patterns in superstring models. Nuclear Phys. B 258 (1985), no. 1, 75–100.

[Yan] C. N. Yang, Selected papers (1945–1980). With commentary. 2005 edition. Reprinted, with two additional papers, from the 1983 edition. World Scientific Series in 20th Century Physics, 36. World Scientific Publishing Co. Pte. Ltd., Singapore, 2005. xvi+603 pp.

[Yan2] C. N. Yang, Condition of self-duality for SU(2) gauge fields on Euclidean four-dimensional space. Phys. Rev. Lett. 38 (1977), no. 24, 1377–1379.

[Yau1] S. T. Yau, A brief history of Kähler geometry. ICCM Notices 3 (2015), no. 2, pp. 1–19.

[Yau2] S. T. Yau, Complex geometry: its brief history and its future. Sci. China Ser. A 48 (2005), suppl., 47–60.

[Yau3] R. Stern, G. Tian, Donaldson and Yau receive Crafoord prize. Notices Amer. Math. Soc. 41 (1994), no. 7, 794–796.

[Yau4] S. T. Yau, Uniformization of geometric structures. The mathematical heritage of Hermann Weyl (Durham, NC, 1987), 265–274, Proc. Sympos. Pure Math., 48, Amer. Math. Soc., Providence, RI, 1988.

[Yau5] S. T. Yau, A survey of Calabi-Yau manifolds. Surveys in differential geometry. Vol. XIII. Geometry, analysis, and algebraic geometry: forty years of the Journal of Differential Geometry, 277–318, Surv. Differ. Geom., 13, Int. Press, Somerville, MA, 2009.

[Yau6] S. T. Yau, Problem section. Seminar on Differential Geometry, pp. 669–706, Ann. of Math. Stud., 102, Princeton Univ. Press, Princeton, N.J., 1982.

Part II
Comments on Yau and His Work

Editor's Comments

Shing-Tung Yau's work is closely related to complex geometry, in particular to Kähler-Einstein manifolds. The brief overview of complex geometry presented in Chapter 1 shows that Yau made an enormous contribution to complex geometry. Since this book was partly motivated by a conference a few years ago for his 65th birthday, it seems appropriate to add some comments on Yau and his work which were collected on that occasion by Shiu-Yuen Cheng, Lizhen Ji, Liping Wang and Hao Xu.

Complex Geometry from Riemann to Kähler-Einstein and Calabi-Yau
ALM 38, pp. 81–106

Calabi-Yau Manifolds at the Interface of Mathematics and Physics

Shiu-Yuen Cheng*, Lizhen Ji†, Liping Wang‡, and Hao Xu§

There were once a young man and an older man. Their mathematics paths crossed and their names became merged into the long-lasting notion of Calabi-Yau manifolds.

How did this happen? Why did Calabi make his famous conjecture? How did Yau prove it? Why are Calabi-Yau manifolds so important? And how are they used? All these and other questions take time to answer and real experts to explain. They can probably be best understood through the interactions between mathematics and physics, and that's where their true value lies. Although these questions have a long history, the future will undoubtedly bring more questions and surprises regarding Calabi-Yau manifolds.

Indeed, one must go back to more than 150 years ago, when Riemann introduced the concepts of Riemann surfaces and Riemannian metrics. His mapping theorem motivated people to endow each Riemann surface with a canonical Riemannian metric, and from there one of the greatest theorems in the history of mathematics, the uniformization theorem for Riemann surfaces and surfaces with Riemannian metrics, took shape.

What happened to higher dimensional Riemannian manifolds? What kind of geometric structures could be built with them? Maybe there was not enough structure. How about adding complex structures? Would they admit canonical metrics? Or perhaps there was still not enough structure. How about Kähler manifolds—structures between complex geometry and algebraic geometry? This

* Mathematical Sciences Center, Tsinghua University, Beijing, China
Email: sycheng@mail.tsinghua.edu.cn
† Department of Mathematics, University of Michigan, Ann Arbor, MI, U.S.A.
Email: lji@umich.edu
‡ Higher Education Press, Beijing, China
Email: wanglp@hep.com.cn
§ Department of Mathematics, University of Pittsburg, PA, U.S.A.
Email: haoxu@pitt.edu

was what Calabi asked 60 years ago, and what Yau answered 24 years later. He cracked a geometric problem by unraveling hard nonlinear differential equations.

The Calabi-Yau metric has been an invaluable tool in differential and algebraic geometry. Here is an incomplete list of long-standing conjectures and problems solved by using the Calabi-Yau metric and Yau's technique of the *a priori* estimate.

1. The Severi conjecture, that every complex surface that is homotopic to the complex projective plane $\mathbb{C}P^2$ is biholomorphic to $\mathbb{C}P^2$.

2. Bogomolov-Miyaoka-Yau Chern number inequalities and their associated uniformization theorems in higher dimensions.

3. Uniqueness of Kähler complex structure on $\mathbb{C}P^n$.

4. Torelli theorem for K3 surfaces.

5. Every K3 surface is Kähler.

6. The existence of Hermitian Yang-Mills connections in stable vector bundles (Donaldson-Uhlenbeck-Yau theorem).

7. The existence of complete Kähler-Einstein metrics on pseudoconvex domains.

8. Calabi-Yau manifolds are used in the context of string theory to make models of particle physics unified with quantum gravity.

9. Construction of non-flat compact simply-connected Kähler manifolds whose Ricci curvature is identically zero.

10. Any compact Kähler manifold with non-negative first Chern class is covered holomorphically by the product of $\mathbb{C}^n$ and a compact simply connected Kähler manifold.

11. The vanishing of c_1 and $c_2 \cup [\omega]^{n-2}$ for a Kähler manifold implies that the manifold admits a metric with zero curvature and is covered by the flat torus.

12. As conjectured by Yau, the existence of Kähler-Einstein metrics on Fano manifolds has intimate connections to the stability of manifolds. Important progress has been made recently.

Such beautiful results are appreciated not only by mathematicians but by physicists as well. For, according to string theory, Calabi-Yau manifolds provide the hidden inner spaces of our universe.

It is true that 60 is a magic number, but 65 is also a milestone in one's life. Here is a poem by Professor Yau reflecting his experience with geometry and physics.

Forty years of geometry,

equation creates structure,

analysis blends perfection and beauty.

Thirty years of physics,

strings and branes drifting through hyperspace,

breaking supersymmetry as they go.

Quantum connects to gravity,

and nature is whole again.

S.-T. Yau

The power of Calabi-Yau manifolds

From the uniformization (Klein, Poincaré, Koebe, Brouwer) to Riemann-Hilbert correspondence, Picard-Fuchs equations, and Kähler-Einstein metrics (Calabi-Yau)

1. Number theory: Calabi-Yau manifolds are natural, higher-dimensional generalizations of elliptic curves—the theory of which dominated number theory in the 20th century. Many believe that Calabi-Yau manifolds will be the main tool in number theory in the 21st century. The related questions of modular forms and quasi-modular forms are now being developed.

2. Algebra: There is a natural category, which appeared in algebra, where the Calabi-Yau condition plays an important role. It is used in classifying topological field theories.

3. Differential geometry: The construction of the Calabi-Yau metric solves the old problem of classification of manifolds with holonomy group $SU(n)$ and $SP(n)$ by reducing the construction of them to algebraic conditions.

4. Algebraic geometry:

(a) The second Chern class of such manifolds is numerically positive in the sense that the cup product of it with the top number of products of any positive (1,1) class is positive unless it is covered by the torus.

(b) Calabi-Yau manifolds are stable and their tangent bundles are also stable with respect to any polarization.

(c) Counting the number of algebraic curves in Calabi-Yau manifolds has become quite successful since the introduction of the concept of mirror symmetry, which has opened up many new directions in mathematics.

(d) It gives a structure theorem for Compact Kähler manifolds with non-negative first Chern class.

5. Nonlinear partial differential equations: The proof of Calabi conjecture gave the first general method to solve the nonlinear complex Ampère equation on a complex manifold. The equation solves a very important inverse problem: given any volume element on any compact Kähler manifold, any Kähler metric can be deformed to a new Kähler metric whose volume form is the given one up to a constant. The methods of estimates have been fundamental to all constructions of Kähler-Einstein metrics, including Kähler metrics with prescribed singularities. The singular metrics were used by Donaldson *et al.* to solve a famous conjecture of Yau on existence of Kähler-Einstein metric on stable Fano manifolds.

Perspectives of experts

When Yau turned 65 this year, we asked experts all over the world, in both mathematics and physics, to comment on the applications and impacts of Calabi-Yau manifolds and on the work of Yau. What followed is the mathematical garden of Yau and Calabi-Yau manifolds as seen through their eyes. We hope you will enjoy a walk through this joyful, colorful, and mind-expanding garden.

In mathematics

> Calabi-Yau manifold is the treasure of human civilization, although many of its mysteries are yet to be explored. In view of its surrounding theorems, tools, questions and conjectures, it has been an extremely active and important research field.

Lo Yang
Academy of Mathematics and System Sciences, Chinese Academy of Sciences
(Hua Loo-Keng Mathematics Prize 1997)

> There is no doubt that Calabi-Yau manifolds have become justly famous. The mere fact that they are now abbreviated to CY manifolds testifies to their importance.
>
> In both geometry and physics the Kummer surface and its relatives have been of central interest since the 19th century. They are the natural successors to elliptic curves, and look how long these curves have fascinated mathematicians. Yau has pioneered many important ideas in global geometric analysis and has influenced much of the mathematics and physics of our time.

Michael Atiyah
University of Edinburgh
(Fields Medal 1966; Abel Prize 2004; King Faisal International Prize 1987)

Calabi-Yau manifolds form a very specific class of mathematical objects, which play an increasingly important role in various branches of mathematics and physics. They are fundamental in algebraic geometry, where they stand somehow in the middle of the complexity hierarchy of algebraic varieties: not too simple, but still accessible. In differential geometry they provide the simplest examples of the so-called Kähler-Einstein manifolds, a very particular and very rich structure. They are fundamental in mathematical physics, in particular in superstring theory: they are the basic objects of mirror symmetry, a surprising duality phenomenon which is one of the major discoveries of the last 30 years in theoretical physics.

Arnaud Beauville
University of Nice, France

I first met Shing-Tung Yau when he joined the Stanford Mathematics Department in 1972. It was immediately clear that he had a remarkable talent and passion for mathematics, not just in his own field of differential geometry, but across the whole spectrum of deep open problems in mathematics. When hearing him talk about mathematics or Chinese civilisation in those days, his many ideas are perhaps best described by the line from Po Chü-i: "Like pearls large and small cascading on a plate of jade."

Yau's whole career since Stanford has borne out the immense promise which was then apparent to all of us who knew him well. His solution of the Calabi conjecture has brought into mathematics a class of complex manifolds which seem likely over the next century to take over the central role played by elliptic curves in the evolution of mathematics in the 19th and 20th century. Indeed, Calabi-Yau manifolds have already been shown to be of great importance in physics, and one can speculate that their arithmetic mysteries, still largely untouched by number-theorists, could some day throw light on even some of the most intractable ancient problems of the subject.

In addition to his own mathematical research, Yau has played an immense role since the middle 1990's in the rebuilding of mathematics in China. His creation and organization of the ICCM's has brought together mathematicians of Chinese origin from all over the world, and built up a new esprit de corps amongst them. He has known how to persuade the highest political voices in China that high-level mathematical research is of great importance for the future of China, and deserves good financial support. Above all, Yau has striven always to foster the very finest mathematical research, and to give gifted young Chinese mathematicians every chance to develop their talents. Just one of many examples of Yau's legacy in this direction, which I enormously appreciate, has been his creation of an outstanding group of number-theorists in the Chinese Academy of Sciences in Beijing. It seems no exaggeration to say that Yau's myriad contributions to all aspects of the mathematical world must place him amongst the greatest scholars in the long and rich intellectual history of China.

John Coates
University of Cambridge
(Senior Whitehead Prize 1997)

The notion of Calabi-Yau manifolds has been generalized to algebras (and categories), and they are used in classifying topological field theories.

Ralph Cohen
Stanford University

Calabi's conjecture from the 1950's, concerning metrics with prescribed Ricci tensor on compact complex manifolds, and Yau's verification of the conjecture from the 1970's are historic landmarks in differential geometry. The verified conjecture yields the first examples of non-flat compact manifolds with vanishing Ricci curvature; answering a fundamental question in global Riemannian geometry. More generally, the nature of the conjecture and its solution-involving a deep interaction between nonlinear elliptic partial differential equations and differential geometry—signalled a radical new development which has pervaded the field in the ensuing 40 years. At a detailed, technical, level, the crucial ideas and estimates for the complex Monge-Ampère equation introduced by Yau at that time are still being refined and exploited in current developments.

The study of elliptic curves goes back to the 18th century origins of algebraic geometry and complex analysis. By the middle of the 20th century, many of the special features of $K3$ surfaces had emerged. Nowadays, with inspiration from string theory in theoretical physics and from developments in algebraic geometry, we see at least a glimpse of the extraordinary and beautiful properties of Calabi-Yau manifolds in all dimensions. A huge amount is known but there are profound questions and mysteries which will surely occupy generations of geometers to come.

Simon Donaldson
Stony Brook University
(Fields Medal 1990; Crafoord Prize 1994; Shaw Prize 2009;
Breakthrough Prize in Mathematics 2014)

For me, the main "application" of Calabi-Yau manifolds is that they are very mysterious and challenging. In fact, this is true already for $K3$ surfaces. (As far as I know, one of the reasons for the name $K3$ was to hint that these surfaces are even less accessible than the famous mountain $K2$ in the Himalayas.) E.g., the proof of the Torelli theorem for $K3$ surfaces is mysterious (I doubt that we know a direct way of constructing a $K3$ surface from its periods). The existence of the Calabi-Yau metric on a $K3$ surface (e.g., on a quartic in $P3$) is again a mystery for me: already in this particular case one needed the power of geometric analysis (and the intellectual power of Prof. Yau) to construct it.

Vladimir Drinfeld
Yale University
(Fields Medal 1990)

Elliptic curves have a long history. In complex geometry, in algebraic complex geometry, their moduli, the j-line is a fundamental invariant. In arithmetic geometry, the Birch and Swinnerton-Dyer conjectures predicts a property of the L-function of elliptic curves over a number field. In positive characteristic, over a finite field, Hasse proved for elliptic curves what later was called the Weil conjectures. $K3$ surfaces have also a long history, even if shorter. In complex geometry, in algebraic complex geometry, their moduli, notably the Torelli theorem (Siu), are fundamental objects and theorems. I do not know about properties of the L-function over number fields. But over finite fields, Deligne proved first the Weil conjectures for $K3$ surfaces before he could prove them in general. Ogus studied the Torelli theorem in positive characteristic, and recently (Madapusi Pera and others) proved the Tate conjecture for divisors in positive characteristic. Those are just a few examples of the role of elliptic curves and $K3$ surfaces in algebraic and arithmetic geometry. As we know, Calabi-Yau varieties generalize to higher dimension elliptic curves and $K3$ surfaces. We do not know much about their moduli or Torelli theorem, nor do we know much about their L-functions. In positive characteristic, their lift to characteristic 0 is not even known, it is also a question of what definition one takes in positive characteristic. So the field is quite open.

Hélène Esnault
Freie Universität Berlin
(Gottfried Wilhelm Leibniz Prize 2003)

I think that the Calabi-Yau theorem and the Donaldson-Uhlenbeck-Yau theorem are the two most important theorems in complex geometry in 20th century after Kodaira's work. Using these two theorems, one can study the moduli spaces of CY manifolds and of stable vector bundles. On the other hand, nowadays these two theorems are still very useful. For example, Demailly and Paun used the CY theorem to study the Kähler cone and got a very deep theorem. Voison used the Donaldson-Uhlenbeck-Yau theorem to prove that the Hodge conjecture is false on the Kähler manifold.

As to the applications of estimates of Yau's paper on Calabi conjecture, I mention only one paper. In 2008, Boucksom, Eyssidieux, Guegj, and Zeriahi wrote a very important paper, "Monge-Ampère equations in big cohomolody classes". In Sections 5 and 6 of that paper, the authors made numerous citations of Yau's estimates.

Jixiang Fu
Fudan University
(Morningside Silver Medal 2010)

The impact of Calabi-Yau manifolds on mathematics—number theory, algebraic and differential geometry, symplectic geometry, and theoretical physics—has been enormous. For example, in number theory, Calabi-Yau manifolds are natural generalizations of abelian varieties and $K3$ surfaces, and there is a lot of work trying to match their zeta functions over number fields with L-functions coming from automorphic forms.

Benedict Gross
Harvard University
(Cole Prize 1987)

Certainly Calabi-Yau manifolds have been widely important. In algebraic geometry, they are one of the key classes of varieties in classification theory, and many mysteries remain. They are crucial in string theory, providing the compactification of space-time to ten dimensions. As a result, there has been a great deal of influence of string theory on algebraic geometry, especially through mirror symmetry. The Strominger-Yau-Zaslow conjecture also introduced differential geometric ideas into mirror symmetry, and has a strong influence on a great deal of work, even though it remains unresolved.

Mark Gross
University of Cambridge

It is hard to overstate the impact that Calabi-Yau manifolds have had in both mathematics and physics. Yau's positive solution of the Calabi conjecture shows the existence of Ricci-flat Kähler metrics on Kähler manifolds with vanishing real first Chern class. This result is what fundamentally enabled the use of Calabi-Yau manifolds for string compactification. Progress in the geometry of Calabi-Yau manifolds and their moduli has been paralleled by similar progress in string theory, leading to the rich cross-fertilization between mathematics and physics that is still going strong 30 years later.

Yau has been a major player in the development of this field over the decades, in mathematics, in physics, and in the hybrid offspring of these two disciplines. His contributions are vast and seminal, extending over a broad swath of the field including work on Gromov-Witten invariants, Picard-Fuchs systems, mirror symmetry, the A-model, elliptic fibred Calabi-Yau manifolds, heterotic string theory, F-theory, and much more. I also mention in this context his celebrated construction with Uhlenbeck of Hermitian Yang-Mills connections on stable bundles, which has also played a fundamental role in the geometry/physics collaboration, extending beyond Calabi-Yau manifolds and string theory.

Yau's work has both informed and inspired my own work and that of many others, and continues to do so. I am happy to join my colleagues in this tribute to Yau on the occasion of his 65th birthday.

Sheldon Katz
University of Illinois at Urbana-Champaign

The most striking applications of Calabi-Yau manifolds are in string theory. In mathematics the CY theorem has inspired the study of special metrics in many other contexts: holomorphic vector bundles, Fano manifolds, special Lagrangian submanifolds, $G2$ and Spin(7) manifolds. Special metrics and algebro-geometric stability enjoy a close and mutually beneficial relationship. Special metrics lead to Hausdorff moduli spaces. Another important feature of CY manifolds is the smoothness of the corresponding moduli spaces.

David Kazhdan
Hebrew University of Jerusalem and Harvard University
(Israel Prize 2002)

Calabi-Yau manifolds are Einstein manifolds (i.e. zero Ricci curvature) with compatible complex structures. Yau's theorem gives a complete characterization of such manifolds in the compact setting. In string theory, they are internal spaces whose geometry determines the physics of our four-dimensional spacetime. Mirror symmetry for Calabi-Yau manifolds is a duality between two seemingly very different types of geometry, namely complex geometry and symplectic geometry. The Strominger-Yau-Zaslow conjecture unveils the mystery of mirror symmetry as a geometric Fourier transformation along special Lagrangian fibrations. It has far reaching implications in almost all branches of geometry, topology and beyond. They include enumerative geometry, gauge theory, calibration theory, toric geometry, affine geometry, hyper-Kähler geometry, geometry of special holonomy, knot theory and even geometric Langlands in number theory.

Conan Leung
The Chinese University of Hong Kong
(Chern Prize 2010)

The beauty of the geometry of Calabi-Yau manifolds lies in its mirror symmetry of the algebraic geometry of the algebraic structures of these manifolds on one hand, and of the differential geometry arising from the Calabi-Yau metrics of these manifolds on the other hand. The study of mirror symmetry of Calabi-Yau manifolds, like a "gold rush", has provided both mathematicians and mathematical physicists a golden opportunity to reveal the inner beauty of the space we imagined ourselves to have lived in.

Jun Li
Stanford University
(Morningside Gold Medal 2001)

Professor Yau has a deep and broad understanding about mathematics as a whole, and he pays a lot of attention to the development of applied mathematics as well. This is clear from the speakers he has invited for important events like ICCM and the inaugural celebration for the Sanya Mathematics Forum. Personally I am grateful to his strong support of number theory, which is not his major field of research. For example, in each ICCM, there are number theorists receiving the Morningside Medals (gold or silver). Behind the scene, Prof. Yau is a strong supporter of Yitang Zhang's various awards, including his introduction to the "Academia Sinica" in Taiwan. Different from many other well-established top mathematicians in the world, Professor Yau pays a lot of attention to mathematics education. He initiated various levels of mathematics competitions, from high school to Ph.D. students, open to students of Chinese origin all over the world. He is the greatest fund raiser among all mathematicians. He has founded several mathematics and applied mathematics centers, raised funds to build new buildings to house many mathematics institutes and centers and to finance their operations, to support the ICCM, and to offer competition awards, etc. He has educated many outstanding mathematicians. His influence and impact on the advancement of mathematics and on the Chinese mathematics community are far-reaching. Some of my Indian friends are very envious of the Chinese, as they feel sorry about the lack of an Indian mathematics leader like Yau.

Winnie Li
Pennsylvania State University
(Chern Prize 2010)

The Estimate of Yau on the complex Monge-Ampère equations is the only global estimate available even now, after its establishment more than three decades ago. Its impact on algebraic geometry is profound, particularly in the classification of minimal models with singularities.

Chang-Shou Lin
Taiwan University
(Morningside Gold Medal 1998)

Calabi-Yau has influenced almost all major areas of mathematics, from number theory, algebra, algebraic geometry, topology, geometric analysis and partial differential equations. Calabi-Yau has become a household name in mathematics and in the real world. Almost every day we hear lectures about Calabi-Yau manifolds, Calabi-Yau algebras, Calabi-Yau categories, Calabo-Yau metrics and Calabi-Yau equations, to mention a few. Calabi-Yau manifolds and their moduli spaces are among the most beautiful and most mysterious mathematical objects I have ever seen.

Kefeng Liu
University of California, Los Angeles and Zhejiang University
(Morningside Gold Medal 2004)

Calabi-Yau manifolds in the strict sense are complex Kähler manifolds, the special unitary group as holonomy. The Calabi conjecture (as proved by Yau) tells us (among other things) that there are plenty of them; for a compact Kähler manifold with trivial canonical bundle we can modify the Kähler metric within its cohomology class such that it becomes a Calabi-Yau manifold. This theorem has (and continues to have) far reaching consequences outside differential geometry (to which it belongs): it permeates algebraic geometry as well as mathematical physics. For instance, this led to a simple and short proof that the Torelli map for Kähler $K3$ surface is surjective. But its impact is being even more strongly felt in the mathematics surrounding mirror symmetry, for it is there the central notion. Indeed, via mirror symmetry this has become so ubiquitous that virtually every person considering him or herself an intellectual has heard it pronounced or seen it written.

Eduard Looijenga
Tsinghua University

Calabi-Yau manifolds belong to the intersection of several geometric worlds (complex, symplectic, and Riemannian geometry) and a world of theoretical physics (quantum string theory). They have an incredible amount of beautiful properties, even if one restricts oneself to the dimension three.

On various scales, our understanding of the physical world is dominated by several basic pictures. We live in Euclidean 3-dim space on a human scale. We live in a (pseudo-)Riemannian Einsteinian 4-dim space-time on a cosmological scale. And on the scales of Planck size and lower, our world is filled with minuscule Calabi-Yau threefolds, determining physics of the observable matter.

Yuri Manin
Max-Planck-Institut für Mathematik, Bonn, Germany and Northwestern University
(Nemmers Prize in Mathematics 1994; Schock Prize 1999; The King Faisal International Prize of Science 2002)

S.-T. Yau's proof of the Calabi conjecture stands as the foundation for much subsequent work, in understanding the structure of Kähler manifolds. These manifolds lie at the crossroads of Riemannian, complex analytic and algebraic geometry. Of singular importance is the subset of Kähler manifolds now known as Calabi-Yau manifolds. These also play a central role in arithmetic geometry, symplectic geometry and string theoretic physics. For example, three-dimensional CY manifolds provide—amongst many other things—the original geometries contributing to the rich field of "mirror symmetry", a field that has profoundly influenced enumerative algebraic geometry, symplectic geometry, and string theory.

Barry Mazur (Veblen Prize 1966; Cole Prize 1982; National Medal of Science 2011)
Peter Kronheimer (Veblen Prize 2007)
Clifford Taubes (Veblen Prize 1991; Shaw Prize 2009)
Harvard University

In algebraic geometry and complex analytic geometry, the theory of Calabi-Yau manifolds is a special and important branch. First of all, Calabi-Yau manifolds constitute one of the five fundamental elements (the remaining four are rationally connected manifolds, complex tori, complex symplectic manifold and manifolds of general type). Second, the manifolds appear as the invisible 6 dimension of our universe in the theory of super strings, which is anticipated to be able to systematically explain the known four forces in physics. Finally, the intrinsic structure of Calabi-Yau manifolds is so mysterious. So far, there are so many examples that we cannot be sure if there are only finitely many deformation types; although counting rational curves on a given Calabi-Yau manifold is an essential recipe in super string theory, we do not know in general if it carries a single rational curve.

Yoichi Miyaoka
University of Tokyo

The existence of Calabi-Yau manifolds, established by Professor Yau in his famous solution of the Calabi conjecture, is one of the most important landmarks in modern mathematics. It is a landmark in the theory of partial differential equations, as it provided the first solution of a full-fledged complex Monge-Ampère equation. It is a landmark in differential geometry, as it is the first significant generalization of the uniformization theorem of nineteenth century mathematics. It is a landmark in algebraic geometry, as it opened the way to the moduli theory of the most important and natural generalizations of elliptic curves. And it is a landmark in mathematical physics, as Calabi-Yau manifolds also arise as vacuum configurations of superstrings. It is hard to think of any other breakthrough which has had so much impact and on so many fields of mathematics.

Duong Phong
Columbia University
(Bergman Prize 2009)

The study of Calabi-Yau manifolds has engaged a vast array of mathematicians and physicists, leading to fantastic new insights in algebraic, differential, and symplectic geometry, category theory, and mathematical physics, over the past three decades. The beautiful properties of their Ricci-flat metrics lead to their primordial role in string theory, motivating the new domain of mirror symmetry. This is now one of the most productive areas of investigation in mathematics, encompassing the theory of pseudo-holomorphic curves and Gromov-Witten invariants, Fukaya categories in symplectic geometry, moduli spaces of Bridgeland stability conditions, character varieties, the study of derived categories, modular forms, and many other things. Major conjectures such as Kontsevich's homological mirror symmetry approach are strongly motivating a wide mathematical community.

Carlos Simpson
University of Nice, France

Before Yau's proof of the Calabi conjecture, almost nothing was known about Calabi-Yau manifolds, and mathematicians had no tools for studying them. His proof has made possible things never dreamt of beforehand. Any result we now know about these manifolds uses Yau's theorem in its proof.

Most importantly, the result created new links between different areas of mathematics, leading to whole new subjects that did not exist before 1980. Many of these new ideas come from "mirror symmetry"—a series of breathtaking predictions from string theory that have dominated large parts of modern geometry in the last 25 years. Yau's theorem provides the links to string theory by solving Einstein's equation on Calabi-Yau manifolds, and has also provided our best approach to solving mirror symmetry, using special Lagrangian submanifolds and the Strominger-Yau-Zaslow conjecture.

Richard Thomas
Imperial College of London

Prof. Yau's proof of the Calabi conjecture is a true gem of mathematics. It introduced the use of Moser iteration to obtain uniform estimates for solutions of geometric PDEs; he then obtained a second-order estimate for the solution which surprisingly did not depend on a gradient bound, but only on the uniform estimate; and then in a true *tour de force*, he derived third-order derivative bounds for the solution, using the second-order bounds. All of these techniques were groundbreaking, and have since then become part of the standard toolbox of any complex geometer and used countless times, to the point that some people nowadays forget that they originated in Prof. Yau's work.

Valentino Tosatti
Northwestern University

What I find amazing about Calabi-Yau manifolds is their impact in physics and the resultant modelling of the universe by ten-dimensional space. At first this seems bizarre but I imagine that early in the last century relativity theory and four-dimensional space-time was also a great conceptual leap for science, so it would be very interesting to look back on Calabi-Yau manifolds in the decades ahead.

Neil Trudinger
Australia National University
(Australian Mathematical Society Medal 1981)

Calabi-Yau manifolds are central in algebraic geometry, because they appear naturally as fibers of the Iitaka fibration of any variety and they sit between two extreme classes, namely varieties of general type, with many holomorphic contravariant tensors, and rationally connected varieties, containing big families of rational curves. Calabi-Yau geometry is beautiful and still mysteriously related to mathematical physics (mirror symmetry) and also to symplectic geometry. Indeed, varieties of general type tend to have trivial Gromov-Witten invariants, while in the rationally connected case, the quantum cohomology corrections tend to be polynomial; in the CY case, the Gromov-Witten potential is actually an infinite series, showing that the study of rational curves in CY manifolds is very rich. The work of Yau establishing the existence of Kähler-Einstein metrics and the stability of their tangent bundle was the starting point for many further progresses, but a number of big questions remain open, both on the analytic geometry and on the algebraic geometry sides. One of them is the relation between the Kodaira dimension zero and the Calabi-Yau property. Another one concerns the classification of CY varieties: Are there infinitely many families in a given dimension? This is unknown in dimension 3, and also for hyper-Kähler manifolds.

Claire Voisin
University of Paris VI and École Polytechnique
(EMS Prize 1992; Sophie Germain Prize 2003;
Satter Prize 2007; Clay Research Award 2008; Shaw Prize 2017)

Being the potential candidate as the inner space in string theory, as well as the building block towards the classification of higher dimensional algebraic varieties, Calabi-Yau manifolds appear to be the most fundamental and influential geometric objects after Yau's resolution of the Calabi conjecture in 1976. Even with many insightful discoveries done by many mathematicians and physicists in the past thirty or more years, it is believed that a significant part of 21st century mathematics will be devoted to uncovering the underlying mysteries of Calabi-Yau spaces.

Chin-Lung Wang
Taiwan University
(Morningside Silver Medal 2001)

Ever since the rise of mirror symmetry some twenty years ago, Calabi-Yau manifolds have been established as one of the most versatile subjects in mathematics and physics.

Spectacular results have been obtained in many different areas such as arithmetic, algebraic, differential and symplectic geometry, differential (Picard-Fuchs) equations, and physics (string theory), with unmistakable focus of research on Calabi-Yau manifolds.

There are numerous examples in which Calabi-Yau manifolds are encountered at the junction of two or more different fields. For instance, the list of relations with physics starts with the unexpected relevance of algebraic varieties as compactification manifolds for string theory, continues with the even less expected enumerative predicts resulting from mirror symmetry, and can be extended all the way to the present day and to the future.

Noriko Yui
University of Toronto

To appreciate the impact of Yau's solution of the Calabi conjecture, we should remember that before the work of Yau, nonlinear PDE was a difficult subject in analysis hardly related to algebraic geometry. Such a link was established only through the insight and technical power of Yau. It gave great inspiration to people working in the field. As an example, Donaldson, in his essay in the book *Fields Medalists' Lectures*, wrote (after Hitchin suggested to him the thesis problem of Hitchin-Kobayashi correspondence): "I soon began to tinker with analytical approaches to Hitchin's conjecture, perhaps motivated by the renowned work of Yau, a few years before, on the Calabi conjecture." Since then, nonlinear analysis has been applied to solve many important problems in disciplines other than algebraic geometry. On the other hand, Calabi conjecture (Yau's theorem) continues to play important roles in recent developments in complex geometry. Here, we should also point out that even in situations where Yau's theorem does not apply directly, the estimates established in his fundamental paper "On the Ricci curvature of a compact Kähler manifold and the complex Monge-Ampère equation" have found numerous applications. A recent example is the work of Chen-Donaldon-Sun verifying a conjecture of Yau relating stability to the existence of KE metrics on a Fano manifold.

Xiping Zhu
Sun Yat-sen University
(Morningside Silver Medal 2004)

In physics

Shing-Tung Yau is one of the most influential people in the international mathematics community. His long-lasting influence on the young generation of Chinese mathematicians and on the Chinese mathematics community is of historical scope.

Chen-Ning Yang
Tsinghua University
(Nobel Prize in Physics 1957; National Medal of Science 1986;
Benjamin Franklin Medal 1993; Einstein Medal 1995)

It is hard to overstate the importance Calabi-Yau manifolds in physics. String compactifications on Calabi-Yau's are the canonical paradigm for connecting string theory to reality, showing that seemingly random and complicated four dimensional particles and interactions of the sort we see in the real world can arise in a simple and striking way from a beautiful higher-dimensional geometry. The study of string theory on Calabi-Yau's has also yielded deep clues into the inner workings of the theory itself, ranging from conifolds to mirror symmetry to twistor string theory. One can only expect this trend to continue in the coming years.

Nima Arkani-Hamed
Institute for Advanced Study, Princeton
(Gribov Medal of the European Physical Society 2003;
Fundamental Physics Prize 2012)

The two pillars on which our understanding of nature is based are Einstein's theory of general relativity, which is used to describe gravity, and quantum field theory, which describes the non-gravitational forces. These theories contradict each other, and a fundamental problem in modern theoretical physics is the incorporation of Einstein's theory into quantum field theory. Such a theory is known as quantum gravity, and the leading theory of quantum gravity is string theory. Calabi-Yau manifolds figure prominently in string theory. It has been possible to show not only that string theory unifies gravity and quantum field theory but that particles like electrons and quarks and known forces can arise. And so it is a candidate as a unified theory of all forces. The needed constructions are based on Calabi-Yau manifolds.

Melanie Becker and Katrin Becker
Texas A&M University

I would say that CY manifolds play an essential role in connecting string theory with the observed 4 dimensional world and with elementary particle physics. Without CY manifolds there would be very little interest in string theory and the many advances in mathematics that are due to connections between mathematics and string theory would not have been made.

Philip Candelas
University of Oxford

Over the past 20 years our understanding of Calabi-Yau manifolds has grown from rudimentary to rich. Shing-Tung Yau's contributions have shaped the study of these important manifolds from the start. His solution to Calabi's conjecture established the existence of a unique Ricci-flat Kähler-Einstein metric in each Kähler class on compact manifolds with trivial canonical bundle. Yau's approach was famously non-constructive; to this day there is no known closed-form expression for this metric on any of the hundreds of millions of (families of) known examples. Fortunately for both physics and mathematics, when string-theorists discovered the importance of Calabi-Yau manifolds in the mid-1980's as part of their quest to create a theory of quantum gravity, the existence of the Calabi-Yau metric proved sufficient for this critical application. It is instead the topological and Hodge-theoretic features of Calabi-Yau manifolds which are crucial for physical application. The interplay between the Hodge-theoretic properties and moduli of Calabi-Yau manifolds, often with string-theoretic motivation (e.g., "B-model" geometry in mirror symmetry), has been a consistent theme of research at the mathematics/strings interface ever since. From his early studies of the mirror map through his recent work on tautological local systems, Shing-Tung Yau has continually led the way. In large part through his efforts, Calabi-Yau manifolds have gone from being little understood curiosities to geometric gems of the first order within a single generation.

Charles Doran
University of Alberta

Superstring theory, the leading candidate for a theory of fundamental interactions, is formulated in ten dimensions. If superstring theory describes our universe, six of these dimensions must be in the form of a very small compact manifold, whose metric must solve Einstein's equation. Calabi-Yau manifolds were the first proposal for such solutions, and though others have been found since, they are still the best motivated and leading proposal. Their topology and geometry leads directly to the spectrum and properties of particle physics in four dimensions, so someday this proposal may find experimental verification.

Michael R. Douglas
Simons Center, Stony brook
(Sackler Prize 2000)

Prof. Yau has suffused the mandate of mathematics, and specifically of algebraic geometry, into the rules of physics to an unprecedented degree. By his contributions to human understanding of our place in nature and reality, he has earned the greatest respect and honor in both fields.

James Gates
University of Maryland
(National Medal of Science 2013; Mendel Medal 2013)

The solution by Prof. Yau of the Calabi conjecture has led to important advances in both algebraic and differential geometry and to a revolution in the picture of the world provided by Super String Theory. Calabi-Yau spaces provide the richest store of Ricci flat manifolds, and manifolds of special holonomy and these special properties are essential for their current role in phenomenologically viable compactification scenarios and in applications to supersymmetric quantum field theories. It is difficult to think what modern theoretical physics would be without them.

Gary Gibbons
University of Cambridge

One of the most amazing developments in theoretical physics in recent decades is the realization that the beautiful geometrical manifolds constructed by E. Calabi and S. T. Yau provided the framework for solutions of 10 dimensional superstring theory wherein 6 of the spatial dimensions are compactified to form a Calabi-Yau manifold. In a manner that would have pleased the Greek philosophers and mathematicians, the nature of matter and the structure of forces is dictated, in Calabi-Yau compactifications, by the geometrical properties of these manifolds. The study of these solutions and these manifolds has deeply influenced developments in both physics and mathematics for the last thirty years.

David Gross
University of California, Santa Barbara
(Dirac Medal 1988; Harvey Prize 2000;
Nobel Prize in Physics 2004)

As a physicist I am always amazed by the great impact of mathematics on physics. There have been at least five profound revolutions in the history of physics, and a sixth is now in progress:

(1) The revolution of *Newtononian mechanics* in the 17th century, for which Newton invented calculus as its mathematical foundation.

(2) The revolution of Einstein's *special relativity* theory in 1905, which unified Newtonian mechanics with the Maxwell equations of electromagnetism, and is formulated by the Lorentz group in mathematics.

(3) The revolution of *quantum mechanics* (from about 1900 to the 1920s), which is mathematically based upon Hilbert space and linear algebra.

(4) The revolution of Einstein's *general relativity* (about 1916) which is deeply rooted in Riemann geometry and opens up the era of modern cosmology.

(5) The revolution, from the 1940s to the 1970s, of *quantum field theory*, especially the *standard model*, which unifies special relativity with quantum mechanics, describing electromagnetic, weak, and strong forces in Nature. Its mathematical foundation includes functional path integrals, gauge groups, and fiber bundles.

(6) The current greatest challenge in physics is to unify gravitational force with the other three forces of nature, and to understand the origin of space-time itself. This definitely calls for a profound new revolution. The leading candidate for such a revolution is the ten-dimensional string theory or eleven-dimensional M-theory, containing hidden 6D Calabi-Yau manifolds, as introduced in the 1980s and 1990s. The impact of Calabi-Yau on modern string theory is far-reaching, similar to the role of Riemann geometry in general relativity. Professor Yau has been standing at the center of these exciting developments since then. He has built up the mathematical foundation, and, in cooperation with prominent physicists, is in the leading position at the interface of mathematics with physics today, at the dawn of the 21st century.

Hong-Jian He
Tsinghua University

Calabi-Yau spaces have had a huge impact in physics. They first gained prominence with the realization that compactifying string theory on one of these geometries gives rise to physics very close to the standard model. It was later shown that singularities in Calabi-Yau spaces can give rise to nonabelian gauge symmetries like the ones seen in physics. Calabi-Yau spaces were also used to give one of the first examples of topology change in quantum gravity. Physicists continue to explore new applications of these remarkable geometries.

Gary Horowitz
University of California, Santa Barbara
(Xanthopoulos Prize 1993)

The investigation of the algebraic topology of Calabi-Yau manifolds has been at the very heart of many of the most striking developments in physical mathematics for the past thirty years. In addition to the important role it has played in string compactification, and in efforts to make realistic models of string theory, investigations into Calabi-Yau compactifications led to the momentous discoveries of mirror symmetry and homological mirror symmetry. The far-reaching ramifications of these discoveries are still being worked out to this day.

Gregory Moore
Rutgers University
(Dannie Heineman Prize 2014)

Einstein's vision of physical laws emerging from the shape of spacetime has been expanded by the higher dimensions of string theory. This vision has transformed not only modern physics, but also modern mathematics. By proving the existence and uniqueness of a large class of solutions to the higher dimensional Einstein equations, Shing-Tung Yau brought these two fields together. This work has had countless implications, from laying out the landscape of string vacua to the discovery of mirror symmetry, in which string theory modifies the basic concepts of geometry.

Joseph Polchinski
University of California, Santa Barbara
(Dirac Medal 2008; Physics Frontier Prize 2013)

Calabi-Yau manifolds were first used in string theory in the context of construction of semi-realistic compactification of heterotic string theory. Since then they have played a central role in many aspects of string theory. Some of the examples I can mention are mirror symmetry, topological string theory, string dualities, and black holes. They continue to play a major role in string theory today and this is likely to continue in the future.

Ashoke Sen
Harish-Chandra Research Institute, Allahabad
(Fundamental Physics Prize 2012; Dirac Medal 2014)

I am sure that many mathematicians and string theorists will comment on the extremely significant role Calabi-Yau manifolds play in mathematics and string theory. Therefore, I would like to focus on a more narrow aspect—their impact on quantum field theories. A two-dimensional non-linear sigma model with (2,2) supersymmetry, whose target space is a Calabi-Yau manifold is a very interesting quantum field theory. In addition to its supersymmetry, it is also conformally invariant. Such theories are extremely rich, exhibiting many important phenomena, some of which are exactly calculable.

Therefore, they are a useful laboratory teaching us about more generic properties of quantum field theories. Lessons extracted from this laboratory have already revolutionized our general understanding of quantum field theory. Examples are the role of exactly marginal deformations, the chiral ring, mirror symmetry/duality, topological field theory, etc. I have no doubt that the study of such quantum field theories will continue to be productive and that it will inspire many more insights.

Nathan Seiberg
Institute for Advanced Study, Princeton
(Heineman Prize 1998; Fundamental Physics Prize 2012)

Over the last thirty years Calabi-Yau manifolds have come to play a central role in the search by physicists for a unified theory of nature. Compactification of ten-dimensional superstring theory on a Calabi-Yau threefold yields a four dimensional universe with properties remarkably like the one we inhabit. A wealth of both mathematical and physical insights have been gained from ongoing analyses of the rich structure of these compactifications.

Andrew Strominger
Harvard University
(Physics Frontiers Prize 2013)

The counterpart of a DNA code in physics is a Calabi-Yau manifold with branes and fluxes—or other data needed to describe a compactification or a background.

Leonard Susskind
Stanford University
(Sakurai Prize 1998)

Calabi-Yau manifolds have been the source of many new and deep ideas in string theory. Not only have they provided us with concrete mathematical objects which are starting points of extracting particle physics data from string theory, but also, in the context of topological strings they have led to new revolutionary symmetry principles such as mirror symmetry and dualities. S.-T. Yau has played a central role in many of these developments.

Cumrun Vafa
Harvard University
(Dirac Medal, 2008; Eisenbud Prize 2014;
Physics Frontiers Prize 2013)

Calabi-Yau manifolds are used in the context of string theory to make models of particle physics unified with quantum gravity. These models, which were originally constructed in the mid-1980's, were the first reasonably realistic models of particle physics unified with gravity, and—together with variants that have been constructed since, usually with Calabi-Yau manifolds as one of the ingredients—they remain the most interesting candidate models that we have today. In addition, compactification on Calabi-Yau manifolds has given important examples in the study of strong coupling or nonperturbative phenomena in string theory. These examples have been remarkably rich and much insight has come from their study.

Edward Witten
Institute for Advanced Study, Princeton
(Fields Medal 1990; Crafoord Prize 2008;
Fundamental Physics Prize 2012; Kyoto Prize 2014)

Inscriptions of experts

丘成桐是当今国際数学界
最有影響的人物之一。他对
中国青年数学家
的影響更是歷
史性的貢獻。

杨振宁
二〇一四年
11月

建筑的美离不开
数学的美。

翁帆
2014年8月

紫气腾茂
福寿安康

张广生
2014.8.4.

事业常青，
松鹤延年。

陈旭
2014.8.4

感谢丘成桐教授
为中国培育一代又一
代数学人材。
陈东升 张东梅
二〇一四年八月.

感谢丘成桐教授：
您对我的好，是对中国的好，是
对数学的好。我无法用言語向您
表示感谢：

张益唐
2014.8.4.

Appendix: Eugenio Calabi

What follows is a brief biography of Calabi with a selective summary of his most important work.

Born 11 May 1923 in Milan, Italy, Calabi earned a Bachelor of Science in chemical engineering from MIT in 1946 and a Master of Arts from the University of Illinois in 1947. He was a Putnam Fellow as an undergraduate at MIT in 1946. He received his Ph.D. degree from Princeton University in 1950 under the supervision of Salomon Bochner. In his dissertation, Calabi developed the theory of isometric complex analytic imbedding of Kähler manifolds.

After graduation, he became an instructor at Princeton University, before joining the faculty of Louisiana State University and later the University of Minnesota. His arrived at the University of Pennsylvania in 1964, was appointed to the Thomas A. Scott Professorship of Mathematics in 1967, and held the post until his retirement in 1993. Calabi's honors include the Guggenheim Fellowship (1962), election to the National Academy of Sciences (1982), the Leroy Steele Prize for Lifetime Achievement in Mathematics (1991) from the AMS, and an Honorary D.Sc. from the University of Pennsylvania (2014).

Calabi made many fundamental contributions to complex differential geometry. In 1950s, he [1] set forth his famous conjecture regarding the existence of Kähler metrics with prescribed Ricci form on Kähler manifolds. The Calabi conjecture was proved by S.-T. Yau two decades later, leading to the creation of Ricci flat compact Kähler manifolds, now known as "Calabi-Yau manifolds". Calabi-Yau manifolds are higher dimensional analogues of elliptic curves and $K3$-surfaces, and provide some of the most important and promising Kähler manifolds. They also provide the foundation for string theory, and many believe the universe to have extra dimensions that are curled up in 6-dimensional Calabi-Yau manifolds.

Calabi pioneered the study of rigidity of locally symmetric space. In a very influential paper [8], Calabi and Vesentini proved the rigidity of the complex structure of a compact quotient of a bounded symmetric domain. Calabi [4] also proved deformation rigidity of compact hyperbolic n-spaceforms for $n \geq 3$.

Calabi [2, 6] also made several pioneering contributions to modern theory of affine differential geometry. Calabi [5] was the first to use ideas of twistor theory in the study of minimal surfaces. In [7], Calabi introduced the important concept of extremal Kähler metrics, which is a natural generalization of Kähler-Einstein metrics. Calabi [3] also constructed some of the most interesting non-Kähler complex manifolds.

References

[1] E. Calabi, The space of Kähler metrics, Proc. Internat. Congress Math. Amsterdam **2**, (1954), 206-207.

[2] E. Calabi, Improper affine hyperspheres of convex type and a generalization of a theorem by K. Jörgens, Michigan Math. J. **5** (1958), 105-126.

[3] E. Calabi, Construction and properties of some 6-dimensional almost complex manifolds, Trans. Amer. Math. Soc. **87** (1958), 407-438.

[4] E. Calabi, On compact, Riemannian manifolds with constant curvature. I, 1961 Proc. Sympos. Pure Math., Vol. III, pp. 155-180. American Mathematical Society, Providence, R.I.

[5] E. Calabi, Minimal immersions of surfaces in Euclidean spheres, J. Differential Geom. **1** (1967), 111-125.

[6] E. Calabi, Complete affine hyperspheres. I, Symposia Mathematica, Vol. X (Convegno di Geometria Differenziale, INDAM, Rome, 1971), pp. 19-38. Academic Press, London, 1972.

[7] E. Calabi, Extremal Kähler metrics, Seminar on Differential Geometry, pp. 259-290, Ann. of Math. Stud., 102, Princeton Univ. Press, Princeton, N.J., 1982.

[8] E. Calabi and E. Vesentini, On compact, locally symmetric Kähler manifolds, Ann. of Math. (2) **71** (1960), 472-507.

Editor's Comments

A foreword to *Selected Expository Works of Shing-Tung Yau with Commentary* seems to complement the comments of many people on Yau, and hence it is included in this book as well.

Complex Geometry from Riemann to Kähler-Einstein and Calabi-Yau
ALM 38, pp. 109–116

Shing-Tung Yau, His Mathematics and Writings

Lizhen Ji

1 Why selected works

There has been a long tradition of publishing collected or selected works of distinguished mathematicians. There are several good reasons for doing this, and it has served many purposes. Probably the most obvious one is that collected and selected works provide an easy access to papers that are scattered in different journals, some of which are not easily accessible to many people. Otherwise, few people, if any, will take the time and trouble to dig up all the papers of their admired mathematicians—especially not those papers that are far away from their interests, and read them. On the other hand, reading papers of a master dealing with different subjects or areas conveys the underlying unity and hence a big picture of mathematics, and it also allows one to gain a historical perspective (or to enter the history). In other words, collected and selected works are more than the simple sum of individual papers.

Indeed, as Abel said famously, we learn "by studying the masters, not their pupils". Even though the world is becoming smaller, few people have many chances to interact with masters who are alive. Of course, the next best way to learn from masters is to read and study their collected works.

Naturally, publishing collected or selected works is also an honor to the authors of these papers. It should be mentioned that collected works of some people can bring honor of the genre of collected works.

Now, with the wide and easy use of e-papers and e-books, most papers in journals can be obtained easily online, and a mere reprinting of papers is probably not as valuable as before. Of course, the value of selected works still stands. For example, holding and reading a beautifully printed book is definitely different from viewing papers online or on e-book to the readers. But they should also provide something else. Several additional things seem to be reasonable: descriptions of how ideas in the paper were formed and time and place the papers were written, relations between papers with the advantage of hindsight, and developments of subjects after the papers were published, and visions for the future. In other words, they should explain the circumstances of the birth of papers and proper, impacts of the papers, and fitting these papers in the grand scheme of mathematics.

These additional things are especially important to beginners, non-experts and even some experts. Most people often concentrate on the best known theorems and most important papers of great mathematicians, but even masters struggled and stumbled sometimes on their mathematical trips. How they found good problems and their ways in their careers, made progress and reached peaks is best described by their own papers, recollections and commentaries, but not textbooks where everything is polished and presented in a streamlined matter, without mentioning that textbooks and research books might not cover some gems in the original papers that are not directly related to the themes of the books. But many people, especially younger ones, often prefer to read polished textbooks. Of course, reading mathematics papers can be difficult (more difficult than reading textbooks), and proper arrangement of related papers and additional guides from the masters are certainly valuable and helpful. Such collected or selected works of distinguished mathematicians often tell good stories of the authors and their mathematics, and browsing through them can be enjoyable and beneficial to people who are not interested in some specific results in the papers.

In these works of expository writings of Shing-Tung Yau, all these things are printed together with his survey papers and papers on open problems. One reason for restricting these volumes to expository papers of Yau is practical. Yau has been very creative and prolific. The collected works including all his papers (both research and expository papers) up to now will occupy too many volumes. Besides, he is also still very active and productive, and the time for collected works may not be ripe yet.

2 Why expository writing

Probably some explanation is needed for publishing these volumes of expository writings of Yau now. Briefly, it is the right time for Yau to share his perspectives and his vision on the broad area of geometric analysis, and his expository writings provide a unique means to this end. They will render a valuable service to the mathematics community.

Colloquium talks have been a common means of communication between mathematicians from different subjects, after they were made successful and popular by Klein and Hilbert in Göttingen about 100 years ago. More expository talks such as *Basic notion seminars* and *What is ...* have also sprung up in many places. They provide effective ways for people to learn and enjoy some beautiful pieces of mathematics, which are outside their fields of speciality. Though there are many books and papers dealing with all kinds of subjects in mathematics, one difficulty is that there are too many of them. It is difficult for people to find the right books and papers, and people may lack the motivation to read mathematics outside their specialities, especially when they involve difficult and technical material. Many people choose to study mathematics not for fame or fortune, but for the beauty and enjoyment of the discipline. To really appreciate the beauty and power of mathematics, one has to roll up one's sleeves and do the work. But not many people can work in many different subjects in mathematics. In the history of mathematics, only a few people have been universal mathematicians. Some obvious names in the recent times include Gauss, Riemann, Poincaré, Hilbert, Weyl, and Hadamard. In spite of the difficulties, one can still enjoy and appreciate many facets of the rich world of mathematics by listening to expository talks and talking to experts. In the abstract world of mathematics, direct interaction and communication is still vital, and the virtual internet is no replacement yet.

Next to listening to talks, one can try to read expository writings and informal comments and notes on technical papers. The former is like colloquium talks, and the latter is like conversations at the colloquium tea or dinner. Expository writings include books, survey papers and descriptions of open problems. It often happens that expository writings are less valuable in a short period than highly technical and original papers, which can give people priorities and more credit. But in the long run, books and expository papers might be read by more people and have a bigger and longer impact. Think of Euler. How many of his papers are still read by people now? But his two elementary books on analysis and algebras are still printed and read by many people. What about Hilbert? His paper on the open problems and his report (or survey) on algebraic number theory are probably most read among his papers. Among contemporary mathematicians, we can think of the expository writings of people such as Atiyah, Milnor and Serre, which have had a huge impact on the modern mathematics. Though not everyone likes every book written by Serge Lang, many people will probably agree that they certainly have had nontrivial impact on mathematics and have been enjoyed by mathematicians from different areas of mathematics.

There are many selected and collected works of famous mathematicians. Usually they consist of their research papers plus some expository writings. But there does not seem to be entire volumes of selected papers that consist of only expository papers by a single mathematician up to now. Writing expository papers takes different skills, and writing it well takes time and broad knowledge, hence it is not easy at all. In this sense, Yau is special among contemporary mathematicians. Besides his many highly original and deep contributions to mathematics, he has also written many expository papers covering the broad areas of geometry and analysis.

3 Writing versus person

There is a Chinese saying that a person's writing resembles the person, i.e., it reflects his or her character. This implies that a mathematician's work and his writing also resemble the person. Instead of commenting on the style of Yau's mathematics and his expository writings, it might be helpful to comment on him as a person.

If one thinks of Yau, the following features stand out. He is powerful and persistent (or stubborn). Once he sets up a goal, he will overcome all obstacles to carry through and achieve it. He strongly and confidently believes in what he is doing. He is full of energy and is always active. It is difficult to imagine him without working on multiple projects or organizing something at any time, while keeping others busy at the same time. For example, he travels extensively and frequently, but the concept of jet lag simply does not exist for him.

He has broad interests, and fine divisions between various things also do not exist for him. For example, the usual gap between pure mathematics and applied mathematics is not visible to him, and the more obvious gap between mathematics and humanities does not matter to him either. He is brave and also delicate (or subtle, as Bacon said that mathematics makes one subtle). He has strong opinions about many things and often express them clearly, both verbally and in writing. He has a curious mind and is full of all kinds of ideas, some of which are rather unusual. His approach to everything is global and massive (the quote below from Armand Borel confirms this).

Yau is highly motivated, or driven, by the ideal to do good for all of mathematic and the mathematics community. By all accounts, he has left his share of permanent marks on the field.

4 A glimpse of the mathematics of Yau

Based on the above brief summary of Yau as a person, we will briefly take a look at his mathematics and his achievements in expository writing.

Shing-Tung Yau is a mathematician of great originality, vision and technical power. His work, together with his dedication to mathematics and to its effective dissemination, has permanently affected many subjects ranging from differential geometry, differential equations, topology, and algebraic geometry to mathematical physics. He is truly a global mathematician and teacher.

Originality and technical power. Yau has proposed and emphasized the philosophy that the subject of nonlinear partial differential equations should be combined with geometry (including differential geometry, complex geometry and algebraic geometry). Many deep results by him and his co-authors have shown that only through geometry, can one thoroughly understand nonlinear differential equations such as the Monge-Ampère equation, and that geometric structures can be constructed effectively by nonlinear partial differential equations coupled with the linear theory of index theorems, which are still the only way to build geometric structures.

Probably the most striking example is the construction of Calabi-Yau metrics, which lead to immediate solution of several longstanding open problems in algebraic geometry and has contributed to the extensive development of mathematical physics, in particular string theory. Other examples of geometric structures constructed in this way include the geometric structures on three manifolds by Hamilton's Ricci flow, the gluing arguments by Taubes and Uhlenbeck for the existence of anti self-dual connections and anti self-dual metrics on four manifolds, and the existence of Hermitian Yang-Mills connections on higher dimensional complex manifolds by Uhlenbeck-Yau.

Before Yau's work in geometric analysis in the 1970s, theories of differential equations were largely separately from differential geometry, and applications of nonlinear differential equations to complex geometry and algebraic geometry could not be imagined. It is fair to say that Yau and his friends such as Rick Schoen, Leon Simon, Shiu-Yuen Cheng, Karen Uhlenbeck, Peter Li, and Richard Hamilton founded the subject of geometric analysis in the 1970s, which was followed by Cliff Taubes, Simon Donaldson and others in the 1980s. Geometric analysis has become the most powerful development in the past 40 years in geometry and topology.

It might not be well-known that the existence of Hermitian Yang-Mills connections is absolutely essential for the construction of supersymmetric heterotic strings and has had a huge impact in algebraic geometry and complex geometry. For example, the proof of Uhlenback and Yau is still the only proof that can be used for applications such as the Higgs field and variations of Hodge structures.

Yau's work on linear differential equations is also very original and influential. For example, the gradient estimate for harmonic functions has become a standard technique in geometric analysis, and the Li-Yau estimate for the heat equation is fundamental to all parabolic equations and has motivated the major estimates of Ricci flow, which was used to prove the Poincaré conjecture.

Another example of the great originality and technique of Yau is the application of harmonic maps and minimal surfaces. Meeks and Yau proved the famous classical problem of embedding of the Douglas solution of the minimal surface equation, which was turned around to solve the equivariant Dehn's Lemma, which was a crucial step in the solution of the famous Smith conjecture for group actions on the three sphere. Schoen and Yau used the theory of harmonic maps to prove the famous Einstein conjecture on the positivity of mass in general relativity—a proof that constitutes one of the most important justifications of relativity theory. (If the mass were negative, the whole system would be unstable and fall apart!) Many important consequences of the solution of positive mass in geometry have been developed, and this led to Yau's recent work with Mu-Tao Wang on the correct definition of quasi-local mass in general relativity.

Vision and open problems. Yau has formulated conjectures and proposed many open problems that have had far reaching consequences. For example, the concept of mirror symmetry in string theory has had many unexpected applications in algebraic geometry, and understanding the mirror symmetry conjecture has been a major challenge in theoretical physics and mathematics. In 1996, Strominger-Yau-Zaslow proposed a new construction of mirror manifolds, called the SYZ conjecture. This conjecture has been the guiding principle for a whole generation of people working on mirror symmetry.

In 1982, Yau proposed a list of 120 open problems when he organized a special year on differential geometry at IAS, Princeton. They covered a broad range of topics. Some of them have been very influential and have changed the subject. For example, Problem 65 called for a notion of rank of nonpositively curved manifolds, which extends the one for locally symmetric spaces, and asked whether higher rank metrics are locally symmetric metrics.

This problem was completely and independently solved by Ballmann, and Burns-Spatzier. The efforts to solve this problem have completely changed the subject of manifolds of nonpositive curvature and attracted many people to work on understanding geometry, topology and the analysis of manifolds of nonpositive curvature. A currently active topic deals with CAT(0)-geodesic metric spaces and their quotients, which are not necessarily smooth manifolds and a natural generalization of nonpositively curved Riemannian manifolds.

After the success of this list of open problems, Yau proposed another 100 open problems in 1992. This list has also had a huge influence on topics covered. For example, one problem can be stated as follows: *Prove that a compact Kähler manifold with positive first Chern class admits a Kähler metric if and only if the manifold is stable in the sense of geometric invariant theory, meaning that the tangent bundle is stable as a bundle and the automorphism group is reductive.*

This problem has been studied intensively by many top experts around the world since the 1980s, and is the focus of current research in algebraic geometry and complex geometry.

Yau also raised many problems in other papers and contexts. All these problems show the global perspective of Yau towards mathematics. More importantly, they teach generations of geometers (in the broadest sense) the importance of understanding and appreciating the connections between different parts of mathematics.

5 Expository writing of Yau and related activities

Besides several hundred research papers, Yau has written many survey papers, as these two volumes show. We will only highlight several surveys here. The paper, *Survey on partial differential equations in differential geometry*, in *Seminar on Differential Geometry*, pp. 3–71, 1982, was the first major survey of geometric analysis. In some sense, it is the record of the firsthand experience of this emerging subject by one of the most important practitioners.

His well-known list of open problems, *Problem section*, in *Seminar on Differential Geometry*, pp. 669–706, Ann. of Math. Stud., 102, Princeton Univ. Press, Princeton, N.J., 1982. was written at the same time for a special year (1979–1980) at the Institute for Advanced Study, Princeton.

This special year was mainly organized by Yau and had a lasting impact on both geometric analysis and also on the Institute for Advanced Study (IAS). It is perhaps best to quote from an article written on the history of the IAS by the distinguished mathematician Armand Borel, who was a longtime permanent member at IAS:

The School of Mathematics at the Institute for Advanced Study, in *A century of mathematics in America*, Part III, 119–147, Hist. Math., 3, Amer. Math. Soc., Providence, RI, 1989.

Borel wrote: "In the sixties, considerable progress was made in ... Algebraic groups, arithmetic groups and automorphic forms, number theory, harmonic analysis on reductive Lie groups. Much of it was done here.... This whole general field had become such an active and important part of 'core mathematics' that it was all to the good. However, that was not matched by activities of similar scope in other areas and created some imbalance, accentuated by Atiyah's resignation in 1972. For reasons already explained, in our view it was not in the best interest of the school in the long run and to correct it by increasing activities in other areas became a concern. There are two obvious means to try to remedy this: the special programs and new faculty appointments... The financial situation of the Institute was worrisome... But we resume both as soon as possible: Enrico Bombieri came to the faculty in 1977 and Shing-Tung Yau in 1980, broadening greatly its coverage. We had also to wait until 1977 for the programs but have had one almost every year from then on... 1979–1980 was the year of the biggest program to date, on differential geometry and analysis, in particular nonlinear PDE. The number of seminars was somewhat overwhelming. Several were concentrated at the end of the week, so as to make it easier for people in neighboring (in a rather wide sense including New York and Philadelphia) institutions to participate. Roughly speaking, the main activities were subdivided in three parts: differential geometry, minimal submanifolds, and mathematical physics, with seminar coordinators L. Simon for the second one, S.-T. Yau for the other two. A remarkable feature of the third one (devoted to relativity, the positive mass conjecture, gauge theories, quantum gravity) was the cooperation between mathematicians and physicists, probably a first here since the early days. Two volume of *notes* resulted from this program."

The book *Seminar on Differential Geometry*, Ann. of Math. Stud., 102, Princeton Univ. Press, 1982, is one of the two *notes* Borel mentioned.

A few years later, Yau wrote another major survey, *Nonlinear analysis in geometry*, Enseign. Math. (2) 33 (1987) 109–158. It gives a broad overview of the role of nonlinear differential equations in differential geometry, algebraic geometry, topology and mathematical physics. It conveys a panoramic view of the rapidly expanding subject of geometric analysis.

With the continuing growth of geometric analysis, other major surveys were needed in the late 1990s. The paper, *Review of geometry and analysis*, Asian J. Math. 4 (2000), no. 1, 235–278, gives a comprehensive overview of areas of mathematics and physics related to Calabi-Yau manifolds and mirror symmetry.

In a more recent survey, *Perspectives on geometric analysis. Surveys in differential geometry.* Vol. X, 275–379, Surv. Differ. Geom., 10, 2006, Yau gave a rather complete historical survey of the field of geometric analysis as one of the major players of this important field. He gave many important observations and philosophical points of views on both problems and methods used to attack them.

One of his most recent substantial surveys is *A survey of Calabi-Yau manifolds*, in *Geometry, analysis, and algebraic geometry: forty years of the Journal of Differential Geometry*, 277–318, Surv. Differ. Geom., 13, 2009. It provides a comprehensive introduction to the history and results of research on Calabi-Yau manifolds, recent developments in this area and their applications.

He has also written several influential books and edited many others.

One particularly influential book is a joint book with Rick Schoen, *Lectures on differential geometry*. It was first published in 1994 and translated into Chinese. An expanded Chinese version was published in 2004. One important feature is that it goes directly to the most important and active topics of the subject and hence guides the reader to get into research quickly. This book has educated generations of differential geometers and will continue to be an indispensable book for many years to come. A closely related book by Schoen and Yau is *Lectures on harmonic maps* published in 1997. It also helps the reader to learn some of the most important recent results on, and applications of, harmonic maps.

As mentioned above, the following book edited by Yau has also been very influential to the development of geometric analysis since the 1980s: *Seminar on Differential Geometry. Papers presented at seminars held during the academic year 1979–1980.* Annals of Mathematics Studies, 102. Princeton University Press, 1982. ix+706 pp.

Five books on mirror symmetry edited (or co-edited) by Yau have been very influential to the emergence of the string theory. The first one, *Essays on mirror manifolds*, published in 1992, was edited by Yau and based on the workshop organized by him in 1991 at MSRI, Berkeley. This is the first book that drew the attention of mathematicians, especially algebraic geometers, to string theory, and has played a decisive role in the trend of mathematical physics.

In some sense, the "expository writing" of Yau might also be interpreted broadly to include many conferences he has organized and various forms of interaction with others. The most prominent one is probably the conference series, *Current Developments in Mathematics* and the resulting book series.

He also founded the book series, *Surveys in Differential Geometry*, each of which consists of substantial surveys on subjects by some of the most active experts. Yau has contributed many papers to them.

There are over 30 volumes in the two series, *Current Developments in Mathematics* and *Surveys in Differential Geometry*, and together they provide a panoramic view of the contemporary mathematics. Their impact on the development of mathematics is huge.

It is important to note that Yau has supervised over 65 Ph.D. students up to now and has reached out to mathematics students and mathematicians at various levels through many mathematics centers and initiatives organized by him.

On top of all of these research, teaching, writing, conferences and mathematics centers, Yau has also launched and edited many journals both in pure and applied mathematics. For example, he turned the *Journal of Differential Geometry* into a leading global journal, which has published the best works of Simon Donaldson, Michael Friedman, Richard Hamilton, Richard Schoen etc.

Part III
Selected Papers in Complex Geometry

Editor's Comments

Riemann is one of the greatest mathematicians in history. Even though he only published a few papers, each of them is a gem or rather a treasure mine. He introduced many new concepts and changed people's perspective on mathematics. Among many of his contributions, the introduction of the notions of Riemann surfaces, Riemannian metric, Riemannian curvature, and the emphasis on using the Cauchy-Riemann equation to study holomorphic functions instead of explicit formulas (or power series) have a huge impact on the subject of complex geometry.

In this book, we include a few pages from his collected works [Rie] to show the originality and depth of his thinking.

Complex Geometry from Riemann
to Kähler-Einstein and Calabi-Yau
ALM 38, pp. 121–123

Foundations for a General Theory of Functions of a Complex Variable*

Bernhard Riemann

If we write the differential quotient $\frac{du+dvi}{dx+dyi}$ in the form

$$\frac{\left(\dfrac{\partial u}{\partial x}+\dfrac{\partial v}{\partial x}i\right)dx+\left(\dfrac{\partial v}{\partial y}-\dfrac{\partial u}{\partial y}i\right)dyi}{dx+dyi},$$

it is plain that it will have the same value for any two values of dx and dy, exactly when

$$\frac{\partial u}{\partial x}=\frac{\partial v}{\partial y},\quad \frac{\partial v}{\partial x}=-\frac{\partial u}{\partial y}.$$

Hence this condition is necessary and sufficient for $w=u+vi$ to be a function of $z=x+yi$. For the individual terms of the function, we deduce the following:

$$\frac{\partial^2 u}{\partial x^2}+\frac{\partial^2 u}{\partial y^2}=0,\quad \frac{\partial^2 v}{\partial x^2}+\frac{\partial^2 v}{\partial y^2}=0.$$

This equation is the basis for the investigation of the properties of the individual terms of such a function. We give the proof of the most important of these properties before undertaking a deeper treatment of the complete function. However, we first establish some points concerning more general matters, in order to smooth the ground of the investigation.†

For the following treatment we permit x, y to vary only over a finite region. The position of the point O is no longer considered as being in the plane A, but in a surface T spread out over the plane. We choose this wording since it is inoffensive to speak of one surface lying on another, to leave open the possibility that the position of O can extend more than once over a given part of the plane. However, in such a case we suppose that the portions of surface lying upon one another do

---

* Excerpted from "*Bernhard Riemann, Collected Papers*", pp.4–5. © Kendrick Press.

† Here Riemann used Cauchy-Riemann equation to define holomorphic functions.

not connect along a line. Thus a folding of the surface, or a splitting of the surface into superimposed parts, does not occur.

Now the number of pieces of surface superimposed in each part of the plane is completely determined, when we give the boundary of the region and its direction (that is, its inner and outer sides). A transit of these pieces can take different forms.

Indeed, if we draw a line through the part of the plane covered by the surface, the number of superimposed surfaces only changes on crossing the boundary. Indeed, the number changes on moving from outside to inside by $+1$, in the opposite case by -1. Thus the number is determined everywhere. Along the edge of the line, each bordering portion of surface continues in a definite way, as long as the line does not meet the contour. For indeterminacy can only occur at an isolated point, and consequently occurs either at a point of the line itself, or at a finite distance from it. If we confine ourselves to part of the line ℓ in the interior of the surface, and both sides of it on a sufficiently narrow strip of surface, we may speak of definite bordering parts of the surface, whose number is equal on either side. Specifying a definite direction of the line, denote the parts of the surface on the left side by $a_1, \ldots, a_n$ and those on the right by $a'_1, \ldots, a'_n$. Each part a will continue into one of the parts a'. Indeed, this will in general be the same part along the length of the line ℓ, except that for certain positions of ℓ it can change at one point. Suppose that above such a point σ (that is, along the preceding part of ℓ) the surface portions $a_1, \ldots, a_n$ are connected respectively with the portions $a'_1, \ldots, a'_n$, while below $\sigma, a_{\alpha_1}, a_{\alpha_2}, \ldots, a_{\alpha_n}$ connect respectively with $a'_1, \ldots, a'_n$. Here $\alpha_1, \ldots, \alpha_n$ is a permutation of $1, \ldots, n$. A point of a_1 that moves into a'_1, above σ, will pass to a_{σ_1} if we go back to the left side below σ. If a point moves around σ from left to right, the index of the part of surface in which it lies takes successively the values

$$1, \alpha_1, \alpha_{\alpha_1}, \ldots, \mu, \alpha_\mu, \ldots.$$

In this sequence, as long as the term 1 does not recur, the terms are necessarily distinct. For an arbitrary intermediate term α_μ is necessarily preceded by μ, and in direct succession by all preceding terms up to 1. But when after a certain number of terms, say m, a number evidently less than n, the term 1 reappears, the other terms must then recur in the same order. The point moving around σ comes back after every m circuits into the same part of the surface and is restricted to m superimposed surface parts, that meet above σ at a unique point. We call this point a branch point of order $m-1$ of the surface T. Repeating the procedure, the $n-m$ remaining surface parts, provided they are not isolated, divide into systems of $m_1, m_2, \ldots$, surface parts. In this case, further branch points of orders $m_1 - 1, m_2 - 1, \ldots$ are located at σ.

When the position and direction of the boundary of T, and the position of the branch points, are given, then T is either completely determined, or restricted to one of a finite number of distinct forms. The latter occurs, in so far as these determining portions can lie on different parts of the superimposed surfaces.

A variable which in general, that is after excluding isolated lines or points,[1] takes a definite value at each point O of the surface T that varies continuously with position, can obviously be considered as a function of x, y. When we refer to functions of x, y in what follows, the term will always be employed in this fashion.[‡]

[1] This restriction does not arise from the definition of a function, but is needed for the application of infinitesimal calculus to it. As an example of a function discontinuous at all points of a surface, take the function whose value is 1 for rational x and y, and otherwise is 2. We cannot apply to the function either differentiation or integration, and so cannot directly use infinitesimal calculus. The arbitrary restriction placed here on the surface T will be justified later (section 15).

[‡] Here Riemann first introduced the notion of Riemann surfaces.

Complex Geometry from Riemann
to Kähler-Einstein and Calabi-Yau
ALM 38, pp. 125–128

The Hypotheses on Which Geometry Is Based*

Bernhard Riemann

Metric relationships which can exist in a manifold of n dimensions, on the assumption that the lines have a length which is independent of their situation, so that every line can be measured by every other line

After having constructed the concept of an n-dimensional variety, and having found as its essential characteristic that defining the position of its elements can be reduced to the determination of n quantities, we turn to the second of the problems posed earlier. This is the study of the metric relationships which can hold in such a manifold and the conditions which suffice to determine these relations. These metric relations can only be studied by using symbols representing abstract quantities and representing their dependence by means of formulae. Under certain hypotheses, however, the relationships can be broken down into simpler relations which can be individually interpreted geometrically: so that the result of the calculation can be expressed in geometric language. In order to remain on *terra firma* we cannot avoid working with abstract formulae, but the results of the investigation can be expressed in a geometric form. The foundations for these two aspects of the question were laid down by Privy Councillor Gauss in his celebrated memoir on curved surfaces.

1.

Measurement requires that the measure of the entities being measured must be independent of their location, and this can be the case in more than one way.

* Excerpted from "*Bernhard Riemann, Collected Papers*", pp.261–264. © Kendrick Press. Here Riemann defined Riemannian metric locally.

The assumption which first suggests itself, and which I intend to pursue here, is that the length of lines is independent of their position, so that every line can be measured by comparing it with any other line. If the determination of the position of a point in a given n-dimensional manifold is reduced to the determination of n variables $x_1, x_2, x_3, \ldots, x_n$, then a line may be defined by the statement that the quantities x are given functions of a single variable. The problem then is to find a mathematical expression for the length of a line, and for this purpose we need to consider the quantities x as expressible in terms of units. I shall handle this problem only under certain restrictions, and confine myself in the first place to lines in which the relations between the quantities dx—the associated variations of the variables x—vary in continuous fashion. We can then visualize the line as being divided up into elements, within which the ratios of the increments dx can be regarded as constant, and the problem reduces to finding a general expression for the line element ds starting from a given point, which will involve the variables x as well as the variables dx. Secondly, I shall assume that the length of the line element, disregarding quantities of the second order of magnitude, remains unchanged if all its points undergo the same infinitesimal displacement. This assumption implies that if all the variables dx are proportionally increased in size, the line element will increase in the same proportion. Under these assumptions, the line element could be an arbitrary homogeneous function of the first degree in the variables dx, that remains unaltered by changing the sign of all dx, and such that the arbitrary constants are continuous functions of the variables x. In order to find the simplest cases, I first seek an expression for the $(n-1)$-dimensional manifolds which are everywhere equidistant from the point of origin of the line element; that is, I look for a continuous function of position which can be used to distinguish one such manifold from another. This function will have the property that it either continually decreases or else continually increases as we move away from the origin in any direction. I shall suppose that it increases in every direction, so that it has a minimum at the origin. It must then, if its first and second derivatives are finite, be a function whose first derivative vanishes and whose second derivative is never negative. I shall suppose that it is always positive. This differential expression of the second order therefore remains constant when ds remains constant and increases in the same proportion as the square of dx if the variables dx and hence ds are all increased proportionately. Thus the function can be written as constant ds^2, and consequently ds is the square root of an everywhere positive quadratic form in the variables dx whose coefficients are continuous functions of the variables x. For space, when the position of a point is expressed in rectangular coordinates, we have

$$ds = \sqrt{\sum (dx)^2}.$$

Thus space falls under this simplest case. The next simplest case would comprise manifolds in which the line element can be expressed by the fourth root of a quartic differential expression. The study of this more general class would not, it is true, require any essentially new principles, but would be time-consuming and probably throw relatively little new light on the theory of space, particularly since the results would not lend themselves to geometric form. I therefore restrict myself to manifolds in which the line element is expressed as the square root of a quadratic differential expression. Such an expression can be transformed into a similar one by substituting, for the n independent variables, functions of n new independent variables. Not every such expression, however, can be transformed into any other in this way; for clearly the expression contains $n(n+1)/2$ coefficients which are arbitrary functions of the independent variables. By introducing n new variables, we can only satisfy n different relations and thus can only make n of the coefficients assume specified values. The other $n(n-1)/2$ coefficients are already completely determined by the nature of the manifold to be represented, and to define its metric properties $n(n-1)/2$ functions of position are required. The manifolds in which, as in the plane and in space, the line element can be expressed in the form $\sqrt{\sum(dx)^2}$ thus constitute a special case of the manifolds under investigation here. They merit a special name, and I designate those manifolds in which the square of the infinitesimal line element can be brought into the form of a sum of the squares of the individual differentials as *flat*. In order to review the essential differences between the diverse flat manifolds, we need to put aside differences which arise from the particular mode of representation. This will be achieved by choosing the variables in accordance with a certain principle.

2.

To this end, imagine that given an arbitrary point of the manifold, a system of shortest lines be constructed emanating from this origin. The position of an indeterminate point will then be defined by the initial direction of the particular shortest line on which it lies, and its distance from the origin measured along this shortest line. This position can therefore be expressed by the ratios of the quantities dx^0, that is, the initial values of dx at the origin of this shortest line, and the length s of the line. We now introduce, instead of dx^0, new linear expressions formed from them, $d\alpha$, such that the initial value of the square of the line element is equal to the sum of the squares of the $d\alpha$. The independent variables are now: the length s, and the ratios of the $d\alpha$. Finally we replace the $d\alpha$ by quantities $x_1, x_2, \ldots, x_n$, proportional to them, which have s^2 as the sum of their squares. If we introduce these quantities as the new variables, then for infinitely small values of x, the square of the line element will be $\sum(dx)^2$, but the term of the next higher order of magnitude in the expression will be a quadratic form in the $n(n-1)/2$ quantities $x_1dx_2 - x_2dx_1, x_1dx_3 - x_3dx_1, \ldots$, that is, an infinitesimal of the fourth dimension. Thus we obtain a finite number if this expression is divided by the square of the area of the infinitesimal triangle at whose vertices the values of the variables are $(0,0,0,\ldots)$, $(x_1, x_2, x_3, \ldots)$ and $(dx_1, dx_2, dx_3, \ldots)$. This number remains unchanged as long as the quantities x and the quantities dx are contained in the same binary linear forms, or in other words, as long as the shortest line from $(0,0,0,\ldots)$ to $(x_1, x_2, x_3, \ldots)$ and the shortest line from $(0,0,0,\ldots)$ to $(dx_1, dx_2, dx_3, \ldots)$ remain in the same surface element, and thus the number depends only on the position and orientation of this element. Its value will obviously be zero when the manifold is flat, that is, when the the square of the infinitesimal line element is reducible to the form $\sum(dx)^2$, and can therefore be regarded as a measure of the departure from flatness of the manifold at the point in question in the surface-direction concerned. When multiplied by $-3/4$, it is equal to the quantity called by Privy Councillor Gauss the curvature [*Krümmungsmass*] of the surface. To determine the metric relations in an n-dimensional manifold susceptible of representation in the assumed form, we found earlier that $n(n-1)/2$ functions of position were needed. If, therefore, the curvature at each point is specified for each of $n(n-1)/2$ surface-directions, then all the metric relations in the manifold can be deduced, unless there are some identical relations between the specified values which, generally speaking, will not be the case. The metric relationships of those manifolds in which the line element is represented by the square root of a quadratic differential expression can thus be expressed in a form which is completely independent of the variables. An entirely analogous method can be employed to achieve the same goal in the case of manifolds in which the line element has a less simple form, for example, the fourth root of a quartic differential expression. Generally speaking, the line element would then no longer be capable of being expressed as the square root of a sum of squares of differential expressions. Consequently in the expression for the square of the line element, the departure from flatness would be an infinitesimal quantity of degree 2, rather than degree 4, as in the case previously considered. This characteristic property of the latter manifolds might well be called *planarity in the smallest parts.* For our present purposes, the most important property of the manifolds which we have considered so far, and indeed the main motivation for this investigation, is that the metric relationships in 2-dimensional manifolds can be interpreted geometrically by surfaces, and those in higher dimensional manifolds can be reduced to relations in the surfaces which are contained in them. This calls for a brief further explanation.

Complex Geometry from Riemann
to Kähler-Einstein and Calabi-Yau
ALM 38, p. 129

A Mathematical Work That Seeks to Answer the Question Posed by the Most Distinguished Academy of Paris*

Bernhard Riemann

On the transformation of the expression $\sum_{i,i'} b_{i,i'} ds_i ds_{i'}$ into a given form $\sum_{i,i'} a_{i,i'}$ $dx_i dx_{i'}$.

Since the question of the most distinguished academy is restricted to homogeneous bodies in which the conductivities are constants, we first examine conditions under which the expression $\sum_{i,i'} b_{i,i'} ds_i ds_{i'}$ can be transformed into the form $\sum_{i,i'} a_{i,i'} dx_i dx_{i'}$, with constant coefficients $a_{i,i'}$, by taking the quantities s to be suitable functions of x. Accordingly we make some remarks about transformation to an arbitrary prescribed form.

It is known that an expression $\sum_{i,i'} a_{i,i'} dx_i dx_{i'}$ can always be brought, to the form $\sum_i dx_i^2$ if, as we suppose here, it is a positive form in the dx. Hence, if $\sum_{i,i'} b_{i,i'} ds_i ds_{i'}$ can be transformed into $\sum_{i,i'} a_{i,i'} dx_i dx_{i'}$, it can be transformed into $\sum_i dx_i^2$, and conversely. We therefore investigate when this expression can be transformed to the form $\sum_i dx_i^2$.

Let B be the determinant $\sum \pm b_{1,1} b_{2,2} \dots b_{n,n}$, and let $\beta_{t,t'}$ be the cofactors. Then

$$\sum_i \beta_{i,i'} b_{i,i'} = B; \quad \sum_i \beta_{i,i'} b_{i,i''} = 0 \text{ if } i' \neq i''.$$

If $\sum_{i,i'} b_{i,i'} ds_i ds_{i'} = \sum_i dx_i^2$ holds for an arbitrary value of dx, then substituting $d + \delta$ for d we can also show that

$$\sum_{i,i'} b_{i,i'} ds_i \delta s_{i'} = \sum_i dx_i \delta x_i$$

for arbitrary values of dx and δx.

* Excerpted from "*Bernhard Riemann, Collected Papers*", p.377. © Kendrick Press. Here Riemann introduced the notion of curvature to characterize flat Riemannian metric locally.

Editor's Comments

Kähler metrics and Kähler manifolds are very basic objects in complex geometry. Any Hermitian metric on a Riemann surface is a Kähler metric, and any projective smooth variety with the induced metric from the Fubini-Study metric of the complex projective space is a Kähler manifold. Though complex manifolds of higher dimensions are natural generalizations of Riemann surfaces, Kähler manifolds are, in some sense, the true and really useful and important generalizations of Riemann surfaces. One reason is that geometric analysis can be carried out efficiently and effectively only on Kähler manifolds, but not on general complex manifolds, and hence the Kähler package establishes a powerful relation between Kähler geometry, in particular algebraic geometry, and algebraic topology.

Kähler was a versatile mathematician and made deep contributions to multiple subjects in mathematics. Though the name Kähler is mainly associated with Kähler manifolds or metrics, it seems that this classical paper of Kähler is not widely read now, and its contents are even surprising to some experts. One reason is that it was written in German, and not many people of the current generation can read German. Because of these consideration, we have this paper translated into English and included it in this book. In some sense, this paper sets the foundation for what to follow in this book.

Complex Geometry from Riemann to Kähler-Einstein and Calabi-Yau
ALM 38, pp. 133–145

About a Remarkable Hermitian Metric*

Erich Kähler
Translated by John Kilgore (University of Michigan) with special thanks to Dan Burns

1.

When examining the invariants of a real, $2n$-dimensional Hermitian metric[1]

$$ds^2 = \sum g_{i\bar{k}}\ dx_i d\bar{x}_k \tag{1}$$

under the "pseudo-conformal" transformations[a]

$$\begin{aligned} x_i' &= \varphi_i(x_1, x_2, \dots, x_n) \\ \bar{x}_i' &= \bar{\varphi}_i(\bar{x}_1, \bar{x}_2, \dots, \bar{x}_n) \end{aligned} \qquad (i = 1, 2, \dots, n), \tag{2}$$

it is obvious that the next thing to consider is the alternating quadratic differential form (forme extérieure)

$$\omega = \sum g_{i\bar{k}}\ d(x_i, \bar{x}_k),$$

where $d(x_i, \bar{x}_k)$ is the so-called exterior product[2] of the differentials $dx_i, d\bar{x}_k$, which means it is a Jacobian determinant $\frac{\partial(x_i, \bar{x}_k)}{\partial(s,t)}\ dsdt$. From this invariant form ω related to (1), more invariants can be constructed by using the machinery of the elegant symbolic calculus of differential forms[3]. For example, the derivative[b]

$$\omega' = \sum d(g_{i\bar{k}}, x_i, \bar{x}_k) = \sum \frac{\partial g_{i\bar{k}}}{\partial x_l}\ d(x_l, x_i, \bar{x}_k) + \frac{\partial g_{i\bar{k}}}{\partial \bar{x}_l}\ d(\bar{x}_l, x_i, \bar{x}_k)$$

---

* Translated from: E. Kähler, Über eine bemerkenswerte Hermitesche Metrik. Abh. Math. Sem. Univ. Hamburg 9 (1933), no. 1, 173–186.

[1] $\bar{x}$ denotes the complex conjugate of x. The indecies corresponding to conjugated variables are always overlined.

[2] To avoid confusion with the ordinary multiplication, the outer product of dx, dy, dz is denoted by $d(x, y, z)$.

[3] E. CARTAN, Invariants intégraux, Chap. VI, VII. Paris (1922); GOURSAT, Lecons sur le probléme de Pfaff, Chap. III Paris (1922).

[a] "Pseudo-conformal transformation" is an archaic term for a multivariable biholomorphism. (Dan Burns)

[b] Kähler calls the exterior derivative just "the derivative."

is a new invariant form, and combining this with ω yields more invariants.

This approach provides the case $\omega' = 0$ as a remarkably interesting case. It turns out that $\omega' = 0$ is equivalent to the metric coming from a "potential" U, which is evidently invariant, in the following way:

$$ds^2 = \sum \frac{\partial^2 U}{\partial x_i \partial \bar{x}_k}\, dx_i d\bar{x}_k. \tag{3}$$

Certain metrics occurring in the theory of automorphic functions are of this type. Namely, projective transformations

$$x'_i = \frac{L_i(x)}{L_0 x} = \frac{a_{i0} + a_{i1}x_1 + \cdots + a_{in}x_n}{a_{00} + a_{01}x_1 + \cdots + a_{0n}x_n} \quad (i = 1, 2, \ldots, n) \tag{4}$$

which map the unit sphere

$$1 - x_1\bar{x}_1 - x_2\bar{x}_2 - \cdots - x_n\bar{x}_n = 0$$

into itself, so because

$$(1 - \sum_v x'_v \bar{x}'_v) = (1 - \sum_v x_v \bar{x}_v)(L_0(x) \cdot \bar{L}_0(\bar{x}))^{-1}$$

the metric

$$ds^2 = \sum \frac{\partial^2 U}{\partial x_i \partial \bar{x}_k}\, dx_i d\bar{x}_k$$

formed from

$$U = k \log(1 - \sum x_v \bar{x}_v) \quad (k \text{ a constant})$$

is invariant under the group of "hyper-Fuchsian" transformations (4). It is analogous for the "hyper-abelian" transformations

$$x'_i = \frac{\alpha_i x_i + \beta_i}{\gamma_i x_i + \delta_i} \quad (i = 1, 2, \ldots, n)$$

which preserve the unit circles

$$1 - x_i \bar{x}_i \quad (i = 1, 2, \ldots, n).$$

In this case, there is an invariant metric derived from the potential

$$U = \sum_{i=1}^{n} k_i \log(1 - \sum x_i \bar{x}_i) \quad (k_i \text{ constants})$$

and it is also clear that all groups of hyper-Fuchschen transformations in variables r and s $(r + s = n)$ lead to metrics of this type.

These correlations will be justified if in the following some formal constraints on the metric (3) are realized because the flexability of these metrics depends heavily on the algebraic studies of automorphic functions.

It should also be noted that

$$ds^2 = \sum \frac{\partial^2 U}{\partial x_i \partial \bar{x}_k}\, dx_i d\bar{x}_k$$

is a solution of Einstein's equations of gravitation

$$R_{\alpha\beta} = \lambda g_{\alpha\beta}$$

if the potential U satisfies the equation

$$\left| \frac{\partial^2 U}{\partial x_i \partial \bar{x}_k} \right| = e^{\lambda U}.$$

2.

The proof that the form of (3) follows from $\omega' = 0$ is easily shown. By setting the coefficients of ω' equal to zero, one obtains

$$\frac{\partial g_{i\bar{k}}}{\partial x_l} - \frac{\partial g_{l\bar{k}}}{\partial x_i} = 0, \quad \frac{\partial g_{i\bar{k}}}{\partial \bar{x}_l} - \frac{\partial g_{l\bar{k}}}{\partial \bar{x}_i} = 0,$$

from which one can set

$$g_{i\bar{k}} = \frac{\partial V_i}{\partial \bar{x}_k} = \frac{\partial V_{\bar{k}}}{\partial x_i}.$$

The integrability conditions

$$\frac{\partial^2 V_{\bar{k}}}{\partial x_i \partial x_l} = \frac{\partial^2 V_i}{\partial \bar{x}_k \partial x_l} = \frac{\partial^2 V_l}{\partial \bar{x}_k \partial x_i},$$

i.e.

$$\frac{\partial}{\partial \bar{x}_k} \left(\frac{\partial V_i}{\partial x_l} - \frac{\partial V_l}{\partial x_i} \right) = 0,$$

show that the expressions $\frac{\partial V_i}{\partial x_l} - \frac{\partial V_l}{\partial x_i}$ do not depend on $\bar{x}$:

$$\frac{\partial V_i}{\partial x_l} - \frac{\partial V_l}{\partial x_i} = \varphi_{il}(x_1, x_2, \ldots, x_n). \tag{5}$$

Now, one can find n functions $V_1', V_2', \ldots, V_n'$ of $x_1, x_2, \ldots, x_n$ such that

$$\frac{\partial V_i'}{\partial x_l} - \frac{\partial V_l'}{\partial x_i} = \varphi_{il}(x_1, x_2, \ldots, x_n),$$

because these equations can be combined into

$$\left(-\sum V_i'\, dx_i \right)' = \sum \varphi_{il} d(x_i, x_l).$$

A necessary and sufficient condition for this, that a differential form $\sum \varphi_{il}\, d(x_i, x_l)$ can be seen as the derivative of another form only containing the variables x, is the vanishing of the derivative

$$\left(\sum \varphi_{il}\, d(x_i, x_l) \right)' = \sum \frac{\partial \varphi_{il}}{\partial x_m}\, d(x_m, x_i, x_l) = 0,$$

which is the case because of the form of the left-hand side of (5).

Obviously, one can replace V_i by $V_i - V_i'$ so that the new V_i satisfy the equations

$$\frac{\partial V_i}{\partial x_l} - \frac{\partial V_l}{\partial x_i} = 0$$

without violating the equations

$$g_{i\bar{k}} = \frac{\partial V_i}{\partial \bar{x}_k}.$$

Then, the V_i can be written in the form

$$V_i = \frac{\partial W}{\partial x_i} + \psi_i(\bar{x}_1, \bar{x}_2, \ldots, \bar{x}_n)$$

and $g_{i\bar{k}}$ is given by

$$g_{i\bar{k}} = \frac{\partial V_i}{\partial \bar{x}_k} = \frac{\partial^2 W}{\partial x_i \partial \bar{x}_k} + \frac{\partial \psi_i}{\partial \bar{x}_k} = \frac{\partial^2 U}{\partial x_i \partial \bar{x}_k},$$

where

$$U = W + x_1\psi_1 + x_2\psi_2 + \cdots + x_n\psi_n.$$

For any metric given by

$$ds^2 = \sum \frac{\partial^2 U}{\partial x_i \partial \bar{x}_k}\, dx_i d\bar{x}_k, \qquad (6)$$

it is immediately seen that the associated exterior differential form

$$\omega = \sum \frac{\partial^2 U}{\partial x_i \partial \bar{x}_k}\, d(x_i, \bar{x}_k)$$

is integrable ($\omega' = 0$).

3.

From the metric (3) we now want to calculate the Riemannian curvature tensor, and, in particular, the Ricci[c] curvature tensor $R_{\alpha\beta}$. It is recommended to add the factor 2 to be consistent with the usual notation:

$$ds^2 = 2\sum \frac{\partial^2 U}{\partial x_i \partial \bar{x}_k}\, dx_i d\bar{x}_k = g_{\alpha\beta}\, dx_\alpha dx_\beta.$$

In the future, the indices taking values in all of $1, 2, \ldots, n, \bar{1}, \bar{2}, \ldots, \bar{n}$ will be denoted by Greek letters, while the Latin letters will denote the numbers $1, 2, \ldots, n$

[c] Kähler does not use the term Ricci curvature. The literal translation is "made youthful looking" curvature tensor. Note this is not the same word used for tensor contraction in German.

and the overlined letters will denote the numbers $\bar{1}, \bar{2}, \ldots, \bar{n}$. Further, the summation symbol will be omitted in the usual manner.

For the quadratic form $g_{\alpha\beta}\, dx_\alpha dx_\beta$ without the terms $g_{ik}, g_{\bar{i}\bar{k}}$, one correspondingly has

$$g^{ik} = 0, \quad g^{\bar{i}\bar{k}} = 0,$$

while

$$g^{i\bar{k}} = \frac{D^{i\bar{k}}}{D},$$

where $D^{i\bar{k}}$ is understood to be the minor of

$$D = D(U) = |U_{i\bar{k}}|$$

corresponding to $U_{i\bar{k}}$[4] with proper sign.

The Christoffel symbols $\left[\begin{smallmatrix} \alpha\ \beta \\ \gamma \end{smallmatrix}\right]$ are calculated the fastest as coefficients of $\dot{x}_\alpha, \dot{x}_\beta$ in the Lagrangian expression

$$P_\gamma = \frac{d}{dt}\left(\frac{\partial T}{\partial \dot{x}_\gamma}\right) - \frac{\partial T}{\partial x_\gamma},$$

where

$$T = \frac{1}{2}\dot{s}^2 = U_{i\bar{k}}\dot{x}_i \dot{\bar{x}}_k$$

is the kinetic energy. One finds

$$\left[\begin{matrix} i\ k \\ \bar{l} \end{matrix}\right] = U_{ik\bar{l}}, \quad \left[\begin{matrix} \bar{i}\ \bar{k} \\ l \end{matrix}\right] = U_{\bar{i}\bar{k}l},$$

and all the other Christoffel symbols are equal to 0. For the curvature tensor, only the components of type $R_{i\bar{k},l\bar{m}}$ are not 0[5], and in fact are given by

$$R_{i\bar{k},l\bar{m}} = -U_{i\bar{k}l\bar{m}} + g^{r\bar{s}} U_{il\bar{s}} U_{\bar{k}\bar{m}r}.$$

In the Ricci tensor

$$R_{\alpha\beta} = R_{\alpha\ \beta\gamma}^{\ \gamma},$$

only the components $R_{i\bar{k}}$ with mixed indices remain because they are, for example, given by

$$R_{i\beta} = R_{i\ \beta\gamma}^{\ \gamma} = g^{\gamma\delta} R_{i\delta\beta\gamma} = -g^{l\bar{m}} R_{i\bar{m},l\beta},$$

which is only non-zero for $\beta = \bar{k}$. In

$$R_{i\bar{k}} = -g^{l\bar{m}} R_{i\bar{m},l\bar{k}} = g^{l\bar{m}} U_{i\bar{k}l\bar{m}} - g^{l\bar{m}} g^{r\bar{s}} U_{\bar{k}\bar{m}r},$$

[4] To simplify the notation, the derivatives of a function will also be denoted by attaching the indices of the derived variables.

[5] And of course also $R_{\bar{i}k,l\bar{m}}$ and so forth.

the expression on the right-hand side very elegantly combines to give

$$\frac{\partial^2}{\partial x_i \partial \bar{x}_k}(\log D(U)),$$

which we want to confirm by recalculating. It is true that

$$\frac{\partial D}{\partial x_i} = D^{r\bar{s}} U_{r\bar{s}i},$$
$$\frac{\partial^2 D}{\partial x_i \partial \bar{x}_k} = D^{r\bar{s}} U_{r\bar{s}i\bar{k}} D^{r\bar{s}}_{l\bar{m}},$$

where $D^{r\bar{s}}_{l\bar{m}}$ denotes the associated 2×2 minor of D, $\begin{vmatrix} U_{r\bar{s}} & U_{r\bar{m}} \\ U_{l\bar{s}} & U_{l\bar{m}} \end{vmatrix}$. From this, it follows that

$$\begin{aligned}\frac{\partial^2}{\partial x_i \partial \bar{x}_k}(\log D(U)) &= \frac{1}{D}\frac{\partial^2 D}{\partial x_i \partial \bar{x}_k} - \frac{1}{D^2}\frac{\partial D}{\partial x_i}\frac{\partial D}{\partial \bar{x}_k} \\ &= \frac{D^{r\bar{s}}}{D} \cdot U_{r\bar{s}i\bar{k}} + U_{r\bar{s}i} U_{l\bar{m}\bar{k}} \left(\frac{D^{r\bar{s}}_{l\bar{m}}}{D} - \frac{D^{r\bar{s}} D^{l\bar{m}}}{D^2} \right),\end{aligned}$$

and from the following easily proven determinantal identity[6]

$$D \cdot D^{r\bar{s}}_{l\bar{m}} - D^{r\bar{s}} D^{l\bar{m}} + D^{r\bar{m}} D^{l\bar{s}} = 0,$$

this is equivalent to

$$g^{l\bar{m}} U_{i\bar{k}l\bar{m}} - g^{l\bar{m}} g^{r\bar{s}} U_{\bar{k}\bar{m}r} = R_{i\bar{k}}.$$

Hence, one has

$$R_{i\bar{k}} = \frac{\partial^2}{\partial x_i \partial \bar{x}_k}(\log D(U)). \tag{7}$$

4.

The tensor $R_{i\bar{k}} = \frac{\partial^2 V}{\partial x_i \partial \bar{x}_k}$ $(V = D(U))$, gives rise to a second alternating[d] differential for

$$\Omega = R_{i\bar{k}}\ d(x_i, \bar{x}_k) = V_{i\bar{k}}\ d(x_i, \bar{x}_k),$$

whose derivative Ω' vanishes because Ω has the same form as

$$\omega = U_{i\bar{k}}\ d(x_i, \bar{x}_k).$$

In fact, both forms can be written directly as derivatives of 1-forms[e]:

$$\begin{aligned} \omega &= -(U_i\ dx_i)' &= (U_{\bar{k}}\ d\bar{x}_k)', \\ \Omega &= -(V_i\ dx_i)' &= (V_{\bar{k}}\ d\bar{x}_k)'. \end{aligned} \tag{8}$$

---

[6] In the expression on the left side one holds $r, \bar{s}, l$ fixed and denotes it by $Z^{\bar{m}}$. The obvious equations $\sum U_{i\bar{m}} Z^{\bar{m}} = 0$ (for $i \neq r$) together with $Z^{\bar{s}} = 0$ give a homogenous system of n equations, from which $Z^{\bar{m}} = 0$ follows.

[d] He uses both "outer/exterior" and "alternating" at different places.

[e] Kähler writes "forms of first degree".

Now, an interesting sequence of new invariant differential forms[7]

$$\Omega^\nu \omega^{n-\nu} \quad (\nu = 0, 1, 2, \ldots, n) \tag{9}$$

that have the form

$$A \cdot d(x_1, \bar{x}_1, x_2, \ldots, \bar{x}_n)$$

can be constructed by multiplication of ω and Ω.

The first,

$$\omega^n = n! \; |U_{i\bar{k}}| \; d(x_1, \bar{x}_1, x_2, \ldots, \bar{x}_n)$$

is equal to the volume form

$$dv = \sqrt{|g_{\alpha\beta}|} \; d(x_1, \bar{x}_1, x_2, \ldots, \bar{x}_n) = |U_{i\bar{k}}| \; d(x_1, \bar{x}_1, x_2, \ldots, \bar{x}_n)$$

up to a constant factor. The next is

$$\begin{aligned} \Omega\omega^{n-1} &= (n-1)! \; R_{i\bar{k}} D^{i\bar{k}} \; d(x_1, \bar{x}_1, x_2, \ldots, \bar{x}_n) \\ &= (n-1)! \; R_{i\bar{k}} g^{i\bar{k}} \cdot D \cdot d(x_1, \bar{x}_1, x_2, \ldots, \bar{x}_n) \\ &= (n-1)! \cdot R \cdot dv, \end{aligned}$$

where R is the scalar curvature.

The integrals $\int \Omega^\nu \omega^{n-\nu}$ can all be written as boundary integrals, where one writes $\Omega^\nu \omega^{n-\nu}$ as the derivative of a form, which is possible in various ways. One has, for example,

$$\omega^n = U_{i\bar{k}} \; d(x_i, \bar{x}_k) \cdot \omega^{n-1} = (U_{\bar{k}} \; d\bar{x}_k)' \cdot \omega^{n-1} = [(U_{\bar{k}} \; d\bar{x}_k)\omega^{n-1}]'$$

because $\omega' = 0$, and also,

$$\omega^n = [(U_i \; dx_i)\omega^{n-1}]'.$$

For a general metric (with $n > 1$), the integral $\int Rdv$ is not convertable to a boundary integral because in more than 2 dimensions, the varation of this integral just provides the Einstein gravitation equations (the old version).

On the other hand, for our metrics, one has

$$(n-1)! \int Rdv = \int \Omega\omega^{n-1} = \int (V_{\bar{k}} \; d\bar{x}_k)' \cdot \omega^{n-1} = \int \; [(V_{\bar{k}} \; d\bar{x}_k) \cdot \omega^{n-1}]',$$

so that by applying the generalized Stokes theorem,

$$\int_C Rdv = \frac{1}{(n-1)!} \int_{\Re(C)} (V_{\bar{k}} \; d\bar{x}_k) \cdot \omega^{n-1}, \tag{10}$$

where C is a $2n$-dimensional cell with boundary $\Re(C)$[f]. It also applies to give

$$\int_C Rdv = -\frac{1}{(n-1)!} \int_{\Re(C)} (V_i \; dx_i) \cdot \omega^{n-1},$$

[7] There is surely no misunderstanding if exterior multiplication of forms is simply written by writing the forms side by side. One must always pay attention to the order, and, if the factors are permuted, possibly reverse the sign.

[f] $\Re$ stands for the German word 'Rand', which means boundary.

and by adding this to (10) one gets

$$\int_C R dv = \frac{1}{2(n-1)!} \int_{\Re(C)} (V_{\bar{k}}\, d\bar{x}_k - V_i\, dx_i) \cdot \omega^{n-1}, \tag{11}$$

which has a real-valued integrand on the right-hand side.

This formula is reminiscent of the Gauss-Bonnet theorem; however, the invariance of the boundary integrands is missing from the complete analogy.

Under a pseudo-conformal transformation (2) is taken to[g]

$$V' = V + \log \Delta + \log \bar{\Delta}, \quad \Delta = \frac{\partial(x_1, x_2, \ldots, x_n)}{\partial(x'_1, x'_2, \ldots, x_n)}.$$

Therefore, $V_{\bar{k}} d\bar{x}_k - V_i dx_i$ is taken to[h]

$$\frac{\partial V}{\partial \bar{x}'_k} d\bar{x}'_k - \frac{\partial V}{\partial x'_i} dx'_i + d \log \frac{\bar{\Delta}}{\Delta},$$

since

$$\frac{\partial \bar{\Delta}}{\partial x'_i} = \frac{\partial \Delta}{\partial x'_i} = 0.$$

Consequently, the entire integrand changes to

$$\left(d \log \frac{\bar{\Delta}}{\Delta} \right) \cdot \omega^{n-1},$$

because

$$\omega = \frac{\partial^2 U}{\partial x_i \partial \bar{x}_k}\, d(x_i \bar{x}_k) = \frac{\partial^2 U}{\partial x'_i \partial \bar{x}'_k}\, d(x'_i \bar{x}'_k)$$

remains unchanged.

If we can find a covariant A associated to the $(2n-1)$-dimensional hypersurfaces[i] such that there is a pseudo-conformal transformation taking A to

$$\log \frac{\Delta}{\bar{\Delta}},$$

one can make the integrand

$$(V_{\bar{k}}\, d\bar{x}_k - V_i\, dx_i) \omega^{n-1}$$

an invariant by adding

$$dA \cdot \omega^{n-1}$$

without essentially changing equation (11). Because $dA \cdot \omega^{n-1}$ is integrable ($(dA \cdot \omega^{n-1})' = 0$), the integral $\int dA \cdot \omega^{n-1}$ over the hypersurface without boundary $\Re(C)$ is invariant under continuous deformations of $\Re(C)$.

---

[g] I left the original typo in the following.

[h] Notice the difference in notation for partial derivatives.

[i] Note that the following expression is only locally defined on the boundary hypersurface.

For $n = 1$, hence a 2 dimensional metric, it is easy to find such a covariant A. $x = x(t), \bar{x} = \bar{x}(t)$ is the equation of a line[j], so

$$A = -\log\left(\frac{\frac{dx}{dt}}{\frac{d\bar{x}}{dt}}\right) = -2\arg.\frac{dx}{dt}$$

has the desired property, and the expression $\frac{\partial V}{\partial \bar{x}}\frac{d\bar{x}}{dt} - \frac{\partial V}{\partial dx}\frac{dx}{dt}$ will be corrected by dA. Hence,

$$\frac{\frac{d^2\bar{x}}{dt^2}\frac{dx}{dt} - \frac{d^2x}{dt^2}\frac{d\bar{x}}{dt}}{\frac{dx}{dt}\frac{d\bar{x}}{dt}} + \frac{U_{x\bar{x}\bar{x}}}{U_{x\bar{x}}}\frac{d\bar{x}}{dt} - \frac{U_{xx\bar{x}}}{U_{x\bar{x}}}\frac{dx}{dt}$$

is essentially the geodesic curvature of the boundary curve $\Re(C)$. The integral $\int dA$ has the value -4π over a simpler, positive encirclement of $\Re(C)$. It should furthermore be remarked that every 2 dimensional metric can be brought to the form

$$ds^2 = \frac{\partial^2 U}{\partial x \partial \bar{x}}\, dx d\bar{x} = \Delta U(dx_1^2 + dx_2^2) \quad (x = x_1 + ix_2).$$

It would be of great importance to the theory of automorphic functions if the curvature integral (11) or some other integral $\Omega^\nu \omega^{n-\nu}$ which is invariant under pseudo-conformal transformations could be converted to a boundary integral of the type just described.

5.

The simple form (7) of the Ricci curvature tensor $R_{i\bar{k}}$ allows one to guess at a solution of the Einstein equations

$$R_{\alpha\beta} = \lambda g_{\alpha\beta}. \tag{12}$$

In the case when the metric is given by

$$ds^2 = 2\frac{\partial^2 U}{\partial x_i \partial \bar{x}_k}\, dx_i d\bar{x}_k, \tag{13}$$

the equations are always satisfied for unmixed indices $(\alpha, \beta) = (i, k)$ or $(\alpha, \beta) = (\bar{i}, \bar{k})$, and the equations

$$R_{i\bar{k}} = \lambda g_{i\bar{k}}$$

require:

$$\frac{\partial^2}{\partial x_i \partial \bar{x}_k}(\log D(U)) = \lambda \frac{\partial^2 U}{\partial x_i \partial \bar{x}_k}.$$

So, the difference $\log D(U) - \lambda U = \varphi$ must be an "n-harmonic" function, that is to say it must satisfy the characteristic differential equations for the real parts of an analytic function:

$$\frac{\partial^2 \varphi}{\partial x_i \partial \bar{x}_k} = 0.$$

---

j Kähler uses the term "line" for curve, as in line integral.

Every such function can be written as

$$\varphi = \psi(x_1, x_2, \dots, x_n) + \bar{\psi}(\bar{x}_1, \bar{x}_2, \dots, \bar{x}_n).$$

It is apparent that the equation

$$\log D(U) - \lambda U = \psi(x_1, x_2, \dots, x_n) + \bar{\psi}(\bar{x}_1, \bar{x}_2, \dots, \bar{x}_n) \tag{14}$$

also implies that (13) satisfies the gravitational equations (12).

The latter equation can be simplified by a pseudo-conformal transformation. Such a transformation changes $\log D(U)$ by

$$\log \frac{\partial(x_1, x_2, \dots, x_n)}{\partial(x'_1, x'_2, \dots, x'_n)} + \frac{\partial(\bar{x}_1, \bar{x}_2, \dots, \bar{x}_n)}{\partial(\bar{x}'_1, \bar{x}'_2, \dots, \bar{x}'_n)}$$

and one can always find a transformation

$$x'_i = x'_i(x_1, x_2, \dots, x_n)$$

such that

$$\frac{\partial(x'_1, x'_2, \dots, x'_n)}{\partial(x_1, x_2, \dots, x_n)} = \psi(x_1, x_2, \dots, x_n).$$

Thus one can achieve $\psi = \bar{\psi} = 0$ in (14).

Since one only searches for essentially different metrics which satisfy (12), one can reduce the problem to finding solutions of the equation

$$D(U) = e^{\lambda U}.$$

The metric from

$$U = -\log(1 - \sum x_\nu \bar{x}_\nu),$$

which is invariant under the hyper-Fuchsian transformations, is a solution of Einstein's equations.

Namely, one has

$$U_{i\bar{k}} = \frac{\bar{x}_i x_k}{S^2} - \frac{\delta_{ik}}{S} \quad \left(\begin{array}{lll} S & = \sum_k x_k \bar{x}_k - 1 & \\ \delta_{ik} & = 0, & i \neq k \\ \delta_{ii} & = 1 & \end{array} \right).$$

The determinant

$$D(U) = S^{-2n} \begin{vmatrix} \bar{x}_1 x_1 - S & \bar{x}_1 x_2 & \cdots & \bar{x}_1 x_n \\ \bar{x}_2 x_1 & \bar{x}_2 x_2 - S & \cdots & \bar{x}_2 x_n \\ \vdots & \vdots & & \vdots \\ \bar{x}_n x_1 & \bar{x}_n x_2 & \cdots & \bar{x}_n x_n - S \end{vmatrix}$$

is easy to calculate by writing it as the determinant of the homogenous system of equations

$$\sum_k \bar{x}_i x_k Y_k = \Lambda \cdot Y_i \tag{15}$$

with unknowns Y_i and with $\Lambda = S$. From (15) it follows that

$$\bar{x}_i \left(\sum_k x_k Y_k \right) = \Lambda Y_i,$$

from which the 1-dimensional[k] solution

$$Y_1 : Y_2 : \cdots : Y_n = \bar{x}_1 : \bar{x}_2 : \cdots : \bar{x}_n \tag{16}$$

can be read. Also,

$$\sum x_k Y_k = 0 \qquad (\Lambda = 0)$$

is at least an $n-1$ dimensional solution. Therefore, the eigenvalue $\Lambda = 0$ must have multiplicity at least $n-1$, and additionally, since the eigenvalue $\Lambda = \sum_k x_k \bar{x}_k$ is related to (16), Λ is exactly of multiplicity $n-1$. So,

$$|\bar{x}_i x_k - \delta_{ik}\Lambda| = (-\Lambda)^{n-1} \cdot \left(\sum x_k \bar{x}_k - \Lambda \right),$$

and for $\Lambda = S = \sum x_k \bar{x}_k - 1$, the value of the determinant in question falls out:

$$(1 - \sum x_k \bar{x}_k)^{n-1},$$

so that

$$D(U) = (1 - \sum x_k \bar{x}_k)^{-n-1} = e^{(n+1)U}. \tag{17}$$

For the hyper-abelian metric, that is, the one constructed from

$$U = -\sum_i \log(1 - x_i \bar{x}_i),$$

there is a similar equation

$$D(U) = e^{2U}.$$

6.

The differential term $D(U)$ can be obtained by variatiaion of integrals

$$\int \mathfrak{C}(U)\; d(x_1, \bar{x}_1, x_2, \ldots, \bar{x}_n),$$

so that the solution of the equation

$$D(U) = e^{kU}$$

can be obtained by a variation problem.

[k] Kähler writes "zero-dimensional".

The integrand $\mathfrak{C}$ is the generalization to n variables by Wirtinger[8] of the Levi invariant integral which occurs in the theory of 3 dimensional singular spaces of analytic functions of 2 variables:

$$\mathfrak{C} = \begin{vmatrix} 0 & U_{\bar{1}} & U_{\bar{2}} & \cdots & U_{\bar{n}} \\ U_1 & U_{1\bar{1}} & U_{1\bar{2}} & \cdots & U_{1\bar{n}} \\ U_2 & U_{2\bar{1}} & U_{2\bar{2}} & \cdots & U_{2\bar{n}} \\ \vdots & \vdots & \vdots & & \vdots \\ U_n & U_{n\bar{1}} & U_{n\bar{2}} & \cdots & U_{n\bar{n}} \end{vmatrix}. \tag{18}$$

In the language of differential forms, one has

$$(n-1)!\ \mathfrak{C}(U)\ d(x_1, \bar{x}_1, x_2, \ldots, \bar{x}_n) = (U_i\ dx_i) \cdot (U_{\bar{k}}\ d\bar{x}_k) \cdot (U_{l\bar{m}}\ d(x_l\bar{x}_m))^{n-1},$$

and the variation of integrals

$$I = \int \varpi = \int (U_i\ dx_i) \cdot (U_{\bar{k}}\ d\bar{x}_k) \cdot (U_{l\bar{m}}\ d(x_l\bar{x}_m))^{n-1}$$

yields (set $\delta U = V$)

$$\delta I = \int \delta\varpi,$$

where

$$\begin{aligned} \delta\varpi =&(v_i\ dx_i) \cdot (U_{\bar{k}}\ d\bar{x}_k) \cdot (U_{l\bar{m}}\ d(x_l\bar{x}_m))^{n-1} \\ &+ (U_i\ dx_i) \cdot (v_{\bar{k}}\ d\bar{x}_k) \cdot (U_{l\bar{m}}\ d(x_l\bar{x}_m))^{n-1} \\ &+ (n-1)(U_i\ dx_i) \cdot (U_{\bar{k}}\ d\bar{x}_k) \cdot (v_{l\bar{m}}\ d(x_l\bar{x}_m))^{n-1} \cdot (U_{r\bar{s}}\ d(x_r\bar{x}_s))^{n-2}. \end{aligned}$$

By calculating using the rules of the differentiation of differential forms, one convinces oneself that

$$\begin{aligned} \delta\varpi =& -(n+1) \cdot v \cdot (U_{i\bar{k}}\ d(x_i\bar{x}_k))^n \\ &+ (n-1) \cdot [(U_i\ dx_i) \cdot (U_{\bar{k}}\ d\bar{x}_k) \cdot (v_{\bar{l}}\ d\bar{x}_l) \cdot (U_{r\bar{s}}\ d(x_r\bar{x}_s))^{n-2}]' \\ &- n[v \cdot (U_i\ dx_i) \cdot (U_{l\bar{m}}\ d(x_l\bar{x}_m))^{n-1}]' \\ &+ [v \cdot (U_{\bar{k}}\ d\bar{x}_k) \cdot (U_{l\bar{m}}\ d(x_l\bar{x}_m))^{n-1}]', \end{aligned} \tag{19}$$

so that δI agrees with

$$-(n+1)! \int D(U) \cdot v \cdot d(x_1, \bar{x}_1, x_2, \ldots, \bar{x}_n)$$

up to a boundary integral.

The equation

$$D(U) = e^{kU}$$

[8] WIRTINGER, Zur formalen Theorie der Funktionen von mehr komplexen Veränderlichen. Math. Ann. 97 (1926), s. 363.

is therefore obtained by variation of

$$\int \left(\mathfrak{C}(U) + \frac{n(n+1)}{k} e^{kU} \right) d(x_1, \bar{x}_1, x_2, \dots, \bar{x}_n). \tag{20}$$

If it behaves like the positive definite metric determined by

$$ds^2 = 2 \frac{\partial^2 U}{\partial x_i \partial \bar{x}_k} \, dx_i d\bar{x}_k,$$

then the integrand of (20) is positive because the quadratic form

$$g^{\bar{i}k} x_i \bar{x}_k$$

is then also positive definite and

$$g^{i\bar{k}} \frac{\partial U}{\partial x_i} \frac{\partial U}{\partial \bar{x}_k} = \frac{D^{i\bar{k}}}{D} \cdot U_i U_{\bar{k}}$$

is just equal to

$$\frac{\mathfrak{C}(U)}{D(U)}.$$

The equation

$$D(U) = 0 \tag{21}$$

also plays a role in the theory of hyper-abelian functions, where it is to be solved under the constraint

$$\mathfrak{C}(U) > 0.$$

By the above, there is an integral

$$\int \mathfrak{C}(U) \; d(x_1, \bar{x}_1, x_2, \dots, \bar{x}_n),$$

whose variation leads to (21).

In a later work, I intend to present the use of the current formal developments to the theory of automorphic functions and the analogy of the equations

$$D(U) = 0, \quad D(U) = e^{kU}$$

with the classical differential equations

$$\Delta U = 0, \quad \Delta U = e^{kU}$$

already pointed out by G. GIRAUD[9] and A. BLOCH[10].

---

9 G. GIRAUD, Sur une équation aux dérivées partielles, on linéaire etc. Comtes Rendus 166, I (1918) p. 893.

10 A. BLOCH, Sur une nouvelle et importante généralisation de l'équation de Laplace. L'Enseignement Mathém. 26 (1927), p. 52.

Editor's Comments

Though the 20th century saw an explosion in mathematics, it is not clear how much of it will remain and will still be used by the mathematicians several centuries later. On the other hand, it seems safe to say that Chern classes will be among those which will remain. This paper of Chern establishes a very fruitful bridge between complex manifolds (in particular algebraic varieties) and topology, which are two central subjects in the 20th century mathematics. It is foreseeable that understanding global properties of algebraic varieties will remain essential in mathematics. After all, we need spaces to study in mathematics, and algebraic varieties are the most natural ones we have or can get. Though this may not be the favorite paper of Chern, it is certainly his most famous and influential paper.

Complex Geometry from Riemann to Kähler-Einstein and Calabi-Yau
ALM 38, pp. 149–185

Characteristic Classes of Hermitian Manifolds*

Shiing-Shen Chern

Introduction

In recent years the works of Stiefel,[1] Whitney,[2] Pontrjagin,[3] Steenrod,[4] Feldbau,[5] Ehresmann,[6] etc. have added considerably to our knowledge of the topology of manifolds with a differentiable structure, by introducing the notion of so-called fibre bundles. The topological invariants thus introduced on a manifold, called the characteristic cohomology classes, are to a certain extent susceptible of characterization, at least in the case of Riemannian manifolds,[7] by means of the local geometry. Of these characterizations the generalized Gauss-Bonnet formula of Allendoerfer-Weil[8] is probably the most notable example.

In the works quoted above, special emphasis has been laid on the sphere bundles, because they are the fibre bundles which arise naturally from manifolds with a differentiable structure. Of equal importance are the manifolds with a complex analytic structure which play an important rôle in the theory of analytic functions of several complex variables and in algebraic geometry. The present paper will be devoted to a study of the fibre bundles of the complex tangent vectors of complex manifolds and their characteristic classes in the sense of Pontrjagin. It will be shown that there are certain basic classes from which all the other characteristic classes can be obtained by operations of the cohomology ring. These basic classes are then identified with the classes obtained by generalizing Stiefel-Whitney's classes to complex vectors. In the sense of de Rham the cohomology classes can be expressed by exact exterior differential forms which are everywhere

* S. S. Chern, Characteristic Classes of Hermitian Manifolds. Annals of Mathematics. Second Series, Ann. of Math. 47 (1946), 85–121. (received July 10, 1945)

1 Stiefel, [24]. The number in the bracket refers to the bibliography at the end of the paper.
2 Whitney, [29], [30].
3 Pontrjagin, [18], [19].
4 Steenrod, [21], [22].
5 Feldbau, [12].
6 Ehresmann, [10].
7 Chern, [5], [6], [7].
8 Allendoerfer-Weil, [2].

regular on the (real) manifold. It is then shown that, in case the manifold carries an Hermitian metric, these differential forms can be constructed from the metric in a simple way. This means that the characteristic classes are completely determined by the local structure of the Hermitian metric. This result also includes the formula of Allendoerfer-Weil and can be regarded as a generalization of that formula.

Concerning the relations between the characteristic classes of a complex manifold and an Hermitian metric defined on it, the problem is completely solved by the above results. It is to be remarked that corresponding questions for Riemannian manifolds remain open. Roughly speaking, the difficulty in the real case lies in the existence of finite homotopy groups of certain real manifolds, namely the manifolds formed by the ordered sets of linearly independent vectors of a finite-dimensional vector space.

The paper is divided into five chapters. In Chapter I we consider the fibre bundles which include the bundles of tangent complex vectors of a complex manifold and which are called complex sphere bundles. To a given base space a complex sphere bundle can be defined by a continuous mapping of the base space into a complex Grassmann manifold and it is shown that this is the most general way of generating a complex sphere bundle. We take the Grassmann manifold to be that in a complex vector space of sufficiently high dimension and define a characteristic cohomology class in the base space to be the inverse image under this mapping of a cohomology class of the Grassmann manifold. We are therefore led to the study of the cocycles or cycles on a complex Grassmann manifold, a problem treated exhaustively by Ehresmann.[9] A close examination of Ehresmann's results is therefore made in Chapter II, in the light of the problems which concern us here. In fact, we are only interested in the cocycles of the Grassmann manifold which are of dimension not greater than the dimension of the base space. If the Grassmann manifold is that of the linear spaces of n (complex) dimensions in a linear vector space of $n + N$ dimensions, there are on it n basic cocycles such that all other cocycles of dimension $\leqq 2n$ can be obtained from them by operations of the cohomology ring. The cycles corresponding to these cocycles are determined and geometrically interpreted. In Chapter III we identify the images of these cocycles in the base space with the cocycles obtained by generalizing the Stiefel-Whitney invariants to complex vectors. A new definition of these cocycles is given, which is important for applications to differential geometry in the large. Chapter IV is devoted to the study of a complex manifold with an Hermitian metric. It is proved that the n basic cocycles in question can be characterized in a simple way in terms of differential forms constructed from the Hermitian metric. These results are then applied in Chapter V to the complex projective space with the elliptic Hermitian metric. Classical formulas of Cartan[10] and Wirtinger[11] are derived from our formulas as particular cases.

---

[9] Ehresmann, [8].
[10] Cartan, [3].
[11] Wirtinger, [31].

Chapter I Complex Sphere Bundles and Their Imbedding

1 The complex sphere

Various definitions have been given of a fibre bundle. For definiteness we shall adopt the one of Steenrod[12] and follow his terminology. We are, however, going to restrict the kind of fibre bundles under consideration.

Let $E(n; C)$ be a complex vector space of n dimensions,[13] whose vectors will be denoted by small German letters. In $E(n; C)$ suppose a positive definite Hermitian form be given, which, in terms of a suitable base, has the expression

$$\mathfrak{z}\bar{\mathfrak{z}} = \sum_{i=1}^{n} z^i \bar{z}^i, \tag{1}$$

where z^i are the components of the vector $\mathfrak{z}$ in terms of the base and the bar denotes the operation of taking the complex conjugate. A vector $\mathfrak{z}$ such that $\mathfrak{z}\bar{\mathfrak{z}} = 1$ is called a unit vector. The group of linear transformations

$$\mathfrak{z}^{*i} = \sum_{j=1}^{n} a^i_j \mathfrak{z}^j, \quad i = 1, \ldots, n, \tag{2}$$

which leaves the form (1) unaltered is the unitary group and will be denoted by $U(n; C)$. We shall call the complex sphere $S(n; C)$ the manifold of all the unit vectors of $E(n; C)$. It is homeomorphic to the real sphere of topological dimension $2n - 1$. The letter C in these notations will be dropped, when there is no danger of confusion.

In this paper we shall be concerned with fibre bundles such that the fibres are homeomorphic to the complex sphere $S(n)$ and that the group in each fibre is the unitary group $U(n)$. Such a fibre bundle is called a *complex sphere bundle.*

The most important complex sphere bundle is obtained from the consideration of the complex tangent vectors of a complex manifold $M(n)$ of complex dimension n and topological dimension $2n$. By a complex manifold $M(n)$ we shall mean a connected Hausdorff space which satisfies the following conditions:

1) It is covered by a finite or denumerable set of neighborhoods each of which is homeomorphic to the interior of the polycylinder

$$|z^i| < 1, \quad i = 1, \ldots, n,$$

in the space of n complex variables, so that z^i can be taken as local coordinates of $M(n)$.

---

12 Steenrod, [22].

13 Throughout this paper we shall mean by dimension the complex dimension. The dimension of a manifold in the sense of topology will be called the topological dimension, which is twice the complex dimension. The dimensions of simplexes, chains, cycles, homology groups, etc., are understood in the sense of topology, so long as there is no danger of confusion.

2) In a region in which two local coordinate systems z^i and z^{*^i} overlap the coordinates of the same point are connected by the relations

$$z^{*^i} = f^i(z^1, \dots, z^n), \tag{3}$$

where f^i are analytic functions.
It follows from this definition that the notions of $M(n)$ which are expressed in terms of local coordinates but which remain invariant under the transformations (3) have an intrinsic meaning in $M(n)$. This is in particular true of a *tangent vector* at a point, which we define in the usual way as an object which has n components Z^i in each local coordinate system and whose components Z^i, Z^{*^i} in two local coordinate systems z^i, z^{*^i} are transformed according to the equations

$$Z^{*^i} = \sum_{k=1}^{n} \frac{\partial z^{*^i}}{\partial z^k} Z^k. \tag{4}$$

It is clear that the space of the tangent vectors at a point is homeomorphic to $E(n)$. We consider the non-zero tangent vectors and call two such vectors equivalent if their components Z^i, W^i with respect to the same local coordinate system satisfy the conditions

$$W^i = \rho Z^i,$$

where ρ is a positive real quantity. This relation remains unchanged under transformation of local coordinates. Also it is an equivalence relation in the sense of algebra, being reflexive, symmetric, and transitive. Hence the non-zero tangent vectors can be divided by means of this equivalence relation into mutually disjoint classes. We call such a class of non-zero tangent vectors a direction. With a natural topology the space of directions at a point is homeomorphic to the complex sphere $S(n)$. Furthermore, by using the so-called unitarian trick in group theory, it is easy to verify that the manifold of all directions at the points of $M(n)$ is a complex sphere bundle with $M(n)$ as the base space. It will be called the tangent bundle of $M(n)$.

Although all the results in this chapter will be formulated for general complex sphere bundles, it is the particular case of the tangent bundle of a complex manifold that justifies the study of complex sphere bundles.

2 The Grassmann manifold and the imbedding theorems

Consider the space $E(n+N;C)$ and the linear subspaces of $E(n+N;C)$ of dimension n. The manifold of all such linear subspaces is called a Grassmann manifold and will be denoted by $H(n,N;C)$ or simply $H(n,N)$. It is of dimension nN. The unit vectors of the complex sphere $S(n+N)$ in $E(n+N)$, which belong to a linear subspace $E(n)$ of dimension n, constitute a complex sphere $S(n)$, to be denoted by $S(n+N) \cap E(n)$.

Now let B be a finite polyhedron in the sense of combinatorial topology and let f be a continuous mapping of B into $H(n,N)$. From the mapping f we can define a complex sphere bundle $\mathfrak{F}$ with B as base space as follows: $\mathfrak{F}$ consists of the points

$(b, \mathfrak{v})$ of the topological product $B \times S(n+N)$ such that $\mathfrak{v} \in f(b) \cap S(n+N)$, and the projection π of $\mathfrak{F}$ onto B is defined by $\pi(b, \mathfrak{v}) = b$. It is easy to verify that $\mathfrak{F}$ is a complex sphere bundle over B, which we shall call the *induced bundle* over B.

The importance of the notion of complex sphere bundles induced by the mapping of the base space into a Grassmann manifold is justified by the following theorems:

Theorem 1. *To every bundle $\mathfrak{F}$ of complex spheres $S(n)$ over a finite polyhedron B of topological dimension d there exists a continuous mapping f of B into $H(n, N)$ with $N \geqq d/2$, such that $\mathfrak{F}$ is equivalent to the bundle induced by f.*

Theorem 2. *Let $\mathfrak{F}_1$ and $\mathfrak{F}_2$ be two bundles of complex spheres $S(n)$ over a finite polyhedron B of topological dimension d induced by the mappings f_1, f_2 respectively of B into $H(n, N)$, $N \geqq d/2$. The bundles $\mathfrak{F}_1$ and $\mathfrak{F}_2$ are equivalent when and only when the mappings f_1 and f_2 are homotopic.*

Similar theorems for real sphere bundles are known.[14] It follows from these theorems that to a class of equivalent bundles of complex spheres $S(n)$ over a finite polyhedron B of topological dimension d corresponds a class of homotopic mappings of B into the Grassmann manifold $H(n, N)$, where N is an integer satisfying $2N \geqq d$. This class of mappings induces a homomorphism h of the cohomology groups of dimension $\leqq d$ of $H(n, N)$ into the cohomology groups of the same dimension of B. A cohomology class of B which is the image under h of a cohomology class of $H(n, N)$ is called a *characteristic cohomology class* or simply a *characteristic class* and each of its cocycles is called a *characteristic cocycle.*

3 Proofs of theorems 1 and 2

The proofs of the Theorems 1 and 2 do not differ essentially from the real case. We shall therefore restrict ourselves to a brief description of the general procedure. We need the following two lemmas:

Lemma 1. (Covering homotopy theorem)[15] *Let $\mathfrak{F}$ be a fibre space over a base space B, which is a compact metric space. Let S be a compact topological space and let I be the unit interval. Suppose a mapping $h(S \times I) \subset \mathfrak{F}$ be given having the property: There exists a mapping $H(S \times 0) \subset \mathfrak{F}$ such that*

$$h(p \times 0) = \pi H(p \times 0), \quad p \in S,$$

where π is the projection of $\mathfrak{F}$ into B. Then there exists a mapping $H(S \times I) \subset \mathfrak{F}$ such that

$$h(p \times t) = \pi H(p \times t), \quad p \in S, \quad 0 \leqq t \leqq 1.$$

From Lemma 1 follows the lemma:

Lemma 2 (Feldbau).[16] *Let $\mathfrak{F}$ be a fibre bundle over a compact metric base space B. If B is contractible to a point, then $\mathfrak{F}$ is equivalent to the topological product of B and one of its fibres F.*

---

14 Whitney, [29]; Steenrod, [22].

15 Hurewicz-Steenrod, [15]. The theorem is given in various papers, in slightly different versions.

16 Feldbau, [12].

To prove Theorem 1 we are going to define the mapping f whose existence was asserted by the theorem. We take a simplicial decomposition of B which is so fine that each simplex lies in a neighborhood, and denote by $\sigma_i^k, i = 1, \ldots, \alpha_k, k = 0, 1, \ldots, d$, its simplexes. We denote as usual by π the projection of $\mathfrak{F}$ onto B. Our purpose is to define a mapping $f(B) \subset H(n,\ H)$ and a mapping $f^*(\mathfrak{F}) \subset B \times S(n+N)$ such that

$$f^*(p) \in \pi(p) \times \{f(\pi(p)) \cap S(n+N)\} \quad . \quad (5)$$

and that for a fixed $\pi(p)$ the mapping $f^*(p)$ is a homeomorphism preserving the scalar product. The definition of these mappings is given by induction on the dimension of the simplexes of B. The images $f(\sigma_i^0) \in H(n, H), i = 1, \ldots, \sigma_0$, are defined in an arbitrary way and it is clear how $f^*(p)$ can be defined for all $p \in \mathfrak{F}$ such that $\pi(p) = \sigma_i^o, i = 1, \ldots, \alpha_0$. We suppose the mappings be defined over the $(k-1)$-dimensional skeleton of B and consider any simplex σ^k of dimension k. We take a neighborhood U which contains σ^k and decompose the set $\pi^{-1}(U)$ into a topological product of U and a complex sphere $S_0(n)$. Then we can define n mappings $\varphi_i(\sigma^k) \subset \mathfrak{F}, i = 1, \ldots, n$, such that: 1) $\pi\varphi_i(p) = p, p \in \sigma^k$; 2) $\varphi_i(p), \varphi_j(p), i \neq j$, are orthogonal vectors on the complex sphere $\pi^{-1}(p)$. We proceed to define by induction $f^*(\varphi_i(p)) = p \times q_i$, which will satisfy the condition that q_i, q_j for $i \neq j$ are orthogonal vectors of $S(n+N)$. By hypothesis, $f^*(\varphi_1(p))$ is defined for all $p \in \partial\sigma^k$.[17] Since $\partial\sigma^k$ is topologically a sphere of topological dimension $k - 1 \leqq d - 1 \leqq 2N - 1 < 2(n+N) - 1$, which is the topological dimension of the complex sphere $S(n+N)$, and since $\pi(p), p \in \varphi_1(\sigma^k)$, is the cell σ^k, it follows that $f^*(\varphi_1(p)), p \in \partial\sigma^k$, is contractible in $B \times S(n+N)$. This means that there is a continuous mapping $g(\partial\sigma^k \times t) \subset B \times S(n+N), 0 \leqq t \leqq 1$, such that the following conditions are satisfied: 1) $g(\partial\sigma^k \times 0)$ is a point; 2) $g(\partial\sigma^k \times 1)$ is identical with $f^*(\varphi_1(p))$. On the other hand, we can introduce in σ^k the "polar coordinates" ρ, p, where $0 \leqq \rho \leqq 1$ and $p \in \partial\sigma^k$. For a point of σ^k having the coordinates ρ, p we define

$$f^*(\varphi_1(\rho, p)) = g(p \times \rho).$$

Suppose now that

$$f^*(\varphi_1(p)) = p \times q_1, \ldots, f^*(\varphi_{i-1}(p)) = p \times q_{i-1}, \quad p \in \sigma^k,$$

are defined, such that $q_k, q_j, k, j = 1, \ldots, i-1,\ k \neq j$, are orthogonal vectors of $S(n+N)$. To define $f^*(\varphi_i(p))$ we consider on $S(n+N)$ the complex spheres $S(n+N-i+1)$ whose vectors are orthogonal to $q_1, \ldots, q_{i-1}$. These complex spheres $S(n+N-i+1)$, depending on p, constitute a complex sphere bundle over the simplex σ^k. By Lemma 2, it is a topological product of σ^k and a complex sphere $S_0(n+N-i+1)$. By induction hypothesis, the boundary $\partial\sigma^k$ is mapped into $S_0(n+N-i+1)$, by means of the vectors $q_i \in S(n+N-i+1)(p)$. Since the topological dimension $k-1$ of $\partial\sigma^k$ is smaller than the topological dimension

[17] We shall make use of the notation $\partial\sigma^k$ to denote both the combinatorial and the settheoretical boundary of the simplex σ^k, as the meaning will be clear by context.

$2(n+N-i)+1$ of $S_0(n+N-i+1)$, the map is contractible and the mapping of $\partial\sigma^k$ can be extended continuously throughout σ^k. It follows that a mapping $h(\sigma^k) \subset S_0(n+N-i+1)$ and hence a mapping $h_1(\sigma^k) \subset S(n+N)$ can be defined such that

$$h_1(p) \in S(n+N-i+1)(p), \quad p \in \sigma^k.$$

We then define $f^*(\varphi_i(p)) = p \times h_1(p) = p \times q_i, p \in \sigma^k$. Clearly the vector q_i is orthogonal to $q_1, \ldots, q_{i-1}$.

To complete the induction on the dimension k, let $p^* \in \mathfrak{F}$ such that $\pi(p^*) = p \in \sigma^k$. Then p^* has n components $u_1, \ldots, u_n$ with respect to $\varphi_1(p), \ldots, \varphi_n(p)$. We define $f(p)$ to be the linear space of n dimensions of $E(n+N)$ which contains the complex sphere determined by $q_1, \ldots, q_n$ and

$$f^*(p^*) = p \times q,$$

where q belongs to $f(p) \cap S(n+N)$ and has the components $u_1, \ldots, u_n$ with respect to $q_1, \ldots, q_n$.

Thus our induction is complete and it is easily seen that the mappings f and f^* fulfill our desired conditions. It is also clear that the complex sphere bundle induced by the mapping $f(B) \subset H(n,N)$ is equivalent to $\mathfrak{F}$. This proves our Theorem 1.

Concerning Theorem 2 it is not difficult to prove that $\mathfrak{F}_1$ and $\mathfrak{F}_2$ are equivalent if f_1 and f_2 are homotopic. The converse is proved by defining a mapping $f(B \times I) \subset H(n,H)$, with $f(B \times 0)$ and $f(B \times 1)$ coinciding with the given mappings f_1 and f_2 respectively. Because of the equivalence of $\mathfrak{F}_1$ and $\mathfrak{F}_2$ a complex sphere bundle can be defined over $B \times I$ in an obvious way. The rest of the argument consists of defining the mapping $f(B \times I)$ by an extension process analogous to the proof of Theorem 1. We shall omit the details here.

Chapter II Study of the Cocycles on a Complex Grassmann Manifold

1 Summary of some known results

Let $H(n,N)$ be the Grassmann manifold of n-dimensional linear subspaces in $E(n,N)$. Our main purpose in this chapter is to give a homology base for the cocycles of dimension $\leqq 2n$ of $H(n,N)$. It is to be remarked that, $H(n,N)$ being a manifold of topological dimension $2nN$, there corresponds to each cycle of dimension s a cocycle of dimension $2nN - s$, and vice versa.

There are two different ways to describe the cocycles of $H(n,N)$, which are both useful to our purpose.

To explain the first method let $0 \leqq \varphi(i) \leqq N$, $1 \leqq i \leqq n$, be a non-decreasing integral-valued function. Let L_i, $1 \leqq i \leqq n$, be a linear vector space of dimension $i + \varphi(i)$ in $E(n+N)$, such that

$$L_1 \subset L_2 \subset \cdots \subset L_n.$$

Let $Z(\varphi(i))$ be the set of all n-dimensional linear spaces $X(n)$ such that

$$\dim(X(n) \cap L_i) \geqq i, \quad i = 1, \dots, n,$$

where the notation in the parenthesis denotes the linear space common to $X(n)$ and L_i. $Z(\varphi(i))$ is called a Schubert variety in algebraic geometry. It is a pseudo-manifold of dimension $s = \sum_{i=1}^{n} \varphi(i)$ and carries an integral cycle of dimension $2s$ of $H(n, N)$. Concerning the significance of the Schubert varieties for the topology of Grassmann manifolds the following theorem was proved by Ehresmann:[18]

Theorem 3. *The Grassmann manifold $H(n, N)$ has no torsion coefficients and has all its Betti numbers of odd dimension equal to zero. Its Betti number of dimension $2s$ is equal to the number of distinct non-decreasing integral-valued functions $\varphi(i), 1 \leqq i \leqq n$, such that $\sum_{i=1}^{n} \varphi(i) = s$. The integral cycles carried by the corresponding Schubert varieties $Z(\varphi(I))$ constitute a homology base for the Betti group of dimension $2s$.*

The second method to describe the cocycles of $H(n, N)$ is by means of differential forms. Let $X(n) \in H(n, H)$ and let $\mathfrak{e}_1, \dots, \mathfrak{e}_n$ be n vectors in $X(n)$ such that

$$\mathfrak{e}_i \cdot \bar{\mathfrak{e}}_j = \delta_{ij}, \quad 1 \leqq i, j \leqq n.$$

To these n vectors we add N further vectors $\mathfrak{e}_{n+1}, \dots, \mathfrak{e}_{n+N}$ satisfying the conditions

$$\mathfrak{e}_A \cdot \bar{\mathfrak{e}}_B = \delta_{AB}, \quad 1 \leqq A, B \leqq n + N. \tag{6}$$

When there is a differentiable family of the vectors $\mathfrak{e}_1, \dots, \mathfrak{e}_{n+N}$, we put

$$\theta_{AB} = d\mathfrak{e}_A \cdot \bar{\mathfrak{e}}_B, \tag{7}$$

which are linear differential forms satisfying the conditions

$$\theta_{AB} + \bar{\theta}_{BA} = 0. \tag{8}$$

Among θ_{AB} the forms

$$\theta_{ir}, \quad \bar{\theta}_{ir}, \quad 1 \leqq i \leqq n, \quad n + 1 \leqq r \leqq n + N,$$

constitute a set of $2nN$ linearly independent forms at each point of $H(n, N)$. Let Θ be a form in $\theta_{ir}, \bar{\theta}_{ir}$ with constant coefficients. Θ is called an invariant form if it remains unchanged under the groups of transformations

$$\theta^*_{ir} = \sum_{j=1}^{n} a_{ij} \theta_{jr}, \tag{9}$$

$$\theta^*_{ir} = \sum_{s=n+1}^{n+N} b_{rs} \theta_{is}, \tag{10}$$

where $(a_{ij}), (b_{rs})$ are arbitrary unitary matrices. It is called exact, if $d\Theta = 0$. It is well-known that on a differentiable manifold of class two a cocycle with rational coefficients can be expressed by an exact differential form, and conversely. For convenience we shall therefore call an exact differential form a cocycle. Then we have the following theorem of E. Cartan:[19]

---

18 Ehresmann, [8], p. 418.

19 Cartan, [3]; Ehresmann, [8], p. 409.

Theorem 4. *Every invariant form of $H(n,N)$ is exact. The Betti number of dimension $2s$ of $H(n,N)$ is equal to the number of linearly independent (With constant coefficients) invariant differential forms of degree $2s$. The set of these forms constitutes a cohomology base of dimension $2s$.*

2 The basic forms

Let r be an integer between 1 and n. For reasons which will be clear later we shall be particularly interested in the n cycles $Z_r, r = 1, \dots, n$, carried by the Schubert varieties defined by the functions

$$\begin{aligned} \varphi_r(i) &= N-1, \quad i = 1, \dots, n-r+1, \\ \varphi_r(i) &= N, \qquad i = n-r+2, \dots, n. \end{aligned} \tag{11}$$

The cycle Z_r is of dimension $2(Nn - n + r - 1)$. We shall find the invariant differential form which gives the cocycle of dimension $2(n-r+1)$ corresponding to Z_r.

For this purpose we put

$$\Theta_{ij} = \sum_{s=n+1}^{n+N} \theta_{is}\theta_{sj}, \quad 1 \leqq i, j \leqq n, \tag{12}$$

and

$$\begin{aligned} \Phi_r = \frac{1}{(2\pi\sqrt{-1})^{n-r+1}(n-r+1)!} \sum \delta(i_1 \cdots i_{n-r+1}; j_1 \cdots j_{n-r+1}) \\ \cdot \Theta_{i_1 j_1} \cdots \Theta_{i_{n-r+1} j_{n-r+1}}, \end{aligned} \tag{13}$$

where $\delta(i_1 \cdots i_{n-r+1}; j_1 \cdots j_{n-r+1})$ is zero except when $j_1, \dots, j_{n-r+1}$ form a permutation of $i_1, \dots, i_{n-r+1}$, in which case it is $+1$ or -1 according as the permutation is even or odd, and where the summation is extended over all indices $i_1, \dots, i_{n-r+1}$ from 1 to n. It is easy to verify that Φ_r is an invariant form on $H(n,N)$. Our problem is then solved by the following theorem:

Theorem 5. *The invariant differential form Φ_r defines a cocycle of dimension $2(n-r+1)$ on $H(n,N)$, which corresponds to the cycle Z_r in the sense that, for any cycle ζ of dimension $2(n-r+1)$, the relation*

$$KI(\zeta_1, Z_r) = \int_\zeta \Phi_r \tag{14}$$

holds,[20] *whenever both sides are defined.*

To prove this theorem we notice that both sides of the equation (14) are linear in ζ, so that it is sufficient to prove (14) for the cycles of a homology base of dimension $2(n-r+1)$. By Theorem 3, these are the cycles carried by the Schubert varieties $Z(\varphi(i))$ such that

$$\sum_{i=1}^{n} \varphi(i) = n - r + 1.$$

[20] The notation KI, due to Lefschetz, means the Kronecker index (or the intersection number).

Since the function $\varphi(i)$ is non-decreasing, we must have

$$\varphi(1) = \cdots = \varphi(r-1) = 0. \tag{15}$$

Let $\zeta_k, k = 1, \ldots, m$, be the cycles of a homology base of dimension $2(n-r+1)$ defined in this way, and let ζ_1 be the cycle defined by

$$\varphi(1) = \cdots = \varphi(r-1) = 0, \quad \varphi(r) = \cdots = \varphi(n) = 1. \tag{16}$$

Then for the cycles $\zeta_k,\ k \neq 1$, we must also have

$$\varphi(r) = 0.$$

It is therefore sufficient to prove that

$$KI(\zeta_k, Z_r) = \int_{\zeta_k} \Phi_r, \quad k = 1, \ldots, m, \tag{17}$$

for the cycles ζ_k which are well chosen so that both sides of (17) are defined.

Let us first assume that $k \neq 1$. By definition, there exists a fixed linear vector space $L(r)$ of dimension r such that any $X(n)$ of the Schubert variety carrying ζ_k satisfies the condition

$$\dim(X(n) \cup L(r)) \geqq r,$$

which means that

$$X(n) \supset L(r).$$

On the other hand, any $X(n)$ of the Schubert variety carrying Z_r has the property that it intersects a fixed linear space $L(N+n-r)$ of dimension $N+n-r$ in a linear space of dimension $\geqq n-r+1$. We take in $E(n+N)$ a frame $\mathfrak{e}_1, \ldots, \mathfrak{e}_{n+N}$ and let $L(r)$ and $L(N+n-r)$ be spanned by the vectors $\mathfrak{e}_1, \ldots, \mathfrak{e}_r$ and $\mathfrak{e}_{r+1}, \ldots, \mathfrak{e}_{n+N}$ respectively. There is clearly no $X(n)$ which contains $\mathfrak{e}_1, \ldots, \mathfrak{e}_r$ and has in common with $L(N+n-r)$ a linear space of dimension $n-r+1$, which means that

$$KI(Z_r, \zeta_k) = 0, \quad k \neq 1.$$

To show that the integral on the right-hand side of (17) is also zero, we choose the frame $\mathfrak{e}_1, \ldots, \mathfrak{e}_{n+N}$ such that $\mathfrak{e}_1, \ldots, \mathfrak{e}_r$ belong to $L(r)$ and $\mathfrak{e}_1, \ldots, \mathfrak{e}_n$ belong to $X(n)$. It is obviously possible to choose $\mathfrak{e}_1, \ldots, \mathfrak{e}_r$ to be fixed. Under this choice we have

$$\Theta_{ij} = 0, \quad 1 \leqq i, j \leqq r,$$

and hence

$$\Phi_r = 0.$$

Thus the relations (17) are proved for $k \neq 1$.

To define ζ_1 we take two linear vector spaces $L(r-1), L(n+1)$ of dimensions $r-1, n+1$ respectively, such that $L(r-1) \subset L(n+1)$. An element $X(n)$ of ζ_1 is then defined by the conditions

$$L(r-1) \subset X(n) \subset L(n+1).$$

Let $\mathfrak{e}_1, \ldots, \mathfrak{e}_{n+N}$ be a frame in $E(n+N)$. We suppose the linear vector spaces in question to be so chosen that

$$\begin{aligned}
&\mathfrak{e}_1, \ldots, \mathfrak{e}_{r-1} \subset L(r-1), \\
&\mathfrak{e}_1, \ldots, \mathfrak{e}_{n+1} \subset L(n+1), \\
&\mathfrak{e}_r, \ldots, \mathfrak{e}_{n-1}, \mathfrak{e}_{n-1}, \ldots, \mathfrak{e}_{n+N} \subset L(n+N-r).
\end{aligned}$$

By this choice $L(n+1)$ and $L(n+N-r)$ have in common a linear vector space of dimension $n-r+1$, namely the one spanned by $\mathfrak{e}_r, \ldots, \mathfrak{e}_{n-1}, \mathfrak{e}_{n+1}$. It follows that Z_r and ζ_1 have in common only one $X(n)$, which is spanned by $\mathfrak{e}_1, \ldots, \mathfrak{e}_{n-1}, \mathfrak{e}_{n+1}$. This intersection is to be counted simply, and we have[21]

$$KI(Z_r, \zeta_1) = 1.$$

It now only remains to evaluate the integral in the right-hand side of (17) for $k=1$. The linear vector spaces $L(r-1)$ and $L(n+1)$ being fixed, we choose a fixed frame $\mathfrak{a}_1, \ldots, \mathfrak{a}_{n+N}$ in $E(n+N)$ such that $\mathfrak{a}_1, \ldots, \mathfrak{a}_{r-1}$ belong to $L(r-1)$ and $\mathfrak{a}_1, \ldots, \mathfrak{a}_{n+1}$ to $L(n+1)$. If $X(n) \in \zeta_1$, we choose the frame $\mathfrak{e}_1, \ldots, \mathfrak{e}_{n+N}$, whose first n vectors belong to $X(n)$ such that

$$\begin{aligned}
&\mathfrak{e}_A = \mathfrak{a}_A, \quad A = 1, \ldots, r-1, n+2, \ldots, n+N, \\
&\mathfrak{e}_B = \sum_C u_{BC}\mathfrak{a}_C, \quad B, C = r, \ldots, n+1,
\end{aligned} \tag{18}$$

where u_{BC} are the elements of a unitary matrix. It follows that

$$\begin{aligned}
\theta_{ki} &= d\mathfrak{e}_k \cdot \bar{\mathfrak{e}}_i = 0, \\
\theta_{ik} &= -\bar{\theta}_{ki} = 0,
\end{aligned} \quad k = n+2, \ldots, n+N, i = 1, \ldots, n+N, \tag{19}$$

and hence that

$$\Theta_{ij} = \theta_{in+1}\theta_{n+1j}, \quad 1 \leqq i, j \leqq n.$$

For simplicity we shall write $\mathfrak{e}$ for $\mathfrak{e}_{n+1}$, u_B for $u_{n+1,B}$, and θ_i for $\theta_{n+1,i}$. We remark that $X(n)$ is completely determined by the vector $\mathfrak{e}$, whose components with respect to $\mathfrak{a}_r, \ldots, \mathfrak{a}_{n+1}$ are $u_r, \ldots, u_{n+1}$. Our purpose therefore is to transform the form Φ_r into an exterior differential form in $u_r, \ldots, u_{n+1}$, from which the integration can be carried out. With this purpose in mind, we have

$$\begin{aligned}
\theta_i = d\mathfrak{e} \cdot \bar{\mathfrak{e}}_i &= \left(\sum_{B=r}^{n+1} du_B \mathfrak{a}_B\right)\left(\sum_{C=r}^{n+1} \bar{u}_{iC}\bar{\mathfrak{a}}_C\right) \\
&= \sum_{B=r}^{n+1} du_B \bar{u}_{iB}, \quad i = r, \ldots, n, \\
\theta_i &= 0, \quad i = 1, \ldots, r-1,
\end{aligned} \tag{20}$$

and

$$\Theta_{ij} = -\bar{\theta}_i\theta_j, \quad i, j = r, \ldots, n,$$

21 Cf. EHRESMANN, [8], p. 421.

all other Θ_{ij} being zero. It follows that Φ_r is equal to

$$\begin{aligned}&\frac{(2\pi\sqrt{-1})^{r-n-1}}{(n-r+1)!}\cdot\sum\delta(i_1\cdots i_{n-r+1};j_1\cdots j_{n-r+1})\theta_{i_1}\bar{\theta}_{j_1}\cdots\theta_{i_{n-r+1}}\bar{\theta}_{j_{n-r+1}}\\&=(-1)^{\frac{1}{2}(n-r)(n-r+1)}\frac{(n-r+1)}{(2\pi\sqrt{-1})^{n-r+1}}\theta_r\cdots\theta_n\bar{\theta}_r\cdots\bar{\theta}_n.\end{aligned}\tag{21}$$

From now on we shall agree on the following ranges of indices:

$$\gamma\leqq\alpha,\beta\leqq n,\quad\gamma\leqq i,j\leqq n+1.$$

From (20) and

$$\theta_{n+1}=\sum_j du_j\bar{u}_j$$

we get, by solving for du_i,

$$du_i=\sum_\alpha u_{\alpha i}\theta_\alpha+u_i\theta_{n+1}.\tag{22}$$

We notice that

$$\theta_{n+1}+\bar{\theta}_{n+1}=0.\tag{23}$$

Consider now the form

$$\Psi_r=\sum_{k=0}^{n-r+1}du_r\cdots du_{r+k-1}du_{r+k+1}\cdots du_{n+1}d\bar{u}_r\cdots d\bar{u}_{r+k-1}d\bar{u}_{t+k+1}\cdots d\bar{u}_{n+1},\tag{24}$$

which we shall prove to differ from Φ_r by a numerical factor. In fact, it is easy to verify, by means of the fact that the matrix (u_{ij}) is unitary, that

$$\Psi_r=\theta_r\cdots\theta_n\bar{\theta}_r\cdots\bar{\theta}_n,$$

and hence that

$$\Phi_r=(-1)^{\frac{1}{2}(n-r)(n-r+1)}\frac{(n-r+1)!}{(2\pi\sqrt{-1})^{n-r+1}}\Psi_r.\tag{25}$$

To integrate Φ_r or Ψ_r over ζ_1 let us notice that $X(n)$ will remain unchanged if $\mathfrak{e}$ is replaced by the vector $e^{\sqrt{-1}\rho}\mathfrak{e}$, ρ being real. We can therefore normalize the coordinates $u_r,\ldots,u_{r+1}$ of $X(n)$ by assuming that u_{n+1} is real and positive. Then we have

$$\Psi_r=du_r\cdots du_n d\bar{u}_r\cdots d\bar{u}_n,$$

which is to be integrated over the domain

$$u_{n+1}^2+u_r\bar{u}_r+\cdots+u_n\bar{u}_n=1,\quad u_{n+1}>0.$$

This integration is then easily achieved. In fact, we put

$$\begin{aligned}u_r&=v_r+\sqrt{-1}w_r,\\\bar{u}_r&=v_r-\sqrt{-1}w_r.\end{aligned}$$

Then

$$\Psi_r = (-1)^{\frac{1}{2}(n-r)(n-r+1)}(-2\sqrt{-1})^{n-r+1} dv_r dw_r \cdots dv_n dw_n,$$

and the integral is over the domain D:

$$u_{n+1}^2 + v_r^2 + w_r^2 + \cdots + v_n^2 + w_n^2 = 1, \quad u_{n+1} > 0.$$

But the integral

$$\int_D dw_r dv_r \cdots dw_n dv_n$$

is the volume of the domain bounded by the unit hypersphere of dimension $2n - 2r + 1$, which is equal to $\pi^{n-r+1}/(n-r+1)!$. Hence we have

$$\int_{\zeta_1} \Psi_r = (-1)^{\frac{1}{2}(n-r)(n-r+1)} \frac{(2\pi\sqrt{-1})^{n-r+1}}{(n-r+1)!},$$

and finally,

$$\int_{\zeta_1} \Phi_r = 1.$$

Our Theorem 5 is therefore completely proved.

3 The basis theorem

The importance of the invariant differential forms Φ_r, $1 \leqq r \leqq n$, lies in a theorem we proceed to prove, which asserts that every invariant differential form of degree $\leqq 2n$ in $H(n, N)$ is a polynomial in Φ_r with constant coefficients. The exact statement of our theorem is as follows:

Theorem 6. *Every invariant differential form of degree $\leqq 2n$ in $H(n, N)$ is a polynomial in Φ_r , $1 \leqq r \leqq n$, with constant coefficients. If the form defines an integral cocycle on $H(n, N)$, the coefficients are integers.*

The theorem follows easily from the so-called first main theorem on vector invariants for the unitary group, which we state as follows:

Lemma 3. *Let $\mathfrak{v}_1, \ldots, \mathfrak{v}_m$ be a set of vectors in $E(n)$ under transformations of the unitary group $U(n)$. Every integral rational invariant in the components of $\mathfrak{v}_k$, $1 \leqq k \leqq m$, is an integral rational function of the scalar products $\mathfrak{v}_i\bar{\mathfrak{v}}_k$, $1 \leqq i, k \leqq m$.*

It is known that[22] under the unimodular unitary group such an invariant is an integral rational function of the scalar products and of determinants of the form $[\mathfrak{v}_1 \cdots \mathfrak{v}_n]$ or $[\bar{\mathfrak{v}}_1 \cdots \bar{\mathfrak{v}}_n]$. But under a general unitary transformation of determinant $e^{\sqrt{-1}\alpha}$ the determinants $[\mathfrak{v}_1 \cdots \mathfrak{v}_n]$ and $[\bar{\mathfrak{v}}_1 \cdots \bar{\mathfrak{v}}_n]$ will be multiplied by $e^{\sqrt{-1}\alpha}$ and $e^{-\sqrt{-1}\alpha}$ respectively. It follows that an invariant will involve the determinants only in products of the form $[\mathfrak{v}_1 \cdots \mathfrak{v}_n] \cdot [\bar{\mathfrak{v}}'_1 \cdots \bar{\mathfrak{v}}'_n]$, which can however be expressed as a determinant of scalar products:

$$[\mathfrak{v}_1 \cdots \mathfrak{v}_n][\bar{\mathfrak{v}}'_1 \cdots \bar{\mathfrak{v}}'_n] = \begin{vmatrix} \mathfrak{v}_1 & \bar{\mathfrak{v}}'_1 & \cdots & \mathfrak{v}_1\bar{\mathfrak{v}}'_n \\ \vdots & \vdots & & \vdots \\ \mathfrak{v}_n & \bar{\mathfrak{v}}'_1 & \cdots & \mathfrak{v}_n\bar{\mathfrak{v}}'_n \end{vmatrix}$$

---

[22] Weyl, [28], p. 45.

Thus the lemma is proved.

To prove Theorem 6 let Ψ be an invariant differential form of degree $2s \leqq 2n$ in $H(n,N)$, which is therefore an exterior form in $\theta_{ir}, \bar{\theta}_{ir}$, $1 \leqq i \leqq n, n+1 \leqq r \leqq n+N$, with constant coefficients. The form Ψ being in particular invariant under the transformation $\theta^*_{ir} = e^{\sqrt{-1}\alpha}\theta_{ir}$, it follows that Ψ, when reduced to its lowest terms, will contain in each term exactly s factors each of θ_{ir} and $\bar{\theta}_{ir}$. Let us fix our attention for the moment to the group (10). We take from Ψ all the terms of the form

$$\text{const.}\theta_{i_1 r_1} \cdots \theta_{i_s r_s} \bar{\theta}_{j_1 t_1} \cdots \bar{\theta}_{j_s t_s},$$

with a fixed set of the indices $i_1, \ldots, i_s, j_1, \ldots, j_s$, and call their sum Ψ_1. Since the indices $i_1, \ldots, i_s, j_1, \ldots, j_s$ are now fixed, we shall drop them for simplicity.

Now it is well-known that there is an isomorphism between the ring of exterior forms and the ring of multilinear forms with alternating coefficients. To Ψ_1 corresponds, in the complex vector space of N dimensions, an alternating multilinear form of degree $2s$. Since Ψ_1 is invariant under the unitary group (10), the same is true of its corresponding alternating multilinear form. By our Lemma the latter is an integral rational function of the scalar products. It follows by the isomorphism that Ψ_1 can be expressed as a polynomial in sums of the form $\sum_r \theta_{ir}\bar{\theta}_{jr} = -\Theta_{ij}$. Consequently, Ψ is a polynomial in Θ_{ij}, $1 \leqq i, j \leqq n$, with constant coefficients.

Let us now put

$$P_r = \sum \Theta_{i_1 i_2}\Theta_{i_2 i_3} \cdots \Theta_{i_{n-r+1} i_1}, \quad r = 1, \ldots, n. \tag{26}$$

By the same argument as above, we can prove that Ψ, being also invariant under the group (9), is a polynomial in $P_r, 1 \leqq r \leqq n$, with constant coefficients. On the other hand, it is easy to show, by induction on r, that P_r is a polynomial in $\Phi_1, \ldots, \Phi_r$, with constant coefficients. Hence the first part of our theorem is proved.

To prove the second part of the theorem consider the products of the form

$$\Phi_n^{\lambda_1} \cdots \Phi_1^{\lambda_n}, \tag{27}$$

such that

$$\lambda_1 + 2\lambda_2 + \cdots + n\lambda_n = s. \tag{27a}$$

These forms constitute a basis for all invariant differential forms of degree $2s \leqq 2n$ on $H(n,N)$. Since $s \leqq n$, their number is equal to the number of partitions of s as a sum of integral summands. By Theorem 3 this is equal to the Betti number of dimension $2s$ of $H(n,N)$. It follows that the products in (27) are linearly independent, and that every invariant differential form of degree $2s$ of $H(n,N)$ representing an integral cocycle is equal to a linear combination of the products (27) with integral coefficients.

Chapter III The Basic Characteristic Classes on a Complex Manifold

1 A second definition of the basic characteristic classes

Let M be a complex manifold of dimension n. We consider the complex sphere bundle defined from the tangent vectors of M and imbed it, according to Theorem 1, in a Grassmann manifold $H(n, N), N \geqq n$, by means of a mapping of M into $H(n, N)$. It follows from Theorem 2 that the inverse image of a cohomology class of dimension $\leqq 2n$ of $H(n, N)$ induced by this mapping is an invariant of M (or rather of the analytic structure of M), which we have called a characteristic cohomology class of M. From Theorem 6 we see that of all the characteristic cohomology classes of M those which are inverse images of the cohomology classes of $H(n, N)$ containing the cocycles $\Phi_r, 1 \leqq r \leqq n$, play a particularly important rôle, because all the others can be obtained from them by operations of the cohomology ring. We therefore call these n classes the basic characteristic classes, the inverse image of the class containing Φ_r being the rth basic class.

Our first aim is to identify these basic classes with the classes obtained by generalizing to complex manifolds the well-known procedure of Stiefel-Whitney.[23]

In order to understand the situation we recall briefly the results of Stiefel-Whitney for real sphere bundles, emphasizing the differences between the real and complex cases. From a bundle of real spheres of dimension $n-1$ over a polyhedron as base space Stiefel and Whitney considered the fibre bundle over the same base space whose fibre at each point is the manifold $V(n, r)$ of r $(1 \leqq r \leqq n)$ linearly independent points of the real sphere at this point. It was proved that all homology groups of dimension $< n-r$ of $V(n, r)$ vanish and that the homology group $H^{n-r}(V(n, r))$ of dimension $n-r$ of $V(n, r)$ is the free cyclic group or the cyclic group of order 2 according as the following condition is satisfied or not: $n-r$ is even or $r = 1$. To define a generator of $H^{n-r}(V(n, r))$ we take an ordered set of $r-1$ mutually perpendicular unit vectors $\mathfrak{e}_1, \ldots, \mathfrak{e}_{r-1}$ in the Euclidean space E^n of dimension n which contains the sphere of dimension $n-1$. The unit vector $\mathfrak{e}_r$ of E^n perpendicular to $\mathfrak{e}_1, \ldots, \mathfrak{e}_{r-1}$ describes a sphere of dimension $n-r$, which, when oriented, defines a cycle ζ_0^{n-r} belonging to one of the generating homology classes of $H^{n-r}(V(n, r))$. We then take a simplicial decomposition K of the base space such that each simplex of K belongs to a neighborhood. Let the simplexes of K be oriented. It is possible to define a continuous mapping of the $(n-r)$-dimensional skeleton of K into the fibre bundle such that each point is mapped into a point belonging to the fibre over it. Let σ^{n-r+1} be a simplex of dimension $n-r+1$ of K. In a neighborhood containing σ^{n-r+1} the fibre bundle can be resolved into a topological product and can therefore be mapped into one fibre. Since the mapping is defined over the boundary $\partial\sigma^{n-r+1}$, we get a mapping of a sphere of dimension $n-r$ into a fibre and hence an element h of $H^{n-r}(V(n, r))$. The delicate point is to get from this element h an integer or a residue class mod. 2. This is possible if a generating element of $H^{n-r}(V(n, r))$ is defined. When $H^{n-r}(V(n, r))$ is cyclic

---

[23] Stiefel, [24]; Whitney, [29].

of order two, it has only one generating element, so that no further assumption is necessary. When $H^{n-r}(V(n,r))$ is a free cyclic group, we assume that a continuous field of generating elements of $H^{n-r}(V(n,r))$ can be defined over the whole M, which is possible if M is orientable. The element h is then equal to the generator of $H^{n-r}(V(n,r))$ so defined, multiplied by an integer or a residue class mod. 2. Taking this integer or the residue class mod. 2 as the value of a cochain for the simplex σ^{n-r+1}, we get an integral cochain or a cochain mod. 2. It was proved that the cochain is a cocycle and that its cohomology class is independent of the choice of the mapping from which it is defined. This cohomology class is the class of Stiefel-Whitney. It is to be remarked that the definition can be given under more general conditions, but we shall be satisfied with the above résume.

The situation is simpler in the case of complex sphere bundles. From a bundle of complex spheres $S(n)$ we consider the fibre bundle $\mathfrak{F}^{(r)}$ over the same base space whose fibre at each point is the manifold $U(n,r)$ of r $(1 \leqq r \leqq n)$ linearly independent vectors in $E(n)$. It can be proved that[24] all homology groups of dimension $< 2n - 2r + 1$ of $U(n,r)$ vanish and that the homology group $H^{2n-2r+1}(U(n,r))$ of dimension $2n - 2r + 1$ is a free cyclic group. To define a generator of $H^{2n-2r+1}(U(n,r))$ we take in $E(n)$ an ordered set of $r - 1$ mutually perpendicular unit vectors $\mathfrak{e}_1, \ldots, \mathfrak{e}_{r-1}$. The unit vector $\mathfrak{e}_r$ in $E(n)$ perpendicular to $\mathfrak{e}_1, \ldots, \mathfrak{e}_{r-1}$ describes a complex sphere in the $E(n - r + 1)$ perpendicular to $\mathfrak{e}_1, \ldots, \mathfrak{e}_{r-1}$. The complex sphere $S(n - r + 1)$ in $E(n - r + 1)$ is topologically a real sphere of topological dimension $2n - 2r + 1$. Its two orientations define two cycles belonging respectively to the two generating classes of $H^{2n-2r+1}(U(n,r))$. The cycle carried by the oriented real sphere $S(n-r+1)$ is completely determined by the orientation of $E(n - r + 1)$ considered as a real Euclidean space of topological dimension $2(n - r + 1)$. This orientation is independent of the order of the vectors $\mathfrak{e}_r, \ldots, \mathfrak{e}_n$ in $E(n-r+1)$. It follows that the fibre bundle $\mathfrak{F}^{(r)}$ is orientable in the sense of Steenrod, which means that there is an isomorphism in the large between the $(2n - 2r + 1)$-dimensional homology groups of the fibres of $\mathfrak{F}^{(r)}$, or that a continuous field of generating elements of $H^{2n-2r+1}(U(n,r))$ can be defined over the whole manifold. The fibre bundle $\mathfrak{F}^{(r)}$ has two opposite orientations and we shall from now on make a definite choice of one of them. Using this "oriented" fibre bundle, we shall be allowed to replace an element of $H^{2n-2r+1}(U(n,r))$ at a point by the integer which, when multiplied by the generating element at this point, is equal to the element in question.

With these explanations understood, we have:

Theorem 7. *The rth basic characteristic class of a complex manifold M of dimension n can be defined as follows: Take a simplicial decomposition of M each of whose simplexes belongs to a neighborhood, and define over its skeleton of dimension $2n - 2r + 1$ a continuous field of ordered sets of r linearly independent complex tangent vectors. To each simplex of dimension $2n - 2r + 2$ take a point in its interior and consider the manifold of the ordered sets of r linearly independent complex tangent vectors at that point. The field on the boundary of the simplex defines a mapping of the boundary into this manifold and hence an element of its*

24 Ehresmann, [9]. This fact is easily proved by making use of the covering homotopy theorem, the complex case being even simpler than the real case.

$(2n-2r+1)$-dimensional homotopy or homology group, which is free cyclic. Attach the corresponding integer to the simplex. The cochain so defined is a cocycle and belongs to the rth basic class.

To prove this theorem we take on $H(n, N)$ a definite Schubert variety V defined by the function in (11), which carries a cycle Z, dual to the cocycle Φ_r. The Schubert variety being an algebraic variety on the algebraic variety $H(n, N)$, it follows from the triangulation theorem of algebraic varieties[25] that $H(n, N)$ can be covered by a complex L such that V is a subcomplex. Let L^* be the dual cellular subdivision of L and let f be a mapping of M into L^*, which, according to Theorem 1, induces over M a complex sphere bundle equivalent to the tangent bundle of M. By the theorem on the simplicial approximation of mappings there exists a subdivision of M and a simplicial mapping f_1 of the subdivision into L^* such that f and f_1 are homotopic. By Theorem 2 the complex sphere bundle over M induced by f_1 is equivalent to the tangent bundle of M. For simplicity of notation we can therefore assume f to be a simplicial mapping of M into L^*.

Let K be the skeleton of M of dimension $2(n-r+1)$. $f(K)$ is a subcomplex of L^* of dimension $\leqq 2(n-r+1)$. The rth basic cocycle of M is by definition a linear function of the integral chains of dimension $2(n-r+1)$ of K such that its value at a simplex σ of K is the intersection number of $f(\sigma)$ and V.

To give a description of the Schubert variety V we take in $E(n+N)$ a linear subspace $L(n+N-r)$ of dimension $n+N-r$, and let $L(r)$ be the linear subspace of dimension r which is totally perpendicular to $L(n+N-r)$. Then V consists of all $X(n)$ of $H(n, N)$ satisfying the condition

$$\dim(X(n) \cap L(n+N-r)) \geqq n-r+1.$$

We take in $L(r) r$ mutually perpendicular unit vectors $\mathfrak{a}_1, \ldots, \mathfrak{a}_r$. To each $X(n)$ not belonging to V the projection of $\mathfrak{a}_1, \ldots \mathfrak{a}_r$ on $X(n)$ will give r vectors which are linearly independent, and the construction fails exactly for the $X(n)$ belonging to V.

With these preparations consider a simplex σ of K. If the intersection number $KI(f(\sigma), V) = 0$, the above construction will give a continuous field of r linearly independent complex vectors at each point of $f(\sigma)$ and hence also at each point of σ, which shows that the integer attached to σ according to the statement of the theorem is also zero. It remains therefore to consider the case that $KI(f(\sigma), V) = \epsilon \neq 0$. In this case $f(\sigma)$ is of dimension $2(n-r+1)$. Since $f(\sigma)$ and V belong to dual subdivisions, we have $\epsilon = +1$. Let $X_0(n)$ be the linear space of $H(n, N)$ common to V and $f(\sigma)$. The orthogonal projection of $\mathfrak{a}_1, \ldots, \mathfrak{a}_r$ defines on each $X(n) \neq X_0(n)$ of $f(\sigma) r$ linearly independent complex vectors, which, by the resolution of all the $X(n)$ belonging to $f(\sigma)$ into a topological product, are mapped into the manifold μ of all the ordered sets of r linearly independent vectors in $X_0(n)$. Our purpose is to prove that by means of the sets of vectors on the boundary $\delta f(\sigma), \delta f(\sigma)$ is mapped into a cycle belonging to the generator of the $(2n-2r+1)$st homology group of μ. For simplicity of language let us call index the integer m obtained by mapping the field on the boundary $\delta f(\sigma)$ into a fibre $U(n, r)$, the image cycle being in the homology class equal to m times the generator of $H^{2n-2r+1}(U(n, r))$.

[25] van der Waerden, [26].

Suppose first that a continuous field of ordered sets of n linearly independent vectors $\mathfrak{e}_1, \ldots, \mathfrak{e}_n$ is defined throughout $f(\sigma)$ and let r linearly independent vectors $\mathfrak{f}_1, \ldots, \mathfrak{f}_r$ be defined over $\delta f(\sigma)$ such that

$$\mathfrak{f}_i = \sum_{k=1}^{n} \mathfrak{f}_{ik} \mathfrak{e}_k, \quad 1 \leqq i \leqq r.$$

Regarding $\mathfrak{e}_1, \ldots, \mathfrak{e}_n$ as fixed, these equations also define a mapping of $\partial f(\sigma)$ into μ. We assert that their indices are equal. In fact, the existence of the field $\mathfrak{e}_1, \ldots, \mathfrak{e}_n$ throughout $f(\sigma)$ provides exactly a deformation of the field $\mathfrak{e}_1, \ldots, \mathfrak{e}_n$ over $\partial f(\sigma)$ into vectors $\mathfrak{e}'_1, \ldots, \mathfrak{e}'_n$ which are constant.

We assume that $X_0(n)$ has the property that the orthogonal projections of $a_1, \ldots, a_{r-1}$ onto it are linearly independent, which is possible, after applying a small deformation if necessary. Since $f(\sigma)$ is a simplex, we can define over $f(\sigma)$ a continuous field of n linearly independent vectors, such that in every $X(n)$ the first $r-1$ of these vectors are the orthogonal projections in $X(n)$ of $\mathfrak{a}_1, \ldots, \mathfrak{a}_{r-1}$ respectively. This continuous field is then deformed into a continuous field $\mathfrak{e}_1, \ldots, \mathfrak{e}_n$ over $f(\sigma)$ such that in each $X(n)$ the vectors $\mathfrak{e}_1, \ldots, \mathfrak{e}_n$ constitute a frame (that is, are mutually perpendicular unit vectors). It is well-known that the deformation can be so chosen that during the deformation the vector subspace determined by the first s vectors $(1 \leqq s \leqq n)$ remains fixed. With these deformations performed, we proceed to study the orthogonal projection $\mathfrak{a}_r^*$ of $\mathfrak{a}_r$ in $X(n)$.

The index of the field of orthogonal projections of $\mathfrak{a}_1, \ldots, \mathfrak{a}_r$ on $\partial f(\sigma)$ is equal to the index of the field $\mathfrak{e}_1, \ldots, \mathfrak{e}_{r-1}, \mathfrak{a}_r^*$ on $\partial f(\sigma)$ and is also equal to the index of the same field, when $\mathfrak{e}_1, \ldots, \mathfrak{e}_{r-1}$ are considered as constant vectors. We also remark that the vector $\mathfrak{a}_r^*$ is linearly independent of $\mathfrak{e}_1, \ldots, \mathfrak{e}_{r-1}$ at every point $\neq X_0(n)$ of $f(\sigma)$.

To show that the index in question is 1 we choose the continuous field of vectors $\mathfrak{e}_{n+1}, \ldots, \mathfrak{e}_{n+N}$ over $f(\sigma)$ such that $\mathfrak{e}_A, 1 \leqq A \leqq n+N$, is a frame in $E(n+N)$. Then we have

$$\mathfrak{a}_r = \sum_{A=1}^{n+N} u_A \mathfrak{e}_A, \quad \sum_{A=1}^{n+N} u_A \bar{u}_A = 1,$$

and

$$\mathfrak{a}_r^* = \sum_{i=1}^{n} u_{ri} \mathfrak{e}_i.$$

According to our previous remark we can regard the vectors $\mathfrak{e}_A, 1 \leqq A \leqq n+N$, as fixed and consider the mapping of $f(\sigma) - X_0(n)$ into $S(n-r)$ defined by the vector whose components with respect to a fixed frame are $u_{r,r+1}, \ldots, u_{rn}$. Thus we see that the index is 1, and Theorem 7 is proved.

It is also possible to introduce from a bundle of complex spheres $S(n)$ the fibre bundles $\mathfrak{F}^{(r)*}$ over the same base space whose fibre at each point is the manifold $U^*(n,r)$ of all ordered sets of r $(1 \leqq r \leqq n)$ mutually perpendicular vectors of $S(n)$. The manifold $U^*(n,r)$ is an absolute retract of $U(n,r)$. Theorem 7 still holds, if we replace everywhere the phrase "ordered sets of r linearly independent complex vectors" by "ordered sets of r mutually perpendicular vectors of $S(n)$" and, naturally also the manifold $U(n,r)$ by $U^*(n,r)$.

2 A third definition of the basic characteristic classes

We suppose in this section that the base space M, which is a complex manifold, is compact. From the tangent bundle $\mathfrak{F}$ over M we construct the fibre bundle $\mathfrak{F}^{(r)*}(1 \leqq r \leqq n)$ as explained at the end of the last section. Then the following theorem gives a third definition of the rth basic characteristic class:

Theorem 8. *The rth basic characteristic class of M is the cohomology class of M, each of whose cocycles γ has the following property: Under the projection of $\mathfrak{F}^{(r)*}$ into M, γ is mapped into a cocycle γ^*. There exists in $\mathfrak{F}^{(r)*}$ a cochain β^*, such that $\delta\beta^* = \gamma^*$ and that β^* reduces to a fundamental cocycle on each fibre of $\mathfrak{F}^{(r)*}$.*

The last statement in the theorem needs some explanation. Let P be a polyhedron and $Q \subset P$ a closed subpolyhedron of P. If γ is a cochain in P, its *reduced cochain* on Q is the cochain γ' such that $\gamma' \cdot \sigma = \epsilon\gamma \cdot \sigma$, where $\epsilon = 1$ or 0 according as the simplex σ belongs to Q or not. Moreover, the integral cohomology group of dimension $2n - 2r + 1$ of a fibre being free cyclic, a *fundamental cocycle* on the fibre is a cocycle of dimension $2n - 2r + 1$ which belongs to a generator of the cohomology group. It is also understood that the cycles and cocycles are defined in terms of simplicial decompositions of M and $\mathfrak{F}^{(r)*}$. To define the inverse mapping of the cocycles of M into the cocycles of $\mathfrak{F}^{(r)*}$ induced by the projection π of $\mathfrak{F}^{(r)*}$ into M, we therefore take a simplicial approximation π' of π. Let σ^* be a simplex of dimension $2n - 2r + 2$ of $\mathfrak{F}^{(r)*}$. Then we define

$$\gamma^* \cdot \sigma^* = \gamma \cdot \pi'(\sigma^*),$$

if $\pi'(\sigma^*)$ is of dimension $2n - 2r + 2$ and $\gamma^* \cdot \sigma^* = 0$ if $\pi'(\sigma^*)$ is of lower dimension. The cocycle γ^* depends on π', but its cohomology class is independent of it. The theorem asserts that any such cocycle γ^* reduces to a fundamental cocycle on a fibre.

In order to prove Theorem 8 we need the following lemma:

Lemma 4. *With the notations of Theorem 8 let U be a neighborhood of M and let $\pi^{-1}(U)$ be its complete inverse image in $\mathfrak{F}^{(r)*}$. Let K be a finite complex and L its skeleton of dimension $\leqq 2n - 2r$. If f and g are two continuous mappings into $\pi^{-1}(U)$ of K and L respectively, there is a continuous mapping f^* of K into $\pi^{-1}(U)$ which is homotopic to f and coincides with g on L.*

We denote by I the unit segment $0 \leqq t \leqq 1$ and consider the topological product $K \times I$. To prove Lemma 4 is to define a continuous mapping $f(K \times I) \subset \pi^{-1}(U)$, with $f(K \times 0)$ and $f(L \times 1)$ given. For this purpose we decompose K simplicially and arrange the simplexes of the decomposition in a sequence that every simplex is preceded by its faces. The mapping is then defined by successive extensions over the prisms constructed on the simplexes of the sequence. We resolve $\pi^{-1}(U)$ into the topological product of U and a fixed fibre F_0 and denote by λ the projection of $\pi^{-1}(U)$ onto F_0. Let σ^0 be a vertex of K. Then $f(\sigma^0 \times 0)$ and $f(\sigma^0 \times 1)$ are both defined in $\mathfrak{F}^{(r)*}$, and can be joined by a segment, on which the prism on σ^0 is mapped. Using mathematical induction we suppose f be defined over all prisms on simplexes preceding σ^m of the sequence, and consider $\sigma^m, m \leqq 2n - 2r$. By hypothesis, the mapping is defined over $\partial(\sigma^m \times I)$, which is topologically a sphere of topological dimension m. The mapping can be extended over $\sigma^m \times I$, if and only if $f(\partial(\sigma^m \times I))$ is homotopic to zero in $\mathfrak{F}^{(r)*}$, that is, by the covering homotopy theorem, if and only if $\lambda f(\partial(\sigma^m \times I))$ is homotopic to zero in F_0. The latter is the case, because the mth homotopy group of F_0 is zero. It follows that f is defined for the subcomplex $L \times I + K \times 0$ of the prism $K \times I$. By a well-known elementary geometric construction[26] f is then extended over $K \times I$. Thus the lemma is proved.

We proceed to prove Theorem 8. Let γ be a cocycle of M belonging to the rth basic class defined by the construction of Theorem 7, with the bundle $\mathfrak{F}^{(r)}$ replaced by $\mathfrak{F}^{(r)*}$. To explain Theorem 7 for this case, we take a simplicial decomposition of M which is so fine that each simplex belongs to a neighborhood of M. Let $K^{2n-2r+1} = K$ be the $(2n-2r+1)$-dimensional skeleton of the simplicial decomposition. There exists a continuous mapping Ψ of K into $\mathfrak{F}^{(r)*}$ such that $\pi\Psi(p) = p$ for every $p \in K$. Let σ be a simplex of dimension $2n-2r+2$. The mapping Ψ defines a mapping of the boundary $\partial\sigma$ of σ into $\mathfrak{F}^{(r)*}$ and the mapping $\lambda\Psi$ defines a mapping of $\partial\sigma$ into F_0, and hence a cycle of dimension $2n-2r+1$ of F_0. Let this cycle be homologous to a multiple $\gamma(\sigma)$ of the generating cycle of dimension $2n-2r+1$ of F_0. According to Theorem 7 the cocycle $\gamma(\psi)$ defined by assigning the integer $\gamma(\sigma)$ to σ belongs to the rth basic class. Moreover, it was also proved[27] that to a given cocycle γ of the rth basic class there exists a mapping ψ such that $\gamma(\psi) = \gamma$. We suppose ψ to be so chosen.

Let π' be a simplicial approximation of π and let γ^* be the inverse image of γ under π' as defined above. We shall show that γ^* is the coboundary of an integral cochain β^* of dimension $2n-2r+1$ of $\mathfrak{F}^{(r)*}$. To define β^* let τ^* be a simplex of dimension $2n-2r+1$ of $\mathfrak{F}^{(r)*}$, and let $\tau = \pi'(\tau^*)$. The discussion will be divided into two cases, according as τ is of dimension equal to or less than $2n-2r+1$.

Suppose τ be of dimension $2n-2r+1$. In this case π' establishes a simplicial and therefore topological mapping between τ and τ^*. By Lemma 4 there exists a mapping $\pi''(r) \subset \mathfrak{F}^{(r)*}$, which is homotopic to $\pi'^{-1}(\tau)$ and coincides with Ψ on the boundary $\partial\tau$. We then take an oriented sphere of topological dimension $2n-2r+1$ and denote by H_1, H_2 its two hemispheres. We map H_1 and H_2 into τ by the mappings h_1 and h_2 of the degrees -1 and $+1$ respectively, such that h_1 and h_2 are identical on the "equator" $H_1 \cap H_2$. A mapping f of the sphere $H_1 + H_2$ into $\mathfrak{F}^{(r)*}$ is then defined by the conditions

$$f(p) = \pi'' h, (p), \quad p \in H_1,$$
$$f(p) = \psi h_2(p), \quad p \in H_2.$$

This mapping f is by construction continuous. Taking its projection λf on F_0, we get an element of the $(2n-2r+1)$-dimensional homotopy group of the fibre and hence an integer, on account of the orientability of the bundle $\mathfrak{F}^{(r)*}$. This integer we define to be $\beta^* \cdot \tau^*$. It is to be remarked that $\beta^* \cdot \tau^*$ in general depends on the deformation which carries π'^{-1} to π'', but only one such deformation will be utilized, and $\beta^* \cdot \tau^*$ is thus well defined.

Next let τ be of dimension $< 2n-2r+1$. We suppose without loss of generality that $p_0 = \pi(F_0) \in \tau$. Let any simplex τ' of dimension $2n-2r+1$ be mapped into τ^*

26 Alexandroff-Hopf, [1], p. 501.
27 Steenrod, [21], p. 124.

by a non-degenerate orientation-preserving simplicial mapping. By Lemma 4 this mapping is homotopic to a mapping $\pi'''(\tau') \subset \mathfrak{F}^{(r)*}$ such that $\pi'''(\partial\tau') = q_0$, where q_0 is a point of F_0. We identify all the points on $\partial\tau'$, thus getting an oriented sphere of topological dimension $2n - 2r + 1$, which is mapped into F_0 by the mapping $\lambda\pi'''(\tau') \subset F_0$. This mapping defines an element of the $(2n - 2r + 1)$-dimensional homotopy group and hence an integer, which is defined to be $\beta^* \cdot \tau^*$.

It remains to show that the coboundary of the cochain β^* so defined is equal to γ^*. For this purpose let σ^* be a simplex of $\mathfrak{F}^{(r)*}$ of dimension $2n - 2r + 2$. It is sufficient to verify that $\gamma^* \cdot \sigma^* = \delta\beta^* \cdot \sigma^* = \beta^* \cdot \partial\sigma^*$.

Suppose first that $\sigma = \pi'(\sigma^*)$ is of dimension $2n - 2r + 2$. The mapping π'' is defined for each simplex of $\partial\sigma$ and hence for $\partial\sigma$ itself, because it coincides with ψ on the $(2n - 2r)$-dimensional skeleton K^{2n-2r}. We therefore have two mappings, $\lambda\psi$ and $\lambda\pi''$ respectively, of $\partial\sigma$ into F_0 such that they are identical on K^{2n-2r}. Our fibre F_0 being $(2n - 2r)$-simple, this is a situation discussed by Eilenberg,[28] who introduced several cochains, denoted in his notation by $c(\lambda\psi), c(\lambda\pi''), d(\lambda\psi, \lambda\pi'')$ respectively. In our notation they are given by

$$\begin{aligned} &c(\lambda\psi) \cdot \sigma = \gamma^* \cdot \sigma^*, \\ &c(\lambda\pi'') = 0, \\ &d(\lambda\psi, \lambda\pi'') \cdot \partial\sigma = \beta^* \cdot \partial\sigma^*, \end{aligned}$$

where the second relation follows from the fact that $\lambda\pi''$ is defined for the simplex σ bounded by $\partial\sigma$. From a theorem of Eilenberg we have

$$\delta d(\lambda\psi, \lambda\pi'') = c(\lambda\psi) - c(\lambda\pi''),$$

or

$$\gamma^* \cdot \sigma^* = \beta^* \cdot \partial\sigma^*,$$

which is to be proved.

Next suppose $\sigma = \pi'(\sigma^*)$ be of dimension $< 2n - 2r + 2$. If each simplex of $\partial\sigma^*$ is mapped by π' into a simplex of dimension $< 2n - 2r + 1, \beta^* \cdot \partial\sigma^*$ is clearly zero. The other possibility is that σ is of dimension $2n - 2r + 1$ and that exactly two of the simplexes of $\partial\sigma^*$, say τ_1^* and τ_2^*, are mapped by π' into σ. If τ_1^* and τ_2^* are coherently oriented with the boundary $\partial\sigma^*$, it follows by definition that $\beta^* \cdot (\tau_1^* + \tau_2^*) = 0$. On the other hand, it is not difficult to see that $\beta^* \cdot (\partial\sigma^* - \tau_1^* - \tau_2^*) = 0$. Hence we have $\beta^* \cdot \partial\sigma^* = 0$. We have thus proved that γ^* is the coboundary of a cochain β^*.

To see that β^* reduces on a fibre F_0, we suppose the simplicial decomposition of $\mathfrak{F}^{(r)*}$ so made that F_0 is a subcomplex. Every simplex of F_0 is mapped by π' into a point, so that we have $\gamma^* \cdot \sigma^* = 0$ for every simplex σ^* of dimension $2n - 2r + 2$ of F_0 , which shows that β^* reduces to a cocycle on F_0. To show that β^* is the fundamental cocycle on F_0, it is sufficient to show that one is the value of its product with a cycle belonging to the generating homology class of dimension $2n - 2r + 1$ of F_0. For this purpose we take an oriented sphere of topological dimension $2n - 2r + 1$ and map it simplicially into F_0 such that the map belongs

[28] Eilenberg, [11], pp. 235–237.

to the generator of the $(2n-2r+1)$-dimensional homotopy group of F_0. The image of this map defines a cycle of F_0 belonging to the generating homology class of the $(2n-2r+1)$-dimensional homology group, and its product with β^* is 1. It follows that β^* reduces to a fundamental cocycle on F_0.

Our Theorem 8 is now completely proved.

3 In terms of differential forms

Consider again the Grassmann manifold $H(n,N)$. We take as point of a new space a linear space $E(n)$ and r vectors $\mathfrak{e}_{n-r+1},\dots,\mathfrak{e}_n$ belonging to $E(n)$ such that $\mathfrak{e}_i\bar{\mathfrak{e}}_j = \delta_{ij}, n-r+1 \leqq i,\ j \leqq n$. This space, to be denoted by $R(r,n,N)$, is clearly a fibre bundle over $H(n,N)$, the projection of a point of $R(r,n,N)$ being the corresponding $E(n)$ and each fibre being homeomorphic to $U^*(n,r)$. This fibre bundle $R(r,n,N)$ is transformed transitively by the unitary group in the space $E(n+N)$. Let W_r be the rth basic class of $H(n,N)$ and $\gamma \in W_r$ be one of its cocycles. γ is mapped by the inverse mapping induced by any simplicial approximation of π into a cocycle having the properties asserted by Theorem 8. In particular, we can take for γ the cocycle defined by the differential form Φ_r. The inverse image of Φ_r in $R(r,n,N)$ under π is a differential form which for convenience we denote by Φ_r. The form Φ_r then defines a cocycle γ^* in $R(r,n,N)$. From Theorem 8 it follows that[29] there exists a cochain β^* in $R(r,n,N)$ such that $\delta\beta^* = \gamma^*$ and such that β^* reduces to a fundamental cocycle on a fibre.

In order to define β^* in $R(r,n,N)$ by means of a suitably chosen differential form, we shall make use of the following lemma proved by de Rham:[30]

Lemma 5. *Let M be a compact differentiable manifold of class $\geqq 2$ and let K be a simplicial decompositi on of M, whose simplexes are $\sigma_i^p, i=1,\dots,\alpha_p, p=0,1,\dots,n$, and whose incidence relations are*

$$\partial\sigma_i^p = \sum_{j=1}^{\alpha_{p-1}} \eta_{ij}^{(p)}\sigma_i^{p-1}.$$

Then there exists a set of differential forms

$$\varphi_i^{(p)}, \quad i=1,\dots,\alpha_p, \quad p=0,1,\dots,n,$$

such that the following conditions are satisfied:

$$1)\ \int_{\sigma_j(p)} \varphi_i^{(p)} = \delta_{ij},$$

$$2)\ d\varphi_i^{(p)} = \sum \eta_{ji}^{(p+1)}\varphi_j^{(p+1)}.$$

29 We have tacitly assumed at this point of our discussion that the Theorem 8, proved by a combinatorial construction for the simplicial approximations of the projection π, holds for π itself, when the cocycles are expressed by means of differential forms. It is, however, possible to avoid this assumption by observing that the cochain β^* exists in $R(r,n,N)$ such that $\delta\beta^* = \gamma^*$. That β^* reduces to a fundamental cocycle on a fibre then follows from the very definition of the characteristic cocycle on $H(n,N)$.

30 de Rahm, [20], p. 178.

We apply this lemma to the manifold $R(r,n,N)$, and write for simplicity $p = 2n-2r+1$. The cochain β^* is defined by definition by a system of equations

$$\beta^* \cdot \sigma_i^p = \lambda_i, \quad i = 1, \ldots, \alpha_p.$$

Let

$$\omega = \sum_{i=1}^{\alpha_p} \lambda_i \varphi_i^{(p)}.$$

Then we have

$$\int_{\sigma_i^p} \omega = \lambda_i$$

which shows that the differential form ω defines the cochain β^*. By construction we have

$$d\omega = \Phi_r.$$

Now the unitary group $U(n+N)$ in $E(n+N)$ transforms transitively the manifold $R(r,n,N)$. Let s be a transformation of $U(n+N)$. If θ is a differential form in $R(r,n,N)$, we shall denote by $s\theta$ its transform by the transformation s. We also use the notation $\theta \sim 0$ to denote that θ is derived.

Lemma 6. *Let β^* be a cochain of dimension $2n-2r+1$, whose coboundary is γ^*. Let ω be a differential form which defines β^*. Then $s\omega - \omega \sim 0$.*

First of all, the differential form $s\omega - \omega$ is exact, since we have

$$d(s\omega - \omega) = s\Phi_r - \Phi_r = 0.$$

Let ζ be a cycle of dimension $2n-2r+1$ of $R(r,n,N)$. It is sufficient to prove that

$$\int_\zeta s\omega = \int_\zeta \omega.$$

The cycle $s\omega$ ω being homologous to zero, let Z^* be the chain it bounds. Then we have

$$\int_\zeta s\omega - \omega = \int_{s\zeta - \zeta} \omega = \int_{\partial Z^*} \omega = \int_{Z^*} \Phi_r.$$

Let Z be the projection in the base space $H(n,N)$ of the chain Z^* in $R(r,n,N)$. The boundary of Z is the projection of $s\zeta - \zeta$. Since the Betti group of dimension $2n-2r+1$ of $H(n,N)$ is zero, the projection of ζ bounds in $H(n,N)$ a chain which we shall call Z_1. Then the projection of $s\zeta$ bounds the chain sZ_1, and we have

$$Z \sim sZ_1 - Z_1.$$

It follows that

$$\int_{Z^*} \Phi_r = \int_Z \Phi_r = \int_{Z_1} s\Phi_r - \int_{Z_1} \Phi_r = 0.$$

Thus Lemma 6 is proved.

Theorem 9. *Under the projection of $R(r,n,N)$ into $H(n,N)$ the differential form Φ_r is mapped by the inverse mapping into $R(r,n,N)$. There exists a differential form π which is invariant under transformations of the unitary group $U(n+N)$ operating in $R(r,n,N)$ and whose exterior derivative $d\pi$ is equal to Φ_r.*

By Lemma 5 we can construct the differential form ω in $R(r,n,N)$ such that

$$d\omega = \Phi_r.$$

For such a differential form ω it follows from Lemma 6 that

$$\delta\omega - \omega \sim 0.$$

Let dv be the invariant volume element of $U(n+N)$ such that the integral of dv over $U(n+N)$ is equal to 1. We put

$$\Pi = \int_{U(n+N)} s\omega dv.$$

Then Π is invariant under $U(n+N)$ and we have

$$\lambda\Pi = \int_{U(n+N)} s \cdot d\omega \cdot dv = \Phi_r \int_{U(n+N)} dv = \Phi_r,$$

which proves our theorem.

To a point of $R(r,n,N)$ we now attach the frames $\mathfrak{e}_1, \ldots, \mathfrak{e}_{n+N}$ in $E(n+N)$ such that $\mathfrak{e}_1, \ldots, \mathfrak{e}_n$ determine the $E(n)$ and that $e_{n-r+1}, \ldots, \mathfrak{e}_n$ are the vectors in question. For clearness let us agree in the remainder of this section on the following ranges of indices:

$$1 \leqq \alpha, \beta, \gamma \leqq n-r, \quad n-r+1 \leqq A, B, C \leqq n,$$
$$n+1 \leqq i, j, k \leqq n+N.$$

In a neighborhood of $R(r,n,N)$ we can choose a differentiable family of such frames, one attached at each point of the neighborhood. By means of the family of frames the forms $\theta_{A\alpha}, \bar{\theta}_{A\alpha}, \theta_{Ai}, \bar{\theta}_{Ai}, \theta_{AB}$ can be constructed according to the equations (7). They constitute a set of linearly independent linear differential forms at each point of $R(r,n,N)$. Our form π, whose existence was asserted by Theorem 9 and which is invariant under $U(n+N)$, is necessarily a polynomial (in the sense of Grassmann algebra) in the forms of this set, with constant coefficients. On the other hand, the form Π, being itself in $R(r,n,N)$, must be invariant under the transformation

$$\theta^*_{Ai} = \sum_j a_{ij}\theta_{A_j}$$

where a_{ij} are the elements of a unitary matrix. We put

$$\Theta_{AB} = \sum_i \theta_{Ai}\theta_{iB}.$$

It follows from the first main theorem on vector invariants of the unitary group that Π is a polynomial in $\theta_{A\alpha}, \bar{\theta}_{A\alpha}, \theta_{AB}, \Theta_{AB}$, with constant coefficients. Moreover, on a fibre, that is, omitting all terms in Θ_{AB}, Π becomes a fundamental cocycle.

All these results can be summarized in the following theorem:

Theorem 10. *There exists in $R(r,n,N)$ a polynomial Π in $\theta_{A\alpha}, \bar{\theta}_{A\alpha}, \theta_{AB}, \Theta_{AB}$, with constant coefficients, such that $d\Pi = \Phi_r$. When all terms involving Θ_{AB} in Π are omitted, the form defines a fundamental cocycle on a fibre.*

Chapter IV Hermitian Manifolds

1 Fundamental formulas of Hermitian geometry

Let M be a compact complex manifold. M is called an Hermitian manifold, if an intrinsic Hermitian differential form is given throughout the manifold. In each local coordinate system z^i the Hermitian differential form is defined by

$$ds^2 = \sum_{i,j=1}^{n} g_{ij}(z,\bar{z})(dz^i d\bar{z}^j), \quad \bar{g}_{ij} = g_{ji}, \tag{28}$$

where, as well as in later formulas, we insert a parenthesis to designate that the multiplication of the differential forms in question is ordinary multiplication. We shall agree, unless otherwise stated, that the indices i, j, k take the values 1 to n.

Our main result in this chapter is to establish that the n basic classes which arise from the analytic structure of a complex manifold, are completely determined by the Hermitian metric, if the manifold in question is an Hermitian manifold. In particular, as we shall see later, the theorem for the class W_1 reduces to the formula of Allendoerfer-Weil, if we interpret the Hermitian metric as a Riemannian metric for the real manifold of $2n$ topological dimensions.

We begin by establishing the fundamental formulas for local Hermitian geometry.

For this purpose we determine in a neighborhood of M n linear differential forms φ_i in the local coordinates z^i such that

$$ds^2 = \sum_{i=1}^{n} (\varphi_i \bar{\varphi}_i). \tag{29}$$

The forms φ_i are determined up to a unitary transformation:

$$\omega_i = \sum_{i=1}^{n} u_{ij}\varphi_j, \quad i = 1, \ldots, n, \tag{30}$$

where u_{ij} are the elements of a unitary matrix

$$U = (u_{ij}),$$

and which we take to be independent variables. Let $\mathfrak{e}_i$ be the dual base corresponding to ω_k, so that

$$\omega_i(\mathfrak{e}_k) = \delta_{ik}.$$

From the Hermitian differential form a scalar product of the contravariant vectors can be defined, and we have

$$\mathfrak{e}_i \cdot \bar{\mathfrak{e}}_k = \delta_{ik}.$$

We shall call a frame the figure formed by a point P and n such vectors $\mathfrak{e}_i$. With a natural topology the set of frames constitutes a fibre bundle over M.

The forms ω_i are intrinsically defined in the fibre bundle. By actual calculation in terms of a local coordinate system, we find that their exterior derivatives are of the form

$$d\omega_i = \sum_j \omega_j \omega_{ji} + \sum_{j,k} a_{ijk}\omega_j\omega_k + \sum_{j,k} b_{ijk}\omega_j\bar{\omega}_k, \quad a_{ijk} + a_{ijk} = 0,$$

where

$$\omega_{ij} = \sum_k \bar{u}_{ik} d\bar{u}_{jk},$$

so that

$$\omega_{ij} + \omega_{ji} = 0.$$

But the forms ω_{ij} are defined up to the transformation

$$\omega_{ij} \to \omega_{ij} + \sum \lambda_{ijk}\omega_k - \sum \bar{\lambda}_{jik}\bar{\omega}_k,$$

where the quantities λ_{ijk} are arbitrary. It is easy to show that there exists one, and only one, set of forms ω_{ij}, such that the following equations are satisfied:

$$d\omega_i = \sum_i \omega_j\omega_{ji} + \sum_{jik} A_{ijk}\omega_j\omega_k, \quad A_{ijk} + A_{ikj} = 0, \quad \omega_{ij} + \bar{\omega}_{ji} = 0. \tag{31}$$

From the uniqueness of this set of forms follows the fact that they are intrinsically defined in the fibre bundle. The forms ω_i, ω_{ij} constitute therefore a set of linearly independent linear differential forms in the fibre bundle.

From equations (31) it is possible to draw all the consequences of local Hermitian geometry. In fact, we put

$$\Omega_i = \sum_{j,k} A_{ijk}\omega_j\omega_k. \tag{32}$$

Exterior differentiation of the first set of equations in (31) will give

$$d\Omega_i - \sum_k \omega_k\Omega_{ki} + \sum_k \Omega_k\omega_{ki} = 0,$$

where we have put

$$\Omega_{ij} = d\omega_{ij} - \sum_k \omega_{ik}\omega_{kj}. \tag{33}$$

We remark that $d\Omega_i + \sum_k \Omega_k\omega_{ki}$ is of the form $\sum_{j,k}\psi_{ijk}\omega_j\omega_k$. It follows that

$$\omega_1 \cdots \omega_n \Omega_{ij} = 0$$

and that we can put

$$\Omega_{ik} = \sum \chi_{ikj}\omega_j.$$

On the other hand, we have

$$\Omega_{ik} + \bar{\Omega}_{ki} = 0$$

or

$$\sum_j \chi_{ikj}\omega_j + \sum_j \bar{\chi}_{kij}\bar{\omega}_j = 0,$$

which shows that χ_{ikj} is a linear combination of $\omega_k, \bar{\omega}_k$:

$$\chi_{ikj} = \sum_{l=1}^{n} a_{ikjl}\omega_l + \sum_{l=1}^{n} b_{ijkl}\bar{\omega}_l.$$

Substituting this expression of χ_{ikj} into the last equation, we see immediately that a_{ikjl} must be symmetric in the last two indices j, l. It follows that Ω_{ik} is of the form

$$\Omega_{ik} = \sum b_{ikjl}\bar{\omega}_l\omega_j.$$

The equations for the exterior derivatives $d\omega_i$, $d\omega_{sj}$ and the equations obtained therefrom by exterior differentiation we shall call the fundamental equations of local Hermitian geometry. These equations will now be summarized as follows:

$$\begin{aligned}
d\omega_i &= \sum_j \omega_j\omega_{ji} + \Omega_i, \\
d\omega_{ij} &= \sum_k \omega_{ik}\omega_{kj} + \Omega_{ij}, \\
d\Omega_i + \sum_j \Omega_j\omega_{ji} &- \sum_j \omega_j\Omega_{ji} = 0, \\
d\Omega_{ij} + \sum_k \Omega_{ik}\omega_{kj} &- \sum_k \omega_{ik}\Omega_{kj} = 0, \\
\Omega_i &= \sum A_{ijk}\omega_j\omega_k, \quad A_{ijk} + A_{ikj} = 0, \\
\Omega_{ij} &= \sum_{l,m=1}^{n} R_{ij,lm}\bar{\omega}_l\omega_m, \quad R_{ij,lm} = \bar{R}_{ji,ml}, \\
\omega_{ij} + \bar{\omega}_{ji} &= 0, \quad \Omega_{ij} + \bar{\Omega}_{ji} = 0.
\end{aligned} \tag{34}$$

In a well-known way the forms ω_i, ω_{ij} can be interpreted as defining an infinitesimal displacement, by means of the equations

$$\begin{aligned}
dp &= \sum_i \omega_i\mathfrak{e}_i, \\
d\mathfrak{e}_i &= \sum_j \omega_{ij}\mathfrak{e}_j.
\end{aligned} \tag{35}$$

Of importance are the Hermitian metrics satisfying the condition

$$\Omega_i = 0, \tag{36}$$

which will be called Hermitian metrics without torsion. An Hermitian metric without torsion can be characterized by the condition

$$d\left(\sum_i \omega_i\bar{\omega}_i\right) = 0, \tag{37}$$

and was studied by E. Kähler.[31] Kähler proved that in this case there exists locally a function $F(z^i, \bar{z}^i)$ such that the metric can be written in the form

$$ds^2 = \sum_{i,k} \frac{\partial^2 F}{\partial z^i \partial \bar{z}^k}(dz^i d\bar{z}^k). \tag{38}$$

Hermitian metrics without torsion play an important rôle in the theory of automorphic functions of several complex variables.

2 Formulas for the basic characteristic classes

As defined in Chapter III there are on M n basic characteristic classes, the rth one ($1 \leqq r \leqq n$) being of dimension $2(n-r+1)$. We shall show that, if M is an Hermitian manifold, these classes are defined by the local properties of the Hermitian metric. For this purpose we put

$$\begin{aligned}\Psi_r = & \frac{1}{(2\pi\sqrt{-1})^{n-r+1}(n-r+1)!} \\ & \sum \delta(i_1 \cdots i_{n-r+1}; j_1 \cdots j_{n-r+1})\Omega_{i_1 j_1} \cdots \Omega_{i_{n-r+1} j_{n-r+1}},\end{aligned} \tag{39}$$

where Ω_{ij} are the forms defined in (34) and where the meaning of the summation has been explained before. Then we have the following theorem, which is the main result of this paper:

Theorem 11. *The form Ψ_r defined by (39) is the form corresponding to the rth basic characteristic class W_r in the sense that the product of any homology class ζ of dimension $2(n-r+1)$ with W_r is equal to the integral of Ψ_r over ζ:*

$$\zeta \cdot W_r = \int_\zeta \Psi_r. \tag{40}$$

We first establish the following lemma:

Lemma 7. *Let Δ be the differential form Π in Theorem 10, with every θ and Θ replaced by the corresponding ω and Ω with the same indices. Then $d\Delta = \Psi_r$.*

We observe that the equations for $d\omega_{ij}, d\Omega_{ij}$ are exactly of the same form as the equations for $d\theta_{ij}, d\Theta_{ij}$, the only difference being that Θ_{ij} is given by the equation (12). It follows that

$$d\Delta - \Psi_r \equiv 0 \text{ mod.} \Omega_{ij} - \sum_{B=n+1}^{n+N} \theta_{iB}\theta_{Bj}.$$

[31] Kähler, [16].

By mathematical induction on the degree of $d\Delta - \Psi_r$ it is easy to show that then

$$d\Delta - \Psi_r = 0.$$

To prove Theorem 11 we make use of the definition of W_r given in Theorem 7, with $\mathfrak{F}^{(r)}$ replaced by $\mathfrak{F}^{(r)*}$. A sufficiently fine simplicial decomposition K of M is taken and a continuous mapping ψ of its $(2n-2r+1)$-dimensional skeleton into $\mathfrak{F}^{(r)*}$ is defined, such that the image of every point belongs to the fibre over it. Let $\sigma_A^{2n-2r+2}$ or $\sigma_A, 1 \leqq A \leqq \alpha_{2n-2r+2}$, be the simplexes of dimension $2n-2r+2$ of K. By the construction of Theorem 7, an integer $\gamma(\sigma_A)$ is defined for every σ_A, and the corresponding cocycle γ belongs to W_r. It is sufficient to prove that

$$\zeta \cdot \gamma = \int_\zeta \Psi_r. \tag{40a}$$

Both sides of the equation (40a) being linear in ζ, equation (40a) will follow from the relation

$$\gamma(\sigma_A) = \int_{\sigma_A} \Psi_r.$$

But, by Lemma 7,

$$\int_{\sigma_A} \Psi_r = \int_{\psi(\sigma_A)} \Psi_r = \int_{\psi(\sigma_A)} d\Delta = \int_{\partial\psi(\sigma)_A} \Delta.$$

The image $\psi(\sigma_A)$ has a singular point and we see that the last integral is precisely the definition of $\gamma(\sigma_A)$ given in integral form. Hence Theorem 11 is proved.

3 The case $r = 1$ and the formula of Allendoerfer-Weil

In the case $r = 1$ we can take for the cycle ζ in (40) one of the fundamental cycles of the manifold M. Then we have $\zeta \cdot W_1 = \chi$, the Euler-Poincaré characteristic of M. On the other hand, the manifold M can be considered as a real differentiable manifold and the Hermitian metric can be used to define a Riemannian metric in the real manifold. It is to be expected that the formula (40) will then reduce to the formula of Allendoerfer-Weil. We shall show that this is actually the case, if the Hermitian metric is without torsion.

To study the Hermitian metric as a Riemannian metric, we decompose each of the forms $\omega_i, \omega_{ij}, \Omega_{ij}$ into its real and imaginary parts, writing

$$\begin{aligned} \omega_i &= \theta_i + \sqrt{-1}\psi_i, \\ \omega_{ij} &= \theta_{ij} + \sqrt{-1}\psi_{ij}, \\ \Omega_{ij} &= \Theta_{ij} + \sqrt{-1}\Psi_{ij}. \end{aligned} \tag{41}$$

From the last two equations of (34) we have

$$\theta_{ij} + \theta_{ji} = 0, \quad \psi_{ij} - \psi_{ji} = 0,$$

$$\Theta_{ij} + \Theta_{ji} = 0, \quad \Psi_{ij} - \Psi_{ji} = 0. \tag{41a}$$

The Hermitian metric can then be written as a Riemannian metric in the form:

$$ds^2 = \sum_i \{(\theta_i)^2 + (\psi_i)^2\}.$$

We get, moreover, by separating the real and imaginary parts of the equations for $d\omega_i, d\omega_{ij}$, the following equations:

$$\begin{aligned} d\theta_i &= \sum_j \theta_j\theta_{ji} - \sum_j \psi_j\psi_{ji}, \\ d\psi_i &= \sum_j \theta_j\psi_{ji} + \sum_j \psi_j\theta_{ji}, \end{aligned} \tag{42}$$

and

$$\begin{aligned} d\theta_{ij} &= \sum_k \theta_{ik}\theta_{kj} - \sum_k \psi_{ik}\psi_{kj} + \Theta_{ij}, \\ d\psi_{ij} &= \sum_k \theta_{ik}\psi_{kj} + \sum_k \psi_{ik}\theta_{kj} + \Psi_{ij}. \end{aligned} \tag{43}$$

It follows that, for the Riemannian Geometry of $2n$ dimensions thus obtained, the curvature forms can be conveniently described by the matrix

$$\begin{pmatrix} \Theta_{ij} & -\Psi_{ij} \\ \Psi_{ij} & \Theta_{ij} \end{pmatrix} \tag{44}$$

or simply by

$$\begin{pmatrix} \Theta & -\Psi \\ \Psi & \Theta \end{pmatrix}.$$

It only remains to compare the integrand of the Allendoerfer-Weil formula calculated from this matrix of curvature forms with the expression Ψ_1 defined in (39).

Let Ω denote the integrand of the Allendoerfer-Weil formula. We observe that $(-2\pi)^n\Omega$ obeys the same expansion rule as a so-called Pfaffian function of the $2n^{\text{th}}$ order,[32] which is an integral rational function in a number of independent variables, whose square is equal to the value of a skew-symmetric determinant of order $2n$. It follows that

$$(2\pi)^{2n}\Omega^2 = \begin{vmatrix} \Theta & -\Psi \\ \Psi & \Theta \end{vmatrix}.$$

On the other hand, we have

$$\Psi_1 = \frac{1}{(2\pi\sqrt{-1})^n}|\Omega_{ij}| = \frac{1}{(2\pi\sqrt{-1})^n}|\Theta + \sqrt{-1}\Psi|,$$

32 Pascal, [17], pp. 60–64.

from which we get, after some reduction,

$$(-1)^n(2\pi)^{2n}\Psi_1^2 = \begin{vmatrix} \Theta+\sqrt{-1}\Psi & 0 \\ 0 & \Theta+\sqrt{-1}\Psi \end{vmatrix}$$

$$= \begin{vmatrix} \Theta+\sqrt{-1}\Psi & 0 \\ \Theta+\sqrt{-1}\Psi & -\Theta+\sqrt{-1}\Psi \end{vmatrix} = (-1)^n \begin{vmatrix} \Theta & -\Psi \\ \Psi & \Theta \end{vmatrix}.$$

Hence we have

$$\Omega^2 = \Psi_1^2,$$

and finally

$$\Omega = \Psi_1,$$

by comparing the coefficients of one of the terms in both sides.

Chapter V Applications to Elliptic Hermitian Geometry

1 Preliminaries

We are going to make use of the above results to derive some consequences for elliptic Hermitian geometry.

The local Hermitian geometry whose fundamental equations are given by (34) is called elliptic Hermitian, if we have

$$\begin{aligned} \Omega_{ii} &= \omega_i\bar{\omega}_i + \sum_k \omega_k\bar{\omega}_k, \\ \Omega_{ij} &= \omega_j\bar{\omega}_i, \quad i \neq j. \end{aligned} \tag{45}$$

This definition owes its origin to the type of geometry studied by G. Fubini,[33] E. Study,[34] and E. Cartan,[35] which we shall prove to satisfy the conditions (45).

The elliptic Hermitian geometry of Fubini-Study is defined as follows: We consider the complex projective space of n dimensions P_n, with the homogeneous coordinates z^α, where, as well as throughout this section, we shall use the following ranges of our indices:

$$0 \leqq \alpha, \beta, \gamma \leqq n, \quad 1 \leqq i, j, k \leqq n.$$

In P_n let a positive definite Hermitian form be given:

$$(z\bar{z}) = \sum_\alpha z^\alpha \bar{z}^{-\alpha}, \tag{46}$$

which will serve to define the scalar product of two vectors in the affine space A_{n+1} of $n+1$ dimensions, with the (non-homogeneous) coordinates z^α. We normalize the coordinates z^α in P_n, such that

$$(z\bar{z}) = 1. \tag{47}$$

---

[33] Fubini, [13].
[34] Study, [25].
[35] Cartan, [4].

An Hermitian metric in P_n is then defined by the Hermitian form

$$ds^2 = (dzd\bar{z}) - (zd\bar{z}) \cdot (\bar{z}dz). \tag{48}$$

The group of linear transformations in A_{n+1} which leaves the form (46) invariant is the unitary group $U(n+1)$. We take in $A_{n+1} n+1$ vectors A_k such that

$$A_\alpha \bar{A}_\beta = \delta_{\alpha\beta}. \tag{49}$$

For a differentiable family of such sets of vectors we have

$$dA_\alpha = \sum_\beta \theta_{\alpha\beta} A_\beta, \tag{50}$$

where

$$\theta_{\alpha\beta} + \bar{\theta}_{\beta\alpha} = 0, \tag{51}$$

and where the forms $\theta_{\alpha\beta}$ satisfy the equations of structure:

$$d\theta_{\alpha\beta} = \sum_\gamma \theta_{\alpha\gamma}\theta_{\gamma\beta}. \tag{52}$$

Let B_α be fixed vectors in A_{n+1}, satisfying the equations

$$B_\alpha \bar{B}_\beta = \delta_{\alpha\beta}.$$

The coordinates of a vector A_0 with respect to B_α are defined by the equation

$$A_0 = \sum_\alpha z^\alpha B_\alpha,$$

from which we find

$$(dA_0 d\bar{A}_0) - (A_0 d\bar{A}_0)(\bar{A}_0 dA_0) = (dz\ d\bar{z}) - (z\ d\bar{z})(\bar{z}\ dz).$$

It follows that, if we regard A_0 as defining the points in P_n, the Hermitian form in (48) can be written as

$$ds^2 = (dA_0 d\bar{A}_0) - (A_0 d\bar{A}_0)(\bar{A}_0 dA_0),$$

and, by (50), as

$$ds^2 = \sum_i (\theta_{0i}\bar{\theta}_{0i}). \tag{53}$$

This proves in particular that the Hermitian form in (48) is positive definite.

To calculate the curvature of the metric (53) we shall make use of the equations (52). Using the notations of local Hermitian geometry in the last chapter, we put

$$\omega_i = \theta_{0i}.$$

Then, by (52), we get

$$d\omega_i = d\theta_{0i} = \sum_{j \neq i} \omega_j \theta_{ji} + \omega_i(\theta_{ii} - \theta_{00}).$$

From the uniqueness of the set of forms ω_{ij} satisfying the first and the ninth equations of (34) it follows that

$$\begin{aligned} \omega_{ij} &= \theta_{ij}, \quad i \neq j, \\ \omega_{ii} &= \theta_{ii} - \theta_{00}. \end{aligned}$$

We find then

$$\begin{aligned} d\omega_{ij} &= \sum_k \omega_{ik}\omega_{kj} + \theta_{i0}\theta_{0j}, \quad i \neq j, \\ d\omega_{ii} &= \sum_k \omega_{ik}\omega_{ki} + \theta_{i0}\theta_{0i} + \sum_k \theta_{k0}\theta_{0k}, \end{aligned}$$

and therefore

$$\begin{aligned} \Omega_{ij} &= -\bar{\omega}_i\omega_j, \quad i \neq j, \\ \Omega_{ii} &= -\bar{\omega}_i\omega_i - \sum_k \bar{\omega}_k\omega_k, \end{aligned}$$

which are exactly the equations (45). Thus we have proved that it is possible to define in the complex projective space an elliptic Hermitian geometry.

2 Formulas of Cartan and Wirtinger

When the Hermitian manifold is locally elliptic, it is possible to calculate the forms Ψ_r in (39) more explicitly. In fact, we are going to prove the following theorem:

Theorem 12. *In a locally elliptic Hermitian manifold let Λ be the exterior differential form corresponding to the Hermitian differential form, that is, $\Lambda = \sum_i \omega_i\bar{\omega}_i$ in case the given Hermitian form is $ds^2 = \sum_i(\omega_i \otimes \bar{\omega}_i)$. Then we have*

$$\Psi_r = \frac{1}{(2\pi\sqrt{-1})^{n-r+1}} \binom{n+1}{r} \Lambda^{n-r+1}, \quad 1 \leqq r \leqq n. \tag{54}$$

It is clear that the construction of Λ from ds^2 is independent of the choice of the base linear differential forms ω_i in terms of which ds^2 is expressed.

The theorem is proved by induction on $n-r$. If $n-r=0$, that is, $r=n$, we have

$$\Psi_n = \frac{1}{2\pi\sqrt{-1}} \sum_i \Lambda_{ii} = \frac{n+1}{2\pi\sqrt{-1}}\Lambda.$$

Suppose the formula (54) be true for $r+1, \ldots, n$. We have

$$\begin{aligned} \Psi_r = \frac{1}{(2\pi\sqrt{-1})^{n-r+1}(n-r+1)!} \Big\{ &\sum \delta(i_1 \cdots i_{n-r}; j_1 \cdots j_{n-r}) \\ &\cdot\Omega_{i_1j_1} \cdots \Omega_{i_{n-r}j_{n-r}} \sum_k \Omega_{kk} - (n-r) \sum \delta(i_1 \cdots i_{n-r}; j_1 \cdots j_{n-r}) \\ &\cdot\Omega_{i_1j_1} \cdots \Omega_{i_{n-r-1}j_{n-r-1}} \cdot \sum_k \Omega_{i_{n-r}k}\Omega_{kj_{n-r}} \Big\}. \end{aligned}$$

Consider the second sum inside the braces. If $i_{n-r} \neq j_{n-r}$, we can replace the sum $\sum_k \Omega_{i_{n-r}k}\Omega_{kj_{n-r}}$ by $\omega_{j_{n-r}}\bar{\omega}_{i_{n-r}}\Lambda = \Omega_{i_{n-r}j_{n-r}}\Lambda$. If $i_{n-r} = j_{n-r}$, the sum can be replaced by $(\omega_{i_{n-r}}\bar{\omega}_{i_{n-r}} + \Lambda)\Lambda = \Omega_{i_{n-r}j_{n-r}}\Lambda$. By our induction hypothesis we get easily the desired formula (54).

As an application let us determine the n basic classes of the complex projective space of n dimensions. The result is given by the following theorem:

Theorem 13. *The rth basic characteristic class $(1 \leqq r \leqq n)$ of a complex projective space of n dimensions is dual to the homology class containing the cycle carried by a linear subspace of dimension $r-1$ multiplied by $\binom{n+1}{r}$.*

To prove this theorem we consider the affine space A_{n+1} of dimension $n+1$ with the coordinates z^α such that the projective space P_n under consideration is the hyperplane at infinity with which A_{n+1} is made into a projective space of $n+1$ dimensions. As before, z^α are homogeneous coordinates in P_n. Suppose that we have defined at each point $\mathfrak{z}$ of A_{n+1} different from the origin r vectors $\mathfrak{v}_i, 1 \leqq i \leqq r$, whose components $v_i^0, \ldots, v_i^n$ are linear forms in the coordinates z^α of $\mathfrak{z}$. Then all the points on the line joining the origin 0 to $\mathfrak{z}$ are projected from 0 into the same point p of P_n and the vectors $\mathfrak{v}_i, 1 \leqq i \leqq r$, at the points of $0_\mathfrak{z}$ are projected from 0 into the same vector of P_n, which we attach to p. It is easy to verify that the r vectors thus defined at each point of P_n are linearly independent when and only when the $r+1$ vectors $\mathfrak{z}, \mathfrak{v}_1, \ldots, \mathfrak{v}_r$ in A_{n+1} are linearly independent.

We take

$$v_i^\alpha = a_i^\alpha z^\alpha, \quad 1 \leqq i \leqq r,$$

where a_i^α are constants. The a_i^α can be so chosen that none of the determinants of order $r+1$ of the matrix

$$\begin{pmatrix} 1 & \cdots & 1 \\ a_1^0 & \cdots & a_1^n \\ & \cdots\cdots & \\ a_r^0 & \cdots & a_r^n \end{pmatrix}$$

will vanish. It then follows that the vectors $\mathfrak{z}, \mathfrak{v}_1, \ldots, \mathfrak{v}_r$ will be linearly dependent when and only when all products

$$z^{\alpha_0} \cdots z^{\alpha_r} = 0, \quad 0 \leqq \alpha_0, \ldots, \alpha_r \leqq n,$$

the indices $\alpha_0, \ldots, \alpha_r$ being distinct from each other. This is possible when and only when $n+1-r$ of the coordinates z^α will vanish. In other words, for this particular field of r vectors the points of P_n at which the r vectors are linearly dependent are the linear spaces of dimension $r-1$ defined by setting $n+1-r$ of the homogeneous coordinates of P_n to zero. The number of such linear spaces is $\binom{n+1}{r}$ and each of them is to be counted simply. Hence our theorem is proved.

Theorem 14 (E. Cartan).[36] *Let M be a closed submanifold of topological dimension $2n-2r+2$ of the complex projective space of n dimensions P_n in*

[36] Cartan, [3], p. 206.

which an elliptic Hermitian geometry is defined. Let m be the number of points of intersection of M with a generic linear subspace of dimension $r-1$ of P_n. Then

$$m = \frac{1}{(2\pi\Gamma - 1^{n-r+1})} \int_m \Lambda^{n-r+1}. \tag{55}$$

In particular, if M is an algebraic variety of dimension $n-r+1$ in P_n, m is its order.

This theorem is an immediate consequence of the Theorems 11, 12, and 13.

Related to these discussions is also a formula due to W. Wirtinger. In Hermitian geometry, as in Riemannian geometry, it is common to define an element of volume of topological dimension $2p$ by means of the equation

$$\varDelta_{2p} = \pm \sum_{(i_1 \cdots i_p)} \omega_{i_1} \cdots \omega_{i_p} \bar{\omega}_{i_1} \cdots \bar{\omega}_{i_p}, \tag{56}$$

the summation being over all the combinations of $i_1, \ldots, i_p$ from 1 to n. Up to a sign Λ_{2p} is equal to $\frac{1}{p!}\Lambda^p$. We define

$$\varDelta_{2p} = \frac{1}{p!}\Lambda^p. \tag{57}$$

From (55) follows the theorem:

Theorem 15 (W. Wirtinger).[37] *In the complex projective space of n-dimensions with the elliptic Hermitian metric let V_{2p} be a p-dimensional algebraic variety of order m and volume V. Then*

$$V = \frac{(2\pi)^p}{p!} m. \tag{58}$$

INSTITUTE FOR ADVANCED STUDY, PRINCETON, AND
TSING HUA UNIVERSITY, CHINA.

References

[1] ALEXANDROFF, P., UND HOPF, H., *Topologie* I, Berlin 1935.

[2] ALLENDOERFER, C. B., AND WEIL, A., *The Gauss-Bonnet theorem for Riemannian polyhedra*, Trans. Amer. Math. Soc., Vol. 53 (1943), pp. 101–129.

[3] CARTAN, E., *Sur les invariants intégraux de certains espaces homogènes clos et les propriétés topologiques de ces espaces*, Annales Soc. pol. Math., Tome 8 (1929), pp. 181–225.

[4] CARTAN, E., *Leçons sur la géométrie projective complexe*, Paris 1931.

---

[37] WIRTINGER, [31].

[5] Chern, S., *A simple intrinsic proof of the Gauss-Bonnet formula for closed Riemannian manifolds*, Annals of Math., Vol. 45 (1944), pp. 747–752.

[6] Chern, S., *Integral formulas for the characteristic classes of sphere bundles*, Proc. Nat. Acad. Sci., Vol. 30 (1944), pp. 269–273.

[7] Chern, S., *Some new viewpoints in differential geometry in the large*, to appear in Bull. Amer. Math. Soc.

[8] Ehresmann, C., *Sur la topologie de certains espaces homogènes*, Annals of Math., Vol. 35 (1934), pp. 396–443.

[9] Ehresmann, C., *Sur la topologie des groupes simples clos*, C. R. Acad. Sci. Paris, Vol. 208 (1939), pp. 1263–1265.

[10] Ehresmann, C., Various notes on fibre spaces in C. R. Acad. Sci. Paris, Vol. 213 (1941), pp. 762–764; Vol. 214 (1942), pp. 144-147; Vol. 216 (1943), pp. 628–630.

[11] Eilenberg, S., *Cohomology and continuous mappings*, Annals of Math., Vol. 41 (1940), pp. 231–251.

[12] Feldbau, J., *Sur la classification des espaces fibrés*, C.R. Acad. Sci. Paris, Vol. 208 (1939), pp. 1621–1623.

[13] Fubini, G., *Sulle metriche definite da una forma Hermitiana*, Instituto Veneto, Vol. 63, 2 (1904), pp. 502–513.

[14] Hopf, H., see Alexandroff, P.

[15] Hurewicz, W., and Steenrod, N., *Homotopy relations in fibre spaces*, Proc. Nat. Acad. Sci., Vol. 27 (1911), pp. 60–64.

[16] Kähler, E., *Über eine bemerkenstwerte Hermitische Metrik*, Abh. Math. Sem. Hamburg, Vol. 9 (1933), pp. 173–186.

[17] Pascal, E., *Die Determinanten*, Leipzig 1900.

[18] Pontrjagin, L., *Characteristic cycles on manifolds*, C. R. (Doklady) Acad. Sci. URSS (N. S), Vol. 35 (1942), pp. 34–37.

[19] Pontrjagin, L., *On some topologic invariants of Riemannian manifolds*, C. R. (Dokiady), Acad. Sci. URSS (N. S.), Vol. 43 (1944), pp. 91–94.

[20] de Rham, G., *Sur l'analysis situs des variétés à n dimensions*, J. Math. pures et appl., Tome 10 (1931), pp. 115–200.

[21] Steenrod, N., *Topological methods for the construction of tensor functions*, Annals of Math., Vol. 43 (1942), pp. 116–131.

[22] Steenrod, N., *The classification of sphere bundles*, Annals of Math., Vol. 45 (1944), pp. 294–311.

[23] STEENROD, N., see HUREWICZ, W.

[24] STIEFEL, E., *Richtungsfelder und Fernparallelismus in n-dimensionalen Mannigfaltigkeiten*, Comm. Math. Helv., Vol. 8 (1936), pp. 305–343.

[25] STUDY, E., *Kürzeste Wege im komplexen Gebiet*, Math. Annalen, Vol. 60 (1905), pp. 321–377.

[26] VAN DER WAERDEN, B. L., *Topologische Begrundung des Kalküls der abzählenden Geometrie*, Math. Annalen, Vol. 102 (1930), pp. 337–362.

[27] WEIL, A., see ALLENDOERFER, C. B.

[28] WEYL, H., *The Classical Groups*, Princeton 1939.

[29] WHITNEY, H., *Topological properties of differentiable manifolds*, Bull. Amer Math. Soc., Vol. 43 (1937), pp. 785–805.

[30] WHITNEY, H., *On the topology of differentiable manifolds*, Lectures in Topology, pp. 101-141, Michigan 1941.

[31] WIRTINGER, W., *Eine Determinantenidentität und ihre Anwendung auf analytische Gebilde in Euklidischer und Hermitischer Massbestimmung*, Monatshefte für Math. u. Physik, Vol. 44 (1936), pp. 343-365.

Editor's Comments

Among Chinese mathematicians, Wei-Liang Chow is probably not as well-known as others, though he was clearly in the same league of other great Chinese mathematicians in the 20th century. Chow is probably best known for his theorem that *every compact complex submanifold of the complex projective space* $\mathbb{C}P^n$ *is an algebraic variety.* This result is now usually included in the famous result of Serre's GAGA theorem as a corollary, but this does not diminish the depth and beauty of Chow's result, which was a classical conjecture in complex analytic geometry before he proved it.

When Riemann wrote his masterpiece in 1857 titled *Theory of abelian functions*, he initiated both the theory of complex analytic geometry and algebraic geometry. Chow's paper is a milestone in the long and fruitful interaction between complex geometry and algebraic geometry, which is the basic theme of this book.

Complex Geometry from Riemann
to Kähler-Einstein and Calabi-Yau
ALM 38, pp. 189–206

On Compact Complex Analytic Varieties*

Wei-Liang Chow

1 Introduction

Let S_n be the projective space of n dimensions over the field k of all complex numbers. For any given choice of the inhomogeneous coordinates $x_1, \ldots, x_n$ in S_n, let $A_n(x)$ be the n-dimensional complex affine space consisting of all the points of S_n which are finite with respect to this coordinate system. A point set U in S_n is said to be analytic in the neighborhood of a point (a) if, choosing the inhomogeneous coordinates so that (a) is a point in $A_n(x)$, there exists a neighborhood R of (a) in $A_n(x)$ such that the intersection $U \cap R$ consists of all the points of R which are common solutions of a finite set of equations,

$$f_i(x_1, \ldots, x_n) = 0 \quad (i = 1, \ldots, s), \tag{1}$$

where the $f_i(x_1, \ldots, x_n), i = 1, \ldots, s$, are holomorphic functions in the region R. A point set in S_n is called an analytic variety if it is analytic in the neighborhood of every one of its points. An analytic variety is called compact if it is a compact point set. A point set in S_n is called an algebraic variety if it consists of all the points whose homogeneous coordinates $x_0, x_1, \ldots, x_n$ satisfy a finite set of equations

$$\phi_i(x_0, x_1, \ldots, x_n) = 0 \quad (i = 1, \ldots, t),$$

where the $\phi_i(x_0, x_1, \ldots, x_n), i = 1, \ldots, t$, are homogeneous polynomials in the $x_0, x_1, \ldots, x_n$. It is easily seen that an algebraic variety is a compact analytic variety. There is a classical conjecture that conversely any compact analytic variety in S_n is also an algebraic variety. So far as we are aware, no proof of this conjecture has ever been offered. It is the purpose of this paper to present a proof of this conjecture.

* Received April 9, 1949. W. L. Chow, On compact complex analytic varieties. Amer. J. Math. 71, (1949), 893–914.

We begin by recalling some well known properties of an analytic variety.[1] Let R be a neighborhood of the origin of $A_n(x)$ defined by the conditions $|x_i| < e, i = 1, \dots, n$. Let $k\{x_1, \dots, x_r\}$ be the ring of all power series in the variables $x_1, \dots, x_r$ which are convergent in the region $|x_i| < e, i = 1, \dots, r$. Consider a set of equations of the following type

$$\begin{gathered} x_{r+1}^{\mu} + B_1 x_{r+1}^{\mu-1} + \cdots + B_{\mu-1} x_{r+1} + B_{\mu} = 0, \\ H x_j = G_j(x_{r+1}) \quad (j = r+2, \dots, n), \end{gathered} \tag{2}$$

where the $B_1, \dots, B_\mu$ and II are elements of the ring $k\{x_1, \dots, x_r\}$ and the $G_j(x_{r+1})$, $j = r+2, \dots, n$, are polynomials in x_{r+1}. with coefficients in $k\{x_1, \dots, x_r\}$; and the left hand side of the first equation is an irreducible distinguished polynomial in x_{r+1} over $k\{x_1, \dots, x_r\}$ and H is its discriminant. Let D be the set of all common solutions of the equations (2) in the region R for which $H \neq 0$, as well as those which are limiting points of such solutions. Such a point set D, or any point set in S_n which can be represented as such a set by a suitable choice of the affine space $A_n(x)$ and its coordinates, is called an analytic element in S_n. The number r is called the dimension of the analytic element D, and we shall write D_r to indicate this. The number μ depends in general upon the choice of the affine space and its coordinates; the smallest possible value of this number for a given D_r is called its order. An analytic element is called regular if it has the order one, otherwise it is called singular. The point (a) is called the center of D_r. It is well known from the theory of functions of several complex variables that every point of an analytic variety W has a neighborhood R such that $R \cap W$ consists of a finite number of analytic elements. In other words, an analytical variety is a topological subspace of S_n which has a system of neighborhoods consisting of analytic elements. An analytic variety is called irreducible if any two analytic elements of it can be obtained from each other by analytic continuations. It is easily seen that all the analytic elements of an irreducible analytic variety W must have the same dimension r; we shall call this number r the dimension of W and we shall write W_r to indicate this. Following a recent practice in algebraic geometry, we shall from now on use the expression " analytic variety " to denote exclusively an irreducible analytic variety; the reducible ones being taken care of later by the more precise concept of an analytic cycle. A point of an analytic variety W_r is called regular if it has a neighborhood consisting of a single regular analytic element; those points of W_r which are not regular are called singular. The set W_r of all singular points of W_r constitutes a finite or enumerably infinite number of analytic varieties of dimensions less than r; we shall call this set W_r the singular part of W$_r$. The set W_r—W_r of all the regular points of W_r is a connected set.

For our present purpose the most important properties of an analytic variety W_r are the following:

(A) *W_r is a topological complex.* This is the triangulation theorem for analytic loci, for the proof of which we shall refer to the literature.[2] In particular, if W_r is

[1] For the results of the theory of functions of several complex variables used in this paper, we refer here once for all to Osgood [7], Bochner and Martin [1].

[2] See Koopman and Brown [4], Lefschetz and Whitehead [6].

compact, it is a finite complex and can be taken as a subcomplex of a simplicial subdivision of the projective space S_n and the singular part $\tilde{W}_r$ can in turn be taken as a subcomplex of W_r.

(B) W_r *can be covered by an enumerable aggregate of regular analytic elements of dimensions* $\leq r$. This can be easily proved by induction, starting with the facts that W_r—$\tilde{W}_r$ can be covered by an enumerable aggregate of analytic elements of dimension r and that $\tilde{W}$ consists of an enumerable aggregate of analytic varieties of dimensions $< r$.

In the following sections we shall show that the main theorem can be proved by using these two properties (A) and (B) alone, without referring directly to the fact that each point of W_r has a neighborhood consisting of a finite number of analytic elements of dimension r. For this reason, we shall give a new (possibly more general) definition of an analytic variety in 2, essentially by means of these two properties (A) and (B) only. In 2 and 3 we shall study the topological properties of analytic varieties and derive from them the Theorem III, which expresses a special property of a regular analytic element D_r contained in a compact analytic variety of r dimensions. In 4 we then proceed to show (Theorem IV) that this property is really a characteristic property of a regular analytic element D_r which is contained in an algebraic variety of r dimensions. The main theorem (Theorem V) then follows immediately from these two results. In the last section (5) we shall indicate some applications of our main theorem.

Finally, we add a few words about the terminology and notations. A topological complex K is a homeomorphic image of an Euclidean complex, and we shall say $2r$-complex if it consists only of simplexes of topological dimension $2r$ and their sides. The boundary of an unoriented complex K (i.e. considered as a chain modulo 2) will be denoted by $\dot{K}$. If the complex K is oriented to become a chain with integral coefficients, then its chain boundary will also be denoted by K. There is no danger of confusion in this, as the meaning will be clear from the context in each case. The word " dimension " without any further qualification will mean the complex dimension, which is twice the topological dimension; subscripts will be used to denote the former, super- scripts the latter.

2 Analytic varieties

An analytic simplex E_r of r dimensions in S_n is a topological $2r$-simplex in S_n, which is a one-to-one analytic image of a topological $2r$-simplex in a complex affine space $A_r(z)$ of r dimensions. More specifically this means that, for a suitable choice of the inhomogeneous coordinates $x_1, \ldots, x_n$ in S_n, the points of E_r can be represented parametrically by a set of n equations

$$x_j = f_j(z_1, \ldots, z_r), \quad j = 1, \ldots, n, \tag{3}$$

where the n functions $f_j(z_1, \ldots, z_r), j = 1, \ldots, n$, are all holomorphic in a region R of the affine space $A_r(z)$ of r complex parameters $z_1, \ldots, z_r$, and the matrix $\|\partial f_j/\partial z_k\|, j = 1, \ldots, n, k = 1, \ldots, r$, has the maximum rank r at every point of the region R; and that this representation induces a homeomorphism between

E_r and a topological $2r$-simplex C^{2r} in the region R. Without loss of generality, we can assume that the coordinate origin $(z) = (0)$ of $A_r(z)$ is contained in the interior of C^{2r}; the corresponding point $(x) = (a)$ in E_r is then called the center of E_r (with respect to the parameters $z_1, \ldots, z_r$). It is obvious that if the analytic simplex E_r is subdivided into a complex, then each simplex of this complex is also an analytic simplex. Since in all our arguments we can always replace any complex by a subdivision in which the simplexes are arbitrarily small, we can therefore assume, without any essential restriction to the concept of an analytic simplex, that for a suitable choice of the coordinates $x_1, \ldots, x_n$, and the parameter space $A_r(z)$, the equations (3) have the form

$$\begin{aligned} &x_1 = z_1, \ldots, x_r = z_r, \\ &x_{r+1} = f_{r+1}(z_1, \ldots, z_r), \ldots, x_n = f_n(z_1, \ldots, z_r); \end{aligned} \tag{4}$$

and that the region R has been chosen so small that the functions $f_{r+1}, \ldots, f_n$ can be represented as power series which are convergent in R. Since the equations (4) map the region R into a regular analytic element D_r in S_n, it follows that an analytic simplex E_r is simply a topological $2r$-simplex imbedded in a regular analytic element D_r. On the other hand, the set of interior points $E_r—\dot{E}_r$ of an analytic simplex is itself a regular analytic element. Hence it is rather indifferent whether a point set is covered by an aggregate of analytic simplexes or an aggregate of regular analytic elements; the use of either one instead of the other is only a matter of convenience.

The concept of an analytic complex in S_n is defined by induction as follows. Any topological 0-complex in S_n is an analytic complex K_0 of zero dimension. A topological $2r$-complex (finite or infinite) in S_n is called an analytic complex K_r of r dimensions, if it contains a subcomplex $\tilde{K}_r$ which consists of an enumerable aggregate of analytic complexes of dimensions less than r, such that any point of $K_r - (\dot{K}_r + \tilde{K}_r)$ has a neighborhood which is a regular analytic element D_r. The points of $K_r - (\dot{K}_r + \tilde{K}_r)$ are called regular points of K. The topological complex $\dot{K}_r$ is also called the boundary of K_r. The aggregate of analytic complexes $\tilde{K}_r$ is called the singular part of K_r, though not every point (or even any point) of it need be a singular point in either the topological or analytic sense. An analytic simplex E_r is of course also an analytic complex of r dimensions, the boundary being the boundary $\dot{E}_r$ and the singular part being the empty set. An analytic variety W_r of r dimensions in S_n is a connected analytic complex with the properties that it has no boundary and its singular part $\tilde{W}_r$ consists of an enumerable aggregate of analytic varieties of dimensions less than r. This definition by induction can be completed by the stipulation that an analytic variety W_0 of zero dimension consists of a single point in S_n. An analytic variety is called compact if it is a compact point set; that is, if it is a finite topological complex.

It is clear that the set $W_r—\tilde{W}_r$ of all the regular points of W_r can be covered by an enumerable aggregate of analytic simplexes of r dimensions. From this and the fact that $\tilde{W}_r$ consists of an enumerable aggregate of analytic varieties of lower dimensions, it follows by induction that the analytic variety W_r can be covered by an enumerable aggregate of analytic simplexes of dimensions $\leq r$. However, it is to be noticed that this covering of W_r by an enumerable aggregate of analytic simplexes is not a simplicial subdivision of W_r as a complex in the topological sense, nor is it a covering by neighborhoods of W_r as a topological subspace of S_n. It is simply a representation of W_r as the set-theoretic sum of an enumerable aggregate of analytic simplexes of various dimensions, and this aggregate will be in general infinite even in case of a compact analytic variety W_r.

Let E_r and E_s be two analytic simplexes in S_n contained in the regular analytic elements D_r and D_s respectively. The intersection $D_r \cap D_s$ of the two analytic elements D_r and D_s, if it is not empty, consists of an enumerable aggregate of connected components, each of which is an analytic variety of dimension not less than $r+s-n$. In case such a component has exactly the dimension $r+s-n$, it is called a proper component of the intersection. It is easily seen that if the join of the tangent spaces of D_r and D_s at some one regular point of such a component has the dimension n, then the component has the dimension $r+s-n$ and is therefore proper. In such a case we shall call the component regular. We shall say that two analytic simplexes E_r and E_s are regular with respect to each other if either (1) the intersection $E_r \cap E_s$ is empty, or (2) each point of $E_r \cap E_s$ is contained in a regular component of $D_r \cap D_s$.

It is clear that for $r+s<n$ the first possibility alone can occur, so that in this case two analytic simplexes E_r and E_s are regular with respect to each other if and only if they are disjoint. In case $r+s=n$, the intersection of two analytic simplexes E_r and E_{n-r}, regular with respect to each other, is either empty or consists of a finite number of points at each of which the tangent spaces of D_r and D_{n-r} are transversal to each other. We are mainly interested in the case $r+s=n$, though the following results hold also for the general case.

Let P be the group of all projective transformations in S_n; it is an analytic manifold of $(n+1)^2-1$ dimensions. It is easily seen that if E_r is an analytic simplex in S_n and T is a projective transformation in S_n, then the transformed set TE_r of E_r is also an analytic simplex of r dimensions. It follows then that if K_r is an analytic complex covered by an aggregate of analytic simplexes $\{E\}$, then TK_r is also an analytic complex covered by the aggregate of the transformed simplexes $\{TE\}$. In the proof of the following theorems, we shall make use of the simple fact about everywhere dense sets which is true for every regular separable topological space: The intersection of a finite or enumerable infinite number of everywhere dense open sets is also an everywhere dense set. Since a closed set is nowhere dense if its complement is everywhere dense, it follows that the sum of a finite or enumerably infinite number of nowhere dense closed sets is a set the complement of which is everywhere dense; hence, if this sum is a closed set, then it is also nowhere dense.

Theorem I. *Let E_r and $E_s(r+s\le n)$ be two analytic simplexes in S_n and let Ω be the set of all elements T of P such that E_r and TE_s are regular with respect to each other, then Ω is an everywhere dense open set in P.*

Proof. We shall choose the inhomogeneous coordinates $x_1,\dots,x_n$ in S_n in such a way that both E_r and E_s are in the affine space $A_n(x)$, which is always possible if these simplexes are sufficiently small. Let E_r be given by the one-to-one analytic mapping

$$x_i = f_i(z_1, \dots, z_r), \quad i = 1, \dots, n, \tag{5}$$

of a region R' in $A_r(z)$ into $A_n(x)$, so that E_r is the image of a topological $2r$-simplex C^{2r} in R'. Similarly, let E_s be given by the one-to-one analytic mapping

$$x_i = g_i(w_1, \dots, w_s), \quad i = 1, \dots, n, \tag{6}$$

of a region R'' in $A_s(w)$ into $A_n(x)$, so that E_s is the image of a topological $2s$-simplex C^{2s} in R''.

We first prove that the set Ω is everywhere dense in P; it is evidently enough to show this for the neighborhood of the identity transformation I. Let $T_{(e)}$ be the affine transformation of $A_n(x)$,

$$x_i' = x_i + e_i, \quad i = 1, \dots, n;$$

it is of course also a projective transformation in S_n and is arbitrarily near to the identity transformation I for points (e) sufficiently near to (0). The transformed analytic simplex $T_{(e)}E_s$ is then given by the one-to-one analytic mapping

$$x_i = g_i(w_1, \dots, w_s) + e_i, \quad i = 1, \dots, n,$$

of the topological simplex C^{2s} in the region R''. Consider the analytic mapping

$$u_i = F_i(z, w) = f_i(z_1, \dots, z_r) - g_i(w_1, \dots, w_s), \quad i = 1, \dots, n, \tag{7}$$

of the region $R = R' \times R''$ in the product space $A_{r+s}(z, w) = A_r(z) \times A_s(w)$ into the space $A_n(u)$. Let C^{2r+2s} be the $2(r+s)$-simplex $C^{2r} \times C^{2s}$ in the region R. It is clear that every common solution of the equations (7) for $(u) = (e)$ in the simplex C^{2r+2s} will correspond to an intersection point of E_r and TE_s and conversely. For $r + s < n$, the image of C^{2r+2s} in $A_n(u)$ under the analytic transformation (7) is a nowhere dense set; hence there exist points (e) in $A_n(u)$ arbitrarily near to the origin such that the equations (7) have no common solution in C^{2r+2s} for $(u) = (e)$. Then the analytic simplexes E_r and $T_{(e)}E_s$ will be disjoint and hence regular with respect to each other. Thus the assertion is proved for this case. For $r+s = n$, we can assume that for all points (e) in a sufficiently small neighborhood of (0) the two analytic simplexes E_r and $T_{(e)}E_s$ have an intersection, for otherwise the assertion is obviously true already. This means that for all points (u) in a sufficiently small neighborhood of the origin in $A_n(u)$ the equations (7) will have a common solution in C^{2r+2s}. It is well known[3] that in this case the Jacobian determinant

$$J(z, w) = |\partial F_i/\partial z_j, \partial F_i/\partial w_k| = |\partial f_i/\partial z_j, -\partial g_i/\partial w_k|,$$

does not vanish identically in the region R. Then the set of all the solutions of the equation $J(z, w) = 0$ in R, if it is not empty, constitutes an enumerable aggregate of analytic varieties of dimension $n - 1$ in R. Let $E^{(i)}, i = 1, 2, \dots$, be

[3] See Knopp and Schmidt [3], p. 379.

a sequence of analytic simplexes covering this set J, and let $E^{(i)^*}, i = 1, 2, \ldots$, be the corresponding images in $A_n(u)$ under the analytic transformation (7). Then the image J^* of J under the analytic transformation (7) is evidently the sum of all the sets $E^{(i)^*}, i = 1, 2, \ldots$. Since each $E^{(i)^*}$ is a nowhere dense set in $A_n(u)$, it follows that the complement of J^* in $A_n(u)$ is an everywhere dense set. This means that there exists in every neighborhood of the origin of $A_n(u)$ at least a point (e) which is not in J^*. Then for this point $(u) = (e)$ the equations (7) will have no common solution in the set J; or, in other words, we have $J(z, w) \neq 0$ at every common solution in R of the equations (7) for $(u) = (e)$. This implies that every intersection point of E_r and $T_{(e)}E_s$ corresponds to a point in C^{2r+2s} at which $J(z, w) \neq 0$; hence E_r and $T_{(e)}E_s$ are regular with respect to each other. Thus our assertion is also proved for the case $r + s = n$.

It remains to show that the set Ω is open in P; that is, we have to show that if E_r and E_s are regular with respect to each other, then E_r and TE_s are also regular with respect to each other for all T in a sufficiently small neighborhood of the identity I. For $r + s < n$, this is obvious; for the E_r and E_s are disjoint closed subsets in S_n, and TE_s varies continuously with T. To prove the assertion for $r + s = n$, let a projective transformation T be given by the equations

$$x'_i = \left(a_{i0} + \sum_{j=1}^{n} a_{ij}x_j\right) \Big/ \left(a_{00} + \sum_{j=1}^{n} a_{0j}x_j\right), \quad i = 1, \ldots, n.$$

If the matrix (a_{ij}) is sufficiently near the identity, then the analytic simplex TE_s will still lie in the affine space $A_n(x)$ and is given by the one-to-one analytic mapping

$$\begin{aligned} x_i = g'_i(w_1, \ldots, w_s) \quad &= \quad \left[a_{i0} + \sum_{j=1}^{n} a_{ij}g_j(w_1, \ldots, w_s)\right] \Big/ \\ &\quad\quad \left[a_{00} + \sum_{j=1}^{n} a_{0j}g_j(w_1, \ldots, w_s)\right], \quad i = 1, \ldots, n, \end{aligned}$$

of the topological simplex C^{2s} in the region R''. Consider the Jacobian determinant

$$J^T(z, w) = |\partial f_i/\partial z_j, \partial g'_i/\partial w_k|,$$

and the n functions

$$F_i^T(z, w) = f_i(z_1, \ldots, z_r) - g'_i(w_1, \ldots, w_s), \quad i = 1, \ldots, n,$$

which are all defined in the region R. It is clear that these functions are also continuous functions of the transformation T in a neighborhood of the identity I. Now, it is easily seen that the two analytic simplexes E_r and TE_s are regular with respect to each other if and only if the following equations

$$\begin{aligned} J^T(z, w) &= 0, \\ F_i^T(z, w) &= 0, \end{aligned} \quad i = 1, \ldots, n, \tag{8}$$

have no common solution in the topological simplex C^{2r+2s} in the region R. Let Δ^T be the set of all the common solutions of the equations (8) in R; then Δ^T is closed in R and varies continuously with T. Since E_r and E_s are assumed to be regular with respect to each other, the intersection $C^{2r+2s} \cap \Delta^I$ is empty. Now, if $C^{2r+2s} \cap \Delta^T$ is not empty for all T in a sufficiently small neighborhood of I, then there is a sequence of $T^{(i)}, i = 1, 2, \ldots$, converging toward I such that $C^{2r+2s} \cap \Delta^{T^{(i)}}$ is not empty. Since C^{2r+2s} is a compact set, the sequence of sets $C^{2r+2s} \cap \Delta^{T^{(i)}}$ has at least one limit point in C^{2r+2s}; and it is easily seen that this limit point is a point of $C^{2r+2s} \cap \Delta^I$, which is a contradiction. Therefore, the intersection $C^{2r+2s} \cap \Delta^T$ is empty for every T in a sufficiently small neighborhood of the identity I; and this means that E_r and TE_s are regular with respect to each other for all such T. This concludes the proof of Theorem I.

From Theorem I we can deduce a similar theorem for analytic varieties. We shall say that a point is a regular intersection of two analytic complexes K_r and K_{n-r} if it is a regular point of both K_r and K_{n-r} and if the tangent spaces of K_r and K_{n-r} at this point are transversal to each other.

Theorem II. *Let W_r and W_{n-r} be two analytic varieties in S_n, and let Ω be the set of all the transformations T in P such that W_r and TW_{n-r} have only regular intersection points. Then the set Ω is everywhere dense in P.*

Proof. Let $\{E\}$ and $\{F\}$ be two enumerable sets of analytic simplexes which cover the analytic varieties W_r and W_{n-r} respectively. Then the set $\{G\} = \{(E, F)\}$ of all the pairs of analytic simplexes, one from each of the two sets $\{E\}$ and $\{F\}$, is also enumerable and hence can be arranged in a sequence $G^{(i)} = (E^{(i)}, F^{(i)}), i = 1, 2, \ldots$. For each $i = 1, 2, \ldots$, let $\Omega\{i\}$ be the set of all transformations T in P such that $E^{(i)}$ and $TF^{(i)}$ are regular with respect to each other. It is easily seen that the intersection of all the sets $\Omega^{(i)}, i = 1, 2, \ldots$, consists of exactly those transformations T such that W_r and TW_{n-r} have only regular intersection points. Since, according to Theorem I, each set $\Omega^{(i)}$ is an everywhere dense open set in P, it follows that Ω is also an everywhere dense set in P.

3 Intersection of analytic varieties

There is a "natural" orientation for an analytic simplex E_r which can be extended to any analytic complex K_r. Let E_r be an analytic simplex in S_n defined by the one-to-one analytic mapping (3) of the topological $2r$-simplex C^{2r} in $A_r(z)$. Let $z_j = z'_j + iz''_j, j = 1, \ldots, r$; then the ordered set of $2r$ real coordinates $z'_1, z''_1, \ldots, z'_r, z''_r$ of the space $A_r(z)$ determines a definite orientation of this space and consequently also an orientation of the simplex C^{2r}. This orientation of C^{2r} determines an orientation of E_r which we shall call its natural orientation. It can be easily shown that the natural orientation is independent of the choice of the analytic parameters and is thus an intrinsic property of an analytic simplex. The set of all regular points $K_r - (\dot{K}_r + \tilde{K}_r)$ of an analytic complex K_r can be covered by a set of analytic simplexes of r dimensions. It can be shown that if each analytic simplex of the set is given its natural orientation, then all these orientations are coherent; and if we orient each topological $2r$-simplex of K_r in concordance with the natural orientation of any one analytic simplex E_r contained in it, we obtain a topological $2r$-chain on K_r with boundary in the sub-complex $\dot{K}_r$, or a $2r$-cycle mod $\dot{K}_r$. Thus an analytic complex K_r is an orientable pseudo-manifold, and it has a natural orientation determined by its analytical structure. This applies in particular also to the projective space S_n itself, since it is also an analytic complex. Furthermore, if K_r is a finite complex, its natural orientation will make it a topological $2r$-chain in S_n with boundary in the subset $\dot{K}_r$. Hence a compact analytic variety W_r, if oriented with the natural orientation, is a topological $2r$-cycle in S_n. We shall from now on assume that all analytic complexes, including the space S_n itself, are oriented with the natural orientation, so that a finite analytic complex K_r corresponds to a uniquely determined topological $2r$-chain in S_n. The significance of the natural orientation is given by the following wellknown result:

Lemma 1. *The topological intersection multiplicity of a regular intersection point of two finite analytic complexes K_r and K_{n-r} is $+1$.*

For the proof of this lemma, which is very simple, we refer to the literature. [4]

As to the topological properties of S_n, it is wellknown that the $2r$-th homology group of S_n is cyclic and the class determined by a linear analytic variety L_r of r dimensions is a generator of this group. From this we can derive at once the following consequences: (1) To each $2r$-cycle in S_n there is associated an integer g called its degree; it is the total topological intersection number of this $2r$-cycle with a linear variety L_{n-r};(2) two $2r$-cycles are homologous if and only if they have the same degree; (3) the total topological intersection number of a $2r$-cycle of degree g and a $2(n-r)$-cycle of degree h is equal to gh. In view of Lemma 1 we can conclude that the degree of a compact analytic variety W_r is always positive; for it is equal to the number of intersection points of W_r with a linear variety L_{n-r} which has only regular intersections with W_r.

We shall need the following lemma concerning the multiplicity of an intersection point. Let B^s and B^{2n-s} be two topological chains in S_n, and let (p) be an isolated point of $B^s \cap B^{2n-s}$ which does not lie in either $\dot{B}^s$ or $\dot{B}^{2n-s}$. Then this point (p) has a uniquely determined multiplicity as an intersection of the two chains B^s and B^{2n-s}. For the determination of this multiplicity we have the following criterion:

Lemma 2. *Let R be a neighborhood of (p) such that the closure of R does not intersect $\dot{B}^s$ or $\dot{B}^{2n-s}$ or any other component of the intersection $B^s \cap B^{2n-s}$, and let A^s and A^{2n-s} be two chains which are d-homologous to the chains B^s and B^{2n-s} respectively, where d is a sufficiently small positive number depending on B^s, B^{2n-s} and R. Then each component of the intersection $A^s \cap A^{2n-s}$ lies either in R or outside of R, and the intersection multiplicity of (p) is equal to the sum of the intersection multiplicities of those components of $A^s \cap A^{2n-s}$ which lie in R.*

In the above lemma the notion of d-homology is defined as follows: Let Σ be a d-neighborhood of B^s and Σ' be a closed d-neighborhood of $\dot{B}^s$, then a chain A^s is said to be d-homologous to B^s if it is contained in Σ and is homologous to B^s mod Σ' on Σ. This includes in particular the case when A^s is a d-deformation of B^s. The proof of Lemma 2 follows easily from the wellknown results about

---

[4] For the topological properties of analytic varieties, we refer once for all to Lefsehetz [5], Ch. VIII, §3 and van der Waerden [9].

intersection of chains,[5] so that we can omit it here.

Let W_r and W_{n-r} be two compact analytic varieties of degrees g and h respectively, and let (p) be an isolated intersection point of them. Let R be a neighborhood of the point (p) in S_n, and let d be the number such that the Lemma 2 holds for all d-deformations of W_r and W_{n-r}. According to Theorem II, there exists a projective transformation such that TW_{n-r} is a d-deformation of W_{n-r} and the two analytic varieties W_r and TW_{n-r} have only regular intersection points. Since according to Lemma 1 each of these intersection points has exactly the multiplicity $+1$, it follows that the intersection $W_r \cap TW_{n-r}$ consists of exactly gh points. If μ of these gh points lie in the neighborhood R, then μ is the multiplicity of (p) as an intersection of W_r and W_{n-r}. Applying this in particular to a regular point of W_r, we obtain the following result: Let E_r be an analytic simplex in S_n and (p) be an interior point of E_r. If E_r is an element of a compact analytic variety W_r, then there exists a positive number N such that if W_{n-r} is a compact analytic variety of degree h and if (p) is an isolated intersection point of E_r with W_{n-r}, then the multiplicity of this intersection is not greater than hN. In fact, we can take N to be any number not less than the degree of W_r. In the next section we shall see that this fact is characteristic of an analytic simplex which is an element of an algebraic variety of r dimensions. However, in order to state this result in the definitive form which we shall use in the next section, we introduce the concept of an analytic cycle. An analytic cycle Z_r of r dimensions in S_n is a topological $2r$-cycle in S_n which can be expressed as a sum of multiples of compact analytic varieties. Thus an analytic cycle Z_r is a finite set of compact analytic varieties $W_r^{(1)}, \dots, W_r^{(s)}$, to each of which is assigned an integer n_i (positive or negative) called its multiplicity. If g_i is the degree of $W_r^{(i)}$ respectively, then the integer $\sum_{i=1}^{s} n_i g_i$ is evidently the degree of the analytic cycle Z_r. An analytic cycle Z_r is called positive if the multiplicities of its component varieties are all positive; it is called algebraic if all of its component varieties are algebraic. It is a simple matter to show that what we have proved above for compact analytic varieties can be generalized to positive analytic cycles. Thus we have the following theorem:

Theorem III. *Let E_r be an analytic simplex in S_n and (p) be an interior point of E_r. If E_r is an element of a compact analytic variety W_r, then there exists a positive number N such that if Z_{n-r} is a positive analytic cycle of degree h and if (p) is an isolated intersection point of E_r with Z_{n-r}, then the multiplicity of this intersection is not greater than hN.*

We have stated the above theorem only for a regular analytic element, as this is the only case we shall need for our purpose. It is however clear that the theorem holds in general for any analytic element, whether regular or not. Incidentally we should like to remark that it is wellknown[6] that an isolated intersection of two analytic varieties, in the usual sense as defined in section 1, has always a positive multiplicity. This property is fundamental for many applications of topological methods to algebraic geometry, but we have no direct use for it in the present paper. In fact, we do not know whether this property is true at all for the ana-

---

[5] See Lefschetz [5], Ch. IV, §3.

[6] See van der Waerden [9], or Lefschetz [5], Ch. VIII, §4.

lytic varieties as we have defined here in the previous section; for the proof of it depends on the fact that the neighborhood of each point of an analytic variety consists of a finite number of analytic elements. What is essential for our present purposes is not to rule out the possibility of a zero multiplicity for an isolated intersection point, but to rule out the possibility of a negative multiplicity for any connected component (of any dimension) of the intersection. That the latter is true for intersections of positive analytic cycles in S_n can be easily deduced from our results, but it is not generally true for intersections of positive analytic cycles in an arbitrary analytic or algebraic manifold.

4 Proof of the main theorem

We begin with two lemmas.

Lemma 3. *Let $f(x_1, \ldots, x_r)$ be a power series in r variables $x_1, \ldots, x_r$. Given any positive integer N, we can always find an integer M with the following property: For every integer $m > M$, there exists a polynomial $F(x_1, \ldots, x_r, x_{r+1})$ in $r+1$ variables $x_1, \ldots, x_{r+1}$ of degree m such that the power series $F(x_1, \ldots, x_r, f(x_1, \ldots, x_r))$ contains no terms of degree $\leq mN$.*

Proof. Let $F(x_1, \ldots, x_{r+1})$ be the general polynomial of degree m in $r+1$ variables $x_1, \ldots, x_{r+1}$. If ${}^hC_j$ denotes the binomial coefficient $h!/j!(h-j)!$, then F contains ${}^{r+1+m}C_{r+1}$ terms and hence that many indeterminate coefficients (c). The power series $F(x_1, \ldots, x_r, f(x_1, \ldots, x_r))$ contains at most ${}^{r+mN}C_r$ terms of degrees $\leq mN$, the coefficients of which are linear forms $L_\lambda(c)$ in the (c). The ${}^{r+mN}C_r$ linear equations $L_\lambda(c) = 0$ will have a non-trivial solution in the (c) if ${}^{r+1+m}C_{r+1} >{}^{r+mN} C_r$. This will be the case if we take m greater than $M = (N+1)^r(1+r)$; for we have then

$$\begin{aligned} (m+1) &> (1+r)m^r(N+1)^r/m^r \\ &> (r+1)[(mN+r)\cdots(mN+1)]/[(m+r+1)\cdots(m+2)], \end{aligned}$$

hence $[(m+r+1)\cdots(m+1)]/[r+1] > (mN+r)\cdots(mN+1)$, hence ${}^{r+1+m}C_{r+1} > {}^{r+mN}C_r$.

Before proceeding to the next lemma we shall recall here some wellknown facts about the intersection of two algebraic cycles Z_r and Z_s in S_n. While in general any two arbitrary cycles C' and C'' in S_n have a "topological intersection" consisting of a homologous family of cycles about their geometric intersection $C' \cap C''$, it is possible in case of the algebraic cycles Z_r and Z_s to define an intersection cycle $Z_r \cdot Z_s$ provided all the components of the geometric intersection $Z_r \cap Z_s$ are proper (i.e. have the dimension $r+s-n$). Let $U^{(1)}, \ldots, U^{(\lambda)}$ be the components of $Z_r \cap Z_s$; they are all algebraic varieties of dimension $r+s-n$. Then we can assign in a unique manner to each component $U^{(i)}$ a positive integer n_i such that the algebraic cycle $\sum_{i=1}^{\lambda} n_i U^{(i)}$ is a member of the homologous family of cycles constituting the "topological intersection" of Z_r and Z_s. This cycle $\sum_{i=1}^{\lambda} n_i U^{(i)}$ is called the intersection cycle $Z_r \cdot Z_s$ of Z_r and Z_s, and the numbers $n_1, \ldots, n_\lambda$ are called the intersection multiplicities of $U^{(1)}, \ldots, U^{(\lambda)}$ respectively. The degree of $Z_r \cdot Z_s$ is the product of the degrees of Z_r and Z_s. Corresponding to the Lemmas

1 and 2 of the preceding section, we have here the following result about the intersection cycle $Z_r \cdot Z_s$: (1) If V_r and V_s are component varieties of multiplicity one of Z_r and Z_s respectively, and if U is a regular component of $V_r \cap V_s$ and is not contained in any other component of Z_r *or* Z_s, then the multiplicity of U in the cycle $Z_r \cdot Z_s$ is equal to $+1$. We shall say the U is a regular component of $V_r \cap V_s$ if it contains a point which is regular for both V_r and V_s and if the join of the tangent spaces of V_r and V_s at this point has the dimension n. (2) Given any positive number d there is a positive number e (depending of course also on Z_r and Z_s) such that if the algebraic cycles Z'_r and Z'_s are e-homologous to Z_r and Z_s respectively, then the cycle $Z'_r \cdot Z'_s$ is d-homologous to the cycle $Z_r \cdot Z_s$. All these hold in fact (with slight modifications) also for the intersection of analytic cycles and complexes, but we shall not need them here.

Lemma 4. *Let E_r be an analytic simplex in S_n, with center at (p) and let L_{n-r+1} be a linear variety containing the point (p) such that the intersection of E_r with L_{n-r+1} is an analytic simplex E_1 with center at (p). If a positive algebraic cycle Z_{n-1} of degree m has an isolated intersection with E_1 at the center (p) with a multiplicity μ, then the intersection cycle of Z_{n-1} and L_{n-r+1} is a positive algebraic cycle Z_{n-r} of degree m which has an isolated intersection of multiplicity μ at the point (p) with E_r.*

Proof. The linear variety L_{n-r+1} does not lie in any component hypersurface of Z_{n-1}; for otherwise the entire analytic simplex E_1 would lie in Z_{n-1}, in contradiction to our assumption that Z_{n-1} has an isolated intersection with E_1 at the point (p). Therefore every component of the intersection $L_{n-r+1} \cap Z_{n-1}$ is an algebraic variety of dimension $n-r$ and hence is proper. Therefore the intersection cycle of L_{n-r+1} and Z_{n-1} is a positive algebraic cycle Z_{n-r} of degree m, and it remains only to show that this cycle Z_{n-r} intersects E_r at the point (p) with the multiplicity μ. It is obviously enough to prove the assertion for the case when Z_{n-1} is an irreducible hypersurface, for otherwise we can apply the same argument to each irreducible component of Z_{n-1}. The assumption that Z_{n-1} intersects E_1 at the center (p) with the multiplicity μ implies that there is a projective transformation arbitrarily near to the identity and such that TZ_{n-1} has exactly μ regular intersections with E_1 which lie in any given sufficiently small neighborhood R of (p). For such a T the components of $TZ_{n-1} \cap L_{n-r+1}$ are all proper and hence there is an intersection cycle $Z^T_{n-r} = TZ_{n-1} \cdot L_{n-r+1}$; and since each of the μ regular intersection points of TZ_{n-1} with E_1 is contained in exactly one regular component of $TZ_{n-1} \cap L_{n-r+1}$, it is contained in exactly one component of multiplicity one of the cycle Z^T_{n-r}. Let the positive number d be chosen with respect to the chains Z_{n-r} and E_r and the neighborhood R of the point (p) as in Lemma 2, and let the positive number e be chosen with respect to the number d and the intersection cycle $Z_{n-r} = Z_{n-1} \cdot L_{n-r+1}$ as in the remark (2) immediately preceding this Lemma. Since for T sufficiently near to the identity the cycle TZ_{n-1} is an e-deformation of Z_{n-1}, it follows that for such a T the intersection cycle Z^T_{n-r} is d-homologous to Z_{n-r}. It is clear that the part of the intersection $Z^T_{n-r} \cap E_r$ which lies in R consists of the same μ intersection points of TZ_{n-1} and E_1. Since each of these points is contained in exactly one component variety of multiplicity one of Z^T_{n-r}, and since it is also a regular intersection of this component variety with

E_r, therefore it is an intersection of multiplicity one of Z_{n-r}^T and E_r. It follows then from Lemma 2 that the multiplicity of the intersection (p) of Z_{n-r} and E_r is equal to μ.

With the help of these two lemmas we can prove the following theorem:

Theorem IV. *Let E_r be an analytic simplex in S_n. Suppose there exists a positive integer N such that if any positive algebraic cycle Z_{n-r} of degree m in S_n has an isolated intersection with E_r at its center, then the multiplicity of this intersection is at most equal to mN. Then E_r is an element of an algebraic variety V_r in S_n.*

Proof. The theorem is evidently equivalent to the following statement: If E_r is not an element of an algebraic variety V_r in S_n, then given any positive integer N there exists a positive algebraic cycle Z_{n-r} of degree m such that Z_{n-r} has an isolated intersection with E_r of multiplicity greater than mN at the center of E_r.

Let D_r be the analytic element containing E_r. Let the affine space $A_n(x)$ be so chosen that the analytic element D_r is given by the set of equations

$$x_{r+1} = f_{r+1}(x_1, \ldots, x_r), \ldots, x_n = f_n(x_1, \ldots, x_r),$$

where the $f_i(x_1, \ldots, x_r), i = r+1, \ldots, n$, are convergent power series in a region in the affine space $A_r(x)$ of the r complex variables $x_1, \ldots, x_r$. We can assume without any loss of generality that E_r is the image of a simplicial $2r$-simplex C^{2r} in $A_r(x)$ and that the center of E_r is the origin of $A_n(x)$.

Since the analytic simplex E_r, hence also the analytic element D_r is not algebraic, at least one of the $n-r$ power series $x_{r+1} = f_{r+1}(x_1, \ldots, x_r), \ldots, x_n = f_n(x_1, \ldots, x_r)$, say the power series $x_{r+1} = f_{r+1}(x_1, \ldots, x_r)$, is not an algebraic element over the field $k(x_1, \ldots, x_r)$. According to Lemma 3, there exists to any given positive integer N a polynomial $F(x) = F(x_1, \ldots, x_{r+1})$ of degree $m(> M)$ in the $r+1$ variables $x_1, \ldots, x_{r+1}$, such that the power series $P(x_1, \ldots, x_r) = F(x_1, \ldots, x_r, f_{r+1}(x_1, \ldots, x_r))$ contains no terms of degree $\leq mN$. Since the element $f_{r+1}(x_1, \ldots, x_r)$ is not algebraic over $k(x_1, \ldots, x_r)$, the power series $P(x_1, \ldots, x_r)$ is not identically zero. Hence, after a suitably chosen linear transformation in $A_r(x)$ if necessary, we can assume that the power series $P(x_1) = P(x_1, 0, \ldots, 0)$ is not identically zero. It is clear that this power series $P(x_1)$ contains no terms of degree $\leq mN$. Let L_{n-r+1} be the linear variety in S_n defined by the $r-1$ linear equations $x_2 = 0, \ldots, x_r = 0$, then the intersection of L_{n-r+1} with D_r is a regular analytic element D_1 given by the equations $x_2 = 0, \ldots, x_r = 0, x_{r+1} = f_{r+1}(x_1, 0, \ldots, 0), \ldots, x_n = f_n(x_1, 0, \ldots, 0)$; and the intersection of L_{n-r+1} with E_r is an analytic simplex E_1 contained in D_1, which is the image of the topological 2-simplex C^2 obtained from C^{2r} by intersection with the subspace $x_2 = x_3 = \cdots = x_r = 0$ in $A_r(x)$. (Thus C^2 is a 2-simplex in the x_1-plane.)

Consider now the homogeneous algebraic equation

$$G(x_0, x_1, \ldots, x_{r+1}) = x_0^m F(x_1/x_0, \ldots, x_{r+1}/x_0) = 0;$$

it defines a positive algebraic cycle Z_{n-1} in S_n consisting of as many hypersurfaces as the distinct irreducible factors on the left side of this equation, each taken with its multiplicity. The fact that $P(x_1) = F(x_1, 0, \ldots, 0, f_{r+1}(x_1, 0, \ldots, 0))$ vanishes for $x_1 = 0$ but not identically shows that the point $(x) = (0)$ is an isolated intersection of Z_{n-1} and E_1. It only remains to prove that the multiplicity of this intersection is greater than mN, for then our theorem will follow from Lemma 4. Here again we can assume that the algebraic cycle Z_{n-1} is an irreducible hypersurface, which means that the form $G(x_0, x_1, \ldots, x_{r+1})$ and hence also the polynomial $F(x_1, \ldots, x_{r+1})$ is irreducible; for otherwise we can apply the same argument to each component hypersurface of Z_{n-1}. According to Lemma 2, it is sufficient to show that given any neighborhood R of the point $(x) = (0)$ in $A_n(x)$ there exists a linear transformation $T_{(e)}$,

$$x'_i = x_i + e_i, \quad i = 1, \ldots, n$$

arbitrarily near to the identity, such that $T_{(e)}Z_{n-1}$ has more than mN regular intersections with E_1 in R.

Consider the power series in x_1 with coefficients in $k(u) = k(u_1, \ldots, u_{r+1})$

$$\Phi(x_1, (u)) = F(x_1 + u_1, u_2, \ldots, u_r, u_{r+1} + f_{r+1}(x_1, 0, \ldots, 0)),$$

which is convergent for all sufficiently small x_1 for any given (u). The intersection of $T_{(e)}Z_{n-1}$ and E_1 is then given by the zeros of the power series $\Phi(x_1, (-e))$ in C^2, and it is well known that a simple zero $x_1 = b_1$ corresponds to a point $(b) = (b_1, 0, \ldots, 0, f_{r+1}(b_1, 0, \ldots, 0), \ldots, f_n(b_1, 0, \ldots, 0))$ on E_1 which is a regular intersection of $T_{(e)}Z_{n-1}$ and E_1. Now we have the fact that $\Phi(x_1, (0)) = P(x_1)$ has a zero of order $\mu > mN$ at the point $x_1 = 0$. Hence, by the Weierstrass preparation theorem, there is a neighborhood R' of $x_1 = 0$ and a neighborhood R'' of $(u) = (0)$ such that

$$\Phi(x_1, (u)) = g(x_1, (u))Q(x_1, (u)),$$

where $Q(x_1, (u))$ is a distinguished polynomial of degree μ in x_1 with center at $x_1 = 0, (u) = (0)$, and $g(x_1, (u))$ is a power series which does not vanish for x_1 in R' and (u) in R''. The distinguished polynomial $Q(x_1, (u))$ might be reducible as a polynomial in x_1 over the field of all quotients of power series in (u), but it cannot have multiple factors. In fact, if any power series $q(x_1, (u))$ with the property $q(0, (0)) = 0$ is a multiple factor of $Q(x_1, (u))$, then the power series $q(u) = q(0, (u))$ is a multiple factor of $Q(0, (u))$ and hence also a multiple factor of the polynomial $\Phi(0, (u)) = F(u)$. One observes that the power series $q(u)$ is not a constant, for $q(u)$ is not identically zero and we have $q(0) = 0$. Hence the equation $q(u) = 0$ defines a finite number of analytic elements of r dimensions in $A_{r+1}(u)$ with center at $(u) = (0)$. If one of the variables (u), say u_1, is actually involved in the polynomial $F(u)$, then $\partial F(u)/\partial u_1$ is also a polynomial in (u). It follows then that both polynomials $F(u)$ and $\partial F(u)/\partial u_1$ are divisible by the power series $q(u)$. This means that the two positive algebraic cycles of r dimensions in $A_{r+1}(u)$ defined by the two equations $F(u) = 0$ and $\partial F(u)/\partial u_1 = 0$ have at least an analytic element of r dimensions in common and consequently also at least one component variety in common. Since $F(u)$ is an irreducible polynomial, the algebraic cycle $F(u) = 0$ is an irreducible algebraic variety and hence must be a component variety of the algebraic cycle $\partial F(u)/\partial u_1 = 0$. This means that the polynomial $F(u)$ is a factor of the polynomial $\partial F(u)/\partial u_1$, both being now considered as elements of the polynomial ring $k[u]$. But this is impossible, for the polynomial $\partial F(u)/\partial u_1$ has a lower degree than $F(u)$. Thus we have shown that the distinguished polynomial $Q(x_1,(u))$ has no multiple factors.[7]

Since $Q(x_1,(u))$ has no multiple factors, its discriminant $H(u)$ is not identically zero in R''. Hence for any point (e) in R'' such that $H(e) \neq 0$, the polynomial $Q(x_1,(e))$ has exactly μ simple zeros and consequently also the power series $\Phi(x_1,(e))$ has exactly μ simple zeros in R'. For all points (e) sufficiently near to the point $(u) = (0)$ with $H(e) \neq 0$, the μ simple zeros will lie in any given neighborhood of the point $x_1 = 0$ and hence the corresponding μ intersection points on E_1 will lie in any given neighborhood of the point $(x) = (0)$. This completes the proof of the theorem.

Combining Theorems III and IV, we obtain immediately the main theorem:

Theorem V. *A compact analytic variety in S_n is an algebraic variety.*

Proof. Let E_r be an analytic simplex contained in the compact algebraic variety W_r. According to Theorems III and IV, there is an algebraic variety V_r which also contains the analytic simplex E_r. By the principle of analytic continuation, any two (irreducible) compact analytic varieties of r dimensions which have an analytic simplex E_r in common must coincide. Hence we have $W_r = V_r$, and the theorem is proved.

It is perhaps not without interest to point out here the analogy between Theorem IV and the wellknown criterion for an algebraic number by means of diophantine approximation: A number ξ is algebraic if and only if there exists a positive integer N such that no form $\sum_{i=0}^{n} z_i\xi^i$ has a proper approximation t^{-N} with respect to the variables $z_0, z_1, \ldots, z_n$. In other words, if $z = \max(|z_0|, |z_1|, \ldots, |z_n|)$, then the number ξ is algebraic if and only if for each n the diophantine inequality

$$\left|\sum_{i=0}^{n} z_i\xi^i\right| < Z^{-N}$$

has only a finite number of solutions for which the left hand side is not zero. Hence, for each n there exists a positive number Γ_n such that the diophantine inequality

$$\left|\sum_{i=0}^{n} z_i\xi^i\right| < (\Gamma_n Z)^{-N}$$

has no solution for which the left hand side is not zero. Taking the negative of the logarithm of both sides, we can write the inequality as follows:

$$-\log\left|\sum_{i=0}^{n} z_i\xi^i\right| > N(\log Z + \log\Gamma_n).$$

In the analogy between algebraic number field and algebraic function field of one variable we can regard the number ξ as an "analytic branch" and the equation

[7] The last part of the argument consists essentially in showing that an irreducible polynomial $F(u_1,\ldots,u_{r+1})$ cannot have multiple factors in the ring of power series $k\{u_1,\ldots,u_{r+1}\}$. This is a special case of a theorem proved by Chevalley, see [2], p. 11, Theorem 1, which asserts that any prime ideal in $k(u_1,\ldots,u_{r+1})$ is the intersection of prime ideals in $k\{u_1,\ldots,u_{r+1}\}$.

$\sum_{i=0}^{n} z_i y^i = 0$, for a given (z), as an "algebraic curve". The number $\log Z + \log \Gamma_n$ can then be regarded as representing the "degree" of the "algebraic curve" and the left hand side of the above inequality as the "intersection multiplicity" of the "analytic branch" with the "algebraic curve". Carrying this over to the algebraic function field of one variable, we obtain the statement that an analytic branch E_1 in S_2 is algebraic if and only if there is a number N such that no algebraic curve of degree m can have an isolated intersection with E_1 with a multiplicity greater than mN. Thus our Theorem IV can be considered as the extension of this criterion to analytic elements in space of higher dimensions.

Finally, we should like to remark that though Theorem IV is stated and proved only for a regular analytic element, it is not difficult to see that it holds in fact for any analytic element, whether regular or singular. Thus this theorem expresses a property which is both necessary and sufficient for any analytic element to be algebraic. Though our proof is partly algebraic and partly topological, as is necessary for our present purpose, the theorem itself can be expressed in purely algebraic terms, using only formal power series. In fact, one could probably obtain a purely algebraic proof of this result by means of the intersection theory for algebroid varieties as developed by Chevalley.[8]

5 Meromorphic transformations

We begin with an almost obvious generalization of Theorem V to a multiply projective space. For the sake of convenience we shall restrict ourselves to the case of a doubly projective space $S_m \times S_n$, though the results hold obviously for the general case. We can define an analytic variety W_r in $S_m \times S_n$ in exactly the same way as we have done in 2 for S_n. Now the space $S_m \times S_n$ can be mapped by a bi-regular birational transformation onto a non-singular algebraic variety V_{m+n} in a suitably chosen projective space S_τ. Under this transformation any analytic or algebraic variety in $S_m \times S_n$ is carried into an analytic or algebraic variety in V_{m+n} respectively, and conversely. Therefore, if W_r is a compact analytic variety in $S_m \times S_n$, then its image W_r^* is also a compact analytic variety in V_{m+n} and hence also in S_τ. By Theorem V, W_r^* is an algebraic variety, hence W_r is also an algebraic variety.

Theorem VI. *A compact analytic variety in a multiply projective space is an algebraic variety.*

In fact, we can extend this theorem to all the so-called "extended spaces" of Osgood,[9] but we shall not go into details here. Instead we shall derive from Theorem VI an important result concerning meromorphic transformation of an analytic variety. A mapping of an analytic variety W_r in S_n into the space S_m is called a meromorphic transformation if in a sufficiently small neighborhood R of any point (a) of W_r and for a suitable choice of the affine spaces $A_n(x)$ and $A_m(y)$ in S_n and S_m respectively, the mapping is defined by a set of equations

$$g(x)y_i = f_i(x), \quad i = 1, \ldots, m, \tag{9}$$

[8] Chevalley [2].

[9] See Osgood [7], Ch. III, §32.

where the functions $g(x), f_1(x), \ldots, f_m(x)$ are holomorphic in R and $g(x)$ does not vanish for all points of $W_r \cap R$. These equations are to be understood with the following stipulations: For a point $(x) = (p)$ in $W_r \cap R$ such that $g(p) \neq 0$, the image under the mapping is the point $y_i = f_i(p)/g(p), i = 1, \ldots, m$, in S_m. We shall call such a point (p) a regular point of the mapping. For a point $(x) = (p)$ such that $g(p) = 0$, the image under the mapping is to be the set of all points in S_n which are limits of the images of a sequence of regular points approaching (p). Thus a meromorphic transformation, just like the rational transformation in case of an algebraic variety, is not a mapping in the strict sense of the word; it assigns only to a sufficiently "general" point of W_r a unique image, while to the others (which correspond to the fundamental points of a rational transformation) it assigns certain subsets as images.

Theorem VII. *A meromorphic transformation of a compact analytic (hence algebraic) variety W_r in S_n into a projective space S_m is a rational transformation of W_r, and the image is an algebraic variety in S_m.*

Proof. The graph of the meromorphic transformation in the product space $S_n \times S_m$ is a point set G, the projection of which in S_n is the variety W_r. Let $((a),(b))$ be any point of G, and let R be a sufficiently small neighborhood of $((a),(b))$ with R' and R'' as its projections in S_n and S_m respectively. The variety W_r being algebraic, it is defined in the neighborhood R' of the point (a) by a set of algebraic equations $\phi_i(x) = 0, i = 1, \ldots, s$. Then the set $G \cap R$ will be defined by the set of equations

$$\begin{aligned} g(x)y_i &= f_i(x), \quad && i = 1, \ldots, m, \\ \phi_j(x) &= 0, && j = 1, \ldots, s, \end{aligned} \tag{10}$$

with the stipulation that any component variety lying entirely in the hypersurface $g(x) = 0$, if it is an isolated one, should be deleted from the set. It is well known[10] from the theory of power series ideals that this can be achieved by the adjunction of a finite number of suitably chosen analytic equations to the equations (10), if the neighborhood R is chosen sufficiently small. Thus the set $G \cap R$ consists of all the common solutions of a set of analytic equations in R. Since $((a),(b))$ is any point of G, this means that G is an analytic variety in $S_n \times S_m$.

It remains to show that G is a compact set. Let $((a^{(i)}),(b^{(i)})), i = 1, 2, \ldots$, be a sequence of points in G converging to a point $((a),(b))$ in $S_n \times S_m$; we have to show that $((a),(b))$ is a point of G. Since W_r is compact and the point (a) is the limit point of the sequence $(a^{(i)}), i = 1, 2, \ldots$, it follows that (a) is a point of W_r. In a sufficiently small neighborhood of (a) the given meromorphic transformation will be represented by a set of equations (9), with the corresponding stipulations. We can assume without loss of generality that for all large i the points $(a^{(i)})$ are regular, i. e. $g(a^{(i)}) \neq 0$. For, since each point $((a^{(i)}),(b^{(i)}))$, with $g(a^{(i)}) = 0$ is by our stipulation the limit of a sequence of regular points, we can always replace each point $((a^{(i)}),(b^{(i)}))$, for sufficiently large i, by a regular point sufficiently near to it so that the resulting sequence will have the same limit point $((a),(b))$. Now,

10 See Rückert [8], or Bochner and Martin [1], Ch. X.

if $g(a) \neq 0$, then we have

$$b_j = \lim_{i\to\infty} b_j^{(i)} = \lim_{i\to\infty} f_j(a^{(i)})/g(a^{(i)}) = f_j(a)/g(a), \quad j = 1, \ldots, m;$$

and hence $((a), (b))$ is a point of G. On the other hand, if $g(a) = 0$, then the point (b) is by our stipulation a point of G. Therefore G is a compact point set.

Thus we have shown that G is a compact analytic variety in $S_n \times S_m$; hence, by Theorem VI, G is an algebraic variety. It follows then that the projection of G in S_m is also an algebraic variety W^* and the given meromorphic transformation is really an (irreducible) algebraic correspondence between W_r and W^*. Since this algebraic correspondence assigns to a generic point of W_r a unique image point in W^* (and since our ground field has the characteristic 0), therefore it is a rational transformation of W_r onto W^*. Thus the proof is completed.

The concept of a meromorphic transformation evidently includes that of a meromorphic function as the special case $m = 1$. Hence we have the corollary:

Corollary. *An everywhere meromorphic function on a compact analytic variety is a rational function.*

So far as we know, even this corollary has been proved up to. now only for a few special cases such as the projective space and the space of analysis (the product of projective lines).

THE JOHNS HOPKINS UNIVERSITY

References

[1] S. Bochner and W. T. Martin, *Several Complex Variables*, Princeton, 1948.

[2] C. Chevalley, "Intersections of algebraic and algebroid varieties," *Transactions of the American Mathematical Society*, Vol. 57 (1945), pp. 1–85.

[3] K. Knopp and R. Schmidt, "Funktionaldeterminanten und Abhaengigkeit von Funktionen," *Mathematische Zeitschrift*, Vol. 25 (1926), pp. 373–381.

[4] B. O. Koopman and A. B. Brown, "On the covering of analytic loci by complexes," *Transactions of the American Mathematical Society*, Vol. 34 (1932), pp. 231–251.

[5] S. Lefschetz, *Topology*, New York, 1930.

[6] ——— and J. H. C. Whitehead, "On analytical complexes," *Transactions of the American Mathematical Society*, Vol. 35 (1933), pp. 510–517.

[7] W. F. Osgood, *Lehrbuch der Funktionentheorie*, Bd. II, 1. Leipzig, 1929.

[8] W. Rückert, "Zum Eliminationsproblem der Potenzreihenideale," *Mathematische Annalen*, Vol. 107 (1932), pp. 259–281.

[9] B. L. van der Waerden, "Topologische Begründung des Kalküls der abzählenden Geometrie," *Mathematische Annalen*, Vol. 102 (1929), pp. 337–362.

Editor's Comments

Probably most people will agree that by any reasonable standard, the uniformization theorem for Riemann surfaces is one of the most useful, important and beautiful theorems in mathematics. One consequence of this theorem is that every Riemann surface admits a canonical metric of constant curvature up to scaling.

There is no direct generalization of the uniformization theorem for higher dimensional complex manifolds. For example, the uniformization theorem is a vast generalization of the Riemann mapping theorem for simply connected domains in $\mathbb{C}$, but there is no analogue of the Riemann mapping theorem even for domains in $\mathbb{C}^2$. Probably the best approximation to the uniformization theorem for complex manifolds in higher dimensions is the existence of Kähler-Einstein metrics.

The optimal results on the existence of Kähler-Einstein metrics were stated by Calabi in two famous papers in the 1950s. Specifically, Calabi first announced his now famous Calabi conjecture as a theorem in a short note titled *The space of Kähler metrics* in the Proceedings of ICM in 1954, pp. 206-207. Later in the paper titled *On Kähler manifolds with vanishing canonical class* in 1957, he proved the uniqueness part of his announced result and turned the result on the existence part into a conjecture, which is usually called the Calabi conjecture.

Complex Geometry from Riemann to Kähler-Einstein and Calabi-Yau
ALM 38, pp. 209–210

The Space of Kähler Metrics*

Eugenio Calabi

Let M^n be a closed, n-dimensional complex manifold. We assume that M^n admits at least one Kähler metric $g_{\alpha\beta^*}$; its associated closed exterior form $\omega = \sqrt{-1}g_{\alpha\beta^*}dz^\alpha \wedge dz^{\beta^*}$ determines a real cohomology class, called the principal class of the metric. Consider the space Ω of all infinitely differentiable Kähler metrics in M^n with the same principal class; the topology of Ω is defined by the L^2 topology of the tensorial components of metrics in Ω in compact subregions of coordinate domains. If $R_{\alpha\beta^*}$ is the Ricci tensor of any metric in Ω, then the Ricci form $\sqrt{-1}R_{\alpha\beta^*}dz^\alpha \wedge dz^{\beta^*}$ is closed and its cohomology class is $2\pi C^{(1)}$ ($C^{(r)} = r$-th Chern class).

Theorem 1. *Given in M^n any real, closed, infinitely differentiable exterior form Σ of type $(1,1)$ and cohomologous to $2\pi C^{(1)}$, there exists exactly one Kähler metric in Ω whose Ricci form equals Σ.*

The proof proceeds by joining the Ricci form of one metric in Ω with Σ by a differentiably parametrized arc in the same linear space of forms (for example by linear interpolation) and constructing a corresponding parametrized path in Ω, then by proving that it is unique, and that its terminal point is independent of the path.

One can introduce in Ω a generalized Riemannian structure by defining the metric form $ds^2 = \left(\frac{\partial\omega(t)}{\partial t}, \frac{\partial\omega(t)}{\partial t}\right) dt^2$, where $\omega(t)$ represents a differentiable path in Ω. One shows that between any two points in Ω there exists exactly one geodesic, that the topology induced by the geodesic distance coincides with the chosen one, and that Ω with this metric is isometric with a geodesically convex subset of a Hilbert sphere.

If $g_{\alpha\beta^*}$ belongs to Ω and Σ is its Ricci form, then (Σ, Σ) is stationary with respect to first order variations in Ω, if and only if the contravariant components of the gradient of the scalar curvature generate a group of complex analytic transformations of M^n. If M^n has the property that each one-parameter complex transformation group of M^n (if any exist) leaves no point fixed, or if it is the

* E. Calabi, The space of Kähler metrics. Proceedings of ICM in 1954, pp. 206–207.

Gaussian sphere, then this implies that the scalar curvature is constant, that Σ is harmonic, and that (Σ, Σ) is minimized.

Theorem 2. *If M^n has the restrictive property described above, then there exists a metric in Ω which minimizes (Σ, Σ), unique but for analytic transformations of M^n, and characterized by the property that the scalar curvature is constant.*

LOUISIANA STATE UNIV.
BATON ROUGE 3, LOUISIANA, U. S. A.

Complex Geometry from Riemann
to Kähler-Einstein and Calabi-Yau
ALM 38, pp. 211–220

On Kähler Manifolds with Vanishing Canonical Class*

Eugenio Calabi

The purpose of this note is to show certain properties of compact Kähler manifolds whose canonical class vanishes; an application of the results obtained leads to a partial classification of all such manifolds, reducing it to the problem of classifying a more restricted class of compact Kähler manifolds, namely, those whose canonical class vanishes and whose first Betti number is zero. A complete classification of manifolds of the latter type is claimed by Severi [9] in the case of algebraic surfaces.

We shall first recall that the canonical class of an algebraic manifold M_n of dimension n is the homology class of the divisor of any non-zero meromorphic n-uple differential. This definition is not suitable for generalization to compact, complex manifolds, since it is not known whether one can always find a non-trivial meromorphic n-uple differential in all manifolds of this wider class. An alternate definition which allows such a generalization is due to Chern [3]. Let M_n be a compact, n-dimensional complex manifold, in which let g be any real, differentiable, positive-valued density of weight 1; g is expressible in any coordinate domain $\{U;(z)\}((z)=(z^1,z^2,\dots,z^n))$ by means of a positive, differentiable function $g_U(z,\overline{z})$, so that, if $\{V;(z')\}$ is another coordinate domain, then at every point of $U\cap V$

$$g_V(z',\overline{z}')=g_U(z,\overline{z})\left|\frac{\partial(z)}{\partial(z')}\right|^2, \tag{1}$$

where $\partial(z)/\partial(z')$ is the Jacobian of the coordinated transformation. It follows from (1) that

$$\log g_V=\log g_U+2\mathrm{Re}\log(\partial(z)/\partial(z')),$$

and hence the tensor components defined in each coordinate domain $\{U;(z)\}$ by

$$R_{\alpha\beta^*}(z,\overline{z})=\frac{\partial^2(\log g_U(z,\overline{z}))}{\partial z^\alpha\partial\overline{z}^\beta}\quad(\alpha,\beta=1,2,\dots,n) \tag{2}$$

* E. Calabi, On Kähler manifolds with vanishing canonical class. Algebraic Geometry and Topology. A Symposium in Honor of S. Lefschetz, Princeton University Press, Princeton, NJ, pp. 78–89. 1957.

define the same hermitian tensor over all of M_n; consequently the exterior form[1]

$$\Sigma = -\frac{1}{2}dCd(\log g) = \sqrt{(-1)}R_{\alpha\beta^*}dz^\alpha \wedge d\overline{z}^\beta \tag{3}$$

is a closed, real-valued exterior differential form of type $(1, 1)^2$ defined over M_n. If a different density g' is chosen, then g'/g is a positive, differentiable scalar; hence

$$\Sigma' - \Sigma = -\frac{1}{2}dCd\left(\log \frac{g'}{g}\right) \tag{4}$$

is a closed form cohomologous to zero. Thus the cohomology class of Σ depends only on the complex analytic structure of M_n, and not on the particular density chosen. It is shown by Chern [3] that whenever a non-trivial meromorphic n-uple differential exists in M_n, then the dual of its divisor cycle is cohomologous to $(1/2\pi)\Sigma$. Thus for any compact, complex manifold one can define the canonical class (or first Chern class) to be the cohomology class of $(1/2\pi)\Sigma$.

From this point we shall assume that the complex manifold M_n admits an infinitely differentiable, positive-definite Kähler metric (for a definition, see [2])

$$ds^2 = 2g_{\alpha\beta^*}(z, \overline{z})dz^\alpha d\overline{z}^\beta; \tag{5}$$

then one can take the density g to be the determinant of the matrix $(g_{\alpha\beta^*})$, i.e. the volume distribution with respect to a local coordinate system. One of the properties of a Kähler metric which we shall use here is that the tensor $R_{\alpha\beta^*}$ defined by (2) from the volume element g is equal to the Ricci tensor of the metric (cf. [1], p. 789); we shall therefore call the form Σ defined by (3) the Ricci form relative to the metric. The property which is characteristic of and was originally used to define a Kähler metric is that the exterior form ω associated with (5) and defined by

$$\omega = \sqrt{(-1)}g_{\alpha\beta^*}dz^\alpha \wedge d\overline{z}^\beta \tag{5a}$$

is, like the Ricci form, closed, differentiable, of type (1, 1), and realvalued (see [6]); we shall call ω the principal form of the metric. Since $\omega^n/n!$ is the volume element of M_n,[3] its integral over the cycle carried by M_n is positive, so that, as a closed differential form of type $(n, n), \omega^n$ cannot be cohomologous to zero; therefore ω belongs to a nontrivial, real cohomology class (see [4]), which we shall call the principal class.

The topology of compact, Kähler manifolds from the standpoint of homology theory has been studied extensively by Hodge [4], who generalized to this class of manifolds the homology properties with respect to real coefficients that Lefschetz [7] had discovered for algebraic varieties. In particular we shall use the following facts. All holomorphic and all antiholomorphic differentials in M_n are harmonic; those of degree 1 (Picard differentials of the first kind and their conjugates) for

[1] The notation here is the same as in Kodaira ([8], pp. 95 et seq.).

[2] An exterior form of type $(p, q) (0 \leqq p, q \leqq n)$ is an exterior form of degree $p + q$, which is p-fold linear in $dz^1, \ldots, dz^n$ and q-fold linear in $d\overline{z}^1, \ldots, d\overline{z}^n$.

[3] An exponent written over an exterior form denotes its power with respect to exterior multiplication.

a complete set of representatives of cohomology classes with complex coefficients for dimension 1; thus the first Betti number B_1 of M_n is twice the irregularity g_1, or number of linearly independent Picard differentials of the first kind in M_n. If $\alpha_1, \ldots, \alpha_{g_1}$ form a basis for these differentials, we call tlie Albanese variety A_M of M_n the complex torus of (real) dimension $2g_1$ whose periods are those of the integrals $\oint \alpha_i (i = 1, \ldots, g_1)$ over the 1-cycles of M_n; we have also a complex analytic mapping of M_n into A_M, which we shall call the Jacobi map, defined up to a translation in A_M, described as follows: if $p_0 \in M_n$ and $a_0 \in A_M$ are fixed points and if p ranges over M_n,

$$J(p) = a_0 + \left(\int_{p_0}^{p} \alpha_i \right)_{i=1}^{g_1} \qquad \text{(modulo periods of } \oint \alpha_i).$$

The map J induces the natural isomorphism of the translation invariant simple differentials of A_M with the Picard differentials of M_n. As in algebraic geometry we shall call M_n *regular*, if $B_1 = 2g_1 = 0$.

We shall now postulate a general result about compact, Kähler manifolds. While the proof of this result is not complete, we feel justified in assuming the truth of the statement for several concurrent, intuitive reasons. A sketch of a heuristic argument for the proposition is given in the Appendix.

Proposition 1. *Let M_n be a compact, complex manifold admitting an infinitely differentiable Kähler metric with principal form ω and Ricci form Σ. If Σ' is any closed, real-valued, infinitely differentiable form of type* (1, 1) *and cohomologous to Σ, then there exists a unique Kähler metric with principal form ω' cohomologous to ω and Ricci form equal to Σ'. This metric is always infinitely differentiable; it is real analytic, if Σ' is analytic.*

We shall now assume that the canonical class of M_n is zero, or, equivalently that the Ricci form Σ is cohomologous to zero. We shall call M_n with this restriction, for short, a *special Kähler manifold*, because (see Chern [3]) this property is equivalent to the reducibility of the structural group of the tangent bundle from the unitary to the special (or unimodular) unitary group.

Examples of special Kähler manifolds of dimension n are given: (a) by the n-dimensional complex tori (group manifolds of compact, abelian, complex Lie groups); (b) by the non-singular hypersurfaces of order $n+2$ in a complex projective $(n+1)$-space; the latter examples are, for $n \geqq 2$, also regular; for $n = 1$ of course the only possible examples are algebraic curves of genus 1.

The following theorem describes the first characteristic property of special Kähler manifolds:

Theorem 1. *If M_n is a special Kähler manifold of dimension n and with irregularity g_1, then $0 \leqq g_1 \leqq n$ (equality holding only for a complex torus); more precisely the Jacobi map J of M_n, into its Albanese variety A_M:*

(i) *is a map onto*;

(ii) *it is the projection map of a complex analytic fibre bundle with base space A_M and bundle space M_n*;

(iii) *the structural group of the bundle is finite and abelian*;

(iv) *the fibre is a connected, special Kähler manifold.*

Proof. Since Σ is cohomologous to zero, we can set $\Sigma' = 0$; then, by applying Proposition 1, we can replace the given Kähler metric by another one with the same principal class, for which the Ricci tensor vanishes identically; this new metric is unique and real analytic. All of the statements that will follow will be referred to this metric.

An immediate application of the theorems of Bochner ([1], Theorems 1, 2, 5, 6) to the case of vanishing Ricci tensor leads to the following statement: a vector function in M_n cannot be harmonic, nor satisfy Killing's equation, nor have holomorphic or antiholomorphic contravariant components, unless it is a parallel vector field (of course if it is parallel, then it trivially satisfies all three other conditions simultaneously). In other words, a real vector function in M_n is harmonic (i.e. the real part of a Picard differential of the first kind), if and only if, its contravariant components are the sum of a holomorphic and an antiholomorphic vector, or if, and only if, it satisfies Killing's equation (i.e. the vector generates a one-parameter group of complex analytic, isometric transformations of M_n). It follows from this that the mapping J induces a one-to-one linear correspondence between the parallel vector fields of M_n and those of A_M; hence the group G' of translations of A_M is locally isomorphic under a map induced by J to the component G of the identity in the group of analytic (and isometric) transformations of M_n. This shows that J maps M_n onto A_M; in the special case where $g_1 = n$, the map is locally one-to-one and without singularities, because of the properties of the group of translations of A_M; therefore M_n is a compact covering of A_M under J, whence it is itself a complex torus, i.e. its own Albanese variety, so that J becomes the identity map.

The inverse image $J^{-1}(a)$ of each point $a \in A_M$ is a complex analytic subvariety of M_n. If U is a geodesically convex, sufficiently small neighborhood of a, the set $\tilde{U}$ of translations of A_M mapping a into U is a neighborhood of the identity in the translation group G', constituting a local group which under the local isomorphism induced by J carries each point of $J^{-1}(a)$ into an analytic cross-section of $J^{-1}(U)$; the image of each element of $\tilde{U}$ under the local isomorphism maps $J^{-1}(a)$ analytically and isometrically onto $J^{-1}(a')$ for some $a' \in U$. One verifies easily that each cross-section over U thus defined is analytically isometric with U and intersects orthogonally the subvariety $J^{-1}(a')$ for each $a' \in U$. Thus $J^{-1}(U)$ is analytically isometric with $U \times J^{-1}(a)$. Since each local cross-section as constructed above is uniquely determined by one of its points but for its extent, we have established the complex fibre bundle structure of M_n over A_M. Furthermore, each component of a fibre is a complex submanifold without singularities, because of the local metric product structure of the bundle space.

It is clear that J induces a homomorphism of G onto G', and that the kemel consists of the analytic isometries of M_n which are analytically deformable to the identity, and map each $J^{-1}(a)$ for each $a \in A_M$ into itself. But it is wellknown that the group of all isometries of a compact Riemannian space is a compact Lie group; thus G is a compact, connected Lie group, locally isomorphic to the compact, abelian group G', so that G is itself abelian. The kernel K of the homomorphism of G onto G' is thus a discrete, and hence finite, abelian subgroup of G. It is easy to see that K is the structural group of the fibre bundle of M_n over A_M, so that the third contention of the theorem is proved.

We have already verified that each component of a fibre is a complex submanifold of M_n without singularities; from the local metric product structure we see that it is totally geodesic, so that its Ricci curvature tensor is the tensor induced from the Ricci tensor of the ambient space. It follows that the Ricci curvature of each component of a fibre vanishes identically, so that it is a special Kähler manifold. In order to prove that each fibre is connected, we identify all points of M_n lying on the same component of any fibre; we obtain thus a complex manifold A', over which M_n is fibered by connected submanifolds, and which is in turn fibered over A_M by fibres consisting of a finite number of points, so that the composition of the two fibre projections is J. The manifold A' thus obtained is a finite covering of A_M; therefore it is a complex torus, in which the group of periods of Picard integrals of the first kind is a subgroup of finite index of the corresponding one for A_M. This would imply, if the covering of A_M by A' were not trivial, that the group of periods of the integrals in M_n would be a proper subgroup of the group of periods of A_M, which would contradict the definition of the Albanese variety. Theorem 1 is thereby proved.

The next theorem is a new interpretation of the first, but it is perhaps more helpful in describing the structure of special Kähler manifolds.

Theorem 2. *If M is a special Kähler manifold, then there exists a finite covering manifold $\tilde{M}$ of M, which is a direct product of its own Albanese variety $A_{\tilde{M}}$ with a special Kähler manifold Y, which is regular, and possibly a point. The relative covering group of $\tilde{M}$ over M is solvable.*

Proof. We construct the principal bundle (cf. Steenrod, [10], pp. 35–9) associated with the fibre bundle $\{M, A_M, J\}$; since the structural group of the bundle is finite, the principal bundle is given by a finite, necessarily abelian, covering A_1 of A_M. Denote the fibre of M by Y_0 and set $M_0 = M$ and $M_1 = Y_0 \times A_1$. Then the principal map P: $M_1 \to M_0$ defines a finite, abelian covering of M_0 by M_1. The complex torus A_1 appearing as a direct factor of M_1 is not necessarily the Albanese variety of M_1, since M_1 may have some Picard differentials of the first kind other than those induced from M under the covering map P; it is obvious that the additional differentials appearing in M_1 are precisely those induced from its other direct factor, namely, Y_0. Thus the Albanese variety of M_1 is the direct product of A_1 with the Albanese variety of Y_0.

Since it is clear that M_1 is again a special Kähler manifold, the process may be repeated with M_0 replaced by its covering manifold M_1, thus obtaining a special Kähler manifold M_2 as a finite, abelian covering of M_1, and continue inductively. At the k'^{th} step one obtains a special Kähler manifold M_k, which is a finite, abelian covering of M_{k-1}, such that the Albanese variety of M_k is the direct product of the corresponding finite covering of the Albanese variety of M_{k-1} with the Albanese variety of the fiber Y_{k-1} of M_{k-1} under J. Since

$$g_1(M_k) - g_1(M_{k-1}) = \dim(Y_{k-1}) - \dim(Y_k) = g_1(Y_{k-1}),$$

either the process stops when the fibre Y_k is reduced to a point, in which case the covering manifold M_k is a complex torus, or one reaches a stage when the irregularity of Y_k is zero, in which case the next inductive step will yield as covering manifold $M_{k+1} = Y_k \times A_{k+1}$ with the properties we set out to prove. Repeated application of the process will then yield only the identity covering. Thus we have obtained a finite covering manifold of M, which is the direct product of a complex torus (its own Albanese variety) with a special Kähler manifold that is regular. The relative covering group is solvable, because it has a normal series consisting of finite, abelian groups. This completes the proof of Theorem 2.

By means of the result just established one could completely classify all special Kähler manifolds in terms of those that are regular. We shall describe briefly how this can be done.

Let A be a complex torus and G a finite soluble group of complex analytic transformations of A such that no element of G except the identity leaves any point of A fixed (for a discussion of these crystallographic groups, cf. H. Weyl [11] and H. Hopf [5]). Let Y be a regular, special Kähler manifold (this may even be one point) and (G_0, ϕ) a (possibly trivial) representation of G as a group of analytic transformations of Y. We then define a group G' of transformations of $Y \times A$, naturally isomorphic to G as follows: if $x \in G$ and x' is the corresponding element of G', and if $y \in Y$ and $a \in A$, then

$$x'(y, a) = (\phi(x)(y), x(a)).$$

It is clear that no element of G' other than the identity leaves any point of $M_0 \times A$ fixed; the identification of points in $y \times A$ equivalent under G' produces a Kähler manifold, which is obviously special. As a result of Theorem 2 we know that every special Kähler manifold can be obtained in this way, so that the whole class of special Kähler manifolds can be constructed from a classification of the regular ones plus the family of all compact Kähler manifolds which are metrically flat.

It might be interesting to know immediately, given a special Kähler manifold M_n with its Ricci flat metric, what the dimension is of the complex torus $A_{\tilde{M}}$ and the complementary dimension of the regular manifold Y such that $Y \times A_{\tilde{M}}$ is a finite covering of M_n. It is clear that an upper bound for the dimension of $A_{\tilde{M}}$ is given by maximum number m such that at each point of M_n there exist m linearly independent (over complex coefficients) parallel vector fields; however, this upper bound is not the best possible, at least for $n \geqq 3$, since one can construct a compact, regular, 3-dimensional Kähler manifold, with Riemann curvature identically zero, admitting a complex torus as a finite covering. Another unanswered question in this connection is whether a special Kähler manifold with no local parallel vector fields with respect to its Ricci flat metric is not only regular, but also simply connected.

One interesting application of the above results is the following one, suggested to the author by G. Washnitzer. Let M_n be a special Kähler manifold with an arbitrary (i.e not necessarily Ricci flat) Kähler metric. We consider on M_n the density g' defined by

$$g' = g \exp(2\Lambda G\Sigma), \tag{6}$$

where g is the volume density, Λ is algebraic operator on exterior p-forms ϕ defined by $\Lambda\phi = (-1)^p * (\omega \wedge *\phi)$ (i.e. the negative of the one so denoted by Kodaira [8]),

G is the Green's operator as in de Rham [8], and Σ is the Ricci form (3). From the wellknown identities in the exterior calculus in Kähler manifolds, and from the assumption that the closed form Σ is cohomologous to 0, we obtain the identity for the functional expression $g'_U(z,\overline{z})$ of g' in a complex coordinate system $\{U;(z)\}$

$$dCd(\log g'_U(z,\overline{z})) = 0.$$

Consequently the density g' is locally equal to the squared absolute value of a complex analytic, nowhere vanishing, n-uple differential Θ, uniquely determined up to constant factors of modulus unity, i.e. a generalization of a Prym differential. The analytical continuations of Θ over closed paths depend on the homology class of the latter only, and hence furnish a multiplicative representation of the first homology group of M_n. We remark that, if M_n is regular, then Θ is single-valued; the same is true if M_n is a complex torus, for then Θ becomes, with a constant factor, the exterior product of a base for the Picard differentials of the first kind.

A direct application of Theorem 2, in connection with the considerations just made, produces the following result.

Corollary to Theorem 2. *Let M_n be a special Kähler manifold and let the smallest finite covering manifold $\tilde{M}_n$ of M_n that is a product of a complex torus with a regular, special Kähler manifold be k-sheeted. Then there exists a divisor d of k such that the m-genera of $M_n (m = 1, 2, \ldots)$ are* 1 *or* 0 *according as to whether d divides m or not.*

Proof. We recall that the m-genus of a complex manifold is the maximum number of linearly independent single-valued, everywhere holomorphic densities of weight m; for $m = 1$ it is also called the geometric genus. If M_n is itself the product of a regular, special Kähler manifold with a complex torus, then $k = d = 1$; in this case we have already verified that there is a single-valued, holomorphic, n-uple differential Θ with no zeros. Each mth tensor power of $\Theta(m = 1, 2, \ldots)$ is a holomorphic density of weight m; it is unique up to constant factors, since its divisor vanishes; thus the m-genera of M_n are all 1.

In the general case there exists a k-sheeted covering $\tilde{M}_n$ of M_n, which is a product of a regular, special Kahler manifold with a complex torus. The real density g' defined by (6) on M_n induces the corresponding one of $\tilde{M}_n$. Hence the generalized Prym differential Θ on M_n has multiplicative periods corresponding to a representation of the relative covering group of M_n by $\tilde{M}_n$, which is finite and of order k; thus the multiplicative periods of Θ are the totality of dth roots of unity for some divisor d of k. For each positive integer m the density Θ^m is an m-uple, holomorphic Prym density on M_n, obviously unique up to a constant factor, since its divisor vanishes, and single-valued, if and only if m is a multiple of d. This concludes the proof.

Appendix

We shall give here a heuristic argument to make the statement of Proposition 1 seem plausible. The proof of the uniqueness of the Kähler metric with prescribed Ricci form (subject to the necessary restrictions stated) will be rigorous;

the existence part will rely on the assumption of the existence of a solution of an integro-differential equation.

We remark first that, if M_n has two distinct metrics whose principal forms ω and ω' belong to the same (principal) class, then there exists a real-valued scalar Φ defined over all of M_n and unique up to an additive real constant, such that

$$\omega' - \omega = dCd\Phi; \tag{7}$$

in fact, it is enough to set $\Phi = \Lambda G(\omega' - \omega)$, where Λ and G are taken relative to either (or any other) Kähler metric, and verify that Φ is a solution. If Φ' is any other solution, then $dCd(\Phi - \Phi') = 0$, meaning that the difference between the two solutions of (7) is the sum of an everywhere holomorphic with an everywhere antiholomorphic scalar, i.e. a constant. Since the conjugate of a solution is clearly a solution, it follows that the imaginary part of any solution of (7) is a constant, so that its real part will do.

We now prove that, if the principal forms ω and ω' are distinct, but cohomologous, then their corresponding Ricci forms Σ and Σ' are distinct. In fact, suppose that $\Sigma = \Sigma'$; from equation (4) it follows that g'/g is the real part of an everywhere holomorphic scalar; therefore it is a constant, namely, the ratio of the volumes of M_n as calculated by each of the two metrics. But the volume elements are given respeotively by the forms $\omega^n/n!$ and $\omega'^n/n!$, which are clearly cohomologous, so that their integrals over M_n are equal. Thus we have $\omega'^n = \omega^n$.

Using the solution Φ of equation (6), we obtain

$$\begin{aligned} 0 &= *(\omega'^n - \omega^n) = * \left[dCd\Phi \wedge \left(\sum_{k=0}^{n-1} \omega^k \wedge \omega'^{n-k-1} \right) \right] \\ &= \sum_{k=0}^{n-1} *(\omega^k \wedge \omega'^{n-k-1} \wedge dCd\Phi). \end{aligned} \tag{8}$$

Each term in the last summation on the above equation can be written in the form

$$H_{(k)}^{\alpha\beta^*} \frac{\partial^2 \Phi}{\partial z^\alpha \partial \bar{z}^\beta},$$

where the coefficient matrices $H_{(k)}^{\alpha\beta^*}$ are, for each value of $k (0 \leqq k \leqq n-1)$, Hermitian contravariant tensors; using the local orthogonal developments of the two metrics, one can easily show that each tensor $H_{(k)}^{\alpha\beta^*}$ is positive-definite. Thus the tensor $H^{\alpha\beta^*} = \sum_{k=0}^{n-1} H_{(k)}^{\alpha\beta^*}$ is positive-definite, and the scalar Φ satisfies the equation

$$H^{\alpha\beta^*} \partial^2 \Phi / \partial z^\alpha \partial \bar{z}^\beta = 0; \tag{9}$$

these two facts imply that, since M_n is compact, Φ is a constant, because no solution of (9) can achieve a maximum value in the interior of its domain of existence, unless it is a constant (cf. Bochner and Yano [2], pp. 26–31). From this it follows trivially that $\omega' - \omega = 0$.

We shall now derive the integro-differential equation whose solution yields the Kähler metric with the prescribed Ricci form Σ', subject to the conditions stated in Proposition 1. Assume that for each such form Σ' there existed *a* corresponding Kähler metric with principal form ω'; suppose further that for each continuous system $\Sigma(t)$ of Ricci forms depending differentiably on a real parameter $t(0 \leqq t \leqq 1)$ the corresponding principal forms $\omega(t)$ depended also differentiably on t. Then, by using equations (3) and (4) we could compute the relationship between the derivatives of $\Sigma(t)$ and $\omega(t)$ with respect to t. In fact

$$\begin{aligned}\frac{\partial\Sigma(t)}{\partial t} &= \lim_{h\to 0} -\frac{1}{2h}dCd\left(\log\frac{g(t+h)}{g(t)}\right) = -\frac{1}{2}dCd\Lambda_t\left(\frac{\partial\omega(t)}{\partial t}\right)\\ &= -\frac{1}{2}\Delta_t\left(\frac{\partial\omega(t)}{\partial t}\right),\end{aligned}$$

where the subscript t under the operators Λ and Δ indicates that they are the ones associated with the metric corresponding to $\omega(t)$. Since both of the derivatives with respect to t of $\omega(t)$ and $\Sigma(t)$ are closed forms cohomologous to zero, using the Green operator G_t relative to $\omega(t)$, we obtain the integro-differential equation

$$\frac{\partial\omega(t)}{\partial t} = -2G_t\left(\frac{\partial\Sigma(t)}{\partial t}\right). \tag{10}$$

It can be shown analytically that, subject to sufficient conditions of differentiability of the function $\partial\Sigma(t)/\partial t$, the integro-differential equation above has a solution $\omega(t)$ with initial value given by the principal form $\omega(0)$ of an arbitrary, infinitely differentiable Kähler metric. In particular, on letting the initial values correspond to the Kähler metric assumed in the statement of the proposition, and choosing $\partial\Sigma(t)/\partial t$ to be the form $\Sigma' - \Sigma$, the solution $w(t)$ of equation (10) such that $\omega(0) = \omega$ formally gives the principal form of the Kähler metric whose Ricci form $\Sigma(t)$ satisfies

$$\Sigma(t) = \Sigma + t(\Sigma' - \Sigma); \tag{11}$$

for $t = 1$ the solution becomes the metric whose existence is asserted.

The first essential step in rigorizing this argument consists in showing that, as t varies, equation (10) preserves its elliptic character. In fact the corresponding equation for the volume element $\omega^n(t)/n!$ is given by

$$\begin{aligned}\frac{\partial(\omega^n(t))}{\partial t} &= n\omega^{n-1}(t)\frac{\partial\omega(t)}{\partial t} = \omega^n(t)\Lambda_t\left(\frac{\partial\omega(t)}{\partial t}\right)\\ &= -2\omega^n(t)\Lambda_t G_t(\Sigma' - \Sigma).\end{aligned} \tag{12}$$

Remembering again that the operator ΛG acting on closed forms of type (1, 1) and cohomologous to zero gives the same scalar modulo additive constants independently of the metric, and that the volume of M_n under any of the metrics considered is constant, we have the explicit solution of equation (12):

$$\omega^n(t) = \omega^n(0)\exp[-2\Lambda G(\Sigma' - \Sigma)]/H\{\exp[-2\Lambda G(\Sigma' - \Sigma)]\}, \tag{13}$$

where the operators Λ, G and H (harmonic projection) are taken relative to the given metric $\omega = \omega(0)$. This fact indicates that, since the determinant $g(t)$ of the metric remains positive for all values of t, the solutions of equation (10) can be continued locally with respect to t.[4] The proof of existence of the solution $\omega(t)$ of (10) for

$$\partial\Sigma/\partial t = \Sigma' - \Sigma(0)$$

can then be completed by evaluating formally from it $\partial^2\omega(t)/\partial t^2$ and obtaining a uniform *a priori* estimate of the latter expression in terms of $\omega(t)$ and $\partial\omega(t)/\partial t$, so that one can then apply the Cauchy-Peano method of polygonal approximations to a solution.

LOUISIANA STATE UNIVERSITY

References

[1] S. BOCHNER, *Vector fiekls and Ricci curvature*, Bull. Amer. Math. Soc., 52 (1946), pp. 776–797.

[2] —— and K. YANO, Curvature and Betti numbers, Annals of Mathematics Studies, no. 32, Princeton, 1953.

[3] SHIING-SHEN CHERN, *Characteristic classes of Hermitian manifolds*, Ann. of Math., 47 (1946), pp. 85–121.

[4] W. V. D. HODGE, Theory and applications of harmonic integrals, Cambridge University Press, 1941.

[5] HEINZ HOPF, *Zum Clifford-Kleinschen Raumproblem*, Math. Ann., 95 (1925), pp. 313–339.

[6] E. KÄHLER, *Über eine bemerkenswerte Hermitesche Metrik*, Abh. Math. Sem. Hamburg Univ., 9 (1933), pp. 173–186.

[7] S. LEFSCHETZ, L'analysis situs et la géométrie algébrique, Paris, Cauthior-Villars, 1950.

[8] G. DE RHAM and K. KODAIRA, Harmonic integrals, Institute for Advanced Study Seminar, 1950.

[9] F. SEVERI, *Sulle superficie algebriche con curva canonica d'ordine zero*, Atti Ist. Venoto, 68 (1908–9), pp. 31–32.

[10] N. E. STEENROD, The topology of fibre bundles, Princeton University Press, 1951.

[11] H. WEYL, Group theory, Lecture Notes, Princeton University, Spring, 1949.

[4] There seems to be some question as to whether the interval in t for which one can solve for $\omega(t)$ is unbounded. This essential gap in the proof of Proposition 1 makes the results of this paper depend on the conjecture that a compact Kähler manifold admits a Kähler metric with any assigned, positive, differentiable volume element.

Editor's Comments

Cohomology groups are essential notions in mathematics. For example, the Riemann-Roch theorem and its generalizations are expressed in terms of cohomology groups. Such theorems are especially useful when all but one cohomology group vanish, and hence desired holomorphic objects (such as forms and sections of vector bundles) are proved to exist from the topological condition imposed by Riemann-Roch theorem or index type theorems. The Kodaira vanishing theorem is probably the most important result of this type. This paper should be viewed together with the next paper by him on embedding compact Kähler manifolds into projective spaces.

Complex Geometry from Riemann to Kähler-Einstein and Calabi-Yau
ALM 38, pp. 223–228

On a Differential-Geometric Method in the Theory of Analytic Stacks*

Kunihiko Kodaira

1 Introduction

Let V be a compact Kähler variety of complex dimension n and let F be a complex line bundle[1] over V whose structure group is the multiplicative group of complex numbers. Moreover let $\Omega^p(F)$ be the stack (faisceau) over V of germs of holomorphic p-forms with coefficients in F and let $H^q(V;\Omega^p(F))$ be the qth cohomology group of V with coefficients in $\Omega^p(F)$. It is important for applications to determine the circumstances under which the cohomology group $H^q(V;\Omega^p(F))$ vanishes. In the present note we shall prove by a differential-geometric method due to Bochner[2] some sufficient conditions for the vanishing of $H^q(V;\Omega^p(F))$ in terms of the characteristic class of the bundle F.

* This work was supported by a research project at Princeton University sponsored by the Office of Ordnance Research, U. S. Army. Department of mathematics, Princeton university, communicated by S. Lefschetz, September 9, 1953. K. Kodaira, On a differential-geometric method in the theory of analytic stacks. Proc. Nat. Acad. Sci. U. S. A. 39, (1953) 1268–1273.

1 By a complex line bundle over V will be meant an analytic fibre bundle over V whose fibre is the complex number field C and whose structure group is the multiplicative group C^* of complex numbers acting on C. Cf. Kodaira, K., and Spencer, D. C., "Groups of Complex Line Bundles Over Compact Kähler Varieties," these PROCEEDINGS, 39, 868-872(1953).

2 Bochner, S., "Curvature and Betti Numbers" (I) and (II), *Ann. Math.*, 49, 379–390 (1948); 50, 77–93 (1949).

2 Harmonic forms with coefficients in complex line bundles[3]

We denote by $ds^2 = 2\sum g_{\alpha\beta^*}(dz^\alpha d\bar{z}^\beta)$ the Kähler metric on V. Take a sufficiently fine finite covering $\mathfrak{U} = \{U_j\}$ of V. Then the bundle F is determined by the system $\{f_{jk}\}$ of non-vanishing holomorphic functions f_{jk} defined, respectively, in $U_j \cap U_k$ and satisfying $f_{jk}f_{kl}f_{lj} = 1$ in $U_j \cap U_k \cap U_l$. Clearly there exists a system $\{a_j\}$ of real positive functions a_j of class C^∞ defined, respectively, in U_j satisfying

$$\frac{a_j}{a_k} = |f_{jk}|^2, \quad \text{in } U_j \cap U_k. \tag{1}$$

A form φ with coefficients in F is, by definition, a system $\{\varphi_j\}$ of exterior differential forms φ_j defined, respectively, in U_j such that

$$\varphi_j = f_{jk} \cdot \varphi_k, \quad \text{in } U_j \cap U_k.$$

For a form $\varphi = \{\varphi_j\}$ with coefficients in $F, \bar{\partial}\varphi = \{(\bar{\partial}\varphi)_j\}$ and $\delta\varphi = \{(\delta_\varphi)_j\}$ are defined, respectively, by

$$(\bar{\partial}\varphi)_j = \bar{\partial}\varphi_j, \quad (\delta\varphi)_j = - * a_j\partial\left(\frac{1}{a_j} * \varphi_j\right),$$

provided that φ is differentiable, where $*\varphi_j$ denotes the dual form of φ_j with respect to the preassigned Kähler metric ds^2 on V. The form φ is called *harmonic* if $\bar{\partial}\varphi = 0$ and $\delta\varphi = 0$.

Let $Q^p(F)$ be the stack (faisceau) over V of germs of holomorphic p-forms with coefficients in F. Then, as was shown by the author,[4] *the cohomology group $H^q(V;\Omega^p(F))$ of V with coefficients in $\Omega^p(F)$ is isomorphic to the space $H^{p,q}(F)$ consisting of all harmonic forms of type (p,q) with coefficients in F*:

$$H^q(V;\Omega^p(F)) \cong H^{p,q}(F). \tag{2}$$

As one readily infers, the mapping

$$\varphi = \{\varphi_j\} \to \varphi^\dagger = \{\varphi_j^\dagger\}, \quad \varphi_j^\dagger = \frac{1}{a_j} * \bar{\varphi}^j$$

maps $H^{p,q}(F)$ isomorphically onto $H^{n-p,n-q}(-F)$, where $-F$ denotes the complex line bundle defined by the system $\{f_{jk}^{-1}\}$. Combining this with (2), we obtain therefore the isomorphism[5]

$$H^q(V;\Omega^p(F)) \cong H^{n-q}(V;\Omega^{n-p}(-F)). \tag{3}$$

---

[3] See Kodaira, K., "On Cohomology Groups of Compact Analytic Varieties with Coefficients in Some Analytic Faisceaux," these PROCEEDINGS, 39, 865–868 (1953).

[4] Kodaira, *loc. cit.*

[5] This result reduces to a special case of Serre's duality theorem.

3 An inequality

The equations $\bar\partial\varphi = 0, \delta\varphi = 0$ for $\varphi = \{\varphi_j\}$,

$$\varphi_j = (1/p!q!)\sum \varphi_{j\tau_1\cdots\tau_p\alpha_1^*\cdots\alpha_q^*} dz^{\tau_1}\cdots dz^{\tau_p} d\bar z^{\alpha_1}\cdots d\bar z^{\alpha_q},$$

can be written explicitly in the forms

$$\sum_{m=0}^{q}(-1)^m \nabla_{\alpha_m^*}\varphi_{j\tau_1\cdots\tau_p\alpha_0^*\cdots\alpha_{m-1}^*}{}_{\alpha_{m+1}^*}\cdots\alpha_q^* = 0, \tag{4$_1$}$$

$$\sum g^{\lambda\mu^*}(\nabla_\lambda + \rho_{j\lambda})\varphi_{j\tau_1\cdots\tau_p\mu^*\alpha_2^*\cdots\alpha_q^*} = 0, \tag{4$_2$}$$

where $\nabla_\lambda, \nabla_{\alpha^*}$ denote the covariant differentiations with respect to the local coordinates $z^\lambda, \bar z^\alpha$ and where

$$\rho_{j\lambda} = -\partial_\lambda \log a_j, \quad \partial_\lambda = \frac{\partial}{\partial z^\lambda}.$$

Let $R^\sigma_{\tau\alpha^*\beta} = \bar\partial_\alpha(\sum g^{\sigma\nu^*}\partial_\beta g_{\tau\nu^*})$ be the curvature tensor and let $R_{\alpha-\beta} = \sum_\tau R^\tau_{\tau\alpha^*\beta}$ be the Ricci tensor.[6] Moreover we set

$$X_{\lambda\alpha^*} = -\bar\partial_\alpha\rho_{j\lambda} = \partial_\lambda\bar\partial_\alpha \log a_j. \tag{5}$$

Then, by a simple computation, we infer from (4) that, for any harmonic form $\varphi = \{\varphi_j\} \in H^{p,q}(F)$, the following relation holds:

$$\sum g^{\lambda\mu^*}(\nabla_\lambda + \rho_{j\lambda})\nabla_\mu * \varphi_{j_{\tau_1\cdots\tau_p}\alpha_1^*\cdots\alpha_q^*} = \sum_m\sum_h\sum R^\sigma_{\tau_h\alpha_m^*}{}^{\mu^*} \times$$
$$\varphi_{j\tau_1\cdots(\sigma)_h\cdots\tau_p\alpha_1^*\cdots(\mu^*)_m\cdots\alpha_q^*} + \sum_m\sum[X^{\mu^*}_{\alpha_m^*} - R^{\mu^*}_{\alpha_m^*}] \times$$
$$\varphi_{j\tau_1\cdots\tau_p\alpha_1^*\cdots(\mu^*)_m\cdots\alpha_{q'}^*}, \tag{6}$$

where $X^{\mu^*}_{\alpha^*} = \sum g^{\lambda\mu^*}X_{\lambda\alpha^*}, R^{\mu^*}_{\alpha^*} = \sum g^{\lambda\mu^*}\bar R_{\lambda^*\alpha}$ and where $(\sigma)_h$ or $(\mu^*)_m$ indicates that τ_h or α_m^* is to be replaced by σ or μ^* respectively. We note here that $\partial\bar\partial\log a_j = \partial\bar\partial\log a_k$ in $U_j \cap U_k$, so

$$\sum X_{\lambda\alpha^*}dz^\lambda d\bar z^\alpha = \partial\bar\partial\log a_j \tag{7}$$

is a well defined 2-form on the whole of V. Now we set

$$\xi_{\mu^*} = \frac{1}{a_j}\sum \nabla_{\mu^*}\varphi_{j\tau_1\cdots\tau_p\alpha_1^*\cdots\alpha_q^*}\cdot\bar\varphi_j^{\tau_1\cdots\tau_p\alpha_1^*\cdots\alpha_q^*}$$

where

$$\bar\varphi_j^{\tau_1\cdots\tau_p\alpha_1^*\cdots\alpha_q^*} = \sum g^{\tau_1\sigma_1^*}\cdots g^{\tau_p\sigma_p^*}g^{\beta_1\alpha_1^*}\cdots g^{\beta_q\alpha_q^*}\overline{\varphi_{j\sigma_1\cdots\sigma_p\beta_1^*\cdots\beta_q^*}}.$$

---

[6] See Garabedian, P. R., and Spencer, D. C., "A Complex Tensor Calculus for Kähler Manifolds," *Acta Math.*, 89, 279–331 (1953).

Then $\xi = \sum \xi_\tau * d\bar{z}^\tau$ is a well defined 1-form on V and

$$\begin{aligned} -\delta\xi &= \sum g^{\lambda\mu^*} \nabla_\lambda \xi_{\mu^*} \\ &= \# + \frac{1}{a_j} \sum g^{\lambda\mu^*} (\nabla_\lambda + \rho_{j\lambda}) \nabla_{\mu^*} \varphi_{j\tau_1\cdots\tau_p\alpha_1^*\cdots\alpha_q^*} \cdot \bar{\varphi}_j^{\tau_1\cdots\tau_p\alpha_1^*\cdots\alpha_q^*}, \end{aligned} \tag{8}$$

where

$$\# = \frac{1}{a_j} \sum g^{\lambda\mu^*} \nabla_{\mu^*} \varphi_{j\tau_1\cdots\tau_p\alpha_1^*\cdots\alpha_q^*} \cdot \overline{\nabla_\lambda^* \varphi_j^{\tau_1\cdots\tau_p\alpha_1^*\cdots\alpha_q^*}} \geqq 0.$$

Denoting by g the determinant $g = |g_{\alpha\beta^*}|$, we have $\int_V \delta\xi \cdot g dV = 0$, where $dV = (i)^n dz^1 d\bar{z}^1 \cdots dz^n d\bar{z}^n$. Combining (6) with (8), we obtain therefore the inequality

$$\begin{aligned} 0 \geqq \int_V \Big\{ & q \sum (\delta_\tau^\sigma [X_{\alpha^*}^{\beta*} - R_{\alpha^*}^{\beta^*}] + p R_{\tau\alpha^*}^{\sigma}{}^{\beta^*}). \\ & \frac{1}{a_j} \varphi_{j\sigma\tau_2\cdots\tau_p\beta^*\alpha_2^*\cdots\alpha_q^*} \bar{\varphi}_j^{\tau\tau_2\cdots\tau_p\alpha^*\alpha_2^*\cdots\alpha_q^*} \Big\} g dV. \end{aligned} \tag{9}$$

By a (p,q)-form we shall mean a $(p+q)$-form of type (p,q). A real $(1,1)$-form ψ on V can be written in the form $\psi = i \sum \psi_{\alpha\beta}^*(z,\bar{z}) dz^\alpha d\bar{z}^\beta$, where $\psi_{\alpha\beta^*} = \psi_{\alpha\beta^*}(z,\bar{z})$ satisfy $\bar{\psi}_{\alpha\beta^*} = \psi_{\beta\alpha^*}$. *We say that such a form ψ is positive and write $\psi > 0$ if the Hermitian form $\sum \psi_{\alpha\beta^*}(z,\bar{z}) \bar{u}^\alpha u^\beta$ in n variables $u^1, u^2, \ldots, u^n$ is positive definite at each point z on V.*

Now, let $c(F)$ be the characteristic class of the principal bundle associated with F.

Lemma. *Let $\{a_j\}$ be a system of functions satisfying* (1). *Then the real d-closed* (1, 1)*-form $\gamma = (i/2\pi)\partial\bar{\partial} \log a_j$ belongs*[7] *to the characteristic class $c(F)$ of the bundle F. Conversely, given a d-closed real* (1,1)*-form γ of class C^∞ belonging to the characteristic class $c(F)$, there exists a system $\{a_j\}$ of positive functions a_j of class C^∞ satisfying* (1) *such that $(i/2\pi)\partial\bar{\partial}\log a_j = \gamma$.*

Proof. Let π be the canonical projection of F onto V and let ζ_j be the linear coordinate on the fibre $\pi^{-1}(z), z \in U_j$, such that $\zeta_j = f_{jk}(z)\zeta_k$ for $z \in U_j \cap U_k$. The principal bundle F^* associated with F is obtained from F by removing the origin $\zeta_j = 0$ of each fibre $\pi^{-1}(z)$. Now

$$\Phi = \frac{1}{2\pi i}(d\log\zeta_j - \partial\log a_j)$$

is a well defined 1-form of class C^∞ on the principal bundle F^* and

$$-d\Phi = \frac{i}{2\pi}\partial\bar{\partial}\log a_j,$$

while the restriction of Φ to the fibre $\pi^{-1}(z)^* = \pi^{-1}(z) - (0)$ of F^* equals the d-closed 1-form $(1/2\pi i)d\log\zeta_j$ which represents the basic cohomology class of $\pi^{-1}(z)^*$. This shows that $\gamma = (i/2\pi)\partial\bar{\partial}\log a_j$ belongs to the characteristic class $c(F)$ of F^*.

[7] In view of de Rham's theorem, any cohomology class may be regarded as a class of d-closed forms. By a d-closed form we shall mean a form which is closed under d.

To prove that every γ belonging to $c(F)$ can be written in the form $\gamma = (i/2\pi)\partial\bar{\partial}\log a_j$, we take a system $\{a'_j\}$ satisfying (1) and consider the difference $\gamma_0 = \gamma - (i/2\pi)\partial\bar{\partial}\log a'_j$. Obviously the harmonic part $H\gamma_0$ of γ_0 vanishes and consequently $\gamma_0 = \Delta G\gamma_0 = 2\bar{\partial}\delta G\gamma_0$, where G denotes the Green's operator. Using the formula[8] $\Lambda\partial - \partial\Lambda = -i\delta$, we get therefore $\gamma_0 = i2\partial\bar{\partial}\Lambda G\gamma_0$. Now $r = 2\Lambda G\gamma_0$ is a real scalar function of class C^∞ defined on the whole of V and, hence, setting $a_j = a'_j \cdot e^{2\pi r}$ we obtain a system $\{a_j\}$ of positive functions a_j of class C^∞ satisfying (1) and also $(i/2\pi)\partial\bar{\partial}\log a_j = \gamma$.

Now, letting $\gamma = (i/2\pi)\sum X_{\alpha\beta^*}dz^\alpha d\bar{z}^\beta$ be an arbitrary real $(1,1)$-form, we introduce the Hermitian form

$$\begin{aligned}\theta^{p,q}(\gamma, u; z) &= \sum(\delta^\sigma_\tau[X^{\beta^*}_{a^*} - R^{\beta^*}_{\alpha^*}] + pR^\sigma_{\tau\alpha^*}{}^{\beta^*}) \times \\ &\qquad u_{\sigma\tau_2\cdots\tau_p\beta^*\alpha^*_2\cdots\alpha^*_q}\bar{u}^{\tau\tau_2\cdots\tau_p\alpha^*\alpha^*_2\cdots\alpha^*_q}\end{aligned}$$

in a skew-symmetric tensor $u_{\tau_1\tau_2\cdots\tau_p\alpha^*_1\cdots\alpha^*_q}$ at each point z on V, provided that $q \geqq 1$. Then the inequality (9) can be written in the following form:

$$\int_V \frac{g}{a_j}\theta^{p,q}(\gamma, \varphi_j; z)dV \leqq 0, \quad \text{for } q \geqq 1. \tag{10}$$

With the help of the above lemma, we derive from (10) the following theorem:

Theorem 1. *If the characteristic class $c(F)$ of F contains a real d-closed* (1,1)-*form $\gamma = (i/2\pi)\sum X_{\alpha\beta^*}dz^\alpha d\bar{z}^\beta$ which is sufficiently large with respect to the order $>$ in the sense that the Hermitian form $\theta^{p,q}(\gamma, u; z)$ is, positive definite at each point z on V, then the cohomology groups $H^q(V, \Omega^p(F))$ and $H^{n-q}(V, \Omega^{n-p}(-F))$ both vanish, provided that $1 \leqq q \leqq n$.*

Proof. In view of (2) and (3), it is sufficient to show that $H^{p,q}(F) = 0$. By virtue of the above lemma, we may choose the system $\{a_j\}$ satisfying (1) such that $(i/2\pi)\partial\bar{\partial}\log a_j = \gamma$. Then the inequality (10) holds for any $\varphi = \{\varphi_j\} \in H^{p,q}(F)$, while, by hypothesis, $\theta^{p,q}(\gamma, \varphi_j; z) > 0$ unless $\varphi_j = 0$. This proves that $H^{p,q}(F) = 0$, q. e. d.

In the case where $p = n$, we have

$$\theta^{n,q}(\gamma, u; z) = n!\sum X^{\beta^*}_{\alpha^*}u_{12\cdots n\beta^*\alpha^*_1\cdots\alpha^*_q}\bar{u}^{12\cdots n\alpha^*\alpha^*_2\cdots\alpha^*_q}.$$

Hence $\theta^{n,q}(\gamma, u; z)$ is positive definite if and only if $\gamma > 0$. As a corollary to the above theorem, we therefore obtain the following:

Theorem 2. *If $c(F)$ contains a real d-closed* (1,1)-*form $\gamma > 0$, then the cohomology groups $H^q(V, \Omega^n(F))$ and $H^{n-q}(V, \Omega^0(-F))$ both vanish for $1 \leqq q \leqq n$.*

By the *canonical bundle* over V we shall mean the complex line bundle K defined by the system $\{J_{jk}\}$ of Jacobians

$$J_{jk} = \frac{\partial(z^1_k, \ldots, z^n_k)}{\partial(z^1_j, \ldots, z^n_j)},$$

[8] Garabedian and Spencer, *loc. cit.*, p. 290.

where $(z_j^1, \dots, z_j^n)$ is the local coordinates in U_j. We note that $-c(K)$ equals the first Chern class of the variety V. Since the isomorphism

$$\Omega^0(F) \cong \Omega^n(F - K)$$

holds in an obvious manner, we get from the above theorem the following:

Theorem 3. *If* $c(F) - c(K)$ *contains a real d-closed* (1,1)*-form* $\gamma > 0$, *then the cohomology group* $H^q(V, \Omega^0(F))$ *vanishes for* $1 \leqq q \leqq n$.

Editor's Comments

When Riemann studied abelian functions and Riemann surfaces in the 1850s, he proved that there is a canonical one-to-one correspondence between plane algebraic curves and compact Riemann surfaces. As mentioned before, every smooth projective subvariety of the complex projective space with the induced metric from the Fubini-Study metric is a Kähler manifold. One important and natural question is when a compact Kähler manifold arises from an algebraic variety in this way. Kodaira's embedding theorem gives a sufficient and necessary condition for a compact Kähler manifold to be embedded as a closed submanifold of a complex projective space, and Chow's theorem implies that the submanifold is a projective algebraic variety. This is a beautiful combination of geometric analysis, topology, and algebraic geometry.

Complex Geometry from Riemann to Kähler-Einstein and Calabi-Yau
ALM 38, pp. 231–250

On Kähler Varieties of Restricted Type*

(An Intrinsic Characterization of Algebraic Varieties)

Kunihiko Kodaira

A compact complex analytic variety V is called a *Kähler variety of restricted type*[1] or a *Hodge variety*[2] if V carries a Kähler metric $ds^2 = 2\sum g_{\alpha\bar{\beta}}(dz_\alpha d\bar{z}_\beta)$ such that the associated exterior form $\omega = i\sum g_{\alpha\bar{\beta}} dz_\alpha d\bar{z}_\beta$ belongs to the cohomology class of an *integral* 2-cocycle on V. In what follows such a metric will be called a *Hodge metric.* The main purpose of the present paper is to prove that *every Hodge variety is (bi-regularly equivalent to) a non-singular algebraic variety imbedded in a projective space.*[3] Since every non-singular algebraic variety in a projective space carries a Hodge metric, our main result may also be stated in the following form:

A compact complex analytic variety V is (bi-regularly equivalent to) a nonsingular algebraic variety imbedded in a projective space if and only if V can carry a Hodge metric.

This paper[4] is divided into four sections. In Section 1 we give a summary of some known results concerning complex line bundles over Kähler varieties. Section 2 is concerned with quadratic transformations. The proof of our main theorem (Theorem 4) is given in the following Section 3. The final Section 4 is devoted to several applications of our main theorem. In particular, we prove that *a compact*

---

* This work was supported by a research project at Princeton University sponsored by the Office of Ordnance Research, U. S. Army. (Received March 18, 1954) K. Kodaira, On Kähler varieties of restricted type (an intrinsic characterization of algebraic varieties). Ann. of Math. 60, (1954) 28–48.

1 W. V. D. Hodge, *A special type of Kähler manifolds*, Proc. London Math. Soc., 1 (1951), pp. 104–117.

2 A. Weil, *On Picard varieties*, Amer. J. Math., 74 (1952), pp. 865–894.

3 This gives, in some sense, an answer to a problem proposed by Hodge. See W. V. D. Hodge, *The topological invariants of algebraic varieties*, Proc. International Congress of Math., 1 (1950), pp. 182-192, §8.

4 The results of the present paper have been announced in Proc. Nat. Acad. Sci., U. S. A., Vol. 40 (1954), pp. 313–316.

complex analytic variety is a non-singular algebraic variety imbedded in a projective space if it carries a Hermitian metric whose Ricci curvature is everywhere negative [or positive] definite (Theorem 5). We also derive the following result: *Let $\mathfrak{B}$ be a bounded domain in the space of n complex variables and let Δ be a discontinuous group of analytic automorphisms without fixed points of $\mathfrak{B}$ such that $\mathfrak{B}/\Delta$ is compact. Then the factor space $\mathfrak{B}/\Delta$ is a non-singular algebraic variety imbedded in a projective space* (Theorem 6).

The author wishes to express his sincere thanks to A. Borel for valuable suggestions given during the period of the preparation of this paper.

1 Preliminaries

Let V be a compact Kähler variety of complex dimension $n \geqq 2$. By a *complex line bundle* F over V we shall mean an analytic fibre bundle whose fibre is a complex line C and whose structure group is the multiplicative group C^* of complex numbers acting on C. The bundle F may be described as follows: Let $\{U_j\}$ be a sufficiently fine finite covering of V and let π be the canonical projection of F onto V. Then the inverse image $\pi^{-1}(U_j)$ has a product representation $\pi^{-1}(U_j) = U_j \times C$, and $(z, \zeta_j) \in U_j \times C$ is identical with $(z, \zeta_k) \in U_k \times C$ if and only if $\zeta_j = f_{jk}(z) \cdot \zeta_k$, where $f_{jk}(z)$ is a non-vanishing holomorphic function defined in $U_j \cap U_k$. Under these circumstances, we say that the bundle F is defined by the system $\{f_{jk}\}$ of the *transition functions* f_{jk}, and we call ζ_j the *fibre coordinate* of the point (z, ζ_j) on F *over the neighborhood* U_j. We identify two complex line bundles which are analytically equivalent. Then the set $\mathfrak{F} = \{F\}$ of all complex line bundles F forms an additive group.[5] By the *characteristic class* $c(F)$ of a complex line bundle F over V we shall mean the characteristic class of the principal bundle associated with F. The characteristic class $c(F)$ is an element of the second cohomology group $H^2(V, Z)$ of V with coefficients in the ring Z of rational integers. Moreover, $F \to c(F)$ is a homomorphism of the additive group $\mathfrak{F}$ into $H^2(V, Z)$. Let R be the real number field. For any element

$$c \in H^2(V, Z)$$

we denote by c_R the element of $H^2(V, R)$ corresponding to c under the homomorphism $H^2(V, Z) \to H^2(V, R)$ induced by the inclusion map $Z \to R$. In view of de Rham's theorem, $c_R \in H^2(V, R)$ may be regarded as a class of closed real 2-forms on V. By a (p, q)-form we shall mean a $(p+q)$-form of type (p, q). In what follows *we assume every form to be of class* C^∞ without mentioning it. In terms of a system of local coordinates $(z_1, \dots, z_n)$ on V, a real (1, 1)-form γ on V is written in the form $\gamma = i \sum \gamma_{\alpha\bar{\beta}}(z, \bar{z}) dz_\alpha d\bar{z}_\beta$, where the $\gamma_{\alpha\bar{\beta}} = \gamma_{\alpha\bar{\beta}}(z, \bar{z})$ satisfy $\bar{\gamma}_{\beta\bar{\alpha}} = \gamma_{\alpha\bar{\beta}}$.

Theorem 1.[6] *Given an element $c \in H^2(V, Z)$, there exists a complex line bundle F over V with $c(F) = c$ if and only if the cohomology class c_R contains a*

[5] K. Kodaira and D. C. Spencer, *Groups of complex line bundles over compact Kähler varieties*, Proc. Nat. Acad. Sci. U. S. A., 39 (1953), pp. 868–872.

[6] Kodaira and Spencer, *loc. cit.*, p. 870; A. Weil, Fibre spaces in algebraic geometry (mimeographed notes), The University of Chicago, 1952, §5.

closed real (1, 1)-*form* γ.

We say that a real (1, 1)-form $\gamma = i\sum \gamma_{\alpha\bar{\beta}}(z,\bar{z})dz_\alpha d\bar{z}_\beta$ on V is *positive* and write $\gamma > 0$ if the Hermitian form $\sum \gamma_{\alpha\bar{\beta}}(z,\bar{z})u_\alpha \bar{u}_\beta$ in n variables $u_1,\ldots,u_n$ is positive definite at each point z on V. Moreover, we say that *a complex line bundle F is positive if its characteristic class $c_R(F)$ contains a closed real* (1, 1)-*form* $\gamma > 0$. By the *canonical bundle* K over V will be meant the complex line bundle of n-forms over V. We note that the characteristic class $c(-K)$ of $-K$ is equal to the first Chern class of V.| Now we denoted by $\Omega(F)$ the *sheaf* (faisceau) of germs of holomorphic sections of the bundle F. Then we have the following

Theorem 2.[7] *The cohomology group $H^q(V,\Omega(F))$ vanishes for $q \geqq 1$ if the bundle $F-K$ is positive in the sense that $c_R(F-K)$ contains a closed real* (1, 1)-*form* $\gamma > 0$.

2 Quadratic transformations[8]

Let p be a point on V and let $(z_1,\ldots,z_n)$ be a system of local coordinates with the center p. We consider the direct product $U_p \times S$ of a neighborhood U_p of p and a projective space S of dimension $n-1$ and, by means of the system of homogenous coordinates $(t_1,t_2,\ldots,t_n)$ on S, we construct the subvariety

$$W_p = \{(z,t)|t_\alpha z_\beta - t_\beta z_\alpha = 0; \quad \alpha,\beta = 1,\ldots,n\}$$

of $U_p \times S$, where $z=(z_1,\ldots,z_n), t=(t_1,\ldots,t_n)$. As one readily infers, W_p is an analytic manifold without singularities and contains the projective space $S = p\times S$. Moreover

$$P'_p; (z,t) \to z$$

is a regular mapping of W_p onto U_p which maps S onto p and $W_p - S$ bi-regularly onto $U_p - p$. Hence, replacing U_p by W_p, we get from V a new analytic variety $\tilde{V} = (V-U_p)\cup W_p$ having the following properties: *There exists a regular mapping P_p of $\tilde{V}$ onto V such that $S = P_p^{-1}(p)$ is a projective space of dimension $n-1$ and P_p maps $\tilde{V} - S$ bi-regularly onto $V - p$.* The inverse mapping $Q_p = P_p^{-1}$ of P_p is called the *quadratic transformation with the center p*. The *quadratic transform* $\tilde{V} = Q_p(V)$ is determined uniquely by p and V and is independent of the choice of the system of local coordinates $(z_1,\ldots,z_n)$ appearing in the above construction. Every complex line bundle F over V induces a complex line bundle $\tilde{F} = Q_pF$ over $\tilde{V} = Q_p(V)$ in a canonical manner.[9] Obviously the restriction $\tilde{F}_s$ of $\tilde{F}$ to S is a trivial bundle $S\times C$ over S. Letting $P_p^* : H^2(V,Z) \to H^2(\tilde{V},Z)$ be the homomorphism induced by the mapping P_p, we have[10] $c(\tilde{F}) = P_p^* c(F)$. Consequently, if a closed real (1,

[7] K. Kodaira, *On a differential-geometric method in the theory of analytic stacks*, Proc. Nat. Acad. Sci., U. S. A., 39 (1953), pp. 1268–1273, Theorem 3.

[8] See F. Hirzebruch, *Über vierdimensionale Riemannsche Flächen mehrdeutiger analytischer Funktionen von zwei komplexen Veränderlichen*, Math. Ann. 126 (1953), pp. 1–22.

[9] See N. Steenrod, The topology of fibre bundles, Princeton University Press, (1951), pp. 47–48.

[10] Steenrod, *loc. cit.*, p. 179.

1)-form $\gamma = \gamma(z)$ belongs to $c_R(F)$, then the induced form $\tilde{\gamma} = \tilde{\gamma}(\tilde{z}) = \gamma(P_p(\tilde{z}))$ belongs to $c_R(\tilde{F})$, where $\tilde{z}$ denotes a "variable" point on $\tilde{V}$.

Now, let $\{S\}$ be the complex line bundle over $\tilde{V}$ determined by the divisor[11] S and let $\{e\}$ be the complex line bundle over S determined by the hyperplane e on S. Then we have

$$\{S\}_S = -\{e\}, \tag{1}$$ [12]

where $\{S\}_S$ denotes the restriction of the bundle $\{S\}$ to S. This relation can be proved as follows: Let $(z, t) = (z_1, \dots, z_n, t_1, \dots, t_n)$ be the coordinates of a variable point $w \in W_p \subset \tilde{V}$ in the product space $U_p \times S$. Then each z_ν or t_λ/t_ν may be regarded as a holomorphic or meromorphic function of w and moreover the identities $z_\lambda/z_\nu = t_\lambda/t_\nu$ hold. Now let e_ν be the hyperplane $t_\nu = 0$ on S, and let S_ν be the open subset $S - e_\nu$ of S. Since the subvariety S of W_p is defined by the local equation $z_\nu = 0$ at each point on S_ν, the bundle $\{S\}_S$ is determined by the system $\{h_{\lambda\nu}\}$ of non-vanishing holomorphic functions $h_{\lambda\nu} = (z_\lambda/z_\nu)_S = t_\lambda/t_\nu$ defined on $S_\lambda \cap S_\nu$, where $()_S$ denotes restriction to S. On the other hand, the local equation of the hyperplane e_1 in S_ν is given by $t_1 t_\nu^{-1} = 0$ and therefore the bundle $\{e_1\}$ is determined by the system $\{h_{\lambda\nu}^{-1}\}$ of functions $h_{\lambda\nu}^{-1} = (t_1 t_\lambda^{-1})/(t_1 t_\nu^{-1})$. Hence we obtain (1).

In this connection we remark that

$$dz_1 dz_2 \cdots dz_n = z_1^{n-1} dz_1 d\left(\frac{t_2}{t_1}\right) \cdots d\left(\frac{t_n}{t_1}\right), \tag{2}$$

and $(z_1, t_2/t_1, \dots, t_n/t_1)$ may be used as the system of local coordinates at each point on S_1. Since the transformation Q_p is bi-regular outside p, we infer from (2) that the canonical bundle $K(\tilde{V})$ over $\tilde{V}$ is related to the canonical bundle K over V by the equality

$$K(\tilde{V}) = \tilde{K} + (n-1)\{S\}, \tag{3}$$

where $\tilde{K} = Q_p K$.

Lemma 1. *There exists a real positive* $(1, 1)$*-form* γ_0 *on* V *with the following property: If a complex line bundle* F *over* V *is sufficiently positive in the sense that* $c_R(F)$ *contains a real* $(1, 1)$*-form* $\gamma > \gamma_0$,[13] *then the bundle* $\tilde{F} - \{S\}$ *over* $\tilde{V} = Q_p(V)$ *is positive for every point* $p \in V$, *where* $\tilde{F} = Q_p F$ *and* $S = Q_p(p)$.

Proof. Let $(z, t) = (z_1, \dots, z_n, t_1, \dots, t_n)$ have the same meaning as above and set $|z|^2 = |z_1|^2 + |z_2|^2 + \cdots + |z_n|^2$. Take a non-increasing function $\rho(r)$ of class C^∞ in $r = |z|^2$ such that

$$\rho(r) = \begin{cases} 1, & \text{for } r \leqq \varepsilon^2, \\ 0, & \text{for } r \geqq (2\varepsilon)^2, \end{cases}$$

ε being a sufficiently small positive number, and set

$$\sigma = \sigma(z) = \frac{i}{2\pi} \partial\bar{\partial}\{\rho(|z|^2) \log |z|^2\}, \quad \text{for } |z| > 0.$$

---

[11] K. Kodaira and D. C. Spencer, *Divisor class groups on algebraic varieties*, Proc. Nat. Acad. Sci., U. S. A., 39 (1953), pp. 872-877.

[12] This formula is due to F. Hirzebruch.

[13] We write $\gamma > \gamma_0$ if $\gamma - \gamma_0 > 0$.

σ is obviously a closed real (1, 1)-form of class C^∞ defined in $U_p - p$ and σ vanishes for $|z| \geqq 2\varepsilon$. Hence we can find a real positive (1, 1)-form γ_0 on V satisfying the condition

$$\gamma_0(z) + \sigma(z) > 0, \quad \text{for } |z| \geqq \varepsilon. \tag{4}$$

Letting $w = (z, t)$, we infer that the induced form $\tilde{\sigma}(w) = \sigma(z)$ on $W_p - S$ has the expression

$$\tilde{\sigma}(w) = \omega(t) \quad \text{for } |z| < \varepsilon,$$

where

$$\omega(t) = \frac{i}{2\pi} \partial\bar{\partial} \log \left(1 + \sum_{\nu=2}^{n} \left| \frac{t_\nu}{t_1} \right|^2 \right)$$

is the fundamental form on S associated with the "standard" Kähler metric. This shows that $\tilde{\sigma}$ is (extended to) a real (1, 1)-form of class C^∞ on the whole of W_p. Moreover, since W_p is a non-singular subvariety of $U_p \times S$, we have

$$\tilde{\gamma}_0(w) + \tilde{\sigma}(w) = \gamma_0(z) + \omega(t) > 0 \quad \text{for } |z| < \varepsilon, \tag{5}$$

where $\tilde{\gamma}_0$ denotes the form on $\tilde{V}$ induced by γ_0. Now we extend $\tilde{\sigma}$ to the whole of $\tilde{V}$ by setting $\tilde{\sigma} = 0$ on $\tilde{V} - W_p$. Then we infer from (4) and (5) that

$$\tilde{\gamma}_0 + \tilde{\sigma} > 0 \quad \text{on } \tilde{V}. \tag{6}$$

In order to show that $\tilde{\sigma}$ belongs to the cohomology class $c_R(-\{S\})$ on $\tilde{V}$, we consider the closed neighborhood $W' = \{w | w = (z, t), |z| \leqq 3\varepsilon\}$ of S and observe that S is a deformation retract of W'. It follows that the inclusion map $S \subset W'$ induces an isomorphism $H^2(W') \cong H^2(S)$, where the cohomology refers in this instance to the reals R. On the other hand, since the boundary $\partial W'$ of W' is a $(2n-1)$-sphere, where $n \geqq 2$, we get the isomorphism

$$H^2(W') \cong H^2(W', \partial W'),$$

while we have the canonical homomorphism $H^2(W', \partial W') \to H^2(\tilde{V})$ induced by the inclusion map $W' \subset \tilde{V}$. Thus we infer that, for any closed 2-form η with support contained in W', the cohomology class of η *on* $\tilde{V}$ is uniquely determined by the cohomology class *on* S of the restriction η_S of η to S. Now, take a closed 2-form $\sigma' \in c_R(\{S\})$ whose support is contained in W'. Then it follows from (1) that $\sigma'_S \in c_R(-\{e\})$ on S, while $\tilde{\sigma}_S = \omega$ is the fundamental form on S and therefore $\tilde{\sigma}_S \in c_R(\{e\})$. Hence $\tilde{\sigma}_S$ is homologous to $-\sigma'_S$ on S and consequently $\tilde{\sigma}$ is homologous to $-\sigma'$ on $\tilde{V}$. Thus we obtain

$$\tilde{\sigma} \in c_R(-\{S\}). \tag{7}$$

Now, since $\gamma \in c_R(F)$ implies $\tilde{\gamma} \in c_R(\tilde{F})$, we infer from (7) that the cohomology class $c_R(\tilde{F} - \{S\})$ contains the closed real (1, 1)-form $\tilde{\gamma} + \tilde{\sigma}$, while (6) proves the inequality $\tilde{\gamma} + \tilde{\sigma} > 0$, provided that $\gamma > \gamma_0$ on V. Consequently *the bundle* $\tilde{F} - \{S\}$ *is positive on* $\tilde{V}$ *if* $\gamma > \gamma_0$ *on* V, *where* γ_0 *is a real positive* (1, 1)-*form on* V *satisfying* (4). This condition (4) for γ_0 depends on the point p, the number ε, and the choice of the system of local coordinates $(z_1, \ldots, z_n)$ with the center p.

Let $\mathfrak{O}$ be a coordinate neighborhood on V with the system of local coordinates $(z_1, \ldots, z_n)$ and let U be an open subset with the closure $[U] \subset \mathfrak{O}$. For any point $p = (a_1, \ldots, a_n)$ in U, we use $(z_1 - a_1, \ldots, z_n - a_n)$ as the system of local coordinates with the center p. Then, for sufficiently small $\varepsilon > 0$, the condition (4) at p in U is written in the *form*

$$\gamma_0(z) + \sigma(z - a) > 0 \quad \text{for } |z - a| \geqq \varepsilon, \tag{4$'$}$$

and V is covered by a finite number of neighborhoods like U. Hence we can find a real positive (1, 1)-form γ_0 satisfying the condition (4)$'$ at each point p on V. This completes our proof of Lemma 1.

For any pair of points $p, q, p \neq q$, on V, we write $Q_p Q_q$ for $Q_{Q_q(p)} Q_q$. Then $Q_p Q_q = Q_q Q_p$. Obviously $S = Q_p Q_q(p)$ and $T = Q_p Q_q(q)$ are both $(n-1)$-dimensional projective spaces lying on $\tilde{V} = Q_p Q_q(V)$. Now, we have

Lemma 2. *Let γ_0 be the positive* (1, 1)*-form on V introduced in the proof of Lemma* 1 *above. If the characteristic class $c_R(F)$ of a complex line bundle F over V contains a closed real* (1, 1)*-form $\gamma > 2\gamma_0$, then the bundle $\tilde{F} - \{S\} - \{T\}$ over $\tilde{V} = Q_p Q_q(V)$ is positive for any pair of points $p, q (p \neq q)$ on V, where $\tilde{F} = Q_p Q_q F$ and $S = Q_p Q_q(p), T = Q_p Q_q(q)$.*

Proof. We set $V' = Q_p(V), S' = Q_p(p)$. Moreover, letting σ have the same meaning as in the proof of Lemma 1, we set

$$\sigma_p(z) = \sigma(z - a), \quad \sigma_q(z) = \sigma(z - b),$$

where $a = (a_1, \ldots, a_n)$ or $b = (b_1, \ldots, b_n)$ denotes the local coordinate of p or q, respectively. Then, as was shown above, the form $\sigma'_p = P^*_p \sigma_p$ induced on $V' - S'$ is (extended to) a form of class C^∞ on the whole of V' and $\gamma'_0 + \sigma'_p$ is positive on V', where $\gamma'_0 = P^*_p \gamma_0$. It follows that $\tilde{\gamma}_0 + \tilde{\sigma}_p = P^*_q(\gamma'_0 + \sigma'_p)$ is positive on $\tilde{V} - T$. Similarly we infer that $\tilde{\gamma}_0 + \tilde{\sigma}_q = P^*_p P^*_q(\gamma_0 + \sigma_q)$ is positive on $\tilde{V} - S$, while $\tilde{\gamma}_0, \tilde{\sigma}_p, \tilde{\sigma}_q$ are forms of class C^∞ on the whole of $\tilde{V}$ and moreover γ_0 is positive on V. Hence we get the inequality

$$2\tilde{\gamma}_0 + \tilde{\sigma}_p + \tilde{\sigma}_q > 0 \quad \text{on} \quad \tilde{V}.$$

On the other hand, since $\sigma'_p \in c_R(-\{S'\})$ on V', we infer that

$$\tilde{\sigma}_p = P^*_q \sigma'_p \in c_R(-\{S\})$$

on $\tilde{V} = Q_q(V')$. Similarly we have $\tilde{\sigma}_q \in c_R(-\{T\})$ on $\tilde{V}$. Consequently we obtain

$$c_R(\tilde{F} - \{S\} - \{T\}) \ni \tilde{\gamma} + \tilde{\sigma}_p + \tilde{\sigma}_q > 0,$$

provided that $\gamma > 2\gamma_0$ on V, q.e.d.

3 Main theorem

In what follows we denote by $\Gamma(F)$ the linear space consisting of all holomorphic sections of the bundle F. Given a complex line bundle F over V such that

$\dim \Gamma(F) \geqq 2$, we associate with F a "meromorphic" mapping Φ_F of V into a projective space $\mathfrak{S}_d$ of dimension $d = \dim \Gamma(F) - 1$ in the following manner: Take a base $\{\varphi_0, \varphi_1, \ldots, \varphi_\lambda, \ldots, \varphi_d\}$ of the linear space $\Gamma(F)$ and, letting $\{U_j\}$ be a finite covering of V, denote by $\varphi_{\lambda j}(z)$ the fibre coordinate of $\varphi_\lambda(z)$ over U_j. Then, considering $(\varphi_{0j}(z), \ldots, \varphi_{\lambda j}(z), \ldots, \varphi_{dj}(z))$ as the homogeneous coordinates of a point in the projective space $\mathfrak{S}_d$, we have

$$(\varphi_{0j}(z), \ldots, \varphi_{dj}(z)) = (\varphi_{0k}(z), \ldots, \varphi_{dk}(z)) \text{ for } z \in U_j \cap U_k.$$

Hence we may define the mapping Φ_F of V into $\mathfrak{S}_d$ by

$$\Phi_F : z \to \Phi_F(z) = (\varphi_{0j}(z), \varphi_{1j}(z), \ldots, \varphi_{dj}(z)).$$

We suppose that the system of coordinates on $\mathfrak{S}_d$ is transformed *cogrediently* with the base $\{\varphi_0, \varphi_1, \ldots, \varphi_d\}$ of $\Gamma(F)$. Then *the mapping Φ_F is determined uniquely by F*. We note that this mapping Φ_F is not necessarily regular; however, *in case the linear space $\Gamma(F)$ is "sufficiently ample" in the sense that, for each point z on V, there exists at least one member $\varphi \in \Gamma(F)$ with $\varphi_j(z) \neq 0$, Φ_F is a regular mapping.*

Now, assuming V to be a Kähler variety whose fundamental form $\omega = i \sum g_{\alpha\bar\beta} dz_\alpha d\bar z_\beta$ belongs to the cohomology class of an integral 2-cocycle, we prove that V can be bi-regularly imbedded in a projective space. Choose a positive integer h_1 such that $h_1\omega > \gamma_0$, where γ_0 is the positive (1, 1)-form introduced in Lemma 1. By virtue of Theorem 1, there exists a complex line bundle F_1 over V such that $h_1\omega \in c_R(F_1)$. Choosing a closed real (1, 1)-form $\kappa \in c_R(K)$ and setting

$$\tau = (n + l)h_1\omega + \kappa, \tag{8}$$

where l is an integer $\geqq 2$, we show that *the mapping Φ_F of V into $\mathfrak{S}_d$ is locally bi-regular if $c_R(F)$ contains a closed real* (1, 1)*-form $\gamma > \tau$.*[14] To do this, take an arbitrary point p on V and form the quadratic transform $\tilde V = Q_p(V)$. Moreover let

$$E_m = \tilde F - m\{S\}, \quad \text{where } \tilde F = Q_p F, \quad S = Q_p(p).$$

Since the restriction $\tilde F_S$ is a trivial bundle over S, we infer from (1) that $(E_m)_S = \{me\}$ on S. Hence we obtain the exact sequence

$$0 \to \Omega(E_{m+1}) \xrightarrow{i} \Omega(E_m) \xrightarrow{r} \Omega(me) \to 0, \tag{9}$$

where $\Omega(me) = \Omega(\{me\})$ is the sheaf over S of germs of holomorphic sections of the bundle $\{me\}$ and where r is the restriction map to S. We note that the mapping i may be described as follows: Let $\{\tilde U_j\}$ be a sufficiently fine finite covering of $\tilde V$ and let $\{\tilde f_{jk}\}$ be the system of non-vanishing holomorphic functions $\tilde f_{jk}$ on $\tilde U_j \cap \tilde U_k$ defining the bundle $\tilde F$. Moreover let $s_j = 0$ be the local equation of S in $\tilde U_j$ (we

[14] Our method of proof is inspired by Castelnuovo-Enriques' argument concerning exceptional curves of the first kind on algebraic surfaces. G. Castelnuovo and F. Enriques, *Sopra alcune questioni fondamentali nella teoria delle superfficie algebriche*, Annali di Math., Ser. III, 6 (1901), pp. 165–225, Chap. III. See also 0. Zariski, Algebraic Surfaces, Berlin (1935), p. 72.

set $s_j = 1$ if $U_j \cap S$ is empty). Then the bundle E_m is defined by the system of functions $\{s_j^{-m} s_k^m \tilde{f}_{jk}\}$, and any germ $\eta \in \Omega(E_m)_p$ is represented by its fibre coordinate η_j on each neighborhood $\tilde{U}_j \ni p$, where

$$\eta_j = s_j^{-m} s_k^m \tilde{f}_{jk} \cdot \eta_k \quad \text{in } \tilde{U}_j \cap \tilde{U}_k.$$

By means of the fibre coordinates η_j of germs $\eta \in \Omega(E_{m+1})_p$ the mapping i is represented in the form

$$i : \eta_j \to s_j \eta_j. \tag{10}$$

Now, setting $\Gamma(E_m) = H^0(\tilde{V}, \Omega(E_m)), \Gamma_S(me) = H^0(S, \Omega(me))$, we obtain from the exact sequence (9) the corresponding exact cohomology sequence

$$0 \to \Gamma(E_{m+1}) \xrightarrow{i^*} \Gamma(E_m) \xrightarrow{r^*} \Gamma_S(me) \xrightarrow{\delta^*} H^1(\tilde{V}, \Omega(E_{m+1})) \to \cdots . \tag{11}$$

With the help of (3) we get

$$E_{m+1} - K(\tilde{V}) = (n+m)(\tilde{F}_1 - \{S\}) + \tilde{F}_2,$$

where

$$\tilde{F}_2 = Q_p F_2, \quad F_2 = F - (n+m)F_1 - K.$$

By Lemma 1, the bundle $\tilde{F}_1 - \{S\}$ over $\tilde{V}$ is positive, i.e., $c_R(\tilde{F}_1 - \{S\})$ contains a closed positive (1, 1)-form α on $\tilde{V}$, while $c_R(F_2)$ contains the closed real (1, 1)-form

$$\gamma_2 = \gamma - (n+m)h_1\omega - \kappa$$

on V. We have

$$\gamma_2 = \gamma - \tau + \ (l \ - m)h_1\omega > 0$$

provided that $m \leqq 1$. Letting $\tilde{\gamma}_2$ be the real (1, 1)-form induced on $\tilde{V}$ by the *positive* (1, 1)-form $\gamma_2 \in c_R(F_2)$ on V, we infer therefore that $c_R(E_{m+1} - K(\tilde{V}))$ contains the real (1, 1)-form $(n+m)\alpha + \tilde{\gamma}_2$, which is *positive* on $\tilde{V}$ if $1-n \leqq m \leqq l$. Thus the bundle $E_{m+1} - K(\tilde{V})$ is positive for $1-n \leqq m \leqq l$ and consequently, by Theorem 2,

$$H^1(\tilde{V}, \Omega(E_{m+1})) = 0 \quad \text{for} \quad 1-n \leqq m \leqq l.$$

Combining this with (11), we obtain

$$0 \to \Gamma(E_{m+1}) \xrightarrow{i^*} \Gamma(E_m) \xrightarrow{r^*} \Gamma_S(me) \xrightarrow{r^*} 0 \quad \text{for } 1-n \leqq m \leqq l. \tag{12}$$

Obviously the mapping $P_p = Q_p^{-1}$ of $\tilde{F} = Q_p F$ onto F induces an isomorphism P_p^* of $\Gamma(F)$ into $\Gamma(\tilde{F})$ in a canonical manner. Since the bundle $\tilde{F}$ is trivial over a neighborhood $W_p = Q_p(U_p)$ of S, we may assume that $\tilde{f}_{jk} = 1$ in $\tilde{U}_j \cap \tilde{U}_k \cap W_p$. Then in U_p the isomorphism $P_p^* : \varphi \to \tilde{\varphi} = P_p^* \varphi$ is given by

$$P_p^* : \varphi_p(Z) \to \tilde{\varphi}_j(\tilde{z}) = \varphi_p(P_p(\tilde{z})), \quad \tilde{z} \in \tilde{U}_j \cap W_p, \tag{13}$$

where $\varphi_p(z)$ or $\tilde{\varphi}_j(\tilde{z})$ is the fibre coordinate of $\varphi(z)$ or $\tilde{\varphi}(\tilde{z})$ over U_p or $\bar{U}_j$, respectively. Now we infer readily that P_p^* *maps* $\Gamma(F)$ *onto* $\Gamma(\tilde{F})$. In fact, let ψ be an arbitrary member of $\Gamma(\tilde{F})$ and let $\psi_j(\tilde{z})$ be the fibre coordinate of $\psi(\tilde{z})$ over $\bar{U}_j \cap W_p$. Then we have $\psi_j(\tilde{z}) = \psi_k(\tilde{z})$ in $\tilde{U}_j \cap \bar{U}_k \cap W_p$ and consequently $\psi_W(\tilde{z}) = \psi_j(\tilde{z}) = \psi_k(\tilde{z})$ is a well defined holomorphic function on W_p. The function $\psi_W(Q_p(z))$ is therefore holomorphic on $U_p - p$ and hence, by a theorem of Hartogs, $\psi_W(Q_p(z))$ *is a holomorphic function on the whole of* U_p. Since P_p maps $\tilde{V} - S$ bi-regularly onto $V - p$, there obviously exists a holomorphic section φ of F over $V - p$ such that $P_p^*\varphi = \psi$ over $\tilde{V} - S$. Letting $\varphi_p(z)$ be the fibre coordinate of $\varphi(z)$ over U_p for $z \in U_p - p$, we infer from (13) that $\varphi_p(z) = \psi_W(Q_p(z))$. Hence, by the above result, $\varphi_p(z)$ is (extended to) a holomorphic function on the whole of U_p, and consequently φ is (extended to) a holomorphic section of F over the whole of V. Thus we see that there exists a $\varphi \in \Gamma(F)$ such that $P_p^*\varphi = \psi$.

Now we have, as particular cases of (12),

$$\begin{cases} 0 \to \Gamma(E_1) \xrightarrow{i^*} \Gamma(E_0) \xrightarrow{r^*} C \to 0, \\ 0 \to \Gamma(E_2) \xrightarrow{i^*} \Gamma(E_1) \xrightarrow{r^*} \Gamma_S(e) \to 0, \end{cases} \tag{14}$$

since $\Gamma_S(0)$ is (isomorphic to) the space C of constants. We note that $\dim \Gamma_S(e) = n$ and that any base $\{t_1, t_2, \ldots, t_n\}$ of $\Gamma_S(e)$ may be used as a system of homogenous coordinates on S. For simplicity's sake we write $\{\psi_0, \psi_1, \ldots\} = \Gamma$ if $\{\psi_0, \psi_1, \ldots\}$ constitutes a base of the linear space Γ. The above formula (14) shows that bases of $\Gamma(E_0), \Gamma(E_1), \ldots$ may be chosen in the following manner:

$$\begin{aligned} \{\psi_0, i^*\psi_1, \ldots, i^*\psi_n, (i^*)^2\psi_{n+1}, \ldots, (i^*)^2\psi_d\} &= \Gamma(E_0), \\ \{\psi_1, \ldots, \psi_n, i^*\psi_{n+1}, \ldots, i^*\psi_d\} &= \Gamma(E_1), \\ \{\psi_{n+1}, \ldots, \psi_d\} &= \Gamma(E_2), \end{aligned} \tag{15}$$

where

$$r^*\psi_0 = 1, \tag{16}$$

$$\{r^*\psi_1, \ldots, r^*\psi_n\} = \Gamma_S(e). \tag{17}$$

In view of the isomorphism $P_p^* : \varphi \to \tilde{\varphi} = P_p^*\varphi$ of $\Gamma(F)$ onto $\Gamma(\tilde{F}) = \Gamma(E_0)$, we find a base $\{\varphi_0, \varphi_1, \ldots, \varphi_d\}$ of $\Gamma(F)$ such that

$$\tilde{\varphi}_0 = \psi_0, \quad \tilde{\varphi}_1 = i^*\psi_1, \ldots, \tilde{\varphi}_n = i^*\psi_n, \quad \tilde{\varphi}_{n+1} = (i^*)^2\psi_{n+1}, \ldots.$$

By means of this base $\{\varphi_0, \ldots, \varphi_d\}$, we show that the mapping $\Phi_F : V \to \mathfrak{S}_d$ is bi-regular in a neighborhood U_p of p. First, we have $\varphi_{0p}(z) = \psi_{0j}(Q_p(z))$ by (13), while (16) shows that $(\psi_{0j})_S = 1$. Hence we get $\varphi_{0p}(p) = 1$. Thus in a sufficiently small neighborhood U_p of p, $\varphi_{0p}(z)$ does not vanish and therefore the mapping

$$z \to \Phi_F(z) = (\varphi_{0p}(z), \varphi_{1p}(z), \ldots, \varphi_{dp}(z))$$

is regular in U_p. Moreover, since $(i^*\psi_\lambda)_j = s_j\psi_{\lambda j}$ by (10), we have

$$\varphi_{\lambda p}(z) = \begin{cases} s_j(\tilde{z})\psi_{\lambda j}(\tilde{z}) & \text{for } \lambda = 1, 2, \ldots, n, \\ s_j^2(\tilde{z})\psi_{\lambda j}(z) & \text{for } \lambda = n+1, n+2, \ldots, d, \end{cases} \tag{18}$$

where $z = P_p(\tilde{z}), \tilde{z} \in W_p$. This implies that

$$\varphi_{\lambda p}(p) = 0 \quad \text{for } \lambda = 1, 2, \ldots, d. \tag{19}$$

Setting $\xi_\lambda = \varphi_{\lambda p}/\varphi_{0p}$, we obtain therefore

$$\Phi_F(z) = (1, \xi_1(z), \ldots, \xi_d(z)),$$

where each $\xi_\lambda(z)$ is holomorphic in U_p and $\xi_\lambda(p) = 0$. To prove that Φ_F is biregular in a neighborhood of p, it is sufficient to show that the Jacobian

$$J(z) = \frac{\partial(\xi_1, \ldots, \xi_n)}{\partial(z_1, \ldots, z_n)}$$

does not vanish at p, where $(z_1, \ldots, z_n)$ is the system of local coordinates in U_p with the center p. Let $(t_1, \ldots, t_n)$ be the system of homogenous coordinates on S corresponding to $(z_1, \ldots, z_n)$. By virtue of (17) we may assume that $(\psi_{\lambda j}/\psi_{\nu j})_S = t_\lambda/t_\nu$ for $\lambda, \nu = 1, 2, \ldots, n$. Moreover (17) implies that $(\psi_{1j})_s$ does not vanish identically on $S \cap \bar{U}_j$. Take a point $\tilde{q}$ on $S \cap \tilde{U}_j$ such that $\psi_{1j}(\bar{q}) \neq 0$ and set $s_1 = s_j\psi_{1j}, w_\lambda = \psi_{\lambda j}/\psi_{1j} (\lambda = 2, 3, \ldots, n)$. Obviously $s_1 = 0$ gives a minimal local equation of S at $\tilde{q}$, while $(w_\lambda)_S = t_\lambda/t_1$. Therefore n holomorphic functions $s_1, w_2, \ldots, w_n$ in $\tilde{U}_j$ are functionally independent at $\tilde{q}$. Moreover we infer from (18) that

$$d\varphi_{1p} d\varphi_{2p} \cdots d\varphi_{np} = s_1^{n-1} ds_1 dw_2 \cdots dw_n.$$

On the other hand, setting $v_\lambda = t_\lambda/t_1 (\lambda = 2, \ldots, n)$, we have

$$dz_1 dz_2 \cdots dz_n = z_1^{n-1} dz_1 dv_2 \cdots dv_n,$$

as (2) shows, where $z_1, v_2, \ldots, v_n$ are also functionally independent at $\tilde{q}$. Using (19) and the equality $\varphi_{0p}(p) = 1$, we obtain therefore

$$J(p) = \left.\frac{\partial(\varphi_{1p}, \ldots, \varphi_{np})}{\partial(z_1, \ldots, z_n)}\right|_p = \left(\frac{s_1}{z_1}\right)_{\tilde{q}}^{n-1} \cdot \left.\frac{\partial(s_1, w_2, \ldots, w_n)}{\partial(z_1, v_2, \ldots, v_n)}\right|_{\tilde{q}} \neq 0.$$

Thus we see that the mapping Φ_F is bi-regular in a neighborhood U_p of p, while p is an arbitrary point on V. Hence *Φ_F is a locally bi-regular mapping of V into the projective space $\mathfrak{S}_d$, provided that $c_R(F)$ contains $\gamma > \tau$.*

Choose a positive integer k such that $k\omega > 2\gamma_0$, where γ_0 is the positive (1, 1)-form introduced in Lemma 2, and take a complex line bundle G_1 over V with $c_R(G_1) \ni k\omega$. The existence of such a bundle G_1 follows from Theorem 1. Now, letting G be a complex line bundle over V whose characteristic class $c_R(G)$ contains $\gamma > nk\omega + \kappa$, we show that the mapping Φ_G of V into a projective space $\mathfrak{S}_e$ is regular and that $\Phi_G(p) \neq \Phi_G(q)$ if $p \neq q$. To do this we form the quadratic transform $\tilde{V} = Q_p Q_q(V)$ and let

$$E' = \tilde{G} - \{S\},$$
$$E'' = \tilde{G} - \{S\} - \{T\},$$

where $\tilde{G} = Q_pQ_qG$, $S = Q_pQ_q(p)$, and $T = Q_pQ_q(q)$. Since $\tilde{G}_S$ or E'_T is a trivial bundle over S or T, respectively, we have the exact sequences

$$0 \to \Omega(E') \xrightarrow{i} \Omega(\tilde{G}) \xrightarrow{r_S} \Omega_S \to 0,$$

$$0 \to \Omega(E'') \xrightarrow{j} \Omega(E') \xrightarrow{r_T} \Omega_T \to 0,$$

where Ω_S or Ω_T is the sheaf of germs of holomorphic functions on S or T and where r_S or r_T denotes the restriction map to S or T, respectively. From these sequences we obtain the corresponding exact cohomology sequences

$$0 \to \Gamma(E') \xrightarrow{i^*} \Gamma(\tilde{G}) \xrightarrow{r_s^*} C_S \xrightarrow{\delta^*} H^1(\tilde{V}, \Omega(E')) \to \dots, \tag{20}$$

$$0 \to \Gamma(E'') \xrightarrow{j^*} \Gamma(E') \xrightarrow{r_T^*} C_T \xrightarrow{\delta^*} H^1(\tilde{V}, \Omega(E''))$$
$$\xrightarrow{j^*} H^1(\tilde{V}, \Omega(E')) \xrightarrow{r_T^*} H^1(T, \Omega_T) \to \dots, \tag{21}$$

where C_S or C_T is the space of constants on S or T, respectively. It follows from (3) that the canonical bundle $K(\tilde{V})$ of $\tilde{V}$ is given by

$$K(\tilde{V}) = \bar{K} + (n-1)(\{S\} + \{T\}).$$

Letting $G_2 = G - nG_1 - K$, we have, by hypothesis,

$$c_R(G_2) \ni \gamma - nk\omega - \kappa > 0 \quad \text{on } V. \tag{22}$$

Hence we infer from Lemma 2 that the bundle

$$E'' - K(\tilde{V}) = n(\tilde{G}_1 - \{S\} - \{T\}) + \tilde{G}_2$$

is positive on $\tilde{V}$. Consequently we have, by Theorem 2, $H^1(\tilde{V}, \Omega(E'')) = 0$. On the other hand, $H^1(T, \Omega_T)$ vanishes[15] since T is a projective space, and therefore, by (21), $H^1(\tilde{V}, \Omega(E')) = 0$. Hence we obtain from (20), (21) the exact sequences

$$\begin{cases} 0 \to \Gamma(E') \xrightarrow{i^*} \Gamma(\tilde{G}) \xrightarrow{r_S^*} C_S \to 0, \\ 0 \to \Gamma(E'') \xrightarrow{j^*} \Gamma(E') \xrightarrow{r_T^*} C_T \to 0. \end{cases} \tag{23}$$

This shows that the bases of $\Gamma(\tilde{G}), \Gamma(E'), \Gamma(E'')$ may be chosen in the following manner:

$$\{\psi_0, i^*\psi_1, i^*j^*\psi_2, i^*j^*\psi_3, \dots\} = \Gamma(\tilde{G}),$$
$$\{\psi_1, j^*\psi_2, j^*\psi_3, \dots\} = \Gamma(E'),$$
$$\{\psi_2, \psi_3, \dots\} = \Gamma(E''),$$

15 This can be easily verified by using a theorem of Dolbeault. See P. Dolbeault, *Sur la cohomologie des variété analytiques complexes*, C. R. Acad. Sci. Paris, 236 (1953), pp. 175–177.

where

$$r_s^*\psi_0 = 1, \quad r_s^* i^* \psi_1 = 0, \quad r_s^* i^* j^* \psi_\lambda = 0 \quad (\lambda = 2, 3, \ldots), \tag{24}$$
$$r_T^*\psi_1 = 1, \quad r_T^* j^* \psi_\lambda = 0 \quad (\lambda = 2, 3, \ldots). \tag{25}$$

The quadratic transformation $Q_p Q_q$ induces the isomorphism $P_p^* Q_q^* : \varphi \to \tilde{\varphi}$ of $\Gamma(G)$ onto $\Gamma(\tilde{G})$. Hence we find a base $\{\varphi_0, \varphi_1, \varphi_2, \ldots\}$ of $\Gamma(G)$ such that

$$\tilde{\varphi}_0 = \psi_0, \quad \tilde{\varphi}_1 = i^*\psi_1, \quad \tilde{\varphi}_2 = i^* j^* \psi_2, \ldots.$$

With the help of (13), we infer from (24) that $\varphi_{0p}(p) = 1, \varphi_{\lambda p}(p) = 0 (\lambda = 1, 2, \ldots)$. Thus the mapping Φ_G is regular in a neighborhood of p, and moreover

$$\Phi_G(p) = (1, 0, 0, \ldots, 0). \tag{26}$$

The mapping i^* is given by $\eta_k \to s_k \eta_k$ over $\tilde{U}_k$, as (10) shows, where we may assume that $s_k = 1$ in a neighborhood of T. Consequently, we infer from (25) that $\varphi_{1q}(q) = 1, \varphi_{\lambda q}(q) = 0 (\lambda = 2, 3, \ldots)$. The mapping Φ_G is therefore regular in a neighborhood of q and

$$\Phi_G(q) = (\varphi_{0q}(q), 1, 0, 0, \ldots, 0).$$

Comparing this with (26), we see that $\Phi_G(p) \neq \Phi_G(q)$, while p, q are arbitrary points on V. Hence we conclude that Φ_G *is a one-to-one regular mapping of* V *into a projective space* $\mathfrak{S}_d$ *if* $c_R(G)$ *contains* $\gamma > nk\omega + \kappa$.

Now, letting β be a real (1, 1)-*form* on V which satisfies both inequalities $\beta > \tau, \beta > nk\omega + \kappa$, we infer from the above results that Φ_F is a one-to-one bi-regular mapping of V into a projective space $\mathfrak{S}_d$ if $c_R(F)$ contains a $\gamma > \beta$. Thus we obtain the following

Theorem 3. *Let* V *be a compact complex analytic variety carrying a Hodge metric. Then there exists on* V *a real* (1, 1)-*form* β *such that, for any complex line bundle* F *whose characteristic class* $c_R(F)$ *contains a closed real* (1, 1)-*form* $\gamma > \beta$, *the mapping* $z \to \Phi_F(z)$ *is a bi-regular mapping of* V *into a projective space.*

Let h be a positive integer such that $h\omega > \beta$ and let F be a complex line bundle over V with $c_R(F) \ni h\omega$. The existence of such a bundle F follows from Theorem 1. By the above Theorem 3, the mapping Φ_F associated with F is a bi-regular mapping of V into a projective space $\mathfrak{S}_d$. Thus V is bi-regularly equivalent to a non-singular sub-variety $\Phi_F(V)$ of $\mathfrak{S}_d$, while, in view of a theorem of Chow,[16] $\Phi_F(V)$ is an algebraic variety. Hence we conclude the following

Theorem 4. (Main Theorem). *Let* V *be a compact complex analytic variety. If* V *carries a Hodge metric, i.e., a Kähler metric* $2\sum g_{\alpha\bar{\beta}}(dz_\alpha d\bar{z}_\beta)$ *such that the associated exterior form* $\omega = i \sum g_{\alpha\bar{\beta}} dz_\alpha d\bar{z}_\beta$ *belongs to the cohomology class of an integral 2-cocycle, then* V *is (bi-regularly equivalent to) a non-singular algebraic variety imbedded in a projective space.*

---

[16] W. L. Chow, *On compact complex analytic varieties*, Amer. J. Math. 71 (1949), pp. 893–914.

4 Applications

(i) *Complex Toruses.* Let C_n be the linear space of n complex variables $(z_1, \dots, z_n)$, Δ the discrete subgroup of C_n generated by $2n$ independent periods $\pi_j = (\pi_{1j}, \dots, \pi_{nj}), j = 1, 2, \dots, 2n$, and let $A_n = C_n/\Delta$. Applying the above theorem to the complex torus A_n, we shall derive a theorem of Lefschetz[17] to the effect that, *if the period matrix* $(\pi_{\lambda j})$ *is a Riemann matrix,* A_n *is a non-singular algebraic variety imbedded in a projective space.* By hypothesis, there exists a positive definite Hermitian matrix $(g_{\lambda\bar{\nu}})$ such that the

$$J_{jk} = i \sum g_{\lambda\bar{\nu}} (\pi_{\lambda j} \bar{\pi}_{\nu k} - \pi_{\lambda k} \bar{\pi}_{\nu j})$$

are rational integers. By means of the matrix $(g_{\lambda\bar{\nu}})$, we define a Kähler metric $ds^2 = 2 \sum g_{\lambda\bar{\nu}} (dz_\lambda d\bar{z}_\nu)$ on A_n. The second homology group $H_2(A_n, Z)$ with coefficients in Z is generated by integral 2-cycles $c_{jk}, 1 \leqq j < k \leqq 2n$, defined respectively by the parametric representations

$$c_{jk} = \{ s\pi_j + t\pi_k | 0 \leqq s, t \leqq 1 \}.$$

The period of the fundamental form $\omega = i \sum g_{\lambda\bar{\nu}}, dz_\lambda d\bar{z}_\nu$ over c_{jk} is given by

$$\int_{c_{jk}} \omega = J_{jk}.$$

Hence ω belongs to the cohomology class of an integral cocycle. By virtue of our main theorem, A_n is therefore a non-singular algebraic variety imbedded in a projective space.

(ii) *Ricci curvature.* First we quote the following

Lemma 3.[18] *Let* F *be a complex line bundle over a compact complex analytic variety* V *and let* $\{f_{jk}\}$ *be the system of transition functions defining the bundle* F *with respect to a finite covering* $\{U_j\}$ *of* V. *Moreover, let* $\{a_j\}$ *be a system of positive functions* a_j *defined respectively on* U_j *satisfying* $a_j = |f_{jk}|^2 a_k$ *in* $U_j \cap U_k$. *Then*

$$\gamma = \frac{i}{2\pi} \partial\bar{\partial} \log a_j$$

is a closed real (1, 1)*-form on* V *belonging to the characteristic class* $c_R(F)$ *of the bundle* F.

Now let V be a compact complex analytic variety with an *Hermitian metric* $\sum g_{\alpha\bar{\beta}} (dz_\alpha d\bar{z}_\beta)$. By the *Ricci curvature* associated with this metric we shall mean the tensor

$$R_{\alpha\bar{\beta}} = -\frac{\partial^2 \log g}{\partial z_\alpha \partial \bar{z}_\beta}, \quad \text{where } g = \det(g_{\alpha\bar{\beta}}).$$

It follows from the above Lemma 3 that the closed real (1, 1)-form

$$\frac{i}{2\pi} \sum R_{\alpha\bar{\beta}} dz_\alpha d\bar{z}_\beta = -\frac{i}{2\pi} \partial\bar{\partial} \log g$$

17 S. Lefschetz, *On certain numerical invariants of algebraic varieties with application to abelian varieties*, Trans. Amer. Math. Soc., 22 (1921), pp. 327-482, Part II, Chap. I, §1.

18 Kodaira, *loc. cit.*, §3, Lemma.

belongs to the characteristic class $c_R(-K)$, where the *first Chern class* $c(-K)$ of V is an element of $H^2(V, Z)$. Consequently the metric

$$ds^2 = -[\text{or}+] \sum \left(\frac{R_{\alpha\bar{\beta}}}{\pi}\right)(dz_\alpha d\bar{z}_\beta)$$

is a Hodge metric on V if it is positive definite. By virtue of our main theorem we obtain therefore the following

Theorem 5. *If a compact complex analytic variety V carries an Hermitian metric such that the associated Ricci curvature is negative [or positive] definite everywhere on V, then V is a non-singular algebraic variety imbedded in a projective space.*

(iii) *Automorphic functions.* Let $\mathfrak{B}$ be a *bounded domain* in the space of n complex variables $(z_1, \ldots, z_n)$ and let Δ be a discontinuous group of analytic automorphisms of $\mathfrak{B}$ such that $\mathfrak{B}$ is compact relative to Δ. We assume further that no element of Δ except the identity has a fixed point in $\mathfrak{B}$. Then the factor space $V = \mathfrak{B}/\Delta$ is a compact complex analytic variety (without singularities). Now we prove

Theorem 6.[19] *The factor space $V = \mathfrak{B}/\Delta$ is a non-singular algebraic variety imbedded in a projective space.*

Proof. Let $\mathfrak{k}(z, \bar{z})$ be Bergmann's kernel function[20] for the domain $\mathfrak{B}$. Then the $2n$-form

$$\varpi = \mathfrak{k}(z, \bar{z}) dz_1 d\bar{z}_1 dz_2 d\bar{z}_2 \cdots dz_n d\bar{z}_n$$

may be regarded as a $2n$-form on $V = \mathfrak{B}/\Delta$, since ϖ is invariant under the group Δ. Therefore we infer from Lemma 3 that

$$\kappa = \frac{i}{2\pi} \partial\bar{\partial} \log \mathfrak{k}(z, \bar{z}) \tag{27}$$

is a closed real (1, 1)-form on V belonging to the characteristic class $c_R(K)$ of the canonical bundle K over V. On the other hand,

$$ds^2 = \sum g_{\alpha\bar{\beta}}(dz_\alpha d\bar{z}_\beta), \quad \text{where } g_{\alpha\bar{\beta}} = \frac{\partial^2 \log \mathfrak{k}(z, \bar{z})}{\partial z_\alpha \partial \bar{z}_\beta}, \tag{28}$$

defines Bergmann's positive definite metric on $\mathfrak{B}$, which is invariant under Δ. Comparing (28) with (27), we see that ds^2/π induces a Hodge metric on V. Hence, by our main theorem, $V = \mathfrak{B}/\Delta$ is a non-singular algebraic variety imbedded in a projective space.

We note that when $\mathfrak{B}$ is *homogenous* the Ricci curvature associated with Bergmann's metric is given by $R_{\alpha\bar{\beta}} = -g_{\alpha\bar{\beta}}$. Hence in this case Theorem 6 can be derived from Theorem 5 as a corollary.[21]

---

[19] A. Borel communicated to the author the fact that a similar result has been obtained by J. P. Serre.

[20] S. Bergmann, Sur les fonctions orthogonales de plusieurs variables complexes, New York, 1941.

[21] This is a remark due to A. Borel. Our proof of Theorem 6 is inspired by this remark.

Theorem 6 above implies *the existence of "sufficiently many" automorphic functions on* $\mathfrak{B}$ *with respect to* Δ.

(iv) *Covering manifolds of algebraic varieties.* It is easy to derive from our main theorem the following

Theorem 7.[22] *Let* V *be a non-singular algebraic variety imbedded in a projective space. Then every finite unramified covering manifold* $\tilde{V}$ *of* V *is also a non-singular algebraic variety imbedded in a projective space.*

In fact, V has a Hodge metric induced by the "standard" Kähler metric of the ambient projective space of V. This Hodge metric induces on $\tilde{V}$ a Hodge metric in a canonical manner. Hence $\tilde{V}$ is also a non-singular algebraic variety imbedded in a projective space.

(v) *Projective bundles over algebraic varieties.* We prove the following

Theorem 8. *Let* B *be an analytic fibre bundle over a complex analytic variety* V *whose fibre is a complex projective space and whose structure group is the group of projective transformations. If the base space* V *is a non-singular algebraic variety imbedded in a projective space, then the bundle* B *is also a non-singular algebraic variety imbedded in a projective space.*

Proof.[23] Let π be the canonical projection of B onto V and let $\{U_j\}$ be a finite covering of V which is differentiably simple.[24] Then $\pi^{-1}(U_j)$ has a product structure: $\pi^{-1}(U_j) = U_j \times S, S$ being a complex projective space, and a point $(z, t_j) \in U_j \times S$ is identical with $(z, t_k) \in U_k \times S$ if and only if $t_j = T_{jk}(z)t_k$, where $T_{jk}(z)$ is a (non-singular) projective transformation of S depending on $z \in U_j \cap U_k$ holomorphically. In terms of the system of homogenous coordinates $(t_0, \ldots, t_\lambda, \ldots, t_m)$ on S, the relation $t_j = T_{jk}(z)t_k$ can be written in the form

$$t_{j\lambda} = \rho(z) \sum_{\nu} M_{jk\lambda\nu}(z) t_{k\nu}, \quad \rho(z) \neq 0.$$

Since, by hypothesis, $U_j \cap U_k$ is simply connected, the factor $\rho(z)$ can be chosen in such a way that the matrix

$$M_{jk}(z) = (M_{jk\lambda\nu}(z))$$

is unimodular and depends on z holomorphically. Then we have

$$\begin{aligned} &M_{jk}(z) M_{kj}(z) = 1 \quad \text{in } U_j \cap U_k, \\ &M_{jk}(z) M_{kl}(z) M_{lj}(z) = \rho_{jkl} \cdot 1 \quad \text{in } U_j \cap U_k \cap U_l, \end{aligned} \tag{29}$$

where ρ_{jkl} is a constant satisfying

$$(\rho_{jkl})^{m+1} = 1, \tag{30}$$

[22] K. Kodaira, *Some results in the transcendental theory of algebraic varieties*, Ann. of Math., 59 (1954), pp. 86–134, Theorem 12.2.

[23] The original proof of the author was incomplete and the present form is a revision following suggestions by A. Borel.

[24] A. Weil, *Sur les théorèmes de de Rham*, Comment. Math. Helv., 26 (1952), pp. 119–145, §1.

m being the dimension of the fibre S. Now, let S_ν be the open subset $\{t|t_\nu \neq 0\}$ of S and let $U_{j\nu} = U_j \times S_\nu$. Obviously the system $\{U_{j\nu}\}$ forms a finite covering of B. We introduce the system $\{f_{j\lambda k\mu}\}$ of non-vanishing holomorphic functions $f_{j\lambda k\mu}$ defined in $U_{j\lambda} \cap U_{k\mu}$ by

$$f_{j\lambda k\mu} = f_{f\lambda k\mu}(z, t_k) = \left(\frac{t_{k\mu}}{\sum_\sigma M_{jk\lambda\sigma}(z) t_{k\sigma}} \right)^{m+1}. \tag{31}$$

By using (29) and (30), we verify that

$$f_{j\lambda k\mu} \cdot f_{k\mu j\lambda} = 1 \quad \text{in } U_{j\lambda} \cap U_{k\mu},$$
$$f_{j\lambda k\mu} \cdot f_{k\mu l\nu} \cdot f_{l\nu j\lambda} = 1 \quad \text{in } U_{j\lambda} \cap U_{k\mu} \cap U_{l\nu}.$$

Hence the system $\{f_{j\lambda k\mu}\}$ defines a complex line bundle F over B. Letting $\{A\}$ be the space of all positive definite Hermitian $(m+1) \times (m+1)$-matrices A, we define a fibre bundle $\mathfrak{U}$ over V with the fibre $\{A\}$ by identifying $(z, A_j) \in U_j \times \{A\}$ with $(z, A_k) \in U_k \times \{A\}$ if any only if

$$A_k = M_{jk}^*(z) A_j M_{jk}(z), \tag{32}$$

where M_{jk}^* denotes the conjugate of the transposed matrix of M_{jk}. The fact that the identifications (32) are compatible in $U_j \cap U_k \cap U_l$ follows from (29) and (30). The bundle $\mathfrak{U}$ has a section $(z, A_j(z))$ of class C^∞, since the fibre $\{A\}$ is contractible. By means of such a section $(z, A_j(z))$, we define the positive function $a_{j\lambda}$ on each $U_{j\lambda}$ by

$$a_{j\lambda} = a_{j\lambda}(z, t_j) = \left(\frac{t_j^* A_j(z) t_j}{|t_{j\lambda}|^2} \right)^{m+1}; \tag{33}$$

t_j^* on the right side denotes the vector $(\bar{t}_{j0}, \ldots, \bar{t}_{jm})$. Then, using (31) and (32), we obtain

$$a_{j\lambda} = |f_{j\lambda k\mu}|^2 a_{k\mu} \quad \text{in } U_{j\lambda} \cap U_{k\mu}.$$

This implies that the closed real (1, 1)-form $\gamma = (i/2\pi)\partial\bar{\partial} \log a_{j\lambda}$ on B belongs to the characteristic class $c_R(F)$ of the bundle F, as Lemma 3 shows. Moreover we infer from (33) that the restriction $\gamma_{z\times S}$ of γ to any fibre $\pi^{-1}(z) = z \times S$ is *positive* on $z \times S$. Now, let E be the complex line bundle over V determined by the hyperplane section of V. Obviously $c_R(E)$ contains the exterior form ω associated with the Hodge metric on V which is induced by the standard Kähler metric of the ambient projective space of V. E or ω induces on B a bundle $\tilde{E}$ or a form $\tilde{\omega}$ in a canonical manner. Obviously we have $\tilde{\omega} \in c_R(\tilde{E})$. Since, as was shown above, $\gamma_{z\times S}$ is positive on each fibre $z \times S$, *the closed real* (1, 1)*-form* $h\tilde{\omega} + \gamma$ *is positive on* B *if the positive integer* h *is sufficiently large*, while we have

$$h\tilde{\omega} + \gamma \in c_R(h\tilde{E} + F),$$

where $c(h\tilde{E} + F)$ is a member of $H^2(B, Z)$. Consequently the metric with the associated form $h\tilde{\omega} + \gamma$ is a Hodge metric on B and therefore B is a non-singular algebraic variety imbedded in a projective space.

Appendix
A theorem concerning quadratic transformations

Let $\tilde{V}$ be a complex analytic variety of complex dimension $n \geqq 2$ and let S be a non-singular subvariety of $\tilde{V}$ of complex dimension $n-1$ which is (biregularly equivalent to) a complex projective space. We infer from formula (1) in Section 2 that *if* $\tilde{V} = Q_p(V)$ *and if* $S = Q_p(p)$, *the relation*

$$c(S)_s = -c(e)$$

holds, where e is a hyperplane on the projective space S *and where* $c(S) = c(\{S\})$, $c(e) = c(\{e\})$. In this appendix we show that the converse of this proposition holds when $\tilde{V}$ is a *Hodge variety.* Namely, we prove the following

Theorem.[25] *Assume* $\tilde{V}$ *to be a Hodge variety. If* S *satisfies the condition*

$$c(S)_S = \quad c(c) \quad \text{on } S, \tag{A}$$

then there exists a non-singular algebraic variety V *embedded in a projective space such that* $Q_p(V) = \tilde{V}$ *and* $Q_p(p) = S$, *where* p *is a point on* V.

We note that the cohomology class $c(S)$ [or $c(e)$] is the dual of the homology class of the cycle S on V [or e on S]. Thus (A) expresses a *topological* condition for S. In particular, in the special case $n = 2$ the condition (A) is equivalent to

$$I(S, S) = -1, \tag{A$_2$}$$

where $I(S, S)$ denotes the intersection multiplicity of the curve S with itself.

Proof of the Theorem. Let F be a complex line bundle on $\tilde{V}$ whose characteristic class $c_R(F)$ contains a closed *positive* (1, 1)-form γ satisfying the inequalities $\gamma > \beta$ and $\gamma > \kappa$, where β is the (1, 1)-form introduced in Theorem 3 and where $\kappa \in c_R(K)$, K being the canonical bundle over $\tilde{V}$. The existence of such a bundle F follows from Theorem 1, since by hypothesis $\tilde{V}$ is a Hodge variety; Theorem 3 asserts that the mapping Φ_F associated with F is a bi-regular mapping of $\tilde{V}$ into a projective space $\mathfrak{S}_d$. Since every complex line bundle on the projective space S is determined uniquely by its characteristic class, the condition (A) is equivalent to

$$\{S\}_S = -\{e\} \quad \text{on } S. \tag{A$'$}$$

The characteristic class $c_R(F_S)$ of the bundle F_S over S is represented by the closed (1, 1)-form γ_S which is obviously *positive* on S, while $c(e)$ is represented by the fundamental form $\omega > 0$ on S. Hence we have

$$F_S = \{l \cdot e\}, \tag{B}$$

where l is a *positive* integer. Moreover, by a suitable choice of F we may assume that $l \geqq 2$. Setting

$$E_m = F + (l - m)\{S\},$$

[25] In the special case where $\tilde{V}$ is an *algebraic surface*, this theorem has been proved by Castelnuovo and Enriques. See footnote 14.

we obtain, by (A)′ and (B),

$$(E_m)_S = \{me\}.$$

Consequently we obtain the exact sequences

$$0 \to \Omega(E_{m+1}) \xrightarrow{i} \Omega(E_m) \xrightarrow{r} \Omega_S(me) \to 0, \quad 0 \leqq m < l, \tag{C}$$

which are similar to (9). Since the canonical bundle over the projective space S of dimension $n-1$ is equal to $-\{ne\}$, we infer from Theorem 2 that

$$H^1(S, \Omega_S(me)) = 0 \quad \text{for } m \geqq 1-n.$$

Therefore we get from (C) the exact cohomology sequences

$$\begin{aligned} 0 \to \Gamma(E_{m+1}) \xrightarrow{i^*} \Gamma(E_m) \xrightarrow{r^*} \Gamma_S(me) \xrightarrow{\delta^*} H^1(V, \Omega(E_{m+1})) \\ \xrightarrow{i^*} H^1(V, \Omega(E_m)) \to 0 \\ \text{for } m \geqq 0, \end{aligned}$$

while, since $c_R(E_l - K) = c_R(F - K)$ contains $\gamma - \kappa > 0$, we have $H^1(V, \Omega(E_l)) = 0$. Hence we obtain successively

$$H^1(V, \Omega(E_m)) = 0 \quad \text{for} \quad m = l-1, l-2, \ldots, 2, 1,$$

and therefore

$$0 \to \Gamma(E_{m+1}) \xrightarrow{i^*} \Gamma(E_m) \xrightarrow{r^*} \Gamma_S(me) \to 0 \quad \text{for } 0 \leqq m < l.$$

This shows that the bases of the linear spaces $\Gamma(E_m)$ can be chosen in the following manner:

$$\begin{gathered} \{\varphi_0, i^*\varphi_1, \ldots, i^*\varphi_n, \ldots, (i^*)^l \varphi_w, \ldots, (i^*)^l \varphi_{w+d}\} = \Gamma(E_0), \\ \{\varphi_1, \ldots, \varphi_n, \ldots, (i^*)^{l-1} \varphi_w, \ldots, (i^*)^{l-1} \varphi_{w+d}\} = \Gamma(E_1), \\ \cdots\cdots\cdots\cdots \\ \cdots\cdots\cdots\cdots \\ \{\varphi_w, \ldots, \varphi_{w+d}\} = \Gamma(E_l), \end{gathered}$$

where

$$r^*\varphi_0 = 1, \tag{D}$$

$$\{r^*\varphi_1, \ldots, r^*\varphi_n\} = \Gamma_S(e). \tag{E}$$

Let $\{U_j\}$ be a finite covering of $\tilde{V}$ and let $\{f_{jk}\}$ be the system of transition functions defining the bundle E_0. Since $(E_0)_S = \{0\}$ is a trivial bundle over S, we may assume that $(f_{jk})_S = 1$. Let $s_j(z) = 0$ be the local equation of S in U_j. Then the bundle E_m is defined by the system $\{(s_j/s_k)^{-m} f_{jk}\}$ and the mapping i^* is represented in the form

$$i^* : \eta_j(z) \to s_j(z)\eta_j(z) \quad \text{on } U_j$$

[see (10)]. Now we consider the mapping

$$\Phi_0 = \Phi_{E_0} : z \to \Phi_0(z) = (\varphi_{0j}(z), (i^*\varphi_1)_j(z), \ldots, ((i^*)^l \varphi_{w+d})_j(z)).$$

It follows from (D) that $(\varphi_{0j})_S = 1$, while $(i^*)^k \varphi_{\nu j} = s_j^k \varphi_{\nu j}$ vanishes on S for $\nu \geqq 1, k \geqq 1$. Hence the mapping $z \to \Phi_0(z)$ is regular at each point on S and maps S onto a single point

$$p = \Phi_0(S) = (1, 0, 0, \ldots, 0).$$

Since $s_j(z)$ does not vanish outside S, we may write

$$\Phi_0(z) = (s_j(z)^{-l}\varphi_{0j}(z), \ldots, \varphi_{wj}(z), \ldots, \varphi_{w+dj}(z))$$

for $z \in \tilde{V} - S$; we have $E_l = F$, and therefore

$$z \to (\varphi_{wj}(z), \ldots, \varphi_{w+dj}(z)) = \Phi_F(z)$$

is a bi-regular mapping of $\tilde{V}$ into $\mathfrak{S}_d$. Consequently the mapping $z \to \Phi_0(z)$ maps $\tilde{V} - S$ bi-regularly into $\mathfrak{S}_{w+d} - p$. Thus, letting $V = \Phi_0(\tilde{V})$, we infer that $z \to \Phi_0(z)$ is a regular mapping of $\tilde{V}$ *onto* the subvariety V of $\mathfrak{S}_{w+d}$ which maps S onto p and $\tilde{V} - S$ bi-regularly onto $V - p$. *Therefore V has no singular point except possibly p.*

We set

$$\xi_\nu(z) = (i^*\varphi_\nu)_j(z)/\varphi_{0j}(z) = s_j(z)\varphi_{\nu j}(z)/\varphi_{0j}(z)$$

for $\nu = 1, 2, \ldots, n$. Since $\varphi_{0j}(z) = 1$ on S, these functions $\xi_\nu(z)$ are holomorphic in a neighborhood $U(S)$ of S and vanish on S. Let $\mathfrak{U}$ be a neighborhood of the origin 0 in the space of n complex variables $(\xi_1, \ldots, \xi_n)$ and let W be the quadratic transform $Q_0(\mathfrak{U})$, i.e., the subvariety of $\mathfrak{U} \times S$ defined by

$$W = \{(\xi, t) | \xi_\lambda t_\nu - \xi_\nu t_\lambda = 0\},$$

where $(t_1, \ldots, t_n)$ denotes the homogenous coordinates of a "variable" point t on S. Now, the formula (E) shows that $(\varphi_{1j}(z), \ldots, \varphi_{nj}(z))$ may be used as the homogenous coordinate of a point z on S;

$$\mu : z \to \mu(z) = (\varphi_{1j}(z), \ldots, \varphi_{nj}(z))$$

is therefore a regular mapping of $U(S)$ onto S satisfying $\mu(z) = z$ for all $z \in S \subset U(S)$. It follows that

$$\Psi : z \to \Psi(z) = (\xi_1(z), \ldots, \xi_n(z), \varphi_{1j}(z), \ldots, \varphi_{nj}(z))$$

is a bi-regular mapping of $U(S)$ onto W and satisfies $\Psi(z) = (0, z)$ for $z \in S$. In fact, the identities $\xi_\lambda(z)\varphi_{\nu j}(z) - \xi_\nu(z)\varphi_{\lambda j}(z) = 0 (\lambda, \nu = 1, 2, \ldots, n)$ show that $\Psi(z)$ lies on W. At each point q on S, one of $\varphi_{\nu j}$, say φ_{1j}, does not vanish, and therefore the n functions $\xi_1(z), \varphi_{2j}(z)/\varphi_{1j}(z), \ldots, \varphi_{nj}(z)/\varphi_{1j}(z)$ constitute a system of local coordinates on $\tilde{V}$ in a neighborhood of q. Hence the mapping Ψ is locally bi-regular at each point on S, while, for $z \in S, z \to \Psi(z) = (0, z)$ is the identity map of S onto $S = 0 \times S \subset W$. Consequently Ψ is a bi-regular mapping of $U(S)$ onto W, and $\Psi(z) = (0, z)$ for $z \in S$. From this we infer that the mapping $\Psi^{-1}Q_0$ maps $\mathfrak{U} - 0$ bi-regularly onto $U(S) - S$. Setting

$$z(\xi) = \Psi^{-1}Q_0(\xi) \quad \text{for } \xi \in \mathfrak{U} - 0,$$

we use $\xi = (\xi_1, \ldots, \xi_n)$ as the coordinate of the point $z(\xi)$ in $U(S) - S$. Now, since the mapping $(\Psi^{-1}Q_0)^{-1} = P_0\Psi$ is given by

$$P_0\Psi : z \to (\xi_1(z), \ldots, \xi_n(z)),$$

we have $\xi_\nu(z(\xi)) = \xi_\nu$ and therefore

$$\Phi_0(z(\xi)) = (1, \xi_1, \ldots, \xi_n, \ldots, s_j^k\varphi_{\nu j}/\varphi_{0j}, \ldots).$$

Each component $s_j^k\varphi_{\nu j}/\varphi_{0j}$ is a holomorphic function of $z \in U(S)$ and $z = z(\xi)$ depends holomorphically on $\xi \in \mathfrak{U} - 0$. Hence $s_j^k\varphi_{\nu j}/\varphi_{0j}$ is a holomorphic function of $\xi = (\xi_1, \ldots, \xi_n)$ in $\mathfrak{U} - 0$ and therefore, by a theorem of Hartogs, it is holomorphic on the whole of $\mathfrak{U}$. This proves that

$$M : \xi \to \Phi_0(z(\xi))$$

is a bi-regular mapping of $\mathfrak{U}$ onto $U(p) = \Phi_0(U(S)) \subset V$. p is therefore a regular point of V and consequently V is a non-singular subvariety of $\mathfrak{S}_{w+d}$.

Moreover, we have the commutative diagram

$$\begin{array}{ccc} U(S) & \xleftarrow{\Psi^{-1}} & W \\ \Phi_0 \downarrow & & \downarrow P_0. \\ U(p) & \xleftarrow{M} & \mathfrak{U} \end{array}$$

Now the mapping Φ_0 is bi-regular between $\tilde{V} - S$ and $V - p$, while the above diagram shows that Φ_0 *is analytically equivalent to* $\mathrm{P}_0 = Q_0^{-1}$ *in a neighborhood* $U(S)$ *of* S. Hence the inverse mapping Φ_0^{-1} coincides with the quadratic transformation Q_p and consequently $\tilde{V} = Q_p(V), S = Q_p(p)$, q.e.d.

PRINCETON UNIVERSITY

Editor's Comments

By the uniformization of Riemann surfaces, any compact Riemann surface diffeomorphic (or even homotopic) to the complex projective space $\mathbb{C}P^1$ is biholomorphic to it. One natural conjecture is that the same result holds in higher dimensions. In two papers in the 1950s (Ann. of Math. (2) 60 (1954), 213–236, problem 28*; Proc. Internat. Congress Math. 1954, Amsterdam, Vol. III, Noordhoff, Groningen, 1956, pp. 457–473; see pp. 468–469), Hirzebruch announced that if X is an n-dimensional compact Kähler manifold which is differeomorphic to the complex projective space $\mathbb{C}P^n$ and n is odd, then X is biholomorphic to $\mathbb{C}P^n$. By combining the Hirzebruch-Riemann-Roch theorem and the Kodaira vanishing theorem, Hirzebruch and Kodaira proved this result.

The case when n is even remained open until Yau solved Calabi conjecture. As mentioned before, one application of the existence of Kähler-Einstein metric implies a positive solution of this rigidity result of $\mathbb{C}P^n$. This is a concrete case showing the power of the integration of complex geometry and algebraic geometry. This also shows the vision of Calabi and the power of Yau.

Complex Geometry from Riemann to Kähler-Einstein and Calabi-Yau
ALM 38, pp. 253–265

On the Complex Projective Spaces*

Friedrich Hirzebruch[1] and Kunihiko Kodaira[2]

Introduction

The principal purpose of this paper is to prove the following theorem announced earlier in [2] and [4].

Theorem. *Let X be an n-dimensional compact Kähler manifold which is* C^∞*-differentiably homeomorphic to the complex projective space* $P_n(\mathbf{C})$ *with its usual differentiable structure. In case the dimension n of X is odd, X is complex-analytically homeomorphic to* $P_n(\mathbf{C})$*, i. e. there exists a bi-regular map of* X *onto* $P_n(\mathbf{C})$*. Let g be the generator of the second cohomology group* $H^2(X, \mathbf{Z})(\cong \mathbf{Z})$ *whose sign is chosen in such a way that g is represented as the fundamental class of a Kähler metric on* X*. In case n is even,* X *is complex-analytically homeomorphic to* $P_n(\mathbf{C})$ *if the first Chern class* c_1 *of* X *is not equal to* $-(n+1)g$.

It is an immediate consequence of a theorem proved in [6] that a compact Kähler manifold X, which is homoomorphic to $P_n(\mathbf{C})$, is an algebraic manifold, i. e. X can be imbedded complex-analytically and without singularities in a complex projective space of sufficiently high dimension. To prove the above theorem, we may assume therefore, without loss of generality, that X is an algebraic manifold. We recall that—c_1 is the cohomology class of the canonical divisor of X.

We note that nothing is known about non-Kähler complex manifolds of dimension $n \geq 2$ which are differentiably homeomorphic to $P_n(\mathbf{C})$. As to the case : $n = 1$, it is a well-known classical theorem that every compact Riemann surface of genus 0 is conformally equivalent to the Riemann 2-sphere (i. e. the complex projective line).

---

1 Received support from a research project at Princeton University sponsored by the National Science Foundation.

2 Received support from a research project at Princeton University sponsored by the Office of Ordnance Research, U. S. Army.

* F. Hirzebruch, K. Kodaira, On the complex projective spaces. J. Math. Pures Appl. (9) 36 (1957), 201–216.

The proof of the above theorem uses essentially the general Riemann-Roch theorem [3] from which we can deduce for odd n that the first Chern class c_1 of X is $(n+1)g$, whereas for even n we can only prove that $c_1 = \pm(n+1)g$. Using the Riemann-Roch theorem together with a theorem on vanishing of cohomology groups [7] we can prove by methods similar to those employed in [6] that X is complex-analytically homeomorphic to $P_n(\mathbf{C})$ provided that $c_1 = +(n+1)g$. In the theorem we made for even n the assumption $c_1 \neq -(n+1)g$. Assuming $c_1 = -(n+1)g$. we can calculate the plurigenera P^r of X which are birational invariants [9]. We obtain $P^r = \binom{nr+r-1}{n}$. Therefore, we can replace the last sentence of our theorem for example by "If the dimension n of X is even and if the bigenus P^2 is different from $\binom{2n+1}{n}$, then X is complex-analytically homeomorphic to $P_n(\mathbf{C})$". We do not know whether the case n even and $c_1 = -(n+1)g$ actually occurs. If $c_1 = -(n+1)g$, then X is not rational.

It is not known whether the theorem remains correct if "C^∞-differentiably homeomorphic" is replaced by "homeomorphic". The reason why we must make the stronger assumption is that the Pontrjagin classes occur in the theorem of Riemann-Roch and that they are known to be invariants of the differentiable manifold X, whereas it is doubtful whether they are invariants of the topological manifold X. Since for $n = 1$ no Pontrjagin classes occur and since the Pontrjagin class of a 4-dimensional manifold is a topological invariant, our theorem is proved for $n = 1$ and 2 under the assumption "X is homeomorphic to $P_n(\mathbf{C})$".

1.

In this paragraph we recall the theorem of Riemann-Roch as given in [3] (*see* in particular Hauptsatz 20.3.2). For various notions and theorems employed in this paragraph we refer the reader to [3] and to the references given there.

By an n-dimensional algebraic manifold we mean a compact complex manifold of (complex) dimension n which can be imbedded complex-analytically and without singularities in some complex projective space. By Chow [1] this definition is equivalent to the usual algebraic definition of non-singular algebraic varieties.

Let X be an n-dimensional algebraic manifold. A divisor D of X can be given with respect to a suitable open covering $\{U_i\}$ of X by a collection of meromorphic functions f_i(f_i is defined and meromorphic in U_i and f_i is not identically o) such that f_i/f_j is holomorphic and different from o in $U_i \cap U_j$. The divisor D is called *holomorphic* if the local functions f_i are all holomorphic (*see* [3], 15.2). To every divisor D of X there is attached a cohomology class $c(D) \in H^2(X, \mathbf{Z})$ which depends only on the divisor class of D and which can be defined as the first Chern class of the complex line bundle (determined up to an isomorphism) belonging to (the divisor class of) D. This complex line bundle has the transition functions $f_{ij} = f_i/f_j$. The divisor D determines a $(2n-2)$-dimensional integral cycle of X, and $c(D)$ is the cohomology class corresponding by Poincaré duality to the homology class of that cycle. Here as always we use the canonical orientation defined by the complex structure (*See* [3], p. 3).

For a divisor D we have the sheaf $\Omega(\mathrm{D})$, i. e. the sheaf belonging to the presheaf (prefaisceau, Garbendatum) which attaches to every open set U of X the **C**-module of those meromorphic functions g (defined in U) for which gf_i is holomorphic in $\mathrm{U} \cap \mathrm{U}_i$. The sheaf $\Omega(\mathrm{D})$ is isomorphic to the sheaf of germs of holomorphic sections of the line bundle belonging to D. Let $\mathrm{H}^k(\mathrm{X}, \mathrm{D})$ be the k-dimensional cohomology group of X with coefficients in $\Omega(\mathrm{D})$. As is wellknown $\mathrm{H}^k(\mathrm{X}, \mathrm{D})$ is a finite dimensional complex vector space. It depends up to an isomorphism only on the divisor class of D. It vanishes if $k > n$. As usual we put

$$\chi(\mathrm{X}, \mathrm{D}) = \sum_{k=0}^{n} (-1)^k \dim \mathrm{H}^k(\mathrm{X}, \mathrm{D}). \tag{1.1}$$

Recall that $\mathrm{H}^0(\mathrm{X}, \mathrm{D})$ equals the vector space L(D) of those (global) meromorphic functions g on X for which gf_i is holomorphic in U_i. Let E be the complex line bundle belonging to D. Then $\mathrm{H}^0(\mathrm{X}, \mathrm{D})$ is isomorphic to the vector space $\Gamma(\mathrm{E})$ of global holomorphic sections of E.

Let c_1 be the first Chern class of $\mathrm{X}(c_1 \in \mathrm{H}^2(\mathrm{X}, \mathbf{Z}))$. We have

$$c_1 = -c(\mathrm{K}), \tag{1.2}$$

where K is a canonical divisor of X.

The algebraic manifold X (of complex dimension n) has to be considered in a natural way as a $2n$-dimensional differentiable manifold. As such it has Pontrjagin classes which we denote by $p_i (p_i \in \mathrm{H}^{4i}(\mathrm{X}, \mathbf{Z}), p_0 = 1)$.

Let $\hat{A}_0, \hat{A}_1, \hat{A}_2, \ldots$ be the multiplicative sequence of polynomials in the p_i belonging to the power series

$$\frac{\frac{1}{2}\sqrt{z}}{\mathrm{sh}\frac{1}{2}\sqrt{z}} \quad (\mathrm{see} [3], 1.2).$$

We have

$$\hat{\mathrm{A}}_0 = 1, \quad \hat{\mathrm{A}}_1 = -\frac{p_1}{24}, \quad \hat{\mathrm{A}}_2 = \frac{1}{2^7 \cdot 45}(-4p_2 + 7p_1^2), \ldots \tag{1.3}$$

In this moment, we consider the $p_i (i > 0)$ as indeterminates. The polynomials $\hat{\mathrm{A}}_k$ are related to the polynomials A_k defined in [3], p. 14, by the equation

$$\mathrm{A}_k = 2^{4k} \hat{\mathrm{A}}_k. \tag{1.4}$$

The theorem of Riemann-Roch asserts that $\chi(\mathrm{X}, \mathrm{D})$ can be calculated by means of the cohomology ring of X, the Pontrjagin classes $p_i \in \mathrm{H}^{4i}(X, \mathbf{R})$ (which only involve the differentiable structure of X) and the class $\left(c(\mathrm{D}) + \frac{1}{2}c_1\right) \in \mathrm{H}^2(\mathrm{X}, \mathbf{R})$. (We consider for the rest of this paragraph all cohomology classes as with real coefficients.) Put $d = c(\mathrm{D})$. The precise formulation of the theorem of Riemann-Roch is as follows (*see* [3], p. 149, formula (6*)).

$$\chi(\mathrm{X}, \mathrm{D}) = \left(\sum \frac{\left(d + \frac{1}{2}c_1\right)^r}{r!} \hat{\mathrm{A}}_s(p_1, \ldots, p_s) \right) [\mathrm{X}]. \tag{1.5}$$

Here the sum has to be extended over all r, s with $r + zs = n$.

For a $2n$-dimensional cohomology class h the value of h on (the $2n$-dimensional fundamental cycle of) X is denoted by $h[\mathrm{X}]$. Since X is canonically oriented by its complex structure, the fundamental cycle is defined without any ambiguity in sign. The formula (1.5) can also be written in the form

$$\chi(\mathrm{X}, \mathrm{D}) = \chi_n \left[e^{d + \frac{1}{2}c_1} \sum_{s=0}^{\infty} \hat{\mathrm{A}}_s(p_1, \ldots, p_s) \right], \tag{1.6}$$

where $\chi_n[\ ]$ stands for the value on X of the $2n$-dimensional component of the expression in $[\ ]$.

The right side of (1.5) or (1.6) is for $d = 0$ the definition of the Todd genus of X which we denote by T(X). If D is the zero divisor, then (1.5) and (1.6) state that the arithmetic genus $\chi(\mathrm{X})$ equals the Todd genus T(X). We recall that

$$\chi(\mathrm{X}) = g_0 - g_1 + g_2 - \cdots + (-1)^n g_n, \tag{1.7}$$

where g_k denotes the dimension of the complex vector space of holomorphic differential forms of degree k on X. In particular, g_0 is the number of connectedness components of X.

2.

Throughout this paragraph let X be a compact complex manifold and assume that X admits a Kähler metric and that X is differentiably homeomorphic to the complex projective space $\mathrm{P}_n(\mathbf{C})(n \geq 1)$. By "differentiably homeomorphic" we mean that there exists a homeomorphism of X onto $\mathrm{P}_n(\mathbf{C})$ which is in both directions differentiable of class C^∞.

Choose a generator g of the infinite cyclic group $\mathrm{H}^2(\mathrm{X}, \mathbf{Z})$ in such a way that for some Kähler metric of X the fundamental class of the metric is g (considered as class with real coefficients). Clearly, the generator g is uniquely determined and depends only on the complex structure of X. After multiplying a Kähler metric of X with a suitable positive scalar, we obtain a Kähler metric of X which has g as its fundamental class. Since g is an integral class, we have a Hodge metric and by the theorem proved in [6] we know that X is an algebraic variety.

We have

$$g^n[\mathrm{X}] = 1. \tag{2.1}$$

In case n is odd, our choice of the generator g can be characterized by (2.1).

Let $h^{p,q}$ be as usual the number of harmonic forms of type (p,q) on X. This number depends only on the complex structure of X. We have $g_k = h^{k,0} = h^{0,k}$. It is well known that $h^{p,p} \geq 1$ (for $0 \leq p \leq n$). By Hodge's theory the r-dimensional Betti number b^r of X is equal to the sum of all $h^{p,q}$ with $p+q=r$. On the other hand, the odd dimensional Betti numbers of X vanish and b^{2p} equals 1 for $0 \leq p \leq n$. From these facts we conclude

$$h^{p,q} = 0 \quad \text{for} \quad p \neq q, \quad h^{p,p} = 1 \quad \text{for} \quad 0 \leq p \leq n. \tag{2.2}$$

In particular, g_k vanishes for $k > 0$. Thus we obtain for the arithmetic genus [*see* (1.7)] that

$$\chi(X) = 1. \tag{2.3}$$

By assumption there exists a differentiable homeomorphism π of X onto $\mathrm{P}_n(\mathbf{C})$. Obviously, the tangential $\mathrm{GL}(2n,\mathbf{R})$-bundle of X is the bundle induced from the tangential $\mathrm{GL}(2n,\mathbf{R})$-bundle of $\mathrm{P}_n(\mathbf{C})$ by π. Therefore the Pontrjagin classes of X are the images under π^* of the Pontrjagin classes of $\mathrm{P}_n(\mathbf{C})$ which are well known (*see* for example [3], Satz 4. 10. 2). The image under π^* of a generator of $\mathrm{H}^2(\mathrm{P}_n(\mathbf{C}),\mathbf{Z})$ is $\pm g$. Thus we get for the $4k$-dimensional Pontrjagin class p_k of X,

$$p_k = \binom{n+1}{k} g^{2k} \in \mathrm{H}^{4k}(\mathrm{X},\mathbf{Z}) \quad (k \geq 0),$$

and for the total Pontrjagin class of X,

$$p = \sum_{k=0}^{\infty} p_k = (1+g^2)^{n+1}. \tag{2.4}$$

From the definition of $\{\hat{\mathrm{A}}_k(p_1,\ldots,p_k)\}$ as a multiplicative sequence (*see* §1) we obtain immediately for the Pontrjagin classes of X,

$$\sum_{k=0}^{\infty} \hat{\mathrm{A}}_k(p_1,\ldots,p_k) = \left(\frac{\frac{g}{2}}{\mathrm{sh}\frac{g}{2}}\right)^{n+1}. \tag{2.5}$$

Here we consider all cohomology classes as classes with real coefficients. Note that $\frac{\frac{g}{2}}{\mathrm{sh}\frac{g}{2}}$ is a power series in g with real (even rational) coefficients and thus an element of the (real) cohomology ring of X.

The Chern class c_1 of the complex manifold X is an element of $\mathrm{H}^2(\mathrm{X},\mathbf{Z})$. The definition of c_1 involves the complex structure of X. However, under the natural homomorphism

$$\rho : \mathrm{H}^2(\mathrm{X},\mathbf{Z}) \to H^2(X,\mathbf{Z}_2) \quad (\cong \mathbf{Z}_2)$$

the class c_1 goes over into the second Stiefel-Whitney class w_2 of X which is a topological invariant of X,

$$\rho(c_1) = w_2. \tag{2.6}$$

The class w_2 vanishes, if n is odd, and is different from 0 for even n. This follows from the wellknown value of the first Chern class of $\mathrm{P}_n(\mathbf{C})$ (with the usual complex structure, *see* for example [3], Satz 4.10.2) and by applying (2.6) on $\mathrm{P}_n(\mathbf{C})$, or we can use the Thom-Wu theory of Steenrod squares in orientable manifolds (*see* for example [5] and the references given there) which yields

$$\mathrm{Sq}^2((\rho g)^{n-1}) = (n-1)(\rho g)^n = w_2(\rho g)^{n-1} \tag{2.7}$$

and thus also the desired result. Now (2.6) gives

$$c_1 = \lambda g \quad \text{with} \quad \lambda \equiv n+1 \pmod 2. \tag{2.8}$$

We interrupt our discussion of the manifold X by a formal algebraic lemma.

Lemma 1. *Let g be an indeterminate. Consider the following power series in g*

$$e^{\left(r+\frac{n+1}{2}\right)g}\left(\frac{\frac{g}{2}}{\mathrm{sh}\frac{g}{2}}\right)^{n+1} = e^{rg}\left(\frac{g}{1-e^{-g}}\right)^{n+1},$$

where r and n are integers ($n \geq 0$). The coefficient of g^n in this power series equals the binomial coefficient $\binom{n+r}{n}$.

Proof. By an easy residue calculation [or by applying the theorem of Riemann-Roch to the complex projective $\mathrm{P}_n(\mathbf{C})$].

We continue the discussion of our manifold X. In virtue of (2.8) we write

$$c_1 = (2s+n+1)g, \tag{2.9}$$

where s is an integer. The Todd genus T(X) equals the arithmetic genus $\chi(\mathrm{X})$. By (2.3) we have $\mathrm{T(X)} = 1$. From the definition of the Todd genus (1.6), from (2.1), (2.5), (2.9), and Lemma 1 we can deduce easily

$$\mathrm{T(X)} = \binom{n+s}{n}.$$

Thus we get

$$1 = \binom{n+s}{n}, \quad \text{i. e.} \quad n! = (s+1)(s+2)\cdots(s+n).$$

From the preceding equation and from (2.9) we readily infer the following.

Lemma 2. *For n odd, we have $s=0$ i. e. $c_1=(n+1)g$. For n even, we have $s=0$ or $s=-(n+1)$, i. e. $c_1=(n+1)g$ or $c_1=-(n+1)g$.*

Remark. *In the case in which $\mathrm{X}=\mathrm{P}_n(\mathbf{C})$ (with the usual complex structure) we always have $c_1=(n+1)g$. It is not known whether the case in which n is even and $c_1=-(n+1)g$ actually occurs.*

We have seen in (2.2) that $h^{0,2} = h^{2,0} = 0$. Thus every element of $\mathrm{H}^2(\mathrm{X}, \mathbf{Z})$ is of type (1, 1) and therefore occurs as cohomology class of a complex line bundle over X. Since X is algebraic, every complex line bundle can be realized by a divisor (*see* [8]). The manifold X is simply connected and therefore has no Picard variety. Hence the group $\mathfrak{D}$ of divisor classes of X is isomorphic to $\mathrm{H}^2(\mathrm{X}, \mathbf{Z})$. The isomorphism

$$\mathfrak{D} \to \mathrm{H}^2(\mathrm{X}, \mathbf{Z})$$

is established by attaching to each divisor D its cohomology class $c(\mathrm{D})$. We write

$$c(\mathrm{D}) = \nu(\mathrm{D})g.$$

Here $v(\mathrm{D})$ is an integer depending only on the divisor class of D, and

$$\mathrm{D} \to \nu(\mathrm{D}) \tag{2.10}$$

induces an isomorphism of $\mathfrak{D}$ onto the integers $\mathbf{Z}$. The hyperplane section of X with respect to some projective imbedding of X is a divisor with positive ν.

Let $\mathrm{H}^k(\mathrm{X}, \mathrm{D})$ be as in paragraph 1 the k-dimensional cohomology group of X with coefficients in $\Omega(\mathrm{D})$. If $\nu = \nu(\mathrm{D})$, we write instead of $\mathrm{H}^k(\mathrm{X}, \mathrm{D})$ simply $\mathrm{H}^k(\mathrm{X}, \nu)$. Because of the isomorphism of $\mathfrak{D}$ onto $\mathbf{Z}$, the finite dimensional complex vector space $\mathrm{H}^k(\mathrm{X}, \nu)$ is defined for every integer ν. A theorem on vanishing of cohomology groups (*see* [7]) yields by virtue of (1.2) and Lemma 2 the following.

Lemma 3. *Assume that the complex dimension n of* X *is odd. Then we have*:

$$\begin{cases} \text{If } \nu > -(n+1), & \text{then } \ \mathrm{H}^k(\mathrm{X}, \nu) = 0 \ \text{ for } \ k \neq 0, \\ \text{If } \nu < 0, & \text{then } \ \mathrm{H}^k(\mathrm{X}, \nu) = 0 \ \text{ for } \ k \neq n. \end{cases} \tag{2.11}$$

If the complex dimension n of X *is even and if $c_1 = (n+1)g$, then* (2.11) *holds. If n is even and if $c_1 = -(n+1)g$, then we have*:

$$\begin{array}{lll} \text{If } \nu > (n+1), & \text{then } \ \mathrm{H}^k(X, \nu) = 0 & \text{for } \ k \neq 0, \\ \text{If } \nu < 0, & \text{then } \ \mathrm{H}^k(X, \nu) = 0 & \text{for } \ k \neq n. \end{array}$$

We write corresponding to (1.1)

$$\chi(X, \nu) = \sum_{j=0}^{n} (-1)^j \dim \mathrm{H}^j(X, \nu).$$

The number $\chi(\mathrm{X}, \nu)$ can be calculated by the theorem of Riemann-Roch (1.5), with the help of (2.1), (2.5), Lemmas 1 and 2. The result is as follows:

Lemma 4. *In case the complex dimension n of* X *is odd, we have*

$$\chi(X, \nu) = \binom{n+\nu}{n}. \tag{2.12}$$

This formula holds also in case n is even if $c_1 = (n+1)g$. In case n is even and $c_1 = -(n+1)g$, we have

$$\chi(X, \nu) = \binom{\nu - 1}{n}.$$

From the preceding discussions we can derive the following theorem.

Theorem 5. *Let* X *be a compact Kähler manifold which is differentiably homeomorphic to the complex projective space* $P_n(\mathbf{C})$. *Then* X *is an algebraic manifold. Let* g *be the generator of* $H^2(X, \mathbf{Z})$ *whose sign is chosen in such a way that* g *is represented as the fundamental class of one of the Kähler metrics on* X. *The mapping* $D \to c(D)$ *induces an isomorphism of the group of divisor classes of* X *onto* $H^2(X, \mathbf{Z})$, *where* D *is an arbitrary divisor on* X *and* c(D) *denotes its cohomology class. Assume either that* n *is odd or that* n *is even and* $c_1 \neq -(n+1)g$, *where* c_1 *denotes the first Chern class of* X. *Then, for a divisor* D *with* $c(D) = \nu g, \nu \geq 0$, *we have*

$$\dim L(D) = \binom{n+\nu}{n}.$$

Thus dim L(D) *is equal to the number of linearly independent homogeneous polynomials of degree* ν *in* $n+1$ *indeterminates. In particular, every divisor* D *with* $c(D) = \nu g, \nu \geq 0$, *is linearly equivalent to a holomorphic divisor.*

Remark. *Let* X *be differentiably homeomorphic to* $P_n(\mathbf{C})$ (n *even*) *and assume that* c_1 *equals* $-(n+1)g$, *then the canonical divisors* K *have the cohomology class* $(n+1)g$. *We know by* (2.2) *that the geometric genus* $g_n = \dim L(K)$ *vanishes. We can calculate the so-called plurigenera* $P^r (r \geq 2)$ *which are defined by* $P^r = \dim L(rK)$. *Lemmas* 3 *and* 4 *yield* $P^r = \binom{nr+r-1}{n}$ *for* $r \geq 2$. *In particular,* $P^2 = \binom{2n+1}{n}$ (*Compare the Introduction.*).

3.

In this paragraph we prove the following:

Theorem 6. *Let* X *be an algebraic manfold of dimension* n *having the following properties*:

(i) *The cohomology ring of* X *has the form*

$$H^*(X, \mathbf{Z}) = \mathbf{Z} + \mathbf{Z}_g + \mathbf{Z}g^2 + \cdots + \mathbf{Z}g^n,$$

where the generator g *belongs to* $H^2(X, \mathbf{Z})$;

(ii) *There exists a Kähler metric on* X *whose fundamental class coincides with the generator* g;

(iii) *There exists a holomorphic divisor* D *on* X *whose cohomology class* c(D) *is equal to* g;

(iv) *The dimension of the linear space* $L(\nu D)$ *is given by*

$$\dim L(\nu D) = \binom{n+\nu}{n} \quad (\nu = 1, 2, 3, \ldots).$$

Then X *is complex analytically homeomorphic to the complex projective space* $P_n(\mathbf{C})$. *Moreover the complex analytic homeomorphism* $X \to P_n(\mathbf{C})$ *is given by*

$$f : z \to f(z) = (1, f_1(z), f_2(z), \ldots, f_n(z)),$$

where $\{1, f_1(z), \dots, f_n(z)\}$ *is a base of the linear space* L(D).

The above assumptions (i)–(iv) will be made throughout this paragraph. The proof of Theorem 6 will depend on various lemmas.

Let E be the complex line bundle belonging to the divisor D and let $\Gamma(\mathrm{E})$ be the vector space of holomorphic sections of E. By our hypothesis (iv) we have

$$\dim \Gamma(\mathrm{E}) = \dim \mathrm{L}(\mathrm{D}) = n + 1.$$

Let $\{\varphi_0, \varphi_1, \dots, \varphi_\nu, \dots, \varphi_n\}$ be a base of $\Gamma(\mathrm{E})$. Take a sufficiently fine covering $\{\mathrm{U}_j\}$ of X. Then each section φ_ν is represented in the form : $z \to \varphi_\nu(z) = (z, \varphi_{\nu j}(z))$, where $\varphi_{\nu j}(z)$ is the fibre coordinate of $\varphi_\nu(z)$ over U_j. These coordinates satisfy

$$\varphi_{\nu j}(z) = e_{jk}(z)\varphi_{\nu k}(z), \quad z \in \mathrm{U}_j \cap \mathrm{U}_k,$$

where $e_{jk}(z)$ are transition functions defining the line bundle E. We introduce a projective space $\mathrm{P}_n^*(\mathbf{C})$ and associate with each point $\lambda \in \mathrm{P}_n^*(\mathbf{C})$ the holomorphic divisor D defined by

$$\mathrm{D}_\lambda = (\varphi_{\lambda j}), \quad \varphi_{\lambda j}(z) = \lambda_0\varphi_{0j}(z) + \cdots + \lambda_\nu\varphi_{\nu j}(z) + \cdots + \lambda_n\varphi_{nj}(z),$$

where $(\lambda_0, \lambda_1, \dots, \lambda_n)$ is the homogeneous coordinate of λ. We note that the set $\{\mathrm{D}_\lambda | \lambda \in \mathrm{P}_n^*(\mathbf{C})\}$ forms the complete linear system $|\mathrm{D}|$ and that $c(\mathrm{D}_\lambda) = c(D) = g$.

Lemma 7. *Let* W *be an algebraic cycle of* X *of dimension* $n - s$ $(0 \le s \le n)$, *which is a finite linear combination of irreducible subvarieties of* X *with non-negative integers as coefficients. If the dual cohomology class of* W *is* g^s, *then* W *is an irreducible subvariety of* X.

Proof. We have

$$\mathrm{W} = \sum_{t=1}^{l} m_l \Theta_t,$$

where m_t are *positive* integers and Θ_t are irreducible subvarieties. The dual cohomology class of Θ_t equals $k_t g^s$ where k_t is a positive integer. Hence we get

$$g^s = \left(\sum_{t=1}^{l} m_l k_t\right) g^s,$$

but this is only possible if $l = 1$ and $\mathrm{W} = \Theta_1$.

The preceding Lemma shows in particular that each divisor D_λ is an irreducible subvariety of X (of dimension $n - 1$).

We denote by B the *base* of the linear system $|\mathrm{D}| = \{\mathrm{D}_\lambda\}$, i. e. the subset of X consisting of all points z such that $\varphi_\nu(z) = 0$ for $\nu = 0, 1, \dots, n$. Clearly, $\mathrm{B} = \bigcap_\lambda \mathrm{D}_\lambda$. If B is not empty, then B is of dimension b with $0 \le b \le n - 2$. For, if $b = n - 1$, we would have $\mathrm{D}_\lambda = \mathrm{B}$ for all λ, since D_λ is irreducible. But this is impossible, because, for an arbitrary point $z \in \mathrm{X}$ we can find λ such that $z \in \mathrm{D}_\lambda$. We shall see later that B is empty.

Lemma 8. *Let* $\lambda^{(1)}, \lambda^{(2)}, \ldots, \lambda^{(r)} (1 \le r \le n)$ *be points of* $P_n^*(\mathbf{C})$ *and let*

$$W^{(r)} = D_{\lambda(1)} \cap D_{\lambda(2)} \cap \cdots \cap D_{\lambda(r)}.$$

Then $W^{(r)}$ *is an irreducible subvariety of* X *containing* B. *Moreover the dual cohomology class of* $W^{(r)}$ *is* g^s *where* $s = n - \dim W^{(r)}$ *and* $1 \le s \le r$.

Proof. The Lemma is already proved for $r = 1$, since every divisor D_λ is an irreducible subvariety with dual cohomology class g. Assume that the lemma is proved for $r = q - 1$. Put

$$W^{(q-1)} = D_{\lambda(1)} \cap D_{\lambda(2)} \cap \cdots \cap D_{\lambda(q-1)}, \quad s' = n - \dim W^{(q-1)}.$$

If $W^{(q-1)} \subset D_{\lambda(q)}$, then

$$W^{(q)} = D_{\lambda(1)} \cap D_{\lambda(2)} \cap \cdots \cap D_{\lambda(q)} = W^{(q-1)}$$

and the assertion of the lemma is proved for $W^{(q)}$. If $W^{(q-1)} \not\subset D_{\lambda(q)}$, then also $D_{\lambda(q)} \not\subset W^{(q-1)}$, because $W^{(q-1)}$ is irreducible. Then the intersection cycle $W^{(q-1)} \cdot D_{\lambda(q)}$ is defined in the sense of the intersection theory of A. Weil [10] and has the dual cohomology class $g^{s'} g$. Thus we get from Lemma 7 that $W^{(q)} = W^{(q-1)} \cap D_{\lambda(q)}$ is irreducible which completes the proof.

Remark. *In the preceding proof and at some other places of this paper we use tacitly the fact that a (compact) complex-analytic subvariety of an algebraic variety is algebraic and other related theorems. (See* [1] *and the recent work of J. P. Serre on Analytic and Algebraic Geometry).*

Now we consider the meromorphic functions

$$f_\nu(z) = \varphi_{\nu j}(z)/\varphi_{0j}(z) \quad (\nu = 1, 2, \ldots, n).$$

Lemma 9. *The meromorphic functions* $f_1(z), \ldots, f_n(z)$ *are algebraically independent.*

Proof. By hypothesis X is a non-singular connected algebraic manifold imbedded in a projective space $P_m(\mathbf{C})$. We denote by $(z_0, z_1, \ldots, z_m)$ the homogeneous coordinate of a point z of X in $P_m(\mathbf{C})$. The meromorphic functions $f(z), \ldots, f_\nu(z), \ldots$ on X are rational functions (*see* Chow [1]), i. e. there exist homogeneous polynomials $\Phi_\nu(z) = \Phi_\nu(z_0, z_1, \ldots, z_m)$ of degree d (not vanishing identically on X) such that

$$f_\nu(z) = \Phi_\nu(z)/\Phi_0(z) \quad \text{for} \quad z \in X.$$

Now, letting $\xi = (1, \xi_1, \xi_2, \ldots, \xi_m)$ be a generic point of X, we denote by V the subvariety of $P_n(\mathbf{C})$ with the generic point

$$(1, f_1(\xi), f_2(\xi), \ldots, f_n(\xi)),$$

where $f_\nu(\xi) = \Phi_\nu(\xi)/\Phi_0(\xi)$. For our purpose it suffices to show that $\dim V = n$.

Let $r = \dim \mathrm{V}$. By our definition of B,

$$f(z) = (1, f_1(z), \ldots, f_n(z)) = (\varphi_{0j}(z), \varphi_{1j}(z), \ldots, \varphi_{nj}(z))$$

is well-defined for each point $z \in \mathrm{X} - \mathrm{B}$ and $z \to f(z)$ is a regular map of $\mathrm{X} - \mathrm{B}$ into V. Let x be a point on $\mathrm{X} - \mathrm{B}$ such that $f(x)$ is a simple point of V. Obviously such a point x exists. V has a tangent space T at $f(x)$ which is a linear subspace of $\mathrm{P}_n(\mathbf{C})$ of dimension r. We shall prove that $f(\mathrm{X} - \mathrm{B}) \subset \mathrm{T}$. Let $y \in \mathrm{X} - \mathrm{B}$ and assume that $f(y) \notin \mathrm{T}$. Then there exists a linear subspace L of $\mathrm{P}_n(\mathbf{C})$ of dimension $n - r$ transversal to T and passing through the points $f(x)$ and $f(y)$. L is defined by r linear equations

$$\lambda_0^{(s)} u_0 + \lambda_1^{(s)} u_1 + \cdots + \lambda_n^{(s)} u_n = 0 \quad (s = 1, \ldots, r),$$

where $(u_0, u_1, \ldots, u_n)$ denotes the homogeneous coordinate of $\mathrm{P}_n(\mathbf{C})$. Consider the intersection

$$\mathrm{W}^{(r)} - \mathrm{D}_{\lambda(1)} \cap \mathrm{D}_{\lambda(2)} \cap \cdots \cap \mathrm{D}_{\lambda(r)}$$

of the subvarieties $\mathrm{D}_{\lambda(s)}$ of X defined by

$$\lambda_0^{(s)} \varphi_{0j}(z) + \cdots + \lambda_n^{(s)} \varphi_{nj}(z) = 0.$$

Clearly $\mathrm{W}^{(r)}$ contains $x, y (x, y \notin \mathrm{B})$ and therefore $\mathrm{W}^{(r)} \supsetneqq \mathrm{B}$, while, by Lemma 8, $\mathrm{W}^{(r)}$ is an irreducible subvariety. Hence $\mathrm{W}^{(r)} - \mathrm{B}$ is a connected open subvariety of $\mathrm{W}^{(r)}$ and consequently $f(\mathrm{W}^{(r)} - \mathrm{B})$ is a connected subset of V combining the points $f(x)$ and $f(y)$. On the other hand we have $f(\mathrm{W}^{(r)} - \mathrm{B}) \subset \mathrm{L} \cap \mathrm{V}$. It follows that $f(x)$ is *not* an isolated point of $\mathrm{L} \cap \mathrm{V}$. This contradicts with L being transversal to T. Thus contrary to our assumption we have $f(y) \in \mathrm{T}$. This proves $f(\mathrm{X} - \mathrm{B}) \subset \mathrm{T}$ which implies that $\mathrm{V} \subset \mathrm{T}$. Therefore V must be equal to the linear subspace T of $\mathrm{P}_n(\mathbf{C})$. Since 1, $f_1(\xi), \ldots, f_n(\xi)$ are *linearly independent*, this is only possible for

$$\dim \mathrm{V} = \dim \mathrm{T} = r = n.$$

This completes the proof of Lemma 9.

Now we come to the proof of Theorem 6.

Let n be a positive integer and consider the vector space $\mathfrak{L}$ consisting of all homogeneous polynomials

$$\Psi(u) = \Psi(u_0, u_1, \ldots, u_n)$$

in $u = (u_0, u_1, \ldots, u_n)$ of degree ν, and let $\{\Psi_0, \Psi_1, \ldots, \Psi_{\mathrm{N}}\}$ be a base of $\mathfrak{L}$. It is well known that

$$\mathrm{N} + 1 = \dim \mathfrak{L} = \binom{n + \nu}{n}$$

and that

$$\mathrm{S} : u \to (\Psi_0(u), \Psi_1(u), \ldots, \Psi_{\mathrm{N}}(u))$$

is a bi-regular map of $\mathrm{P}_n(\mathbf{C})$ *into* $\mathrm{P}_{\mathrm{N}}(\mathbf{C})$. Setting

$$\psi_{\nu j}(z) = \Psi_\nu(\varphi_{0j}(z), \varphi_{1j}(z), \ldots, \varphi_{nj}(z)),$$

we associate with each Ψ_ν the holomorphic section

$$\psi_\nu : z \to \psi_\nu(z) = (z, \psi_{\nu j}(z))$$

of the complex line bundle $\mathrm{E}^\nu = \underbrace{\mathrm{E} \otimes \mathrm{E} \otimes \cdots \otimes \mathrm{E}}_{\nu}$ which belongs to the divisor $\nu\mathrm{D}$.

Since, by Lemma 9, $f_1(z), \ldots, f_n(z)$ are *algebraically independent*, the meromophic functions

$$\mathrm{F}_\nu(z) = \Psi_\nu(1, f_1(z), \ldots, f_n(z)) \quad (\nu = 0, 1, 2, \ldots, \mathrm{N})$$

are *linearly independent*,. while we have

$$\psi_{\nu j}(z)/\psi_{0j}(z) = \mathrm{F}_\nu(z)/\mathrm{F}_0(z).$$

Consequently the $\mathrm{N}+1$ sections $\psi_0, \psi_1, \ldots, \psi_\mathrm{N}$ of E^ν are linearly independent. On the other hand, we have, by our hypothesis (iv),

$$\dim \Gamma(\mathrm{E}^\nu) = \dim \mathrm{L}(\nu\mathrm{D}) = \binom{n+\nu}{n} = \mathrm{N}+1.$$

Hence *these sections* $\psi_0, \psi_1, \ldots, \psi_\mathrm{N}$ *form a base of* $\Gamma(\mathrm{E}^\nu)$. By hypothesis, the characteristic class $c(\mathrm{E})$ of E is *positive*. Therefore

$$\psi : z \to \psi(z) = (\psi_{0j}(z), \psi_{1j}(z), \ldots, \psi_{\mathrm{N}j}(z))$$

is a bi-regular map of X *into* $\mathrm{P}_\mathrm{N}(\mathbf{C})$, provided that ν is sufficiently large (*see* [6], p. 39, Theorem 3). In particular, for each point $z \in \mathrm{X}$, at least one of coordinates $\psi_{\nu j}(z)$ does not vanish. This implies that at least one of $\varphi_{\nu j}(z)$ does not vanish. Thus the base B of $|\mathrm{D}| = \{\mathrm{D}_\lambda\}$ is empty and the map

$$f : z \to f(z) = (1, f_1(z), \ldots, f_n(z)) = (\varphi_{0j}(z), \varphi_{1j}(z), \ldots, \varphi_{nj}(z))$$

is a regular map of X into $\mathrm{P}_n(\mathbf{C})$. Now it follows from the commutative diagram

$$\begin{array}{ccc} \mathrm{X} & \xrightarrow{f} & \mathrm{P}_n(\mathbf{C}) \\ \psi \searrow & & \swarrow \mathrm{s} \\ & \mathrm{P}_\mathrm{N}(\mathbf{C}) & \end{array}$$

that f is a bi-regular map of X into $\mathrm{P}_n(\mathbf{C})$, since ψ is bi-regular.

On the other hand we have $\dim \mathrm{X} = \dim f(\mathrm{X}) = \dim \mathrm{P}_n(\mathbf{C}) = n$. Hence f is a bi-regular map of X *onto* $\mathrm{P}_n(\mathbf{C})$. This completes the proof of Theorem 6.

By combining Theorem 6 with Theorem 5 we obtain immediately our main theorem mentioned in the Introduction.

References

[1] W. L. Chow *On compact complex analytic varieties* (*Amer. J. Math.*, t. 71, 1949, p. 893–914).

[2] F. HIRZEBRUCH, *Some problems on differentiable and complex manifolds* (*Ann. Math.*, t. 60; 1954, p. 213–236).

[3] F. HIRZEBRUCH, *Neue topologische Methoden in der algebraischen Geometrie* (*Ergebnisse der Mathematik und ihrer Grenzgebiete*, N. F. *Heft* 9, Berlin-Göttingen-Heidelberg, Springer Verlag).

[4] F. HIRZEBRUCH, *Der Satz von Riemann-Roch in faisceau-theoretischer Formulierung*, (*Proc. Int. Cong. Math.*, 1954, p. 457–472.

[5] F. HIRZEBRUCH, *On Steenrod's reduced powers, the index of inertia, and the Todd genus* (*Proc. Nat. Acad. Sc., U. S. A.*, t. 39, 1953, p. 951–956).

[6] K. KODAIRA, *On Kähler varieties of restricted type* (*Ann. .Math.*, t. 60, 1954, p. 28-48).

[7] K. KODAIRA. *On a differential-geometric method in the theorie of analytic stacks* (*Proc. Nat. Acad. Sc. U. S. A.*, t. 39, 1953, p. 1268–1273).

[8] K. KODAIRA and D. C. SPENCER, *Groups of complex line bundles over compact Kähler varieties. Divisor class groups on algebraic varieties* (*Proc. Nat. Acad. Sc., U. S. A.*, t. 39, 1953, p. 868–877).

[9] B. L. VAN DER WAERDEN, *Birationale Transformationen von linearen Scharen auf algebraischen Mannigfaltigkeiten* (*Math. Z.*, t. 51, 1948, p. 502–523).

[10] A. WEIL, *Foundations of algebraic geometry* (*Amer Math., Soc. Coll. Publ.*, t. 24, 1946).

Editor's Comments

There are many generalizations of Riemann-Roch theorem and they usually go under the name of index theorems. Though they have different expressions, the basic unifying idea is the same: relating data from analysis on the manifolds and bundles above them to the topological data of the spaces under consideration. Though the spaces are not necessarily complex manifolds, some of the most important results and applications are adapted to algebraic varieties and complex manifolds.

It is difficult to pick one paper which gives a taste of this vast and important subject. This short announcement of a major result fits the general idea of index theory relating analytic data to topological data and also shows connections to several other subjects. For example, it is related to the basic Lefschetz fixed formula in algebraic topology and trace formulas in automorphic forms and number theory.

Complex Geometry from Riemann to Kähler-Einstein and Calabi-Yau
ALM 38, pp. 269–274

A Lefschetz Fixed Point Formula for Elliptic Differential Operators[1,*]

Michael Atiyah and Raoul Bott

Introduction

The classical Lefschetz fixed point formula expresses, under suitable circumstances, the number of fixed points of a continuous map $f : X \to X$ in terms of the transformation induced by f on the cohomology of X. If X is not just a topological space but has some further structure, and if this structure is preserved by f, one would expect to be able to refine the Lefschetz formula and to say more about the nature of the fixed points. The purpose of this note is to present such a refinement (Theorem 1) when X is a compact differentiable manifold endowed with an elliptic differential operator (or more generally an elliptic complex). Taking essentially the classical operators of complex and Riemannian geometry we obtain a number of important special cases (Theorems 2, 3). The first of these was conjectured to us by Shimura and was proved by Eichler for dimension one.

1 The main theorem

Let X be a smooth compact manifold and let E, F be smooth complex vector bundles over X. A differential operator from E to F means a linear map $d : \Gamma(E) \to \Gamma(F)$ on the spaces of smooth sections which is given in local coordinates by a matrix of partial differential operators with smooth coefficients. By an *elliptic complex* E on X we mean a sequence $E_0, E_1, \ldots, E_n$ of smooth vector bundles over X and a sequence of differential operators $d_i : \Gamma(E_i) \to \Gamma(E_{i+1})$ so that

(i) $d_{i+1}d_i = 0$ for $i = 1, \ldots, n-1$, and

[1] This work was supported in part by the National Science Foundation.

* Communicated by E. Spanier, August 20, 1965. Atiyah, M. F.; Bott, R., A Lefschetz fixed point formula for elliptic differential operators. Bull. Amer. Math. Soc. 72 (1966) 245–250.

(ii) the sequence of symbols

$$\cdots \to E_{i,x} \xrightarrow{\sigma_i(x,\xi)} E_{i+1,x} \to \cdots$$

is *exact* for all $x \in X$ and all nonzero cotangent vectors ξ to X at x. Here σ_i denotes the symbol of d_i, and d_i as well as σ_i is formally put equal to zero for $i < 0$ and $i \geqq n$.

In view of (i) the cohomology groups $H^i(E)$ =Kernel d_i/Image d_{i-1}, are well defined and it is a consequence of (ii) that $H^i(E)$ is finite dimensional. Note that, if $n = 1$, we are dealing with a single elliptic operator d_0. H^0 is then the space of solutions of $d_0 u = 0$ and H^1 can be shown to be isomorphic to the space of solutions of the adjoint equation $d_0^* v = 0$.

By an endomorphism T of an elliptic complex E we mean a sequence of linear maps $T_i : \Gamma(E_i) \to \Gamma(E_i)$ such that $d_i T_i = T_{i+1} d_i$. Such an endomorphism induces endomorphisms $H^i T$ of $H^i(E)$ and we define the *Lefschetz number* of T by the formula

$$L(T) = \sum_{i=0}^{n} (-1)^i \text{ Trace } H^i T.$$

For instance if T is the identity endomorphism I the Lefschetz number $L(I)$ is just the Euler-characteristic $\chi(E) = \sum(-1)^i \dim H^i(E)$, and in particular if $n = 1$, so that we are dealing with a single elliptic operator d_0, then $L(I) = \dim \operatorname{Ker} d_0 - \dim \operatorname{Coker} d_0 = \operatorname{index} d_0$. The problem of computing index d_0 (and more generally $\chi(E)$) has been solved in [1]. In this note we are concerned with endomorphisms which are, in a sense, at the opposite extreme from the identity.

Suppose then that $f : X \to X$ is a smooth map and that $\phi_i : f^* E_i \to E_i$ are smooth vector bundle homomorphisms. We define linear maps $T_i : \Gamma(E_i) \to \Gamma(E_i)$ as the composition

$$\Gamma(E_i) \xrightarrow{f^*} \Gamma(f^* E_i) \xrightarrow{\phi_i} \Gamma(E_i).$$

If further $d_i T_i = T_{i+1} d_i$ then the T_i define an endomorphism T of the elliptic complex E. An endomorphism of this type, associated to $\{f, \phi_i\}$ we call a *geometric* endomorphism of E.

Finally we define a fixed point P of a map $f : X \to X$ to be *simple* if $\det(1 - df_P) \neq 0$, where df_P is the induced map on the tangent space to X at P. This implies that P is an isolated fixed point. Hence if all fixed points of f are simple it follows, since X is compact, that they are finite in number. Let us observe that ϕ_i may be interpreted as a family of linear maps $\phi_{i,P} : E_{i,f(P)} \to E_{i,P}$, parameterized by $P \in X$. Hence if P is a fixed point, $f(P) = P$, then $\phi_{i,P}$ is an endomorphism of the vector space $E_{i,P}$ and so Trace $\phi_{i,P}$ is defined.

Theorem 1. *Let E be an elliptic complex on X and let T be a geometric endomorphism of E associated to (f, ϕ_i) where $f : X \to X$ has only simple fixed points. Then the Lefschetz number $L(T)$ is given by the formula*

$$L(T) = \sum_P \nu(P)$$

where the summation is over the set of fixed points of f and $\nu(P)$ is given by

$$\nu(P) = \frac{\sum(-1)^i \text{Trace } \phi_{i,P}}{|\det(1 - df_P)|}$$

Remark.

(1) *Note that the formula for $\nu(P)$ does not explicitly involve the differential operators d_i. This is in marked contrast to the index formula* [1, *Theorem* 1]. *Of course the d_i are implicitly involved by the condition $T_i d_i = T_{i+1} d_i$.*

(2) *If we take E to be the de Rham complex* (*exterior differential forms with the exterior derivative*), *and ϕ_i to be the natural map on i-forms induced by f, we find*

$$\nu(P) = \frac{\det(1 - df_P)}{|\det(1 - df_P)|} = \pm 1$$

and we recover the original Lefschetz theorem, since $L(T)$ is now the usual Lefschetz number of the map f.

(3) *Note that in general $\nu(P)$ is a complex number and not an integer. The classical Lefschetz formula,* where $\nu(P) = \pm 1$, *is highly special in this direction. Note on the other hand that the Lefschetz number $L(T)$ is a linear combination of traces and so, if T is of finite order, $L(T)$ will be an algebraic integer. In these cases Theorem 1 leads to "integrality theorems" analogous to the integrality theorems obtainable from the index theorem.*

2 The holomorphic case

Using the $\bar{\partial}$-complex on a complex manifold and the Dolbeault isomorphism Theorem 1 leads easily to:

Theorem 2. *Let X be a compact complex manifold, V a holomorphic vector bundle over $X, f : X \to X$ a holomorphic map with simple fixed points and $\phi : f^*V \to V$ a holomorphic vector bundle homomorphism. Let $H^i(f, \phi)$ denote the composite homomorphism*

$$H^i(X, v) \xrightarrow{f^*} H^i(X, f^*v) \xrightarrow{\phi^*} H^i(X, v),$$

where v is the sheaf of germs of holomorphic sections of V. Then

$$\sum(-1)^i \text{ Trace } H^i(f, \phi) = \sum_P \nu(P),$$

where the summation is over the fixed points of f,

$$\nu(P) = \frac{\text{Trace } \phi_P}{\det_C(1 - df_P)}$$

and df_P denotes the C-endomorphism [2] *of the tangent space to X at P induced by f.*

[2] For a complex linear endomorphism we must distinguish carefully between $\det_C$ and $\det_R$ the determinant of the underlying real transformation: they are related by $\det_R = |\det_C|^2$.

Like the Riemann-Roch theorem, Theorem 2 makes sense and is also true in the case of abstract algebraic geometry. As a simple corollary of Theorem 2 we have:

Corollary 1. *Any rational projective algebraic manifold has the fixed point property for holomorphic maps.*

3 The Riemannian case

Let X be a compact oriented Riemann manifold of dimension $2l$. Let H^l denote the space of (complex-valued) harmonic l-forms. The usual dualizing operator $*: H^l \to H^l$ has $(*)^2 = (-1)^l$. Hence $\alpha = i^{l^2} *$ has $\alpha^2 = 1$. Let H^l_+ and H^l_- denote the $+1, -1$ eigenspaces of α. Now there is an elliptic operator (cf. [1, (3.1) (ii)]), canonically associated to the structure of X, whose index is $\dim H^l_+ - \dim H^l_-$. If l is even this is the Hirzebruch index of X. More generally therefore if $f: X \to X$ is an oriented isometry we may define the Hirzebruch number

$$\tau(f) = \text{ Trace } f^l_+ - \text{Trace } f^l_-,$$

where f^l_+ is the endomorphism of H^l_+ induced by f and similarly for f^l_-. It is a special Lefschetz number. Its interest lies in the fact that it depends only on the section of f on the cohomology of X and so it is quite easily computed. Applying Theorem 1 to this special operator we obtain

Theorem 3. *Let X be a compact oriented smooth manifold of dimension $2l$. Let $f: X \to X$ be an oriented isometry with isolated fixed points. For each fixed point P let $T_P = \oplus_{k=1}^l T_{P,k}$ be a decomposition of the oriented tangent space T_P into oriented 2-planes $T_{P,k}$ invariant under df_P, and let df_P induce a rotation through an angle $\theta_{P,k}$ on $T_{P,k}$. Then the Hirzebruch number $\tau(f)$ is given by the formula*

$$\tau(f) = \sum_P \nu(P),$$

where

$$\nu(P) = i^l \prod_{k=1}^{l} \cot\left(\frac{\theta_{P,k}}{2}\right).$$

If G is a compact Lie group acting differentiably on X then by averaging over G we can find a Riemannian metric invariant under G. Theorem 3 can then be used to obtain results concerning such group actions. For example Theorem 3 together with some delicate number theory (*for which we are indebted to J. Milnor*) gives:

Corollary 2. *Let G be a compact Lie group acting differentiably on a (homology) sphere. Assume that G has just two fixed points P, Q and that elsewhere the action is free. Then the representations of G on the tangent spaces at P, Q are equivalent.*

This result is particularly interesting when G is a finite cyclic group since it implies that two lens spaces which are h-cobordant are isometric (cf. [3, §12]).

4 Outline of proof

If T is a geometric endomorphism of an elliptic complex E the operators T_i : $\Gamma(E_i) \to \Gamma(E_i)$ do not unfortunately have traces in any accepted functional analytic sense. If, however, T is defined by a map f with only simple fixed points it turns out that one can define a trace for T_i which we call the flat trace and denote bv $\text{Trace}^{\flat}\, T_i$. It is defined as a limit

$$\text{Trace}^{\flat}\, T_i = \lim_{n\to\infty} \text{ Trace } (T_i \circ I_n), \tag{1}$$

where I_n is a sequence of operators with smooth kernels which approximate the identity operator in a certain rather precise way (referred to as flat approximation). Note that the composition $T_i \circ I_n$ has smooth kernel and therefore has a trace in an accepted sense, namely as $\int k(x,x)dx$, where $k(x,y)$ is the kernel.

From its definition as the limit of a sequence of integrals it is not difficult to evaluate $\text{Trace}^{b}\, T_i$ explicitly, and we obtain

$$\text{Trace}^{b}\, T_i = \sum_P \frac{\text{Trace } \phi_{i,P}}{|\det(1 - df_P)|},$$

where the summation is over the fixed points of f. To prove Theorem 1 therefore, we have to show that

$$\sum (-1)^i \text{ Trace } T_i = \sum (-1)^i \text{ Trace } H^i T. \tag{2}$$

For this purpose consider the "Zeta-function" [4]

$$\zeta_i(s) = \text{Trace}\{(1+\Delta)^{-s} \circ T_i\},$$

where Δ is the operator[3] $dd^* + d^*d$ relative to Riemannian structures on the E_i and X. For $\text{Re}(s)$ large this trace exists in the Hilbert Space sense and there, by the Hodge theory:

$$\sum (-1)^i \zeta_i(s) = \sum (-1)^i \text{ Trace } H^i T. \tag{3}$$

Using more delicate properties of Δ, as given in [2], and more recently further refined by Hormander and Seeley, one then shows that $\text{Trace}^{b}\, \{(1+\Delta)^{-s} \circ T_i\}$ defined as in (1), furnishes an analytic continuation of $\zeta_i(s)$ to the whole plane. The result then follows by putting $s = 0$.

References

[1] M. F. Atiyah and I. M. Singer, *The index of elliptic operators on compact manifolds*, Bull. Amer. Math. Soc. 69 (1963), 422–433.

[2] J. Kotake and M. Narasimhan, *Regularity theorems for fractional powers of a linear elliptic operator*, Bull. Soc. Math. France 90 (1962), 449–471.

[3] For simplicity of exposition we assume all d_i are of the same order.

[3] J. Milnor, Whitehead torsion, (to appear).

[4] S. Minakshisundaram and A. Pleijel, *Some properties of the eigenfunctions of the Laplace-operator on Riemann manifolds*, Canad. J. Math. 1 (1949), 242–256.

Oxford University,
Harvard University.

Editor's Comments

If one wants to learn how Kähler-Einstein metrics are used and why they are important, this is the paper to read. It shows how several major and longstanding conjectures in algebraic geometry and complex analytic geometry can be solved with ease with the help of Kähler-Einstein metrics. The rigidity of the complex projective space $\mathbb{C}P^n$ mentioned above is one such example. The basic point is that algebraic varieties enhanced with canonical Kähler-Einstein metrics are very special.

This paper is short and written by the world leading figure on Kähler-Einstein metrics and the interaction between complex geometry and algebraic geometry. The best recommendation one can offer is to read it and see for yourself.

Complex Geometry from Riemann to Kähler-Einstein and Calabi-Yau
ALM 38, pp. 277–281

Calabi's Conjecture and Some New Results in Algebraic Geometry*

Shing-Tung Yau

Abstract

We announce a proof of Calabi's conjectures on the Ricci curvature of a compact Kähler manifold and then apply it to prove some new results in algebraic geometry and differential geometry. For example, we prove that the only Kähler structure on a complex projective space is the standard one.

Let M be a compact Kähler manifold with Kähler metric $\Sigma g_{i\bar{j}}\, dz^i \otimes d\bar{z}^j$. Let $\Sigma R_{i\bar{j}}\, dz^i \otimes d\bar{z}^j$ be the Ricci tensor of M. Then, according to Chern (ref. 1), the (1,1) form $\sqrt{-1}/2\pi\Sigma R_{i\bar{j}}\, dz^i \Lambda d\bar{z}^j$ is closed and represents the first Chern class of M. Hence, there is a natural obstruction for a tensor to be the Ricci tensor of some Kähler metric. In refs. 2 and 3, Calabi asked whether this is the only obstruction. Namely, given a closed (1,1) form $\sqrt{-1}/2\pi\Sigma\tilde{R}_{i\bar{j}}\, dz^i \Lambda d\bar{z}^j$, which represents the first Chern class of M, can one find a Kähler metric $\Sigma\tilde{g}_{i\bar{j}}\, dz^i \otimes d\bar{z}^j$ such that $\Sigma\tilde{R}_{i\bar{j}}\, dz^i \otimes d\bar{z}^j$ is the Ricci tensor of this metric and that $\sqrt{-1}/2\Sigma\tilde{g}_{i\bar{j}}\, dz^i \Lambda d\bar{z}^j$ determines the same cohomology class as $\sqrt{-1}/2\Sigma g_{i\bar{j}}\, dz^i \Lambda d\bar{z}^j$. Calabi (ref. 3) establishes the uniqueness of such a metric $\Sigma\tilde{g}_{i\bar{j}}\, dz^i \otimes d\bar{z}^j$ and proved the existence under the assumption that $\tilde{R}_{i\bar{j}}$ is close to R_{ij}. The existence of $\Sigma\tilde{g}_{i\bar{j}}\, dz^i \otimes d\bar{z}^j$, without any assumption on M or $\tilde{R}_{i\bar{j}}$, is known as Calabi's conjecture.

Since $\sqrt{-1}/2\pi\Sigma\tilde{R}_{i\bar{j}}\, dz^i \Lambda d\bar{z}^j = -\sqrt{-1}/2\pi\partial\bar{\partial}\log\det[g_{i\bar{j}} + (\partial^2\varphi/\partial z^i\partial\bar{z}^j)]$, the condition that $\sqrt{-1}/2\pi\Sigma\tilde{R}_{i\bar{j}}\, dz^i \Lambda d\bar{z}^j$ is cohomologous to $\sqrt{-1}/2\pi\Sigma R_{i\bar{j}}\, dz^i \Lambda d\bar{z}^j$

---

* Mathematics Department, Stanford University, Stanford, California 94305. *Communicated by S. S. Chern, January 31, 1977.* S.-T. Yau, Calabi's conjecture and some new results in algebraic geometry. Proc. Natl. Acad. Sci. USA, 74 (1977) 1798–1799.

implies that we can find a smooth function F on M so that

$$-\frac{\sqrt{-1}}{2\pi}\partial\bar{\partial}\log\det\left(g_{i\bar{j}}+\frac{\partial^2\varphi}{\partial z^i\partial\bar{z}^j}\right)$$
$$+\frac{\sqrt{-1}}{2\pi}\partial\bar{\partial}\log\det(g_{i\bar{j}})=\frac{\sqrt{-1}}{2\pi}\partial\bar{\partial}F. \tag{1}$$

The conjecture of Calabi is therefore equivalent to solving [1] with F and $g_{i\bar{j}}$ given. Here φ is required to be smooth so that $\Sigma_{r,j}[g_{i\bar{j}}+(\partial^2\varphi/\partial z^i\partial\bar{z}^j)]dz^i\otimes d\bar{z}^j$ defines a metric tensor. Adding F by a known constant, we see that it is sufficient to solve the following equation

$$\det\left(g_{i\bar{j}}+\frac{\partial^2\varphi}{\partial z^i\partial\bar{z}^j}\right)\det(g_{i\bar{j}})^{-1}=\exp(F), \tag{2}$$

where

$$\int_M \exp(F)=\mathrm{Vol}\,(M). \tag{3}$$

In ref. 2, Calabi also conjectured the existence of Kähler Einstein metric on a Kähler manifold whose first Chern class is negative, zero, or positive and which does not admit any holomorphic vector field. (A theorem of Nakano says that negativity of the first Chern class implies the nonexistence of holomorphic vector field.) Kähler Einstein metrics are the Kähler metrics whose Ricci form is proportional to the Kähler form. If the first Chern class is negative, then by definition we can choose Kähler metric so that $\sqrt{-1}/2\Sigma g_{i\bar{j}}\,dz^i\Lambda d\bar{z}^j$ represents $-c_1(M)$. If $c_1(M)$ is zero, we choose $\sqrt{-1}/2\pi\Sigma g_{i\bar{j}}\,dz^i\Lambda d\bar{z}^j$ to be arbitrary Kähler form. If $c_1(M)$ is positive, we choose $\sqrt{-1}/2\Sigma g_{i\bar{j}}\,dz^i\Lambda d\bar{z}^j$ to represent $c_1(M)$. In these cases, the existence of Kähler Einstein metric is equivalent to solving the following equation

$$\det\left(g_{i\bar{j}}+\frac{\partial^2\varphi}{\partial z^i\partial\bar{z}^j}\right)\det(g_{i\bar{j}})^{-1}=\exp(c\varphi+F), \tag{4}$$

where $c=+1,0$ or -1, and F is a smooth function defined on M that satisfies [3] when $c=0$.

In this note, we announce the following solution of Calabi's conjecture.

Theorem 1. *Let M be a compact Kähler manifold with Kähler metric $\Sigma g_{i\bar{j}}\,dz^i\otimes d\bar{z}^j$. Then when $c\geq 0$ eqs. 2 and 4 can be solved with a smooth φ. In particular, any $(1,1)$ form on M that represents $c_1(M)$ can be realized as the Ricci form of a unique Kähler metric whose Kähler class is that of $\sqrt{-1}/2\Sigma g_{i\bar{j}}\,dz^i\Lambda d\bar{z}^j$. If the first Chern class of M is zero and negative, then there exists a Kähler Einstein metric on M. In the latter case, if we require the Ricci curvature to be -1, the metric is unique, depending only on the complex structure of M.*

We shall discuss eq. 4 with $c<0$ on a latter occasion. We should mention that if the bisectional curvature of M is nonnegative, then T. Aubin (ref. 4) has solved eq. 2. However, the class of such manifolds is rather restrictive. Very recently he also announced (ref. 5) a solution of a special case of [4] when $c>0$.

Instead of giving details of how to prove *Theorem 1*, which will appear elsewhere, we shall now give some applications of *Theorem 1*. Some other applications will be published elsewhere also.

Theorem 2. *There are compact simply-connected Kähler manifolds whose Ricci curvature is identically zero and whose curvature tensor is not identically zero.*

Proof. It is wellknown that for $n > 1$, every complex hypersurface of degree $n + 2$ in the complex projective space Cp^{n+1} is simply connected and has zero first Chern class. Hence by *Theorem 1*, we can represent the zero (1,1) form as the Ricci tensor of some Kähler metric.

Theorem 3. *There are compact simply-connected Kähler manifolds whose Ricci curvature is negative everywhere.*

Proof. For $n > 1$, any complex hypersurface of CP^{n+1} with degree greater than $n+2$ is simply connected and has negative first Chern class. Using *Theorem 1*, we can find a Kähler metric with negative Ricci curvature on them.

Theorem 4. *Let M be a Kähler surface with ample canonical bundle. Then $3c_2(M) \geq c_1(M)^2$ and the equality holds if and only if M is covered biholomorphically by the ball in C^2.*

Proof. According to *Theorem* 1, we can find a Kähler Einstein metric on M whose Ricci curvature is a negative constant. By Chern's theorem (ref. 1), we can write $3c_2(M) - c_1^2(M)$ as the integral of a function that depends only on the curvature tensor of the above Einstein metric. This function can be described as follows. Let $\{e_1, e_2 = Je_1, e_3, e_4 = Je_3\}$ be a unitary frame at a point p. Then the value of the function at p is given by $1/4\pi^2\{[R_{1212} - 2(R_{1313} + R_{1414})]^2 + 3(R_{1313} - R_{1414})^2 + 6(R_{1312}^2 + R_{1412}^2 + R_{1413}^2)\}$. Since this function is non-negative, we have established the inequality $3c_2(M) \geq c_1^2(M)$. Furthermore, the equality holds iff the function is zero identically. As the Ricci curvature $R_{1212} + R_{1313} + R_{1414}$ is a negative constant, we see that in this case, the holomorphic sectional curvature of M is a negative constant. Therefore, M is covered by the ball holomorphically.

Remarks. *(i) Under the assumption that M is a surface of general type, Van de Ven (ref. 6) proved the inequality $8c_2(M) \geq c_1^2(M)$. Recently D. Mumford informed us that F. Bogomolv improved the inequality to $4c_2(M) \geq c_1^2(M)$ while Y. Miyaoka improved Bogomolov's argument to obtain $2c_2(M) \geq c_1^2(M)$. However, Miyaoka's argument does not give information for the case $3c_2(M) = c_1^2(M)$. We believe our method can be generalized to cover the case of general type. This will be discussed later.*

(ii) Assuming the surface is Kähler Eistein, we knew the inequality $3c_2(M) \geq c_1^2(M)$ about 4 years ago. However, it was pointed out by D. Mumford that more than 20 years ago, Guggenheimer (ref. 7) had already found the inequality, assuming the existence of Kähler Einstein metric.

(iii) We can generalize the above inequality to other inequalities for higher dimensional manifolds with ample canonical class. We can prove, for example, the inequality $(-1)^n c_1^{n-2} c_2(M) \geq (-1)^n n/2(n+1) c_1^n(M)$, and the equality holds only if M is covered by the ball holomorphically.

An important application of *Theorem* 4 is the following resolution of a conjecture of Severi (ref. 8)

Theorem 5. *Every complex surface that is homotopic to the complex projective plane* $\mathrm{C}P^2$ *is biholomorphic to* $\mathrm{C}P^2$.

Proof. Since the index of a four-dimensional (real) manifold is invariant (up to sign) under homotopic equivalence, we conclude that $1/3[c_1^2(M) - 2c_2(M)] = \pm 1$. As $c_2(M) = 3$, this shows immediately that $c_1^2(M) > 0$. By a theorem of Kodaira (ref. 9) this implies that M is algebraic.

Let $\mathcal{O}$ be the structure sheaf of M. Then the exact sequence

$$1 \to Z \to \mathcal{O} \to \mathcal{O}^* \to 1 \tag{5}$$

induces the long exact sequence

$$H^1(M, Z) \to H^1(M, \mathcal{O}) \to H^1(M, \mathcal{O}^*) \to H^2(M, Z) \to \tag{6}$$

Since $H^1(M, Z) = 0$ and $H^2(M, Z) = Z$, we have $H^1(M, \mathcal{O}) = 0$ and $H^1(M, \mathcal{O}^*)$ is infinitely cyclic. Therefore, line bundles of M are multiples of each other. As M is algebraic, the canonical line bundle of M is a multiple of some positive line bundle. Therefore, the canonical line bundle is either negative or positive. In the previous case, Hirzebruch and Kodaira (ref. 9) already established that M is biholomorphic to $\mathrm{C}P^2$. In both cases, they also established $c_1(M)^2 = 9$. Therefore, in the latter case $3c_2(M) = c_1^2(M)$ and we can apply *Theorem* 4 to conclude that M is covered by the ball. As M is simply connected, we conclude that the latter case cannot happen and M is biholomorphic to $\mathrm{C}p^2$.

Remark: We can apply the remark in *Theorem* 4 to prove that any Kähler manifold that is homeomorphic to $\mathrm{C}P^2$ is biholomorphic to $\mathrm{C}P^n$. One can also prove that $\mathrm{C}P^2$ is the only simply connected algebraic surface with positive definite index form (see ref. 6).

Theorem 6. *Let N be a compact complex surface that is covered by the ball in C_2. Then any complex surface M that is oriented homotopic to N is biholomorphic to N.*

Proof. Since N is Kähler, the first Betti number of N, and hence M, is even. According to a result of Y. Miyaoka (ref. 10), our complex surface that is not covered by the $K - 3$ surface is Kähler. We claim that the first Chern class of M is negative.

To see this, we notice the obvious fact that every compact holomorphic curve in a Kähler manifold is not homologous to zero in that manifold. We use this to show that M admits no rational curves or elliptic curves. In fact, it is wellknown that topologically, M is a $K(\pi, 1)$ whose fundamental group contains no nontrivial abelian subgroup besides the integers. Hence topologically, every continuous map from the rational curve or the elliptic curve into M must be homotopic to a map whose image is either a point or a circle. As a consequence, no nontrivial holomorphic map from a rational curve or elliptic curve into M is possible. According to the classification of complex surfaces, we conclude that M must be an algebraic surface of general type. On the other hand, for an algebraic surface of general type that contains no rational curves, the first Chern class is negative (see ref. 9).

Since M is oriented homotopic to N, the index of M is equal to that of N. One concludes immediately $c_1^2(M) = c_1^2(N)$ and $c_2(M) = c_2(N)$. Therefore, $3c_2(M) =$

$c_1^2(M)$. *Theorem* 4 then shows that M is covered by the ball in C^2. By Mostow's rigidity theorem (ref. 11), M is in fact biholomorphic to N.

This research was partially supported by a Sloan fellowship.

The costs of publication of this article were defrayed in part by the payment of page charges from funds made available to support the research which is the subject of the article. This article must therefore be hereby marked "advertisement" in accordance with 18 U. S. C. §1734 solely to indicate this fact.

References

[1] Chern, S. S. (1946) "Characteristic classes of Hermitian manifolds," *Ann. Math.* 47, 85–121.

[2] Calabi, E. (1954) "The space of Kähler metrics," *Proc. Int. Congr. Math. Amsterdam* 2, 206–207.

[3] Calabi, E. (1955) *On Kähler Manifolds with Vanishing Canonical Class, Algebraic Geometric and Topology, A Symposium in Honor of S. Lefschitz* (Princeton Univ. Press, Princeton, NJ), pp. 78–89.

[4] Aubin, T., (1970) "Métriques riemanniennes et courbure," *J. Diff. Geom.* 4, 383-424.

[5] Aubin, T. (1976) "Equations du type Mongi-Ampère sur les variétés kähleriennes compactes," *C. R. Acad. Sci. Hebd. Seances* 283, 119–121.

[6] Van de Ven, A. (1976) "On the Chern numbers of surfaces of general type," *Invent. Math.* 36, 285–293.

[7] Guggenheimer, H. (1952) "Über vierdimensionale Einsteinräume," *Experientia* **8**, 420–421.

[8] Severi, F. (1954) *Some remarks on the topological characterization of algebraic surfaces, in Studies presented to R. von Mises* (Academic Press, New York), pp. 54–61; MR 16, p. 397.

[9] Kodaira, K. (1975) *Collected Works* (Princeton Univ. Press, Princeton, NJ), Vols. II and III.

[10] Miyaoka, Y. (1974) "Kähler metrics on elliptic surfaces," *Proc. Jpn. Acad. Sci.*, 50.

[11] Mostow, G. D. (1973) *Strong Rigidity of Locally Symmetric Spaces* (Princeton Univ. Press, Princeton, NJ).

Editor's Comments

This is the first major paper which combines nonlinear partial differential equations and differential geometry. It is a modern masterpiece and a tour de force. It creates methods and solves the Calabi conjecture, which is the higher dimensional generalization of the uniformization of Riemann surfaces as explained above.

Briefly, the Calabi conjecture was reduced to questions in nonlinear partial differential equations of Monge-Ampère type, which are solved by using the method of continuity after establishing some extremely difficult a priori estimates. Besides this application to the Calabi conjecture, the methods and results on partial differential equations in this paper are of great interest in themselves and can be applied to other problems.

Though textbook expositions of this important result might polish the proof in some ways, reading the original always conveys the spirit and courage that the originator built roads and bridges to reach the final destination. It is an awe-inspiring and encouraging process.

The greatest work of a writer, artist, or composer is called his or her magnum opus. If mathematicians are considered artists (they should), then this paper is the magnum opus of Yau. It is the culmination of many years of innovative and hard work by Yau. He invented various methods and tools in geometric analysis in order to solve the Calabi conjecture.

This paper was carefully and well written, and it is also direct and down to earth. It is a must-read for anyone who is serious about geometric analysis.

Let us quote a few sentences from the paper. Yau started the paper by writing:

> Let M be a compact Kähler manifold with Kähler metric $\Sigma_{i,j}\ g_{i,\bar{j}} dz^i \otimes d\bar{z}^j$. Then it is wellknown that the Ricci curvature of M can be computed in the following way....
>
> Therefore a necessary condition for a $(1,1)$ form ... to be the Ricci form of some Kähler metric is that it must be closed and its cohomology class must represent the first Chern class of M.
>
> More than twenty years ago, E. Calabi [3] conjectured that the above necessary condition is in fact sufficient. This conjecture of Calabi can be reduced to a problem in non-linear partial differential equation.

On the second page, Yau wrote:

> In this paper, we shall solve equation (0.4) with the (necessary) compatibility condition (0.5).

On the fourth page, Yau wrote:

> My proof of Calabi's conjecture was completed in the middle of 1976. A large amount of work was done while I was working on the real Monge-Ampère equation (cf. [8], [9]). For example, Pogorelov's second order estimate in the real Monge-Ampère equation had direct bearing on my work here. Indeed, besides the uniform estimate, the higher order estimates were discovered at that time.

Yau also put his heart and soul into this paper. For example, on the fourth page, he wrote:

> I am very grateful to all my teachers and my friends for their constant encouragement during this work. Among them I should mention E. Calabi, S. S. Chern, S. Y. Cheng, S. Kobayashi, J. J. Kohn, B. Lawson, L. Nirenberg and Y. T. Siu....
>
> I dedicate this paper to my mother Leung Yeuk Lam and to the memory of my father Chiou Chen Ying. My father worked in both philosophy, economics and Chinese literature. The depth of his insight had a tremendous influence on me. He died before he was wellknown. My mother took over the most difficult job of teaching and feeding seven children in a poor family. Without my parents' persistent teaching, I would never become a mathematician.

After reading this paper, the reader can appreciate Yau both as a mathematician and a man.

Complex Geometry from Riemann to Kähler-Einstein and Calabi-Yau
ALM 38, pp. 285–345

On the Ricci Curvature of a Compact Kähler Manifold and the Complex Monge-Ampère Equation, I*

Shing-Tung Yau

0 Introduction

Let M be a compact Kähler manifold with Kähler metric $\sum_{i,j} g_{i\bar{j}} dz^i \otimes d\bar{z}^j$. Then it is wellknown that the Ricci curvature of M can be computed in the following way. If $\sum_{i,j} R_{i\bar{j}} dz^j \otimes d\bar{z}^j$ is the Ricci tensor, then

$$R_{i\bar{j}} = -\frac{\partial^2}{\partial z^i \partial \bar{z}^j}[\log \det(g_{s\bar{t}})].$$

In particular, the (1, 1) form $(\sqrt{-1}/2\pi)\sum_{i,j} R_{i\bar{j}} dz^i \wedge d\bar{z}^j$ is a closed (1, 1) form which is equal to $(-\sqrt{-1}/2\pi)\partial\bar{\partial}[\log \det(g_{s\bar{t}})]$. A wellknown theorem of S. S. Chern [6] shows that the cohomology class of this (1, 1) form depends only on the complex structure of M and is equal to the first Chern class of M. Therefore a necessary condition for a (1, 1) form $(\sqrt{-1}/\alpha\pi)\sum_{i,j} \tilde{R}_{i\bar{j}} dz^i \wedge d\bar{z}^j$ to be the Ricci form of some Kähler metric is that it must be closed and its cohomology class must represent the first Chern class of M.

More than twenty years ago, E. Calabi [3] conjectured that the above necessary condition is in fact sufficient. This conjecture of Calabi can be reduced to a problem in non-linear partial differential equation.

* The research for this paper was done while the author was a visiting member at the Courant Institute of Mathematical Sciences. Reproduction in whole or in part is permitted for any purpose of the United States Government. Received June, 1977. S.-T. Yau, On the Ricci curvature of a compact Kähler manifold and the complex Monge-Ampère equation I. Comm. Pure Appl. Math. 31 (1978), no. 3, 339–411. Stanford university.

In fact, if $(\sqrt{-1}/2\pi)\sum_{i,j}\tilde{R}_{i\bar{j}}dz^i\wedge d\bar{z}^j$ represents the first Chern class of M, then we can find a smooth function F defined on M such that

$$\tilde{R}_{i\bar{j}}-R_{i\bar{j}}=-\frac{\partial^2 F}{\partial z^i\partial\bar{z}^j}. \tag{0.1}$$

If $\sum_{i,j}\tilde{R}_{i\bar{j}}dz^i\otimes d\bar{z}^j$ is the Ricci tensor of some Kähler metric $\sum_{i,j}\tilde{g}_{ij}dz^i\otimes d\bar{z}^j$, then

$$\tilde{R}_{i\bar{j}}=-\frac{\partial^2}{\partial z^i\partial\bar{z}^j}[\log\det(\tilde{g}_{s\bar{t}})]$$

and the above equation shows that

$$\partial\bar{\partial}\log\frac{\det(\tilde{g}_{s\bar{t}})}{\det(g_{s\bar{t}})}=\partial\bar{\partial}F. \tag{0.2}$$

Since M is compact and det $(\tilde{g}_{s\bar{t}})/\det(g_{s\bar{t}})$ is a globally defined function, the maximal principle shows that log det $(g_{s\bar{t}})/\det(g_{s\bar{t}})-F$ is a constant. Hence, for some constant $C>0$,

$$\det(\tilde{g}_{s\bar{t}})=C\exp\{F\}\det(g_{s\bar{t}}). \tag{0.3}$$

If $\frac{1}{2}\sqrt{-1}\sum_{i,j}\tilde{g}_{i\bar{j}}dz^i\wedge d\bar{z}^j$ is cohomologous to $\frac{1}{2}\sqrt{-1}\sum_{i,j}g_{i\bar{j}}dz^i\wedge d\bar{z}^j$, then for some smooth function φ we have $\tilde{g}_{i\bar{j}}=g_{i\bar{j}}+\partial^2\varphi/\partial z^i\partial\bar{z}^j$. In this case, the above equation is equivalent to the following equation:

$$\det\left(g_{i\bar{j}}+\frac{\partial^2\varphi}{\partial z^i\partial\bar{z}^j}\right)=C\exp\{F\}\det(g_{s\bar{t}}). \tag{0.4}$$

Conversely, if for some constant $C>0$, we can solve (0.4) so that the solution φ is smooth and $\sum_{i,j}(g_{i\bar{j}}+\partial^2\varphi/\partial z^i\partial\bar{z}^i)dz^i\otimes d\bar{z}^j$ defines a Kähler metric, then we can solve the conjecture of Calabi. It turns out, by integrating (0.4), that there is only one choice of C:

$$C\int_M\exp\{F\}=\mathrm{Vol}(M). \tag{0.5}$$

In this paper, we shall solve equation (0.4) with the (necessary) compatibility condition (0.5). In view of its application in Kähler geometry, we generalize (0.4) to the following more general case.

Let L_1 and L_2 be holomorphic Hermitian line bundles over M. Let $s_1,\ldots,s_p$ be holomorphic sections of L_1 and $t_1,\ldots,t_q$ be holomorphic sections of L_2. Then for $h_1\geqq 0,\ldots,h_p\geqq 0,\ldots,k_q\geqq 0$, we study the equation

$$\begin{aligned}\det\left(g_{i\bar{j}}+\frac{\partial^2\varphi}{\partial z^i\partial\bar{z}^j}\right)&=(|s_1|^{2h_1}+\cdots+|s_p|^{2h_p})(|t_1|^{2k_1}+\cdots+|t_q|^{2k_q})^{-1}\\&\quad\times\exp\{F(x,\varphi)\}\det(g_{i\bar{j}}),\end{aligned} \tag{0.6}$$

where F is a smooth function defined on $M\times R$ such that $\partial F/\partial\varphi\geqq 0$.

We assume that we can find a smooth function ψ defined on M such that

$$\int_M(|s_1|^{2h_1}+\cdots+|s_p|^{2h_p})(|t_1|^{2k_1}+\cdots+|t_q|^{2k_q})^{-1}\exp\{F(x,\psi)\}=\mathrm{Vol}(M). \tag{0.7}$$

It is also necessary to assume that the right-hand side of (0.6) cannot tend to infinity too fast. Hence we assume that, for $m = \dim M$,

$$\int_M (|t_1|^{2k_1} + \cdots + |t_q|^{2k_q})^{-m} < \infty \tag{0.8}$$

and

for some $q > 0$, the function $(|t_1|^{2k_1} + \cdots + |t_q|^{2k_q})^{-q-1/m}$ $|\Delta \log(|t_1|^{2k_1} + \cdots + |t_q|^{2k_q})|^{(m-1)/m}$ is integrable over M (0.9)
and over every analytic disc of M.

We shall show that, under the conditions (0.7)–(0.9), equation (0.6) has a bounded solution φ which is smooth and defines a Kähler metric

$$\sum_{i,j} (g_{i\bar{j}} + \partial^2\varphi/\partial z^i \partial \bar{z}^j) dz^i \otimes d\bar{z}^j$$

outside the variety defined by the s_i and the variety defined by the t_i. Near the variety defined by the s_i, the Kähler metric is uniformly bounded from above and near the variety defined by the t_i the Kähler metric grows upward at most like a "pole". Furthermore, the function φ is unique up to a constant.

Equation (0.6) is considerably more complicated than equation (0.4) and will be proved only at the end of the paper. However, when the right-hand side of (0.6) specializes to the form exp $\{\varphi + F(x)\}$, then equation (0.6) is much easier than (0.4) and can be used to deduce the existence and uniqueness of a canonical Kähler-Einstein metric on a compact Kähler manifold with ample canonical line bundle. This was also conjectured by E. Calabi.

It should be mentioned that Calabi proved the uniqueness of the solution of (0.4) a long time ago. He also proved that when the right-hand side of (0.4) is close enough to a constant, then (0.4) has a smooth solution. (This part of Calabi's work was also made precise by T. Ochiai.) His later paper [5] on affine hyperspheres is very important for the study of the Monge-Ampère equation. The computation of the Laplacian of the third derivatives originated from there and was shown to us by L. Nirenberg. Furthermore, in [1], T. Aubin published a proof of Calabi's conjecture assuming the original Kähler manifold has non-negative holomorphic bisectional curvature. This class of manifolds is rather restrictive. Moreover, he used the variational method and his procedure is rather difficult to comprehend.

My proof of Calabi's conjecture was completed in the middle of 1976. A large amount of work was done while I was working on the real Monge- Ampère equation (cf. [8], [9]). For example, Pogorelov's second order estimate in the real Monge-Ampère equation had direct bearing on my work here. Indeed, besides the uniform estimate, the higher order estimates were discovered at that time. A special case of equation (0.6) when the right side is the function exp $\{\varphi + F(x)\}$ had been studied independently by Aubin [2].

The plan of this paper consists in starting from special cases of the theorem to make it easier for the reader to comprehend what is going on before treating the more involved equation (0.6). We use the continuity method and the required estimates are worked out for equation (0.4) in Sections 2 and 3. We shall give applications in the second part of the paper.

I am very grateful to all my teachers and my friends for their constant encouragement during this work. Among them I should mention E. Calabi, S. S. Chern, S. Y. Cheng, S. Kobayashi, J. J. Kohn, B. Lawson, L. Nirenberg and Y. T. Siu. Professor Chern kindly lectured on this paper and made significant improvements on its presentation. His help and his encouragement can hardly be exaggerated. My joint papers with Cheng have definite influence on the present paper. I am also very grateful to L. Nirenberg for many helpful conversations on the complex Monge-Ampère equation, especially on the complex analogue of Calabi's computation.

I dedicate this paper to my mother Leung Yeuk Lam and to the memory of my father Chiou Chen Ying. My father worked in both philosophy, economics and Chinese literature. The depth of his insight had a tremendous influence on me. He died before he was wellknown. My mother took over the most difficult job of teaching and feeding seven children in a poor family. Without my parents' persistent teaching, I would never have been able to become a mathematician.

Most of the results of this paper and their applications were announced in [15].

1 Notations and local formulae

Let M be an m-dimensional Kähler manifold with Kähler metric $ds^2 = \sum_{i,j} g_{i\bar{j}} dz^i \otimes d\bar{z}^j$. The Hermitian condition of the Kähler metric shows that, for all i, j,

$$\bar{g}_{i\bar{j}} = g_{\bar{i}j} = g_{j\bar{i}}. \tag{1.1}$$

If we let

$$H = \frac{1}{2}\sqrt{-1}\sum_{i,j} g_{i\bar{j}} dz^i \wedge d\bar{z}^j \tag{1.2}$$

be the Kähler form, then the Kähler condition is that

$$dH = 0, \tag{1.3}$$

or, for all i, j, k,

$$\frac{\partial g_{i\bar{j}}}{\partial z^k} = \frac{\partial g_{k\bar{j}}}{\partial z^i}. \tag{1.4}$$

The connection form ω_i^j of the Kähler metric is characterized by the fact that it is of type (1, 0) and it satisfies the following equation (of preserving scalar product):

$$dg_{i\bar{j}} = \omega_{i\bar{j}} + \omega_{\bar{j}i}, \tag{1.5}$$

where

$$\omega_{i\bar{j}} = \sum_k g_{k\bar{j}} \omega_i^k, \tag{1.6}$$

$$\omega_{i\bar{j}} = \bar{\omega}_{\bar{j}i}. \tag{1.7}$$

The matrix $(g^{i\bar{j}})$ satisfies the equation

$$\sum_i g^{i\bar{j}} g_{i\bar{k}} = \delta^{\bar{j}}_{\bar{k}}. \tag{1.8}$$

Hence

$$\omega_i^j = \sum_k g^{ik} \partial g_{ik}. \tag{1.9}$$

The curvature form is given by

$$\begin{aligned} \Omega_i^j &= d\omega_i^j - \sum_k \omega_i^k \wedge \omega_k^j \\ &= \sum_k dg^{j\bar{k}} \wedge \partial g_{i\bar{k}} + \sum_k g^{j\bar{k}} \bar{\partial}\partial g_{i\bar{k}} - \sum_{k,p,q} g^{k\bar{p}} \partial g_{i\bar{p}} \wedge g^{j\bar{q}} \partial g_{k\bar{q}}. \end{aligned} \tag{1.10}$$

Differentiating (1.8), we have

$$dg^{i\bar{j}} = -\sum_{k,p} g^{k\bar{j}} dg_{k\bar{p}} g^{i\bar{p}}. \tag{1.11}$$

Substituting (1.11) into (1.10), we obtain

$$\Omega_i^j = \sum_k g^{i\bar{k}} \bar{\partial}\partial g_{i\bar{k}} + \sum_{k,p,q} g^{k\bar{p}} g^{j\bar{q}} \partial g_{i\bar{p}} \wedge \bar{\partial} g_{k\bar{q}}. \tag{1.12}$$

Hence Ω_i^j is of type (1.1) and we can set

$$\Omega_i^j = \sum_{k,l} R^j_{ik\bar{l}} dz^k \wedge d\bar{z}^l, \tag{1.13}$$

where

$$R^j_{ik\bar{l}} = -\sum g^{j\bar{p}} \frac{\partial^2 g_{i\bar{p}}}{\partial z^k \partial \bar{z}^l} + \sum_{t,p,q} g^{t\bar{p}} g^{j\bar{q}} \frac{\partial g_{i\bar{p}}}{\partial z^k} \frac{\partial g_{t\bar{q}}}{\partial \bar{z}^l} \tag{1.14}$$

or

$$R_{i\bar{j}k\bar{l}} = -\frac{\partial^2 g_{i\bar{j}}}{\partial z^k \partial \bar{z}^l} + \sum_{p,q} g^{p\bar{q}} \frac{\partial g_{p\bar{j}}}{\partial \bar{z}^l}. \tag{1.15}$$

By Bianchi's identity

$$\sum_j dz^i \wedge \Omega_i^j = 0, \tag{1.16}$$

we have

$$R^j_{ik\bar{l}} = R^j_{ki\bar{l}} \tag{1.17}$$

and

$$R_{i\bar{j}k\bar{l}} = R_{k\bar{j}i\bar{l}}. \tag{1.18}$$

On the other hand,

$$\frac{\partial g_{i\bar{j}}}{\partial \bar{z}^l} = \frac{\partial g_{i\bar{l}}}{\partial \bar{z}^l} \tag{1.19}$$

so that

$$R_{i\bar{j}k\bar{l}} = R_{i\bar{l}k\bar{j}}. \tag{1.20}$$

Therefore the tensor $R_{i\bar{j}k\bar{l}}$ is symmetric in the first and third indices, as well as in the second and fourth indices.

The Ricci tensor is

$$\begin{aligned} R_{i\bar{j}} &= \sum_{k,l} g^{k\bar{l}} R_{k\bar{l}i\bar{j}} \\ &= -\sum_{k,l} g^{k\bar{l}} \frac{\partial^2 g_{k\bar{l}}}{\partial z^i \partial \bar{z}^j} + \sum_{p,q,k,l} g^{k\bar{l}} g^{p\bar{q}} \frac{\partial g_{k\bar{q}}}{\partial z^i} \frac{\partial g_{p\bar{l}}}{\partial \bar{z}^j}. \end{aligned} \tag{1.21}$$

By direct computation, we get

$$\bar{\partial} \log \det(g_{i\bar{j}}) = \sum_{i,j} g^{i\bar{j}} \bar{\partial} g_{i\bar{j}}. \tag{1.22}$$

Hence,

$$\partial\bar{\partial} \log \det(g_{i\bar{j}}) = \sum_{i,j} g^{i\bar{j}} \partial\bar{\partial} g_{i\bar{j}} - \sum_{i} g^{p\bar{j}} g^{i\bar{q}} \partial g_{p\bar{q}} \wedge \bar{\partial} g_{i\bar{j}}.$$

Therefore, the Ricci tensor is given by

$$R_{i\bar{j}} = -\frac{\partial^2 \log \det(g_{i\bar{j}})}{\partial z^i \partial \bar{z}^j}. \tag{1.23}$$

The curvature tensor will be an obstruction for the commuting of covariant differentiations. Thus let us set up the terminology for covariant differentiation.

Let f be any real-valued function defined on M. Then we define

$$df = \sum_i f_i \omega^i + \sum_i f_{\bar{i}} \bar{\omega}^i, \tag{1.24}$$

$$df_i - \sum_j f_j \omega_i^j = \sum_j f_{ij} \omega^j + \sum_j f_{i\bar{j}} \bar{\omega}^j, \tag{1.25}$$

$$df_{\bar{i}} - \sum_j f_{\bar{j}} \bar{\omega}_i^j = \sum_j f_{\bar{i}j} \omega^j + \sum_j f_{\bar{i}\bar{j}} \bar{\omega}^j. \tag{1.26}$$

By exterior differentiating (1.24), one verifies that

$$\begin{aligned} f_{ij} &= f_{ji}, \\ f_{i\bar{j}} &= f_{\bar{i}j}, \\ f_{\bar{i}\bar{j}} &= f_{\bar{j}\bar{i}}, \end{aligned} \tag{1.27}$$

For convenience, we shall write Δf as $\sum_i f_{i\bar{i}}$. Then, as is wellknown, 2Δ is the Laplace-Beltrami operator of the Kähler manifold M. In local coordinates, it is given by $\sum_{i,j} g^{i\bar{j}} \partial^2 f / \partial z^i \partial \bar{z}^j$.

Exterior differentiating (1.25), we have

$$\sum_j \left(df_{ij} - \sum_k f_{ik}\omega_j^k - \sum_k f_{kj}\omega_i^k \right) \wedge \omega_j^k$$
$$+ \sum_j \left(df_{i\bar{j}} - \sum_k f_{i\bar{k}}\bar{\omega}_j^k - \sum_k f_{k\bar{j}\omega_i^k} \right) \wedge \bar{\omega}^j = - \sum_j f_j \Omega_i^j. \tag{1.28}$$

Hence we define

$$df_{ij} - \sum_k f_{ik}\omega_j^k - \sum_k f_{kj}\omega_i^k = \sum_k f_{ijk}\omega^k + \sum_k f_{ij\bar{k}}\bar{\omega}^k, \tag{1.29}$$

$$df_{i\bar{j}} - \sum_k f_{i\bar{k}}\bar{\omega}_j^k - \sum_k f_{k\bar{j}}\omega_i^k = \sum_k f_{i\bar{j}k}\omega^k + \sum_k f_{i\bar{j}\bar{k}}\bar{\omega}^k. \tag{1.30}$$

By (1.28), we have

$$f_{ijk} = f_{ikj}, \tag{1.31}$$

$$f_{i\bar{j}\bar{k}} = f_{i\bar{k}\bar{j}}, \tag{1.32}$$

$$f_{i\bar{k}j} - f_{ij\bar{k}} = - \sum_l f_l R^l_{ij\bar{k}}. \tag{1.33}$$

Continuing in this way, we can find the commutation formulae of the higher covariant derivatives of f. For example,

$$f_{i\bar{j}\bar{k}l} - f_{i\bar{j}l\bar{k}} = \sum_n f_{i\bar{n}} R \frac{\bar{n}}{ikl} - \sum_n f_{n\bar{j}} R^n_{il\bar{k}}. \tag{1.34}$$

The main fact that we shall use is that, when we apply covariant differentiation successively with respect to two indices without bar, we can interchange these two indices. The same principle applies when the two indices have a bar at the same time. If one of the indices has a bar and the other does not, then after commuting them, we still have terms which are contraction of a curvature tensor with the covariant derivatives of f whose order is the order of the original one minus two.

2 Estimates up to second order

From now on, we assume that M is a compact Kähler manifold. We shall study the equation

$$\det \left(g_{i\bar{j}} + \frac{\partial^2 \varphi}{\partial z_i \partial z_{\bar{j}}} \right) (\det(g_{i\bar{j}}))^{-1} = \exp\{F\}, \tag{2.1}$$

where F is assumed to be in $C^3(M)$.

We are going to look for solutions φ of (2.1) such that $(g_{i\bar{j}} + \partial^2\varphi/\partial z_i \partial z_{\bar{j}})$ is a positive definite Hermitian matrix. The tensor $\sum_{i,j}(g_{i\bar{j}} + \partial^2\varphi/\partial z_i \partial \bar{z}_i) dz^i \otimes d\bar{z}^j$

will define another Kähler metric on M. We shall denote it by $\sum_{i,j} g'_{i\bar{j}} dz' \otimes d\bar{z}^j$ so that

$$g'_{i\bar{j}} = g_{i\bar{j}} + \frac{\partial^2 \varphi}{\partial z_i \partial \bar{z}_j}. \tag{2.2}$$

Before proving the existence of φ, we need *a priori* estimates of φ. Therefore we assume that $\varphi \in C^5(M)$. Our first aim will be to obtain the estimates of φ up to second derivatives. This will be achieved under the normalization

$$\int_M \varphi = 0. \tag{2.3}$$

Differentiating (2.1), we have

$$\sum_{i,j} g'^{i\bar{j}} \left(\frac{\partial g_{i\bar{j}}}{\partial z^k} + \frac{\partial^3 \varphi}{\partial z^i \partial \bar{z}^j \partial z^k} \right) - \sum_{i,j} g^{i\bar{j}} \frac{\partial g_{i\bar{j}}}{\partial z^k} = \frac{\partial F}{\partial z^k}, \tag{2.4}$$

where $(g'^{i\bar{j}})$ is the inverse matrix of $(g_{i\bar{j}})$. Now we differentiate (2.4) again and obtain

$$\begin{aligned} &- \sum_{i,j,t,n} g'^{t\bar{j}} g'^{i\bar{n}} \left(\frac{\partial g_{t\bar{n}}}{\partial \bar{z}^l} + \frac{\partial^3 \varphi}{\partial z^t \partial \bar{z}^n \partial \bar{z}^l} \right) \left(\frac{\partial g_{i\bar{j}}}{\partial z^k} + \frac{\partial^3 \varphi}{\partial z^i \partial \bar{z}^j \partial z^k} \right) \\ &+ \sum_{i,j} g'^{i\bar{j}} \left(\frac{\partial^2 g_{i\bar{j}}}{\partial z^k \partial \bar{z}^l} + \frac{\partial^4 \varphi}{\partial z^i \partial \bar{z}^j \partial z^k \partial \bar{z}^l} \right) + \sum_{i,j,k,n} g^{t\bar{j}} g^{i\bar{n}} \frac{\partial g_{t\bar{n}}}{\partial \bar{z}^l} \frac{\partial g_{i\bar{j}}}{\partial z^k} \\ &- \sum_{i,j} g^{i\bar{j}} \frac{\partial^2 g_{i\bar{j}}}{\partial z^k \partial \bar{z}^l} = \frac{\partial^2 F}{\partial z^k \partial \bar{z}^l}. \end{aligned} \tag{2.5}$$

Let Δ' be the normalized Laplacian associated with the metric $\sum_{ij} g^*_{i\bar{j}} dz^i \otimes d\bar{z}^j$. Then

$$\begin{aligned} \Delta'(\Delta\varphi) &= \sum_{k,l} g'^{k\bar{l}} \frac{\partial^2}{\partial z^k \partial \bar{z}^l} \left(\sum_{i,j} g^{i\bar{j}} \frac{\partial^2 \varphi}{\partial z^i \partial \bar{z}^j} \right) \\ &= \sum_{i,j,k,l} g'^{k\bar{l}} g^{i\bar{j}} \frac{\partial^4 \varphi}{\partial z^i \partial \bar{z}^j \partial z^k \partial \bar{z}^l} + \sum_{i,j,k,l} g'^{k\bar{l}} \frac{\partial^2 g^{i\bar{j}}}{\partial z^k \partial \bar{z}^l} \frac{\partial^2 \varphi}{\partial z^i \partial \bar{z}^j} \\ &+ \sum_{i,j,k,l} g'^{kl} \frac{\partial g^{i\bar{j}}}{\partial z^k} \frac{\partial^3 \varphi}{\partial z^i \partial \bar{z}^j \partial \bar{z}^l} + \sum_{i,j,k,l} g'^{kl} \frac{\partial g_{i\bar{j}}}{\partial \bar{z}^l} \frac{\partial^3 \varphi}{\partial z^i \partial \bar{z}^j \partial z^k}. \end{aligned} \tag{2.6}$$

Since the left-hand side of (2.6) is independent of the choice of coordinate system, we can compute the right-hand side by picking a coordinate system at one point so that $g_{i\bar{j}} = \delta_{ij}, \partial g_{i\bar{j}}/\partial \bar{z}^k = \partial g_{i\bar{j}}/\partial \bar{z}^l = 0$. Then inserting (2.5) and (1.9) into (2.6), we have

$$\begin{aligned} \Delta'(\Delta\varphi) &= \Delta F + \sum_{i,j,k,n,l} g'^{k\bar{j}} g'^{i\bar{n}} \varphi_{k\bar{n}\bar{l}} \varphi_{i\bar{j}l} + \sum_{i,j,l} g'^{ij} R_{i\bar{j}l\bar{l}} - \sum_{i,l} R_{i\bar{i}l\bar{l}} \\ &+ \sum_{i,j,k,l} g'^{k\bar{l}} R_{i\bar{j}k\bar{l}} \varphi_{i\bar{j}}. \end{aligned} \tag{2.7}$$

Note that each term on the right-hand side is a scalar function and is independent of the choice of the coordinate system if we contract the terms suitably with the tensor $\{g^{i\bar{j}}\}$.

If we choose another coordinate system so that $g_{i\bar{j}} = \delta_{ij}$ and $\varphi_{i\bar{j}} = \delta_{ij}\varphi_{i\bar{i}}$, then we have

$$g'^{i\bar{j}} = \delta_{ij}(1+\varphi_{i\bar{i}})^{-1} \tag{2.8}$$

and

$$\begin{aligned}
&\sum_{i,j,l} g'^{ij} R_{i\bar{j}l\bar{l}} - \sum_{i,l} R_{i\bar{i}l\bar{l}} + \sum_{i,j,k,l} g'^{k\bar{l}} R_{i\bar{j}k\bar{l}}\varphi_{i\bar{j}} \\
&= -\sum_{\bar{i},l} R_{i\bar{i}l\bar{l}} \frac{\varphi_{i\bar{i}}}{1+\varphi_{i\bar{i}}} + \sum_{i,l} R_{i\bar{i}l\bar{l}} \frac{\varphi_{i\bar{i}}}{1+\varphi_{l\bar{l}}} \\
&= -\sum_{\bar{i},l} R_{i\bar{i}l\bar{l}} \frac{\varphi_{i\bar{i}}(\varphi_{l\bar{l}}-\varphi_{i\bar{i}})}{(1+\varphi_{i\bar{i}})(1+\varphi_{l\bar{l}})} \\
&= \frac{1}{2}\left\{-\sum_{i,l} R_{i\bar{i}l\bar{l}} \frac{\varphi_{i\bar{i}}(\varphi_{l\bar{l}}-\varphi_{i\bar{i}})}{(1+\varphi_{i\bar{i}})(1+\varphi_{l\bar{l}})} - \sum_{i,j} R_{i\bar{i}l\bar{l}} \frac{\varphi_{l\bar{l}}(\varphi_{i\bar{i}}-\varphi_{l\bar{l}})}{(1+\varphi_{i\bar{i}})(1+\varphi_{l\bar{l}})}\right\} \\
&= \frac{1}{2}\sum_{i,l} R_{i\bar{i}l\bar{l}} \frac{(\varphi_{l\bar{l}}-\varphi_{i\bar{i}})^2}{(1+\varphi_{i\bar{i}})(1+\varphi_{l\bar{l}})}.
\end{aligned} \tag{2.9}$$

Combining (2.7) and (2.9), we see that

$$\Delta'(\Delta\varphi) \geqq \Delta F + \sum g'^{k\bar{j}} g'^{i\bar{n}} \varphi_{k\bar{n}\bar{l}}\varphi_{i\bar{j}l} + \left(\inf_{i\neq l} R_{i\bar{i}l\bar{l}}\right)\left[\sum_{i,l} \frac{1+\varphi_{i\bar{i}}}{1+\varphi_{l\bar{l}}} - m^2\right]. \tag{2.10}$$

On the other hand,

$$\begin{aligned}
\Delta'\varphi &= \sum_i \frac{\varphi_{i\bar{i}}}{1+\varphi_{i\bar{i}}} \\
&= m - \sum_i \frac{1}{1+\varphi_{i\bar{i}}}.
\end{aligned} \tag{2.11}$$

Let C be a positive constant. Then we have

$$\begin{aligned}
\Delta'(\exp\{-C\varphi\}(m+\Delta\varphi)) &= C^2\exp\{-C\varphi\}\left(\sum_{i,j} g'^{i\bar{j}}\varphi_i\varphi_{\bar{j}}\right)(m+\Delta\varphi) \\
&\quad -C\exp\{-C\varphi\}\sum_{i,j} g'^{i\bar{j}}\varphi_i(\Delta\varphi)_{\bar{j}} \\
&\quad -C\exp\{-C\varphi\}\sum_{i,j} g'^{i\bar{j}}(\Delta\varphi)_i\varphi_{\bar{j}} \\
&\quad -C\exp\{-C\varphi\}\Delta'\varphi(m+\Delta\varphi) + \exp\{-C\varphi\}\Delta'(\Delta\varphi).
\end{aligned} \tag{2.12}$$

Applying Schwarz' inequality to the first three terms of (2.13), we have

$$\begin{aligned}\Delta'(\exp\{-C\varphi\}(m+\Delta\varphi)) &\geqq -(m+\Delta\varphi)^{-1}\exp\{-C\varphi\}\sum_{k,j} g'^{i\bar{j}}(\Delta\varphi)_i(\Delta\varphi)_{\bar{j}} \\ &\quad -C\exp\{-C\varphi\}(\Delta'\varphi)(m+\Delta\varphi)+\exp\{-C\varphi\}\Delta'(\Delta\varphi). \qquad (2.13)\end{aligned}$$

In computing the right-hand side of (2.13), we assume that $g_{i\bar{j}} = \delta_{ij}$ and $\varphi_{i\bar{j}} = \varphi_{i\bar{i}}\delta_{ij}$ at a point. Then, using (2.10), we have

$$\begin{aligned}&-(m+\Delta\varphi)^{-1}\sum_{i,j} g'^{i\bar{j}}(\Delta\varphi)_i(\Delta\varphi)_{\bar{j}} + \Delta'(\Delta\varphi) \\ &\geqq -(m+\Delta\varphi)^{-1}\sum_i (1+\varphi_{i\bar{i}})^{-1}\left|\sum_k \varphi_{k\bar{k}i}\right|^2 \\ &\quad +\Delta F + \sum_{i,j,k}(1+\varphi_{k\bar{k}})^{-1}(1+\varphi_{i\bar{i}})^{-1}\varphi_{k\bar{i}\bar{j}}\varphi_{i\bar{k}j} \\ &\quad + \sum_{i,l}\left(\inf_{i\neq l} R_{i\bar{i}l\bar{l}}\right)\left[\sum_{i,l}\frac{1+\varphi_{l\bar{l}}}{1+\varphi_{i\bar{i}}} - m^2\right]. \qquad (2.14)\end{aligned}$$

But

$$\begin{aligned}&(m+\Delta\varphi)^{-1}\sum_i (1+\varphi_{i\bar{i}})^{-1}\left|\sum_k \varphi_{k\bar{k}i}\right|^2 \\ &= (m+\Delta\varphi)^{-1}\sum_i (1+\varphi_{i\bar{i}})^{-1}\left|\sum_k \frac{\varphi_{k\bar{k}i}}{(1+\varphi_{k\bar{k}})^{1/2}}(1+\varphi_{k\bar{k}})^{1/2}\right|^2 \\ &\leqq (m+\Delta\varphi)^{-1}\left(\sum_{i,k}(1+\varphi_{i\bar{i}})^{-1}(1+\varphi_{k\bar{k}})^{-1}\varphi_{k\bar{k}i}\varphi_{\bar{k}k\bar{i}}\right)\left(\sum_k (1+\varphi_{k\bar{k}})\right) \\ &= \sum_{i,k}(1+\varphi_{i\bar{i}})^{-1}(1+\varphi_{k\bar{k}})^{-1}\varphi_{k\bar{k}i}\varphi_{\bar{k}k\bar{i}} \\ &= \sum_{i,k}(1+\varphi_{i\bar{i}})^{-1}(1+\varphi_{k\bar{k}})^{-1}\varphi_{i\bar{k}k'}\varphi_{k\bar{i}\bar{k}} \\ &\leqq \sum_{i,j,k}(1+\varphi_{k\bar{k}})^{-1}(1+\varphi_{i\bar{i}})^{-1}\varphi_{k\bar{i}\bar{j}}\varphi_{i\bar{k}j}. \qquad (2.15)\end{aligned}$$

(Here we use the fact that $\varphi_{k\bar{k}i} = \varphi_{\bar{k}ki} = \varphi_{\bar{k}ik} = \varphi_{i\bar{k}k}$ and $\varphi_{\bar{k}k\bar{i}} = \varphi_{k\bar{k}\bar{i}} = \varphi_{k\bar{i}\bar{k}}$, see Section 1). Hence (2.14) and (2.15) imply

$$\begin{aligned}&-(m+\Delta\varphi)^{-1}\sum_{i,j} g'^{i\bar{j}}(\Delta\varphi)_i(\Delta\varphi)_{\bar{j}} + \Delta'(\Delta\varphi) \\ &\geqq \Delta F + \left(\inf_{i\neq l} R_{i\bar{i}l\bar{l}}\right)\left[\sum_{i,l}\frac{1+\varphi_{i\bar{i}}}{1+\varphi_{l\bar{l}}} - m^2\right]. \qquad (2.16)\end{aligned}$$

Inserting (2.16) into (2.13), we obtain

$$\Delta'(\exp\{-C\varphi\}(m+\Delta\varphi)) \geqq \exp\{-C\varphi\}\left[\Delta F + \inf_{i\neq l}(R_{i\bar{i}l\bar{l}})\left(\sum_{i,l}\frac{1+\varphi_{i\bar{i}}}{1+\varphi_{l\bar{l}}} - m^2\right)\right]$$
$$-C\exp\{-C\varphi\}(\Delta'\varphi)(m+\Delta\varphi). \tag{2.17}$$

Inserting (2.11) into (2.17), we get

$$\begin{aligned}
&\Delta'(\exp\{-C\varphi\}(m+\Delta\varphi))\\
&\geqq \exp\{-C\varphi\}\left(\Delta F - m^2\inf_{i\neq l}(R_{i\bar{i}l\bar{l}})\right)\\
&\quad + \exp\{-C\varphi\}\inf_{i\neq l}(R_{i\bar{i}l\bar{l}})\left(\sum_{i,l}\frac{1+\varphi_{i\bar{i}}}{1+\varphi_{l\bar{l}}}\right)\\
&\quad -C\exp\{-C\varphi\}m(m+\Delta\varphi) + C\exp\{-C\varphi\}(m+\Delta\varphi)\sum_i\frac{1}{1+\varphi_{i\bar{i}}}\\
&= \exp\{-C\varphi\}\left(\Delta F - m^2\inf_{i\neq l}(R_{i\bar{i}l\bar{l}})\right) - C\exp\{-C\varphi\}m(m+\Delta\varphi)\\
&\quad + \left(C + \inf_{i\neq l}(R_{i\bar{i}l\bar{l}})\right)\exp\{-C\varphi\}(m+\Delta\varphi)\left(\sum_i\frac{1}{1+\varphi_{i\bar{i}}}\right).
\end{aligned} \tag{2.18}$$

Now let us notice the following inequality:

$$\sum_i\frac{1}{1+\varphi_{i\bar{i}}} \geqq \left(\frac{\sum_i(1+\varphi_{i\bar{i}})}{\prod_i(1+\varphi_{i\bar{i}})}\right)^{1/(m-1)} \tag{2.19}$$

which can be verified by taking the $(m-1)$-th power of both sides. Therefore, by (2.1),

$$\sum_i\frac{1}{1+\varphi_{i\bar{i}}} \geqq (m+\Delta\varphi)^{1/(m-1)}\exp\left\{\frac{-F}{m-1}\right\}. \tag{2.20}$$

Choose C so that

$$C + \inf_{i\neq l}(R_{i\bar{i}l\bar{l}}) > 1. \tag{2.21}$$

Then it follows from (2.18) and (2.20) that

$$\begin{aligned}
&\Delta'(\exp\{-C\varphi\}(m+\Delta\varphi))\\
&\geqq \exp\{-C\varphi\}\left(\Delta F - m^2\inf_{i\neq l}(R_{i\bar{i}l\bar{l}})\right)\\
&\quad -C\exp\{-C\varphi\}m(m+\Delta\varphi)\\
&\quad + \left(C + \inf_{i\neq l}(R_{i\bar{i}l\bar{l}})\right)\exp\{-C\varphi\}\exp\left\{\frac{-F}{m-1}\right\}(m+\Delta\varphi)^{1+1/(m-1)}.
\end{aligned} \tag{2.22}$$

Using (2.22), we can now estimate $\exp\{-C\varphi\}(m+\Delta\varphi)$. In fact, it must achieve its maximum at some point p so that the right-hand side of (2.22) is nonpositive. At this point,

$$\begin{aligned}0 \geqq \Delta F - m^2 \inf_{i\neq l}(R_{i\bar{i}l\bar{l}}) - Cm(m+\Delta\varphi)\\ + \left(C + \inf_{i\neq l}(R_{i\bar{i}l\bar{l}})\right)\exp\left\{\frac{-F}{m-1}\right\}(m+\Delta\varphi)^{1+1/(m-1)}.\end{aligned} \quad (2.23)$$

Hence $(m+\Delta\varphi)(p)$ has an upper bound C_1 depending only on $\sup_M(-\Delta F)$, $\sup_M|\inf_{i\neq l}(R_{i\bar{i}l\bar{l}})|, C\cdot m$ and $\sup_M F$. (It is clear that the inequality $y^{1+1/(m-1)} \leqq ay+b$ implies either $y^{1+1/(m-1)} \leqq 2ay$ or $y^{1+1/(m-1)} \leqq 2b$. In either case, we have an upper estimate of y in terms of a and b.)

Since $\exp\{-C\varphi\}(m+\Delta\varphi)$ achieves its maximum at p, we have the following inequality

$$0 < m + \Delta\varphi \leqq C_1 \exp\left\{C\left(\varphi - \inf_M \varphi\right)\right\}. \quad (2.24)$$

We are going to make use of (2.24) to estimate $\sup_M|\varphi|$. First of all, since $\Delta\varphi + m = \sum_i(1+\varphi_{i\bar{i}}) = \sum_{i,j} g^{i\bar{j}}g'_{i\bar{j}} > 0$, we can estimate $\sup_M \varphi$ as follows.

Let $G(p,q)$ be the Green's function (cf. [10]) of the operator Δ on M. Let K be a constant (depending only on M) such that

$$G(p,q) + K \geqq 0. \quad (2.25)$$

Then (2.3) shows that

$$\begin{aligned}\varphi(p) &= -\int_M G(p,q)\Delta\varphi(q)dq\\ &= -\int_M (G(p,q)+K)\Delta\varphi(q)dq.\end{aligned} \quad (2.26)$$

Therefore,

$$\sup_M \varphi \leqq m \sup_{p\in M}\int_M (G(p,q)+K)dq. \quad (2.27)$$

The inequalities (2.3) and (2.27) also imply

$$\begin{aligned}\int_M |\varphi| &\leqq \int_M \left|\sup_M \varphi - \varphi\right| + \int_M |\sup \varphi|\\ &\leqq \left(\sup_M \varphi\right)\mathrm{Vol}(M) - \int_M \varphi + \left(\sup_M \varphi\right)\mathrm{Vol}(M)\\ &= 2m\mathrm{Vol}(M)\sup_{p\in M}\int_M (G(p,q)+K)dq.\end{aligned} \quad (2.28)$$

To estimate $\inf_M \varphi$, we offer two different proofs. One works for $m=2$ and the other works in general.

For $m=2$, we renormalize φ so that $\sup_M \varphi \leqq -1$. (This is possible because we have estimated $\sup_M \varphi$ already so that we still have an estimate on $\int_M |\varphi|$.)

Let p be any positive number greater than or equal to 1. Then

$$\begin{aligned}\Delta'(-\varphi)^p &= p(p-1)(-\varphi)^{p-2}\sum_i \frac{|\varphi_i|^2}{1+\varphi_{i\bar{i}}} - p(-\varphi)^{p-1}\Delta'\varphi \\ &= p(p-1)(-\varphi)^{p-2}\sum_i \frac{|\varphi_i|^2}{1+\varphi_{i\bar{i}}} - p(-\varphi)^{p-1}\left(m - \sum_i \frac{1}{1+\varphi_{i\bar{i}}}\right).\end{aligned} \tag{2.29}$$

When $m = 2$, we find

$$\begin{aligned}\Delta'(-\varphi)^p &= p(p-1)(-\varphi)^{p-2}\sum_i \frac{|\varphi_i|^2}{1+\varphi_{i\bar{i}}} \\ &\quad -2p(-\varphi)^{p-1} + p(-\varphi)^{p-1}(1+\varphi_{1\bar{1}})^{-1}(1+\varphi_{2\bar{2}})^{-1}(2+\Delta\varphi) \\ &= p(p-1)(-\varphi)^{p-2}\sum_i \frac{|\varphi_i|^2}{1+\varphi_{i\bar{i}}} - 2p(-\varphi)^{p-1} \\ &\quad +2p(-\varphi)^{p-1}\exp\{-F\} \\ &\quad +\exp\{-F\}[-\Delta(-\varphi)^p + p(p-1)(-\varphi)^{p-2}|\nabla\varphi|^2].\end{aligned} \tag{2.30}$$

Multiplying (2.30) by exp $\{F\}$ and integrating with respect to the volume form of the original metric $\sum_{i,j} g_{i\bar{j}} dz^i \otimes d\bar{z}^j$, we have

$$\begin{aligned}2p\int_M (-\varphi)^{p-1}(1-\exp\{F\}) &= p(p-1)\int_M (-\varphi)^{p-2}\left(\sum_i \frac{|\varphi_i|^2}{1+\varphi_{i\bar{i}}}\right)\exp\{F\} \\ &\quad -\int_M \Delta(-\varphi)^p + p(p-1)\int_M (-\varphi)^{p-2}|\nabla\varphi|^2 \\ &= p(p-1)\int_M (-\varphi)^{p-2}\left(\sum_i \frac{|\varphi_i|^2}{1+\varphi_{i\bar{i}}}\right)\exp\{F\} \\ &\quad +\frac{4(p-1)}{p}\int_M |\nabla(-\varphi)^{p/2}|^2.\end{aligned} \tag{2.31}$$

Since φ is negative, (2.31) implies that, for $p \geqq 2$,

$$\begin{aligned}\int_M |\nabla(-\varphi)^{p/2}|^2 &\leqq \frac{p^2}{2(p-1)}\left|\int_M (1-\exp\{F\})(-\varphi)^{p-1}\right| \\ &\leqq pC_2\int_M |\varphi|^{p-1},\end{aligned} \tag{2.32}$$

where C_2 depends only on sup F.

Now applying the Sobolev inequality (this is true on a compact manifold because we can cut the manifold into a finite number of pieces which are diffeomorphic to a domain in Euclidean space and we can apply the standard Sobolev inequality in Euclidean space), we can find a constant C_3 depending only on M such that

$$\begin{aligned}\left(\int_M |\varphi|^{2p}\right)^{1/2} &\leqq C_3 \int_M |\varphi|^p + C_3 \int_M |\nabla(-\varphi)^{p/2}|^2 \\ &\leqq C_3 \int_M |\varphi|^p + pC_1C_3 \int_M |\varphi|^{p-1}. \end{aligned} \tag{2.33}$$

Since $\varphi \leqq -1$, we derive from (2.33) another constant C_4 depending only on M such that

$$\int_M |\varphi|^{2p} \leqq C_4 p^2 \left(\int_M |\varphi|^p\right)^2. \tag{2.34}$$

In order to use (2.34), we need to find an estimate of $\int_M |\varphi|^2$. But this follows from the estimate of $\int_M |\varphi|$ (see (2.28)), the Poincarè inequality and (2.32).

We now claim that we can find a constant C_5 depending only on M such that

$$\int_M |\varphi|^p \leqq C_5^p \left(\frac{p}{2}\right)^{p/2} \tag{2.35}$$

for all integers $p \geqq 1$. In fact, let p_0 be the first integer such that, for $p \geqq p_0$,

$$C_4(1 + \mathrm{Vol}(M)) \left(\frac{p+2}{4}\right)^{(p+1)/2} \leqq \left(\frac{p+1}{2}\right)^{(p-3)/2}. \tag{2.36}$$

Then we choose C_5 so that (2.34) is valid for $1 \leqq p \leqq 2p_0$. (This is possible by (2.34), the L^2-estimate of φ and the Hölder inequality.) We are going to prove (2.34) for all $p \geqq 1$ by induction.

There are two cases. If $p+1$ is divisible by two, then from (2.34) we have

$$\begin{aligned}\int_M |\varphi|^{p+1} &\leqq C_4 \left(\frac{p+1}{2}\right)^2 \left(\int_M |\varphi|^{(p+1)/2}\right)^2 \\ &\leqq C_4 \left(\frac{p+1}{2}\right)^2 C_5^{p+1} \left(\frac{p+1}{4}\right)^{(p+1)/2}, \end{aligned} \tag{2.37}$$

where the last inequality follows by the induction hypothesis. If we apply (2.36) to (2.37), then (2.35) is valid with p replaced by $p+1$.

If $p+1$ is not divisible by two, then, by (2.34), the Schwarz inequality and the induction hypothesis,

$$\begin{aligned}\int_M |\varphi|^{p+1} &\leqq C_4 \left(\frac{p+1}{2}\right)^2 \left(\int_M |\varphi|^{(p+1)/2}\right)^2 \\ &\leqq C_4 \left(\frac{p+1}{2}\right)^2 \left[\int_M |\varphi|^{(p+2)/2}\right]^{2(p+1)/(p+2)} \mathrm{Vol}(M)^{2/(p+2)} \\ &\leqq C_4 \left(\frac{p+1}{2}\right)^2 C_5^{p+1} \left(\frac{p+2}{4}\right)^{(p+1)/2} \mathrm{Vol}(M)^{2(p+2)}. \end{aligned} \tag{2.38}$$

Applying (2.36) to (2.38), we verify (2.35) with p replaced by $p+1$.

In conclusion, (2.35) is valid for all p. Therefore,

$$\int_M \exp\{k\varphi^2\} \leqq \sum_{p=0}^{\infty} \frac{k^p}{p!} \int_M |\varphi|^{2p}$$
$$\leqq \sum_{p=0}^{\infty} \frac{k^p}{p!} (C_5^2)^p p^p. \tag{2.39}$$

Since the Stirling's formula tells us that

$$p! \geqq \sqrt{2\pi} p^{p+1/2} e^{-p} \tag{2.40}$$

for all positive integers p, we find

$$\int_M \exp\{k\varphi^2\} \leqq \sum_{p=0}^{\infty} (2\pi)^{-1/2} (kC_5^2 e)^p p^{-1/2}. \tag{2.41}$$

When

$$k < C_5^{-2} e^{-1}, \tag{2.42}$$

the right-hand side of (2.41) is finite and we have obtained an estimate of $\int_M \exp\{k\varphi^2\}$ with any k satisfying (2.42).

We are now in a position to use (2.24) to estimate $\sup_M |\varphi|$. Rewrite (2.24) as

$$\Delta\varphi = f, \tag{2.43}$$

where

$$-m \leqq f \leqq C_1 \exp\{C \sup \varphi\} \exp\{-\inf_M \varphi\}. \tag{2.44}$$

Then by the Schauder estimate (see [13] p. 156 inequality (5.5.23)), there is a constant C_6 depending only on M and $C_1 \exp\{C \sup \varphi\}$ such that

$$\sup_M |\nabla\varphi| \leqq C_6 \left(\exp\{-C \inf \varphi\} + \int_M |\varphi| \right). \tag{2.45}$$

Since (2.27) and (2.28) already provide estimates for $\sup \varphi$ and $\int_M |\varphi|$, we have a constant C_7 depending only on M and C such that

$$\sup_M |\nabla\varphi| \leqq C_7 (\exp\{-C \inf \varphi\} + 1). \tag{2.46}$$

Let q be a point in M where $\varphi(q) = \inf \varphi$. Then in the geodesic ball, with center q and radius $-\frac{1}{2}(\inf \varphi) C_7^{-1} (\exp\{-C \inf \varphi\} + 1)^{-1}$, φ is not greater than $\frac{1}{2} \inf \varphi$. Since we may assume $-\inf \varphi$ to be large, we may assume that the radius of this geodesic ball is smaller than the injectivity radius of M. Therefore, the integral of $\exp\{k\varphi^2\}$ in this ball is not less than $C_8 \exp\{\frac{1}{4} k (\inf \varphi)^2\} \left(-\frac{1}{2} \inf \varphi\right)^{2m} C_7^{-2m}$ $(\exp\{-C \inf \varphi\} + l)^{-2m}$, where C_8 is a positive constant depending only on M. Since (2.40) gives an estimate of the integral of $\exp\{k\varphi^2\}$ over M, the last quantity is estimated. This of course implies an estimate of $|\inf_M \varphi|$.

Let us now give an estimate of $|\inf_M \varphi|$ without assuming $m = 2$. Let N be any positive number. Then (2.18) shows that

$$\begin{aligned}\Delta'(\exp\{-N\varphi\}(m+\Delta\varphi)) \geqq \exp\{-N\varphi\}\left(\Delta F - m^2 \inf_{i\neq l}(R_{i\bar{i}l\bar{l}})\right) \\ -N\exp\{-N\varphi\}m(m+\Delta\varphi) + \left(N + \inf_{i\neq l}(R_{i\bar{i}l\bar{l}})\right) \\ \times \exp\{-N\varphi\}(m+\Delta\varphi)\left(\sum_i \frac{1}{1+\varphi_{i\bar{i}}}\right). \qquad (2.47)\end{aligned}$$

Choose N so that

$$N + \inf_{i\neq l}(R_{i\bar{i}l\bar{l}}) \geqq \frac{1}{2}N. \qquad (2.48)$$

Then, by (2.20),

$$\left(N + \inf_{i\neq l}(R_{i\bar{i}l\bar{l}})\right)(m+\Delta\varphi)\left(\sum_i \frac{1}{1+\varphi_{i\bar{i}}}\right) \geqq \frac{1}{2}N\exp\left\{\frac{-F}{m-1}\right\}(m+\Delta\varphi)^{m(m-1)}. \qquad (2.49)$$

There is a constant C_9, depending only on $\sup F$ and m such that

$$\frac{1}{2}N\exp\left\{\frac{-F}{m-1}\right\}(m+\Delta\varphi)^{m/(m-1)} \geqq 2Nm(m+\Delta\varphi) - NC_9. \qquad (2.50)$$

Inserting (2.48), (2.49) and (2.50) into (2.47), we find

$$\begin{aligned}\Delta'(\exp\{-N\varphi\}(m+\Delta\varphi)) \geqq \exp\{-N\varphi\}\left(\Delta F - m^2 \inf_{i\neq l}(R_{i\bar{i}l\bar{l}}) - NC_9\right) \\ +N\exp\{-N\varphi\}m(m+\Delta\varphi). \qquad (2.51)\end{aligned}$$

Therefore,

$$\begin{aligned}&\exp\{F\}\Delta'(\exp\{-N\varphi\}(m+\Delta\varphi)) \\ &\geqq \exp\{-N\varphi\}\exp\{F\}\left(\Delta F - m^2 \inf_{i\neq l}(R_{i\bar{i}l\bar{l}}) - NC_9\right) \\ &\quad +N\exp\{-N\varphi\}\exp\{\inf F\}m(m+\Delta\varphi) \\ &= \exp\{-N\varphi\}\left\{\exp\{F\}\left[\left(\Delta F - m^2 \inf_{1\neq l}(R_{i\bar{i}l\bar{l}}) - NC_9\right) + m^2N\exp\{\inf F\}\right]\right\} \\ &\quad -mN\exp\{\inf F\}\exp\{-N\varphi\}\Delta\varphi \\ &= \exp\{-N\varphi\}\left\{\exp\{F\}\left[\left(\Delta F - m^2 \inf_{i\neq l}(R_{i\bar{i}l\bar{l}}) - NC_9\right) + m^2N\exp\{\inf F\}\right]\right\} \\ &\quad +m\exp\{\inf F\}(-\Delta\exp\{-N\varphi\} + N^2\exp\{-N\varphi\}|\nabla\varphi|^2) \\ &\geqq -C_{10}\exp\{-N\varphi\} + m\exp\{\inf F\}(-\Delta\exp\{-N\varphi\} + N^2\exp\{-N\varphi\}|\nabla\varphi|^2), \qquad (2.52)\end{aligned}$$

where C_{10} depends only on N, F and M. Integrating (2.52), we obtain

$$\int_M \left|\nabla \exp\left\{-\frac{1}{2}N\varphi\right\}\right|^2 = \frac{1}{4}N^2 \int_M \exp\{-N\varphi\}|\nabla\varphi|^2$$
$$\leqq \frac{1}{4}C_{10}m^{-1}\exp\{-\inf F\}\int_M \exp\{-N\varphi\}. \quad (2.53)$$

We claim that for each N satisfying (2.48), the inequalities (2.53) and (2.28) furnish an estimate of $\int_M \exp\{-N\varphi\}$ (depending on N, F and M). We are going to prove this statement by contradiction.

Suppose there exists a sequence $\{\varphi_i\}$ satisfying (2.28) and (2.53) such that $\lim_{i\to\infty}\int_M \exp\{-N\varphi_i\} = \infty$. Then we define

$$\exp\{-N\tilde{\varphi}_i\} = \exp\{-N\varphi_i\}\left\{\int_M \exp\{-N\varphi_i\}\right)^{-1}. \quad (2.54)$$

It follows from (2.53) that the sequence $\int_M \left|\nabla \exp\left\{-\frac{1}{2}N\tilde{\varphi}\right\}\right|^2$ is uniformly bounded from above by a constant depending only on N, F and M. Since $\int_M \exp\{-N\tilde{\varphi}_i\} = 1$ for all i, this last fact implies (see [13]) that a subsequence of $\left\{\exp\left\{-\frac{1}{2}N\tilde{\varphi}_i\right\}\right\}$ converges in $L^2(M)$ to some function f in $L^2(M)$. We assume this subsequence is $\left\{\exp\left\{-\frac{1}{2}N\tilde{\varphi}_i\right\}\right\}$ itself.

On the other hand, we know that, for any $\lambda > 0$,

$$\mathrm{Vol}\{x|\lambda \leqq |\varphi|(x)\} \leqq \frac{1}{\lambda}\int_M |\varphi|. \quad (2.55)$$

Furthermore,

$$\mathrm{Vol}\left\{x|\lambda \leqq \exp\left\{-\frac{1}{2}N\tilde{\varphi}_i\right\}\right\} = \mathrm{Vol}\left\{x\Big|\frac{2}{N}\log\lambda + \frac{1}{N}\log\int_M \exp\{-N\varphi_i\} \leqq -\varphi_i\right\}. \quad (2.56)$$

Since $\lim_{i\to\infty}\int_M \exp\left\{-\frac{1}{2}N\varphi_i\right\} = \infty$, we conclude that, for i large enough,

$$\mathrm{Vol}\left\{x|\lambda \leqq \exp\left\{-\frac{1}{2}N\tilde{\varphi}_i\right\}\right\}$$
$$\leqq \mathrm{Vol}\left\{x|0 < \frac{2}{N}\log\lambda + \frac{1}{N}\log\int_M \exp\{-N\varphi_i\} \leqq |\varphi_i|\right\}$$
$$\leqq \left(\frac{2}{N}\log\lambda + \frac{1}{N}\log\int_M \exp\{-N\varphi_i\}\right)^{-1}\int_M |\varphi_i|. \quad (2.57)$$

By (2.28) $\left\{\int_M |\varphi_i|\right\}$ is uniformly bounded and the inequality (2.57) then implies that

$$\lim_{i\to\infty}\mathrm{Vol}\left\{x|\lambda \leqq \exp\left\{-\frac{1}{2}N\tilde{\varphi}_i\right\}\right\} = 0 \quad (2.58)$$

for all $\lambda > 0$. Clearly,

$$\mathrm{Vol}\{x|\lambda \leqq f\} \leqq \mathrm{Vol}\left\{x\Big|\frac{1}{2}\lambda \leqq \left|f - \exp\left\{-\frac{1}{2}N\tilde{\varphi}_i\right\}\right|\right\}$$
$$+\mathrm{Vol}\left\{x\Big|\frac{1}{2}\lambda \leqq \exp\left\{-\frac{1}{2}N\tilde{\varphi}_i\right\}\right\}$$

$$\leq \frac{4}{\lambda^2}\int_M \left| f - \exp\left\{\frac{1}{2}N\tilde{\varphi}_i\right\}\right|^2$$
$$+\mathrm{Vol}\left\{x \middle| \frac{1}{2}\lambda \leq \exp\left\{-\frac{1}{2}N\tilde{\varphi}_i\right\}\right\}. \tag{2.59}$$

Since $\exp\left\{-\frac{1}{2}N\tilde{\varphi}_i\right\}$ converges to f in $L^2(M)$, (2.58) and (2.59) shows that, for all $\lambda > 0$,

$$\mathrm{Vol}\{x|\lambda \leqq f\} = 0. \tag{2.60}$$

Since f is the L^2-limit of $\exp\left\{-\frac{1}{2}N\tilde{\varphi}_i\right\}$, (2.59) implies that f is zero almost everywhere. This is a contradiction because $\int_M f^2 = 1$ which is a consequence of (2.54) and the L^2 convergence of $\exp\left\{-\frac{1}{2}N\tilde{\varphi}_i\right\}$ to f. Therefore we have arrived at the conclusion that whenever N verifies (2.53), (2.28) and (2.48) $\int_M \exp\{-N\varphi\}$ has an estimate from above depending only on N, F and M.

We can now repeat the previous argument to find an estimate of $|\inf_M \varphi|$. Inequality (2.46) is valid for arbitrary dimension. As before, we find a geodesic ball with radius $-\frac{1}{2}\inf\varphi C_7^{-1}(\exp\{-C\inf\varphi\}+1)^{-1}$ (which is not greater than the injectivity radius) such that φ is not greater than $\frac{1}{2}\inf\varphi$ in this ball. Then we choose N so large that (2.48) is satisfied and that N is greater than $4mC$. Since the integral of $\exp\{-N\varphi\}$ in the above geodesic ball is not less than

$$C_{12}\exp\left\{-\frac{1}{2}N\inf\varphi\right\}\left(-\frac{1}{2}\inf\varphi\right)^{2m} C_7^{-1}(\exp\{-C\inf\varphi\}+1)^{-2m},$$

where C_{12} is a positive constant depending only on M, we have an estimate of $-\inf_M \varphi$.

Since we have already found an estimate of $\sup\varphi$, this gives an estimate of $\sup_M|\varphi|$. The inequalities (2.46) and (2.24) then give estimates of $\sup_M|\nabla\varphi|$ and $\sup_M(m+\Delta\varphi)$. On the other hand, since $(\delta_{ij}+\varphi_{ij})$ is a positive definite Hermitian matrix, we can find upper estimates of $1+\varphi_{i\bar{i}}$ for each i. The equation $\prod_{i=1}^m(1+\varphi_{i\bar{i}}) = \exp\{F\}$ then gives a positive lower estimate of $(1+\varphi_{i\bar{i}})$ for each i. In particular, the metric tensor $\sum_{i,j}(g_{i\bar{j}}+\partial^2\varphi/\partial z^i\partial\bar{z}^j)dz^i\otimes d\bar{z}^j$ is uniformly equivalent to $\sum_{i,j} g_{i\bar{j}}dz^i\otimes d\bar{z}^j$.

In conclusion, we have proved the following

Proposition 2.1. *Let M be a compact Kähler manifold with metric tensor $2\sum_{i,j} g_{i\bar{j}}dz^i\otimes d\bar{z}^j$. Let φ be a real-valued function in $C^4(M)$ such that $\int_M\varphi = 0$ and $\sum_{i,j}(g_{i\bar{j}}+\partial^2\varphi/\partial z^i\partial\bar{z}^j)dz^i\otimes d\bar{z}^j$ defines another metric tensor on M. Suppose $\det(g_{i\bar{j}}+\partial^2\varphi/\partial z^i\partial\bar{z}^j)\det(g_{i\bar{j}})^{-1} = \exp\{F\}$. Then there are positive constants C_1, C_2, C_3, and C_4, depending on $\inf_M F, \sup_M F, \inf\Delta F$ and M such that $\sup_M|\varphi| \leqq C_1, \sup_M|\nabla\varphi| \leqq C_2, 0 < C_3 \leqq 1+\varphi_{i\bar{i}} \leqq C_4$ for all i.*

3 Third-Order estimates

In this section, we estimate the third derivatives of $\varphi_{i\bar{j}k}$ assuming φ solves the equation (2.1) and F is $C^3(M)$.

Following E. Calabi [5], we consider the function

$$S = \sum g'^{i\bar{r}} g'^{\bar{j}s} g'^{k\bar{t}} \varphi_{i\bar{j}k} \varphi_{\bar{r}s\bar{t}}. \tag{3.1}$$

We are going to compute the Laplacian of S. For convenience, we shall introduce the following convention. We say that $A \simeq B$ if $|A - B| \leqq C_1\sqrt{S} + C_2$, where C_1 and C_2 are constants that can be estimated. We also say that $A \cong B$ if $|A - B| \leqq C_3 S + C_4\sqrt{S} + C_5$, where C_3, C_4, and C_5, are constants that can be estimated.

As in the last section, we shall diagonalize our metric tensor and the Hessian $(\varphi_{i\bar{j}})$ at a point that is under consideration. Then by the computation which will be carried out in the Appendix A, we have

$$\begin{aligned} \Delta' S \cong \sum &(1+\varphi_{i\bar{i}})^{-1}(1+\varphi_{j\bar{j}})^{-1}(1+\varphi_{k\bar{k}})^{-1}(1+\varphi_{\alpha\bar{\alpha}})^{-1} \\ &\times \Bigg\{ \Big| \varphi_{\bar{i}j\bar{k}\alpha} - \sum_p \varphi_{\bar{i}p\bar{k}} \varphi_{\bar{p}j\alpha} (1+\varphi_{p\bar{p}})^{-1} \Big|^2 \\ &+ \Big| \varphi_{i\bar{j}k\alpha} - \sum_p (\varphi_{\bar{p}i\alpha}\varphi_{p\bar{j}k} + \varphi_{\bar{p}ik}\varphi_{p\bar{j}\alpha})(1+\varphi_{p\bar{p}})^{-1} \Big|^2 \Bigg\}. \end{aligned} \tag{3.2}$$

On the other hand, by (2.7)

$$\Delta'(\Delta\varphi) \geqq \sum (1+\varphi_{k\bar{k}})^{-1}(1+\varphi_{i\bar{i}})^{-1} |\varphi_{k\bar{i}\bar{j}}|^2 - C_6, \tag{3.3}$$

where C_6 is a constant that can be estimated.

Therefore by letting C_7, be a large positive constant, we find

$$\Delta'(S + C_7\Delta\varphi) \geqq C_8 S - C_9, \tag{3.4}$$

where C_7, C_8, C_9 are positive constants that can be estimated.

At the point where $S + C_7\Delta\varphi$ achieves its maximum, (3.4) shows that

$$C_8(S + C_7\Delta\varphi) \leqq C_9 + C_8 C_7 \Delta\varphi. \tag{3.5}$$

Since we have estimated $\Delta\varphi$ already, (3.5) gives an estimate of the quantity $\sup_M(S + C_7\Delta\varphi)$ and hence of $\sup_M S$. This in turn gives estimates of $\varphi_{i\bar{j}k}$ for all i, j, k.

Proposition 3.1. *Let M be a compact Kähler manifold with metric tensor $\sum_{i,j} g_{i\bar{j}} dz^i \otimes d\bar{z}^j$. Let φ be a real-valued function in $C^5(M)$ such that $\int_M \varphi = 0$ and*

$$\sum_{i,j} \left(g_{i\bar{j}} + \frac{\partial^2\varphi}{\partial z^i \partial \bar{z}^j} \right) dz^i \otimes d\bar{z}^j$$

defines another metric tensor on M. Suppose

$$\det\left(g_{i\bar{j}} + \frac{\partial^2\varphi}{\partial z^i \partial \bar{z}^j} \right) \det(g_{i\bar{j}})^{-1} = \exp\{F\}.$$

Then there is an estimate of the derivatives $\varphi_{i\bar{j}k}$ in terms of $\sum_{i,j} g_{i\bar{j}} dz^i \otimes d\bar{z}^j$, $\sup|F|, \sup|\nabla F|, \sup_M \sup_i |F_{i\bar{i}}|$ and $\sup_M \sup_{i,j,k} |F_{i\bar{j}k}|$.

4 Solution of the equation

With the estimates of Sections 1 and 3, we can now solve the equation $\det[g_{i\bar{j}} + (\partial^2\varphi/\partial z^i\partial\bar{z}^j)]\det(g_{i\bar{j}})^{-1} = \exp\{F\}$, where $F \in C^3(M)$.

If Ω is the Kähler form of M, then the above equation is equivalent to

$$(\Omega + \partial\bar{\partial}\varphi)^m = (\exp\{F\})\Omega^m. \tag{4.1}$$

Hence integrating (4.1), we see immediately that

$$\int \exp\{F\} = \mathrm{Vol}(M). \tag{4.2}$$

Conversely, we shall now prove that if $F \in C^k(M)$ with $k \geqq 3$ and F satisfies (4.2), then we can find a solution φ of (4.1) where $\varphi \in C^{k+1,\alpha}(M)$ for any $0 \leqq \alpha < 1$. ($C^{k-1,\alpha}(M)$ are the functions whose $(k-1)$-derivatives are Hölder continuous with exponent α.)

We are going to use the continuity method. We consider the set

$$\begin{aligned} S = \Big\{t \in [0,1] \Big| \text{ the equation } \det\left(g_{i\bar{j}} + \frac{\partial^2\varphi}{\partial z^i\partial\bar{z}^j}\right)\det(g_{i\bar{j}})^{-1} \\ = \mathrm{Vol}(M)\left[\int \exp\{tF\}\right]^{-1} \exp\{tF\} \text{ has a solution in } C^{k+1,\alpha}(M)\Big\}. \end{aligned} \tag{4.3}$$

Obviously $0 \in S$. Hence we need only to show that S is both closed and open in $[0, 1]$. This will imply that $1 \in S$ and that our original equation has a solution in $C^{k+1,\alpha}(M)$.

To see that the set S is open, we use the standard inverse function theorem (cf. [14]). Let $\theta = \{\varphi \in C^{k+1,\alpha}(M) | 1 + \varphi_{i\bar{i}} > 0$ for all i and $\int_M \varphi = 0\}$ and $B = \{f \in C^{k-1,\alpha}(M) | \int_M f = \mathrm{Vol}(M)\}$. Then θ is an open set in the Banach space $C^{k+1,\alpha}(M)$ and B is a hyperplane in the Banach space $C^{k-1,\alpha}(M)$. We have a map G mapping θ into B:

$$G(\varphi) = \det\left(g_{i\bar{j}} + \frac{\partial^2\varphi}{\partial z^i\partial\bar{z}^j}\right)\det(g_{i\bar{j}})^{-1}. \tag{4.4}$$

The differential of G at a point φ_0 is given by $\det(g_{i\bar{j}} + \partial^2\varphi_0/\partial z^i\partial\bar{z}^j) \cdot \det(g_{i\bar{j}})^{-1}\Delta_{\varphi_0}$, where Δ_{φ_0} is the (normalized) Laplacian of the metric $\sum(g_{i\bar{j}} + \partial^2\varphi_0/\partial z^i\partial\bar{z}^j)dz^i \otimes d\bar{z}^j$. The tangent space of B is the space of $C^{k-1,\alpha}$-functions such that $\int_M f = 0$.

It is wellknown that the condition for $\Delta_{\varphi_0}\varphi = g$ to have a weak solution on M is that $\int_M g dV_{\varphi_0} = 0$. Hence the condition for

$$\det\left(g_{i\bar{j}} + \frac{\partial^2\varphi_0}{\partial z^i\partial\bar{z}^j}\right)\det(g_{i\bar{j}})^{-1}\Delta_{\varphi_0}\varphi = f$$

to have a weak solution is that $\int_M f = 0$. The Schauder theory (cf. [13]) makes sure that $\varphi \in C^{k+1,\alpha}(M)$ when $f \in C^{k-1,\alpha}(M)$. The solution is clearly unique if we require that $\int_M \varphi = 0$. Hence the differential dG of G at φ_0 is invertible and G maps an open neighborhood of φ_0 to an open neighborhood of $G(\varphi_0)$ in B. this proves that the set S is open.

It remains to prove that the set S is closed, If $\{t_q\}$ is a sequence in S, then we have a sequence $\varphi_q \in C^{k+1,\alpha}(M)$ such that

$$\det\left(g_{i\bar{j}} + \frac{\partial\varphi_q}{\partial z^i \partial \bar{z}^j}\right)\det(g_{i\bar{j}})^{-1} = \mathrm{Vol}(M)\left(\int_M \exp\{t_q F\}\right)^{-1}\exp\{t_q F\}.$$

Normalizing, we can assume $\int_M \varphi_q = 0$.

Differentiating the above equation, we have

$$\det\left(g_{i\bar{j}} + \frac{\partial^2\varphi_q}{\partial z^i \partial \bar{z}^j}\right)\sum_{i,j} g_q^{\prime i\bar{j}} \frac{\partial^2}{\partial z^i \partial \bar{z}^j}\left(\frac{\partial\varphi_q}{\partial z^p}\right)$$

$$- \mathrm{Vol}(M)\left(\int_M \exp\{t_q F\}\right)^{-1}\frac{\partial}{\partial z^p}[\exp\{t_q F\}\det(g_{i\bar{j}})], \tag{4.5}$$

where $(g_q^{\prime i\bar{j}})$ is the inverse matrix of $(g_{i\bar{j}} + \partial^2\varphi_q/\partial z^i \partial\bar{z}^j)$ for each q.

Proposition 2.1 shows that the operator on the left-hand side of (4.5) is uniformly elliptic. Proposition 3.1 shows that the coefficients are Hölder continuous with exponent α for any $0 \leqq \alpha \leqq 1$. The Schauder estimate then gives an estimate on the $C^{2,\alpha}$-norm of $\partial\varphi_q/\partial z^p$. Similarly we can find the $C^{2,\alpha}$-norm of $\partial\varphi_q/\partial\bar{z}^p$. From this information we know that the coefficients of the elliptic operator on the left-hand side of (4.5) have better differentiability. The Schauder estimate again provides better differentiability of $\partial\varphi_q/\partial z^p$ and $\partial\varphi_q/\partial\bar{z}^p$. Iterating, one finds $C^{k+1,\alpha}$-estimates of φ_q. Therefore the sequence $\{\varphi_q\}$ converges in the $C^{k+1,\alpha}$-norm to a solution of the equation

$$\det\left(g_{i\bar{j}} + \frac{\partial^2\varphi}{\partial z^i \partial \bar{z}^j}\right)\det(g_{i\bar{j}})^{-1} = \mathrm{Vol}(M)\left(\int_M \exp\{t_0 F\}\right)^{-1}\exp\{t_0 F\},$$

where $t_0 - \lim_{q\to\infty} t_q$. Hence S is closed.

Theorem 1. *Assume that M is a compact Kähler manifold with metric $\sum g_{i\bar{j}} dz^i \otimes d\bar{z}^j$. Let F be $C^k(M)$ with $k \geqq 3$ and $\int_M \exp\{F\} = \mathrm{Vol}(M)$. Then there is a function φ in $C^{k+1,\alpha}(M)$ for any $0 \leqq \alpha < 1$ such that $2\sum(g_{i\bar{j}} + \partial^2\varphi/\partial z^i \partial z^j) dz^i \otimes d\bar{x}^j$ defines a Kähler metric and*

$$\det\left(g_{i\bar{j}} + \frac{\partial^2\varphi}{\partial z^i \partial \bar{z}^j}\right) = \exp\{F\}\det(g_{i\bar{j}}).$$

As a consequence of Theorem 1, we can prove the conjecture of Calabi [3]. Let $\sum g_{\alpha\bar{\beta}} dx^\alpha \otimes d\bar{z}^\beta$ be a Kähler metric on a complex manifold M. Let $\sum R_{\alpha\bar{\beta}} dz^\alpha \otimes d\bar{z}^\beta$ be the corresponding Ricci tensor. Then Chern [6] proved that the (1, 1) form $(i/2\pi)\sum R_{\alpha\bar{\beta}} dz^\alpha \wedge d\bar{z}^\beta$ represents the first Chern class of M. Calabi asked the following question: Given a (1, 1) form $(i/2\pi)\sum \tilde{R}_{\alpha\bar{\beta}} dz^\alpha \wedge d\bar{z}^\beta$ which represents the first Chern class of M, can one find a Kähler metric on M so that $\sum \tilde{R}_{\alpha\bar{\beta}} dz^\alpha \otimes d\bar{z}^\beta$

is the Ricci tensor of this metric. Calabi himself observed the uniqueness of this problem. Namely, there is at most one Kähler metric in each cohomology class which does this job. He also observed the local version of this problem. He proved that, if the tensor $\sum R_{\alpha\bar{\beta}} dz^\alpha \otimes d\bar{z}^\beta$ can be realized, then all tensors which are close to this one and which represent the first Chern class can also be realized.

To see how Calabi's conjecture follows from Theorem 1, we notice the following wellknown fact (see (1.10)):

$$R_{\alpha\bar{\beta}} = -\frac{\partial^2}{\partial z^\alpha \partial \bar{z}^\beta} \log \det(g_{i\bar{j}}). \tag{4.6}$$

Since we assume that $(i/2\pi)\sum \tilde{R}_{\alpha\bar{\beta}} dz_\alpha \wedge d\bar{z}_\beta$ represents the first Chern class, we conclude immediately from Chern's theorem that

$$\tilde{R}_{\alpha\bar{\beta}} = R_{\alpha\bar{\beta}} - \frac{\partial^2}{\partial z^\alpha \partial \bar{z}^\beta} f \tag{4.7}$$

for some smooth real-valued function f.

According to Theorem 1, we can find a smooth function φ so that $\sum(g_{\alpha\bar{\beta}} + \partial^2\varphi/\partial z^\alpha \partial \bar{z}^\beta) dz^\alpha \otimes d\bar{z}^\beta$ defines a Kähler metric and that

$$\det\left(g_{\alpha\bar{\beta}} + \frac{\partial^2 \varphi}{\partial z^\alpha \partial \bar{z}^\beta}\right) \det(g_{\alpha\bar{\beta}})^{-1} = C \exp\{f\}, \tag{4.8}$$

where C is a constant chosen to satisfy the equation

$$\int_M C \exp\{f\} = \mathrm{Vol}(M). \tag{4.9}$$

From (4.6), (4.7) and (4.8), it is easy to see that $\sum \tilde{R}_{\alpha\bar{\beta}} dz^\alpha \otimes d\bar{z}^\beta$ is the Ricci tensor of $\sum(g_{\alpha\bar{\beta}} + \partial^2\varphi/\partial z^\alpha \partial \bar{z}^\beta) dz^\alpha \otimes d\bar{z}^\beta$.

Theorem 2. *Let M be a compact Kähler manifold with Kähler metric $\sum g_{\alpha\bar{\beta}} dz^\alpha \otimes d\bar{z}^\beta$. Let $\sum \tilde{R}_{\alpha\bar{\beta}} dz^\alpha \otimes d\bar{z}^\beta$ be a tensor whose associated* (1, 1) *form $(i/2\pi)\sum_{\alpha,\beta} \tilde{R}_{\alpha\bar{\beta}} dz^\alpha \wedge d\bar{z}^\beta$ represents the first Chern class of M. Then we can find a Kähler metric $\sum \bar{g}_{\alpha\bar{\beta}} dz^\alpha \otimes d\bar{z}^\beta$ whose Ricci tensor is given by $\sum \tilde{R}_{\alpha\bar{\beta}} dz^\alpha \otimes d\bar{z}^\beta$. Furthermore, we can require that this Kähler metric has the same Kähler class as the original one. In this case, the required Kähler metric is unique.*

Remark. *The uniqueness was proved by Calabi* [4] *and will also be indicated and proved in Theorem* 3.

5 Complex Monge-Ampère equation with degenerate right-hand side

Let L be a line bundle over a compact Kähler manifold M. Let s be a (non-trivial) holomorphic section of L. Suppose L is equipped with an Hermitian metric so that we have a globally defined function $|s|^2$ on M. Then, for $k \geqq 0$, we shall study an equation of the form

$$\det\left(g_{i\bar{j}} + \frac{\partial^2\varphi}{\partial z^i \partial \bar{z}^j}\right)\det(g_{i\bar{j}})^{-1} = |s|^{2k}\exp\{F\}, \tag{5.1}$$

where $\sum_{i,j} g_{i\bar{j}} dz^i \otimes d\bar{z}^j$ is a Kähler metric and F is a smooth function such that

$$\int_M |s|^{2k}\exp\{F\} = \mathrm{Vol}(M). \tag{5.2}$$

In order to solve (5.1), we approximate the equation by the following non-degenerate elliptic equation:

$$\det\left(g_{i\bar{j}} + \frac{\partial^2\varphi}{\partial z^i \partial \bar{z}^j}\right)\det(g_{i\bar{j}})^{-1} = C_\varepsilon(|s|^2+\varepsilon)^k\exp\{F\}, \tag{5.3}$$

where $\varepsilon > 0$ is a small constant and

$$C_\varepsilon = \mathrm{Vol}(M)\left(\int_M (|s|^2+\varepsilon)^k\exp\{F\}\right)^{-1}. \tag{5.4}$$

It was proved in Section 4 that (5.3) has a smooth solution φ_ε, such that $(g_{i\bar{j}} + \partial^2\varphi_\varepsilon/\partial z^i\partial\bar{z}^j)$ is positive definite and

$$\int_M \varphi_\varepsilon = 0. \tag{5.5}$$

We are going to prove that when ε approaches zero, φ_ε tends to a solution of (5.1). This depends on estimates of φ_ε which are independent of ε.

Since the right-hand side of (2.28) depends only on the Green's function of the fixed Kähler metric, we can find a constant C_1 independent of ε such that

$$\sup_M \varphi_\varepsilon \leqq C_1. \tag{5.6}$$

In order to estimate $\inf_M \varphi_\varepsilon$ and $\Delta\varphi_\varepsilon$ we notice that, when $s \neq 0$,

$$\Delta\log(|s|^2+\varepsilon) \geqq \frac{|s|^2}{|s|^2+\varepsilon}(\Delta\log|s|^2). \tag{5.7}$$

By direct computation, one knows that, when $s \neq 0, \Delta\log|s|^2$ depends only on the fixed Hermitian metric of L. In fact, it is the trace of the first Chern form of L with respect to the tensor $\sum_{ij} g_{i\bar{j}} dz^i \otimes d\bar{z}^j$. Hence $\Delta\log(|s|^2+\varepsilon)$ is uniformly bounded from below by a constant independent of ε.

Let Δ be the normalized Laplacian of the metric $\sum_{i,j}(g_{i\bar{j}} + (\partial^2\varphi_\varepsilon/\partial z^i\partial\bar{z}^j))dz^i \otimes d\bar{z}^j$. Then according to (2.22), we have

$$\begin{aligned}
&\Delta'_\varepsilon(\exp\{-C\varphi_\varepsilon\}(m+\Delta\varphi_\varepsilon))\\
&\geqq \exp\{-C\varphi_\varepsilon\}\left(k\Delta\log(|s|^2+\varepsilon) + \Delta F - m^2\inf_{i\neq l}(R_{i\bar{i}l\bar{l}})\right)\\
&\quad -mC\exp\{-C\varphi_\varepsilon\}(m+\Delta\varphi_\varepsilon)
\end{aligned}$$

$$+\left(C+\inf_{i\neq l}(R_{i\bar{i}l\bar{l}})\right) C_\varepsilon^{-1/(m-1)} \exp\{-C\varphi_\varepsilon\}(|s|^2+\varepsilon)^{-k/(m-1)}$$
$$\times \exp\left\{\frac{-F}{m-1}\right\}(m+\Delta\varphi_\varepsilon)^{1+1/(m-1)}. \tag{5.8}$$

By (5.7), $\Delta \log(|s|^2+\varepsilon)$ is bounded from below. Hence one can conclude as in Section 2 that the following inequality holds:

$$\exp\{-C\varphi_\varepsilon\}(m+\Delta\varphi_\varepsilon) \leqq C_2 \exp\{-C \inf \varepsilon_\varepsilon\}, \tag{5.9}$$

where C_2 is a constant independent of ε.

If $m=2$, we multiply (5.8) by $(|s|^2+\varepsilon)^k \exp\{F\}$ and integrate the resulting inequality. We find

$$C_\varepsilon^{-1}\left(C+\inf_{i\neq l}(R_{i\bar{i}l\bar{l}})\right)\int_M \exp\{-C\varphi_\varepsilon\}(m+\Delta\varphi_\varepsilon)^2$$
$$-mC\int_M \exp\{-C\varphi_\varepsilon\}(m+\Delta\varphi_\varepsilon)(|s|^2+\varepsilon)^k \exp\{F\} \tag{5.10}$$
$$\leqq \int_M \exp\{-C\varphi_\varepsilon\}(|s|^2+\varepsilon)^k \exp\{F\}\left(k|C_L|-\Delta F+m^2\inf_{i\neq l}(R_{i\bar{i}l\bar{l}})\right)$$

Here C_L is the trace of the first Chern form of L with respect to the tensor $\sum_{i,j} g_{i\bar{j}} dz^i \otimes d\bar{z}^j$.

For any $\delta > 0$, we can find a constant C_δ, such that

$$m+\Delta\varphi_\varepsilon \leqq \delta(m+\Delta\varphi_\varepsilon)^2 + C_\delta. \tag{5.11}$$

Hence by choosing $C+\inf_{i\neq l}(R_{i\bar{i}l\bar{l}}) \geqq 1$, we can obtain an estimate of the form

$$\int_M \exp\{-C\varphi_\varepsilon\}(m+\Delta\varphi_\varepsilon) \leqq C_3 \int_M \exp\{-C\varphi_\varepsilon\}, \tag{5.12}$$

where C_3 is a positive constant independent of ε.

One can now proceed as in Section 2 to find an estimate of $\inf_M \varphi_\varepsilon$, independent of ε and hence also an estimate of $\sup_M \Delta\varphi_\varepsilon$ independent of ε.

For $m > 2$, we proceed as follows. Let p be any non-negative number. Then by Schwarz' inequality, we have

$$\begin{aligned}\Delta'_\varepsilon[(|s|^2+\varepsilon)^p \exp\{-C\varphi_\varepsilon\}] &= [\Delta'_\varepsilon(|s|^2+\varepsilon)^p]\exp\{-C\varphi_\varepsilon\}\\ &\quad +2\nabla'_\varepsilon(|s|^2+\varepsilon)^p \cdot \nabla'_\varepsilon \exp\{-C\varphi_\varepsilon\}\\ &\quad +(|s|^2+\varepsilon)^p[-C\exp\{-C\varphi_\varepsilon\}\Delta'_\varepsilon\varphi_\varepsilon\\ &\quad +|\nabla'_\varepsilon \exp\{-C\varphi_\varepsilon\}|^2 \exp\{C\varphi_\varepsilon\}]\\ &\geqq [\Delta'_\varepsilon(|s|^2+\varepsilon)^p]\exp\{-C\varphi_\varepsilon\}\\ &\quad -\frac{|\nabla'_\varepsilon(|s|^2+\varepsilon)^p|^2}{(|s|^2+\varepsilon)^p}\exp\{-C\varphi_\varepsilon\}\\ &\quad -C(|s|^2+\varepsilon)^p \exp\{-C\varphi_\varepsilon\}\Delta'_\varepsilon\varphi_\varepsilon.\end{aligned} \tag{5.13}$$

For $s \neq 0$, $\Delta'_\varepsilon \log |s|^2$ is dominated from below by the trace of the first Chern form of L with respect to $\sum_{i,j}(g_{i\bar{j}} + \partial^2\varphi_\varepsilon/\partial z^i \partial \bar{z}^j) dz^i \otimes d\bar{z}^j$. Hence there is a positive constant C_4, independent of ε such that

$$\Delta'_\varepsilon \log |s|^2 \geqq -C_4 \sum_i \frac{1}{1 + (\varphi_\varepsilon)_{i\bar{i}}}. \tag{5.14}$$

As in (5.7), we have

$$\Delta'_\varepsilon \log(|s|^2 + \varepsilon)^p \geqq -pC_4 \sum_i \frac{1}{1 + (\varphi_\varepsilon)_{i\bar{i}}}. \tag{5.15}$$

Hence

$$\Delta'_\varepsilon (|s|^2 + \varepsilon)^p \geqq \frac{|\nabla'_\varepsilon (|s|^2 + \varepsilon)^p|^2}{(|s|^2 + \varepsilon)^p} - pC_4 (|s|^2 + \varepsilon)^p \sum_i \frac{1}{1 + (\varphi_\varepsilon)_{i\bar{i}}}. \tag{5.16}$$

Inserting (5.16) into (5.13), we obtain

$$\begin{aligned}
&\Delta'_\varepsilon [(|s|^2 + \varepsilon)^p \exp\{-C\varphi_\varepsilon\}] \\
&\geqq -pC_4 (|s|^2 + \varepsilon)^p \sum_i \frac{1}{1 + (\varphi_\varepsilon)_{i\bar{i}}} \\
&\quad -C(|s|^2 + \varepsilon)^p \exp\{-C\varphi_\varepsilon\} \left[\sum_i \frac{1 + (\varphi_\varepsilon)_{i\bar{i}} - 1}{1 + (\varphi_\varepsilon)_{i\bar{i}}}\right] \\
&= -mC(|s|^2 + \varepsilon)^p \exp\{-C\varphi_\varepsilon\} \\
&\quad +(C - pC_4)(|s|^2 + \varepsilon)^p \left(\sum_i \frac{1}{1 + (\varphi_\varepsilon)_{i\bar{i}}}\right) \exp\{-C\varphi_\varepsilon\}.
\end{aligned} \tag{5.17}$$

Using the arithmetic-geometric mean inequality and (5.3), we deduce that

$$\begin{aligned}
&\Delta'_\varepsilon [(|s|^2 + \varepsilon)^p \exp\{-C\varphi_\varepsilon\}] \\
&\geq -mC(|s|^2 + \varepsilon)^p \exp\{-C\varphi_\varepsilon\} + m(C - pC_3)(|s|^2 + \varepsilon)^p \\
&\quad \times \left[\prod_{i=1}^m (1 + (\varphi_\varepsilon)_{i\bar{i}})^{-1/m}\right] \exp\{-C\varphi_\varepsilon\} \\
&= -mC(|s|^2 + \varepsilon)^p \exp\{-C\varphi_\varepsilon\} \\
&\quad +m(C - pC_3)(|s|^2 + \varepsilon)^p C_\varepsilon^{-1/m} (|s|^2 + \varepsilon)^{-k/m} \\
&\quad \times \exp\left\{-\frac{F}{m}\right\} \exp\{-C\varphi_\varepsilon\}.
\end{aligned} \tag{5.18}$$

Multiplying (5.18) by $(|s|^2 + \varepsilon)^k \exp\{F\}$ and integrating, we find a positive constant C_5 independent of ε such that

$$\begin{aligned}
&(C - pC_3) \int_M (|s|^2 + \varepsilon)^{(m-1)k/m + p} \exp\{-C\varphi_\varepsilon\} \\
&\leqq C_5 \int_M (|s|^2 + \varepsilon)^{k+p} \exp\{-C\varphi_\varepsilon\}.
\end{aligned} \tag{5.19}$$

In (5.19), C and p are arbitrary non-negative constants. By iterating (5.19), we see that, for all fixed $p \geqq 0$, we can find (large) constants C and C_6, such that

$$\int_M (|s|^2+\varepsilon)^k \exp\{-C\varphi_\varepsilon\} \leqq C_6 \int_M (|s|^2+\varepsilon)^{k+p} \exp\{-C\varphi_\varepsilon\}. \tag{5.20}$$

On the other hand, by (5.8), we can find positive constants C_7, and C_8, independent of ε such that

$$\begin{aligned} &\Delta'_\varepsilon(\exp\{-C\varphi_\varepsilon\}(m+\Delta\varphi_\varepsilon)) \\ &\geqq -C_7\exp\{-C\varphi_\varepsilon\} + C_8\exp\{-C\varphi_\varepsilon\}(m+\Delta\varphi_\varepsilon). \end{aligned} \tag{5.21}$$

Multiplying (5.21) by $(|s|^2+\varepsilon)^k \exp\{F\}$ and integrating, we obtain

$$\int_M (|s|+\varepsilon)^k \exp\{-C\varphi_\varepsilon\}(m+\Delta_\varepsilon) \leqq C_9 \int_M (|s|^2+\varepsilon)^k \exp\{-C\varphi\}, \tag{5.22}$$

where C_9 is independent of ε. Since $m+\Delta\varphi_\varepsilon > 0$, it follows from (5.22) that we can find a positive constant C_{10} independent of ε such that

$$\int_M (|s|^2+\varepsilon)^{k+1}(m+\Delta\varphi)\exp\{-C\varphi_\varepsilon\} \leqq C_{10}\int_M (|s|^2+\varepsilon)^k \exp\{-C\varphi_\varepsilon\}. \tag{5.23}$$

Integrating by part in (5.23), we derive

$$\begin{aligned} &C\int_M (|s|^2+\varepsilon)^{k+1}\exp\{-C\varphi_\varepsilon\}|\nabla\varphi_\varepsilon|^2 \leqq (k+1)\int_M (|s|^2+\varepsilon)^k \cdot \\ &(\nabla|s|^2\cdot\nabla\varphi_\varepsilon)\exp\{-C\varphi_\varepsilon\} \\ &+C_{10}\int_M (|s|^2+\varepsilon)^k\exp\{-C\varphi_\varepsilon\} \\ \leqq\ &\frac{(k+1)^2}{C}\int_M (|s|^2+\varepsilon)^{k-1}|\nabla|s|^2|^2\exp\{-C\varphi_\varepsilon\} \\ &+\frac{1}{4}C\int_M (|s|^2+\varepsilon)^{k+1}\exp\{-C\varphi_\varepsilon\}|\nabla\varphi_\varepsilon|^2 \\ &+C_{10}\int_M (|s|^2+\varepsilon)^k\exp\{-C\varphi_\varepsilon\}. \end{aligned} \tag{5.24}$$

Hence

$$\begin{aligned} &\frac{3}{4}C\int_M (|s|^2+\varepsilon)^{k+1}\exp\{-C\varphi_\varepsilon\}|\nabla\varphi_\varepsilon|^2 \\ \leqq\ &\frac{(k+1)^2}{C}\int_M (|s|^2+\varepsilon)^{k-1}|\nabla|s|^2|^2\exp\{-C\varphi_\varepsilon\} \\ &+C_{10}\int_M (|s|^2+\varepsilon)^k\exp\{-C\varphi_\varepsilon\}, \end{aligned} \tag{5.25}$$

$$\begin{aligned} &\int_M \left|\nabla\left[(|s|^2+\varepsilon)^{(k+1)/2}\exp\left\{-\frac{1}{2}C\varphi_\varepsilon\right\}\right]\right|^2 \\ \leqq\ &\frac{1}{2}(k+1)^2\int_M (|s|^2+\varepsilon)^{k-1}|\nabla|s|^2|^2\exp\{-C\varphi_\varepsilon\} \end{aligned}$$

$$
\begin{aligned}
&+\frac{1}{2}C^2\int_M (|s|^2+\varepsilon)^{k+1}\exp\{-C\varphi_\varepsilon\}|\nabla\varphi_\varepsilon|^2\\
&\leqq \frac{5}{6}(k+1)^2\int_M (|s|^2+\varepsilon)^{k-1}|\nabla|s|^2|^2\exp\{-C\varphi_\varepsilon\}\\
&+\frac{2}{3}CC_{10}\int_M (|s|^2+\varepsilon)^{k}\exp\{-C\varphi_\varepsilon\}\\
&\leqq C_{11}\int_M (|s|^2+\varepsilon)^{k}\exp\{-C\varphi_\varepsilon\},
\end{aligned}
\tag{5.26}
$$

where C_{11} is independent of ε. (We use the fact that $|\nabla|s|^2|^2$ is dominated by $|s|^2$.)

By making use of (5.20), $\int_N (|s|^2+\varepsilon)^k \exp\{-C\varphi_\varepsilon\}$ is dominated by

$$\int_M (|s|^2+\varepsilon)^{k+1}\exp\{-C\varphi_\varepsilon\}.$$

Therefore we can find a constant C_{12}, independent of ε such that C_{12} independent of ε such that

$$\int_M \left|\nabla\left[(|s|^2+\varepsilon)^{(k+1)/2}\exp\left\{-\frac{1}{2}C\varphi_\varepsilon\right\}\right]\right|^2 \leqq C_{12}\int_M (|s|^2+\varepsilon)^{k+1}\exp\{-C\varphi_\varepsilon\}. \tag{5.27}$$

Since (5.5) and (5.6) provide an estimate of $\int_M |\varphi_\varepsilon|$ independent of ε, we see applying the arguments of Section 2 that (5.27) gives an estimate of $\int_M (|s|^2+\varepsilon)^{k+1}\exp\{-C\varphi_\varepsilon\}$ independent of ε. (Suppose there is no estimate. Then one can find a sequence $\varepsilon_j \to 0$ such that $\int_M (|s|^2+\varepsilon_j)^{k+1}\exp\{-C\varphi_{\varepsilon_j}\}$ tends to infinity. Then one defines

$$\exp\{-C\tilde{\varphi}_j\} = \exp\{-C\varphi_{\varepsilon_j}\}\left(\int_M (|s|^2+\varepsilon_j)^{k+1}\exp\{-C\varphi_{\varepsilon_j}\}\right)^{-1}.$$

By (5.27), $(|s|^2+\varepsilon_j)^{(k+1)/2}\exp\left\{-\frac{1}{2}C\tilde{\varphi}_j\right\}$ converges in $L^2(M)$ to a non-negative function f. Using the estimate of $\int_M |\varphi_\varepsilon|$ and restricting to sets of the form $|s| \geqq 1/n$, one shows that f is zero almost everywhere and hence obtains a contradiction.)

As in Section 2, inequality (5.9) and an upper estimate of φ_ε, give an estimate of $|\nabla\varphi_\varepsilon|(\exp\{-C\inf\varphi_\varepsilon\}+1)^{-1}$ independent of ε. Hence, for some geodesic ball $B(x,R)$ of radius $R = C_{13}(-\inf\varphi_\varepsilon)(\exp\{-C\inf\varphi_\varepsilon\}+1)^{-1}$, φ_ε is not greater then $\frac{1}{2}\inf\varphi_\varepsilon$. (Here C_{13} is a positive constant independent of ε, and R is less than the injectivity radius of M.) Over this geodesic ball, the integral of $(|s|^2+\varepsilon)^{k+1}\exp\{-N\varphi_\varepsilon\}$ is not less than

$$
\begin{aligned}
&C_{14}\exp\left\{-\frac{1}{2}N\inf\varphi_\varepsilon\right\}\int_0^R r^{a(k+1)}\,dr\\
&\geqq \frac{C_{14}}{ak+a}\exp\left\{-\frac{1}{2}N\inf\varphi_\varepsilon\right\}[C_{13}(-\inf\varphi_\varepsilon)(\exp\{-C\inf\varphi_\varepsilon\}+1)^{-1}]^{ak+a+1},
\end{aligned}
\tag{5.28}
$$

where C_{14} and a are positive constants independent of ε.

By choosing N large enough, inequality (5.28) and the estimate of $\int_M(|s|^2+\varepsilon)^{k+1}\exp\{-N\varphi_\varepsilon\}$ provide an estimate of $-\inf\varphi_\varepsilon$ independent of ε and inequality (5.9) gives an upper estimate of $m+\Delta\varphi_\varepsilon$ independent of ε.

Let us now show that $|(\varphi_\varepsilon)_{i\bar{j}k}|^2$ has an upper bound independent of ε on compact subsets of the complement of the divisor of s. In fact, we can follow the arguments of Section 2 to define S_ε to be $\sum g_\varepsilon'^{i\bar{r}} g_\varepsilon'^{\bar{j}s} g'^{k\bar{t}} (\varphi_\varepsilon)_{i\bar{j}k} (\varphi_\varepsilon)_{\bar{r}s\bar{t}}$. Let ρ be a non-negative smooth function in M whose support is disjoint from the divisor of s. Then over the support of ρ, the metric tensor $\sum_{i,j}(g_{i\bar{j}}+\partial^2\varphi_\varepsilon/\partial z^i\partial\bar{z}^j)dz^i\otimes d\bar{z}^j$ is uniformly bounded from above and below by some positive multiple of $\sum_{i,j} g_{i\bar{j}}dz^i\otimes d\bar{z}^j$, the constants being independent of ε. (This is because we have an upper estimate of $m+\Delta\varphi_\varepsilon$ and, over the compact set under consideration, the volume form has a fixed positive volume form as a lower bound.)

From inequality (2.10), we can find positive constants C_{15} and C_{16} independent of ε such that

$$\rho\Delta_\varepsilon'(\Delta\varphi_\varepsilon) \geqq C_{15}\rho S_\varepsilon - C_{16}\rho. \tag{5.29}$$

Integrating (5.29) with respect to the volume form $(|s|^2+\varepsilon)^k\exp\{F\}dV$, we see that

$$\begin{aligned} C_{15}\int_M \rho S_\varepsilon(|s|^2+\varepsilon)^k\exp\{F\} &\leqq C_{16}\int_M \rho(|s|^2+\varepsilon)^k\exp\{F\} \\ &\quad + \int_M \Delta_\varepsilon'(\rho)\cdot\Delta\varphi_\varepsilon\cdot(|s|^2+\varepsilon)^k\exp\{F\}. \end{aligned} \tag{5.30}$$

Because the new metric is uniformly equivalent to the old one on the support of ρ, the right-hand side of (5.30) can be estimated by a constant independent of ε. Since $|s|>0$ on the support of ρ, we can therefore find an estimate of $\int_M \rho S_\varepsilon$, independent of ε. Since the function ρ is chosen quite arbitrary, we see that we have found an L_1-estimate of S_ε over any compact subset of M which is disjoint from the divisor of s.

Let $B(R)=\{(z_1,\dots,z_m)|\sum_i|z_i|^2\leqq R\}$ be a coordinate chart in the complement of the divisor of s. Then we claim that we can estimate $S_\varepsilon(0)$ in terms of the L_1-norm of S_ε over the coordinate ball. First of all, we note that, using the computations of Section 3, we know that there are positive constants C_{17} and C_{18} independent of ε and φ_ε, such that the inequality

$$\Delta_\varepsilon'\left(S_\varepsilon + C_{17}\Delta\varphi_\varepsilon + C_{18}\sum_i|z_i|^2\right) > 0 \tag{5.31}$$

holds on $B(R)$.

By choosing C_{18} larger if necessary, we may assume that the function $\tilde{S}_\varepsilon = S_\varepsilon + C_{17}\Delta\varphi_\varepsilon + C_{18}\left(\sum_i|z_i|^2+1\right)$ is positive. We can now solve the Dirichlet problem to find a smooth function $\tilde{\tilde{S}}_\varepsilon=\tilde{S}_\varepsilon$ on the boundary of $B(R)$ and $\Delta_\varepsilon'\tilde{\tilde{S}}_2=0$. By the maximum principle, $\tilde{\tilde{S}}_\varepsilon \geqq \tilde{S}_\varepsilon > 0$ in $B(R)$.

By using the fact that Δ'_ε is the normalized Laplacian of a Kähler metric, we see that Δ'_ε can be written as

$$\det(g'_{\varepsilon p\bar{q}})^{-1}\sum_{i,j}\frac{\partial}{\partial z^i}\left[\det(g'_{\varepsilon p\bar{q}})g'^{i\bar{j}}_{\varepsilon}\frac{\partial}{\partial \bar{z}^j}\right],$$

where $g'_{\varepsilon i\bar{j}} = g_{i\bar{j}} + \partial^2\varphi_\varepsilon/\partial z^i\partial\bar{z}^j$ and $(g'^{i\bar{j}}_\varepsilon)$ is the inverse matrix of $(g'_{\varepsilon i\bar{j}})$. Since the metric $\sum_{i,j} g'_{\varepsilon i\bar{j}}dz^i\otimes d\bar{z}^j$ is uniformly equivalent to $\sum_{i,j} g_{i\bar{j}}dz^i\otimes d\bar{z}^j$ on $B(R)$, we know that $\tilde{\tilde{S}}_\varepsilon$ is a solution of a uniform elliptic equation of divergence form whose ellipticity is estimated.

Using Moser's Harnack inequality [ ], we conclude that there is a positive constant C_{19} independent of φ_ε, and ε such that

$$\tilde{\tilde{S}}_\varepsilon(0) \leqq C_{19}\int_{B(R/2)}\tilde{\tilde{S}}_\varepsilon. \tag{5.31'}$$

We are going to prove that the right-hand side of (5.31′) can in turn be estimated by $\int_{\partial B(R)}\tilde{\tilde{S}}$. To achieve this, let σ be a non-decreasing C^1-function defined on the real line such that $\sigma(t) = 0$ for $t \leqq 0$. Then, for $\tau < R$, we can define ψ on the real line to be $\psi(s) = \int_s^\infty t\sigma(\tau - t)dt$. It is easy to check that $\psi(r) = \psi\left(\left(\sum_i |z_i|^2\right)^{1/2}\right)$ is in $C^{1,1}(B(R))$ and that $\psi(r)$ vanishes outside a compact subset of the interior of $B(R)$.

By direct computation, we have

$$\begin{aligned}\Delta'\psi(r) &= \sum_{i,j} g'^{i\bar{j}}_\varepsilon\frac{\partial^2\psi(r)}{\partial z^i\partial\bar{z}^j}\\ &= -\frac{1}{2}\sigma(\tau - r)\sum_{i,j} g'^{i\bar{j}}_\varepsilon\frac{\partial^2 r^2}{\partial z^i\partial\bar{z}^j} + r\sigma'(\tau - r)\sum_{i,j} g'^{i\bar{j}}_\varepsilon\frac{\partial r}{\partial z^i}\frac{\partial r}{\partial\bar{z}^j}\\ &= -\frac{1}{2}\sigma(\tau - r)\sum_{i} g'^{i\bar{i}}_\varepsilon + r\sigma'(\tau - r)\sum_{i,j} g'^{i\bar{j}}_\varepsilon\frac{\partial r}{\partial z^i}\frac{\partial r}{\partial\bar{z}^j}.\end{aligned} \tag{5.32}$$

Since $\psi(r)$ has compact support in the interior of $B(R)$, we can multiply (5.32) by $\tilde{\tilde{S}}_\varepsilon\det(g'_{\varepsilon p\bar{q}})$ and integrate with respect to the Euclidean volume form dE to obtain

$$\begin{aligned}0 &\leqq \int_{B(R)}\psi(r)(\Delta'\tilde{\tilde{S}}_\varepsilon)\det(g'_{\varepsilon p\bar{q}})dE\\ &= \int_{B(R)}\tilde{\tilde{S}}_\varepsilon(\Delta'\psi(r))\det(g'_{\varepsilon p\bar{q}})dE\\ &= -\frac{1}{2}\int_{B(R)}\tilde{\tilde{S}}_\varepsilon\sigma(\tau - r)\left(\sum_i g'^{i\bar{i}}_\varepsilon\right)\det(g'_{\varepsilon p\bar{q}})dE\\ &\quad + \int_{B(R)}\tilde{\tilde{S}}_\varepsilon r\sigma'(\tau - r)\left(\sum_{i,j} g'^{i\bar{j}}_\varepsilon\frac{\partial r}{\partial z^i}\frac{\partial r}{\partial\bar{z}^j}\right)\det(g'_{\varepsilon p\bar{q}})dE.\end{aligned} \tag{5.33}$$

Since $\sigma \geqq 0$, and $\sigma' \geqq 0$, it follows immediately from (5.33) that

$$\frac{1}{2}\inf_{B(R)}\left[\left(\sum_i g_\varepsilon'^{i\bar{i}}\right)\det(g'_{\varepsilon p\bar{q}})\right]\int_{B(R)}\tilde{\tilde{S}}_\varepsilon\sigma(\tau-r)dE$$

$$\leqq \sup_{B(R)}\left[\left(r\sum_{i,j} g_\varepsilon'^{i\bar{j}}\frac{\partial r}{\partial z^i}\frac{\partial r}{\partial \bar{z}^j}\right)\det(g'_{\varepsilon p\bar{q}})\right]\int_{B(R)}\tilde{\tilde{S}}_\varepsilon\sigma'(\tau-r)dE. \qquad (5.34)$$

Therefore, by the uniform ellipticity of Δ', we can find a positive constant C_{20} independent of $\sigma, \tau, \varphi_\varepsilon$ such that

$$\int_{B(R)}\tilde{\tilde{S}}_\varepsilon\sigma(\tau-r)dE \leqq C_{20}\int_{B(R)}\tilde{\tilde{S}}_\varepsilon\sigma'(\tau-r)dE. \qquad (5.35)$$

Letting τ tend to R in (5.35), we see that (5.35) is also valid when τ is replaced by R. Now choose σ so that $\sigma(t) = 1$ when $t \geqq \varepsilon$ and $\sigma'(t) \leqq 1/\varepsilon$ for all t. Then (5.35) implies

$$\int_{B(R-\varepsilon)}\tilde{\tilde{S}}_\varepsilon dE \leqq C_{20}\varepsilon^{-1}\int_{B(R)\setminus B(R-\varepsilon)}\tilde{\tilde{S}}_\varepsilon dE. \qquad (5.36)$$

Letting $\varepsilon \to 0$ in (5.36), we see that $\int_{B(R)}\tilde{\tilde{S}}_\varepsilon dE$ can be estimated from above by $\int_{\partial B(R)}\tilde{S}_\varepsilon$. Since $\tilde{S}_\varepsilon|\partial B(R) = \tilde{S}_\varepsilon|\partial B(R)$ and $\tilde{\tilde{S}}_\varepsilon > 0$, we conclude from (5.31′) that there is a positive constant C_{21} independent of φ_ε and ε such that

$$\tilde{S}_\varepsilon(0) \leqq \tilde{\tilde{S}}_\varepsilon(0) \leqq C_{21}\int_{\partial B(R)}\tilde{S}_\varepsilon. \qquad (5.37)$$

Since C_{21} can be chosen to be independent of R when $B(R)$ lies in a fixed coordinate chart in the complement of the divisor of s, we can integrate (5.37) to find an estimate of $\tilde{S}_\varepsilon(0)$ in terms of the L^1-norm of $\tilde{S}_\varepsilon$ over a ball in the coordinate chart. Combining this with the previous L^1-estimate of $\tilde{S}_\varepsilon$, we conclude that we have an estimate of $\tilde{S}_\varepsilon$ on every compact subset of the complement of the divisor of s.

Therefore we have a uniform Lipschitz estimate of $(\varphi_\varepsilon)_{i\bar{j}}$ over any compact subset of the complement of the divisor of s. It is now easy to estimate the higher derivatives of φ_ε. In fact, we can fix a local coordinate system in M so that (5.3) holds with φ replaced by φ_ε. Then we differentiate (5.3) with respect to $\partial/\partial z^k$. We see that $\partial\varphi_\varepsilon/\partial z^k$ satisfies an equation of the form

$$\Delta'_\varepsilon\left(\frac{\partial\varphi_\varepsilon}{\partial z^k}\right) = \frac{\partial}{\partial z^k}[\log(C_\varepsilon(|s|^2+\varepsilon)^k\exp\{F\}\det(g_{i\bar{j}}))]. \qquad (5.38)$$

In our coordinate system, the operator Δ_ε is an elliptic operator whose coefficients are the coefficients of the inverse matrix of $(g_{i\bar{j}} + \partial\varphi_\varepsilon/\partial z^i\partial\bar{z}^j)$. Since we have Lipschitz estimates of these coefficients over compact subsets of M which are disjoint from the divisor of s, the Schauder estimate (see [13]) shows that all higher derivatives of φ_ε can be estimated over these sets.

By letting $\varepsilon \to 0$, we can now conclude that $\{\varphi_\varepsilon\}$ has a subsequence converging to a solution φ of (5.11) such that φ is smooth outside of the divisor of s and $\{|\varphi_{i\bar{j}}|\}$ is bounded for all i, j.

Theorem 3. *Let L be a holomorphic line bundle over a compact Kähler manifold M. Let s be a holomorphic section of L. Let $\sum_{ij} g_{i\bar{j}} dz^i \otimes d\bar{z}^j$ be the Kähler metric of M. Then, for any $k \geqq 0$ and any smooth function F with $\int_M |s|^{2k} \exp\{F\} = \operatorname{Vol}(M)$, we can find a solution φ of the equation*

$$\det\left(g_{i\bar{j}} + \frac{\partial^2 \varphi}{\partial z^i \partial \bar{z}^j}\right) = |s|^{2k} \exp\{F\} \det(g_{i\bar{j}})$$

with the following properties: Outside the divisor of s, φ is smooth and, over M, $\{|\varphi_{i\bar{j}}|\}$ is bounded for all i, j. Furthermore, any function ψ satisfying the above properties must be equal to φ plus a constant.

Proof. We have only to prove the last statement. Let Δ' be the Laplacian of the metric $\sum_{i,j}(g_{i\bar{j}} + \partial^2\varphi/\partial z^i \partial \bar{z}^j) dz^i \otimes d\bar{z}^j$. We claim that, if f is a function such that $\{|f_{i\bar{j}}|\}$ is bounded over M for all i, j, then

$$\int_M (\Delta' f)|s|^{2k} \exp\{F\} = 0. \tag{5.39}$$

Indeed, using our previous notation, we have

$$\int_M (\Delta'_\varepsilon f)(|s|^2 + \varepsilon)^k \exp\{F\} = 0. \tag{5.40}$$

If we let $c(g_\varepsilon)_{i\bar{j}}$ be the $(i, \bar{j})$-th cofactor of the matrix $(g_{i\bar{j}} + \partial^2\varphi_\varepsilon/\partial z^i \partial \bar{z}^j)$, then (5.40) is equivalent to

$$\int_M \sum_{i,j} (g'_\varepsilon)_{i\bar{j}} \frac{\partial^2 f}{\partial z^i \partial \bar{z}^j} dx^1 \wedge \cdots \wedge dz^m \wedge d\bar{z}^1 \wedge \cdots \wedge d\bar{z}^m = 0. \tag{5.41}$$

Since $c(g'_\varepsilon)_{i\bar{j}}$ and $\partial^2 f/\partial z^i \partial \bar{z}^j$ are bounded independent of ε, we can use the Lebesgue dominated convergence theorem to obtain (5.39) from (5.41).

Now let ψ be another solution of (5.1) satisfying the properties mentioned in the theorem. Then we have

$$\det(g_{i\bar{j}} + \varphi_{i\bar{j}}) = \det(g_{i\bar{j}} + \psi_{i\bar{j}}). \tag{5.42}$$

Rewriting (5.42), we have

$$\det(g_{i\bar{j}} + \varphi_{i\bar{j}} + (\psi - \varphi)_{i\bar{j}}) \det(g_{i\bar{j}} + \varphi_{i\bar{j}})^{-1} = 1. \tag{5.43}$$

Using the arithmetic-geometric mean inequality, we have

$$\frac{1}{m}[m + \Delta'(\psi - \varphi)] \geqq 1. \tag{5.44}$$

Therefore, $\psi - \varphi$ is subharmonic with respect to the operator Δ'. Since $|\psi_{i\bar{j}}|$ and $|\varphi_{i\bar{j}}|$ are both bounded, $|(\psi-\varphi)^2_{i\bar{j}}|$ is also bounded over M and $\psi - \varphi$ is a C^1-function over M. By adding a constant to $\psi - \varphi$, we may assume that $\psi - \varphi \geqq 0$. Then applying (5.39) to $(\psi - \varphi)^2$, we obtain

$$2\int_M (\psi-\varphi)\Delta'(\psi-\varphi) + 2\int_M |\nabla'(\psi-\varphi)|^2 = \int_M \Delta'(\psi-\varphi)^2 = 0. \tag{5.45}$$

Since $(\psi - \varphi) \geqq 0$ and $\Delta'(\psi - \varphi) \geqq 0$, we conclude from (5.45) that $\nabla'(\psi - \varphi) = 0$ and $\psi - \varphi$ is a constant.

6 Complex Monge-Ampère equation with more general right-hand side

In this section, we study equations of the form

$$\det\left(g_{i\bar{j}} + \frac{\partial^2\varphi}{\partial z^i \partial \bar{z}^j}\right) = \exp\{F(x,\varphi)\}\det(g_{i\bar{j}}), \tag{6.1}$$

where $\sum_{i,j} g_{i\bar{j}} dz^i \otimes d\bar{z}^j$ is a Kähler metric defined on M, and F is a smooth function defined on $M \times R$ with $\partial F/\partial t \geqq 0$.

We shall look for real-valued smooth functions φ satisfying (6.1) such that $\sum(g_{i\bar{j}} + \partial^2\varphi/\partial z^i\partial\bar{z}^j)dz^i \otimes d\bar{z}^j$ defines another Kähler metric. If such a function φ exists, then integrating (6.1), the integral of the right-hand side is equal to the volume of M. Therefore, a certain compatibility condition must be satisfied for the solvability of (6.1). The condition is that there exists a smooth function ψ defined on M such that

$$\int_M \exp\{F(x,\psi)\} = \mathrm{Vol}(M). \tag{6.2}$$

Relying on (6.2), we are going to use an iteration method to solve (6.1).

First of all, let us prove the uniqueness of the solution of (6.1).

Lemma 1. *Let φ and $\bar{\varphi}$ be two smooth solutions of* (6.1) *such that both*

$$\sum_{i,j}\left(g_{i\bar{j}} + \frac{\partial^2\varphi}{\partial z^i \partial \bar{z}^j}\right) dz^i \otimes d\bar{z}^j$$

and

$$\sum_{i,j}\left(g_{i\bar{j}} + \frac{\partial^2\bar{\varphi}}{\partial z^i \partial \bar{z}^j}\right) dz^i \otimes d\bar{z}^j$$

define Kähler metrics on M. Then $\varphi - \bar{\varphi}$ is a constant.

Proof of Lemma 1. Using the fact that both φ and $\bar{\varphi}$ satisfy (6.1), we can immediately derive the following equation:

$$\det\left(g_{i\bar{j}} + \frac{\partial^2\bar{\varphi}}{\partial z^i \partial \bar{z}^j} + \frac{\partial^2(\varphi-\bar{\varphi})}{\partial z^i \partial \bar{z}^j}\right)\det\left(g_{i\bar{j}} + \frac{\partial^2\varphi}{\partial z^i \partial \bar{z}^j}\right)^{-1} = \exp\{F(x,\varphi) - F(x,\bar{\varphi})\}. \tag{6.3}$$

Let Δ_φ be the normalized Laplacian of the metric

$$\sum_{i,j}\left(g_{i\bar{j}}+\frac{\partial^2\varphi}{\partial z^i\partial\bar{z}^j}\right)dz^i\otimes d\bar{z}^j.$$

Then it follows from the arithmetic-geometric mean inequality and (6.3) that we have the inequality

$$m+\Delta_\varphi(\varphi-\bar{\varphi})\geqq m\exp\left\{\frac{1}{m}[F(x,\varphi)-F(x,\bar{\varphi})]\right\}. \tag{6.4}$$

By the mean valued theorem we have

$$F(x,\varphi)-F(x,\bar{\varphi})=\int_{\bar{\varphi}(x)}^{\varphi(x)}\frac{\partial F}{\partial t}(x,\tau)d\tau=\frac{\partial F}{\partial t}(x,\bar{t}(x))(\varphi(x)-\bar{\varphi}(x)), \tag{6.5}$$

where $\bar{t}(x)$ is a number between $\inf(\varphi(x),\bar{\varphi}(x))$ and $\sup(\varphi(x),\bar{\varphi}(x))$.

Since $\partial F/\partial t\geqq 0$, we can combine (6.4) with (6.5) to conclude that whenever $\varphi(x)-\bar{\varphi}(x)$ is strictly positive, $\Delta_\varphi(\varphi-\bar{\varphi})(x)$ is non-negative.

If $\sup_{x\in M}(\varphi-\bar{\varphi})(x)>0$, then, since M is compact and $\varphi-\bar{\varphi}$ is smooth, the standard proof of the maximal principle shows that $\varphi-\bar{\varphi}$ is a constant function. Therefore we conclude that either $\varphi-\bar{\varphi}$ is a constant function or $\varphi(x)\leqq\bar{\varphi}(x)$ for all $x\in M$. Clearly we can interchange the roles of φ and $\bar{\varphi}$ in the above assertion. Combining these two inequalities, we can deduce Lemma 1.

After this preparation, we proceed to solve (6.1). By Theorem 1, we can find a smooth function φ_0 such that $\sum_{i,j}(g_{i\bar{j}}+\partial^2\varphi_0/\partial z^i\partial\bar{z}^j)dz^i\otimes d\bar{z}^j$ defines a Kähler metric and

$$\det\left(g_{i\bar{j}}+\frac{\partial^2\varphi_0}{\partial z^i\partial\bar{z}^j}\right)=\exp\{F(x,\psi)\}\det(g_{i\bar{j}}). \tag{6.6}$$

If we define

$$\bar{\varphi}_0=\varphi_0+\sup_M|\varphi_0-\psi| \tag{6.7}$$

and

$$\bar{\bar{\varphi}}_0=\varphi_0-\sup_M|\varphi_0-\psi|, \tag{6.8}$$

then clearly both $\bar{\varphi}$ and $\bar{\bar{\varphi}}_0$ satisfy equation (6.6) with φ_0 replaced by $\bar{\varphi}_0$ and $\bar{\bar{\varphi}}_0$, respectively. Furthermore,

$$\bar{\varphi}_0(x)\geqq\psi(x)\geqq\bar{\bar{\varphi}}_0(x) \tag{6.9}$$

for all x on M.

The set $A=\{(x,t)|x\in M,\bar{\varphi}_0(x)\geqq t\geqq\bar{\bar{\varphi}}_0(x)\}$ is a compact subset of $M\times R$. Hence we can define

$$k=\sup_{(x,t)\in A}\frac{\partial F}{\partial t}(x,t)+1>0. \tag{6.10}$$

Using this definition, we are going to prescribe our iteration scheme as follows.

For each $i \geqq 1$, we define $\bar{\varphi}_i$ and $\bar{\bar{\varphi}}_i$ as the smooth solutions of the following equations:

$$\det\left(g_{\alpha\bar{\beta}} + \frac{\partial^2 \bar{\varphi}_i}{\partial z^\alpha \partial \bar{z}^\beta}\right) = \exp\{k\bar{\varphi}_i + F(x, \bar{\varphi}_{i-1}) - k\bar{\varphi}_{i-1}\} \det(g_{\alpha\bar{\beta}}), \quad (6.11)$$

$$\det\left(g_{\alpha\bar{\beta}} + \frac{\partial^2 \bar{\bar{\varphi}}_i}{\partial z^i \partial \bar{z}^j}\right) = \exp\{k\bar{\bar{\varphi}}_i + F(x, \bar{\bar{\varphi}}_{i-1}) - k\bar{\bar{\varphi}}_{i-1}\} \det(g_{\alpha\bar{\beta}}), \quad (6.12)$$

so that both

$$\sum_{\alpha,\beta}\left(g_{\alpha\bar{\beta}} + \frac{\partial^2 \bar{\varphi}_i}{\partial z^\alpha \partial \bar{z}^\beta}\right) dz^\alpha \otimes d\bar{z}^\beta \quad \text{and} \quad \sum_{\alpha\beta}\left(g_{\alpha,\bar{\beta}} + \frac{\partial^2 \bar{\bar{\varphi}}_i}{\partial z^\alpha \partial \bar{z}^\beta}\right) dz^\alpha \otimes d\bar{z}^\beta$$

define Kähler metrics.

To make sure that such functions $\bar{\varphi}_i$ and $\bar{\bar{\varphi}}_i$ exist, we need the following

Lemma 2. *Let M be a compact Kähler manifold with Kähler metric $\sum_{i,j} g_{i\bar{j}} dz^i \otimes d\bar{z}^j$. Let $F(x)$ be any smooth function defined on M. Then, for any constant $k > 0$, there exists a unique smooth function φ such that*

$$\det\left(g_{i\bar{j}} + \frac{\partial^2 \varphi}{\partial z^i \partial \bar{z}^j}\right) = \exp\{k\varphi + F\} \det(g_{i\bar{j}})$$

and $\sum_{i,j} g_{i\bar{j}} dz^i \otimes d\bar{z}^j$ defines a Kähler metric.

Proof of Lemma 2. As in Theorem 1, we can use the continuation method where the one parameter family of equations is

$$\det\left(g_{i\bar{j}} + \frac{\partial^2 \varphi}{\partial z^i \partial \bar{z}^j}\right) + \exp\{k\varphi + tF\} \det(g_{i\bar{j}}).$$

The fact that $k > 0$ makes the proof of Lemma 2 much easier. In fact, the estimate of $\sup_M |\varphi|$ can be derived easily by the maximum principle as follows. When φ achieves its maximum at a point x_0, we must have

$$\det\left(g_{i\bar{j}} + \frac{\partial^2 \varphi}{\partial z^i \partial \bar{z}^j}\right) \det(g_{i\bar{j}})^{-1} \leqq 1.$$

Hence $\exp\{k\varphi(x_0) + tF(x_0)\} \leqq 1$. This implies immediately $\sup \varphi = \varphi(x_0) \leqq -(t/k)F(x_0)$. Similarly one can draw an estimate of $\inf_M \varphi$. The uniqueness part of Lemma 2 follows easily from Lemma 1 noting that $k \neq 0$.

Lemma 2 justifies our construction of the iteration scheme. We claim that, for all $i \geqq 0$,

$$\bar{\bar{\varphi}}_0 \leqq \bar{\bar{\varphi}}_i \leqq \bar{\bar{\varphi}}_{i+1} \leqq \bar{\varphi}_{i+1} \leqq \bar{\varphi}_i \leqq \bar{\varphi}_0. \quad (6.13)$$

For $i = 0$, we use the fact that $\bar{\varphi}_0$ satisfies equation (6.6) and inequality (6.9) to conclude that

$$\begin{aligned}\det\left(g_{\alpha\bar{\beta}} + \frac{\partial^2 \bar{\varphi}_1}{\partial z^\alpha \partial \bar{z}^\beta}\right) &= \exp\{k\bar{\varphi}_1 - k\bar{\varphi}_0\} \exp\{F(x, \bar{\varphi}_0)\} \det(g_{\alpha\bar{\beta}}) \\ &\geqq \exp\{k(\bar{\varphi}_1 - \bar{\varphi}_0)\} \exp\{F(x, \psi)\} \det(g_{\alpha\bar{\beta}}) \\ &= \exp\{k(\bar{\varphi}_1 - \bar{\varphi}_0)\} \det\left(g_{\alpha\bar{\beta}} + \frac{\partial^2 \bar{\varphi}_0}{\partial z^\alpha \partial \bar{z}^\beta}\right). \quad (6.14)\end{aligned}$$

At the point where $\bar{\varphi}_1 - \bar{\varphi}_0$ achieves its maximum,

$$\det\left(g_{\alpha\bar{\beta}} + \frac{\partial^2 \bar{\varphi}_1}{\partial z^\alpha \partial \bar{z}^\beta}\right) \leqq \det\left(g_{\alpha\bar{\beta}} + \frac{\partial^2 \bar{\varphi}_0}{\partial z^\alpha \partial \bar{z}^\beta}\right).$$

Hence it follows from (6.14) that $\sup_M(\bar{\varphi}_1 - \bar{\varphi}_0) \leqq 0$. Similarly one can also prove that $\sup_M(\bar{\bar{\varphi}}_0 - \bar{\bar{\varphi}}_1) \leqq 0$.

To show that $\bar{\bar{\varphi}}_1 \leqq \bar{\varphi}_1$, we divide (6.11) by (6.12) to obtain

$$\begin{aligned} &\det\left(g_{\alpha\bar{\beta}} + \frac{\partial^2 \bar{\varphi}_1}{\partial z^\alpha \partial \bar{z}^\beta}\right) \det\left(g_{\alpha\bar{\beta}} + \frac{\partial^2 \bar{\bar{\varphi}}_1}{\partial z^\alpha \partial \bar{z}^\beta}\right)^{-1} \\ &= \exp\{k\bar{\varphi}_1 - k\bar{\bar{\varphi}}_1 + F(x, \bar{\varphi}_0) - F(x, \bar{\bar{\varphi}}_0) - k(\bar{\varphi}_0 - \bar{\bar{\varphi}}_0)\}. \end{aligned} \tag{6.15}$$

Since $\bar{\varphi}_0 \geqq \bar{\bar{\varphi}}_0$ and k is defined by (6.10), it follows that

$$F(x, \bar{\varphi}_0) \quad F(x, \bar{\bar{\varphi}}_0) \quad k(\bar{\varphi}_0 \quad \bar{\bar{\varphi}}_0) \leqq 0. \tag{6.16}$$

Therefore,

$$\det\left(g_{\alpha\bar{\beta}} + \frac{\partial^2 \bar{\varphi}_1}{\partial z^\alpha \partial \bar{z}^\beta}\right) \det\left(g_{\alpha\bar{\beta}} + \frac{\partial^2 \bar{\bar{\varphi}}_1}{\partial z^\alpha \partial \bar{z}^\beta}\right)^{-1} \leqq \exp\{k(\bar{\varphi}_1 - \bar{\bar{\varphi}}_1)\}. \tag{6.17}$$

At the point where $\bar{\varphi}_1 - \bar{\bar{\varphi}}_1$ achieves its minimum, the right-hand side of (6.17) is non-negative and hence $\bar{\varphi}_1 \geqq \bar{\bar{\varphi}}_1$.

Suppose now that we have proved

$$\bar{\bar{\varphi}}_0 \leqq \bar{\bar{\varphi}}_{j-1} \leqq \bar{\bar{\varphi}}_j \leqq \bar{\varphi}_j \leqq \bar{\varphi}_{j-1} \leqq \bar{\varphi}_0 \tag{6.18}$$

for all $j < i$. We claim that (6.18) remains valid when j is replaced by i. In fact, applying (6.11) twice, we have

$$\begin{aligned} &\det\left(g_{\alpha\bar{\beta}} + \frac{\partial^2 \bar{\varphi}_i}{\partial z^\alpha \partial \bar{z}^\beta}\right) \det\left(g_{\alpha\bar{\beta}} + \frac{\partial^2 \bar{\varphi}_{i-1}}{\partial z^\alpha \partial \bar{z}^\beta}\right)^{-1} \\ &= \exp\{k\bar{\varphi}_i - k\bar{\varphi}_{i-1} + F(x, \bar{\varphi}_{i-1}) - F(x, \bar{\varphi}_{i-2}) - k(\bar{\varphi}_{i-1} - \bar{\varphi}_{i-2})\}. \end{aligned} \tag{6.19}$$

Since $\bar{\varphi}_0 \geqq \bar{\varphi}_{i-2} \geqq \bar{\varphi}_{i-1} \geqq \bar{\bar{\varphi}}_0$ and k is defined by (6.10), we see that

$$\det\left(g_{\alpha\bar{\beta}} + \frac{\partial^2 \bar{\varphi}_i}{\partial z^\alpha \partial \bar{z}^\beta}\right) \det\left(g_{\alpha\bar{\beta}} + \frac{\partial^2 \bar{\varphi}_{i-1}}{\partial z^\alpha \partial \bar{z}^\beta}\right)^{-1} \geqq \exp\{k\bar{\varphi}_i - k\bar{\varphi}_{i-1}\}. \tag{6.20}$$

Hence the maximal principle shows that $\bar{\varphi}_{i-1} \geqq \bar{\varphi}_i$. Similarly one can show that $\bar{\bar{\varphi}}_{i-1} \leqq \bar{\bar{\varphi}}_i$.

To prove that $\bar{\varphi}_i \geqq \bar{\bar{\varphi}}_i$, one just divides (6.11) by (6.12) and obtains

$$\begin{aligned} &\det\left(g_{\alpha\bar{\beta}} + \frac{\partial^2 \bar{\varphi}_i}{\partial z^\alpha \partial \bar{z}^\beta}\right) \det\left(g_{\alpha\bar{\beta}} + \frac{\partial^2 \bar{\bar{\varphi}}_i}{\partial z^\alpha \partial \bar{z}^\beta}\right) \\ &= \exp\{k\bar{\varphi}_i - k\bar{\bar{\varphi}}_i + F(x, \bar{\varphi}_{i-1}) - F(x, \bar{\bar{\varphi}}_{i-1}) - k\bar{\varphi}_{i-1} + k\bar{\bar{\varphi}}_{i-1}\}. \end{aligned} \tag{6.21}$$

Using $\bar{\varphi}_{i-1} \geqq \bar{\bar{\varphi}}_{i-1}$ one can repeat the above argument to show that $\bar{\varphi}_i \geqq \bar{\bar{\varphi}}_i$. Therefore the inequality (6.13) is valid in general and both $\bar{\varphi}_i$ and $\bar{\bar{\varphi}}_i$ are uniformly bounded.

Our next step is to find a uniform estimate of $\partial^2\bar{\varphi}_i/\partial z^\alpha \partial \bar{z}^\alpha$. As in Section 2, it suffices to estimate $\Delta\bar{\varphi}_i$.

Let Δ_i be the normalized Laplacian operator associated with the metric $\sum_{\alpha,\beta}(g_{\alpha\bar{\beta}} + \partial^2\bar{\varphi}_i/\partial z^\alpha \partial \bar{z}^\beta) dz^\alpha \otimes d\bar{z}^\beta$. Let C be any positive constant such that $C + \inf_{i\neq l}(R_{i\bar{i}l\bar{l}}) > 1$. Then by the same computation as in (2.18), we have

$$\begin{aligned}
&\Delta_i[\exp\{-C\bar{\varphi}_i\}(m + \Delta\bar{\varphi}_i)\\
&= \exp\{-C\bar{\varphi}_i\}\Big[k(\Delta\bar{\varphi}_i - \Delta\bar{\varphi}_{i-1}) + \sum_{\alpha,\beta} g^{\alpha\bar{\beta}} \frac{\partial F(x, \bar{\varphi}_{i-1})}{\partial z^\alpha \partial \bar{z}^\beta}\\
&\quad + \sum_{\alpha,\beta} g^{\alpha\bar{\beta}} \frac{\partial^2 F(x, \bar{\varphi}_{i-1})}{\partial t \partial z^\alpha} \frac{\partial \bar{\varphi}_{i-1}}{\partial \bar{z}^\beta} + \sum_{\alpha,\beta} g^{\alpha\bar{\beta}} \frac{\partial^2 F(x, \bar{\varphi}_{i-1})}{\partial t \partial \bar{z}^\beta} \frac{\partial \bar{\varphi}_{i-1}}{\partial z^\alpha}\\
&\quad + \frac{\partial^2}{\partial t^2}(x, \bar{\varphi}_{i-1}) \sum_{\alpha,\beta} g^{\alpha\bar{\beta}} \frac{\partial \bar{\varphi}_{i-1}}{\partial z^\alpha} \frac{\partial \bar{\varphi}_{i-1}}{\partial \bar{z}^\beta}\\
&\quad + \frac{\partial F}{\partial t}(x, \bar{\varphi}_{i-1}) \Delta\bar{\varphi}_{i-1}\Big] - C \exp\{-C\bar{\varphi}_i\} m(m + \Delta\bar{\varphi}_i)\\
&\quad + (C + \inf_{i\neq l}(R_{i\bar{i}l\bar{l}})) \exp\{-C\bar{\varphi}_i\}(m + \Delta\bar{\varphi}_i) \left(\sum_\alpha \frac{1}{1 + (\bar{\varphi}_i)_{\alpha\bar{\alpha}}}\right).
\end{aligned} \tag{6.22}$$

Since $\sup_M |\bar{\varphi}_i|$ has been estimated, it follows from Schauder's estimate (cf. [13]) that, for some constant C_1 independent of i,

$$\sup_M |\nabla\bar{\varphi}_i| \leqq C_1 \left(\sup_M |\Delta\bar{\varphi}_i| + 1\right). \tag{6.23}$$

As in (2.20),

$$\begin{aligned}
&\sum_\alpha \frac{1}{1 + (\bar{\varphi}_i)_{\alpha\bar{\alpha}}}\\
&\geqq (m + \Delta\bar{\varphi}_i)^{1/(m-1)} \exp\left\{\frac{-k}{m-1}(\bar{\varphi}_i - \bar{\varphi}_{i-1}) + \frac{1}{m-1} F(x, \bar{\varphi}_{i-1})\right\}\\
&\geqq C_2(m + \Delta\bar{\varphi}_i)^{1/(m-1)},
\end{aligned} \tag{6.24}$$

where C_2 is a positive constant independent of i.

Noting again that $\sup_M |\bar{\varphi}_i|$ has been estimated, it follows from (6.22), (6.23) and (6.24) that there are positive constants C_3, C_4, C_5 and C_6, independent of i, such that

$$\begin{aligned}
&\Delta_i[\exp\{-C\bar{\varphi}_i\}(m + \Delta\bar{\varphi}_i)]\\
&\geqq -C_3(m + \Delta\bar{\varphi}_i) - C_4(m + \Delta\bar{\varphi}_{i-1}) - C_5 - \sup_M(m + \Delta\bar{\varphi}_{i-1})\\
&\quad + C_6(m + \Delta\bar{\varphi}_i)^{1+1/(m-1)} - C_3.
\end{aligned} \tag{6.25}$$

At the point where $\exp\{-C\bar{\varphi}_i\}(m+\Delta\bar{\varphi}_i)$ achieves its maximum, the right-hand side must be non-positive and so

$$\begin{aligned} &C_6\left(m+\sup_M \Delta\bar{\varphi}_i\right)^{1+1/(m-1)} \\ &\leqq \exp\left\{\frac{m}{m-1}\sup_M \bar{\varphi}_i\right\}\times\left[C_3\left(m+\sup_M\Delta\bar{\varphi}_i\right)\right. \\ &\left.+C_4\left(m+\sup_M\Delta\bar{\varphi}_{i-1}\right)+C_5\left(m+\sup_M\Delta\bar{\varphi}_{i-1}\right)C_3\right]. \end{aligned} \tag{6.26}$$

Choose a constant C_7, such that, for all $x \geqq 0$,

$$x^{1+1/(m-1)} \geqq (C_3+2C_4+2C_5)C_6^{-1}\exp\left\{\frac{m}{m-1}\sup_M\bar{\varphi}_i\right\}x - C_7. \tag{6.27}$$

Then (6.26) implies that we can find a positive constant C_8, independent of i, such that

$$\left(m+\sup_M\Delta\bar{\varphi}_i\right) \leqq \frac{1}{2}\left(m+\sup_M\Delta\bar{\varphi}_{i-1}\right)+C_8. \tag{6.28}$$

By iteration, (6.28) gives

$$m+\sup_M\Delta\bar{\varphi}_i \leqq \left(\frac{1}{2}\right)^i\left(m+\sup_M\Delta\bar{\varphi}_0\right)+2C_8. \tag{6.29}$$

Therefore we have found uniform estimates for $\partial^2\bar{\varphi}_i/\partial z^\alpha\partial\bar{z}^\beta$. To find uniform estimate of $\partial^2\bar{\varphi}_i/\partial z^\alpha\partial\bar{z}^\beta\partial z^\gamma$, let

$$S_i = \sum g_i^{\prime\alpha\bar{l}}g_i^{\prime\bar{\beta}p}g_i^{\prime\gamma\bar{q}}\frac{\partial^2\bar{\varphi}_i}{\partial z^\alpha\partial\bar{z}^\beta\partial z^\gamma}\frac{\partial^2\bar{\varphi}_i}{\partial\bar{z}^l\partial z^p\partial\bar{z}^q}, \tag{6.30}$$

where $(g_i^{\prime\alpha\bar{l}})$ is the inverse matrix of $(g_{\alpha\bar{l}}+\partial^2\bar{\varphi}_i/\partial z^\alpha\partial\bar{z}^l)$. By a computation similar to that of (3.12), we have

$$\Delta_i(S_i+C_9\Delta\bar{\varphi}_i) \geqq C_{10}S_i - C_{11}\sqrt{S_i}\sqrt{S_{i-1}}-C_{12}, \tag{6.30$'$}$$

where C_9, C_{10}, C_{11} and C_{12} are positive constants independent of i.

Since $|\Delta\bar{\varphi}_i|$ has been estimated, it follows from the maximum principle that

$$\sup_M S_i \leqq \frac{C_{11}}{C_{10}}\sqrt{\sup_M S_i}\sqrt{\sup_M S_{i-1}}+\frac{C_{12}}{C_{10}}+C_9\sup_M|\Delta\bar{\varphi}_i|. \tag{6.31}$$

It should be noted that in (6.30), we can choose C_{10} to be arbitrarily large if we are allowed to increase C_9, and C_{12}. In particular, we may assume that $C_{11}/C_{10} \leqq \frac{1}{2}$. Then (6.31) shows that

$$\sup_M S_i \leqq \frac{1}{3}\sup_M S_{i-1}+\frac{4C_{12}}{C_{10}}+4C_9\sup_M|\Delta\bar{\varphi}_i|. \tag{6.32}$$

By iteration, we can find a uniform estimate of S_i and hence a uniform estimate of $\partial^2\bar{\varphi}_i/\partial z^\alpha \partial \bar{z}^\beta \partial z^\gamma$.

Letting i tend to infinity in (6.11), we can then obtain a solution of (6.1). The Schauder estimate guarantees the solution to be smooth.

Theorem 4. *Let M be a compact Kähler manifold with Kähler metric $\sum_{i,j} g_{i\bar{j}} dz^i \otimes d\bar{z}^j$. Let F be a smooth function defined on $M \times R$ with $\partial F/\partial t \geqq 0$. Suppose that, for some smooth function ψ, defined on M, $\int_M \exp\{F(x,\psi(x))\} = \mathrm{Vol}(M)$. Then there exists a smooth function φ on M such that*

$$\det\left(g_{i\bar{j}} + \frac{\partial^2\varphi}{\partial z^i \partial \bar{z}^j}\right) = \exp\{F(x,\varphi(x))\}\det(g_{i\bar{j}})$$

and $\sum_{i,j}(g_{i\bar{j}} + \partial^2\varphi/\partial z^i\partial\bar{z}^j) dz^i \otimes d\bar{z}^j$ defines a Kähler metric. Furthermore, any other smooth function satisfying the same property differs from φ by only a constant.

As a consequence of Theorem 4, one deduces that, on a Kähler manifold with ample canonical line bundle, one can find a (unique) Kähler-Einstein metric whose Ricci tensor is the negative of the metric tensor.

In fact, by hypothesis, the negative of the first Chern class of M is represented by some positive $(1,1)$ form $\sqrt{-1}\sum_{i,j} g_{i\bar{j}} dz^i \wedge d\bar{z}^j$. Take this form as our Kähler form. Then by Chern's theorem [6], the closed $(1,1)$ form $-\partial\bar{\partial}\log\det(g_{i\bar{j}})$ also represents the first Chern class of M. Hence we can find a smooth function f such that

$$\partial\bar{\partial}\log\det(g_{i\bar{j}}) = \sqrt{-1}\sum_{i,j} g_{i\bar{j}} dz^i \wedge d\bar{z}^j + \partial\bar{\partial} f. \tag{6.33}$$

Now by Theorem 4, we can solve the equation

$$\det\left(g_{i\bar{j}} + \frac{\partial^2\varphi}{\partial z^i\partial\bar{z}^j}\right) = \exp\{\varphi - f\}\det(g_{i\bar{j}}) \tag{6.34}$$

so that $\sum_{i,j}(g_{i\bar{j}} + \partial^2\varphi/\partial z^i\partial\bar{z}^j) dz^i \otimes d\bar{z}^j$ defines a Kähler metric. By (6.33) and (6.35) we have

$$\begin{aligned} -\partial\bar{\partial}\log\det\left(g_{i\bar{j}} + \frac{\partial^2\varphi}{\partial z^i\partial\bar{z}^j}\right) &= -\partial\bar{\partial}\varphi + \partial\bar{\partial} f - \sqrt{-1}\sum_{i,j} g_{i\bar{j}} dz^i \wedge d\bar{z}^j - \partial\bar{\partial} f \\ &= -\sqrt{-1}\sum_{i,j}\left(g_{i\bar{j}} + \frac{\partial^2\varphi}{\partial z^i\partial\bar{z}^j}\right) dz^i \wedge d\bar{z}^j. \end{aligned} \tag{6.35}$$

This last equation simply says that the Ricci tensor of the metric $\sum_{i,j}(g_{i\bar{j}} + \partial^2\varphi/\partial z^i\partial\bar{z}^j) dz^i \otimes d\bar{z}^j$ is the negative of the metric tensor itself. Hence we have found the metric that we are looking for.

To see that the metric is unique (that is to say, that the metric depends only on the complex structure of M) we notice that if $\sum_{i,j}\tilde{g}_{i\bar{j}} dz^i \otimes d\bar{z}^j$ is another such metric, then its Kähler form must represent the negative of the first Chern class of M. Hence we can find a smooth function ψ defined on M such that

$\tilde{g}_{i\bar{j}} = g_{i\bar{j}} + \partial^2\psi/\partial z^i \partial \bar{z}^j$. On the other hand, the fact that the Ricci tensor of $\sum_{i,j} \tilde{g}_{i\bar{j}} dz^i \otimes d\bar{z}^j$ is the negative of the metric tensor can be written as

$$-\partial\bar{\partial} \log \det \left(g_{i\bar{j}} + \frac{\partial^2 \psi}{\partial z^i \partial \bar{z}^j} \right) = -\sqrt{-1} \sum_{i,j} g_{i\bar{j}} dz^i \wedge d\bar{z}^j - \partial\bar{\partial}\psi$$
$$= -\partial\bar{\partial} \log \det(g_{i\bar{j}}) + \partial\bar{\partial} f - \partial\bar{\partial}\psi, \qquad (6.36)$$

where the last equation follows from (6.33).

Equation (6.36) shows that the function

$$\log \left[\det \left(g_{i\bar{j}} + \frac{\partial^2 \psi}{\partial z^i \partial \bar{z}^j} \right) \det(g_{i\bar{j}})^{-1} \exp\{-\psi + f\} \right]$$

is a pluriharmonic function. Since we are on a compact manifold, this function must reduce to a constant $\exp\{c\}$. Therefore,

$$\det \left(g_{i\bar{j}} + \frac{\partial^2 \psi}{\partial z^i \partial \bar{z}^j} \right) = \exp\{\psi + c - f\} \det(g_{i\bar{j}}). \qquad (6.37)$$

The function $\psi + c$ then satisfies an equation of the form of (6.34). Lemma 1 shows that $\varphi - \psi$ is a constant. Hence,

$$\sum_{i,j} \tilde{g}_{i\bar{j}} dz^i \otimes d\bar{z}^j = \sum_{ij} \left(g_{i\bar{j}} + \frac{\partial^2 \psi}{\partial z^i \partial \bar{z}^j} \right) dz^i \otimes d\bar{z}^j = \sum_{i,j} \left(g_{i\bar{j}} + \frac{\partial^2 \varphi}{\partial z^i \partial \bar{z}^j} \right) dz^i \otimes d\bar{z}^j.$$

Theorem 5. *Let M be a Kähler manifold with ample canonical line bundle. Then there is a Kähler-Einstein metric whose Ricci tensor is the negative of the metric tensor. Furthermore, a metric of this form is unique and depends only on the complex structure of M.*

7 Degenerate complex Monge-Ampère equation with general right-hand side

In this section, we combine the main results of the last two sections. Let L be a line bundle over a compact Kähler manifold M. Let s be a (non-trivial) holomorphic section of L. Suppose L is equipped with an Hermitian metric so that the function $|s|^2$ is globally defined on M. Then, for $k \geqq 0$, we study equations of the form

$$\det \left(g_{i\bar{j}} + \frac{\partial^2 \varphi}{\partial z^i \partial \bar{z}^j} \right) = |s|^{2k} \exp\{F(x, \varphi)\} \det(g_{i\bar{j}}), \qquad (7.1)$$

where F is a smooth function defined on $M \times R$ with $\partial F/\partial t \geqq 0$.

As in Section 6, we look for solutions φ of (7.1) such that $|\varphi_{i\bar{j}}|$ is uniformly bounded on M and φ is smooth outside the divisor of s. (In fact, φ is going to be the uniform limit of a sequence of smooth functions $\{\varphi_\alpha\}$ so that $\det(g_{i\bar{j}} +$

$\partial^2\varphi_\alpha/\partial z^i\partial z^j)\det(g_{i\bar{j}})^{-1}$ converges uniformly to $|s|^{2k}\exp\{F(x,\varphi)\}$. Clearly a necessary condition is obtained by integrating (7.1). The result is that there is a function ψ whose partial derivatives $|\psi_{i\bar{j}}|$ are uniformly bounded on M so that

$$\int_M |s|^{2k}\exp\{F(x,\psi)\} = \mathrm{Vol}(M). \tag{7.2}$$

With this sole assumption on the existence of ψ, we proceed to solve (7.1). We approximate (7.1) by the following non-degenerate elliptic equation:

$$\det\left(g_{i\bar{j}} + \frac{\partial^2\varphi}{\partial z^i\partial\bar{z}^j}\right) = C_\varepsilon(|s|^2+\varepsilon)^k\exp\{F(x,\varphi)\}\det(g_{i\bar{j}}), \tag{7.3}$$

where $\varepsilon > 0$ is a smooth constant and

$$C_\varepsilon = \mathrm{Vol}(M)\left[\int_M (|s|^2+\varepsilon)^k\exp\{F(x,\psi_\varepsilon)\}\right]^{-1}. \tag{7.4}$$

Here we choose $\{\psi_\varepsilon\}$ to be a sequence of smooth functions such that ψ_ε converges to ψ uniformly on M and that $\sup_M|\partial^2\psi_\varepsilon/\partial z^i\partial\bar{z}^j|$ is uniformly bounded on every coordinate chart.

According to Theorem 4, we can find smooth solutions φ_ε of (7.3) such that $\sum_{i,j}(g_{i\bar{j}} + \partial^2\varphi_\varepsilon/\partial z^i\partial\bar{z}^j)dz^i\otimes d\bar{z}^j$ defines a metric. By examining the proof of Theorem 4, we can also assume that we have an estimate of $\sup_M|\varphi_\varepsilon|$ in the following way.

Let $\bar{\varphi}_\varepsilon$ and $\bar{\bar{\varphi}}_\varepsilon$ be two smooth solutions of the equation

$$\det\left(g_{i\bar{j}} + \frac{\partial^2\varphi}{\partial z^i\partial\bar{z}^j}\right) = C_\varepsilon(|s|^2+\varepsilon)^k\exp\{F(x,\psi_\varepsilon)\}\det(g_{i\bar{j}}) \tag{7.5}$$

such that

$$\bar{\varphi}_\varepsilon \geqq \psi_\varepsilon \geqq \bar{\bar{\varphi}}_\varepsilon. \tag{7.6}$$

Then the arguments of Theorem 4 show that

$$\bar{\varphi}_\varepsilon \geqq \varphi_\varepsilon \geqq \bar{\bar{\varphi}}_\varepsilon. \tag{7.7}$$

On the other hand, for the unique solution of (7.5) with $\int_M\varphi = 0$, we can find an estimate of $\sup_M|\varphi|$ which is independent of ε. (This is seen in the proof of Theorem 3. Note that boundedness of $\Delta\psi_\varepsilon$, is needed.) In particular, we can choose $\bar{\varphi}_\varepsilon$ to be bounded from above by a constant independent of ε, and $\bar{\bar{\varphi}}_\varepsilon$ to be bounded from below by a constant independent of ε.

Hence according to (7.7), we can find a smooth solution φ_ε of (7.3) whose supremum norm is bounded from above by a constant independent of ε and whose matrix $(g_{i\bar{j}} + \partial^2\varphi_\varepsilon/\partial z^i\partial\bar{z}^j)$ is positive definite.

Let us now proceed to estimate $\Delta\varphi_\varepsilon$ from above. Let Δ_ε be the normalized Laplacian of the Kähler metric $\sum_{i,j}(g_{i\bar{j}} - \partial^2\varphi_\varepsilon/\partial z^i\partial\bar{z}^j)dz^i\otimes d\bar{z}^j$.

Then, as in (5.8), we have

$$\begin{aligned}
&\Delta_\varepsilon(\exp\{-C\varphi_\varepsilon\}(m+\Delta\varphi_\varepsilon))\\
&\geqq \exp\{-C\varphi_\varepsilon\}\Big[(k\Delta(\log|s|^2+\varepsilon)+\sum_{i,j} g^{i\bar{j}}\frac{\partial^2 F}{\partial z^i\partial\bar{z}^j}(x,\varphi_\varepsilon(x))\\
&\quad+\sum_{i,j} g^{i\bar{j}}\frac{\partial^2 F}{\partial t\partial z^i}\frac{\partial\varphi_\varepsilon}{\partial\bar{z}^j}+\sum_{i,j} g^{i\bar{j}}\frac{\partial^2 F}{\partial t\partial\bar{z}^j}\frac{\partial\varphi_\varepsilon}{\partial z^i}+\sum_{i,j} g^{i\bar{j}}\frac{\partial\varphi_\varepsilon}{\partial z^i}\frac{\partial\varphi_\varepsilon}{\partial\bar{z}^j}\\
&\quad-m^2\inf_{i\neq l}(R_{i\bar{i}l\bar{l}})\Big]-mC\exp\{-C\varphi_\varepsilon\}(m+\Delta\varphi_E)\\
&\quad+\left(C+\inf_{i\neq l}(R_{i\bar{i}l\bar{l}})\right)C_\varepsilon^{-1/(m-1)}\exp\{-C\varphi_\varepsilon\}(|s|^2+\varepsilon)^{-k/(m-1)}\\
&\quad\times\exp\left(\frac{-F}{m-1}\right)(m+\Delta\varphi_\varepsilon)^{1+1/(m-1)}.
\end{aligned}\tag{7.8}$$

Choose C so that $C+\inf_{i\neq l}(R_{i\bar{i}l\bar{l}})\geqq\frac{1}{2}C\geqq 1$. Then noting (5.7) and the fact that $\sup_M|\varphi_\varepsilon|$ is uniformly bounded, we can find positive constants C_1, C_2, C_3 and C_4 independent of ε such that

$$\begin{aligned}
\Delta'(\exp\{-C\varphi_\varepsilon\}(m+\Delta\varphi_\varepsilon))\geqq& -C_1-C_2(m+\Delta\varphi_\varepsilon)\\
&-C_3|\nabla\varphi_\varepsilon|+C_4(m+\Delta\varphi_\varepsilon)^{m/(m-1)}.
\end{aligned}\tag{7.9}$$

On the other hand, by Schauder's estimate and the estimate of $\sup_M|\varphi_\varepsilon|$, there are positive constants C_5, and C_6, such that

$$\sup_M|\nabla\varphi_\varepsilon|\leqq C_5\left(\sup_M|\Delta\varphi_\varepsilon|+\sup|\varphi_\varepsilon|\right)\leqq C_6\left(\sup_M(m+\Delta\varphi_\varepsilon)+1\right).\tag{7.10}$$

Combining (7.9) and (7.10) and the estimate of $\sup_M|\varphi_\varepsilon|$, it becomes clear that we can apply the maximum principle to find an upper estimate of $m+\Delta\varphi_\varepsilon$. Therefore, since $\sum_{i,j}(g_{i\bar{j}}+\partial^2\varphi_\varepsilon/\partial z^i\partial\bar{z}^j)dz^i\otimes d\bar{z}^j$ defines a Kähler metric, we have uniform estimates of $|\partial^2\varphi_\varepsilon/\partial z^i\partial\bar{z}^j|$ on every coordinate chart of M.

Using the uniform estimate of $\partial^2\varphi_\varepsilon/\partial z^i\partial\bar{z}^j$, one can now follow the arguments of Section 5 to provide higher derivative estimates of φ_ε on compact subsets of the complement of the divisor of s. Letting $\varepsilon\to 0$, we have then proved the following theorem.

Theorem 6. *Let L be a holomorphic line bundle over a compact Kähler manifold M whose Kähler metric is given by $\sum_{i,j} g_{i\bar{j}}dz^i\otimes d\bar{z}^j$. Let s be a holomorphic section of L. Let F be a smooth function defined on $M\times R$ such that $\partial F/\partial t\geqq 0$. Suppose, for some function ψ whose partial derivatives $|\partial^2\psi/\partial z^i\partial\bar{z}^j|$ are bounded on every coordinate chart of M, we have $\int_M|s|^{2k}\exp\{F(x,\psi(x)\}=\mathrm{Vol}(M)$. Then we can find a solution φ of the equation $\det(g_{i\bar{j}}+\partial^2\varphi/\partial z^i\partial\bar{z}^j)=|s|^{2k}\exp\{F(x,\varphi(x)\}$ with the following properties. Outside the divisor of s,φ is smooth and $\sum_{i,j}(g_{i\bar{j}}+\partial^2\varphi/\partial z^i\partial\bar{z}^j)dz^i\otimes d\bar{z}^j$ defines a Kähler metric. Over M,*

$\{|\varphi_{i\bar{j}}|\}$ is bounded for all i, j. Furthermore, any solution satisfying the above properties must be equal to φ plus a constant.

Proof. We have only to prove the last statement. Let Δ' be the normalized Laplacian of the metric $\sum_{i,j}(g_{i\bar{j}} + \partial^2\varphi/\partial z^i\partial\bar{z}^j)dz^i \otimes d\bar{z}^j$. Then we claim that if f is a C^1-function on M such that, for all i, j, $|\partial^2 f/\partial z^i\partial\bar{z}^j|$ is bounded on every coordinate chart of M, then

$$\int_{\{x|f(x)>0\}} \Delta'(f^2)|s|^{2k} \exp\{F(x,\varphi(x)\} = 0. \tag{7.11}$$

Indeed, let Δ'_ε be the normalized Laplacian of the Kähler metric $\sum_{i,j}(g_{i\bar{j}} + \partial^2\varphi_\varepsilon/\partial z^i\partial\bar{z}^j)dz^i \otimes d\bar{z}^j$. Then, for all $\delta > 0$ such that the boundary of $\{x|f(x) \geqq \delta\}$ is a C^1-manifold, we know that by Stoke's theorem

$$\int_{\{x|f(x)\geqq\delta\}} \Delta'_\varepsilon(f^2)(|s|^2+\varepsilon)^k \exp\{F(x,\varphi_\varepsilon(x)\}$$

can be expressed in terms of the boundary integral of $2f\partial f/\partial n$. Here $\partial/\partial n$ is the normal of the sets $\{x|f(x) = \delta\}$ taken with respect to our metric $\sum_{i,j}(g_{i\bar{j}} + \partial^2\varphi_\varepsilon/\partial z^i\partial\bar{z}^j)dz^i \otimes d\bar{z}^j$. Since $\{z|f(x) > 0\}$ is an increasing union of the sets $\{x|f(x) \geqq \delta\}$, we know that the measure of $\{x|\delta > f(x) > 0\}$ taken with respect to $(|s|^2+\varepsilon)^k \exp\{F(x,\varphi_\varepsilon(x)\}dV$ tends to zero as δ tends to zero. Applying the co-area formula (see [11]) to the set $\{x|\delta > f(x) > \frac{1}{2}\delta\}$, it is easy to see that we can find a sequence $\{\delta_i\}$ such that the product of δ_i with the codimension one volume of $\{x|f(x) = \delta_i\}$ tends to zero as δ_i tends to zero. Considering this together with the previous fact, we conclude that

$$\int_{\{z|f(x)\geqq\delta_i\}} \Delta'_\varepsilon(f^2)(|s|^2+\varepsilon)^k \exp\{F(x,\varphi_\varepsilon(x)\}$$

tends to zero as δ_i tends to zero. Hence we have

$$\int_{\{x|f(x)>0\}} \Delta'_\varepsilon(f^2)(|s|^2+\varepsilon)^k \exp\{F(x,\varphi_\varepsilon(x)\} = 0. \tag{7.12}$$

Letting $\varepsilon \to 0$ as in Theorem 3, we see that (7.11) follows from (7.12).

Suppose now that $\bar{\varphi}$ is another solution of (7.1) with all the properties described in the theorem. Then

$$\det(g_{i\bar{j}} + \varphi_{i\bar{j}} + (\bar{\varphi}-\varphi)_{i\bar{j}})\det(g_{i\bar{j}} + \varphi_{i\bar{j}})^{-1} = \exp\{F(x,\bar{\varphi}) - F(x,\varphi)\}. \tag{7.13}$$

Consider the set $\Omega = \{x|\bar{\varphi}(x) - \varphi(x) > 0\}$; if it is nonempty, then the arithmetic-geometric mean inequality shows that the inequality

$$\Delta'(\bar{\varphi}-\varphi) \geqq m\exp\left\{\frac{1}{m}[F(x,\bar{\varphi}(x)) - F(x,\varphi(x))]\right\} - m \geqq 0 \tag{7.14}$$

holds on Ω. (Note that $\partial F/\partial t$ is used here.)

Applying (7.11) to the function $(\bar{\varphi}-\varphi)^2$, we conclude immediately that

$$\begin{aligned} 2 & \int_{\Omega}(\bar{\varphi}-\varphi)\Delta'(\bar{\varphi}-\varphi)|s|^{2k}\exp\{F(x,\varphi(x))\} \\ +2 & \int_{\Omega}|\nabla'(\bar{\varphi}-\varphi)|^2|s|^{2k}\exp\{F(x,\varphi(x)\} \\ = & \int_{\Omega}\Delta'(\bar{\varphi}-\varphi)^2|s|^{2k}\exp\{F(x,\varphi(x)\} = 0. \end{aligned} \tag{7.15}$$

Combining (7.14) and (7.15), we see that $\nabla'(\bar{\varphi}-\varphi)=0$ on Ω and $\bar{\varphi}-\varphi$ is a constant on each component of $\Omega=\{x|\bar{\varphi}(x)-\varphi(x)>0\}$. Since $\bar{\varphi}-\varphi$ is continuous, this is possible only if Ω is empty or $\Omega=M$. In the first case, $\bar{\varphi}(x)\leqq\varphi(x)$ for all $x\in M$. In the second case, $\bar{\varphi}-\varphi$ is a constant. By reversing the role of $\bar{\varphi}$ and φ, we conclude easily that, in any case, $\bar{\varphi}-\varphi$ is a constant.

8 Complex Monge-Ampère equations with meromorphic right-hand side

Let L_1 and L_2 be two holomorphic line bundles over a compact Kähler manifold M. Let s_1 and s_2 be two (non-trivial) holomorphic sections of L_1 and L_2 that are equipped with Hermitian metrics so that we have globally defined functions $|s_1|^2$ and $|s_2|^2$ on M. Then, for $k_1\geqq 0$ and $k_2\geqq 0$, we shall study equations of the form

$$\det\left(g_{i\bar{j}}+\frac{\partial^2\varphi}{\partial z^i\partial\bar{z}^j}\right)\det(g_{i\bar{j}})^{-1}=|s_1|^{2k_1}|s_2|^{-2k_2}\exp\{F\}, \tag{8.1}$$

where $\sum_{i,j}g_{i\bar{j}}dz^i\otimes d\bar{z}^j$ is a Kähler metric and F is a smooth function such that

$$\int_M|s_1|^{2k_1}|s_2|^{-2k_2}\exp\{F\}=\mathrm{Vol}(M). \tag{8.2}$$

As before we approximate (8.1) by the equation

$$\det\left(g_{i\bar{j}}+\frac{\partial^2\varphi}{\partial z^i\partial\bar{z}^j}\right)\det(g_{i\bar{j}})^{-1}=C_\varepsilon(|s_1|^2+\varepsilon)^k(|s_2|^2+\varepsilon)^{-k_2}\exp\{F\}, \tag{8.3}$$

where

$$C_\varepsilon=\mathrm{Vol}(M)\left[\int_M(|s_1|^2+\varepsilon)^{k_1}(|s_2|^2+\varepsilon)^{-k_2}\exp\{F\}\right]^{-1}. \tag{8.4}$$

In order to prove that the solutions φ_ε of (8.3) converge on the complement of the divisors of s_1 and s_2, we consider the expression $(|s_2|^2+\varepsilon)^p\exp\{-C\varphi_\varepsilon\}(m+\Delta_\varepsilon)$ with $p\geqq 0$.

Letting Δ'_ε be the normalized Laplacian of the metric $\sum_{i,j}(g_{i\bar{j}}+\partial^2\varphi_\varepsilon/\partial z^i\partial\bar{z}^j)$ $dz^i\otimes d\bar{z}^j$, we compute the Laplacian of the above expression as follows:

$$\begin{aligned} & \Delta'_\varepsilon[(|s_2|^2+\varepsilon)^p\exp\{-C\varphi_\varepsilon\}(m+\Delta\varphi_\varepsilon)] \\ = & [\Delta'_\varepsilon\exp\{p\log(|s_2|^2+\varepsilon)-C\varphi_\varepsilon\}](m+\Delta\varphi_\varepsilon) \end{aligned}$$

$$
\begin{aligned}
&+2\nabla'_\varepsilon \exp\{p\log(|s_2|^2+\varepsilon)-C\varphi_\varepsilon\}\cdot\nabla'_\varepsilon(\Delta\varphi_\varepsilon)\\
&+\exp\{p\log(|s_2|^2+\varepsilon)-C\varphi_\varepsilon\}\Delta'_\varepsilon(\Delta\varphi_\varepsilon)\\
\geqq{}& [\Delta'_\varepsilon \exp\{p\log(|s_2|^2+\varepsilon)-C\varphi\}](m+\Delta\varphi_\varepsilon)\\
&-[\exp\{p\log(|s_2|^2+\varepsilon)-C\varphi_\varepsilon\}]\\
&\times|\nabla'_\varepsilon(p\log(|s_2|^2+\varepsilon)-C\varphi_2)|^2(m+\Delta\varphi_\varepsilon)\\
&-[\exp\{p\log(|s_2|^2+\varepsilon)-C\varphi_\varepsilon\}]\frac{|\nabla'_\varepsilon(\Delta\varphi_\varepsilon)|^2}{m+\Delta\varphi_\varepsilon}\\
&+\exp\{p\log(|s_2|^2+\varepsilon)\}-C\varphi_\varepsilon)\Delta'_\varepsilon(\Delta\varphi_\varepsilon)\\
\geqq{}& [\exp\{p\log(|s_2|^2+\varepsilon)-C\varphi_\varepsilon\}]\\
&\times(m+\Delta\varphi_\varepsilon)(p\Delta'_\varepsilon\log(|s_2|^2+\varepsilon)-C\Delta'_\varepsilon\varphi_\varepsilon)\\
&-[\exp\{p\log(|s_2|^2+\varepsilon)-C\varphi_\varepsilon\}]\frac{|\nabla'_\varepsilon(\Delta\varphi_\varepsilon)|^2}{m+\Delta\varphi_\varepsilon}\\
&+\exp\{p\log(|s_2|^2+\varepsilon)-C\varphi_\varepsilon\}\Delta'_\varepsilon(\Delta\varphi_\varepsilon). \qquad (8.5)
\end{aligned}
$$

By applying the same reasoning as in (2.15), we have

$$
\begin{aligned}
&\Delta'_\varepsilon[(|s_2|^2+\varepsilon)^p\exp\{-C\varphi_\varepsilon\}(m+\Delta\varphi_\varepsilon)]\\
&\geqq (|s_2|^2+\varepsilon)^p\exp\{-C\varphi_\varepsilon\}(m+\Delta\varphi_\varepsilon)(p\Delta'_\varepsilon\log(|s_2|^2+\varepsilon)-C\Delta'_\varepsilon\varphi_\varepsilon)\\
&\quad+(|s_2|^2+\varepsilon)^p\exp\{-C\varphi_\varepsilon\}\Big[\Delta F+k_1\Delta\log(|s_1|^2+\varepsilon)-k_2\Delta\log(|s_2|^2+\varepsilon)\\
&\quad+\left(\inf_{i\neq l}R_{i\bar{i}l\bar{l}}\right)\left(\sum_{i,l}\frac{1+(\varphi_\varepsilon)_{i\bar{i}}}{1+(\varphi_\varepsilon)_{l\bar{l}}}-m^2\right)\Big]. \qquad (8.6)
\end{aligned}
$$

As in (5.15), we have a positive constant C_1 which is independent of ε such that

$$
p\Delta'_\varepsilon\log(|s_2|^2+\varepsilon)\geqq -pC_1\sum_i\frac{1}{1+(\varphi_\varepsilon)_{i\bar{i}}}. \qquad (8.7)
$$

Combining (8.6) and (8.7) and computing as before, we can find positive constants C_2 and C_3 which are independent of ε such that

$$
\begin{aligned}
&\Delta'_\varepsilon[(|s_2|^2+\varepsilon)^p\exp\{-C\varphi_\varepsilon\}(m+\Delta\varphi_\varepsilon)]\\
&\geqq -C_2(|s_2|^2+\varepsilon)^p\exp\{-C\varphi_\varepsilon\}-mC(|s_2|^2+\varepsilon)^p\exp\{-C\varphi_\varepsilon\}(m+\Delta\varphi_\varepsilon)\\
&\quad-k_2(|s_2|^2+\varepsilon)^p\exp\{-C\varphi_\varepsilon\}\Delta\log(|s_2|^2+\varepsilon)\\
&\quad+\left(C-pC_1+\inf_{i\neq l}R_{i\bar{i}l\bar{l}}\right)(|s_2|^2+\varepsilon)^p\exp\{-C\varphi_\varepsilon\}(m+\Delta\varphi_\varepsilon)\sum_i\frac{1}{1+(\varphi_\varepsilon)_{i\bar{i}}}\\
&\geqq -C_2(|s_2|^2+\varepsilon)^p\exp\{-C\varphi_\varepsilon\}-mC(|s_2|+\varepsilon)^p\exp\{-C\varphi_\varepsilon\}(m+\Delta\varphi_\varepsilon)\\
&\quad-k_2(|s_2|^2+\varepsilon)^p\exp\{-C\varphi_\varepsilon\}\Delta\log(|s_2|^2+\varepsilon)\\
&\quad+C_3\left(C-pC_1+\inf_{i\neq l}R_{i\bar{i}l\bar{l}}\right)(|s_2|^2+\varepsilon)^p\exp\{-C\varphi_\varepsilon\}\\
&\quad\times(|s_1|^2+\varepsilon)^{-k_1/(m-1)}(|s_2|^2+\varepsilon)^{k_2/(m-1)}(m+\Delta\varphi_\varepsilon)^{m/(m-1)}. \qquad (8.8)
\end{aligned}
$$

Clearly, for any fixed p, we can choose C large enough so that

$$C_3\left(C - pC_1 + \inf_{i\neq l} R_{i\bar{i}l\bar{l}}\right)(|s_1|^2+\varepsilon)^{-k_1/(m-1)} \geqq 1.$$

With this choice of C, we consider the point where $(|s_2|^2+\varepsilon)^p \exp\{-C\varphi_\varepsilon\} \times (m+\Delta\varphi_\varepsilon)$ achieves its maximum. At such a point, the left-hand side of (8.8) is non-positive and so is the right-hand side of (8.8). Therefore, at this point,

$$\begin{aligned}&(|s_2|^2+\varepsilon)^{k_2/(m-1)}(m+\Delta\varphi_\varepsilon)^{m/(m-1)}\\ &\leqq \frac{1}{3}\max\{C_2, mC(m+\Delta\varphi_\varepsilon), k_2\Delta|\log(|s_2|^2+\varepsilon)|\}. \end{aligned} \tag{8.9}$$

It follows easily from (8.9) that, for some positive constant C_4 independent of ε. we have

$$\begin{aligned}&\sup_M(|s_2|^2+\varepsilon)^p \exp\{-C\varphi_\varepsilon\}(m+\Delta\varphi_\varepsilon)\\ &\leqq C_4(C^{m-1}+1)\max\Big\{\sup_M(|s_2|^2+\varepsilon)^{p-(k_2/m)}\exp\{-C\varphi_\varepsilon\}\\ &\sup_M(|s_2|^2+\varepsilon)^{p-k_2}\exp\{-C\varphi_\varepsilon\}, \sup(|s_2|^2+\varepsilon)^{p-(m-1)/m-(k_2/m)}\exp\{-C\varphi_\varepsilon\}\Big\}. \end{aligned} \tag{8.10}$$

It is easy to see from (8.2) that $k_2 < 1$. Hence the third term in the right-hand side of (8.10) will be the dominating term. If we choose p to be $(m-1)/m + (k_2/m) + Cq$ with $q \geqq 0$ in (8.10), we see that, for some positive constant C_5 independent of ε,

$$\begin{aligned}&\sup_M(|s_2|^2+\varepsilon)^{(m-1)/m+(k_2/m)+Cq}\exp\{-C\varphi_\varepsilon\}(m+\Delta\phi_\varepsilon)\\ &\leqq C_5(C^{m+1}+1)[\sup(|s_2|^2+\varepsilon)^q\exp\{-\varphi_\varepsilon\}]^C. \end{aligned} \tag{8.11}$$

We are going to estimate the supremum norm of $|\varphi_\varepsilon|$. First of all, we recall that when we solve (8.3), we can normalize our solution φ_ε so that $\int_M \varphi_\varepsilon = 0$. Then, using the inequality $\Delta\varphi_\varepsilon + m > 0$, we have an upper estimate of φ_ε. Hence it remains to find a lower estimate of φ_ε.

Integrating (8.8) with respect to the volume from $(|s_2|^2+\varepsilon)^{-k_2}(|s_1|^2+\varepsilon)^{k_1} \times \exp\{F\}dV$, we have

$$\begin{aligned}&C_3\left(C - pC_1 + \inf_{i\neq l} R_{i\bar{i}l\bar{l}}\right)\exp\{\inf F\}\\ &\times\int_M (|s_1|^2+\varepsilon)^{(m-2)k_1/(m-1)}(|s_2|^2+\varepsilon)^{p-(m-2)k_2/(m-1)}\\ &\times(m+\Delta\varphi_\varepsilon)^{m/(m-1)}\exp\{-C\varphi_\varepsilon\}\\ &-k_2\int_M(|s_2|^2+\varepsilon)^{p-k_2}(|s_1|^2+\varepsilon)^{k_1}\exp\{-C\varphi_\varepsilon\}\exp\{F\}\Delta\log(|s_2|^2+\varepsilon)\\ &\leqq C_2\exp\{\sup F\}\int_M(|s_2|^2+\varepsilon)^{p-k_2}(|s_1|^2+\varepsilon)^{k_1}\exp\{-C\varphi_\varepsilon\}\end{aligned}$$

$$+mC\exp\{\sup F\}\int_M(|s_2|^2+\varepsilon)^{p-k_2}(|s_1|^2+\varepsilon)^{k_1}\exp\{-C\varphi_\varepsilon\}(m+\Delta\varphi_\varepsilon). \tag{8.12}$$

Using (5.7), we can find a positive constant C_6 which is independent of ε such that

$$\Delta\log(|s_2|^2+\varepsilon)\geqq -C_6. \tag{8.13}$$

Hence, for any positive constant L,

$$\begin{aligned}
&L\int_M(|s_2|^2+\varepsilon)^{p-k_2}(|s_1|^2+\varepsilon)^{k_1}\exp\{-C\varphi_\varepsilon\}\exp\{F\}\Delta\log(|s_2|^2+\varepsilon)\\
&=L\int_M(|s_2|^2+\varepsilon)^{p-k_2}(|s_1|^2+\varepsilon)^{k_1}\exp\{-C\varphi_\varepsilon\}\exp\{F\}[\Delta\log(|s|^2+\varepsilon)+C_6]\\
&\quad -LC_6\int_M(|s_2|^2+\varepsilon)^{p-k_2}(|s_1|^2+\varepsilon)^{k_1}\exp\{-C\varphi_\varepsilon\}\exp\{F\}\\
&\leqq L\int_M(|s_2|^2+\varepsilon)^{p-k_2}(|s_1|^2+\varepsilon)^{k_1}\exp\{-C\varphi_\varepsilon\}\exp\{\sup F\}\\
&\quad \times[\Delta\log(|s_2|^2+\varepsilon)+C_6]\\
&=L\exp\{\sup F\}\int_M(|s_2|^2+\varepsilon)^{p-k_2}(|s_1|^2+\varepsilon)^{k_1}\exp\{-C\varphi_\varepsilon\}\Delta\log(|s_2|^2+\varepsilon)\\
&\quad +LC_6\exp\{\sup F\}\int_M(|s_2|^2+\varepsilon)^{p-k_2}(|s_1|^2+\varepsilon)^{k_1}\exp\{-C\varphi_\varepsilon\}.
\end{aligned}\tag{8.14}$$

By Hölder's inequality, we know that, for any $\delta>0$,

$$\begin{aligned}
&\int_M(|s_2|^2+\varepsilon)^{p-k_2}(|s_1|^2+\varepsilon)^{k_1}\exp\{-C\varphi_\varepsilon\}(m+\Delta\varphi_\varepsilon)\\
&=\Big[\int_M(|s_2|^2+\varepsilon)^{p-(m-2)k_2/(m-1)}(|s_1|^2+\varepsilon)^{(m-2)k_1/(m-1)}\exp\{-C\varphi_\varepsilon\}\\
&\quad \times(m+\Delta\varphi_\varepsilon)^{m/(m-1)}\Big]^{(m-1)/m}\\
&\quad \times\Big[\int_M(|s_2|^2+\varepsilon)^{p-2k_2}(|s_1|^2+\varepsilon)^{2k_1}\exp\{-C\varphi_\varepsilon\}\Big]^{1/m}\\
&\leqq \frac{m-1}{m}\delta^{m/(m-1)}\int_M(|s_2|^2+\varepsilon)^{p-(m-2)k_2/(m-1)}\\
&\quad \times(|s_1|^2+\varepsilon)^{(m-2)k_1/(m-1)}\exp\{-C\varphi_\varepsilon\}(m+\Delta\varphi_\varepsilon)^{m/(m-1)}\\
&\quad +\frac{\delta^{-m}}{m}\int_M(|s_2|^2+\varepsilon)^{p-2k_2}(|s_1|^2+\varepsilon)^{2k_1}\exp\{-C\varphi_\varepsilon\}.
\end{aligned}\tag{8.15}$$

For any $p\geqq 0$, we choose a constant C large enough so that $C\geqq 1, C_3(C-pC_1+\inf_{i\neq l}R_{i\bar{i}l\bar{l}})\geqq\frac{1}{2}C_3C$. Then we choose δ so that

$$(m-1)\delta^{m/(m-1)}CC_3^{-1}\left(C-pC_1+\inf_{i\neq l}R_{i\bar{i}l\bar{l}}\right)^{-1}\exp\{\sup F-\inf F\}=\frac{1}{2}.$$

Substituting (8.15) into (8.12) and keeping (8.14) in mind, we see that we can find positive constants C_7 and C_8 which are independent of ε and C for which

$$\begin{aligned}
&\int_M (|s_2|^2+\varepsilon)^{p-k_2}(|s_1|^2+\varepsilon)^{k_1}\exp\{-C\varphi_\varepsilon\}(m+\Delta\varphi_\varepsilon)\log(|s_2|^2+\varepsilon)\\
&\quad -\frac{C_7}{C}\int_M (|s_2|^2+\varepsilon)^{p-k_2}(|s_1|^2+\varepsilon)^{k_1}\exp\{-C\varphi_\varepsilon\}\Delta\log(|s_2|^2+\varepsilon)\\
&\leqq C_8\int_M (|s_2|^2+\varepsilon)^{p-k_2}(|s_1|^2+\varepsilon)^{k_1}\exp\{-C\varphi_\varepsilon\}\\
&\quad +C_8\int_M (|s_2|^2+\varepsilon)^{p-2k_2}(|s_1|^2+\varepsilon)^{2k_1}\exp\{-C\varphi_\varepsilon\} \qquad (8.16)
\end{aligned}$$

In order to make use of (8.16) to derive an integral estimate $\exp\{-C\varphi_\varepsilon\}$, we shall assume that the integral $\int_M |s_2|^{-2mk_2}$ is finite. Then by choosing p so that $C_7(p-k_2+C_7)^{-1}=\frac{1}{2}$, we shall prove in the Appendix B that, for some positive constant C_{11} independent of ε and C,

$$\begin{aligned}
&\int_M \left|\nabla\left[(|s_2|^2+\varepsilon)^{(p-k_2)/2}(|s_1|^2+\varepsilon)^{k_1/2}\exp\left\{-\frac{1}{2}C\varphi_\varepsilon\right\}\right]\right|^2\\
&\leqq CC_{11}\left(\int_M [(|s_2|^2+\varepsilon)^{p-k_2}(|s_1|^2+\varepsilon)^{k_1}\exp\{-C\varphi_\varepsilon\}]^{m/(m-1)}\right)^{(m-1)/m}.
\end{aligned} \qquad (8.17)$$

Since we have an estimate of $\int_M |\varphi_\varepsilon|$, we can use the method of Section 5 to find an estimate of

$$\int_M \left[\left(\int |s_2|^2+\varepsilon\right)^{p-k_2}(|s_1|^2+\varepsilon)^{k_1}\exp\{-C\varphi_\varepsilon\}\right]^{m/(m-1)}$$

which is independent of ε.

Note that in (8.11) and (8.17) we may choose two different values of C as long as they are larger than some fixed constant. From (8.11) and the supremum estimate of φ_ε, we derive that, for any $q \geqq 0$,

$$\begin{aligned}
&\sup_M(|s_2|^2+\varepsilon)^{(m-1)/m+(k_2/m)+Cq}(m+\Delta\varphi_\varepsilon)\\
&\leqq C_5(C^{m-1}+1)\exp\{C\sup\varphi_\varepsilon\}(\sup(|s_2|^2+\varepsilon)^q\exp\{-\varphi_\varepsilon\})^C, \qquad (8.18)
\end{aligned}$$

where C is any positive constant so that

$$C_3\left[C-\left(\frac{m-1}{m}+\frac{k_2}{m}+Cq\right)C_1+\inf_{i\neq l}R_{i\bar{i}l\bar{l}}\right]\geqq\sup(|s_1|^2+\varepsilon)^{k_1/(m-1)}.$$

From (8.17), we conclude that when $(p-k_2)(p-k_2+C_7)=\frac{1}{2}$ and N is a large constant, we can find a positive constant C_{12}, independent of ε such that

$$\begin{aligned}
&\int_M (|s_2|^2+\varepsilon)^{p-k_2}(|s_1|^2+\varepsilon)^{k_1}\exp\{-N\varphi_\varepsilon\}\\
&=\int_M (|s_2|^2+\varepsilon)^{C_7}(|s_1|^2+\varepsilon)^{k_1}\exp\{-N\varphi_\varepsilon\}\leqq C_{12}. \qquad (8.19)
\end{aligned}$$

Recall that C_7 is independent of N also.

Using (8.18) and (8.19) we shall show that, for any $q > 0$, $(|s_2|^2+\varepsilon)^q \exp\{-\varphi_\varepsilon\}$ has an upper bound which is independent of ε when ε tends to zero.

In fact, if this assertion were not true, we could find a sequence of numbers $\{\varepsilon_i\}$ and a sequence of points $\{x_i\}$ in M such that $\varepsilon \to 0$ and $(|s_2|^2(x_i) + \varepsilon_i)^q \exp\{-\varphi_{\varepsilon_i}\}(x_i) = \sup_M(|s_2|^2 + \varepsilon_i)^q \exp\{-\varphi_{\varepsilon_i}\}$ tends to infinity. Without loss of generality, we may assume that $\{x_i\}$ converges to a point x_0 in M.

Suppose the sequence $\{\varepsilon_i^{-1}|s_i|^2(x_i)\}$ is bounded independent of i. Then, clearly, $\varepsilon_i^q \exp\{-\varphi_{\varepsilon_i}\}(x_i)$ tends to infinity when x_i tends to x_0. On the other hand, using inequality (8.18) and the L^1-estimate of φ_ε, we can apply the Schauder estimate to find a constant C_{13} independent of i such that

$$\sup_M |\nabla\varphi_{\varepsilon_i}| \leqq C_{13}\left[\varepsilon_i^{-(m-1)/m-(k_2/m)-Cq}\left(\sup_M(|s_2|^2+\varepsilon_i)^q \exp\{-\varphi_{\varepsilon_i}\}\right)^C + 1\right]. \tag{8.20}$$

It follows from (8.20) that there is a constant C_{14} independent of i such that

$$\begin{aligned}&\sup_M |\nabla[\log(|s_2|^2+\varepsilon_i)^q - \varphi_{\varepsilon_i}]| \\ &\leqq C_{14}\left[\varepsilon^{-(m-1)/m-(k_2/m)-Cq}\left(\sup_M(|s_2|^2+\varepsilon_i)^q \exp\{-\varphi_{\varepsilon_i}\}\right)^C + \varepsilon_i^{-1/2}\right].\end{aligned} \tag{8.21}$$

Clearly we may assume that $\sup_M[q\log(|s_2|^2+\varepsilon_i)-\varphi_{\varepsilon_i}] \geqq 0$. Then proceeding as in Section 2, we can now conclude that for some constant $\tilde{C}_{14}$, independent of i,

$$\begin{aligned}&\varepsilon_i^{2(m-1)+2k_2+2mCq}\sup_M[\log(|s_2|^2+\varepsilon_i)^q - \varphi_{\varepsilon_i}]^{2m} \\ &\times\left[\left(\sup_M(|s_2|^2+\varepsilon_i)^q\exp\{-\varphi_{\varepsilon_i}\}\right)^C + 1\right]^{-2m} \\ &\times\varepsilon_i^{C_7+k_1}\varepsilon_i^{-Nq/2}\left[\sup_M(|s_2|^2+\varepsilon_i)^q\exp\{-\varphi_{\varepsilon_i}\}\right]^{N/2} \\ &\leqq \tilde{C}_{14}\int_M(|s_2|^2+\varepsilon_i)^{C_7}(|s_1|^2+\varepsilon_i)^{k_1}\exp\{-N\varphi_{\varepsilon_i}\}.\end{aligned} \tag{8.22}$$

In (8.19), the number N can be made as large as we want (where C_{12} depends on N). Therefore, (8.22) shows that $\sup_M(|s_2|^2+\varepsilon_i)^q\exp\{-\varphi_{\varepsilon_i}\}$ is bounded from above independently of i.

The above contradiction shows that, by passing to a subsequence, we may assume that $\lim_{i\to\infty}\varepsilon_i^{-1}|s_2|^2(x_i) = \infty$.

For each x_i, let $B(x_i, \delta_i)$ be a geodesic ball around x_i such that, for each $x \in B(x_i, \delta_i)$,

$$\frac{1}{2}|s_2|^2(x_i) \geqq |s_2|^2(x) \geqq \frac{1}{2}|s_2|^2(x_i). \tag{8.23}$$

Let C_{15} be the supremum of $|\nabla|s_2|^2|$ on M. Then one can easily prove that we may assume

$$\delta_i \geqq \frac{1}{2C_{15}}|s_2|^2(x_i). \tag{8.24}$$

We only need inequalities (8.23) and (8.24) with suitably large C_{15}. Hence we may assume that δ_i is smaller than the injectivity radius of M.

It is easy to derive from (8.18) that, over the ball $B(x_i, \delta_i)$,

$$\begin{aligned} &0 < m + \Delta\varphi_{\varepsilon_i} \\ &\leqq C_5\left[\frac{1}{2}|s_2|^2(x_i)\right]^{-(m-1)/m-(k_2/m)-C_q}(\sup(|s_2|^2+\varepsilon_i)^q\exp\{-\varphi_{\varepsilon_i}\})^C. \end{aligned} \tag{8.25}$$

By applying the Schauder estimate on the ball $B(x_i, \delta_i)$, we can find a constant C_{16} which is independent of i such that

$$\begin{aligned} &\sup_{x\in B(x_i,\delta_i)} |\nabla\varphi_{\varepsilon_i}|(x) \leqq C_{16}\Big[(|s_2|^2(x_i))^{-(m-1)/m-(k_2/m)-Cq} \\ &\times\left(\sup_M(|s_2|^2+\varepsilon_i)^q\exp\{-\varphi_{\varepsilon_i}\}\right)^C + \delta_i^{-2m-1}\int_{B(x_i,\delta_i)}|\varphi_{\varepsilon_i}|\Big] \\ &\leqq C_{16}\Big[(a|s_2|^2(x_i))^{-(m-1)/m-(k_2/m)-Cq}\left(\sup_M(|s_2|^2+\varepsilon_i)^q\exp\{-\varphi_{\varepsilon_i}\}\right)^C \\ &+(2C_{15})^{2m+1}(|s_2|^2(x_i))^{-2m-1}\int_{B(x_i,\delta_i)}|\varphi_{\varepsilon_i}|\Big]. \end{aligned} \tag{8.26}$$

Since we have an L^1-estimate of $\int_M|\varphi_{\varepsilon_i}|$, it follows from (8.26) that there is a constant C_{17}, independent of i, such that

$$\begin{aligned} &\sup_{x\in B(x_i,\delta_i)} |\nabla(q\log(|s_2|^2+\varepsilon_i)-\varphi_{\varepsilon_i})| \\ &\leqq C_{17}[(|s_2|^2(x_i))^{-(m-1)/m-(k_2/m)-Cq}(\sup_M(|s_2|^2+\varepsilon_i)^q\exp\{-\varphi_{\varepsilon_i}\})^C \\ &+(|s_2|^2(x_i))^{-2m-1}]. \end{aligned} \tag{8.27}$$

Since s_1 is holomorphic, one can find positive constants l and C_{18} independent of i, such that, for any (small) $r > 0$ and $x \in M$,

$$\int_{B(x,r)}(|s_1|^2)^{k_1} \geqq C_{18}r^l. \tag{8.28}$$

As before, we may assume that $\sup_M[q\log(|s_2|^2+\varepsilon_i)-\varphi_{\varepsilon_i}] \geqq 0$. Then proceeding as in Section 2, we can now conclude from (8.23), (8.27) and (8.28) that there is a constant C_{19} independent of i, such that

$$\begin{aligned} &(|s_2|^2(x_i))^{C_7}\left(\frac{3}{2}|s_2|^2(x_i)+\varepsilon_i\right)^{-qN/2}\left[\sup_M(|s_2|^2+\varepsilon_i)^q\exp\{-\varphi_{\varepsilon_i}\}\right]^{N/2} \\ &\times\sup_M(\log(|s_2|^2+\varepsilon_i)^q-\varphi_{\varepsilon_i}) \end{aligned}$$

$$\times[(|s_2|^2(x_i))^{-(m-1)/m-(k_2/m)-Cq}(\sup_M(|s_2|^2+\varepsilon_i)^q\exp\{-\varphi_{\varepsilon_i}\})^C$$
$$+(|s_2|^2(x_i))^{-2m-1}]^{-1}$$
$$\leqq C_{19}\int_M(|s_2|^2+\varepsilon_i)^{C_7}(|s_1|^2+\varepsilon_i)^{k_1}\exp\{-N\varphi_{\varepsilon_i}\}. \tag{8.29}$$

Since $\lim_{i\to\infty}\varepsilon_i^{-1}|s_2|^2(x_i)=\infty$ and the right-hand side of (8.29) can be estimated even when N is large, we see that the quantity $\sup_M(|s_2|^2+\varepsilon_i)^q\times\exp\{-\varphi_{\varepsilon_i}\}$ can be estimated by a constant independent of i.

In conclusion we have proved that, for any $q>0, -\varphi_\varepsilon+q\log(|s_2|^2+\varepsilon)$ is bounded from above by a constant independent of ε. In particular, $-\varphi_\varepsilon$ is uniformly bounded over compact subsets of the complement of the divisor of s_2. From (8.18) and the supremum estimate of φ_ε, we see that both $|\varphi_\varepsilon|$ and $|\Delta\varphi_\varepsilon|$ are uniformly bounded over compact subsets of the complement of the divisor of s_2. The arguments of Section 5 now show that one can find uniform estimates of $\{(\varphi_\varepsilon)_{i\bar{j}k}\}$ over compact subsets of the complement of the divisors of s_1 and s_2. Hence we have proved the following theorem.

Theorem 7. *Let L_1 and L_2 be two holomorphic line bundles over a compact Kähler manifold M whose Kähler metric is given by $\sum_{i,j}g_{i\bar{j}}dz^i\otimes d\bar{z}^j$. Let s_1 and s_2 be two holomorphic sections of L_1 and L_2, respectively. Let F be a smooth function defined on M such that $\int_M|s_1|^{2k_1}|s_2|^{-2k_2}\exp\{F\}=\mathrm{Vol}(M)$, where k_1 and k_2, are two non-negative integers. Suppose that, for $m=\dim M$, $\int_M|s_2|^{2mk_2}<\infty$. Then we can solve the equation*

$$\det\left(g_{i\bar{j}}+\frac{\partial^2\varphi}{\partial z^i\partial\bar{z}^j}\right)\det(g_{i\bar{j}})^{-1}=|s_1|^{2k_1}|s_2|^{-2k_2}\exp\{F\}$$

so that

(i) *φ is smooth outside the divisors of s_1 and s_2 with $\sup_M\varphi<\infty$,*

(ii) *$(\varphi_{i\bar{j}})$ is a bounded matrix outside the divisor of s_2 and, for any $q>0$, $|s_2|^{2(m-1)m^{-1}+2k_2m^{-1}+q}\Delta\varphi$ is bounded on M,*

(iii) *for any $q>0$, the function $\varphi-q\log|s_2|^2$ is bounded from below,*

(iv) *the matrix $(g_{i\bar{j}}+\partial^2\varphi/\partial z^i\partial\bar{z}^j)$ is positive definite outside the complement of the divisors of s_1 and s_2.*

Furthermore, if we assume that $\int_M(|s_2|^2)^{(1-m)m^{-1}-k_2m^{-1}-q}<\infty$ for some $q>0$, then any two solutions of the equation which has the above properties (i), (ii) *and* (iv) *must differ from each other by a constant. If we also know that $(|s_2|^2)^{(1-m)m-k_2m^{-1}-q}$ is integrable over every analytic disc of M, then the unique solution φ is bounded from below on M.*

Proof. We have only to prove the last sentence. Suppose ψ is another solution of the equation which possess properties (i), (ii) and (iv) stated in the theorem. Then clearly we have

$$\det\left(g_{i\bar{j}}+\frac{\partial^2\psi}{\partial z^i\partial\bar{z}^j}\right)\det\left(g_{i\bar{j}}+\frac{\partial^2\varphi_\varepsilon}{\partial z^i\partial\bar{z}^j}\right)^{-1}$$
$$=C_\varepsilon^{-1}|s_1|^{2k_1}(|s_1|^2+\varepsilon)^{-k_1}|s_2|^{-2k_2}(|s_2|^2+\varepsilon)^{k_2}. \tag{8.30}$$

On the other hand, we can rewrite $\det(g_{i\bar{j}} + \partial^2\psi/\partial z^i \partial \bar{z}^j)$ as

$$\det\left(g_{i\bar{j}} + \frac{\partial^2 \varphi_\varepsilon}{\partial z^i \partial \bar{z}^j} + \frac{\partial^2(\psi - \varphi_\varepsilon)}{\partial z^i \partial \bar{z}^j}\right).$$

Therefore the arithmetic-geometric mean inequality shows that

$$m + \Delta'_\varepsilon(\psi - \varphi_\varepsilon) \geqq m[C_\varepsilon^{-1}|s_1|^{2k_1}(|s_1|^2 + \varepsilon)^{-k_1}|s_2|^{-2k_2}(|s_2|^2 + \varepsilon)^{k_2}]^{1/m}, \tag{8.31}$$

where Δ'_ε is the normalized Laplacian of the metric

$$\sum_{i,j}\left(g_{i\bar{j}} + \frac{\partial^2 \varphi_\varepsilon}{\partial z^i \partial \bar{z}^j}\right) dz^i \otimes d\bar{z}^j.$$

Let k be any constant. We claim that, over the domain $\Omega_{\varepsilon,k} = \{x \in M | \psi(x) - \varphi_\varepsilon(x) > k\}$,

$$\int_{\Omega_{\varepsilon,k}} (\psi - \varphi_\varepsilon - k)\Delta'_\varepsilon(\psi - \varphi_\varepsilon) \leqq 0. \tag{8.32}$$

In fact, let $\delta > 0$ be a positive constant. Then the domain $\Omega_{\varepsilon,k,\delta} = \{x|\psi(x) - \varphi_\varepsilon(x) > k - \delta \log |s_2|^2\}$ is disjoint from the divisor of s_2 by property (i). Hence both $(\psi_{i\bar{j}})$ and $((\varphi_\varepsilon)_{i\bar{j}})$ are bounded on $\Omega_{\varepsilon,k,\delta}$ and we can apply the divergence theorem on $\Omega_{\varepsilon,k,\delta}$ to find

$$\begin{aligned}&\int_{\Omega_{\varepsilon,k,\delta}} (\psi - \varphi_\varepsilon - k + \delta \log |s_2|^2)\Delta'_\varepsilon(\psi - \varphi_\varepsilon - k + \delta \log |s_2|^2) \\ &= -\int_{\Omega_{\varepsilon,k,\delta}} |\nabla'_\varepsilon(\psi - \varphi_\varepsilon - k + \delta \log |s_2|^2)|^2.\end{aligned} \tag{8.33}$$

Outside the divisor of s_2, $\Delta'_\varepsilon \log |s_2|^2$ is simply the trace of the first Chern form of L_2 with respect to the metric

$$\sum_{i,j}\left(g_{i\bar{j}} + \frac{\partial^2 \varphi_\varepsilon}{\partial z^i \partial \bar{z}^j}\right) dz^i \otimes d\bar{z}^j.$$

Hence for some constant C_{20}, independent of δ, we have

$$|\Delta'_\varepsilon(\delta \log |s_2|^2)| \leqq \delta C_{20}. \tag{8.34}$$

Using property (ii) and the assumption that $\int_M (|s_2|^2)^{(1-m)m^{-1} - k_2 m^{-1} - q} < \infty$, we can find a constant C_{21} independent of δ, such that

$$\delta \int_{\Omega_{\varepsilon,k,\delta}} \log |s_2|^2 |\Delta'_\varepsilon(\psi - \varphi_\varepsilon)| \leqq \delta C_{21}. \tag{8.35}$$

It follows easily from (8.33), (8.34) and (8.35) that

$$\lim_{\delta \to 0} \int_{\Omega_{\varepsilon,k,\delta}} (\psi - \varphi_\varepsilon - k)\Delta'_\varepsilon(\psi - \varphi_\varepsilon) \leqq 0. \tag{8.36}$$

Using the definition of $\Omega_{\varepsilon,k,\delta}$, we see that, over $\Omega_{\varepsilon,k,\delta}$, $\psi-\varphi_\varepsilon-k$ is bounded by a constant independent of δ when δ is small. The function $(\psi-\varphi_\varepsilon-k)\times\Delta'_\varepsilon(\psi-\varphi_\varepsilon)$ is therefore uniformly integrable and we can apply Lebesgue's dominated convergence theorem to prove (8.32).

Applying (8.31) and (8.32), we can now prove the following inequality:

$$\begin{aligned}
&\int_{\Omega_{\varepsilon,k}} (\psi-\varphi_\varepsilon-k)\{m-m[C_\varepsilon^{-1}|s_1|^{2k_1}(|s_1|^2+\varepsilon)^{-k_1}|s_2|^{-2k_2}(|s_2|^2+\varepsilon)^{k_2}\}^{1/m}\\
&\geqq \int_{\Omega_{\varepsilon,k}} (\psi-\varphi_\varepsilon-k)\{m-m[C_\varepsilon^{-1}|s_1|^{2k_1}(|s_1|^2+\varepsilon)^{-k_1}|s_2|^{-2k_2}\\
&\quad\times(|s_2|^2+\varepsilon)^{k_2}]^{1/m}+\Delta'_\varepsilon(\psi-\varphi_\varepsilon)\}\\
&\geqq 0.
\end{aligned}\tag{8.37}$$

When ε tends to zero, the integral on the left-hand side tends to zero. Let K be any compact set of the complement of the divisor of s_2, then by (8.31) and (8.37), we have

$$\begin{aligned}
&\lim_{\varepsilon\to 0}\int_{K\cap\Omega_{\varepsilon,k}} (\psi-\varphi_\varepsilon-k)\{m-m[C_\varepsilon^{-1}|s_1|^{2k_1}(|s_1|^2+\varepsilon)^{-k_1}\\
&\times|s_2|^{-2k_2}(|s_2|^2+\varepsilon)^{k_2}]^{1/m}+\Delta'_\varepsilon(\psi-\varphi_\varepsilon)\}=0.
\end{aligned}\tag{8.38}$$

(Note that (8.31) is used strongly here.) Let $\Omega_k=\{x|\psi(x)-\varphi(x)>k\}$. Then (8.38) gives

$$\int_{K\cap\Omega_k}(\psi-\varphi-k)\Delta'(\psi-\varphi)=0,\tag{8.39}$$

where Δ' is the normalized Laplacian of the metric

$$\sum_{i,j}\left(g_{i\bar{j}}+\frac{\partial^2\varphi}{\partial z^i\partial\bar{z}^j}\right)dz^i\otimes d\bar{z}^j.$$

As in (8.31), we know that $\Delta'(\psi-\varphi)\geqq 0$ and hence we conclude from (8.39) that $\Delta'(\psi-\varphi)=0$ on Ω_k. When the arithmetic-geometric mean inequality becomes equality, the eigenvalues of the corresponding matrix are equal. Therefore, $\psi_{i\bar{j}}=\varphi_{i\bar{j}}$ everywhere on Ω_k. Since k is arbitrary, $\psi_{i\bar{j}}=\varphi_{i\bar{j}}$ everywhere on the complement of the divisor of s_2. One can now prove that $\varphi-\psi$ =constant by simply letting first ε tend to zero and then δ tend to zero in (8.33). This shows that

$$\int_{K\cap\Omega_k}|\nabla'(\psi-\varphi)|^2<-\lim_{\varepsilon\to 0}\int_{\Omega_{\varepsilon,k}}(\psi-\varphi_\varepsilon-k)\Delta'_\varepsilon(\psi-\varphi_\varepsilon)$$

which is equal to zero by (8.37).

It remains to prove that the unique solution φ is bounded from below.

From (8.11) and the estimate of $(|s_2|^2+\varepsilon)^q\exp\{-\varphi_\varepsilon\}$ for any $q>0$, we know that

$$\sup_M[(m+\Delta\varphi_\varepsilon)(|s_2|^2+\varepsilon)^{q+(m-1)m^{-1}+k_2m^{-1}}]\leqq C_{22}\tag{8.40}$$

for any $q > 0$, where C_{22} is a constant independent of ε.

Let x be any point on the divisor of s_2. Let D_x be an analytic disc passing through x such that s_2 is not zero on ∂D_x. Then $|\varphi_\varepsilon|$ is uniformly bounded on ∂D_x when ε tends to zero.

Since $[g_{i\bar{j}} + (\varphi_\varepsilon)_{i\bar{j}}]$ is positive definite, it follows from (8.40) that when we restrict φ_ε to D_x, the absolute value of its Laplacian is estimated by $(|s_2|^2 + \varepsilon)^{-q-(m-1)m^{-1}-k_2m^{-1}}$. On the other hand, the assumption guarantees the integrability of $(|s_2|^2+\varepsilon)^{-\bar{q}-(m-1)m^{-1}-k_2m^{-1}}$ over D_x for some $\bar{q} > 0$.

We can now apply Green's formula to D_x, to prove that $\sup_{D_x} |\varphi_\varepsilon|$ can be estimated from above by a quantity involving $\sup_{\partial D_x} |\varphi_\varepsilon|$ and

$$\int_{D_x} |\log(|s_2|^2+\varepsilon)|(|s_2|^2+\varepsilon)^{-q-(m-1)m^{-1}-k_2m^{-1}}.$$

Therefore, by choosing $q < \bar{q}$, we obtain an estimate of $|\varphi_\varepsilon|$ on D_x. Since x is arbitrary, we can conclude the boundedness of φ.

9 The general case

In this section, we write down a general theorem.

Let M be a compact Kähler manifold with Kähler metric $\sum_{i,j} g_{i\bar{j}} dz^i \otimes d\bar{z}^j$. Let $t_1, t_2, \ldots, t_{n_1+n_2}$ be non-zero non-negative functions defined on M with the following properties. For each t_i, we can find holomorphic sections $s_1, s_2, \ldots, s_l$ of some Hermitian holomorphic line bundles such that $t_i = \sum_{j=1}^{l} |s_j|^{2k_j}$, where $k_j \geqq 0$ for each j.

Then we consider the equation

$$\det\left(g_{i\bar{j}} + \frac{\partial^2\varphi}{\partial z^i \partial \bar{z}^j}\right)\det(g_{i\bar{j}})^{-1} = t_1 t_2 \cdots t_n t_{n_1+1}^{-1} t_{n_1+n_2}^{-1} F(x,\varphi), \tag{9.1}$$

where F is a smooth function defined on $M \times R$ with $\partial F/\partial t \geqq 0$.

As in the previous section, we assume that we can find a smooth function ψ defined on M such that

$$\int_M t_1 t_2 \cdots t_n t_{n_1+1}^{-1} \cdots t_{n_1+n_2}^{-1} F(x,\psi) = \mathrm{Vol}(M). \tag{9.2}$$

If $m = \dim M$, we also assume that

$$\int_M (t_{n_1+1} \cdots t_{n_1+n_2})^{-m} < \infty \tag{9.3}$$

and, for some $q > 0$,

$$\int_M (t_{n_1+1} \cdot t_{n_1+n_2})^{-q/m} |\Delta \log t_{n_1+1} \cdots t_{n_1+n_2}|^{(m-1)/m} < \infty \tag{9.4}$$

is integrable over M and over every analytic disk of M.

The arguments of Section 8 can easily be modified to prove the solvability of (9.1) when either F is independent of φ or $F(x,\varphi) = \exp\{k\varphi\}$ with $k > 0$. The arguments of Section 6 can then be modified to solve (9.1) in general. (The required comparison theorem can be treated as in the end of Section 8.)

Theorem 8. *Let M be a compact Kähler manifold. Suppose that, in equation (9.1), the t_i are functions satisfying the above mentioned properties. Then we can find a solution φ of (9.1) such that*

(i) *φ is smooth outside the zeros of the t_i and $\sup_M |\varphi| < \infty$,*

(ii) $\sup_M(\Delta\varphi)(t_{n_1+1}\cdots t_{n_1+n_2})^{q+1/m}(|\Delta \log t_{n_1+1}\cdots t_{n_1+n_2}| + 1)^{(1-m)/m}$ *is bounded,*

(iii) *outside the zeros of the t_i, $\sum_{i,j}(g_{i\bar{j}} + \partial^2\varphi/\partial z^i\partial\bar{z}^j)dz^i \otimes d\bar{z}^j$ defines a Kähler metric.*

Furthermore, any solution of (9.1) satisfying the above three properties differs from φ by a constant.

Appendix A

In this appendix, we carry out the computation needed in Section 3. In other words, we prove formula (3.2) here.

From the commutation formula of Section 1, we have

$$\begin{aligned}\varphi_{i\bar{j}k\bar{\beta}\alpha} &= \varphi_{i\bar{j}\bar{\beta}k\alpha} + \left(\sum_p \varphi_{i\bar{p}} R^{\bar{p}}_{\bar{j}\bar{\beta}k} - \sum_p \varphi_{p\bar{j}} R^{p}_{ik\bar{\beta}}\right)_\alpha \\ &= \varphi_{i\bar{\beta}\alpha\bar{j}k} + \left(\sum_p \varphi_{i\bar{p}} R^{\bar{p}}_{\bar{\beta}\bar{j}\alpha} - \sum_p \varphi_{p\bar{\beta}} R^{p}_{i\alpha\bar{j}}\right)_k \\ &\quad + \left(\sum_p \varphi_{i\bar{p}} R^{\bar{p}}_{\bar{j}\bar{\beta}k} - \sum_p \varphi_{p\bar{j}} R^{p}_{ik\bar{\beta}}\right)_\alpha .\end{aligned} \tag{A.1}$$

Since Proposition 2.1 shows that the metric tensor $\sum_{\alpha,\beta} g'_{\alpha\bar{\beta}} dz^\alpha \otimes d\bar{z}^\beta$ is uniformly equivalent to $\sum_{\alpha,\beta} g_{\alpha\bar{\beta}} dz^\alpha \otimes d\bar{z}^\beta$, we see from (3.2) that

$$\varphi_{i\bar{j}k\bar{\beta}\alpha} \simeq \varphi_{i\bar{\beta}\alpha\bar{j}k}. \tag{A.2}$$

Similarly, one can also prove that

$$\varphi_{i\bar{j}k\bar{\beta}\alpha} \simeq \varphi_{i\bar{j}k\alpha\bar{\beta}}. \tag{A.3}$$

Let us now write (2.5) as

$$\sum_{i,j} g'^{i\bar{j}} \varphi_{i\bar{j}k\bar{l}} = F_{k\bar{l}} + \sum g'^{t\bar{j}} g'^{i\bar{n}} \varphi_{t\bar{n}\bar{l}} \varphi_{i\bar{j}k}. \tag{A.4}$$

Then differentiating, we have

$$\sum_{i,j} g'^{i\bar{j}} \varphi_{i\bar{j}k\bar{l}s} = \sum g'^{i\bar{t}} g'^{n\bar{j}} \varphi_{n\bar{t}s} \varphi_{i\bar{j}k\bar{l}} + F_{k\bar{l}s} + \left(\sum g'^{t\bar{j}} g'^{i\bar{n}} \varphi_{t\bar{n}\bar{l}} \varphi_{i\bar{j}k}\right)_s . \tag{A.5}$$

Now we compute the Laplacian of S:

$$
\begin{aligned}
\Delta' S &= \sum g'^{\alpha\bar\beta} S_{\bar\beta\alpha}\\
&= \sum g'^{\alpha\bar\beta}[-g'^{i\bar p} g'^{q\bar r} g'^{\bar j s} g'^{k\bar t} \varphi_{q\bar p\bar\beta}\varphi_{i\bar j k}\varphi_{\bar r s\bar t}\\
&\quad -g'^{i\bar r} g'^{\bar j p} g'^{\bar q s} g'^{k\bar t} \varphi_{p\bar q\bar\beta}\varphi_{i\bar j k}\varphi_{\bar r s\bar t}\\
&\quad -g'^{i\bar r} g'^{\bar j s} g'^{k\bar p} g'^{q\bar t} \varphi_{\bar p q\bar\beta}\varphi_{i\bar j k}\varphi_{\bar r s\bar t}\\
&\quad +g'^{i\bar r} g'^{\bar j s} g'^{k\bar t} \varphi_{i\bar j k\bar\beta}\varphi_{\bar r s t} + g'^{i\bar r} g'^{\bar j s} g'^{k\bar t}\varphi_{i\bar j k}\varphi_{\bar r s\bar t\bar\beta}]_\alpha\\
&\simeq \sum g'^{\alpha\bar\beta}[(-2g'^{i\bar p} g'^{q\bar r} g'^{\bar j s} g'^{k\bar t} \varphi_{q\bar p\bar\beta}\varphi_{i\bar j k}\varphi_{\bar r s\bar t}\\
&\quad -g'^{i\bar r} g'^{\bar j p} g'^{\bar q s} g'^{k\bar t} \varphi_{p\bar q\beta}\varphi_{i\bar j k}\varphi_{\bar r s\bar t})\\
&\quad +g'^{i\bar r} g'^{\bar j s} g'^{k\bar t}(\varphi_{i\bar j k\bar\beta}\varphi_{\bar r s\bar t} + \varphi_{i\bar j k}\varphi_{\bar r s\bar t\bar\beta})]_\alpha\\
&\simeq \sum g'^{\alpha\bar\beta}[2g'^{i\bar\alpha} g'^{b\bar p} g'^{q\bar r} g'^{\bar j s} g'^{k\bar t}\varphi_{\bar a b\alpha}\varphi_{q\bar p\bar\beta}\varphi_{i\bar j k}\varphi_{\bar r s t}\\
&\quad +2g'^{i\bar p} g'^{q\bar a} g'^{b\bar r} g'^{\bar j s} g'^{k\bar t}\varphi_{\bar a b\alpha}\varphi_{q\bar p\bar\beta}\varphi_{i\bar j k}\varphi_{\bar r s\bar t}\\
&\quad +2g'^{i\bar p} g'^{q\bar r} g'^{\bar j b} g'^{\bar a s} g'^{k\bar t}\varphi_{\bar a b\alpha}\varphi_{q\bar p\bar\beta}\varphi_{i\bar j k}\varphi_{\bar r s\bar t}\\
&\quad +2g'^{i\bar p} g'^{q\bar r} g'^{\bar j s} g'^{k\bar a} g'^{b\bar t}\varphi_{\bar a b\alpha}\varphi_{q\bar p\bar\beta}\varphi_{i\bar j k}\varphi_{\bar r s\bar t}\\
&\quad -2g'^{i\bar p} g'^{q\bar r} g'^{\bar j s} g'^{k\bar t}(\varphi_{q\bar p\bar\beta\alpha}\varphi_{i\bar j k}\varphi_{\bar r s\bar t}\\
&\quad +\varphi_{q\bar p\bar\beta}\varphi_{i\bar j k\alpha}\varphi_{\bar r s\bar t} + \varphi_{q\bar p\bar\beta}\varphi_{i\bar j k}\varphi_{\bar r s\bar t\alpha})\\
&\quad +g'^{i\bar a} g'^{b\bar r} g'^{\bar j p} g'^{\bar q s} g'^{k\bar t}\varphi_{\bar a b\alpha}\varphi_{p\bar q\bar\beta}\varphi_{i\bar j k}\varphi_{\bar r s\bar t}\\
&\quad +g'^{i\bar r} g'^{\bar j b} g'^{\bar a p} g'^{\bar q s} g'^{k\bar t}\varphi_{\bar a b\alpha}\varphi_{p\bar q\bar\beta}\varphi_{i\bar j k}\varphi_{\bar r s t}\\
&\quad +g'^{i\bar r} g'^{\bar j p} g'^{\bar q b} g'^{\bar a s} g'^{k\bar t}\varphi_{\bar a b\alpha}\varphi_{p\bar q\bar\beta}\varphi_{i\bar j k}\varphi_{\bar r s\bar t}\\
&\quad +g'^{i\bar r} g'^{\bar j p} g'^{\bar q s} g'^{k\bar a} g'^{b\bar t}\varphi_{\bar a b\alpha}\varphi_{p\bar q\bar\beta}\varphi_{i\bar j k}\varphi_{\bar r s\bar t}\\
&\quad -g'^{i\bar r} g'^{\bar j p} g'^{\bar q s} g'^{k\bar t}(\varphi_{p\bar q\bar\beta\alpha}\varphi_{i\bar j k}\varphi_{\bar r s\bar t} + \varphi_{p\bar q\bar\beta}\varphi_{i\bar j k\alpha}\varphi_{\bar r s\bar t}\\
&\quad -g'^{i\bar r} g'^{\bar j p} g'^{\bar q s} g'^{k\bar t}(\varphi_{p\bar q\bar\beta\alpha}\varphi_{i\bar j k}\varphi_{\bar r s\bar t} + \varphi_{p\bar q\bar\beta}\varphi_{i\bar j k\alpha}\varphi_{\bar r s\bar t} + \varphi_{p\bar q\bar\beta}\varphi_{i\bar j k}\varphi_{\bar r s t\bar\alpha})\\
&\quad -(2g'^{i\bar a} g'^{b\bar r} g'^{\bar j s} g'^{k\bar t}\varphi_{\bar\alpha b\alpha} + g'^{i\bar r} g'^{\bar j b} g'^{\bar a s} g'^{k\bar t}\varphi_{\bar a b\alpha})(\varphi_{i\bar j k\bar\beta}\varphi_{\bar r s t} + \varphi_{i\bar j k}\varphi_{\bar r s t\bar\beta})\\
&\quad +g'^{i\bar r} g'^{\bar j s} g'^{k\bar t}(\varphi_{i\bar j k\bar\beta\alpha}\varphi_{\bar r s\bar t} + \varphi_{i\bar j k\bar\beta}\varphi_{\bar r s\bar t\alpha} + \varphi_{i\bar j k\alpha}\varphi_{\bar r s\bar t\bar\beta} + \varphi_{i\bar j k}\varphi_{r s\bar t\bar\beta\alpha})].
\end{aligned}
\tag{A.6}
$$

By (A.1),

$$
\begin{aligned}
&\sum g'^{\alpha\bar\beta}\varphi_{i\bar j k\bar\beta\alpha}\\
&\simeq \sum g'^{\alpha\bar\beta}\varphi_{\alpha\bar\beta i\bar j k}\\
&= \sum g'^{\alpha\bar p} g'^{q\bar\beta}\varphi_{q\bar p k}\varphi_{\alpha\bar\beta i\bar j} + F_{i\bar j k} + \left(\sum g'^{p\bar\beta} g'^{\alpha\bar q}\varphi_{p\bar q\bar j}\varphi_{\alpha\bar\beta i}\right)_k\\
&= \sum g'^{\alpha\bar p} g'^{q\bar\beta}\varphi_{q\bar p k}\varphi_{\alpha\bar\beta i\bar j} + F_{i\bar j k} - \sum g'^{p\bar a} g'^{b\bar\beta} g'^{\alpha\bar q}\varphi_{\bar a b k}\varphi_{p\bar q\bar j}\varphi_{\alpha\bar\beta i}\\
&\quad -\sum g'^{p\bar\beta} g'^{\alpha\bar a} g'^{b\bar q}\varphi_{\bar a b k}\varphi_{p\bar q\bar j}\varphi_{\alpha\bar\beta i} + \sum g'^{p\bar\beta} g'^{\alpha\bar q}\varphi_{p\bar q\bar j k}\varphi_{\alpha\bar\beta i}\\
&\quad +\sum g'^{p\bar\beta} g'^{\alpha\bar q}\varphi_{p\bar q\bar j}\varphi_{\alpha\bar\beta i k}.
\end{aligned}
\tag{A.7}
$$

Inserting (A.4) and (A.7) into (A.6), we find

$$
\begin{aligned}
\Delta' S \simeq \sum & g'^{\alpha\bar\beta}[2g'^{i\bar a}g'^{b\bar p}g'^{q\bar r}g'^{\bar j s}g'^{k\bar t}\varphi_{\bar a b\alpha}\varphi_{q\bar p\bar\beta}\varphi_{i\bar j k}\varphi_{\bar r s\bar t}\\
&+2g'^{i\bar p}g'^{q\bar a}g'^{b\bar r}g'^{\bar j s}g'^{k\bar t}\varphi_{\bar a b\alpha}\varphi_{q\bar p\bar\beta}\varphi_{i\bar j k}\varphi_{\bar r s\bar t}\\
&+2g'^{i\bar p}g'^{q\bar r}g'^{\bar j b}g'^{\bar a s}g'^{k\bar t}\varphi_{\bar a b\alpha}\varphi_{q\bar p\beta}\varphi_{ij\bar k}\varphi_{\bar r s\bar t}\\
&+2g'^{i\bar p}g'^{q\bar r}g'^{\bar j s}g'^{k\bar a}g'^{b\bar t}\varphi_{\bar a b\alpha}\varphi_{q\bar p\bar\beta}\varphi_{i\bar j k}\varphi_{\bar r s\bar t}]\\
&-2\sum g'^{i\bar p}g'^{q\bar r}g'^{\bar j s}g'^{k\bar t}\\
&\times[F_{q\bar p}\varphi_{i\bar j k}\varphi_{\bar r s\bar t}+g'^{t\bar\beta}g'^{\alpha\bar n}\varphi_{t\bar n\bar p}\varphi_{\alpha\bar\beta q}\varphi_{i\bar j k}\varphi_{\bar r s\bar t}\\
&+g'^{\alpha\bar\beta}\varphi_{q\bar p\bar\beta}\varphi_{i\bar j k\alpha}\varphi_{\bar r st}+g'^{\alpha\bar\beta}\varphi_{q\bar p\bar\beta}\varphi_{i\bar j k}\varphi_{\bar r s\bar t\alpha}]\\
&+\sum g'^{\alpha\bar\beta}g'^{i\bar a}g'^{b\bar r}g'^{\bar j p}g'^{\bar q s}g'^{k\bar t}\varphi_{\bar a b\alpha}\varphi_{p\bar q\bar\beta}\varphi_{i\bar j k}\varphi_{\bar r s\bar t}\\
&+\sum g'^{\alpha\bar\beta}g'^{i\bar r}g'^{\bar j b}g'^{\bar a p}g'^{\bar q s}g'^{k\bar t}\varphi_{\bar a b\alpha}\varphi_{p\bar q\bar\beta}\varphi_{i\bar j k}\varphi_{\bar r s\bar t}\\
&+\sum g'^{\alpha\bar\beta}g'^{i\bar r}g'^{\bar j p}g'^{\bar q b}g'^{\bar a s}g'^{k\bar t}\varphi_{\bar a b\alpha}\varphi_{p\bar q\bar\beta}\varphi_{i\bar j k}\varphi_{\bar r s\bar t}\\
&+\sum g'^{\alpha\bar\beta}g'^{i\bar r}g'^{\bar j p}g'^{\bar q s}g'^{k\bar a}g'^{b\bar t}\varphi_{\bar a b\alpha}\varphi_{p\bar q\bar\beta}\varphi_{i\bar j k}\varphi_{\bar r s\bar t}\\
&-\sum g'^{i\bar r}g'^{\bar j p}g'^{\bar q s}g'^{k\bar t}(F_{p\bar q}\varphi_{i\bar j k}\varphi_{\bar r s\bar t}+g'^{t\bar\beta}g'^{\alpha\bar n}\varphi_{t\bar n\bar q}\varphi_{\alpha\bar\beta p}\varphi_{i\bar j k}\varphi r_{st}\\
&+g'^{\alpha\bar\beta}\varphi_{p\bar q\bar\beta}\varphi_{i\bar j k\alpha}\varphi_{\bar r s\bar t}+g'^{\alpha\bar\beta}\varphi_{p\bar q\bar\beta}\varphi_{i\bar j k}\varphi_{\bar r s\bar t\alpha})\\
&-\sum g'^{\alpha\bar\beta}(2g'^{i\bar a}g'^{b\bar r}g'^{\bar j s}g'^{k\bar t}\varphi_{\bar a b\alpha}\\
&+g'^{i\bar r}g'^{\bar j b}g'^{\bar a s}g'^{k\bar t}\varphi_{\bar a b\alpha})(\varphi_{i\bar j k\bar\beta}\varphi_{\bar r s\bar t}+\varphi_{i\bar j k}\varphi_{\bar r s\bar t\bar\beta})\\
&+2\mathrm{Re}\sum g'^{i\bar r}g'^{\bar j s}g'^{k\bar t}\varphi_{\bar r s\bar t}(g'^{\alpha\bar p}g'^{q\bar\beta}\varphi_{q\bar p k}\varphi_{\alpha\bar\beta i\bar j}+F_{i\bar j k}\\
&-g'^{p\bar a}g'^{b\bar\beta}g'^{\alpha\bar q}\varphi_{\bar a b k}\varphi_{p\bar q\bar j}\varphi_{\alpha\bar\beta i}\\
&-g'^{p\bar\beta}g'^{\alpha\bar a}g'^{b\bar q}\varphi_{\bar a b k}\varphi_{p\bar q\bar j}\varphi_{\alpha\bar\beta i}\\
&+g'^{p\bar\beta}g'^{\alpha\bar q}\varphi_{p\bar q\bar j k}\varphi_{\alpha\bar\beta i}+g'^{p\bar\beta}g'^{\alpha\bar q}\varphi_{p\bar q\bar j}\varphi_{\alpha\bar\beta ik})\\
&+\sum g'^{\alpha\bar\beta}g'^{i\bar r}g'^{\bar j s}g'^{k\bar t}(\varphi_{i\bar j k\bar\beta}\varphi_{\bar r s\bar t\alpha}+\varphi_{i\bar j k\alpha}\varphi_{\bar r s\bar t\bar\beta}).
\end{aligned}
\tag{A.8}
$$

Diagonalizing and simplifying, we obtain

$$
\begin{aligned}
\Delta' S \cong & \sum(1+\varphi_{\alpha\bar\alpha})^{-1}(1+\varphi_{i\bar i})^{-1}(1+\varphi_{j\bar j})^{-1}\\
&\times(1+\varphi_{k\bar k})^{-1}(1+\varphi_{p\bar p})^{-1}(1+\varphi_{q\bar q})^{-1}\\
&\times(2\varphi_{\bar i p\alpha}\varphi_{q\bar p\bar\alpha}\varphi_{i\bar j k}\varphi_{\bar q j\bar k}+2\varphi_{\bar k p\alpha}\varphi_{q\bar i\bar\alpha}\varphi_{i\bar j k}\varphi_{\bar q j\bar p}+\varphi_{\bar p q\alpha}\varphi_{j\bar q\bar\alpha}\varphi_{i\bar j k}\varphi_{\bar i p\bar k})\\
&-2\mathrm{Re}\sum(1+\varphi_{\alpha\bar\alpha})^{-1}(1+\varphi_{i\bar i})^{-1}(1+\varphi_{j\bar j})^{-1}(1+\varphi_{k\bar k})^{-1}\\
&\times(1+\varphi_{p\bar p})^{-1}(\varphi_{p\bar i\bar\alpha}\varphi_{i\bar j k\alpha}\varphi_{\bar p j\bar k}+\varphi_{j\bar p\bar\alpha}\varphi_{i\bar j k}\varphi_{\bar i p\bar k\alpha}+\varphi_{\bar i p\alpha}\varphi_{i\bar j k}\varphi_{\bar p j\bar k\bar\alpha})\\
&+\sum(1+\varphi_{\alpha\bar\alpha})^{-1}(1+\varphi_{i\bar i})^{-1}(1+\varphi_{j\bar j})^{-1}(1+\varphi_{k\bar k})^{-1}(|\varphi_{i\bar j k\alpha}|^2+|\varphi_{i\bar j k\bar\alpha}|^2)\\
= & \sum(1+\varphi_{i\bar i})^{-1}(1+\varphi_{j\bar j})^{-1}(1+\varphi_{k\bar k})^{-1}(1+\varphi_{\alpha\bar\alpha})^{-1}
\end{aligned}
$$

$$\times\Big\{\Big|\varphi_{\bar{i}j\bar{k}\alpha} - \sum_p \varphi_{\bar{i}p\bar{k}}\varphi_{\bar{p}j\alpha}(1+\varphi_{p\bar{p}})^{-1}\Big|^2$$

$$+\Big|\varphi_{i\bar{j}k\alpha} - \sum_p (\varphi_{\bar{p}i\alpha}\varphi_{p\bar{j}k} + \varphi_{\bar{p}ik}\varphi_{p\bar{j}\alpha})(1+\varphi_{p\bar{p}})^{-1}\Big|^2\Big\}. \tag{A.9}$$

Appendix B

In this appendix, we carry out the proof of (8.17) promised in Section 8. Integrating by parts in (8.16), we have

$$\begin{aligned}
&-(p-k_2)\int_M (|s_2|^2+\varepsilon)^{p-k_2-1}(|s_1|^2+\varepsilon)^{k_1}\exp\{-C\varphi_\varepsilon\}\nabla|s_2|^2\cdot\nabla\varphi_\varepsilon\\
&-k_1\int_M (|s_2|^2+\varepsilon)^{p-k_2}(|s_1|^2+\varepsilon)^{k_1-1}\exp\{-C\varphi_\varepsilon\}\nabla|s_1|^2\cdot\nabla\varphi_\varepsilon\\
&+C\int_M (|s_2|^2+\varepsilon)^{p-k_2}(|s_1|^2+\varepsilon)^{k_1}\exp\{-C\varphi_\varepsilon\}|\nabla\varphi_\varepsilon|^2\\
&+(p-k_2)C_7C^{-1}\int_M (|s_2|^2+\varepsilon)^{p-k_2-2}(|s_1|^2+\varepsilon)^{k_1}\exp\{-C\varphi_\varepsilon\}|\nabla|s_2|^2|^2\\
&+k_1C_7C^{-1}\int_M (|s_2|^2+\varepsilon)^{p-k_2-1}(|s_1|^2+\varepsilon)^{k_1-1}\exp\{-C\varphi_\varepsilon\}\nabla|s_1|^2\cdot\nabla|s_2|^2\\
&-C_2\int_M (|s_2|^2+\varepsilon)^{p-k_2-1}(|s_1|^2+\varepsilon)^{k_1}\exp\{-C\varphi_\varepsilon\}\nabla\varphi_\varepsilon\cdot\nabla|s_2|^2\\
&\leqq C_8\int_M (|s_2|^2+\varepsilon)^{p-k_2}(|s_1|^2+\varepsilon)^{k_1}\exp\{-C\varphi_\varepsilon\}\\
&\quad+C_8\int_M (|s_2|^2+\varepsilon)^{p-2k_2}(|s_1|^2+\varepsilon)^{2k_1}\exp\{-C\varphi_\varepsilon\}.
\end{aligned} \tag{B.1}$$

Hence, estimating the term

$$\int_M (|s_2|^2+\varepsilon)^{p-k_2-1}(|s_1|^2+\varepsilon)^{k_1}\exp\{-C\varphi_\varepsilon\}\nabla|s_2|^2\cdot\nabla\varphi_\varepsilon$$

from (B.1),

$$\begin{aligned}
&\int_M \Big|\nabla\Big[(|s_2|^2+\varepsilon)^{(p-k_2)/2}(|s_1|^2+\varepsilon)^{k_1/2}\exp\Big\{-\frac{1}{2}C\varphi_\varepsilon\Big\}\Big]\Big|^2\\
&=\frac{1}{4}(p-k_2)^2\int_M (|s_2|^2+\varepsilon)^{p-k_2-2}(|s_1|^2+\varepsilon)^{k_1}\exp\{-C\varphi_\varepsilon\}|\nabla|s_2|^2|^2\\
&-\frac{1}{4}k_1^2\int_M (|s_2|^2+\varepsilon)^{p-k_2}(|s_1|^2+\varepsilon)^{k_1-2}\exp\{-C\varphi_\varepsilon\}|\nabla|s_1|^2|^2\\
&+\frac{1}{4}C^2\int_M (|s_2|^2+\varepsilon)^{p-k_2}(|s_1|^2+\varepsilon)^{k_1}\exp\{-C\varphi_\varepsilon\}|\nabla\varphi_\varepsilon|^2\\
&+\frac{1}{2}k_1(p-k_2)\int_M (|s_2|^2+\varepsilon)^{p-k_2-1}(|s_1|^2+\varepsilon)^{k_1-1}\exp\{-C\varphi_\varepsilon\}\nabla|s_2|^2\cdot\nabla|s_1|^2
\end{aligned}$$

$$
\begin{aligned}
&-\frac{1}{2}C(p-k_2)\int_M (|s_2|^2+\varepsilon)^{p-k_2-1}(|s_1|^2+\varepsilon)^{k_1}\exp\{-C\varphi_\varepsilon\}\nabla\varphi_\varepsilon\cdot\nabla|s_2|^2\\
&-\frac{1}{2}k_1C\int_M (|s_2|^2+\varepsilon)^{p-k_2}(|s_1|^2+\varepsilon)^{k_1-1}\exp\{-C\varphi_\varepsilon\}\nabla\varphi_\varepsilon\cdot\nabla|s_1|^2\\
\leqq{}& \frac{1}{4}(p-k_2)^2\int_M (|s_2|^2+\varepsilon)^{p-k_2-2}(|s_1|^2+\varepsilon)^{k_1}\exp\{-C\varphi_\varepsilon\}|\nabla|s_2|^2|^2\\
&+\frac{1}{4}k_1^2\int_M (|s_2|^2+\varepsilon)^{p-k_2}(|s_1|^2+\varepsilon)^{k_1-2}\exp\{-C\varphi_\varepsilon\}|\nabla|s_1|^2|^2\\
&+\frac{1}{4}C^2\int_M (|s_2|^2+\varepsilon)^{p-k_2}(|s_1|^2+\varepsilon)^{k_1}\exp\{-C\varepsilon_\varepsilon\}|\nabla\varphi_\varepsilon|^2\\
&+\frac{1}{2}k_1(p-k_2)\int_M (|s_2|^2+\varepsilon)^{p-k_2-1}(|s_1|^2+\varepsilon)^{k_1-1}\exp\{-C\varphi_\varepsilon\}\nabla|s_2|^2\cdot\nabla|s_1|^2\\
&+\frac{1}{2}C(p-k_2)\Big[k_1(p-k_2+C_7)^{-1}\int_M (|s_2|^2+\varepsilon)^{p-k_2}(|s_1|^2+\varepsilon)^{k_1-1}\\
&\times\exp\{-C\varphi_\varepsilon\}\nabla|s_1|^2\cdot\nabla\varphi_\varepsilon\\
&-C(p-k_2+C_7)^{-1}\int_M (|s_2|^2+\varepsilon)^{p-k_2}(|s_1|^2+\varepsilon)^{k_1}\\
&\times\exp\{-C\varphi_\varepsilon\}|\nabla\varphi_\varepsilon|^2-(p-k_2)C_7C^{-1}(p-k_2+C_7)^{-1}\int_M (|s_2|^2+\varepsilon)^{p-k_2-2}\\
&\times(|s_1|^2+\varepsilon)^{k_1}\exp\{-C\varphi_\varepsilon\}|\nabla|s_2|^2|^2-k_1C_7C^{-1}(p-k_2+C_7)^{-1}\\
&\times\int_M (|s_2|^2+\varepsilon)^{p-k_2-1}(|s_1|^2+\varepsilon)^{k_1-1}\exp\{-C\varphi_\varepsilon\}\nabla|s_1|^2\cdot\nabla|s_2|^2\\
&+C_8(p-k_2+C_7)^{-1}\int_M (|s_2|^2+\varepsilon)^{p-k_2}(|s_1|^2+\varepsilon)^{k_1}\exp\{-C\varphi_\varepsilon\}\\
&+C_8(p-k_2+C_7)^{-1}\int_M (|s_2|^2+\varepsilon)^{p-2k_2}(|s_1|^2+\varepsilon)^{2k_1}\exp\{-C\varphi_\varepsilon\}\Big]\\
&-\frac{1}{2}k_1C\int_M (|s_2|^2+\varepsilon)^{p-k_2}(|s_1|^2+\varepsilon)^{k_1-1}\exp\{-C\varphi_\varepsilon\}\nabla\varphi_\varepsilon\cdot\nabla|s_1|^2. \qquad \text{(B.2)}
\end{aligned}
$$

Since we have assumed that $C_7(p-k_2+C_7)^{-1}=\frac{1}{2}$, we know that $(p-k_2)(p-k_2+C_7)^{-1}=\frac{1}{2}$ and (B.2) shows that

$$
\begin{aligned}
&\int_M \nabla|(|s_2|^2+\varepsilon)|^{(p-k_2)/2}(|s_1|^2+\varepsilon)^{k_1-2}\exp\left\{-\frac{1}{2}C\varphi_\varepsilon\right\}\Big]\Big|^2\\
\leqq{}& \frac{1}{4}k_1^2\int_M (|s_2|^2+\varepsilon)^{p-k_2}(|s_1|^2)^{k_1-2}\exp\{-C\varphi_\varepsilon\}|\nabla|s_1|^2|^2\\
&+\frac{1}{4}k_1(p-k_2)\int_M (|s_2|^2+\varepsilon)^{p-k_2-1}(|s_1|^2+\varepsilon)^{k_1-1}\exp\{-C\varphi_\varepsilon\}\nabla|s_2|^2\cdot\nabla|s_1|^2\\
&-\frac{1}{4}k_1C\int_M (|s_2|^2+\varepsilon)^{p-k_2}(|s_1|^2+\varepsilon)^{k_1-1}\exp\{-C\varphi_\varepsilon\}\nabla\varphi_\varepsilon\cdot\nabla|s_1|^2\\
&+\frac{1}{4}CC_8\int_M (|s_2|^2+\varepsilon)^{p-k_2}(|s_1|^2+\varepsilon)^{k_1}\exp\{-C\varphi_\varepsilon\}
\end{aligned}
$$

$$+\frac{1}{4}CC_8\int_M(|s_2|^2+\varepsilon)^{p-2k_2}(|s_1|^2+\varepsilon)^{2k_1}\exp\{-C\varphi_\varepsilon\} \tag{B.3}$$

To estimate the right-hand side of (B.3), we use (5.7) again to find a positive constant C_9 independent of ε such that

$$\int_M(|s_2|^2+\varepsilon)^{p-k_2}(|s_1|^2+\varepsilon)^{k_1}\exp\{-C\varphi_\varepsilon\}\Delta\log(|s_1|^2+\varepsilon)$$
$$\geqq -C_9\int_M(|s_2|^2+\varepsilon)^{p-k_2}(|s_1|^2+\varepsilon)^{k_1}\exp\{-C\varphi_\varepsilon\}. \tag{B.4}$$

Integrating by parts in (B.4), we have

$$(p-k_2)\int_M(|s_2|^2+\varepsilon)^{p-k_2-1}(|s_1|^2+\varepsilon)^{k_1-1}\exp\{-C\varphi_\varepsilon\}\nabla|s_1|^2\cdot\nabla|s_2|^2$$
$$+k_1\int_M(|s_2|^2+\varepsilon)^{p-k_2}(|s_1|^2+\varepsilon)^{k_1-2}\exp\{-C\varphi_\varepsilon\}|\nabla|s_1|^2|^2$$
$$-C\int_M(|s_2|^2+\varepsilon)^{p-k_2}(|s_1|^2+\varepsilon)^{k_1-1}\exp\{-C\varphi_\varepsilon\}\nabla\varphi_\varepsilon\cdot\nabla|s_1|^2$$
$$\leqq C_9\int_M(|s_2|^2+\varepsilon)^{p-k_2}(|s_1|^2+\varepsilon)^{k_1}\exp\{-C\varphi_\varepsilon\}. \tag{B.5}$$

Multiplying (B.5) by $\frac{1}{4}k_1$ and sub stituting we obtain the resulting inequality into (B.3), we obtain

$$\int_M\left|\nabla\left[(|s_2|^2+\varepsilon)^{(p-k_2)/2}(|s_1|^2+\varepsilon)^{k_1/2}\exp\left\{-\frac{1}{2}C\varphi_\varepsilon\right\}\right]\right|^2$$
$$\leqq\left(\frac{1}{4}k_1C_9+\frac{1}{4}CC_8\right)\int_M(|s_2|^2+\varepsilon)^{p-k_2}(|s_1|^2+\varepsilon)^{k_1}\exp\{-C\varphi_\varepsilon\}$$
$$+\frac{1}{4}CC_8\int_M(|s_2|^2+\varepsilon)^{p-2k_2}(|s_1|^2+\varepsilon)^{2k_1}\exp\{-C\varphi_\varepsilon\}. \tag{B.6}$$

Since we assume $\int_M|s_2|^{-2mk_2}<\infty$, we can find a constant C_{10} independent of ε such that

$$\int_M(|s_2|^2+\varepsilon)^{-mk_2}\leqq C_{10}. \tag{B.7}$$

Then by Hölder's inequality, we have

$$\int_M(|s_2|^2+\varepsilon)^{p-2k_2}(|s_1|^2+\varepsilon)^{2k_1}\exp\{-C\varphi_\varepsilon\}$$
$$\leqq\left(\int_M(|s_2|^2+\varepsilon)^{-mk_2}\right)^{1/m}$$
$$\times\left(\int_M[(|s_2|^2+\varepsilon)^{p-k_2}(|s_1|^2+\varepsilon)^{2k_1}\exp\{-C\varphi_\varepsilon\}]^{m/(m-1)}\right)^{(m-1)/m} \tag{B.8}$$

Inserting (B.7) and (B.8) into (B.6), we see that, for some positive constant C_{11} independent of ε and C, we have

$$\int_M \left|\nabla\left[(|s_2|^2+\varepsilon)^{(p-k_2)/2}(|s_1|^2+\varepsilon)^{k_1/2}\exp\left\{-\frac{1}{2}-C\varphi_\varepsilon\right\}\right]\right|^2$$
$$\leqq CC_{11}\left(\int_M[(|s_2|^2+\varepsilon)^{p-k_2}(|s_1|^2+\varepsilon)^{k_1}\exp\{-C\varphi_\varepsilon\}]^{m/(m-1)}\right)^{(m-1)/m} \quad \text{(B.9)}$$

Added in proof (February 21, 1978): Some developments on the problem after the completion of this paper may be recorded here. In Section 2, I offered two proofs of Calabi's conjecture, one for complex dimension two and one for the general case. After I finished my proof, I was kindly invited by Professor E. Calabi to Philadelphia to present a talk in 1976 in which Professor J. Kazdan was present. After a few months, J. Kazdan was able to simplify my two-dimensional proof. This simplification will also appear in this issue. Then Professor J. Kazdan was invited by Professor T. Aubin to present in Paris my two-dimensional proof and his simplification. At the end of 1977, Professor Aubin and Professor J. P. Bourguignon, who was kindly giving a talk on my paper in the Bourbaki seminar, independently generalized my two-dimensional proof to higher dimensions. I wish to thank these colleagues for their interest in my work and for giving another complete proof of the general case.

References

[1] Aubin, T., *Métriques Riemanniennes et courbure*, J. Diff. Geom. 4, 1970, pp. 383-424.

[2] Aubin, T., *Equations du type Monge-Ampère sur les variétés Kähleriennes compactes*, C.R. Acad. Sc. Paris, 283, 1976, pp. 119-121.

[3] Calabi, E., *The space of Kähler metrics*, Proc. Internat. Congress Math. Amsterdam, 1954, Vol. 2, pp. 206-207.

[4] Calabi, E., *On Kähler manifolds with vanishing canonical class*, Algebraic Geometry and Topology, A symposium in honor of S. Lefschatz, Princeton Univ. Press, Princeton, 1955, pp. 78-89.

[5] Calabi, E., *Improper affine hyperspheres and generalization of a theorem of K. Jörgens*, Mich. Math. J. 5, 1958, pp. 105-126.

[6] Chern, S. S., *Characteristic classes of Hermitian manifolds*, Ann. of Math. 47, 1946, pp. 85-121.

[7] Chern, S. S., *Complex Manifold Without Potential Theory*, Princeton, N.J., van Nostrand, 1967.

[8] Cheng, S. Y. and Yau, S. T., *On the regularity of the Monge-Ampère equation* $\det(\partial^2 u/\partial x_i \partial x_j) = F(x, u)$, Comm. Pure Appl. Math., Vol. 30, 1977, pp. 41-68.

[9] Cheng, S . Y. and Yau, S. T., *On the regularity of the solution of the ndimensional Minkowski problem*, Comm. Pure Appl. Math., Vol. 29, 1976, pp. 495-516.

[10] de Rham,G., *Variétés Differentiables; Formes, Courants, Formes Harmoniques*, Paris, Hermann, 1960.

[11] Federer, H., *Geometric Measure Theory*, Springer-Verlag, 1969.

[12] Moser, J., *A new proof of De Giorgi's theorem concerning the regularity problem for elliptic differential equations*, Comm. Pure Appl. Math., Vol. 13, 1963, pp. 457-468.

[13] Morrey, C. B., *Multiple Integrals in the Calculus of Variation*, Springer-Verlag, 1966.

[14] Schwartz, J., *Nonlinear Functional Analysis*, N.Y., Gordon and Breach, 1968.

[15] Yau, S. T., *On Calabi's conjecture and some new results in algebraic geometry*, Nat. Acad. Sci. U.S.A., 74, 1977, pp. 1798-1799.

Editor's Comments

This paper is an important example showing that differential geometry methods can be used effectively to solve a major conjecture in complex geometry and algebraic geometry, the conjecture of Frankel, which states that every compact Kähler manifold M of positive bisectional curvature is biholomorphic to the complex projective space. It follows the tradition set up by Riemann.

Complex Geometry from Riemann to Kähler-Einstein and Calabi-Yau
ALM 38, pp. 349–364

Compact Kähler Manifolds of Positive Bisectional Curvature*

Yum-Tong Siu and Shing-Tung Yau

In this paper we prove the following conjecture of Frankel [7].

Main Theorem. *Every compact Kähler manifold of positive bisectional curvature is biholomorphic to the complex projective space.*

The case of dimension two was proved by Andreotti-Frankel [7] and the case of dimension three by Mabuchi [18] using the result of Kobayashi-Ochiai [12].

Our method of proof uses harmonic maps and the characterization of the complex projective space obtained by Kobayashi-Ochiai [15]. According to the result of Kobayashi-Ochiai the complex projective space of dimension n is characterized by the fact that its first Chern class equals $\lambda c_1(F)$ for some $\lambda \geqq n+1$ and some positive holomorphic line bundle F over it. Since by the result of Bishop-Goldberg [2] the second Betti number of a compact Kähler manifold M of positive bisectional curvature is 1, for the Main Theorem it suffices to show that $c_1(M)$ is λ times a generator of $H^2(M,\mathbb{Z})$ for some $\lambda \geqq 1+\dim M$. This can be done by proving that a generator of the free part of $H_2(M,\mathbb{Z})$ can be represented by a rational curve, because the tangent bundle of M restricted to the rational curve splits into a direct sum of holomorphic line bundles over the rational curve according to the result of Grothendieck [11]. The existence of the rational curve is obtained in the following way. According to the result of Sacks-Uhlenbeck [22] and its improved formulation by Meeks-Yau [19], the infimum of the energies of maps from S^2 to M representing the generator of $\pi_2(M)$ can be achieved by a sum of stable harmonic maps f_i from S^2 to $M (1 \leqq i \leqq m)$. The key step in our proof is to show that each f_i is either holomorphic or conjugate holomorphic. The known methods of proving the complex-analyticity of a harmonic map use the formula for the Laplacian of the energy function [23, 25, 26] or a variation of it [24]. Here we use instead the second variation formula of the energy function. In this second

* Research partially supported by NSF grants. Department of Mathematics, Stanford University, Stanford, CA, 94305, USA. Siu, Yum Tong; Yau, Shing Tung, Compact Kähler manifolds of positive bisectional curvature. Invent. Math. 59 (1980), no. 2, 189–204. Received March 29, 1979.

variation formula a 2-parameter variation has to be used to imitate the situation of holomorphic deformation. After this key step we use holomorphic deformations of rational curves in M to show that $m = 1$. This is done by proving that in case $m > 1$ we can holomorphically deform the images of some holomorphic f_i and some conjugate holomorphic f_j so that they are tangential to each other at some point. By removing a disc centered at the point of contact from each and joining the two disc boundaries by a suitable surface, we obtain a map from S^2 to M with energy smaller than the minimum energy. Thus $m = 1$ and the image of f_1 is a rational curve representing a generator of the free part of $H_2(M, \mathbb{Z})$.

In our proof the existence of a rational curve plays a very important role. This important role of a rational curve has already been observed earlier by Hartshorne and Kobayashi-Ochiai [13].

We learn that very recently Mori [20] has also given a proof of Frankel's conjecture by using the method of algebraic geometry of characteristic $p > 0$. Mori's result is stronger than ours. He needs only the assumption that the tangent bundle of the manifold is ample, whereas we have to assume that the manifold has positive holomorphic bisectional curvature. Though our result is weaker, our proof has the advantage that it uses only methods of Kähler geometry to answer a question in Kähler geometry. On the other hand, even in the case of complex manifolds Mori's proof involves the use of methods of algebraic geometry of characteristic $p > 0$.

Contents

1 Second variation formula

Suppose M, N are compact Kähler manifolds whose Kähler metrics are respectively

$$ds_M^2 = 2\text{Re} \sum_{\alpha,\beta=1}^{m} h_{\alpha\bar{\beta}} dz^\alpha \overline{dz^\beta}$$

and

$$ds_N^2 = 2\text{Re} \sum_{i,j=1}^{n} g_{i\bar{j}} dw^i \overline{dw^j}.$$

Let $f(t) : N \to M, t \in \mathbb{C}, |t| < \varepsilon$, be a family of smooth maps parametrized by an open disc in $\mathbb{C}$. The pointwise $\bar{\partial}$-energy of f is defined by

$$|\bar{\partial} f|^2 = g^{\bar{i}j} f^{\alpha}_{\bar{i}} \overline{f^{\beta}_{\bar{j}}} h_{\alpha\bar{\beta}},$$

where the summation convention is used and

$$f^{\alpha}_{\bar{i}} = \frac{\partial f^{\alpha}}{\partial \overline{w^{i}}}.$$

In order to compute $\frac{\partial^2}{\partial t \partial \bar{t}} |\bar{\partial} f|^2$ at a point P of N, we choose local holomorphic coordinate systems at P and $Q = f(P)$ such that

$$dg_{i\bar{j}} = 0 \text{ at } P$$

and

$$\begin{aligned} dh_{\alpha\bar{\beta}} &= 0, \\ \partial_{\gamma}\partial_{\delta} h_{\alpha\bar{\beta}} &= 0 \text{ at } Q. \end{aligned}$$

This is possible because the metrics of M and N are both Kähler. Direct computation yields

$$\begin{aligned} \frac{\partial^2}{\partial t \partial \bar{t}} |\bar{\partial} f|^2 = {} & 2\mathrm{Re}\, g^{\bar{i}j} \left(\frac{\partial^2}{\partial t \partial \bar{t}} f^{\alpha}_{\bar{i}} \right) \overline{f^{\beta}_{\bar{j}}} h_{\alpha\bar{\beta}} + g^{\bar{i}j} \left(\frac{\partial}{\partial t} f^{\alpha}_{\bar{i}} \right) \overline{\left(\frac{\partial}{\partial t} f^{\beta}_{\bar{j}} \right)} h_{\alpha\bar{\beta}} \\ & + g^{\bar{i}j} \left(\frac{\partial}{\partial \bar{t}} f^{\alpha}_{\bar{i}} \right) \overline{\left(\frac{\partial}{\partial \bar{t}} f^{\beta}_{\bar{j}} \right)} h_{\alpha\bar{\beta}} + g^{\bar{i}j} f^{\alpha}_{\bar{i}} \overline{f^{\beta}_{\bar{j}}} (\partial_{\mu}\partial_{\bar{\nu}} h_{\alpha\bar{\beta}}) \frac{\partial f^{\mu}}{\partial \bar{t}} \overline{\frac{\partial f^{\nu}}{\partial \bar{t}}} \\ & + g^{\bar{i}j} f^{\alpha}_{\bar{i}} \overline{f^{\beta}_{\bar{j}}} (\partial_{\mu}\partial_{\bar{\nu}} h_{\alpha\bar{\beta}}) \frac{\partial f^{\mu}}{\partial t} \overline{\frac{\partial f^{\nu}}{\partial t}}. \end{aligned}$$

Consider the vector field ξ on N defined by

$$\xi^{\bar{i}} = g^{\bar{i}j} \left(\frac{D}{\partial t} \frac{\partial}{\partial \bar{t}} f^{\alpha} \right) \overline{f^{\beta}_{\bar{j}}} h_{\alpha\bar{\beta}},$$

where $\frac{D}{\partial t}$ is the covariant differentiation with respect to the connection of the tangent bundle of M, i.e.

$$\frac{D}{\partial t} \frac{\partial}{\partial \bar{t}} f^{\alpha} = \frac{\partial^2}{\partial t \partial \bar{t}} f^{\alpha} + {}^{M}\Gamma^{\alpha}_{\beta\gamma} \frac{\partial f^{\beta}}{\partial t} \frac{\partial f^{\gamma}}{\partial \bar{t}},$$

${}^{M}\Gamma^{\alpha}_{\beta\gamma}$ being the Christoffel symbol of M. The divergence of ξ at P is

$$\begin{aligned} \nabla_{\bar{i}} \xi^{\bar{i}} = {} & g^{\bar{i}j} \left(\frac{\partial^2}{\partial t \partial \bar{t}} f^{\alpha}_{\bar{i}} \right) \overline{f^{\beta}_{\bar{j}}} h_{\alpha\bar{\beta}} + g^{\bar{i}j} \left(\frac{\partial^2}{\partial t \partial \bar{t}} f^{\alpha} \right) \overline{\partial_i f^{\beta}_{\bar{j}}} h_{\alpha\bar{\beta}} \\ & + g^{\bar{i}j} (\partial_{\mu}\partial_{\bar{\nu}} h_{\alpha\bar{\beta}}) \overline{f^{\nu}_{i}} \overline{f^{\beta}_{\bar{j}}} \frac{\partial f^{\mu}}{\partial t} \frac{\partial f^{\alpha}}{\partial \bar{t}}. \end{aligned}$$

Now assume that f is harmonic at 0. Then at P and $t=0$

$$g^{\bar{i}j}\overline{\partial_i f_{\bar{j}}^{\beta}}h_{\alpha\bar{\beta}}=0$$

and

$$\begin{aligned}\frac{\partial^2}{\partial t\partial\bar{t}}|\bar{\partial}f|^2-2\mathrm{Re}\,\mathrm{div}\,\xi &= g^{\bar{i}j}\left(\frac{D}{\partial t}f_{\bar{i}}^{\alpha}\right)\left(\overline{\frac{D}{\partial t}f_{\bar{j}}^{\beta}}\right)h_{\alpha\bar{\beta}}\\ &+g^{\bar{i}j}\left(\frac{D}{\partial\bar{t}}f_{\bar{i}}^{\alpha}\right)\left(\overline{\frac{D}{\partial\bar{t}}f_{\bar{j}}^{\beta}}\right)h_{\alpha\bar{\beta}}+g^{\bar{i}j}f_{\bar{i}}^{\alpha}\overline{f_{\bar{j}}^{\beta}}R_{\mu\bar{\nu}\alpha\bar{\beta}}\frac{\partial f^{\mu}}{\partial\bar{t}}\overline{\frac{\partial f^{\nu}}{\partial\bar{t}}}\\ &+g^{\bar{i}j}f_{\bar{i}}^{\alpha}\overline{f_{\bar{j}}^{\beta}}R_{\mu\bar{\nu}\alpha\bar{\beta}}\frac{\partial f^{\mu}}{\partial t}\overline{\frac{\partial f^{\nu}}{\partial t}}-2\mathrm{Re}\,g^{\bar{i}j}R_{\mu\bar{\nu}\alpha\bar{\beta}}\overline{f_i^{\nu}}\overline{f_{\bar{j}}^{\beta}}\frac{\partial f^{\mu}}{\partial\bar{t}}\frac{\partial f^{\alpha}}{\partial t}.\end{aligned}$$

Hence at $t=0$

$$\begin{aligned}\frac{\partial^2}{\partial t\partial\bar{t}}\int_N|\bar{\partial}f|^2 &= \int_N g^{\bar{i}j}\left(\frac{D}{\partial t}f_{\bar{i}}^{\alpha}\right)\left(\overline{\frac{D}{\partial t}f_{\bar{j}}^{\beta}}\right)h_{\alpha\bar{\beta}}+\int_N g^{\bar{i}j}\left(\frac{D}{\partial\bar{t}}f_{\bar{i}}^{\alpha}\right)\left(\overline{\frac{D}{\partial\bar{t}}f_{\bar{j}}^{\beta}}\right)h_{\alpha\bar{\beta}}\\ &+\int_N g^{\bar{i}j}f_{\bar{i}}^{\alpha}\overline{f_{\bar{j}}^{\beta}}R_{\mu\bar{\nu}\alpha\bar{\beta}}\frac{\partial f^{\mu}}{\partial\bar{t}}\overline{\frac{\partial f^{\nu}}{\partial\bar{t}}}\\ &+\int_N g^{\bar{i}j}f_{\bar{i}}^{\alpha}\overline{f_{\bar{j}}^{\beta}}R_{\mu\bar{\nu}\alpha\bar{\beta}}\frac{\partial f^{\mu}}{\partial t}\overline{\frac{\partial f^{\nu}}{\partial t}}\\ &-2\mathrm{Re}\int_N g^{\bar{i}j}R_{\mu\bar{\nu}\alpha\bar{\beta}}\overline{f_i^{\nu}}\overline{f_{\bar{j}}^{\beta}}\frac{\partial f^{\mu}}{\partial\bar{t}}\frac{\partial f^{\alpha}}{\partial t},\end{aligned}$$

where

$$R_{\mu\bar{\nu}\alpha\bar{\beta}}=\partial_{\mu}\partial_{\bar{\nu}}h_{\alpha\bar{\beta}}-h^{\bar{\gamma}\delta}\partial_{\mu}h_{\alpha\bar{\gamma}}\overline{\partial_{\nu}h_{\beta\bar{\delta}}}$$

is the curvature tensor of M.

2 Relation of energy and ∂-energy

Suppose M,N are as in §1. Let $f:N\to M$ be a smooth map. Similar to $|\bar{\partial}f|^2$, the pointwise ∂-energy $|\partial f|^2$ of f is defined by

$$|\partial f|^2=g^{\bar{i}j}f_j^{\alpha}\overline{f_i^{\beta}}h_{\alpha\bar{\beta}},$$

where

$$f_j^{\alpha}=\frac{\partial f^{\alpha}}{\partial w^j}.$$

The pointwise energy $e(f)$ of f, which is defined as the trace of $f^*(ds_M^2)$ with respect to ds_N^2, is therefore equal to $|\bar{\partial}f|^2+|\partial f|^2$. Now assume $\dim_{\mathbb{C}}N=1$. The pullback of the Kähler form of M under f is

$$\sqrt{-1}h_{\alpha\bar{\beta}}df^{\alpha}\wedge d\overline{f^{\beta}}=\sqrt{-1}h_{\alpha\bar{\beta}}\left(\frac{\partial f^{\alpha}}{\partial w}\overline{\frac{\partial f^{\beta}}{\partial w}}-\frac{\partial f^{\alpha}}{\partial\bar{w}}\overline{\frac{\partial f^{\beta}}{\partial\bar{w}}}\right)dw\wedge d\bar{w}.$$

Hence

$$\int_N |\partial f|^2 - \int_N |\bar{\partial} f|^2 = \int_N \sqrt{-1} h_{\alpha\bar{\beta}} df^\alpha \wedge d\overline{f^\beta}$$

which is equal to the Kähler class $\omega(M)$ of M evaluated at the homology class $[f(N)]$ defined by $f : N \to M$. It follows that

$$\int_N |\partial f|^2 = \frac{1}{2}\int_N e(f) + \frac{1}{2}\omega(M)[f(N)],$$

$$\int_N |\bar{\partial} f|^2 = \frac{1}{2}\int_N e(f) - \frac{1}{2}\omega(M)[f(N)].$$

As a consequence, the energy-minimizing maps from N to M are precisely the same as the $\bar{\partial}$-energy-minimizing maps.

Remark. *This last fact was first observed by Lichnerowicz* [16].

3 Complex-analyticity of energy-minimizing maps

Suppose M is a compact Kähler manifold with positive holomorphic bisectional curvature. Let $f_0 : \mathbb{P}_1 \to M$ be an energy-minimizing map.

Proposition 1. *If $f_0^* c_1(M)$ evaluated at $\mathbb{P}_1$ is nonnegative (respectively nonpositive), then f_0 is holomorphic (respectively conjugate holomorphic).*

Proof. Since the proof of the other case is similar, we prove here only the holomorphic case.

Let T_M be the holomorphic tangent bundle of M, let w be a local coordinate of $\mathbb{P}_1$, and let $\frac{D}{\partial \bar{w}}$ be the covariant differentiation, in the antiholomorphic direction, of local cross sections of $f_0^* T_M$ with respect to the connection of T_M. Let $\mathscr{F}$ be the sheaf of germs of local cross sections s of $f_0^* T_M$ with $\frac{D}{\partial \bar{w}} s = 0$. Clearly $\mathscr{F}$ is an analytic sheaf over $\mathbb{P}_1$. We claim that $\mathscr{F}$ is locally free and the holomorphic vector bundle associated to $\mathscr{F}$ is topologically isomorphic to $f_0^* T_M$.

It suffices to show that for an arbitrary point P of $\mathbb{P}_1$ there exist local cross sections $s_1, \ldots, s_m$ at P such that $\frac{D}{\partial \bar{w}} s_i \equiv 0, 1 \leqq i \leqq m$, and $s_1(P), \ldots, s_m(P)$ form a basis of the fiber $f_0^* T_M$ at P, where $m = \dim_{\mathbb{C}} M$.

These local sections s_i can be constructed as follows. Choose the local coordinate (z^α) of M at $Q = f_0(P)$ such that the metric tensor $h_{\alpha\bar{\beta}}$ of M satisfies $dh_{\alpha\bar{\beta}} = 0$ at Q. Every local cross section s of $f_0^* T_M$ at P can be written as

$$s = \sum_\alpha s^\alpha \frac{\partial}{\partial z^\alpha}.$$

Then at P

$$\frac{D}{\partial \bar{w}} s = \sum_\alpha \frac{\partial s^\alpha}{\partial \bar{w}} \frac{\partial}{\partial z^\alpha}.$$

Choose smooth local cross sections

$$t_i = \sum_\alpha t_i^\alpha \frac{\partial}{\partial z^\alpha}$$

of $f_0^* T_M$ at $P, 1 \leqq i \leqq m$, such that the matrix

$$(t_i^\alpha)_{1 \leqq \alpha, i \leqq m}$$

is nonsingular at P and each t_i^α is holomorphic at P as a function of w. Then $\frac{D}{\partial \bar{w}} t_i$ vanishes at P. Solve the equations

$$\frac{D}{\partial \bar{w}} u_i = \frac{1}{w - w(P)} \frac{D}{\partial \bar{w}} t_i \qquad (*)$$

for u_i locally at P by using the Cauchy kernel

$$\frac{1}{2\pi\sqrt{-1}} \frac{dw \wedge d\bar{w}}{w - \zeta}$$

and the standard classical iteration process of Korn-Lichtenstein (see e.g. p. 394 of [21]; cf. pp. 254–267 of [1]). Since the right-hand side of $(*)$ is bounded, it follows that u_i is ε-Hölder continuous for any $0 < \varepsilon < 1$. Define s_i by

$$s_i = t_i - (w - w(P)) u_i.$$

Then s_i satisfies the requirements.

Now we can regard $f_0^* T_M$ as a holomorphic vector bundle over $\mathbb{P}_1$ (after its identification with the holomorphic vector bundle associated to $\mathcal{F}$). By the theorem of Grothendieck [11], $f_0^* T_M$ is a direct sum of holomorphic line bundles $L_1, \ldots, L_m$ over $\mathbb{P}_1$. Since the first Chern class of $f_0^* T_M$ evaluated at $\mathbb{P}_1$ is nonnegative, it follows that for some $i, c_1(L_i)$ evaluated at $\mathbb{P}_1$ is nonnegative. By the theorem of Riemann-Roch, we can find a nontrivial global holomorphic section

$$s = \sum s^\alpha \frac{\partial}{\partial z^\alpha}$$

of L_i (and hence of $f_0^* T_M$) over $\mathbb{P}_1$.

Construct a smooth family of smooth maps $f(t) : \mathbb{P}_1 \to M, t \in \mathbb{C}, |t| < \varepsilon$, such that $f(0) = f_0$ and at $t = 0$

$$\frac{\partial}{\partial \bar{t}} f^\alpha(t) = 0$$

and

$$\frac{\partial}{\partial t} f^\alpha(t) = s^\alpha.$$

Since s is a holomorphic cross section of $f_0^* T_M$, it follows that at $t = 0$

$$\frac{D}{\partial t} \frac{\partial}{\partial \bar{w}} f^\alpha = \frac{D}{\partial \bar{w}} \frac{\partial f^\alpha}{\partial t} = \frac{D}{\partial \bar{w}} s^\alpha = 0.$$

Moreover,

$$\frac{D}{\partial \bar{t}} \frac{\partial}{\partial \bar{w}} f^\alpha = \frac{D}{\partial \bar{w}} \frac{\partial f^\alpha}{\partial \bar{t}} = 0$$

at $t = 0$.

From the second variation formula derived in §1, it follows that at $t = 0$

$$\frac{\partial^2}{\partial t \partial \bar{t}} \int_{\mathbb{P}_1} |\bar{\partial} f|^2 = \int_{\mathbb{P}_1} \frac{\partial f^\alpha}{\partial \bar{w}} \overline{\frac{\partial f^\beta}{\partial \overline{w}}} R_{\mu\bar{\nu}\alpha\bar{\beta}} \frac{\partial f^\mu}{\partial t} \overline{\frac{\partial f^\nu}{\partial t}} (\sqrt{-1} dw \wedge d\bar{w}), \qquad (\dagger)$$

where $R_{\mu\bar{\nu}\alpha\bar{\beta}}$ is the curvature tensor of M. By §2, f_0 is $\bar{\partial}$-energy minimizing. Hence at $t = 0$ we have

$$\frac{\partial^2}{\partial t \partial \bar{t}} \int_{\mathbb{P}_1} |\bar{\partial} f|^2 \geqq 0.$$

Since the holomorphic bisectional curvature of M is positive, i.e.

$$R_{\mu\bar{\nu}\alpha\bar{\beta}} \xi^\mu \overline{\xi^\nu} \eta^\alpha \overline{\eta^\beta} < 0$$

for $(\xi^\mu), (\eta^\alpha) \in \mathbb{C}^m - 0$, it follows from $(\dagger)$ that

$$\frac{\partial f_0^\alpha}{\partial \bar{w}} = 0$$

on $\mathbb{P}_1 - Z$ for all α, where Z is the zero-set of s. The map f_0 is holomorphic, because Z is a finite subset of $\mathbb{P}_1$. Q.E.D.

4 Existence of energy-minimizing maps

Let M be a compact Riemannian manifold and S^2 be the 2-sphere. For each C^1 map $f : S^2 \to M$, we let $E(f)$ be the energy of f and let $E([f])$ be the infimum of the sum of the energies of maps whose sum is homotopic to f. The following proposition is proved by using the method of Sacks-Uhlenbeck [22] (see also [19] for results related to the present situation).

Proposition 2. *For every C^1 map $f : S^2 \to M$ there exist energy-minimizing maps $f_i : S^2 \to M, 1 \leqq i \leqq m$, such that the sum of f_i is homotopic to f and $E([f]) = \sum_{i=1}^m E(f_i)$.*

Proof. It is proved in [22] that any nonconstant harmonic map or any homotopically nontrivial C^1 map from S^2 into M has energy $> c$ for some fixed positive constant c. Let k be the smallest nonnegative integer such that

$$E(f) \leqq \frac{kc}{2}.$$

We are going to prove the proposition by induction on k. The case $k = 0$ is true. We want to prove the case $k = n + 1$ under the assumption that the case $k = n$ is true.

By [22] for $\alpha > 1$ we can find smooth maps $f_\alpha : S^2 \to M$ which minimize the functional

$$E_\alpha(g) = \int_{S^2} (1 + |dg|^2)^\alpha$$

over the space of all C^1 maps homotopic to f. Let x_α be a point of S^2 so that

$$|df_\alpha|^2(x_\alpha) = \sup_{S^2} |df_\alpha|^2.$$

If $\sup_{S^2} |df_\alpha|^2$ is uniformly bounded in α, the f_α converges to an energy-minimizing map and the assertion is proved. If $\sup_{S^2} |df_\alpha|^2$ is unbounded in α, then for some disc D_α with center at x_α the map $f_\alpha|D_\alpha$, after identifying D_α with an open disc in $\mathbb{R}^2$ by a suitable homothetic map, converges on compact subsets to a harmonic map $g\colon \mathbb{R}^2 \to M$. It is proved in [22] that g can be extended to a harmonic map $\tilde{g} : S^2 \to M$ and we can assume that $f_\alpha|\partial D_\alpha$ has arbitrarily small length.

As in [19], we can construct a map $f_\alpha^1 : S^2 \to M$ so that $E(f_\alpha^1)$ is arbitrarily close to $E(f_\alpha) - E(\tilde{g})$ and f is homotopic to the sum of f_α^1 and $\tilde{g}$. For later purpose, we outline the argument as follows. Fix a disc D so that $f_\alpha|D$ converges smoothly on D to $\tilde{g}|D$ and the energy of $\tilde{g}$ over D is close to the energy of $\tilde{g}$. We define maps $\tilde{f}_\alpha$ from S^2 into M by extending $f_\alpha|D$ to S^2 in the following way. Since $f_\alpha|\partial D$ is close to $\tilde{g}|\partial D$, we can join each point $f_\alpha(x)$ to $\tilde{g}(x)$ (for $x \in \partial D$) by a short geodesic and obtain a map h_α from an annulus into M whose image has small area. Then we obtain a map $\tilde{f}_\alpha$ from S^2 into M by putting together the maps $f_\alpha|S^2 - D, h_\alpha$ and $\tilde{g}|S^2 - D$.

By approximation, we may assume that the map $\tilde{f}_\alpha$ is a smooth immersion from S^2 into M and the area of the image of $\tilde{f}_\alpha$ is close to the sum of the areas of the images of $f_\alpha|S^2 - D, h_\alpha$, and $\tilde{g}|S^2 - D$. The map $\tilde{f}_\alpha$ pulls back the metric from M to a metric on S^2 which defines a conformal structure on S^2. Since there is only one conformal structure on S^2, there is an orientation preserving diffeomorphism from S^2 into S^2 which pulls this conformal structure back to the standard conformal structure on S^2. We define f_α^1 to be $\tilde{f}_\alpha$ composed with this diffeomorphism so that f_α^1 is conformal. Observe that the area of the image of a map from a real surface to a Riemannian manifold is the same as the energy when the map is conformal. Since $\tilde{g}$ is conformal (see for example [19]) and $E(\tilde{g}|S^2 - D)$ is small, the area of the image of the map $\tilde{g}|S^2 - D$ is small. This together with the smallness of the area of the image of h_α implies that $E(f_\alpha^1)$ is close to $E(f_\alpha) - E(\tilde{g})$.

Since $\tilde{g}$ is harmonic, $E(\tilde{g}) > c$ and $E(f_\alpha^1) \leqq \frac{nc}{2}$ for α sufficiently close to 1. By induction hypothesis, for α sufficiently close to 1, we can find energy-minimizing maps $f_i : S^2 \to M, 1 \leqq i \leqq m-1$, such that the sum of $f_i (1 \leqq i \leqq m-1)$ is homotopic to f_α^1 and $E([f_\alpha^1]) = \sum_{i=1}^{m-1} E(f_i)$.

Set $f_m = \tilde{g}$, then

$$\begin{aligned} E([f]) &= \lim_{\alpha \to 1} E(f_\alpha) = \lim_{\alpha \to 1} E(f_\alpha^1) + E(\tilde{g}) \\ &\geqq E([f_\alpha^1]) + E(\tilde{g}) \\ &= \sum_{i=1}^{m} E(f_i). \end{aligned}$$

On the other hand, from the definition of $E([f])$, it is clear that

$$E([f]) \leqq \sum_{i=1}^{m} E(f_i).$$

Hence $E([f]) = \sum_{i=1}^{m} E(f_i)$. It follows from the definition of $E([f])$ that f_m is energy-minimizing. Q.E.D.

5 Holomorphic deformation of rational curves

Let M be a compact complex manifold of complex dimension m whose tangent bundle T_M is positive in the sense of Griffiths [10]. Let C_0 be a rational curve in M, possibly with singularities, that is, C_0 is the image of a holomorphic map $f_0 : \mathbb{P}_1 \to M$, which is the normalization of C_0.

Let $V \subset \mathbb{P}_1 \times M$ be the graph of f_0. Let

$$\pi : \mathbb{P}_1 \times M \to \mathbb{P}_1,$$
$$\sigma : \mathbb{P}_1 \times M \to M$$

be the natural projections. Since $T_{\mathbb{P}_1 \times M}$ is isomorphic to $\pi^* T_{\mathbb{P}_1} \oplus \sigma^* T_M$ and T_V is isomorphic to $\pi^* T_{\mathbb{P}_1}$, it follows that $\sigma^* T_M | V$ is isomorphic to the normal bundle N_V of V in $\mathbb{P}_1 \times M$, that is, $f_0^* T_M$ is isomorphic to N_V.

Let $\tilde{D}$ be the moduli space when V is deformed as a subspace of $\mathbb{P}_1 \times M$ (see [3]). Let D be the irreducible component of $\tilde{D}$ which contains the point x_0 of $\tilde{D}$ corresponding to V. The infinitesimal deformation of V is given by $\Gamma(V, N_V)$. We claim that $H^1(V, N_V)$ vanishes so that all infinitesimal deformations are realized as actual deformations and x_0 is a regular point of $\tilde{D}$ of dimension equal to the dimension of $\Gamma(V, N_V)$.

Since T_M is positive, the μth symmetric tensor product $T_M^{(\mu)}$ of T_M is ample in the sense of Griffiths [10] for $\mu \geqq$ some positive integer μ_0. Take $v \in V$ such that at $y = \pi(v)$ the map f_0 is an immersion. Then for $\mu \geqq \mu_0$, we have an exact sequence

$$0 \to F \to \Gamma(V, N_V^{(\mu)}) \to (N_V^{(\mu)})_v \to 0$$

so that the natural map $F \to (N_V^{(\mu)})_v \otimes (T_V^*)_v$ is surjective. By a theorem of Grothendieck [11], N_V splits into a sum of holomorphic line bundles $L_1, \ldots, L_m$ over V. Then for $\mu \geqq \mu_0$ and $1 \leqq i \leqq m$ we have an exact sequence

$$0 \to F_i \to \Gamma(V, L_i^{\mu}) \to (L_i^{\mu})_v \to 0$$

so that the natural map $F_i \to (L_i^{\mu})_v \otimes (T_V^*)_v$ is surjective. Hence each L_i is a positive holomorphic line bundle over V. By the theorem of Riemann-Roch, $H^1(V, L_i) = 0$ for each i. It follows that $H^1(V, N_V) = 0$.

We have a complex subspace $\mathscr{C}$ of $D \times \mathbb{P}_1 \times M$ with the following property. Let

$$\alpha : \mathscr{C} \to D,$$
$$\beta : \mathscr{C} \to \mathbb{P}_1 \times M$$

be the natural projections. Then $\mathscr{C}$ is α-flat and for every $x \in D, \beta$ maps $\alpha^{-1}(x)$ biholomorphically onto the complex subspace of $\mathbb{P}_1 \times M$ corresponding to the point x. In particular, β maps $\alpha^{-1}(x_0)$ biholomorphically onto V.

Let $\mathscr{C}'$ be the subset of $\mathscr{C}$ consisting of all $w \in \mathscr{C}$ such that

i) the structure sheaf of $\mathscr{C}$ is reduced at w,

ii) D is regular at $\alpha(w)$, and

iii) the map α is a submersion at w.

Clearly $\mathscr{C} - \mathscr{C}'$ is a subvariety of $\mathscr{C}$. The fiber $\alpha^{-1}(x_0)$ is contained in $\mathscr{C}'$. Let $D_1 = \alpha(\mathscr{C} - \mathscr{C}')$. Then D_1 is a proper subvariety of D.

Let Ω_D^1 (respectively $\Omega_{\mathscr{C}}^1$) be the sheaf of germs of holomorphic 1-forms on D (respectively $\mathscr{C}$) in the sense that it is locally the sheaf of germs of holomorphic 1-forms on an ambient manifold modulo the defining functions and their differentials. Let $\mathscr{L}$ be the linear space over $\mathscr{C}$ associated to the coherent analytic sheaf $\Omega_{\mathscr{C}}^1/\alpha^*\Omega_D^1$ on $\mathscr{C}$, i.e., the sheaf of germs of holomorphic functions on $\mathscr{L}$ which are linear forms along the fibers of $\mathscr{L}$ is isomorphic to $\Omega_{\mathscr{C}}^1/\alpha^*\Omega_D^1$ (see [6]). When restricted to $\mathscr{C}'$, $\mathscr{L}$ is isomorphic to the vector bundle of tangent vectors along the fibers of $\mathscr{C}$. Let $\mathbb{P}(\mathscr{L})$ be obtained by replacing each fiber of $\mathscr{L}$ by the projective space of all complex lines, i.e. $\mathbb{P}(\mathscr{L})$ is the orbit space of $\mathscr{L}$ under the $\mathbb{C}^*$ action on the fibers. The projection $\sigma\beta : \mathscr{C} \to M$ induces a holomorphic map

$$(\sigma\beta)_* : \mathbb{P}(\mathscr{L}) \to \mathbb{P}(T_M),$$

where $\mathbb{P}(T_M)$ is the projectivization of the tangent bundle T_M. We claim that $(\sigma\beta)_*$ is surjective.

Since D is compact (see [8, 17]), it follows that $\mathscr{C}$ is compact and $\mathbb{P}(\mathscr{L})$ is compact. Hence the image of $(\sigma\beta)_*$ is a subvariety of $\mathbb{P}(T_M)$. To prove the surjectivity of $(\sigma\beta)_*$, it suffices to show that $(\sigma\beta)_*$ is open at some point. Take a regular point y_0 of C_0. Then it corresponds to a point w_0 in $\mathscr{C}$, i.e. $\alpha(w_0) = x_0$ and $(\sigma\beta)(w_0) = y_0$. Let $\tilde{y}_0 \in \mathbb{P}_1$ be the point such that $f_0(\tilde{y}_0) = y_0$. Let ζ be a local coordinate of $\mathbb{P}_1$ at $\tilde{y}_0$ vanishing at y_0. Then w_0 and $\left(\frac{\partial}{\partial\zeta}\right)_{\tilde{y}_0}$ determine a point θ_0 in $\mathbb{P}(\mathscr{L})$. We are going to show that $(\sigma\beta)_*$ is open at θ_0.

Let E be the divisor of the differential df_0 of f_0 and let $[E]$ be the line bundle over $\mathbb{P}_1$ associated to the divisor E. Then $T_{\mathbb{P}_1} \otimes [E]$ is a line subbundle of $f_0^*T_M$. Let Q be the quotient bundle $f_0^*T_M/T_{\mathbb{P}_1} \otimes [E]$. By the theorem of Grothendieck [11], Q splits into a direct sum of holomorphic line bundles $Q_2, \ldots, Q_m$ over $\mathbb{P}_1$. Since $f_0^*T_M$ is positive, each $Q_\nu, 2 \leqq \nu \leqq m$, is a positive line bundle over $\mathbb{P}_1$. Let $[\tilde{y}_0]$ be the line bundle over $\mathbb{P}_1$ associated to the divisor $\tilde{y}_0$. For $2 \leqq \nu \leqq m$ there exists a holomorphic cross section s_ν of $Q_\nu \otimes [\tilde{y}_0]^{-1}$ over $\mathbb{P}_1$ which does not vanish at $\tilde{y}_0$. We can regard s_ν as a holomorphic cross section of $Q \otimes [\tilde{y}_0]^{-1}$. Consider the following exact sequence

$$\Gamma(\mathbb{P}_1, (f_0^*T_M) \otimes [\tilde{y}_0]^{-1}) \to \Gamma(\mathbb{P}_1, Q \otimes [\tilde{y}_0]^{-1}) \to H^1(\mathbb{P}_1, T_{\mathbb{P}_1} \otimes [E] \otimes [\tilde{y}_0]^{-1})$$

coming from the exact sequence

$$0 \to T_{\mathbb{P}_1} \otimes [E] \otimes [\tilde{y}_0]^{-1} \to (f_0^*T_M) \otimes [\tilde{y}_0]^{-1} \to Q \otimes [\tilde{y}_0]^{-1} \to 0.$$

Since

$$H^1(\mathbb{P}_1, T_{\mathbb{P}_1} \otimes [E] \otimes [\tilde{y}_0]^{-1}) = 0$$

by the vanishing theorem of Kodaira, the cross sections $s_\nu, 2 \leqq \nu \leqq m$, can be lifted to holomorphic cross sections $\tilde{s}_\nu$ of $f_0^*T_M \otimes [\tilde{y}_0]^{-1}$ over $\mathbb{P}_1$.

Let t be the holomorphic cross section of $[\tilde{y}_0]$ over $\mathbb{P}_1$ whose divisor is $\tilde{y}_0$. Let $u_\nu = \tilde{s}_\nu t$. Then u_ν is a holomorphic cross section of $f_0^*T_M$ over $\mathbb{P}_1, 2 \leqq \nu \leqq$

m, and these sections have the property that $\frac{1}{\zeta}u_\nu, 2 \leqq \nu \leqq m$, form a basis of $(f_0^* T_M / T_M)_{\tilde{y}_0}$.

For $1 \leqq \nu \leqq m-1$ there exists a holomorphic cross section $s_{m+\nu}$ of $Q_{\nu+1}$ over $\mathbb{P}_1$ which does not vanish at $\tilde{y}_0$. We can regard $s_{m+\nu}, 1 \leqq \nu \leqq m-1$, as a holomorphic cross section of Q. Consider the following exact sequence

$$\Gamma(\mathbb{P}_1, f_0^* T_M) \to \Gamma(\mathbb{P}_1, Q) \to H^1(\mathbb{P}_1, T_{\mathbb{P}_1}, \otimes [E])$$

coming from the exact sequence

$$0 \to T_{\mathbb{P}_1} \otimes [E] \to f_0^* T_M \to Q \to 0.$$

Since

$$H^1(\mathbb{P}_1, T_{\mathbb{P}_1} \otimes [E]) = 0$$

by the vanishing theorem of Kodaira, the sections $s_{m+\nu}, 1 \leqq \nu \leqq m-1$, can be lifted up to a holomorphic cross section $u_{m+\nu}, 1 \leqq \nu \leqq m-1$, of $f_0^* T_M$ over $\mathbb{P}_1$.

Each $u_\nu, 2 \leqq \nu \leqq 2m-1$, defines a tangent vector v_ν of D at x_0 and defines $\xi_\nu \in \Gamma(\alpha^{-1}(x_0), T_{\mathscr{C}})$ such that

i) $(d\alpha)(\xi_\nu) \equiv v_\nu$ at every point of $\alpha^{-1}(x_0)$, and

ii) $(d(\sigma\beta))(\xi_\nu)$ at $w \in \alpha^{-1}(x_0)$ equals u_ν at $\pi\beta(w)$.

Choose a local submanifold R of D at x_0 such that the tangent space of R at x_0 is spanned by $v_2, \dots, v_{2m-1}$. At w_0, we choose a local coordinate system $t_1, \dots, t_{2n-1}$ of $\alpha^{-1}(R)$ such that

i) t_ν is of the form $t'_\nu \circ \alpha$ for $2 \leqq \nu \leqq 2m-1$,

ii) $t_1(w_0) = \cdots = t_{2m-1}(w_0) = 0$,

iii) $\frac{\partial}{\partial t_\nu} = \xi_\nu (2 \leqq \nu \leqq 2m-1)$ at $\alpha^{-1}(x_0)$, and

iv) $t_1 = \zeta \circ f_0^{-1} \circ (\sigma\beta)$ on $\alpha^{-1}(x_0)$.

Choose a local coordinate system $z_1, \dots, z_m$ of M at y_0 such that

i) C_0 is defined by $z_2 = \cdots = z_m = 0$, and

ii) $z_1 = \zeta \circ f_0^{-1}$ on C_0.

Let

$$u_\nu = \sum_{\mu=1}^{m} u_{\nu\mu} \frac{\partial}{\partial z_\mu} \quad (2 \leqq \nu \leqq m-1)$$

where $u_{\nu\mu}$ is a holomorphic function on $\mathbb{P}_1$ near $\tilde{y}_0$. Since $\frac{1}{\zeta}u_\nu, 2 \leqq \nu \leqq m$, form a basis of $(f_0^* T_M / T_M)_{\tilde{y}_0}$, it follows that the matrix

$$\left(\frac{1}{\zeta} u_{\nu\mu}\right)_{2 \leqq \nu, \mu \leqq m}$$

is nonsingular at $\tilde{y}_0$. Let

$$a_{\nu\mu} = \left(\frac{1}{\zeta} u_{\nu\mu}\right) \circ f_0^{-1}$$

on C_0 near y_0. Then on C_0 near y_0 we have

$$(d(\sigma\beta)) \frac{\partial}{\partial t_1} = \frac{\partial}{\partial z_1},$$

$$(d(\sigma\beta)) \frac{\partial}{\partial t_\nu} = z_1 \sum_{\mu=1}^{m} a_{\nu\mu} \frac{\partial}{\partial z_\mu} \quad (2 \leqq \nu \leqq m).$$

Let

$$(d(\sigma\beta))\frac{\partial}{\partial t_\nu} = \sum_{\mu=1}^{m} b_{\nu\mu}\frac{\partial}{\partial z_\mu} \quad (m < \nu < 2m).$$

Since $u_\nu, m < \nu < 2m$, form a basis of $(f_0^* T_M/T_M)_{\bar{y}_0}$, it follows that the matrix

$$(b_{\nu\mu})_{m<\nu<2m, 2\leqq\mu\leqq m}$$

is nonsingular at y_0.

Now we calculate the Jacobian matrix of the map $(\sigma\beta)_*$, restricted to $\mathbb{P}(\mathscr{L})|R$, with respect to local coordinate systems we are going to describe and verify that the Jacobian matrix has rank $2m-1$ over $\mathbb{C}$ at θ_0.

Since $\alpha^{-1}(x)$ is of complex dimension 1 for every $x \in D$, for $w \in \mathscr{C}'$ the fiber of $\mathbb{P}(\mathscr{L})$ at w consists only of a single point. Hence we can identify $\mathbb{P}(\mathscr{L})|\mathscr{C}'$ with $\mathscr{C}'$ and use $t_1, \ldots, t_{2m-1}$ as local coordinates at w_0 for $\mathbb{P}(\mathscr{L})$ (after identification of $\mathbb{P}(\mathscr{L})|\mathscr{C}'$ with $\mathscr{C}'$).

Every element η of T_M can be written as

$$\sum_{\mu=1}^{m} \eta_\nu \frac{\partial}{\partial z_\nu}$$

and we use as local coordinates of $\mathbb{P}(T_M)$ near $(y_0, \frac{\partial}{\partial z_1})$ the functions

$$z_1, \ldots, z_m, \frac{\eta_2}{\eta_1}, \ldots, \frac{\eta_m}{\eta_1}.$$

The map $\sigma\beta$ makes $z_1, \ldots, z_m$ functions of $t_1, \ldots, t_{2m-1}$. The Jacobian matrix of $(\sigma\beta)_*$, restricted to $\mathbb{P}(\mathscr{L})|R$, with respect to the coordinate systems

$$t_1, \ldots, t_{2m-1},$$

and

$$z_1, \ldots, z_m, \frac{\eta_2}{\eta_1}, \ldots, \frac{\eta_m}{\eta_1},$$

equals to the $(2m-1) \times (2m-1)$ matrix (A, B), where

$$A = \left(\frac{\partial z_\mu}{\partial t_\nu}\right)_{1\leqq\nu\leqq 2m-1, 1\leqq\mu\leqq m},$$

$$B = \left(\frac{\partial}{\partial t_\nu}\left(\frac{\frac{\partial z_\mu}{\partial t_1}}{\frac{\partial z_1}{\partial t_1}}\right)\right)_{1\leqq\nu\leqq 2m-1, 2\leqq\mu\leqq m}.$$

We have

$$\frac{\partial z_\mu}{\partial t_1} = \delta_{\mu 1} \quad \text{(the Kronecker delta) for } 1 \leqq \mu \leqq m,$$

$$\frac{\partial z_\mu}{\partial t_\nu} = t_1 a_{\nu\mu} \quad \text{for } 2 \leqq \nu \leqq m, 1 \leqq \mu \leqq m,$$

$$\frac{\partial z_\mu}{\partial t_\nu} = b_{\nu\mu} \quad \text{for } m < \nu < 2m, 1 \leqq \mu \leqq m$$

on $\alpha^{-1}(x_0)$ near w_0. Now

$$\frac{\partial}{\partial t_\nu}\left(\frac{\frac{\partial z_\mu}{\partial t_1}}{\frac{\partial z_1}{\partial t_1}}\right) = \frac{\frac{\partial z_\mu}{\partial t_1}\frac{\partial z_\mu}{\partial t_\nu}}{\frac{\partial z_1}{\partial t_1}} \frac{\frac{\partial}{\partial t_1}\frac{\partial z_1}{\partial t_\nu}}{\frac{\partial z_1}{\partial t_1}} \frac{\partial z_\mu}{\partial t_1} = a_{\nu\mu}$$

at w_0 for $2 \leqq \nu \leqq m$, $2 \leqq \mu \leqq m$. Since the matrices

$$(a_{\nu\mu})_{2\leqq\nu,\mu\leqq m}$$

and

$$(b_{\nu\mu})_{m<\nu<2m,2\leqq\mu\leqq m}$$

are nonsingular at y_0, it follows that (A, B) is nonsingular at w_0. This concludes the proof that $(\sigma\beta)_*$ is open at θ_0.

Let G be the subset of $\mathbb{P}(T_M)$ consisting of all points $\hat{y}$ of $\mathbb{P}(T_M)$ such that $(\sigma\beta)_*^{-1}(\hat{y})$ is entirely contained in the restriction of $\mathbb{P}(\mathscr{L})$ to $\mathscr{C}' - \alpha^{-1}(D_1)$. Let $Z = \mathbb{P}(T_M) - G$. Then Z is a proper subvariety of $\mathbb{P}(T_M)$. We thus have proved the following.

Proposition 3. *Let M be a compact complex manifold whose tangent bundle T_M is positive. Let C_0 be a rational curve in M and $f_0 : \mathbb{P}_1 \to C_0$ be its normalization. Then there exists a proper subvariety Z of $\mathbb{P}(T_M)$ with the following property. If $y \in M$ and $\xi \in (T_M)_y - 0$ define an element of $\mathbb{P}(T_M) - Z$, then there exists a holomorphic map $f : \mathbb{P} \to M$ homotopic to f_0 (when f_0 is regarded as a map from $\mathbb{P}_1$ to M) such that y is a regular point of $f(\mathbb{P}_1)$ and the tangent vector of $f(\mathbb{P}_1)$ at y is a nonzero multiple of ξ.*

6 Proof of the main theorem

Suppose M is an m-dimensional compact Kähler manifold of positive bisectional curvature. Since the Ricci curvature of M is positive, it follows from the theorem of Bonnet-Myers that the universal covering of M is compact. Since $\mathbb{P}_m$ has no fixed-point-free automorphism, in order to prove the Main Theorem, by replacing M by its universal covering, we can assume without loss of generality that M is simply connected. It follows that $\pi_2(M)$ is isomorphic to $H_2(M, \mathbb{Z})$.

Since the holomorphic bisectional curvature of M is positive, by the result of Bishop-Goldberg [2] (see also [9]) the second Betti number of M is 1. From the universal coefficient theorem it follows that $H^2(M, \mathbb{Z}) \approx \mathbb{Z}$. There exists a positive holomorphic line bundle F over M whose first Chern class $c_1(F)$ is a generator of $H^2(M, \mathbb{Z})$. Let g be a generator of the free part of $H_2(M, \mathbb{Z})$ such that the value of $c_1(F)$ at g is 1. Let $f : \mathbb{P}_1 \to M$ be a smooth map so that the element in $\pi_2(M)$ defined by f corresponds to g in the isomorphism between $\pi_2(M)$ and $H_2(M, \mathbb{Z})$.

By Proposition 2, there exist energy-minimizing maps $f_i : \mathbb{P}_1 \to M, 0 \leqq i \leqq k$, such that the sum of $f_i(0 \leqq i \leqq k)$ is homotopic to f and $E([f]) = \sum_{i=0}^k E(f_i)$.

By Proposition 1, each $f_i(0 \leqq i \leqq k)$ is either holomorphic or conjugate holomorphic. So each $f_i(\mathbb{P}_1)(0 \leqq i \leqq k)$ is a rational curve. Since $c_1(T_M)$ is a positive integral multiple of $c_1(F)$ and the value of $c_1(F)$ at g is 1, it follows that at least one f_i is holomorphic. If $k > 0$, then at least one f_j is conjugate holomorphic. We distinguish now between two cases.

Case 1. $k = 0$.

We use the notations of §5. The line bundle $T_M \otimes [E]$ is a subbundle of $f_0^* T_M$ and the quotient bundle $(f_0^* T_M)/(T_M \otimes [E])$ splits into a direct sum of line bundles $Q_2, \ldots, Q_m$. Each $Q_i (2 \leqq i \leqq m)$ is a positive line bundle. It follows that

$$c_1(f_0^* T_M) = c_1(T_M) + c_1([E]) + \sum_{i=2}^{m} c_1(Q_i).$$

Hence $c_1(T_M)$ evaluated at g is $\geqq n+1$. that is, $c_1(T_M) = \lambda c_1(F)$ for some integer $\lambda \geqq n+1$. By the result of Kobayashi-Ochiai [14], M is biholomorphic to $\mathbb{P}_m$.

Case 2. $k > 0$.

Without loss of generality we can assume that f_0 is holomorphic and f_1 is conjugate holomorphic. From Proposition 3 we obtain a proper subvariety Z of $\mathbb{P}(T_M)$ for the rational curve $f_0(\mathbb{P}_1)$ of M. Similarly we obtain a proper subvariety Z' of $\mathbb{P}(T_M)$ for the rational curve $f_1(\mathbb{P}_1)$ of M. Take a point $y \in M$ and a nonzero tangent vector ξ of M at y so that the point of $\mathbb{P}(T_M)$ defined by y and ξ does not belong to $Z \cup Z'$. The map f_0 (respectively f_1) is homotopic to a holomorphic map f_0' (respectively conjugate holomorphic map f_1') From $\mathbb{P}_1$ to M such that y is a regular point of $f_0'(\mathbb{P}_1)$ (respectively $f_1'(\mathbb{P}_1)$) whose tangent vector at y is a multiple of ξ. Since for $\nu = 0, 1$ both $E(f_\nu)$ and $E(f_\nu')$ are equal to the absolute value of the Kähler class of M at $[f_\nu(\mathbb{P}_1)]$, by replacing f_ν by $f_\nu' (\nu = 0, 1)$ we can assume without loss of generality $f_\nu = f_\nu' (\nu = 0, 1)$.

Choose a local coordinate system $z_1, \ldots, z_n$ at y with y as the origin such that $\xi = \frac{\partial}{\partial z_1}$. Choose a local coordinate system ζ of $\mathbb{P}_1$ so that both $f_0^{-1}(y)$ and $f_1^{-1}(y)$ correspond to $\zeta = 0$ and $f_\nu (\nu = 0, 1)$ is of the form

$$\begin{aligned} z_1 &= \zeta \\ z_\mu &= f_{\nu\mu}(\zeta) \quad (2 \leqq \mu \leqq m) \end{aligned}$$

near $\zeta = 0$.

For $\delta > 0$ let Δ_δ be the closed disc in $\mathbb{C}$ of radius δ centered at 0. For δ sufficiently small, we remove $f_\nu(\Delta_\delta)$ from $f_\nu(\mathbb{P}_1) (\nu = 1, 2)$ and replace these two discs by the surface S_δ defined by

$$\begin{aligned} z_1 &= \delta e^{i\theta}, \\ z_\mu &= t f_0(\delta e^{i\theta}) + (1-t) f_1(\delta e^{i\theta}) \end{aligned}$$

for $0 \leqq \theta \leqq 2\pi$ and $0 \leqq t \leqq 1$. The surface

$$S_\delta \cup \left(\bigcup_{\nu=1,2} f_\nu(\mathbb{P}_1 - \Delta_\delta) \right)$$

oriented in the obvious way is the image of a map $\tilde{f}$ from S^2 to M which is homotopic to the sum of f_0 and f_1. Moreover, for δ sufficiently small the area of this surface is strictly less than the combined area of $f_0(\mathbb{P}_1)$ and $f_1(\mathbb{P}_1)$. By smoothing out $\tilde{f}$ we obtain a smooth map $\hat{f}$ from S^2 to M such that the area of $\hat{f}(S^2)$ is strictly less than the combined area of $f_0(\mathbb{P}_1)$ and $f_1(\mathbb{P}_1)$. We make $\hat{f}$ conformal by replacing it by its composite with a suitable diffeomorphism of S^2. Then $E(\hat{f}) < E(f_0) + E(f_1)$. The sum of $\hat{f}$ and $f_2, \ldots, f_k$ is homotopic to f and yet

$$E(\hat{f}) + \sum_{i=2}^{k} E(f_i) < \sum_{i=0}^{k} E(f_i) = E([f]),$$

which is a contradiction. Hence $k = 0$ and M is biholomorphic to $\mathbb{P}_m$.

References

[1] Bers, L., Fohn, F., Schechter, M.: Partial Differential Equations. New York: Interscience Publishers 1964

[2] Bishop, R.L., Goldberg, S.I.: On the other second cohomology group of a Kähler manifold of positive curvature. Proc. Amer. Math. Soc. **16**, 119–122 (1965)

[3] Douady, A.: Le problème de modules pour les sous-espaces analytiques compacts d'un espace analytique donné Ann. Inst. Fourier (Grenoble) **16**, 1–95 (1966)

[4] Eells, J., Lemaire, L.: A report on harmonic maps. Bull. London Math. Soc. **10**, 1–68 (1978)

[5] Eells, J., Sampson, J.H.: Harmonic mappings of Riemannian manifolds. Amer. J. Math. **86**, 109–160 (1964)

[6] Fischer, G.: Lineare Faserräume und kohärente Modulgarben über komplexen Räumen. Archiv der Math. **18**, 609–617 (1967)

[7] Frankel, T.: Manifolds with positive curvature. Pacific J. Math. **11**, 165–174 (1961)

[8] Fujiki, A.: Closedeness of the Douady spaces of compact Kähler manifolds. Publ. Math. R.I.M.S. Kyoto Univ. **14**, 1-52 (1978)

[9] Goldberg, S.I., Kobayashi, S.: Holomorphic bisectional curvature. J. Diff. Geom. **1**, 225-234 (1967)

[10] Griffiths, P.: Hermitian differential geometry, Chern classes, and positive vector bundles. In: Global Analysis, (D.C. Spencer and S. Iyanaga, eds.) pp. 185-251. Princeton Univ. Press 1969

[11] Grothendieck, A.: Sur la classification des fibrés holomorphes sur la sphère de Riemann. Amer. J. Math. **79**, 121–138 (1957)

[12] Kobayashi, S., Ochiai, T.: On complex manifolds with positive tangent bundles. J. Math. Soc. Japan **22**, 499–525 (1970)

[13] Kobayashi, S., Ochiai, T.: Compact homogeneous complex manifolds with positive tangent bundle. In: Differential Geometry, in honor of K. Yano, pp. 233-242, (S. Kobayashi, M. Obata, and T. Takahashi, eds.) pp. 233-242. Tokyo: Kinokuniya 1972

[14] Kobayashi, S., Ochiai, T.: Three-dimensional compact Kähler manifolds with positive holomorphic bisectional curvature. J. Math. Soc. Japan **24**, 465–480 (1972)

[15] Kobayashi, S., Ochiai, T.: Characterizations of complex projective spaces and hyperquadrics. J. Math. Kyoto Univ. **13**, 31-47 (1973)

[16] Lichnerowicz, A.: Applications harmoniques et variétés kählériennes, Symp. Math. **3**, 341-402 (1970)

[17] Lieberman, D.: Compactness of the Chow scheme: applications to automorphisms and deformations of Kähler manifolds. Séminaire F. Norguet pp.140–186, 1976

[18] Mabuchi, T.: $\mathbb{C}$-actions and algebraic threefolds with ample tangent bundle. Nagoya Math. J. 69, 33–64 (1978)

[19] Meeks, W., Yau, S.-T.: Topology of three dimensional manifolds and the embedding problems in minimal surface theory

[20] Mori, S.: Projective manifolds with ample tangent bundles. Ann. of Math. **110**, 593-606 (1979)

[21] Newlander, A., Nirenberg. L.: Complex-analytic coordinates in almost complex manifolds. Ann. of Math. **65**, 391-404 (1957)

[22] Sacks, J., Uhlenbeck, K.: The existence of minimal immersion of 2-spheres

[23] Schoen, R., Yau S.-T.: On univalent harmonic maps between surfaces. Invent. Math. **44**, 265-278 (1978)

[24] Siu, Y.-T.: The complex-analyticity of harmonic maps and the strong rigidity of compact Kähler manifolds. Ann. of Math. in press (1980)

[25] Toledo, D.: On the Schwarz lemma for harmonic maps and characteristic numbers of flat bundles

[26] Wood, J.C.: Holomorphicity of certain harmonic maps from a surface to complex projective n-space

Editor's Comments

This paper is yet another important example showing that geometric analysis can be used to prove a major rigidity result in complex geometry. It solves a variant of a conjecture by Yau that two compact Kähler manifolds of complex dimension at least 2 of negative sectional curvature are biholomorphic or conjugate biholomorphic if they have the same homotopy type. It was motivated by and is an analogue of the celebrated Mostow strong rigidity for locally symmetric spaces. Besides making the conjecture, Yau also suggested the method of using harmonic maps to the author.

Complex Geometry from Riemann to Kähler-Einstein and Calabi-Yau
ALM 38, pp. 367–408

The Complex-Analyticity of Harmonic Maps and the Strong Rigidity of Compact Kähler Manifolds[1,*]

Yum-Tong Siu

In 1960 Calabi and Vesentini [3] proved that compact quotients of bounded symmetric domains are rigid in the sense that they do not admit any nontrivial infinitesimal holomorphic deformation. In 1970 Mostow discovered the phenomenon of strong rigidity [7]. He proved that the fundamental group of a compact locally symmetric Riemannian manifold of nonpositive curvature determines the manifold up to an isometry and a choice of normalizing constants if the manifold admits no closed one or two dimensional geodesic submanifolds which are locally direct factors. In particular, two compact quotients of the ball of complex dimension $\geqq 2$ with isomorphic fundamental groups are either biholomorphic or conjugate biholomorphic. Yau conjectured that this phenomenon of strong rigidity should hold also for compact Kähler manifolds of complex dimension $\geqq 2$ with negative sectional curvature. That is, two compact Kähler manifolds of complex dimension $\geqq 2$ with negative sectional curvature are biholomorphic or conjugate biholomorphic if they are of the same homotopy type. In this paper we prove that Yau's conjecture is true when the curvature tensor of one of the two compact Kähler manifolds is strongly negative in the sense defined in Section 2 with no curvature assumption on the other manifold. The strong negativity of the curvature tensor is a condition stronger than the negativity of the sectional curvature. This strong negative curvature condition is satisfied by quotients of the ball and also by the compact Kähler surface recently constructed by Mostow and Siu [8] which has negative sectional curvature and whose universal covering is not biholomorphic to the ball. Until now there is no known example of a compact Kähler manifold of

1 Research partially supported by an NSF grant. Siu, Yum Tong, The complex-analyticity of harmonic maps and the strong rigidity of compact Kähler manifolds. Ann. of Math. (2) 112 (1980), no. 1, 73–111. (Received March 2, 1979).

negative sectional curvature which does not admit a Kähler metric with strongly negative curvature tensor.

Our result is proved by showing that a harmonic map of compact Kähler manifolds is either holomorphic or conjugate holomorphic if the rank over R of the differential of the map is $\geqq 4$ at some point and if the curvature tensor of the image manifold is strongly negative. This result on the complex-analyticity of harmonic maps is obtained by a Bochner type argument. The usual technique of proving properties of a harmonic map f is to obtain a Bochner type formula by considering the Laplacian of the pointwise square norm of df. For the complex manifold case the pointwise square norm of df is replaced by the pointwise square norm of $\bar{\partial} f$ in this technique (see [9]). The metric tensor of the image manifold and the inverse matrix of the metric tensor of the domain manifold appear in the pointwise square norm of $\bar{\partial} f$. Hence in this Bochner type formula an expression involving the difference of the curvature tensors of the domain manifold and the image manifold appears. This prevents one from drawing any conclusion when both manifolds have negative curvature. In our proof we overcome this difficulty by replacing the pointwise square norm of $\bar{\partial} f$ by the contraction of $\bar{\partial} f \wedge \overline{\bar{\partial} f}$ with the metric tensor of the image manifold and by replacing the Laplacian by $\partial\bar{\partial}$. This method enables us to get rid of the curvature tensor of the domain manifold in the Bochner type formula. Our result on the complex-analyticity of harmonic maps can be applied to the problem of representing homology classes by complex-analytic subvarieties. Unfortunately it is still far from being able to prove the Hodge conjecture even for the case of the compact quotients of the ball.

Our method of proving the complex-analyticity of harmonic maps and strong rigidity can be applied to a much wider class of compact Kähler manifolds than the class of those with strongly negative curvature tensor. As examples, we apply our method to compact quotients of the four types of classical bounded symmetric domains and obtain the following result. Any harmonic map from a compact Kähler manifold to a compact quotient of an irreducible classical bounded symmetric domain of dimension $\geqq 2$ is either holomorphic or conjugate holomorphic if the map is a submersion at some point. In particular, any compact Kähler manifold which is of the same homotopy type as a compact quotient of an irreducible classical bounded symmetric domain of dimension $\geqq 2$ is either biholomorphic or conjugate biholomorphic to it. This rigidity result is stronger than the corresponding strong rigidity theorem of Mostow [7], because here only one manifold is assumed to be locally symmetric whereas in Mostow's theorem both manifolds have to be assumed locally symmetric.

These results were announced in [10].

I would like to thank S.-T. Yau for introducing me to harmonic maps and his conjecture and for many conversations in connection with his conjecture. E. Bedford drew my attention to the inequality that the second elementary symmetric function of a finite number of real numbers is dominated by the square of their sum. He showed me how to use such an inequality and the complex Bernstein formula [2, p. 378] to obtain, in the pseudoconvex case, a new proof of Bochner's theorem on extending holomorphic functions from boundaries. This simple inequality is used in our proof of the complex-analyticity of harmonic maps and its use is inspired by

Bedford's proof of the special case of Bochner's theorem. I would like to express my indebtedness to him.

Contents

1 Harmonic maps

Let $f : N \to M$ be a map of Riemannian manifolds whose metrics are respectively

$$ds^2_M = \sum_{\alpha,\beta} g_{\alpha\beta} dx^\alpha dx^\beta,$$

$$ds^2_N = \sum_{i,j} h_{ij} dy^i dy^j.$$

The energy $E(f)$ of f is defined to be

$$\frac{1}{2}\int_N \mathrm{trace}_{ds^2_N}(f^* ds^2_M),$$

that is,

$$E(f) = \frac{1}{2}\int \sum_{i,j.\alpha,\beta} g_{\alpha\beta} h^{ij} \frac{\partial f^\alpha}{\partial y^i}\frac{\partial f^\beta}{\partial y^j}$$

in terms of local coordinate charts. The Euler-Lagrange equation for the energy functional E is

$$\Delta_N f^\alpha + \sum_{i,j,\beta,\gamma} {}^M\Gamma^\alpha_{\beta\gamma} \frac{\partial f^\beta}{\partial y^i}\frac{\partial f^\gamma}{\partial y^j} h^{ij} = 0$$

for all α, where Δ_N is the Laplace-Beltrami operator of N and ${}^M\Gamma^\alpha_{\beta\gamma}$ is the Christoffel symbol of M. The map f is said to be *harmonic* if f satisfies the Euler-Lagrange equation for the energy functional.

Eells-Sampson [4] proved that when M and N are both compact and M has nonpositive sectional curvature, every continuous map from N to M is homotopic to a harmonic map. Hartman [6] proved that the harmonic map is unique in each homotopy class if M has strictly negative curvature. Since the Euler-Lagrange equation for the energy functional is a second-order quasilinear elliptic system of partial differential equations, it follows that any harmonic map between smooth Riemannian manifolds is smooth. The reader is referred to [4] for a survey of the theory of harmonic maps.

2 Curvature conditions and statement of results

Suppose M is a complex manifold with Kähler metric

$$ds^2_M = 2\mathrm{Re}\sum_{\alpha,\beta} g_{\alpha\bar\beta} dz^\alpha dz^{\bar\beta}.$$

The curvature tensor is given by

$$R_{\alpha\bar\beta\gamma\bar\delta} = \partial_\gamma\partial_{\bar\delta} g_{\alpha\bar\beta} - \sum_{\lambda,\mu} g^{\lambda\mu}\partial_\alpha g_{\gamma\bar\mu}\partial_{\bar\beta} g_{\lambda\bar\delta}.$$

The sectional curvature at the 2-plane spanned by the two tangent vectors

$$p = 2\mathrm{Re}\sum_\alpha \xi^\alpha \frac{\partial}{\partial z_\alpha},$$

$$q = 2\mathrm{Re}\sum_\alpha \eta^\alpha \frac{\partial}{\partial z_\alpha}$$

is given by

$$-||p\wedge q||^{-2} \sum_{\alpha,\beta,\gamma,\delta} R_{\alpha\bar\beta\gamma\bar\delta}(\xi^\alpha\overline{\eta^\beta} - \eta^\alpha\overline{\xi^\beta})(\overline{\xi^\delta\overline{\eta^\gamma} - \eta^\delta\overline{\xi^\gamma}}),$$

where

$$\begin{aligned}\|p\wedge q\|^2 = \sum_{\alpha,\beta,\gamma,\delta} g_{\alpha\bar\delta} g_{\gamma\bar\beta}[&(\xi^\alpha\overline{\eta^\beta} - \eta^\alpha\overline{\xi^\beta})(\overline{\xi^\delta\overline{\eta^\gamma} - \eta^\delta\overline{\xi^\gamma}}) \\ &+(\xi^\alpha\eta^\gamma - \eta^\alpha\xi^\gamma)\overline{(\xi^\beta\eta^\delta - \eta^\beta\xi^\delta)}].\end{aligned}$$

Nonpositivity of the sectional curvature is equivalent to

$$\sum_{\alpha.\beta,\gamma,\delta} R_{\alpha\bar\beta\gamma\bar\delta}(\xi^\alpha\overline{\eta^\beta} - \eta^\alpha\overline{\xi^\beta})(\overline{\xi^\delta\overline{\eta^\gamma} - \eta^\delta\overline{\xi^\gamma}}) \geq 0$$

for all complex numbers ξ^α, η^α. The sectional curvature is negative if, in addition in the above inequality, equality holds if and only if

$$\xi^\alpha \overline{\eta^\beta} - \eta^\alpha \overline{\xi^\beta} = 0 \quad \text{for all } \alpha, \beta,$$

because the vanishing of all $\xi^\alpha \overline{\eta^\beta} - \eta^\alpha \overline{\xi^\beta}$ is equivalent to the vanishing of $\|p \wedge q\|^2$ as one can easily verify by diagonalizing the matrix $(g_{\alpha\bar\beta})$ and making use of the identity

$$|\xi^\alpha \eta^\beta - \eta^\alpha \xi^\beta|^2 = |\xi^\alpha \overline{\eta^\beta} - \eta^\alpha \overline{\xi^\beta}|^2 - (\xi^\alpha \overline{\eta^\alpha} - \eta^\alpha \overline{\xi^\alpha})\overline{(\xi^\beta \overline{\eta^\beta} - \eta^\beta \overline{\xi^\beta})}.$$

We now introduce some notions of negative curvature stronger than the negativity of sectional curvature.

Definition. The curvature $R_{\alpha\bar\beta\gamma\bar\delta}$ is said to be *strongly negative* (respectively *strongly seminegative*) if

$$\sum_{\alpha,\beta,\gamma,\delta} R_{\alpha\bar\beta\gamma\bar\delta}(A^\alpha \overline{B^\beta} - C^\alpha \overline{D^\beta})\overline{(A^\delta \overline{B^\gamma} - C^\delta \overline{D^\gamma})}$$

is positive (respectively nonnegative) for arbitrary complex numbers $A^\alpha, B^\alpha, C^\alpha, D^\alpha$ when $A^\alpha \overline{B^\beta} - C^\alpha \overline{D^\beta} \neq 0$ for at least one pair of indices (α, β).

Definition. The curvature tensor $R_{\alpha\bar\beta\gamma\bar\delta}$ is said to be *very strongly negative* (respectively *very strongly seminegative*) if

$$\sum_{\alpha,\beta,\gamma,\delta} R_{\alpha\bar\beta\gamma\bar\delta}\xi^{\alpha\bar\beta}\overline{\xi^{\delta\bar\gamma}}$$

is positive (respectively nonnegative) for arbitrary complex numbers $\xi^{\alpha\bar\beta}$ when $\xi^{\alpha\bar\beta} \neq 0$ for at least one pair of indices (α, β).

Clearly very strong negativity of the curvature tensor implies strong negativity of the curvature tensor. Strong negativity of the curvature tensor implies negativity of the sectional curvature.

Theorem 1. *Suppose M and N are compact Kähler manifolds and the curvature tensor of M is strongly negative. Suppose $f : N \to M$ is a harmonic map and the rank over $\mathbf{R}$ of the differential df of f is at least 4 at some point of N. Then f is either holomorphic or conjugate holomorphic.*

In Theorem 1 the curvature condition on M can be weakened to the following. The curvature tensor of M is strongly seminegative everywhere and is strongly negative at $f(P)$ for some $P \in N$ with $\operatorname{rank}_{\mathbf{R}} df \geqq 4$ at P.

The following strong rigidity theorem is a consequence of Theorem 1.

Theorem 2. *Let M be a compact Kähler manifold of complex dimension at least two whose curvature tensor is strongly negative. Then a compact Kähler manifold of the same homotopy type as M must be either biholomorphic or conjugate biholomorphic to M.*

Though most of the bounded symmetric domains do not have strongly negative curvature tensor, our method can also yield the strong rigidity of compact quotients of classical bounded symmetric domains and the complex-analyticity of harmonic maps into them. The precise statements are contained in the following two theorems.

Theorem 3. *Suppose N is a compact Kähler manifold and M is a compact quotient of a bounded symmetric domain of type* $\mathrm{I}_{m,n}(mn \geqq 2), \mathrm{II}_n(n \geqq 3), \mathrm{III}_n(n \geqq 2)$, *or* $\mathrm{IV}_n(n \geqq 3)$ *(whose precise descriptions are given in §5). Suppose $f : N \to M$ is a harmonic map which is a submersion at some point of N. Then f is either holomorphic or conjugate holomorphic.*

Theorem 4. *Let M be a compact quotient of a bounded symmetric domain of type* $\mathrm{I}_{m,n}(mn \geqq 2), \mathrm{II}_n(n \geqq 3), \mathrm{III}_n(n \geqq 2)$, *or* $\mathrm{IV}_n(n \geqq 3)$. *Then any compact Kähler manifold of the same homotopy type as M must be either biholomorphic or conjugate biholomorphic to M.*

The main part of this paper will devote to the proofs of these theorems.

As examples of compact Kähler manifolds with very strongly negative curvature tensor, we show that any compact quotient of the ball has very strongly negative curvature tensor.

Proposition 1. *The curvature tensor of the invariant metric of the ball is very strongly negative.*

Proof. The invariant metric $2\ \mathrm{Re} \sum_{\alpha\cdot\beta} g_{\alpha\bar\beta} dz^\alpha \overline{dz^\beta}$ of the ball B of C^n is given by

$$g_{\alpha\bar\beta} = \partial_\alpha \partial_{\bar\beta}(-\log(1-|z|^2)).$$

Since

$$-\log(1-|z|^2) = |z|^2 + \frac{|z|^4}{2} + \frac{|z|^6}{3} + \dots,$$

it follows that at the origin all the components of the curvature tensor

$$R_{\alpha\bar\beta\gamma\bar\delta} = \partial_\alpha \partial_{\bar\beta} \partial_\gamma \partial_{\bar\delta}(-\log(1-|z|^2)) = \partial_\alpha \partial_{\bar\beta} \partial_\gamma \partial_{\bar\delta}\left(\frac{|z|^4}{2}\right)$$

are zero except the following

$$\begin{aligned} R_{\alpha\bar\alpha\alpha\bar\alpha} &= 2, \\ R_{\alpha\bar\alpha\beta\bar\beta} &= R_{\alpha\bar\beta\beta\bar\alpha} = 1 \quad \text{for } \alpha \neq \beta. \end{aligned}$$

Hence at the origin

$$\begin{aligned} \sum_{\alpha,\beta,\gamma,\delta} R_{\alpha\bar\beta\gamma\bar\delta} \xi^{\alpha\bar\beta} \overline{\xi^{\delta\bar\gamma}} &= \sum_\alpha \left(\sum_{\gamma\delta} R_{\alpha\bar\alpha\gamma\bar\delta} \xi^{\alpha\bar\alpha} \overline{\xi^{\delta\bar\gamma}} \right) + \sum_{\alpha\neq\beta} \left(\sum_{\gamma,\delta} R_{\alpha\bar\beta\gamma\bar\delta} \xi^{\alpha\bar\beta} \overline{\xi^{\delta\bar\gamma}} \right) \\ &= \sum_\alpha \left(\sum_\gamma R_{\alpha\bar\alpha\gamma\bar\gamma} \xi^{\alpha\bar\alpha} \overline{\xi^{\gamma\bar\gamma}} \right) + \sum_{\alpha\neq\beta} R_{\alpha\bar\beta\beta\bar\alpha} \xi^{\alpha\bar\beta} \overline{\xi^{\alpha\bar\beta}} \\ &= 2\sum_\alpha |\xi^{\alpha\bar\alpha}|^2 + \sum_{\alpha\neq\gamma} \xi^{\alpha\bar\alpha} \overline{\xi^{\gamma\bar\gamma}} + \sum_{\alpha\neq\beta} |\xi^{\alpha\bar\beta}|^2 \end{aligned}$$

$$= \sum_{\alpha} |\xi^{\alpha\bar{\alpha}}|^2 + \sum_{\alpha,\gamma} \xi^{\alpha\bar{\alpha}}\xi^{\gamma\bar{\gamma}} + \sum_{\alpha\neq\beta} |\xi^{\alpha\bar{\beta}}|^2$$

$$= \sum_{\alpha} |\xi^{\alpha\bar{\alpha}}|^2 + \left|\sum_{\alpha} \xi^{\alpha\bar{\alpha}}\right|^2 + \sum_{\alpha\neq\beta} |\xi^{\alpha\bar{\beta}}|^2$$

which is $\geqq 0$ and is zero if and only if all $\xi^{\alpha\bar{\beta}} = 0$. Q.E.D

Compact quotients of the ball are not the only examples in complex dimension $\geqq 2$ of compact Kähler manifolds with very strongly negative curvature tensor. Recently Mostow-Siu [8] constructed a compact Kähler surface with very strongly negative curvature tensor whose universal covering is not biholomorphic to the ball of complex dimension two.

3 A Bochner type identity

Suppose M is a Kähler manifold with metric tensor

$$g = 2\text{Re} \sum_{\alpha,\beta} g_{\alpha\bar{\beta}} dz^{\alpha} d\overline{z^{\beta}}$$

and suppose N is a complex manifold and $f : N \to M$ is a smooth (i.e., C^{∞}) map. Let T_M denote the real tangent bundle of M when M is regarded as a real manifold. The complex structure of M gives a decomposition of $T_M \otimes \mathbf{C}$ into tangent vectors of type (1, 0) and type (0, 1),

$$T_M \otimes \mathbf{C} = T_M^{1,0} \oplus T_M^{0,1}.$$

The differential $df : T_N \to T_M$ of f given rise to a map

$$df \otimes \mathbf{C} : T_N \otimes \mathbf{C} \to T_M \otimes \mathbf{C}.$$

Composing this with the projection map

$$\Pi_{1,0} : T_M \otimes \mathbf{C} \to T_M^{1,0},$$

we obtain

$$\Pi_{1,0} \circ (df \otimes \mathbf{C}) : T_N \otimes \mathbf{C} \to T_M^{1,0}.$$

This is equivalent to a bundle map

$$T_N \otimes \mathbf{C} \to f^* T_M^{1,0}$$

of C-vector bundles over N. Composing this bundle map with the inclusion map

$$T_N^{0,1} \hookrightarrow T_N \otimes \mathbf{C},$$

we obtain a bundle map from $T_N^{0,1} \to f^* T_M^{1,0}$ which we denote by $\bar{\partial} f$. Hence $\bar{\partial} f$ is a smooth section of the C-vector bundle

$$\text{Hom}_{\mathbf{C}}(T_N^{0,1}, f^* T_M^{1,0}) = (T_N^{1,0})^* \otimes f^* T_M^{1,0}$$

over N. In other words $\bar{\partial} f$ is an $f^*T^{1,0}_M$-valued $(0,1)$-form on N.

Let (w^i) be a local holomorphic coordinate chart of N. Then in terms of the local coordinate charts (z^α) and (w^i) of M and N, $\bar{\partial} f$ is simply represented by $(\partial_{\bar{i}} f^\alpha)$, where $\partial_{\bar{i}} = \partial/\partial\overline{w^i}$. In Sections 1—5 we will use the notation $\partial_{\bar{i}} = \partial/\partial\overline{w^i}$ and also the notations $\partial_i = \partial/\partial w^i, \partial_\alpha = \partial/\partial z^\alpha$, and $\partial_{\bar{\alpha}} = \partial/\partial\overline{z^\alpha}$. In these notations we may substitute another lower case italic (respectively Greek) for i (respectively α).

Likewise we define $\partial f, \partial\bar{f}$, and $\bar{\partial}\bar{f}$. ∂f is an $f^*T^{1,0}$-valued (1, 0)-form on N represented by $(\partial_i f^\alpha)$.$\partial\bar{f}$ is an $f^*T^{0,1}$-valued (1, 0)-form on N represented by $(\partial_i\overline{f^\alpha})$. $\bar{\partial} f$ is an $f^*T^{0,1}$-valued $(0,1)$-form on N represented by $(\partial_{\bar{i}}\overline{f^\alpha})$. It is clear that $\partial\bar{f}$ is the complex conjugate of $\bar{\partial}\bar{f}$ and $\bar{\partial}\bar{f}$ is the complex conjugate of ∂f.

Let ∇ be the Riemannian connection of M defined by its Kähler metric. It induces a connection $f^*\nabla$ on the vector bundle $f^*T^{1,0}$. This connection together with the ∂ operator of N enable us to define, for any $f^*T^{1,0}$-valued (0, 1)-form ω on N, the ∂ exterior derivative of ω, which we denote by $D\omega$. It is an $f^*T^{1,0}$-valued (1, 1)-form on N. In terms of local coordinates, if $\omega = (\omega^\alpha_{\bar{j}})$, then

$$D\omega = \left(\sum_{i,j} \eta^\alpha_{i\bar{j}} dw^i \wedge d\overline{w^j} \right)$$

with

$$\eta^\alpha_{i\bar{j}} = \partial_i \omega^\alpha_{\bar{j}} + \sum_{\beta,\gamma} {}^M\Gamma^\alpha_{\beta\gamma} \omega^\beta_{\bar{j}} \partial_i f^\gamma,$$

where ${}^M\Gamma^\alpha_{\beta\gamma}$ is the Christoffel symbol of M (evaluated at $f(Q)$ when the equation is considered at the point Q of N).

Likewise we define, for any $f^*T^{0,1}$-valued (1, 0)-form ω' on N, the $\bar{\partial}$ exterior derivative $\bar{D}\omega'$ of ω' which is an $f^*T^{0,1}$-valued (1, 1)-form on N.

Let A^r_s be the bundle over M of (complex-valued) tensors of contra-variant order r and covariant order s. Let σ be a section of A^r_s on M and let τ be an $f^*A^s_r$-valued p-form on N. We denote by $\langle\sigma, \tau\rangle$ the p-form on N obtained by the contraction which contracts elements of A^r_s with elements of A^s_r to form scalars.

Let $R = (R_{\alpha\bar{\beta}\gamma\bar{\delta}})$ denote the curvature tensor of M. We are now ready to state our Bochner type identity.

Proposition 2.

$$\partial\bar{\partial}\langle g, \bar{\partial} f \wedge \partial\bar{f}\rangle = \langle R, \bar{\partial} f \wedge \partial\bar{f} \wedge \partial f \wedge \overline{\partial f}\rangle - \langle g, D\bar{\partial} f \wedge \bar{D}\partial\bar{f}\rangle.$$

In local coordinates

$$\partial\bar{\partial} \sum_{\alpha,\beta} g_{\alpha\bar{\beta}} \bar{\partial} f^\alpha \wedge \partial\bar{f}^\beta = \sum_{\alpha,\beta,\gamma,\delta} R_{\alpha\bar{\beta}\gamma\bar{\delta}} \bar{\partial} f^\alpha \wedge \partial\bar{f}^\beta \wedge \partial f^\gamma \wedge \partial\bar{f}^\delta$$
$$- \sum_{\alpha,\beta} g_{\alpha\bar{\beta}} D\bar{\partial} f^\alpha \wedge \bar{D}\partial\overline{f^\beta},$$

where

$$D\bar{\partial} f^\alpha = \partial\bar{\partial} f^\alpha + \sum_{\beta,\gamma} {}^M\Gamma^\alpha_{\beta\gamma} \partial f^\beta \wedge \bar{\partial} f^\gamma,$$

$$\bar{D}\partial\overline{f^\beta} = \bar{\partial}\partial\overline{f^\beta} + \sum_{\alpha,\gamma} {}^M\overline{\Gamma^\beta_{\alpha\gamma}}\bar{\partial}\overline{f^\alpha} \wedge \partial\overline{f^\gamma}.$$

Proof. Fix a point Q of N where we want to verify the equation. Let $P = f(Q)$. We choose a holomorphic local coordinate system (z^α) of M at P such that $dg_{\alpha\bar{\beta}} = 0$ at P. Then at P all the Christoffel symbols ${}^M\Gamma^\alpha_{\beta\gamma}$ vanish. It follows that at P,

$$R_{\alpha\bar{\beta}\gamma\bar{\delta}} = \partial_\gamma\partial_{\bar{\delta}}g_{\alpha\bar{\beta}}$$

and at Q

$$D\bar{\partial}f^\alpha = \partial\bar{\partial}f^\alpha,$$
$$\bar{D}\partial\overline{f^\alpha} = \bar{\partial}\partial\overline{f^\alpha}.$$

Using $\partial_\gamma g_{\alpha\bar{\beta}} = 0$ at P and recalling that in the equation $g_{\alpha\bar{\beta}}$ stands for $g_{\alpha\bar{\beta}} \circ f$, we have at Q

$$\begin{aligned}\partial\bar{\partial}\sum_{\alpha\beta} g_{\alpha\bar{\beta}}\bar{\partial}f^\alpha \wedge \partial\overline{f^\beta} = & \sum_{\alpha,\beta,\gamma,\delta} \partial_\delta\partial_\gamma g_{\alpha\bar{\beta}}\partial f^\delta \wedge \bar{\partial}f^\gamma \wedge \bar{\partial}f^\alpha \wedge \partial\overline{f^\beta} \\ & + \sum_{\alpha,\beta,\gamma,\delta} \partial_{\bar{\delta}}\partial_\gamma g_{\alpha\bar{\beta}}\partial\overline{f^\delta} \wedge \bar{\partial}f^\gamma \wedge \bar{\partial}f^\alpha \wedge \partial\overline{f^\beta} \\ & + \sum_{\alpha,\beta,\gamma,\delta} \partial_\delta\partial_{\bar{\gamma}} g_{\alpha\bar{\beta}}\partial f^\delta \wedge \bar{\partial}\overline{f^\gamma} \wedge \bar{\partial}f^\alpha \wedge \partial\overline{f^\beta} \\ & + \sum_{\alpha,\beta,\gamma,\delta} \partial_{\bar{\delta}}\partial_{\bar{\gamma}} g_{\alpha\bar{\beta}}\partial\overline{f^\delta} \wedge \bar{\partial}\overline{f^\gamma} \wedge \bar{\partial}f^\alpha \wedge \partial\overline{f^\beta} \\ & - \sum_{\alpha,\beta} g_{\alpha\bar{\beta}}\partial\bar{\partial}f^\alpha \wedge \bar{\partial}\partial\overline{f^\beta}\end{aligned}$$

Since $\partial_\delta\partial_\gamma g_{\alpha\bar{\beta}}$ is symmetric in α, γ (due to the fact that the metric $2\mathrm{Re}\sum g_{\alpha\bar{\beta}}dz^\alpha \overline{dz^\beta}$ is Kähler) and since $\partial f^\delta \wedge \bar{\partial}f^\gamma \wedge \bar{\partial}f^\alpha \wedge \partial\overline{f^\beta}$ is skew-symmetric in α, γ, it follows that the first term on the right-hand side of the preceding equation is zero. Likewise, it follows from the symmetry of $\partial_{\bar{\delta}}\partial_\gamma g_{\alpha\bar{\beta}}$ in α, γ and the skew-symmetry of $\partial\overline{f^\delta} \wedge \bar{\partial}f^\gamma \wedge \bar{\partial}f^\alpha \wedge \partial\overline{f^\beta}$ in α, γ that the second term on the right-hand side is zero. And it follows from the symmetry of $\partial_{\bar{\delta}}\partial_{\bar{\gamma}} g_{\alpha\bar{\beta}}$ in β, δ and the skew-symmetry of $\partial\overline{f^\delta} \wedge \bar{\partial}\overline{f^\gamma} \wedge \bar{\partial}f^\alpha \wedge \partial\overline{f^\beta}$ that the last term on the right-hand side is zero. Hence

$$\begin{aligned}& \partial\bar{\partial}\sum_{\alpha,\beta} g_{\alpha\bar{\beta}}\bar{\partial}f^\alpha \wedge \partial\overline{f^\beta} \\ = & \sum_{\alpha,\beta,\gamma,\delta} R_{\alpha\bar{\beta}\delta\bar{\gamma}}\partial f^\delta \wedge \overline{\partial f^\gamma} \wedge \bar{\partial}f^\alpha \wedge \partial\overline{f^\beta} \\ & - \sum_{\alpha,\beta} g_{\alpha\bar{\beta}}D\bar{\partial}f^\alpha \wedge \bar{D}\partial\overline{f^\beta}.\end{aligned}$$

Q.E.D.

4 Complex-Analyticity of harmonic maps

Proposition 3. *Let M, N be compact Kähler manifolds and $f : N \to M$ be a harmonic map. Let $m = \dim_C M, n = \dim_C N$, and $f^\alpha (1 \leqq \alpha \leqq m)$ be the components of f with respect to a local coordinate system of M. Let $w^i (1 \leqq i \leqq n)$ be a local coordinate system of N and let*

$$\xi_{\bar{i}\bar{j}}^{\alpha\bar{\beta}} = (\partial_{\bar{i}} f^\alpha)(\overline{\partial_j f^\beta}) - (\partial_{\bar{j}} f^\alpha)(\overline{\partial_i f^\beta}).$$

If the curvature tensor $R_{\alpha\bar{\beta}\gamma\bar{\delta}}$ of M is strongly seminegative, then for $1 \leqq i, j \leqq n$,

$$\sum_{\alpha,\beta,\gamma,\delta} R_{\alpha\bar{\beta}\gamma\bar{\delta}} \xi_{\bar{i}\bar{j}}^{\alpha\bar{\beta}} \overline{\xi_{\bar{i}\bar{j}}^{\delta\bar{\gamma}}} = 0.$$

Proof. Let g=2Re$\sum_{\alpha,\beta} g_{\alpha\bar{\beta}} dz^\alpha \overline{dz^\beta}$ and h=2Re$\sum_{i,j} h_{i\bar{j}} dw^i \overline{dw^j}$ be respectively the Kähler metrics of M and N. Denote the Kähler form $\sqrt{-1}\sum_{i,j} h_{i\bar{j}} dw^i \overline{dw^j}$ of N by ω. We use the notations of Section 3. First we show that

$$\langle g, D\bar{\partial} f \wedge \bar{D}\partial \bar{f}\rangle \wedge \omega^{n-2} = \chi\omega^n \tag{4.1}$$

for some nonnegative function χ on N.

Fix a point Q of N and let $P = f(Q)$. To prove (4.1) at Q, we choose local holomorphic coordinate systems (z^α) at P and (w^i) at Q such that with respect to these coordinate systems

$g_{\alpha\bar{\beta}} = \delta_{\alpha\beta}$	(the Kronecker delta) at P,
$dg_{\alpha\bar{\beta}} = 0$	at P,
$h_{i\bar{j}} = \delta_{ij}$	at Q,
$dh_{i\bar{j}} = 0$	at Q.

Let u^α and v^α be respectively the real and imaginary parts of f^α. Let $\lambda_1^{(\alpha)}, \ldots, \lambda_n^{(\alpha)}$ (respectively $\mu_1^{(\alpha)}, \ldots, \mu_n^{(\alpha)}$) be the eigenvalues of the complex Hessian of u^α (respectively v^α) at Q with respect to the coordinates (w^i). We have

$$\begin{aligned}\langle g, D\bar{\partial} f \wedge \bar{D}\partial \bar{f}\rangle \wedge \omega^{n-2} &= \sum_\alpha \partial\bar{\partial} f^\alpha \wedge \bar{\partial}\partial\overline{f^\alpha} \wedge \omega^{n-2} \\ &= \sum_\alpha (\partial\bar{\partial} u^\alpha \wedge \bar{\partial}\partial u^\alpha + \partial\bar{\partial} v^\alpha \wedge \bar{\partial}\partial v^\alpha) \wedge \omega^{n-2} \\ &= \frac{-1}{n(n-1)} \sum_\alpha \sum_{i\neq j} (\lambda_i^{(\alpha)}\lambda_j^{(\alpha)} + \mu_i^{(\alpha)}\mu_j^{(\alpha)})\omega^n,\end{aligned}$$

where the last identity is obtained by diagonalizing the complex Hessians of u^α and v^α (nonsimultaneously). Now (4.1) follows from the identity

$$\sum_{i\neq j} \lambda_i \lambda_j = \left(\sum_i \lambda_i\right)^2 - \sum_i \lambda_i^2$$

and the fact that the harmonicity of f implies

$$\sum_i \lambda_i^{(\alpha)} = \sum_i \mu_i^{(\alpha)} = 0.$$

This proof of (4.1) yields also the statement that $\chi = 0$ at Q if and only if $D\bar{\partial} f^\alpha = 0$ at Q for all α.

We now express

$$\langle R, \bar{\partial} f \wedge \partial \bar{f} \wedge \partial f \wedge \overline{\partial f} \rangle \wedge \omega^{n-2}$$

in terms of local coordinates. Fix $Q \in N$. Choose an arbitrary local holomorphic coordinate system (w^i) at Q such that $h_{i\bar{j}} = \delta_{ij}$ at Q. Then at Q

$$\begin{aligned}
&\langle R, \bar{\partial} f \wedge \partial \bar{f} \wedge \partial f \wedge \overline{\partial f} \rangle \wedge \omega^{n-2} \\
&= \sum_{\alpha,\beta,\gamma,\delta} R_{\alpha\bar{\beta}\gamma\bar{\delta}} \bar{\partial} f^\alpha \wedge \partial \overline{f^\beta} \wedge \partial f^\gamma \wedge \overline{\partial f^\delta} \wedge (n-2)!(\sqrt{-1})^{n-2} \\
&\quad \times \sum_{1 \leqq i < j \leqq n} \wedge_{1 \leqq k \leqq n, k \neq i,j} (dw^k \wedge \overline{dw^k}) \\
&= (n-2)!(\sqrt{-1})^{n-2} \sum_{\alpha,\beta,\gamma,\delta,1 \leqq i<j \leqq n} R_{\alpha\bar{\beta}\gamma\bar{\delta}} (-(\partial_{\bar{i}} f^\alpha)(\partial_i \overline{f^\beta})(\partial_j f^\lambda)(\partial_{\bar{j}} f^\delta) \\
&\quad +(\partial_{\bar{i}} f^\alpha)(\partial_j \overline{f^\beta})(\partial_i f^\gamma)(\partial_{\bar{j}} f^{\bar{\delta}}) + (\partial_{\bar{j}} f^\alpha)(\partial_i \overline{f^\beta})(\partial_j f^\gamma)(\partial_{\bar{i}} f^{\bar{\delta}}) \\
&\quad -(\partial_{\bar{j}} f^\alpha)(\partial_j \overline{f^\beta})(\partial_i f^\gamma)(\partial_{\bar{i}} f^{\bar{\delta}})) \wedge \left(\bigwedge_{1 \leqq k \leqq n} (dw^k \wedge \overline{dw^k}) \right) \\
&= \frac{-1}{n(n-1)} \sum_{\alpha,\beta,\gamma,\delta,1 \leqq i<j \leqq n} R_{\alpha\bar{\beta}\gamma\bar{\delta}} ((\partial_{\bar{i}} f^\alpha)(\overline{\partial_{\bar{j}} f^\delta}) - (\partial_{\bar{j}} f^\alpha)(\overline{\partial_{\bar{i}} f^\delta})) \\
&\quad \times \overline{((\partial_{\bar{i}} f^\beta)(\partial_j f^\gamma) - (\partial_{\bar{j}} f^\beta)(\partial_i f^\gamma))} \omega^n.
\end{aligned}$$

Since $R_{\alpha\bar{\beta}\gamma\bar{\delta}}$ is symmetric in β, δ, it follows that

$$\langle R, \bar{\partial} f \wedge \partial \bar{f} \wedge \partial f \wedge \overline{\partial f} \rangle \wedge \omega^{n-2} = \frac{4}{n(n-1)} \sum_{\alpha,\beta,\gamma,\delta,1 \leqq i<j \leqq n} R_{\alpha\bar{\beta}\gamma\bar{\delta}} \xi_{\bar{i}\bar{j}}^{\alpha\bar{\beta}} \overline{\xi_{\bar{i}\bar{j}}^{\delta\bar{\gamma}}} \omega^n$$

$$\text{at any point } Q \text{ of } N \text{ when } h_{i\bar{j}} = \delta_{ij} \text{ at } Q. \tag{4.2}$$

The strong seminegativity of the curvature tensor $R_{\alpha\bar{\beta}\gamma\bar{\delta}}$ implies that

$$\langle R, \bar{\partial} f \wedge \partial \bar{f} \wedge \partial f \wedge \overline{\partial f} \rangle \wedge \omega^{n-2} = \sigma \omega^n \tag{4.3}$$

for some nonpositive function σ on N.

By Proposition 2,

$$\begin{aligned}
\partial\bar{\partial} \langle g, \bar{\partial} f \wedge \partial \bar{f} \rangle \wedge \omega^{n-2} &= \langle R, \bar{\partial} f \wedge \partial \bar{f} \wedge \partial f \wedge \overline{\partial f} \rangle \wedge \omega^{n-2} \\
&\quad - \langle g, D\bar{\partial} f \wedge \bar{D} \partial f \rangle \wedge \omega^{n-2}.
\end{aligned}$$

Since the left-hand side is an exact form, it follows that

$$\int_N \langle R, \bar{\partial} f \wedge \partial \bar{f} \wedge \partial f \wedge \overline{\partial f} \rangle \wedge \omega^{n-2} - \langle g, D\bar{\partial} f \wedge \bar{D}\partial f \rangle \wedge \omega^{n-2} = 0.$$

From (4.1) and (4.3) we conclude that χ and σ are identically zero. By (4.2),

$$\sum_{\alpha,\beta,\gamma,\delta} R_{\alpha\bar{\beta}\gamma\bar{\delta}} \xi^{\alpha\bar{\beta}}_{\bar{i}\bar{j}} \xi^{\overline{\delta\bar{\gamma}}}_{\bar{i}\bar{j}} = 0. \tag{4.4}$$

To finish the proof, we take an arbitrary local holomorphic coordinate system (ζ^i) at Q. Let

$$\eta^{\alpha\bar{\beta}}_{\bar{i}\bar{j}} = \frac{\partial f^\alpha}{\partial \overline{\zeta^i}} \frac{\overline{\partial f^\beta}}{\partial \zeta^j} - \frac{\partial f^\alpha}{\partial \overline{\zeta^j}} \frac{\overline{\partial f^\beta}}{\partial \zeta^i}$$

evaluated at Q. We have to show that

$$\sum_{\alpha,\beta,\gamma,\delta} R_{\alpha\bar{\beta}\gamma\bar{\delta}} \eta^{\alpha\bar{\beta}}_{\bar{i}\bar{j}} \eta^{\overline{\delta\bar{\gamma}}}_{\bar{i}\bar{j}} = 0. \tag{4.5}$$

Fix $1 \leqq i, j \leqq n$. Since $\eta^{\alpha\bar{\beta}}_{\bar{i}\bar{j}} = 0$ when $i = j$, we can assume that $i \neq j$. At Q let

$$\frac{\partial}{\partial \zeta^i} = \sum_k a_k \frac{\partial}{\partial w^k},$$
$$\frac{\partial}{\partial \zeta^j} = \sum_k b_k \frac{\partial}{\partial w^k}.$$

There exist uniquely two unit orthogonal n-vectors $(A_1^{(\nu)}, \ldots, A_n^{(\nu)}), \nu = 1, 2$, such that $a_k = cA_k^{(1)}$ and $b_k = c'A_k^{(1)} + c''A_k^{(2)}, 1 \leqq k \leqq n$, for some complex numbers c, c', c''. Construct a unitary matrix $(A_k^{(\nu)}), 1 \leqq \nu, k \leqq n$ whose first two rows are these two unit orthogonal n-vectors. Define a local holomorphic coordinate system (τ^k) at Q so that

$$\frac{\partial}{\partial \tau^\nu} = \sum_k A_k^{(\nu)} \frac{\partial}{\partial w^k}$$

at Q. Let

$$\theta^{\alpha\bar{\beta}} = \frac{\partial f^\alpha}{\partial \overline{\tau^1}} \frac{\overline{\partial f^\beta}}{\partial \tau^2} - \frac{\partial f^\alpha}{\partial \overline{\tau^1}} \frac{\overline{\partial f^\beta}}{\partial \tau^2}$$

evaluated at Q. By replacing the coordinate system (w^k) by the coordinate system (τ^k), corresponding to (4.4) we obtain the following:

$$\sum_{\alpha,\beta,\gamma,\delta} R_{\alpha\bar{\beta}\gamma\bar{\delta}} \theta^{\alpha\bar{\beta}} \theta^{\overline{\delta\bar{\gamma}}} = 0. \tag{4.6}$$

Now

$$\begin{aligned}\eta^{\alpha\bar\beta}_{\bar i\bar j} &= \sum_{k,l}\bar a_k\bar b_l\xi^{\alpha\bar\beta}_{\bar k\bar l}\\ &= \bar c\bar c'\sum_{k,l}\overline{A^{(1)}_k}\,\overline{A^{(1)}_l}\xi^{\alpha\bar\beta}_{\bar k\bar l}+\bar c\bar c''\sum_{k,l}\overline{A^{(1)}_k}\,\overline{A^{(2)}_l}\xi^{\alpha\bar\beta}_{\bar k\bar l}\ .\\ &= \bar c\bar c''\theta^{\alpha\bar\beta},\end{aligned}$$

where the vanishing of $\sum_{k,l}\overline{A^{(1)}_k}\,\overline{A^{(1)}_l}\xi^{\alpha\bar\beta}_{\bar k\bar l}$ is due to the skew-symmetry of $\xi^{\alpha\bar\beta}_{\bar k\bar l}$ in k,l. Hence (4.5) follows from (4.6). Q.E.D.

In order to avoid repeating part of the argument in the case of classical bounded symmetric domains, instead of proving directly Theorem 1, we prove a more general result. To state this more general result, we have to introduce the following definition.

Definition. The curvature tensor $R_{\alpha\bar\beta\gamma\bar\delta}$ of a Kähler manifold of complex dimension m is called *negative of order* k if it is strongly seminegative and it enjoys the following property. If $A=(A^\alpha_{\bar i}),B=(B^\alpha_i)$ are any two $m\times k$ matrices $(1\leqq\alpha\leqq m,1\leqq i\leqq k)$ with

$$\operatorname{rank}\begin{pmatrix}A & B\\ \bar B & \bar A\end{pmatrix}=2k$$

and if

$$\sum_{\alpha,\beta,\gamma,\delta}R_{\alpha\bar\beta\gamma\bar\delta}\xi^{\alpha\bar\beta}_{\bar i\bar j}\overline{\xi^{\delta\bar\gamma}_{\bar i\bar j}}=0$$

for all $1\leqq i,j\leqq k$, where

$$\xi^{\alpha\bar\beta}_{\bar i\bar j}=A^\alpha_{\bar i}\overline{B^\beta_j}-A^\alpha_{\bar j}\overline{B^\beta_i},$$

then either $A=0$ or $B=0$.

The curvature tensor $R_{\alpha\bar\beta\gamma\bar\delta}$ is called *adequately negative* if it is negative of order m.

The above definition is motivated by the following equivalent definition which is clumsier to state but which renders more transparent the motivation behind the definition.

Definition. The curvature tensor $R_{\alpha\bar\beta\gamma\bar\delta}$ of a Kähler manifold M is said to be *negative of order* k at a point P of M if it is strongly seminegative at P and if it enjoys the following property. If $f:U\to M$ is a smooth map from an open neighborhood U of 0 in C^k to M with $f(0)=P$ and $\operatorname{rank}_{\mathbf{R}}df=2k$ at 0 and if at P

$$\sum_{\alpha,\beta,\gamma,\delta}R_{\alpha\bar\beta\gamma\bar\delta}\xi^{\alpha\bar\beta}_{\bar i\bar j}\overline{\xi^{\delta\bar\gamma}_{\bar i\bar j}}=0$$

for $1\leqq i,j\leqq k$, where

$$\xi^{\alpha\bar\beta}_{\bar i\bar j}=(\partial_{\bar i}f^\alpha)(0)\overline{(\partial_j f^\beta)(0)}-(\partial_{\bar j}f^\alpha)(0)\overline{(\partial_i f^\beta)(0)}$$

(the coordinates of $\mathbf{C}^k$ being $w^i, 1 \leqq i \leqq k$), then either $\partial f = 0$ at 0 or $\bar{\partial} f = 0$ at 0.

The curvature tensor $R_{\alpha\bar{\beta}\gamma\bar{\delta}}$ is said to be *adequately negative* at P if the above holds with the condition $\operatorname{rank}_{\mathbf{R}} df = 2k$ at 0 replaced by the condition that f is locally diffeomorphic at 0.

We will use only the latter definition, because there the indices have transparent meanings and are easier to keep track of.

Theorem 5. *Let $k \geqq 2$. Suppose M and N are compact Kähler manifolds and the curvature tensor of M is negative of order k. Suppose $f : N \to M$ is a harmonic map and the rank over $\mathbf{R}$ of the differential df of f is at least $2k$ at some point of N. Then f is either holomorphic or conjugate holomorphic.*

Before we prove Theorem 5, we have to prove first the following very simple lemma in linear algebra.

Lemma 1. *Suppose V is a vector space of dimension n over $\mathbf{C}$ and W is an $\mathbf{R}$-vector subspace of V and the dimension of W over $\mathbf{R}$ is $\leqq 2n - 2k$. Let E be the set of all bases of V over $\mathbf{C}$ and let F be the subset of E consisting of all bases $(e_1, \ldots, e_n)$ of V over $\mathbf{C}$ such that $(\sum_{\nu=1}^k \mathbf{C} e_{i_\nu}) \cap W = 0$ for all $1 \leqq i_1 < \cdots < i_k \leqq n$. Then F is a dense open subset of E.*

Proof. For $1 \leqq i_1 < \cdots < i_k \leqq n$ let $F_{i_1 \cdots i_k}$ be the subset of E consisting of all $(e_1, \ldots, e_n)$ with $(\sum_{\nu=1}^k \mathbf{C} e_{i_\nu}) \cap W = 0$. It suffices to show that each $F_{i_1 \cdots i_k}$ is a dense open subset of E. Let H be the set of all $\mathbf{C}$-vector subspaces L of complex dimension k in V with $L \cap W = 0$. It suffices to show that H is a dense open subset of the Grassmannian $G_k(V)$ of all $\mathbf{C}$-vector subspaces of complex dimension k in V. Clearly H is open in $G_k(V)$.

Let $K = W \cap \sqrt{-1} W$ be the maximum $\mathbf{C}$-vector subspace of V contained in W. We can choose a basis $e_1, \ldots, e_n$ of V over $\mathbf{C}$ such that $e_1, \ldots, e_p$ is a basis of K over $\mathbf{C}$ and

$$e_1, \ldots, e_p, \sqrt{-1} e_1, \ldots, \sqrt{-1} e_p, e_{p+1}, \ldots, e_l$$

is a basis of W over $\mathbf{R}$. We have $l + p = \dim_{\mathbf{R}} W \leqq 2n - 2k$. First we show that H is nonempty. Let q be the largest integer $\leqq (l - p/2)$. Let Q be the $\mathbf{C}$-vector subspace of V spanned by

$$e_{p+1} + \sqrt{-1} e_{p+2}, e_{p+3} + \sqrt{-1} e_{p+4}, \ldots, e_{p+2q-1} + \sqrt{-1} e_{p+2q}, e_{l+1}, \ldots, e_n$$

over $\mathbf{C}$. Clearly $Q \cap W = 0$. We claim that $\dim_{\mathbf{C}} Q \geqq k$. Since $2q \geqq l - p - 1$ and $2n - 2k \geqq l + p$, it follows that $2n - 2k + 2q \geqq 2l - 1$, which implies $n - k + q \geqq 1$, because n, k, q, l are integers. Hence $\dim_{\mathbf{C}} Q = q + (n - l) \geqq k$. Choose a $\mathbf{C}$-vector subspace L' of complex dimension k in Q. Then $L' \in H$.

Take $L \in G_k(V)$ and let $g_1, \ldots, g_k$ be a basis of L over $\mathbf{C}$. Let

$$g_i = \sum_{j=1}^n a_{ji} e_j \quad (1 \leqq i \leqq k).$$

Let $A = (a_{ji})_{p<j\leqq l, 1\leqq i\leqq k}$ and $B = (a_{ji})_{l<j\leqq n, 1\leqq i\leqq k}$. Then $L \cap W \neq 0$ if and only if for some nonzero column c of k complex numbers Im $Ac = 0$ and $Bc = 0$, i. e.,

$$\begin{cases} Ac - \bar{A}\bar{c} = 0, \\ Bc = 0, \\ \bar{B}\bar{c} = 0, \end{cases}$$

which is equivalent to

$$\operatorname{rank} \begin{pmatrix} A & -\bar{A} \\ B & 0 \\ 0 & \bar{B} \end{pmatrix} < 2k,$$

because, if

$$\begin{cases} Ac - \bar{A}d = 0, \\ Bc = 0, \\ \bar{B}d = 0 \end{cases}$$

for some c, d not both zero, then by suitably adding the equations to and subtracting the equations from their complex conjugates one obtains

$$\begin{cases} A\xi - \bar{A}\bar{\xi} = 0, \\ B\xi = 0, \\ \bar{B}\bar{\xi} = 0 \end{cases}$$

for $\xi = c + \bar{d}, \sqrt{-1}(c - \bar{d})$, one of which is nonzero. By replacing L by L' we obtain g_i', a_{ji}', A', and B'. For $0 \leqq t \leqq 1$, let

$$\begin{aligned} g_i(t) &= (1-t)g_i + tg_i', \\ a_{ji}(t) &= (1-t)a_{ji} + ta_{ji}', \\ A(t) &= (1-t)A + tA', \\ B(t) &= (1-t)B + tB'. \end{aligned}$$

The following two inequalities

$$\operatorname{rank}\,(a_{ji}(t))_{1\leqq j\leqq n, 1\leqq i\leqq k} \geqq k,$$
$$\operatorname{rank} \begin{pmatrix} A(t) & -\overline{A(t)} \\ B(t) & 0 \\ 0 & \overline{B(t)} \end{pmatrix} \geqq 2k.$$

hold at $t = 1$ and hence hold for all $t \in [0,1] - J$, where J is some finite set. For $t \in [0,1] - J$ let $L(t)$ be the C-vector subspace of V spanned by $g_1(t), \ldots, g_k(t)$ over C. Then $L(t) \in H$ and $L(t)$ approaches L in $G_k(V)$ as t approaches 0. Hence H is dense in $G_k(V)$. Q.E.D.

Proof of Theorem 5. Now $\operatorname{rank}_{\mathbf{R}} df \geqq 2k$ at some point of N and hence at every point of some nonempty connected open subset U of N. We first prove that

$$\text{either } \partial f \equiv 0 \text{ on } U \text{ or } \bar{\partial} f \equiv 0 \text{ on } U. \tag{4.7}$$

It suffices to show that for every point Q of U

$$\text{either } \partial f = 0 \text{ at } Q \text{ or } \bar{\partial} f = 0 \text{ at } Q, \tag{4.8}$$

because the nowhere-vanishing of df on U implies that the two closed subsets $U \cap \{\partial f = 0\}$ and $U \cap \{\bar{\partial} f = 0\}$ are disjoint and since by (4.8) their union is the connected set U, one of them is equal to U.

Since the curvature tensor $R_{\alpha\bar{\beta}\gamma\bar{\delta}}$ is strongly seminegative, it follows from Proposition 3 that for $1 \leqq i, j \leqq n$ and $1 \leqq \alpha, \beta \leqq m$

$$\sum_{\alpha,\beta,\gamma,\delta} R_{\alpha\bar{\beta}\gamma\bar{\delta}} \xi_{\bar{i}\bar{j}}^{\alpha\bar{\beta}} \xi_{\bar{i}\bar{j}}^{\overline{\delta\bar{\gamma}}} = 0 \text{ at } P, \tag{4.9}$$

where

$$\xi_{\bar{i}\bar{j}}^{\alpha\bar{\beta}} = (\partial_{\bar{i}} f^\alpha)(Q)\overline{(\partial_j f^\beta)(Q)} - (\partial_{\bar{j}} f^\alpha)(Q)\overline{(\partial_i f^\beta)(Q)}$$

($w^i, 1 \leqq i \leqq m$, being local coordinates of N and $f^\alpha, 1 \leqq \alpha \leqq n$, being components of f with respect to a local coordinate system of M).

Let $P = f(Q)$ and let $T_{M,P}$ (respectively $T_{N,Q}$) be the tangent space of M at P (respectively N at Q) when M (respectively N) is regarded as a real manifold. Let K be the kernel of $df : T_{N,Q} \to T_{M,P}$. The complex structure of M makes $T_{M,P}$ a vector space over C. Since $\text{rank}_{\mathbf{R}} df \geqq 2k$, it follows that $\dim_{\mathbf{R}} K \leqq 2n - 2k$. By Lemma 1, there exists a basis $g_1, \ldots, g_n$ of $T_{N,Q}$ over C such that for $1 \leqq i_1 < \cdots < i_k \leqq n$ the intersection of K with the C-vector subspace of $T_{N,Q}$ spanned by $g_{i_1}, \ldots, g_{i_k}$ is the zero vector subspace. Choose a holomorphic coordinate system (w^i) of an open neighborhood W of Q in U such that $g_i = 2\text{Re}(\partial/\partial w^i)$. For $1 \leqq i_1 < \cdots < i_k \leqq n$ let $\Phi_{i_1 \cdots i_k}$ be the restriction of f to

$$W \cap \{w_\gamma = w_\gamma(Q) \quad \text{for} \quad 1 \leqq \gamma \leqq n, \gamma \neq i_1, \ldots, i_k\}.$$

Then $\text{rank}_{\mathbf{R}} d\Phi_{i_1 \cdots i_k} = 2k$ at Q. Since $R_{\alpha\bar{\beta}\gamma\bar{\delta}}$ is negative of order k at every point of M, it follows from (4.9) that for $1 \leqq i_1 < \cdots < i_k \leqq n$ either $\partial_{\bar{j}} f^\alpha = 0$ at Q for $1 \leqq \alpha \leqq m$ and $j = i_1, \ldots, i_k$ or $\partial_j f^\alpha = 0$ at Q for $1 \leqq \alpha \leqq m$ and $j = i_1, \ldots, i_k$. Now (4.8) follows from $k \geqq 2$, because, if $\partial_{\bar{j}} f^\alpha \neq 0$ at Q for some $1 \leqq \alpha \leqq m$ and some $1 \leqq j \leqq n$, and $\partial_l f^\beta \neq 0$ at Q for some $1 \leqq \beta \leqq m$ and some $1 \leqq l \leqq n$, then we can select $1 \leqq i_1 < \cdots < i_k \leqq n$ such that both j and l belong to the set $\{i_1, \ldots, i_k\}$. The theorem now follows from (4.8) and the following proposition.

Q.E.D.

Proposition 4. *Suppose M, N are compact Kähler manifolds and $f : N \to M$ is a harmonic map. Let U be a nonempty open subset of N. If f is holomorphic (respectively conjugate holomorphic) on U, then f is holomorphic (respectively conjugate holomorphic) on N.*

Proof. Since the proofs of the holomorphic case and the conjugate holomorphic case are similar, we prove only the holomorphic case.

Let Ω be the largest connected open subset of N containing U such that $\bar{\partial} f$ vanishes identically on Ω. It suffices to show that Ω is closed in N. Suppose the contrary Ω has a boundary point Q. Let W be a connected open neighborhood of Q in N such that

i) there exists a holomorphic coordinate system (w^i) on some open neighborhood of the closure of W and

ii) there exists a holomorphic coordinate system (z^α) on some open neighborhood of the closure of $f(W)$.

The harmonicity of f is given by the equation

$$\Delta_N f^\alpha + \sum_{i,j,\beta,\gamma} {}^M\Gamma^\alpha_{\beta\gamma}(\partial_i f^\beta)(\partial_{\bar{j}} f^\gamma) h^{\bar{j}\,i} = 0,$$

where $(h^{\bar{j}\,i})$ is the inverse matrix of the matrix $(h_{i\bar{j}})$ of the Kähler metric of N, Δ_N is the Laplace-Beltrami operator of N, and ${}^M\Gamma^\alpha_{\beta\gamma}$ is the Christoffel symbol of M. Applying $\partial_{\bar{k}}$ to this equation and recalling $\Delta_N = 2\sum_{ij} h^{\bar{j}\,i}\partial_i\partial_{\bar{j}}$, we obtain

$$\begin{aligned}\Delta_N(\partial_{\bar{k}} f^\alpha) + \sum_{i,j}(2\partial_{\bar{k}} h^{\bar{j}\,\bar{i}})\partial_i\partial_{\bar{j}} f^\alpha \quad &+ \sum_j \left(\sum_{i,\beta,\gamma} \partial_{\bar{k}}({}^M\Gamma^\alpha_{\beta\gamma}(\partial_i f^\beta) h^{\bar{j}\,i})\right)\partial_{\bar{j}} f^\gamma \\ &+ \sum_j \left(\sum_{i,\beta,\gamma} {}^M\Gamma^\alpha_{\beta r}(\partial_i f^\beta) h^{\bar{j}\,i}\right)\partial_{\bar{k}}\partial_{\bar{j}} f^\gamma = 0.\end{aligned}$$

Hence for some positive number C

$$|\Delta_N(\partial_{\bar{k}} f^\alpha)| \leqq C\left(\sum_{i,j}|\partial_i\partial_{\bar{j}} f^\alpha| + \sum_{j,\gamma}|\partial_{\bar{j}} f^\gamma| + \sum_{j,\gamma}|\partial_{\bar{k}}\partial_{\bar{j}} f^\gamma|\right)$$

on W. Let $u^\alpha_{\bar{k}}$ and $v^\alpha_{\bar{k}}$ be respectively the real and imaginary parts of $\partial_{\bar{k}} f^\alpha$. Since Δ_N is a real operator (i.e., it maps real functions to real functions), it follows that for some positive number C'

$$\begin{aligned}|\Delta_N u^\alpha_{\bar{k}}|^2 &\leqq C' \sum_{\beta,l}(|\mathrm{grad}\, u^\beta_{\bar{l}}|^2 + |\mathrm{grad}\, v^\beta_{\bar{l}}|^2 + |u^\beta_{\bar{l}}|^2 + |v^\beta_{\bar{l}}|^2), \\ |\Delta_N v^\alpha_{\bar{k}}|^2 &\leqq C' \sum_{\beta,l}(|\mathrm{grad} u^\beta_{\bar{l}}|^2 + |\mathrm{grad} v^\beta_{\bar{l}}|^2 + |u^\beta_{\bar{l}}| + |v^\beta_{\bar{l}}|^2)\end{aligned}$$

on W, where grad is the gradient with respect to the coordinate system whose coordinate functions are the real and imaginary parts of w^i. By applying Aronszajn's unique continuation theorem [1, p. 248] to the system of functions $u^\alpha_{\bar{k}}, v^\alpha_{\bar{k}} (1 \leqq \alpha \leqq m, 1 \leqq k \leqq n)$ and to the elliptic operator Δ_N, we conclude from the identical vanishing of $u^\alpha_{\bar{k}}, v^\alpha_{\bar{k}}$ on $W \cap \Omega$ that $u^\alpha_{\bar{k}}, v^\alpha_{\bar{k}}$ vanish identically on W. This contradicts the fact that Q is a boundary point of Ω. Hence $\Omega = N$ and $\bar{\partial} f \equiv 0$ on N. Q.E.D.

Theorem 1 follows from Theorem 5 and the following lemma.

Lemma 2. *If the curvature tensor $R_{\alpha\bar{\beta}\gamma\bar{\delta}}$ of a Kähler manifold M is strongly negative, then it is negative of order* 2.

Proof. Let U be an open neighborhood of 0 in $\mathbf{C}^2$ with coordinates w^1, w^2. Suppose $f : U \to M$ is a smooth map which is a local immersion at 0 such that for $1 \leqq i, j \leqq 2$

$$\sum_{\alpha,\beta,\gamma,\delta} R_{\alpha\bar{\beta}\gamma\bar{\delta}} \xi_{\bar{i}\bar{j}}^{\alpha\bar{\beta}} \xi_{\bar{i}\bar{j}}^{\overline{\delta\bar{\gamma}}} = 0$$

at $P = f(0)$, where

$$\xi_{\bar{i}\bar{j}}^{\alpha\bar{\beta}} = (\partial_{\bar{i}} f^\alpha)(0)\overline{(\partial_j f^\beta)(0)} - (\partial_{\bar{j}} f^\alpha)(0)\overline{(\partial_i f^\beta)(0)}.$$

We have to show that either $\partial f = 0$ at 0 or $\bar{\partial} f = 0$ at 0.

Let $m = \dim_{\mathbf{C}} M$. Let $T_{M,P}$ be the tangent space of M at P when M is regarded as a real manifold. The complex structure of M makes $T_{M,P}$ a vector space over C. Since f is a local immersion at 0, the image L of df is an R-vector subspace of real dimension 4 in $T_{M,P}$. By Lemma 1, we can find a basis $e_1, \ldots, e_n$ of T_M over C such that for $1 \leqq \alpha < \beta \leqq m$ the inter section of L with the C-vector subspace of $T_{M,P}$ spanned by $e_\gamma, 1 \leqq \gamma \leqq n, \gamma \neq \alpha, \beta$, is the zero vector subspace. Choose a local coordinate system (z^α) of M at P such that $e_\alpha = 2\mathrm{Re}(\partial/\partial z^\alpha)$ at P. For $1 \leqq \alpha < \beta \leqq m$ the map

$$\Phi_{\alpha\beta} : (w^1, w^2) \mapsto (f^\alpha(w^1, w^2), f^\beta(w^1, w^2))$$

is locally diffeomorphic at 0.

Since $R_{\alpha\bar{\beta}\gamma\bar{\delta}}$ is strongly negative, it follows that $\xi_{\bar{i}\bar{j}}^{\alpha\bar{\beta}} = 0$ for $1 < \alpha, \beta \leqq m$ and $1 \leqq i, j \leqq 2$. Assume $\partial f \neq 0$ at 0. We want to prove that $\bar{\partial} f = 0$ at 0. Since $\partial f^\alpha \neq 0$ at 0 for some $1 \leqq \alpha \leqq m$, without loss of generality we can assume that $\partial f^1 \neq 0$ at 0. Let p be the number of C-linearly independent (1, 0)-forms among $\partial f^1, \ldots, \partial f^m$ at 0. We distinguish between two cases.

Case 1. $p = 2$. Without loss of generality we can assume that ∂f^1 and ∂f^2 are C-linearly independent at 0. For $1 \leqq \alpha \leqq m$ and $\beta = 1, 2$, it follows from $\xi_{\bar{1}\bar{2}}^{\alpha\bar{\beta}} = 0$ that $\bar{\partial} f^\alpha = c_{\alpha\beta}\overline{\partial f^\beta}$ at 0 for some $c_{\alpha\beta} \in \mathbf{C}$. Hence, if $\bar{\partial} f^\alpha \neq 0$ at 0 for some $1 \leqq \alpha \leqq n$, then $\partial f^1 = (c_{\alpha 2}/c_{\alpha 1})\partial f^2$ at 0, contradicting the C-linear independence of ∂f^1 and ∂f^2 at 0.

Case 2. $p = 1$. For $1 < \alpha \leqq m$ there exist complex numbers r_α such that $\partial f^\alpha = r_\alpha \partial f^1$ at 0. For $1 \leqq \alpha \leqq m$ it follows from $\xi_{\bar{1}\bar{2}}^{\alpha\bar{1}} = 0$ that $\bar{\partial} f^\alpha = c_\alpha \overline{\partial f^1}$ at 0 for some $c_\alpha \in \mathbf{C}$. For $1 \leqq \alpha < \beta \leqq m$, $df^\alpha \wedge d\overline{f^\alpha} \wedge df^\beta \wedge d\overline{f^\beta}$ is a linear combination of exterior products of ∂f^1 and $\overline{\partial f^1}$ at 0 and hence vanishes at 0, contradicting the fact that $\Phi_{\alpha\beta}$ is locally diffeomorphic at 0. Q.E.D.

5 Bounded symmetric domains

We recall the four types of classical bounded symmetric domains which we denote by $D_{m\ n}^{\mathrm{I}}, D_n^{\mathrm{II}}, D_n^{\mathrm{III}}, D_n^{\mathrm{IV}}$ and compute their curvature tensors.

Type $\mathrm{I}_{m,n}$. The domain $D_{m,n}^{\mathrm{I}}$ is an open subset of $\mathbf{C}^{mn}$ and is the set of all $m \times n$ matrices $Z = (z_{\alpha\beta})$ with complex entries such that $\mathrm{I}_n - {}^t Z \bar{Z}$ is positive definite, where I_n is the identity matrix of order n and ${}^t\bar{Z}$ is the transpose of the complex conjugate of Z. An invariant Kähler metric has the potential function

$$\begin{aligned}\Phi &= \log\det(\mathrm{I}_n - {}^t\bar{Z}Z)^{-1}\\ &= \sum |z_{\alpha\beta}|^2 + \frac{1}{2}\sum \bar{z}_{\alpha\gamma}z_{\alpha\delta}\bar{z}_{\beta\delta}z_{\beta\gamma} + \text{higher order terms.}\end{aligned}$$

At $Z = 0$ the coordinates $(Z_{\alpha\beta})$ are normal coordinates in the sense that the first order derivatives of the coefficients of the metric tensor with respect to these coordinates vanish at $Z = 0$. Hence the curvature tensor at $Z = 0$ is given by

$$R_{\alpha\gamma\overline{\beta\rho}\lambda\sigma,\overline{\mu\tau}} = \frac{\partial^4\Phi}{\partial z_{\alpha\gamma}\partial\bar{z}_{\beta\rho}\partial z_{\lambda\sigma}\partial\bar{z}_{u\tau}} = \delta_{\alpha\beta}\delta_{\lambda\mu}\delta_{\gamma\tau}\delta_{\rho\sigma} + \delta_{\alpha\mu}\delta_{\beta\lambda}\delta_{\gamma\rho}\delta_{\sigma\tau}.$$

It follows that at $Z = 0$

$$\begin{aligned}&\sum R_{\alpha\gamma,\overline{\beta\rho},\lambda\sigma,\overline{\mu\tau}}\xi^{\alpha\gamma,\overline{\beta\rho}}\overline{\xi^{\mu\tau,\overline{\lambda\sigma}}} \qquad (5.1)\\ &= \sum \delta_{\alpha\beta}\delta_{\lambda\mu}\delta_{\gamma\tau}\delta_{\rho\sigma}\xi^{\alpha\gamma,\overline{\beta\rho}}\overline{\xi^{\mu\tau,\overline{\lambda\sigma}}} + \sum \delta_{\alpha\mu}\delta_{\beta\lambda}\delta_{\gamma\rho}\delta_{\sigma\tau}\xi^{\alpha\gamma,\overline{\beta\rho}}\overline{\xi^{\mu\tau,\overline{\lambda\sigma}}}\\ &= \sum_{\alpha,\lambda,\gamma,\rho} \xi^{\alpha\gamma,\overline{\alpha\rho}}\overline{\xi^{\lambda\gamma,\overline{\lambda\rho}}} + \sum_{\alpha,\beta,\gamma,\sigma} \xi^{\alpha\gamma,\overline{\beta\gamma}}\overline{\xi^{\alpha\sigma,\overline{\beta\sigma}}}\\ &= \sum_{\gamma,\rho}\left|\sum_{\alpha}\xi^{\alpha\gamma,\overline{\alpha\rho}}\right|^2 + \sum_{\alpha\beta}\left|\sum_{\gamma}\xi^{\alpha\gamma,\overline{\beta\gamma}}\right|^2.\end{aligned}$$

Type II_n. The domain D_n^{II} is an open subset of $\mathrm{C}^{n(n-1)/2}$ and is the set of all skew-symmetric $n \times n$ matrices $Z = (z_{\alpha\beta})$ with complex entries such that $\mathrm{I}_n - {}^t\bar{Z}Z$ is positive definite. It is a complex submanifold of $D_{n,n}^{\mathrm{I}}$. The invariant Kähler metric of $D_{n,n}^{\mathrm{I}}$ induces an invariant Kähler metric of D_n^{II}. At $Z = 0$, the coordinates $(z_{\alpha\beta})$ for $\alpha < \beta$ are normal coordinates. Hence the second fundamental form of D_n^{II} in $D_{n,n}^{\mathrm{I}}$ vanishes at $Z = 0$ and the curvature tensor of D_n^{II} at $Z = 0$ is the restriction of the curvature of $D_{n,n}^{\mathrm{I}}$. It follows that at $Z = 0$

$$\begin{aligned}&\sum_{\alpha<\gamma\ \beta<\rho\ \lambda<\sigma\ \mu<\tau} R_{\alpha\gamma\overline{\beta\rho}\lambda\sigma,\overline{\mu\tau}}\xi^{\alpha\gamma,\overline{\beta\rho}}\overline{\xi^{\mu\tau,\overline{\lambda\sigma}}} \qquad (5.2)\\ &= \sum_{\gamma,\rho=1}^{n}\left|\sum_{\alpha=1}^{n}\xi^{\alpha\gamma,\overline{\alpha\rho}}\right|^2 + \sum_{\alpha,\beta=1}^{n}\left|\sum_{\gamma=1}^{n}\xi^{\alpha\gamma,\overline{\beta\gamma}}\right|^2,\end{aligned}$$

where $\xi^{\alpha\gamma,\overline{\beta\rho}}$ is skew-symmetric in α, γ and skew-symmetric in β, ρ.

Type III_n. The domain D_n^{III} is an open subset of $\mathrm{C}^{n(n+1)/2}$ and is the set of all symmetric $n \times n$ matrices $Z = (z_{\alpha\beta})$ with complex entries such that $\mathrm{I}_n - {}^t\bar{Z}Z$ is positive definite. It is a complex submanifold of $D_{n,n}^{\mathrm{I}}$. The invariant Kähler metric of $D_{n,n}^{\mathrm{I}}$ induces an invariant Kähler metric of D_n^{III}. At $Z = 0$, the coordinates $(z_{\alpha\beta})$ for $\alpha \leqq \beta$ are normal coordinates. Hence the second fundamental form of

D_n^{III} in $D_{n,n}^{\mathrm{I}}$ vanishes at $Z=0$ and the curvature tensor of D_n^{III} at $Z=0$ is the restriction of the curvature tensor of $D_{n,n}^{\mathrm{I}}$. It follows that at $Z=0$

$$\sum_{\alpha\leqq\gamma,\beta\leqq\rho,\lambda\leqq\sigma,\mu\leqq\tau} R_{\alpha\gamma,\overline{\beta\rho}\lambda\sigma,\overline{\mu\tau}}\xi^{\alpha\gamma,\overline{\beta\rho}}\overline{\xi^{\mu\tau,\overline{\lambda\sigma}}} \tag{5.3}$$
$$= \sum_{\gamma,\rho=1}^{n}\left|\sum_{\alpha=1}^{n}\xi^{\alpha\gamma,\overline{\alpha\rho}}\right|^2 + \sum_{\alpha,\beta=1}^{n}\left|\sum_{\gamma=1}^{n}\xi^{\alpha\gamma,\overline{\beta\gamma}}\right|^2,$$

where $\xi^{\alpha\gamma,\overline{\beta\rho}}$ is symmetric in α,γ and symmetric in β,ρ.

Type IV_n. The domain D_n^{IV} is an open subset of $\mathbf{C}^n$ and is the set of all $z=(z_1,\ldots,z_n)$ in $\mathbf{C}^n$ satisfying

$$\begin{cases} 1+\left|\sum_\alpha z_\alpha^2\right|^2 - 2\sum_\alpha |z_\alpha|^2 > 0, \\ \sum_\alpha |z_\alpha|^2 < 1. \end{cases}$$

An invariant Kähler metric is given by the potential function

$$\Phi = -\log\left(1-2\sum_\alpha|z_\alpha|^2+\left|\sum_\alpha z_\alpha^2\right|^2\right)$$
$$= 2\sum_\alpha|z_\alpha|^2 - \left|\sum_\alpha z_\alpha^2\right|^2 + 2\left(\sum_\alpha|z_\alpha|^2\right)^2 + \text{higher order terms.}$$

At $z=0$ the coordinates $z_\alpha (1\leqq\alpha\leqq n)$ are normal coordinates. Hence the curvature tensor is given by

$$R_{\alpha\bar\rho\beta\bar\sigma} = \frac{\partial^4\Phi}{\partial z_\alpha\partial\bar z_\rho\partial z_\beta\partial\bar z_\sigma} = 4(\delta_{\alpha\rho}\delta_{\beta\sigma}+\delta_{\alpha\sigma}\delta_{\beta\rho}-\delta_{\alpha\beta}\delta_{\rho\sigma}).$$

It follows that at $z=0$

$$\sum R_{\alpha\bar\rho\beta\bar\sigma}\xi^{\alpha\bar\rho}\overline{\xi^{\alpha\bar\beta}} \tag{5.4}$$
$$= 4\sum\delta_{\alpha\rho}\delta_{\beta\sigma}\xi^{\alpha\bar\rho}\overline{\xi^{\sigma\bar\beta}} + 4\sum\delta_{\alpha\sigma}\delta_{\beta\rho}\xi^{\alpha\bar\rho}\overline{\xi^{\sigma\bar\beta}} - 4\sum\delta_{\alpha\beta}\delta_{\rho\sigma}\xi^{\alpha\bar\rho}\overline{\xi^{\sigma\bar\beta}}$$
$$= 4\sum_{\alpha,\beta}\xi^{\alpha\bar\alpha}\overline{\xi^{\beta\bar\beta}} + 4\sum\xi^{\alpha\bar\beta}_{\alpha,\beta}\overline{\xi^{\alpha\bar\beta}} - 4\sum_{\alpha,\rho}\xi^{\overline{\alpha\rho}}\overline{\xi^{\bar\rho\alpha}}$$
$$= 4\left|\sum_\alpha\xi^{\alpha\bar\alpha}\right|^2 + 2\sum_{\alpha\neq\beta}|\xi^{\alpha\bar\beta}-\xi^{\beta\bar\alpha}|^2.$$

The curvature tensors of these four types of classical bounded symmetric domains are strongly seminegative, but are not strongly negative. In the following four Sections 6–9, we will show that these curvature tensors are adequately negative so that Theorem 3 follows from Theorem 5. For the proof of the adequate negativity of the curvature tensors, we will need the following very simple lemma in linear algebra.

Lemma 3. *Suppose φ is a smooth map from an open neighborhood of 0 in C^m to an open neighborhood of 0 in C^n with $\varphi(0) = 0$. Let $w_1, \ldots, w_m$ (respectively $z_1, \ldots, z_n$) be the coordinates of C^m (respectively C^n).*

a) *If the rank of the $n \times m$ matrix $(\partial z_\alpha / \partial w_i)$ at 0 is n, then there exists a nonsingular linear transformation*

$$w_i = \sum_{j=1}^{m} b_{ij} w'_j \quad (1 \leqq i \leqq m) \tag{5.5}$$

such that at 0,

$$\frac{\partial z_\alpha}{\partial w'_i} = \delta_{\alpha i} \quad \textit{for} \quad 1 \leqq \alpha \leqq n, \quad 1 \leqq i \leqq m.$$

b) *If the rank of the $n \times m$ matrix $(\partial z_\alpha / \partial w_i)$ at 0 is $p < n$, then there exist a nonsingular linear transformation (5.5) and a unitary transformation*

$$z'_\alpha = \sum_{\beta=1}^{n} a_{\alpha\beta} z_\beta \quad (1 \leqq \alpha \leqq n)$$

such that at 0,

$$\frac{\partial z'_\alpha}{\partial w'_i} = \delta_{\alpha i} \quad \textit{for} \quad 1 \leqq \alpha \leqq p, \quad 1 \leqq i \leqq m$$

and

$$\frac{\partial z'_\alpha}{\partial w'_i} = 0 \quad \textit{for} \quad p < \alpha \leqq n, \quad 1 \leqq i \leqq m.$$

c) *If the rank of the $n \times m$ matrix $(\partial z_\alpha / \partial w_i)$ at 0 is $p < n$ and $\partial z_1 / \partial w_i \neq 0$ at 0 for some $1 \leqq i \leqq m$, then there exist a nonsingular linear transformation (5.5) and a unitary transformation of the form*

$$z'_1 = z_1,$$
$$z'_\alpha = \sum_{\beta=2}^{n} a_{\alpha\beta} z_\beta \quad (2 \leqq \alpha \leqq n)$$

such that at 0

$$\frac{\partial z'_\alpha}{\partial w'_i} = \delta_{\alpha i} \quad \textit{for} \quad 1 \leqq \alpha \leqq p, \quad 1 \leqq i \leqq m$$

and

$$\frac{\partial z'_\alpha}{\partial w'_i} = 0 \quad \textit{for} \quad p < \alpha \leqq n, \quad 2 \leqq i \leqq m.$$

Proof. Let $\Phi : C^m \to C^n$ be the C-linear map defined by the $n \times m$ matrix $(\partial z_\alpha / \partial w_i)$ at 0. Let $e_1, \ldots, e_n$ (respectively $u_1, \ldots, u_m$) be the standard unit basis vectors of C^n (respectively C^m).

a) Take a basis $v_{n+1}, \ldots, v_m$ of Ker Φ. Let v_i be a vector in $\mathbf{C}^m$ with $\Phi(v_i) = e_i (1 \leqq i \leqq n)$. Let

$$v_i = \sum_{j=1}^{m} b_{ij} u_j \quad (1 \leqq i \leqq m).$$

Then (b_{ij}) satisfies the requirement.

b) Take a unitary basis $d_1, \ldots, d_n$ of $\mathbf{C}^n$ such that $d_\alpha \in \mathrm{Im}\Phi$ for $1 \leqq \alpha \leqq p$. Take a basis $v_{p+1}, \ldots, v_m$ of Ker Φ. Let v_i be a vector in $\mathbf{C}^m$ with $\Phi(v_i) = e_i (1 \leqq i \leqq p)$. Let

$$v_i = \sum_{j=1}^{m} b_{ij} u_j \quad (1 \leqq i \leqq m)$$

and

$$d_\alpha = \sum_{\beta=1}^{n} \alpha_{\alpha\beta} e_\beta \quad (1 \leqq \alpha \leqq n).$$

Then (b_{ij}) and $(a_{\alpha\beta})$ satisfy the requirement.

c) Let $\pi : \mathbf{C}^n \to \mathbf{C}e_1$ denote the orthogonal projection. Since $\partial z_1 / \partial w_i \neq 0$ at 0 for some $1 \leqq i \leqq m, \pi \circ \Phi$ is surjective. Let $K = \mathrm{Ker}(\pi \circ \Phi)$. Let $\Psi : K \to \mathrm{Ker}\, \pi = \oplus_{\alpha=2}^{n} Ce_\alpha$ be induced by Φ. Then rank $\Psi = p - 1$. Take a unitary basis $d_2, \ldots, d_n$ in $\oplus_{\alpha=2}^{n} Ce_\alpha$ such that $d_\alpha \in \mathrm{Im}\Psi$ for $2 \leqq \alpha \leqq p$. Let $d_1 = e_1$. Let π' denote the orthogonal projection from $\mathbf{C}^n$ onto the linear subspace V spanned by $d_1, \ldots, d_p$. Then $\pi' \circ \Phi : \mathbf{C}^m \to V$ is surjective. Choose a basis $v_{p+1}, \ldots, v_m$ of $\mathrm{Ker}(\pi' \circ \Phi)$. Let v_i be a vector in $\mathbf{C}^m$ with $(\pi' \circ \Phi)(v_i) = d_i$ for $1 \leqq i \leqq p$. Let

$$v_i = \sum_{j=1}^{m} b_{ij} u_j \quad (1 \leqq i \leqq m)$$

and

$$d_\alpha = \sum_{\beta=2}^{n} a_{\alpha\beta} e_\beta \quad (2 \leqq \alpha \leqq p).$$

Then (b_{ij}) and $(a_{\alpha\beta})$ satisfy the requirements, because $\Phi(v_i) = d_i$ for $2 \leqq i \leqq p$ which follows from the fact that $\Phi(v_i) - d_i$ belongs both to Im Φ and to the linear subspace spanned by $d_{p+1}, \ldots, d_n$. Q.E.D.

6 Adequate negativity of the curvature of D^{I}_{mn}

Proposition 5. *For $mn \geqq 2$ the curvature tensor $R_{\alpha\gamma,\overline{\beta\rho},\lambda\sigma,\overline{\mu\tau}} (1 \leqq \alpha, \beta, \lambda, \mu \leqq m, 1 \leqq \gamma, \rho, \sigma, \tau \leqq n)$ of D^{I}_{mn} is adequately negative.*

Proof. Let U be an open neighborhood of 0 in $\mathbf{C}^{mn}$. Let $(w_{ij})_{1 \leqq i \leqq m, 1 \leqq j \leqq n}$ be the coordinates of $\mathbf{C}^{mn}$. Denote $\partial / \partial \overline{w_{ij}}, \partial / \partial \overline{w_{ij}}$ by $\partial_{ij}, \partial_{\overline{ij}}$ respectively. Let $f : U \to D^{\mathrm{I}}_{m\,n}$ be a smooth map which maps the origin to the zero matrix and which is locally diffeomorphic at 0. Let $f^{\alpha\beta}(1 \leqq \alpha \leqq m, 1 \leqq \beta \leqq n)$ be the components and let

$$\xi_{\overline{ij},\overline{kl}}^{\alpha\beta,\overline{\gamma\delta}} = (\partial_{\overline{ij}} f^{\alpha\beta})(0)\overline{(\partial_{kl} f^{\gamma\delta})(0)} - (\delta_{\overline{kl}} f^{\alpha\beta})(0)\overline{(\partial_{ij} f^{\gamma\delta})(0)}.$$

Assume that

$$\sum_{\alpha,\beta,\gamma,\rho,\lambda,\mu,\sigma,\tau} R_{\alpha\gamma,\overline{\beta\rho},\lambda\sigma,\overline{\mu\tau}} \xi_{\overline{ij},\overline{kl}}^{\alpha\gamma,\overline{\beta\delta}} \overline{\xi_{\overline{ij},\overline{kl}}^{\mu\tau,\overline{\lambda\sigma}}} = 0 \tag{6.1}$$

at $Z = 0$ for all $(i,j),(k,l)$. We have to prove that

$$\begin{cases} \text{either } \partial_{\overline{ij}} f^{\alpha\beta} = 0 \text{ at } 0 \text{ for all } (i,j) \text{ and } (\alpha,\beta) \\ \text{or } \partial_{ij} f^{\alpha\beta} = 0 \text{ at } 0 \text{ for all } (i,j) \text{ and } (\alpha,\beta). \end{cases} \tag{6.2}$$

The following three conventions will be used in this proof and also in the proofs of Propositions 6 and 7. Moreover, in the proofs of Propositions 6 and 7 the notations $\partial_{ij}, \partial_{\overline{ij}}, f^{\alpha\beta}$, and $\xi_{\overline{ij},\overline{kl}}^{\alpha\beta,\overline{\gamma\sigma}}$ will carry analogous meanings.

i) For notational simplicity we will denote $(\partial_{ij} f^{\alpha\beta})(0), (\partial_{\overline{ij}} f^{\alpha\beta})(0), (\partial_{ij} f^{\overline{\alpha\beta}})$ $(0), (\partial_{\overline{ij}} f^{\overline{\alpha\beta}})(0)$ simply by $\partial_{ij} f^{\alpha\beta}, \partial_{\overline{ij}} f^{\alpha\beta}, \partial_{ij} f^{\overline{\alpha\beta}}, \partial_{\overline{ij}} f^{\overline{\alpha\beta}}$ respectively.

ii) After we apply linear transformations to the coordinates (w_{ij}) and $(z_{\alpha\beta})$, we will use the same symbols for the new coordinates.

iii) We use the lexicographical ordering for the double indices (i,j). By $(i,j) > (k,l)$ we mean either $i = k$ and $j > l$ or $i > k$. By $(i,j) \geqq (k,l)$ we mean $(i,j) > (k,l)$ or $(i,j) = (k,l)$.

From (6.1) and (5.1) it follows that

$$\begin{aligned} \sum_{\alpha} \xi_{\overline{ij},\overline{kl}}^{\alpha\gamma,\overline{\alpha\rho}} &= 0 \quad \text{for all } (\gamma,\rho), \\ \sum_{\gamma} \xi_{\overline{ij},\overline{kl}}^{\alpha\gamma,\overline{\beta\gamma}} &= 0 \quad \text{for all } (\alpha,\beta). \end{aligned} \tag{6.3}$$

We will prove (6.2) from the equations (6.3). The equations (6.3) are invariant under the following transformations of $D_{m,n}^{\mathrm{I}}$.

$$\begin{aligned} Z &\mapsto \bar{Z}, \\ Z &\mapsto {}^tZ, \\ Z &\mapsto AZB, \end{aligned} \tag{6.4}$$

where A and B are fixed unitary matrices respectively of orders m and n.

To prove the proposition from (6.3) we can assume without loss of generality (after the transformation $Z \to {}^tZ$ if necessary) that $m \geqq 2$.

Assume that $\partial_{ij} f^{\alpha\beta} \neq 0$ for some (i,j) and some (α,β). We want to prove that $\partial_{\overline{ij}} f^{\alpha\beta} = 0$ for all (i,j) and (α,β). By applying a unitary transformation to the rows of Z (i.e., a transformation of the form (6.4) with $B = \mathrm{I}_n$), we can assume without loss of generality that $\partial_{ij} f^{1\beta} \neq 0$ for some β and some (i,j). Let the rank of the $n \times (mn)$ matrix

$$(\partial_{ij} f^{1\beta})_{1 \leqq \beta \leqq n, 1 \leqq i \leqq m, 1 \leqq j \leqq n}$$

be p. By applying Lemma 3b) to the smooth map

$$(f^{11}, \ldots, f^{1n}); U \to \mathrm{C}^n,$$

we conclude that, after we apply a linear transformation to the coordinates (w_{ij}) and apply a unitary transformation to the columns of Z (i.e., a transformation of the form (6.4) with $A = \mathrm{I}_m$), we can assume without loss of generality that for $1 \leqq \beta \leqq n$,

$$\begin{cases} \partial_{1j} f^{1\beta} = \delta_{j\beta} & \text{for } 1 \leqq j \leqq p, \\ \partial_{ij} f^{1\beta} = 0 & \text{for } (i,j) > (1,p). \end{cases} \tag{6.5}$$

For $1 \leqq i \leqq p$ and $(k,l) > (1,p)$, consider

$$\sum_{\beta=1}^{n} \xi_{\overline{1i},kl}^{1\beta,\overline{1\beta}} = 0;$$

that is,

$$\sum_{\beta=1}^{n} (\partial_{\overline{1i}} f^{1\beta} \overline{\partial_{kl} f^{1\beta}} - \partial_{\overline{kl}} f^{1\beta} \overline{\partial_{1i} f^{1\beta}}) = 0.$$

Since $\partial_{kl} f^{1\beta} = 0$ and $\partial_{1i} f^{1\beta} = \delta_{i\beta}$, it follows that

$$\partial_{\overline{kl}} f^{1i} = 0 \text{ for } 1 \leqq i \leqq p \text{ and } (k,l) > (1,p). \tag{6.6}$$

We claim that the rank of the $(n-p) \times (mn-p)$ matrix

$$P := (\partial_{\overline{kl}} f^{1\gamma})_{p<\gamma \leqq n, (k,l)>(1,p)}$$

is $n-p$. Suppose the contrary. Then at 0,

$$\bar{\partial} f^{1,p+1} \wedge \cdots \wedge \bar{\partial} f^{1n} = 0 \quad \mathrm{mod}\ (d\bar{w}_{11}, \ldots, d\bar{w}_{1\mathrm{p}}).$$

By (6.5), $\partial f^{1\beta} = 0$ for $p < \beta \leqq n$. Hence at 0,

$$df^{1,p+1} \wedge \cdots \wedge df^{1n} = 0 \quad \mathrm{mod}\ (d\bar{w}_{11}, \ldots, d\bar{w}_{1p}). \tag{6.7}$$

By (6.5) and (6.6), $df^{1\beta}, 1 \leqq \beta \leqq p$, is a linear combination of $dw_{11}, \ldots, dw_{1p}$, $d\bar{w}_{11}, \ldots, d\bar{w}_{1p}$ at 0. It follows that at 0,

$$\bigwedge_{\beta=1}^{p} (df^{1\beta} \wedge df^{\overline{1\beta}}) = c \bigwedge_{\beta=1}^{p} (dw_{1\beta} \wedge d\bar{w}_{1\beta})$$

for some constant c. This together with (6.7) implies that at 0,

$$\bigwedge_{\beta=1}^{n} (df^{1\beta} \wedge df^{\overline{1\beta}}) = 0,$$

contradicting the assumption that f is locally diffeomorphic at 0. Hence the rank of P is $n-p$. By applying Lemma 3 a) to the map $(\overline{f^{1,p+1}}, \ldots, \overline{f^{1n}})$ whose domain variables are $(w_{kl}), (k,l) > (1,p)$ (the other variables being fixed), we conclude that after a linear coordinate change in the variables $(w_{kl}), (k,l) > (1,p)$, we have for $p < \gamma \leqq n$,

$$\begin{cases} \partial_{\overline{1j}} f^{1\gamma} = \delta_{j\gamma} & \text{for } p < j \leqq n, \\ \partial_{\overline{kl}} f^{1\gamma} = 0 & \text{for } 2 \leqq k \leqq m, 1 \leqq l \leqq n. \end{cases} \tag{6.8}$$

Since this last coordinate change involves only w_{kl} with $(k,l) > (1,p)$, the validity of (6.5) is not affected.

Take any $1 < \alpha \leqq m$. Let $1 < k \leqq m, 1 \leqq l \leqq n$, and $p < i \leqq n$. Consider

$$\sum_{\beta=1}^{n} \xi_{\overline{1ikl}}^{1\beta\overline{\alpha\beta}} = 0;$$

that is,

$$\sum_{\beta=1}^{n} (\partial_{\overline{1i}} f^{1\beta} \overline{\partial_{kl} f^{\alpha\beta}} - \partial_{\overline{kl}} f^{1\beta} \overline{\partial_{1i} f^{\alpha\beta}}) = 0.$$

Since $\partial_{\overline{kl}} f^{1\beta} = 0$ and $\partial_{\overline{1i}} f^{1\beta} = \delta_{i\beta}$ by (6.6) and (6.8), it follows that

$$\partial_{kl} f^{\alpha i} = 0 \text{ for } 1 < \alpha \leqq m, 1 < k \leqq m, 1 \leqq l \leqq n, \text{ and } p < i \leqq n. \tag{6.9}$$

Consider now

$$\sum_{\beta=1}^{n} \xi_{\overline{kl},\, \overline{1j}}^{\alpha\beta, \overline{1\beta}} = 0,$$

where $1 < \alpha \leqq m, 1 < k \leqq m, 1 \leqq l \leqq m$, and $1 \leqq j \leqq p$. That is,

$$\sum_{\beta=1}^{n} (\partial_{\overline{kl}} f^{\alpha\beta} \overline{\partial_{1j} f^{1\beta}} - \partial_{\overline{1j}} f^{\alpha\beta} \overline{\partial_{kl} f^{1\beta}}) = 0.$$

Since $\partial_{kl} f^{1\beta} = 0$ and $\partial_{1j} f^{1\beta} = \delta_{j\beta}$ by (6.5), it follows that

$$\partial_{\overline{kl}} f^{\alpha j} = 0 \text{ for } 1 < \alpha \leqq m, 1 < k \leqq m, 1 \leqq l \leqq n, \text{ and } 1 \leqq j \leqq p. \tag{6.10}$$

We claim that, after a linear coordinate change in (w_{kl}) for $k > 1$ the following hold for $1 \leqq i \leqq m$:

$$\begin{cases} \partial_{il} f^{i\beta} = \delta_{l\beta} & \text{for } 1 \leqq l \leqq p \text{ and } 1 \leqq \beta \leqq n. \\ \partial_{kl} f^{i\beta} = 0 & \text{for } (k,l) > (i,p) \text{ and } 1 \leqq \beta \leqq n. \\ \partial_{\overline{il}} f^{i\beta} = \delta_{l\beta} & \text{for } p < l, \beta \leqq n. \\ \partial_{\overline{kl}} f^{i\beta} = 0 & \text{for } k > i, 1 \leqq l, \beta \leqq n. \end{cases} \tag{6.11$_i$}$$

We prove this by induction on i. When $i = 1$, $(6.11)_i$ follows from (6.5) and (6.8). Suppose $(6.11)_i$ holds for $1 \leqq i < j$ and $j \leqq m$ and we want to show that $(6.11)_j$ holds. From $(6.11)_i, 1 \leqq i < j$, it follows that

$$\begin{cases} df^{\alpha\beta}, 1 \leqq \alpha < j, 1 \leqq \beta \leqq n, \text{is a linear combination of} \\ dw_{kl}, d\bar{w}_{kl}, 1 \leqq k < j, 1 \leqq l \leqq n, \text{ at } 0. \end{cases} \tag{6.12}$$

From (6.10) it follows that

$$\begin{cases} df^{j\beta}, 1 \leqq \beta \leqq p, \text{is a linear combination of } dw_{kl}, 1 \leqq k \leqq m, \\ 1 \leqq l \leqq n, \text{and } d\bar{w}_{1r}, 1 \leqq r \leqq n \text{ at } 0. \end{cases} \tag{6.13}$$

The $p \times (m-j+1)n$ matrix

$$(\partial_{kl} f^{j\beta})_{1 \leqq \beta \leqq p, j \leqq l \leqq m, l \leqq l \leqq n}$$

must have rank p, otherwise from (6.12) and (6.13) it follows that

$$\left(\bigwedge_{1 \leqq \alpha < j, 1 \leqq \beta \leqq n} (df^{\alpha\beta} \wedge d\overline{f^{\alpha\beta}}) \right) \wedge \left(\bigwedge_{1 \leqq \beta \leqq p} (df^{j\beta} \wedge d\overline{f^{j\beta}}) \right)$$

vanishes at 0, contradicting the assumption that f is locally diffeomorphic at 0. By applying Lemma 3 a) to the map $(f^{j1}, \ldots, f^{jp})$ whose domain variables are $(w_{kl}), j \leqq k \leqq m, 1 \leqq l \leqq n$ (the other variables being fixed), we conclude that, after a linear coordinate change in $(w_{kl}), j \leqq k \leqq m, 1 \leqq l \leqq n$, we have the first two equations of $(6.11)_j$. This implies that

$$\begin{cases} df^{j\beta}, 1 \leqq \beta \leqq p, \text{is a linear combination of } dw_{kl}, d\bar{w}_{kl}, \\ (j,p) \geqq (k,l), \text{at } 0. \end{cases} \tag{6.14}$$

From (6.9) it follows that

$$\begin{cases} df^{j\beta}, p < \beta \leqq n, \text{is a linear combination of } d\bar{w}_{kl}, 1 \leqq k \leqq m, \\ 1 \leqq l \leqq n, \text{and } dw_{1r}, 1 \leqq \gamma \leqq n, \text{at } 0. \end{cases} \tag{6.15}$$

The rank of the $(n-p) \times ((m-j+1)n-p)$ matrix

$$(\partial_{\overline{kl}} f^{j\beta})_{p < \beta \leqq n, (k,l) > (j,p)}$$

must be $n-p$, otherwise it follows from (6.12), (6.14), and (6.15) that

$$\bigwedge_{1 \leqq \alpha \leqq j, 1 \leqq \beta \leqq n} (df^{\alpha\beta} \wedge d\overline{f^{\alpha\beta}})$$

vanishes at 0, contradicting the assumption that f is locally diffeomorphic at 0. By applying Lemma 3 a) to the map $(\overline{f^{j,p+1}}, \ldots, \overline{f^{jn}})$ whose domain variables are (w_{kl}) with $(k,1) > (j,p)$ (the other variables being fixed), we conclude that, after a linear coordinate change in (w_{kl}) with $(k,l) > (j,p)$, we have the last two equations of $(6.11)_j$.

We distinguish between two cases.

Case 1. $p = n$. By the first two equations of $(6.11)_i$ we have for all (α, β)

$$\begin{cases} \partial_{\alpha\beta} f^{\alpha\beta} = 1 \\ \partial_{kl} f^{\alpha\beta} = 0 \qquad \text{for } (k,l) > (\alpha,\beta). \end{cases}$$

Hence the $(mn) \times (mn)$ matrix

$$(\partial_{ij} f^{\alpha\beta})_{1 \leqq i, \alpha \leqq m, 1 \leqq j, \beta \leqq n}$$

is nonsingular. By Lemma 3 a) we can apply a linear change of coordinates to $(w_{ij}), 1 \leqq i \leqq m, 1 \leqq j \leqq n$ so that after this linear coordinate change we have

$$\partial_{ij} f^{\alpha\beta} = \delta_{i\alpha}\delta_{j\beta}; \tag{6.16}$$

but statements derived above concerning $\partial_{\overline{ij}} f^{\alpha\beta}$ may no longer remain valid, i.e., (6.6), (6.8), (6.10), and the last two equations of $(6.11)_i$ may no longer hold.

We are now ready to show that $\partial_{\overline{ij}} f^{\alpha\beta} = 0$ for all α, β, i, j. Fix arbitrarily (α, β) and (i, j). Consider first the case $i \neq \alpha$. We have

$$\sum_{\gamma=1}^{n} \xi_{\overline{ij}, \alpha\beta}^{\alpha\gamma, \overline{\alpha\gamma}} = 0,$$

i. e.,

$$\sum_{\gamma=1}^{n} (\partial_{\overline{ij}} f^{\alpha\gamma} \overline{\partial_{\alpha\beta} f^{\alpha\gamma}} - \partial_{\overline{\alpha\beta}} f^{\alpha\gamma} \overline{\partial_{ij} f^{\alpha\gamma}}) = 0.$$

By (6.16), $\partial_{ij} f^{\alpha\gamma} = 0$ and $\partial_{\alpha\beta} f^{\alpha\gamma} = \delta_{\beta\gamma}$. Hence $\partial_{\overline{ij}} f^{\alpha\beta} = 0$. Consider now the remaining case $i = \alpha$. Since $m \geqq 2$, there exists $1 \leqq k \leqq m$ with $k \neq \alpha$. We have

$$\sum_{\gamma=1}^{n} \xi_{\overline{\alpha j}, k\beta}^{\alpha\gamma, \overline{k\gamma}} = 0,$$

that is,

$$\sum_{\gamma=1}^{n} (\partial_{\overline{\alpha j}} f^{\alpha\gamma} \overline{\partial_{k\beta} f^{k\gamma}} - \partial_{\overline{k\beta}} f^{\alpha\gamma} \overline{\partial_{\alpha j} f^{k\gamma}}) = 0.$$

By (6.16), $\partial_{\alpha j} f^{k\gamma} = 0$ and $\partial_{k\beta} f^{k\gamma} = \delta_{\beta\gamma}$. Hence $\partial_{\overline{\alpha j}} f^{\alpha\beta} = 0$.

Case 2. $p < n$. From the last two equations of $(6.11)_i$, we have

$$\begin{cases} \partial_{\overline{in}} f^{in} = 1, & \\ \partial_{\overline{in}} f^{jn} = 0 & \text{for } i > j. \end{cases}$$

It follows that the rank of the $m \times (mn)$ matrix

$$(\partial_{\overline{ij}} f^{\alpha n})_{1 \leqq i \leqq m, 1 \leqq j \leqq n, 1 \leqq \alpha \leqq m}$$

is m. Since $p < n$, we must have $n \geqq 2$. After we apply the transformations

$$Z \mapsto A^t \bar{Z},$$

where A is the matrix obtained from I_m by interchanging the first and last rows, we reduce this case to Case 1 and we therefore conclude, by the arguments of Case 1, that $\partial_{ij} f^{\alpha\beta} = 0$ for all $1 \leqq i, \alpha \leqq m$ and $1 \leqq j, \beta \leqq n$, which is a contradiction. Hence Case 2 cannot occur and $\partial_{\overline{ij}} f^{\alpha\beta} = 0$ for all $1 \leqq i, \alpha \leqq m$ and $1 \leqq j, \beta \leqq n$.

Q.E.D.

7 Adequate negativity of the curvature of D_n^{II}

Proposition 6. *For $n \geqq 3$ the curvature tensor $R_{\alpha\gamma,\overline{\beta\rho},\lambda\sigma,\overline{\mu\tau}}(1 \leqq \alpha < \gamma \leqq n, 1 \leqq \beta < \rho \leqq n, 1 \leqq \lambda < \sigma \leqq n, 1 \leqq \mu < \tau \leqq n)$ of D_n^{II} is adequately negative.*

Proof. Let U be an open neighborhood of 0 in $\mathbf{C}^{(1/2)n(n-1)}$. Let $(w_{ij})_{1\leqq i<j\leqq n}$ be the coordinates of $\mathbf{C}^{(1/2)n(n-1)}$. Let $f : U \to D_n^{\mathrm{II}}$ be a smooth map which maps the origin to the zero matrix and which is locally diffeomorphic at 0. The components $f^{\alpha\beta}$ of the map f satisfy $f^{\alpha\beta} = -f^{\beta\alpha}$ for $1 \leqq \alpha, \beta \leqq n$. Assume

$$\sum_{\alpha<\gamma,\beta<\rho,\lambda<\sigma,\mu<\tau} R_{\alpha\gamma,\overline{\beta\sigma},\lambda\sigma,\overline{\mu\tau}} \xi^{\alpha\gamma,\overline{\beta\rho}}_{\overline{ij},kl} \overline{\xi^{\mu\tau,\overline{\lambda\sigma}}_{\overline{ij},kl}} = 0 \tag{7.1}$$

at $Z = 0$. We have to prove that

$$\begin{cases} \text{either } \bar\partial\, f^{\alpha\beta} = 0 \text{ at } 0 \text{ for all } 1 \leqq \alpha < \beta \leqq n \\ \text{or } \partial f^{\alpha\beta} = 0 \text{ at } 0 \text{ for all } 1 \leqq \alpha < \beta \leqq n. \end{cases} \tag{7.2}$$

It follows from (7.1) and (5.2) that

$$\sum_{\gamma=1}^{n} \xi^{\alpha\gamma,\overline{\beta\gamma}}_{\overline{ij},\,\overline{kl}} = 0 \quad \text{for } 1 \leqq \alpha, \beta \leqq n. \tag{7.3}$$

We will prove (7.2) from the equations (7.3). The equations (7.3) are invariant under the following transformations of D_n^{II}:

$$\begin{aligned} Z &\mapsto \bar Z, \\ Z &\mapsto {}^tAZA, \end{aligned} \tag{7.4}$$

where A is a fixed unitary matrix of order n. Since f is locally diffeomorphic at 0, we have either

$$\partial_{ij} f^{1\beta} \neq 0 \quad \text{for some } i < j \text{ and some } \beta$$

or

$$\partial_{\overline{ij}} f^{1\beta} \neq 0 \text{ for some } i < j \text{ and some } \beta$$

(otherwise $d\,f^{1\beta} = 0$ at 0 for all β, contradicting $\bigwedge_{\alpha<\beta}(df^{\alpha\beta} \wedge df^{\overline{\alpha\beta}}) \neq 0$ at 0). Since the equations (7.3) are invariant under the transformation $Z \to \bar Z$, we can assume without loss of generality that $\partial_{ij} f^{1\beta} \neq 0$ for some $i < j$ and some β. Let the rank of the $(n-1) \times (1/2)n(n-1)$ matrix

$$(\partial_{ij} f^{1\beta})_{1<\beta\leqq n, 1\leqq i<j\leqq n}$$

be $p - 1$. By applying Lemma 3 b) to the smooth map

$$(f^{12}, \ldots, f^{1n}); U \mapsto \mathbf{C}^{n-1},$$

we conclude that, after we apply a linear transformation to the coordinates (w_{ij}) and apply a transformation to Z of the form (7.4) with

$$A = \begin{pmatrix} 1 & 0 \\ 0 & B \end{pmatrix},$$

where B is a unitary matrix of order $n-1$, we can assume without loss of generality that for $1 < \beta \leqq n$

$$\begin{cases} \partial_{1j} f^{1\beta} = \delta_{j\beta} & \text{for } 1 < j \leqq p, \\ \partial_{ij} f^{1\beta} = 0 & \text{for } (i,j) > (1,p). \end{cases} \tag{7.5}$$

For $1 < i \leqq p$ and $(k,l) > (1,p)$, consider

$$\sum_{\beta=1}^{n} \xi^{1\beta,\overline{1\beta}}_{\overline{1i},\ \overline{kl}} = 0.$$

Since $\partial_{kl} f^{1\beta} = 0$ and $\partial_{1i} f^{1\beta} = \delta_{i\beta}$, it follows that

$$\partial_{\overline{kl}} f^{1i} = 0 \quad \text{for} \quad 1 < i \leqq p \text{ and } (k,l) > (1,p). \tag{7.6}$$

The rank of the $(n-p) \times ((1/2)n(n-1) - p + 1)$ matrix

$$(\partial_{\overline{kl}} f^{1\gamma})_{p<\gamma\leqq n,(k,l)>(1,p)}$$

must be $n-p$, otherwise it follows from (7.5) and (7.6) that

$$\bigwedge_{\beta=1}^{n} (df^{1\beta} \wedge d\overline{f^{1\beta}})$$

vanishes at 0, contradicting the assumption that f is locally diffeomorphic at 0. By applying Lemma 3 a) to the map $(f^{1,p+1}, \dots, f^{1n})$ whose domain variables are $(w_{k,l})$ with $(k,l) > (1,p)$ (the other variables being fixed), we conclude that, after a linear change in the coordinates (w_{kl}), $(k,l) > (1,p)$, we have for $p < \gamma \leqq n$,

$$\begin{cases} \partial_{\overline{1j}} f^{1\gamma} = \delta_{j\gamma} & \text{for } p < j \leqq n, \\ \partial_{\overline{kl}} f^{1\gamma} = 0 & \text{for } 2 \leqq k < l \leqq n. \end{cases} \tag{7.7}$$

This does not affect the validity of (7.5).

For $1 < \alpha \leqq n, 1 < k < l \leqq n$, and $p < i \leqq n$, consider

$$\sum_{\beta=1}^{n} \xi^{1\beta,\overline{\alpha\beta}}_{\overline{1i},\ \overline{kl}} = 0.$$

Since $\partial_{\overline{kl}} f^{1\beta} = 0$ by (7.7) and $\partial_{\overline{1i}} f^{1\beta} = \delta_{i\beta}$ by (7.6) and (7.7), it follows that

$$\partial_{kl} f^{\alpha i} = 0 \quad \text{for} \quad 1 < \alpha \leqq n, 1 < k < l \leqq n \text{ and } p < i \leqq n. \tag{7.8}$$

For $1 < \alpha \leqq n, 1 < k < l \leqq n$, and $1 < j \leqq p$, consider

$$\sum_{\beta=1}^{n} \xi^{\alpha\beta,\overline{1\beta}}_{\overline{kl},\ \overline{1j}} = 0.$$

Since $\partial_{kl} f^{1\beta} = 0$ and $\partial_{1j} f^{1\beta} = \delta_{j\beta}$ by (7.5), it follows that

$$\partial_{\overline{kl}} f^{\alpha j} = 0 \quad \text{for} \quad 1 < \alpha \leqq n, 1 < k < l \leqq n \text{ and } 1 < j \leqq p. \tag{7.9}$$

We claim that, after a linear coordinate change in (w_{kl}) for $1 < k < l \leqq n$ the following holds for $1 \leqq i \leqq n$:

$$\begin{cases} \partial_{il} f^{i\beta} = \delta_{l\beta} & \text{for } i < l \leqq p \text{ and } i < \beta \leqq n, \\ \partial_{kl} f^{i\beta} = 0 & \text{for } (k,l) > (i,p), \\ \partial_{\overline{il}} f^{i\beta} = \delta_{l\beta} & \text{for } \max(i,p) < l \leqq n \text{ and } i < \beta \leqq n, \\ \partial_{\overline{kl}} f^{i\beta} = 0 & \text{for } i < k < l \leqq n \text{ and } i < \beta \leqq n. \end{cases} \tag{7.10$_i$}$$

We prove it by induction on i. When $i = 1$, $(7.10)_i$ follows from (7.5) and (7.7). Suppose $(7.10)_i$ holds for $1 \leqq i < j$ and $j \leqq n$ and we want to show that $(7.10)_j$ holds. From $(7.10)_i, 1 \leqq i < j$, it follows that

$$\begin{cases} df^{\alpha\beta}, 1 \leqq \alpha < j, \alpha < \beta \leqq n, \text{is a linear combination of} \\ dw_{kl}, d\bar{w}_{kl}, 1 \leqq k < j, k < l \leqq n, \text{at } 0. \end{cases} \tag{7.11}$$

From (7.9) it follows that

$$\begin{cases} df^{j\beta}, j < \beta \leqq p, \text{is a linear combination of } dw_{kl}, 1 \leqq k < l \leqq n, \\ \text{and } d\bar{w}_{1\gamma}, 1 < r \leqq n. \end{cases} \tag{7.12}$$

When $p > j$, the $(p-j) \times (1/2)(n-j+1)(n-j)$ matrix

$$(\partial_{kl} f^{j\beta})_{j<\beta\leqq p, j\leqq k<l\leqq n}$$

must have rank $p - j$, otherwise from (7.11) and (7.12) it follows that

$$\left(\bigwedge_{1\leqq\alpha<j,\alpha<\beta\leqq n} (df^{\alpha\beta} \wedge d\overline{f^{\alpha\beta}}) \right) \wedge \left(\bigwedge_{j<\beta\leqq p} (df^{j\beta} \wedge d\overline{f^{j\beta}}) \right)$$

vanishes at 0, contradicting the assumption that f is locally diffeomorphic at 0. By applying Lemma 3 a) to the map $(f^{j,j+1}, \ldots, f^{jn})$ whose domain variables are $(w_{kl}), j \leqq k < l \leqq n$ (the other variables being fixed), we conclude that, after a linear coordinate change in $(w_{kl}), j \leqq k < l \leqq n$, we have the first two equations of $(7.10)_j$. This implies that

$$\begin{aligned} &df^{j\beta}, j < \beta \leqq p, \text{ is a linear combination of } dw_{kl}, d\bar{w}_{kl}, \\ &(j,p) \geqq (k,l) \text{ at } 0. \end{aligned} \tag{7.13}$$

From (7.8) it follows that

$$\begin{aligned} &df^{j\beta}, \max(j,p) < \beta \leqq n, \text{ is a linear combination of} \\ &d\bar{w}_{kl}, 1 \leqq k < l \leqq n, \text{ and } dw_{1r}, 1 < r \leqq n, \text{at } 0. \end{aligned} \tag{7.14}$$

The rank of the $(n - \max(j,p)) \times (1/2(n-j+1)(n-j) - \max(0, p-j))$ matrix

$$(\partial_{\overline{kl}} f^{j\beta})_{\max(j,p)<\beta\leqq n,(k,l)>(j,p)}$$

must be $n - \max(j,p)$, otherwise it follows from (7.11), (7.13), and (7.14) that

$$\bigwedge_{1 \leqq \alpha \leqq j, \alpha < \beta \leqq n} (df^{\alpha\beta} \wedge d\overline{f^{\alpha\beta}})$$

vanishes at 0, contradicting the assumption that f is locally diffeomorphic at 0. By applying Lemma 3 a) to the map $(f^{j,\max(j,p)+1}, \ldots, f^{jn})$ whose domain variables are (w_{kl}) with $(k,l) > (j,p)$ (the other variables being fixed), we conclude that, after a linear coordinate change in (w_{kl}) with $(k,l) > (j,p)$, we have the last two equations of $(7.10)_j$.

We distinguish between the following two cases.

Case 1. $p = n$. By the first two equations of $(7.10)_i$ the $((1/2)n(n-1)) \times ((1/2)n(n-1))$ matrix

$$(\partial_{ij} f^{\alpha\beta})_{1 \leqq i < j \leqq n, 1 \leqq \alpha < \beta \leqq n}$$

is nonsingular. By Lemma 3 a) we can apply a linear coordinate change to (w_{ij}), $1 \leqq i < j \leqq n$, so that after this coordinate change we have

$$\partial_{ij} f^{\alpha\beta} = \delta_{i\alpha}\delta_{j\beta} \text{ for } 1 \leqq i < j \leqq n \text{ and } 1 \leqq \alpha < \beta \leqq n, \tag{7.15}$$

at the expense of possibly sacrificing the validity of (7.6), (7.7), (7.9), and the last two equations of $(7.10)_i$.

We want to show that $\partial_{\overline{ij}} f^{\alpha\beta} = 0$ for $1 \leqq i < j \leqq n$ and $1 \leqq \alpha < \beta \leqq n$. For notational convenience, we define, for $1 \leqq i < j \leqq n$, $\partial_{ji} = -\partial_{ij}$ and $\partial_{\overline{ji}} = -\partial_{\overline{ij}}$. First we show that

$$\partial_{\overline{\alpha j}} f^{\alpha\beta} = 0 \text{ for } 1 \leqq \alpha, \beta, j \leqq n \text{ with } j \neq \alpha \text{ and } \beta \neq \alpha. \tag{7.16}$$

Since $n \geqq 3$, we can choose $1 \leqq k \leqq n$ with $k \neq \alpha$ and $k \neq \beta$. Consider

$$\sum_{\gamma=1}^{n} \xi^{\alpha\gamma, \overline{k\gamma}}_{\overline{\alpha j}, k\beta} = 0;$$

that is,

$$\sum_{\gamma=1}^{n} (\partial_{\overline{\alpha j}} f^{\alpha\gamma} \overline{\partial_{k\beta} f^{k\gamma}} - \partial_{\overline{k\beta}} f^{\alpha\gamma} \overline{\partial_{\alpha j} f^{k\gamma}}) = 0. \tag{7.17}$$

Since $\alpha \neq k$, it follows from (7.15) that $\partial_{\alpha j} f^{k\gamma} = 0$ if $\gamma \neq \alpha$. On the other hand, if $\gamma = \alpha$, then $f^{\alpha\gamma} \equiv 0$. Hence in any case $\partial_{\overline{k\beta}} f^{\alpha\gamma} \overline{\partial_{\alpha j} f^{k\gamma}} = 0$. Since $\partial_{k\beta} f^{k\gamma} = \delta_{\beta\gamma}$ by (7.15), it follows from (7.17) that $\partial_{\overline{\alpha j}} f^{\alpha\beta} = 0$.

Now fix $1 \leqq \alpha < \beta \leqq n$ and $1 \leqq i, j \leqq n$ with $i \neq j$. We want to prove that $\partial_{\overline{ij}} f^{\alpha\beta} = 0$. Because of (7.16), we can assume without loss of generality that α, β, i, j are distinct. Consider

$$\sum_{\gamma=1}^{n} \xi^{\alpha\gamma, \overline{\alpha\gamma}}_{\overline{ij}, \alpha\beta} = 0;$$

that is,

$$\sum_{\gamma=1}^{n}(\partial_{\overline{ij}} f^{\alpha\gamma}\overline{\partial_{\alpha\beta} f^{\alpha\gamma}} - \partial_{\overline{\alpha\beta}} f^{\alpha\gamma}\overline{\partial_{ij} f^{\alpha\gamma}}) = 0. \tag{7.18}$$

It follows from (7.16) (and the fact that $f^{\alpha\alpha} \equiv 0$) that $\partial_{\overline{\alpha\beta}} f^{\alpha\gamma} = 0$. Since $\partial_{\alpha\beta} f^{\alpha\gamma} = \delta_{\beta\gamma}$ by (7.15), we have $\partial_{\bar{i}\bar{j}} f^{\alpha\beta} = 0$ from (7.18).

Case 2. $p < n$. From the last two equations of $(7.10)_i$, we have

$$\partial_{\overline{jn}} f^{in} = 1 \text{ for } 1 \leqq i \leqq j < n.$$

Since $f^{ni} = -f^{in}$, it follows that the rank of the $(n-1) \times (1/2)n(n-1)$ matrix

$$(\partial_{\overline{ij}} f^{n\alpha})_{1 \leqq i < j \leqq n, 1 \leqq \alpha < n}$$

is $n-1$. After we apply the transformation

$$Z \mapsto {}^t A \bar{Z} A,$$

where A is the matrix obtained from I_n by interchanging the first and last columns, we reduce this case to Case 1 and we therefore conclude, by the arguments of Case 1, that $\partial_{ij} f^{\alpha\beta} = 0$ for all $1 \leqq i < j \leqq n$ and $1 \leqq \alpha < \beta \leqq n$, which is a contradiction. Hence Case 2 cannot occur and $\partial_{\overline{ij}} f^{\alpha\beta} = 0$ for all $1 \leqq i < j \leqq n$ and $1 \leqq \alpha < \beta \leqq n$. Q.E.D.

8 Adequate negativity of the curvature of D_n^{III}

Proposition 7. *For $n \geqq 2$ the curvature tensor $R_{\alpha\gamma,\overline{\beta\rho}\lambda\sigma\overline{\mu\tau}} (1 \leqq \alpha \leqq \gamma \leqq n, 1 \leqq \beta \leqq \rho \leqq n, 1 \leqq \lambda \leqq \sigma \leqq n, 1 \leqq \mu \leqq \tau \leqq n)$ of D_n^{III} is adequately negative.*

Proof. Let U be an open neighborhood of 0 in $\mathbf{C}^{(1/2)n(n+1)}$. Let $(w_{ij})_{1 \leqq i \leqq j \leqq n}$ be the coordinates of $\mathbf{C}^{(1/2)n(n+1)}$. Let $f : U \to D_n^{\mathrm{III}}$ be a smooth map which maps the origin to the zero matrix and which is locally diffeomorphic at 0. The components $f^{\alpha\beta}$ of f satisfy $f^{\alpha\beta} = f^{\beta\alpha}$ for $1 \leqq \alpha, \beta \leqq n$. Assume

$$\sum_{\alpha \leqq \gamma, \beta \leqq \rho, \lambda \leqq \sigma, \mu \leqq \tau} R_{\alpha\gamma,\overline{\beta\rho},\lambda\sigma,\overline{\mu\tau}} \xi^{\alpha\gamma,\overline{\beta\rho}}_{\overline{ij},\ \overline{kl}} \overline{\xi^{\mu\tau,\overline{\lambda\sigma}}_{\overline{ij},\ \overline{kl}}} = 0 \tag{8.1}$$

at $Z = 0$. We have to prove that

$$\begin{aligned} &\text{either } \bar{\partial} f^{\alpha\beta} = 0 \text{ at } 0 \text{ for } 1 \leqq \alpha \leqq \beta \leqq n \\ &\text{or } \quad \partial f^{\alpha\beta} = 0 \text{ at } 0 \text{ for } 1 \leqq \alpha \leqq \beta \leqq n. \end{aligned} \tag{8.2}$$

From (8.1) and (5.3) it follows that

$$\sum_{\gamma=1}^{n} \xi^{\alpha\gamma,\overline{\beta\gamma}}_{\overline{ij},\ \overline{kl}} = 0 \quad \text{for} \quad 1 \leqq \alpha, \beta \leqq n. \tag{8.3}$$

We will prove (8.2) from the equations (8.3). The equations (8.3) are invariant under the following transformations of D_n^{III}:

$$Z \mapsto \bar{Z}, \qquad Z \mapsto {}^t AZA, \tag{8.4}$$

where A is a fixed unitary matrix of order n.

Since f is locally diffeomorphic at 0, we have either

$$\partial_{ij} f^{11} \neq 0 \quad \text{for some } i \leqq j$$

or

$$\partial_{\overline{ij}} f^{11} \neq 0 \quad \text{for some } i \leqq j$$

(otherwise $d\,f^{11} = 0$ at 0, contradicting $\bigwedge_{\alpha \leqq \beta}(d\,f^{\alpha\beta} \wedge d\,\overline{f^{\alpha\beta}}) \neq 0$ at 0). Since the equations (8.3) are invariant under the transformation $Z \to \bar{Z}$, we can assume without loss of generality that $\partial_{ij} f^{11} \neq 0$ for some $i < j$. Let the rank of the $n \times (1/2)n(n+1)$ matrix

$$(\partial_{ij} f^{1\beta})_{1 \leqq \beta \leqq n, 1 \leqq i \leqq j \leqq n}$$

be p. By applying Lemma 3 c) to the smooth map

$$(f^{11}, \ldots, f^{1n}) : U \to \mathbf{C}^n,$$

we conclude that, after we apply a linear transformation to the coordinates (w_{ij}) and apply a transformation to Z of the form (8.4) with

$$A = \begin{pmatrix} 1 & 0 \\ 0 & B \end{pmatrix},$$

where B is a unitary matrix of order $n-1$, we can assume without loss of generality that

$$\begin{cases} \partial_{1j} f^{1\beta} = \delta_{j\beta} & \text{for } 1 \leqq j \leqq p, 1 \leqq \beta \leqq p, \\ \partial_{1j} f^{1\beta} = 0 & \text{for } 1 < j \leqq p, p < \beta \leqq n, \\ \partial_{ij} f^{1\beta} = 0 & \text{for } (i,j) > (1,p), 1 \leqq \beta \leqq n. \end{cases} \tag{8.5}$$

For $1 < i \leqq p$ and $(k,l) > (1,p)$, consider

$$\sum_{\beta=1}^{n} \xi_{\overline{1i},\, \overline{kl}}^{1\beta, \overline{1\beta}} = 0;$$

that is,

$$\sum_{\beta=1}^{n} (\partial_{\overline{1i}} f^{1\beta} \overline{\partial_{kl} f^{1\beta}} - \partial_{\overline{kl}} f^{1\beta} \overline{\partial_{1i} f^{1\beta}}) = 0.$$

Since $\partial_{1i} f^{1\beta} = \delta_{i\beta}$ and $\partial_{kl} f^{1\beta} = 0$ by (8.5), it follows that

$$\partial_{\overline{kl}} f^{1i} = 0, \quad 1 < i \leqq p \quad \text{and} \quad (k,l) > (1,p). \tag{8.6}$$

Let q be the rank of the $(n-p)\times((1/2)n(n+1)-p)$ matrix

$$P := (\partial_{\overline{kl}} f^{1i})_{p<i\leqq n,(k,l)>(1,p)}.$$

Select $p < i_1 < \cdots < i_q \leqq n$ such that the rank of the $q \times ((1/2)n(n+1)-p)$ matrix

$$(\partial_{\overline{kl}} f^{1i_\nu})_{1\leqq\nu\leqq q,(k,l)>(1,p)}$$

is q. By applying Lemma 3 a) to the map $(f^{1i_1},\ldots,f^{1i_q})$ whose domain variables are $(w_{kl}),(k,l)>(1,p)$ (the other variables being fixed), we conclude that, after a linear change in the variables $(w_{kl}),(k,l)>(1,p)$, we can assume without loss of generality that

$$\begin{cases} \partial_{\overline{1,j+p}} f^{1i_\nu} = \delta_{j\nu} & \text{for } 1 \leqq \nu, j \leqq q, \\ \partial_{\overline{kl}} f^{1i_\nu} = 0 & \text{for } (k,l) > (1,p+q), \leqq \nu \leqq q. \end{cases} \tag{8.7}$$

We must have

$$\partial_{\overline{kl}} f^{1i} = 0 \quad \text{for} \quad (k,l) > (1,p+q), p < i \leqq n, \tag{8.8}$$

otherwise the rank of the matrix P is $> q$. For $(k,l) > (1,p+q)$, consider

$$\sum_{\gamma=1}^{n} \xi_{\overline{kl},11}^{\overline{1\gamma},\overline{1\gamma}} = 0;$$

this is,

$$\sum_{\gamma=1}^{n} (\partial_{\overline{kl}} f^{1\gamma} \overline{\partial_{11} f^{1\gamma}} - \partial_{\overline{11}} f^{1\gamma} \overline{\partial_{kl} f^{1\gamma}}) = 0.$$

Since $\partial_{kl} f^{1\gamma} = 0$ and $\partial_{11} f^{1\gamma} = \delta_{1\gamma}$ for $1 \leqq \gamma \leqq p$ by (8.5), it follows from (8.8) that

$$\partial_{\overline{kl}} f^{11} = 0 \quad \text{for } (k,l) > (1,p+q). \tag{8.9}$$

When $p+q<n$, this implies together with (8.5), (8.6), (8.7), and (8.8) that at 0,

$$df^{11} \wedge \cdots \wedge df^{1n} \wedge df^{\overline{11}} \wedge \cdots \wedge df^{\overline{1n}}$$

is a linear combination of exterior products of $dw_{1l}, d\bar{w}_{1l}$ for $1 \leqq l \leqq p+q$ and hence must be zero, which contradicts the fact that f is locally diffeomorphic to 0. Therefore we must have $q = n-p$. Thus $i_\nu = p+\nu$ and (8.7), (8.8), (8.9) read

$$\begin{cases} \partial_{\overline{1l}} f^{1\beta} = \delta_{l\beta} & \text{for } p < l, \beta \leqq n, \\ \partial_{\overline{kl}} f^{1\beta} = 0 & \text{for } 1 < k \leqq l \leqq n, p < \beta \leqq n, \\ \partial_{\overline{kl}} f^{11} = 0 & \text{for } 1 < k \leqq l \leqq n. \end{cases} \tag{8.10}$$

Take any $1 < \alpha \leqq n$. Let $1 < k \leqq l \leqq n$ and $p < i \leqq n$. Consider

$$\sum_{\beta=1}^{n} \xi_{\overline{1i},\,\overline{kl}}^{1\beta,\overline{\alpha\beta}} = 0;$$

that is,

$$\sum_{\beta=1}^{n}(\partial_{\overline{1i}}f^{1\beta}\overline{\partial_{kl}f^{\alpha\beta}}-\partial_{\overline{kl}}f^{1\beta}\overline{\partial_{1i}f^{\alpha\beta}})=0.$$

Since $\partial_{\overline{kl}}f^{1\beta}=0$ by (8.6) and (8.10) and since $\partial_{\overline{1i}}f^{1\beta}=\delta_{i\beta}$ for $1<\beta\leqq n$ by (8.6) and (8.10), it follows that

$$\partial_{\overline{1i}}f^{11}\overline{\partial_{kl}f^{\alpha 1}}+\partial_{kl}f^{\alpha i}=0.$$

Because $f^{\alpha 1}=f^{1\alpha}$, by (8.5) we have $\partial_{kl}f^{\alpha 1}=0$. Hence

$$\partial_{kl}f^{\alpha i}=0 \quad \text{for } 1<\alpha\leqq n, p<i\leqq n, 1<k\leqq l\leqq n. \tag{8.11}$$

Consider now

$$\sum_{\beta=1}^{n}\xi^{\alpha\beta,\overline{1\beta}}_{\overline{kl},\ \overline{1j}}=0,$$

where $1<\alpha\leqq n, 1<k\leqq l\leqq n$, and $1<j\leqq p$. that is,

$$\sum_{\beta=1}^{n}(\partial_{\overline{kl}}f^{\alpha\beta}\overline{\partial_{1j}f^{1\beta}}-\partial_{\overline{1j}}f^{\alpha\beta}\overline{\partial_{kl}f^{1\beta}})=0.$$

Since $\partial_{kl}f^{1\beta}=0$ and $\partial_{1j}f^{1\beta}=\delta_{j\beta}$ by (8.5), it follows that

$$\partial_{\overline{kl}}f^{\alpha j}=0 \quad \text{for } 1<\alpha\leqq n, 1<k\leqq l\leqq n, \text{and } 1<j\leqq p. \tag{8.12}$$

We are going to prove by induction on i for $1<i\leqq n$ that after a linear coordinate change in (w_{kl}) for $1<k\leqq l\leqq n$, the following hold:

$$\begin{cases} \partial_{il}f^{i\beta}=\delta_{l\beta} & \text{for } i\leqq\beta\leqq n \text{ and } i\leqq l\leqq p,\\ \partial_{kl}f^{i\beta}=0 & \text{for } i\leqq\beta\leqq n \text{ and } (k,l)>(i,p),\\ \partial_{\overline{il}}f^{i\beta}=\delta_{l\beta} & \text{for } \max(i,p+1)\leqq\beta,l\leqq n,\\ \partial_{\overline{kl}}f^{i\beta}=0 & \text{for } i<k\leqq l\leqq n \text{ and } i\leqq\beta\leqq n. \end{cases} \tag{8.13$_i$}$$

To avoid a repetitious initial step of the induction process, we agree to mean by $(8.13)_1$ the vacuous statement and prove $(8.13)_i$ by induction on i for $1\leqq i\leqq n$. Now we prove $(8.13)_{i-1}\Rightarrow(8.13)_i$ for $1<i\leqq n$. From (8.5), (8.6), (8.10), it follows that

$$df^{11},\dots,df^{1n} \text{ is a linear combination of } dw_{11},\dots,dw_{1n}, \tag{8.14}$$
$$d\bar{w}_{11},\dots,d\bar{w}_{1n} \text{ at } 0.$$

From $(8.13)_\nu$ $(1<\nu<i)$ it follows that for $1<\nu<i$

$$df^{\nu\nu},\dots,df^{\nu n} \text{ is a linear combination of } dw_{kl}, d\bar{w}_{kl}, \tag{8.15$_\nu$}$$

$$1\leqq k\leqq\nu, \quad k\leqq l\leqq n.$$

From (8.11) and (8.12) it follows that

$$df^{ii},\dots,df^{ip} \text{ is a linear combination of } dw_{kl}, 1 \leqq k \leqq l \leqq n, \text{ and } d\bar{w}_{1j}, 1 \leqq j \leqq n, \text{ at } 0, \tag{8.16}$$

and

$$df^{i\beta}, \max(i,p+1) \leqq \beta \leqq n, \text{ can be expressed in terms of } d\bar{w}_{kl}, 1 \leqq k \leqq l \leqq n, \text{ and } dw_{1j}, 1 \leqq j \leqq n \text{ at } 0. \tag{8.17}$$

When $p \geqq i$, the $(p-i+1) \times (1/2)(n-i+1)(n-i+2)$ matrix

$$Q := (\partial_{kl} f^{i\beta})_{i \leqq \beta \leqq p, i \leqq k < l \leqq n},$$

must have rank $p-i+1$; otherwise from (8.14), $(8.15)_\nu, 1 < \nu < i$, and (8.16) it follows that

$$\left(\bigwedge_{1 \leqq \alpha < i, \alpha \leqq \beta \leqq n} (df^{\alpha\beta} \wedge d\overline{f^{\alpha\beta}}) \right) \wedge \left(\bigwedge_{\beta=i}^{p} (df^{i\beta} \wedge d\overline{f^{i\beta}}) \right)$$

vanishes at 0, contradicting the assumption that f is locally diffeomorphic at 0. By applying Lemma 3a) to the map $(f^{ii},\dots,f^{ip})$ whose domain variables are $w_{kl}, i \leqq k < l \leqq n$ (the other variables being fixed), we conclude that, after a linear coordinate change in $w_{kl}, i \leqq k < l \leqq n$, the first two equations of $(8.13)_i$ are satisfied. This implies that

$$df^{ii},\dots,df^{ip} \text{ is a linear combination of } dw_{kl}, d\bar{w}_{kl}, (i,p) \geqq (k,l). \tag{8.18}$$

The $(n-\max(i,p+1)+1) \times ((1/2)(n-i+1)(n-i+2) - \max(0,p-i+1))$ matrix

$$L := (\partial_{\overline{kl}} f^{i\beta})_{\max(i,p+1) \leqq \beta \leqq n, (k,l) \geqq (i,\max(i,p+1))}$$

must have rank $n - \max(i,p+1)+1$; otherwise from (8.14), $(8.15)_\nu$, $1 < \nu < i$, (8.18), and (8.17) it follows that

$$\bigwedge_{1 \leqq \alpha \leqq i, \alpha \leqq \beta \leqq n} (df^{\alpha\beta} \wedge d\overline{f^{\alpha\beta}})$$

vanishes at 0, contradicting the assumption that f is locally diffeomorphic at 0. By applying Lemma 3a) to the map $(f^{i,\max(i,p+1)},\dots,f^{in})$ whose domain variables are $w_{kl}, (k,l) \geqq (i,\max(i,p+1))$ (the other variables being fixed), we conclude that, after a linear coordinate change in $w_{kl}, (k,l) \geqq (\max(i,p+1))$, the last two equations of $(8.13)_i$ are satisfied.

We distinguish between two cases.

Case 1. $p = n$. By (8.5) and $(8.13)_i, 1 < i \leqq n$, the $(1/2)n(n+1) \times (1/2)n(n+1)$ matrix

$$(\partial_{ij} f^{\alpha\beta})_{1 \leqq i \leqq j \leqq n, 1 \leqq \alpha \leqq \beta \leqq n}$$

is nonsingular. By Lemma 3a) we can apply a linear change of coordinates to $(w_{ij}), 1 \leqq i \leqq j \leqq n$ so that after this linear coordinate change we have

$$\partial_{ij} f^{\alpha\beta} = \delta_{i\alpha}\delta_{j\beta} \text{ for } 1 \leqq i \leqq j \leqq n \text{ and } 1 \leqq \alpha \leqq \beta \leqq n, \tag{8.19}$$

at the expense of possibly sacrificing the validity of (8.6), (8.10), (8.12), and the last two equations of $(8.13)_i$.

We want to show that $\partial_{\overline{ij}} f^{a\beta} = 0$ for $1 \leqq i \leqq j \leqq n$ and $1 \leqq \alpha \leqq \beta \leqq n$. For notational convenience, we define, for $1 \leqq i \leqq j \leqq n, \partial_{ji} = \partial_{ij}$ and $\partial_{\overline{ji}} = \partial_{\overline{ij}}$.

Take $1 \leqq \alpha, \beta, j, k, l \leqq n$ with $\beta \neq k, l$. Consider

$$\sum_{\gamma=1}^{n} \xi^{\alpha\gamma\overline{\beta\gamma}}_{\overline{kl},\overline{\beta j}} = 0;$$

i. e.,

$$\sum_{\gamma=1}^{n} (\partial_{\overline{kl}} f^{\alpha\gamma} \overline{\partial_{\beta j} f^{\beta\gamma}} \quad \partial_{\overline{\beta j}} f^{\alpha\gamma} \overline{\partial_{kl} f^{\beta\gamma}}) = 0.$$

It follows from $\beta \neq k, l$ and (8.19) that $\partial_{kl} f^{\beta\gamma} = 0$. From (8.19) we also have $\partial_{\beta j} f^{\beta\gamma} = \delta_{j\gamma}$. Hence

$$\partial_{\overline{kl}} f^{\alpha j} = 0 \quad \text{for } 1 \leqq \alpha, j, k, l \leqq n, \tag{8.20}$$

if there exists $1 \leqq \beta \leqq n$ with $\beta \neq k, l$. When $n \geqq 3$, for any given $1 \leqq k, l \leqq n$, we can always find $1 \leqq \beta \leqq n$ with $\beta \neq k, l$. It remains to prove the case $n = 2$. Consider

$$\sum_{\gamma=1}^{n} \xi^{\alpha\gamma,\overline{\beta\gamma}}_{\overline{ij},\ \overline{kl}} = 0$$

for $(\alpha, \beta, i, j, k, l) = (1, 1, 1, 2, 1, 1), (1,\ 2,\ 1,\ 2,\ 2,\ 2), (2,\ 2,\ 1,\ 2,\ 2,\ 2)$. That is,

$$\begin{cases} \partial_{\overline{12}} f^{11} \overline{\partial_{11} f^{11}} - \partial_{\overline{11}} f^{11} \overline{\partial_{12} f^{11}} + \partial_{\overline{12}} f^{12} \overline{\partial_{11} f^{12}} - \partial_{\overline{11}} f^{12} \overline{\partial_{12} f^{12}} = 0, \\ \partial_{\overline{12}} f^{11} \overline{\partial_{22} f^{21}} - \partial_{\overline{22}} f^{11} \overline{\partial_{12} f^{21}} + \partial_{\overline{12}} f^{12} \overline{\partial_{22} f^{22}} - \partial_{\overline{22}} f^{12} \overline{\partial_{12} f^{12}} = 0, \\ \partial_{\overline{12}} f^{21} \overline{\partial_{22} f^{21}} - \partial_{\overline{22}} f^{21} \overline{\partial_{12} f^{21}} + \partial_{\overline{12}} f^{22} \overline{\partial_{22} f^{22}} - \partial_{\overline{22}} f^{22} \overline{\partial_{12} f^{12}} = 0. \end{cases} \tag{8.21}$$

By (8.20),

$$\partial_{\overline{11}} f^{\alpha\beta} = \partial_{\overline{22}} f^{\alpha\beta} = 0 \quad \text{for } 1 \leqq \alpha, \beta \leqq 2.$$

By (8.19)

$$\begin{aligned} \partial_{11} f^{11} &= \partial_{22} f^{22} = \partial_{22} f^{22} = 1, \\ \partial_{11} f^{12} &= \partial_{22} f^{21} = 0. \end{aligned}$$

It follows from (8.21) that

$$\partial_{\overline{12}} f^{11} = \partial_{\overline{12}} f^{12} = \partial_{\overline{12}} f^{22} = 0.$$

Case 2. $p < n$. It follows from (8.10) and $(8.13)_i$, $1 < i \leqq n$, that

$$\partial_{\overline{jn}} f^{in} = \delta_{ij} \quad \text{for } 1 \leqq i \leqq j \leqq n.$$

Since $f^{ni} = f^{ni}$, it follows that the rank of the $n \times (1/2)n(n+1)$ matrix

$$(\partial_{\bar{i}\bar{j}} f^{n\alpha})_{1 \leqq \alpha \leqq n, 1 \leqq i \leqq j \leqq n}$$

is n. After we apply the transformation

$$Z \longmapsto {}^t A \bar{Z} A,$$

where A is the matrix obtained from I_n by interchanging the first and last columns, we reduce this case to Case 1 and we therefore conclude, by the arguments of Case 1, that $\partial_{ij} f^{\alpha\beta} = 0$ for all $1 \leqq i \leqq j \leqq n$ and $1 \leqq \alpha \leqq \beta \leqq n$, which is a contradiction. Hence Case 2 cannot occur and $\partial_{\bar{i}\bar{j}} f^{\alpha\beta} = 0$ for all $1 \leqq i \leqq j \leqq n$ and $1 \leqq \alpha \leqq \beta \leqq n$. Q.E.D.

9 Adequate negativity of the curvature of D_n^{IV}

Proposition 8. *For $n \geqq 3$ the curvature tensor $R_{\alpha\bar{\beta}\gamma\bar{\delta}}(1 \leqq \alpha, \rho, \beta, \sigma \leqq n)$ of D_n^{IV} is adequately negative.*

Proof. Let U be an open neighborhood of 0 in $\mathbf{C}^n$. Let $w_i (1 \leqq i \leqq n)$ be the coordinates of $\mathbf{C}^n$. Denote $\partial/\partial w^i, \partial/\partial \overline{w^i}$ by $\partial_i, \partial_{\bar{i}}$. Let $f : U \to D_n^{\mathrm{IV}}$ be a smooth map which sends the origin to the origin and which is locally diffeomorphic at 0. Let $f^\alpha (1 \leqq \alpha \leqq n)$ be the components of f and let

$$\xi_{\bar{i}\bar{j}}^{\alpha\bar{\beta}} = (\partial_{\bar{i}} f^\alpha)(0)(\overline{\partial_j f^\beta})(0) - (\partial_{\bar{j}} f^\alpha)(0)(\overline{\partial_i f^\beta})(0).$$

Assume

$$\sum_{\alpha,\beta,\rho,\sigma} , R_{\alpha\bar{\rho}\beta\bar{\sigma}} \xi_{\bar{i}\bar{j}}^{\alpha\bar{\rho}} \overline{\xi_{\bar{i}\bar{j}}^{\sigma\bar{\beta}}} = 0 \tag{9.1}$$

at 0 for all $1 \leqq i, j \leqq n$. We have to show that

$$\text{either } \bar{\partial} f^\alpha = 0 \text{ at } 0 \text{ for } 1 \leqq \alpha \leqq n \text{ or } \partial f^\alpha = 0 \text{ at } 0 \text{ for } 1 \leqq \alpha \leqq n. \tag{9.2}$$

For notational simplicity we will denote $(\partial_i f^\alpha)(0), (\partial_{\bar{i}} f^\alpha)(0), (\partial_i \overline{f^\alpha})(0), (\partial_{\bar{i}} \overline{f^\alpha})(0)$ simply by $\partial_i f^\alpha, \partial_{\bar{i}} f^\alpha, \partial_i \overline{f^\alpha}, \partial_{\bar{i}} \overline{f^\alpha}$ respectively. From (9.1) and (5.4) it follows that

$$\begin{cases} \displaystyle\sum_{\alpha=1}^n \xi_{\bar{i}\,\bar{j}}^{\alpha\bar{\alpha}} = 0 & \text{for all } i, j \\ \xi_{\bar{i}\,\bar{j}}^{\alpha\bar{\beta}} - \xi_{\bar{i}\,\bar{j}}^{\beta\bar{\alpha}} = 0 & \text{for all } i, j, \alpha, \beta. \end{cases} \tag{9.3}$$

We will prove (9.2) only from the second equation of (9.3). The second equation of (9.3) is invariant under the automorphism $z \mapsto \bar{z}$ of D_n^{IV}.

For $1 \leqq \alpha_1 < \alpha_2 < \alpha_3 \leqq n$ let $V_{\alpha_1\alpha_2\alpha_3}$ be the vector subspace of the tangent space $T_{U,0}$ of U at 0 consisting all tangent vectors whose images under *df* are annihilated by $dz_{\alpha_\nu}, d\bar{z}_{\alpha_\nu}, 1 \leqq \nu \leqq 3$. Since f is locally diffeomorphic at 0, $V_{\alpha_1\alpha_2\alpha_3}$

is of real codimension 6 in $T_{U,0}$. By Lemma 1, we can choose a basis $e_1, \dots, e_n$ of $T_{U,0}$ over $\mathbf{C}$ such that

$$V_{\alpha_1\alpha_2\alpha_3} \cap \left(\bigoplus_{\nu=1}^{n-3} \mathbf{C} e_{i_\nu} \right) = 0$$

for all $1 \leqq \alpha_1 < \alpha_2 < \alpha_3 \leqq n$ and all $1 \leqq i_1 < \cdots < i_{n-3} \leqq n$. We denote again by $w_i, 1 \leqq i \leqq n$, the linear coordinate system of U such that $e_i = 2\mathrm{Re}(\partial/\partial w_i)$ at $0, 1 \leqq i \leqq n$. Then for any $1 \leqq \alpha_1 < \alpha_2 < \alpha_3 \leqq n$ and any $1 \leqq i_1 < i_2 < i_3 \leqq n$, the map $(f^{\alpha_1}, f^{\alpha_2}, f^{\alpha_3})$ whose domain variables are $w_{i_1}, w_{i_2}, w_{i_3}$ (the other variables being fixed at 0) is locally diffeomorphic at 0. For the conclusion of the proposition, clearly it suffices to show that for any $1 \leqq \alpha_1 < \alpha_2 < \alpha_3 \leqq n$ and any $1 \leqq i_1 < i_2 < i_3 \leqq n$ either $\partial_{i_\mu} f^{\alpha_\lambda} = 0$ for $1 \leqq \mu, \lambda \leqq 3$ or $\partial_{\bar{i}_\mu} f^{\alpha_\lambda} = 0$ for $1 \leqq \mu, \lambda \leqq 3$. Hence in order to derive the conclusion of the proposition from the second equation of (9.3), it suffices to consider the case $n = 3$. We assume now $n = 3$.

We write out below in full the nine equations $\xi^{\alpha\bar{\beta}}_{i\bar{j}} = \xi^{\beta\bar{\alpha}}_{i\bar{j}}, \alpha \neq \beta, i \neq j$ from the second equation of (3.9):

$$\partial_{\bar{1}} f^1 \overline{\partial_2 f^2} - \partial_{\bar{2}} f^1 \overline{\partial_1 f^2} = \partial_{\bar{1}} f^2 \overline{\partial_2 f^1} - \partial_{\bar{2}} f^2 \overline{\partial_1 f^1}, \tag{9.4}$$
$$\partial_{\bar{2}} f^1 \overline{\partial_3 f^2} - \partial_{\bar{3}} f^1 \overline{\partial_2 f^2} = \partial_{\bar{2}} f^2 \overline{\partial_3 f^1} - \partial_{\bar{3}} f^2 \overline{\partial_2 f^1}, \tag{9.5}$$
$$\partial_{\bar{3}} f^1 \overline{\partial_1 f^2} - \partial_{\bar{1}} f^1 \overline{\partial_3 f^2} = \partial_{\bar{3}} f^2 \overline{\partial_1 f^1} - \partial_{\bar{1}} f^2 \overline{\partial_3 f^1}, \tag{9.6}$$
$$\partial_{\bar{1}} f^3 \overline{\partial_2 f^2} - \partial_{\bar{2}} f^3 \overline{\partial_1 f^2} = \partial_{\bar{1}} f^2 \overline{\partial_2 f^3} - \partial_{\bar{2}} f^2 \overline{\partial_1 f^3}, \tag{9.7}$$
$$\partial_{\bar{2}} f^3 \overline{\partial_3 f^2} - \partial_{\bar{3}} f^3 \overline{\partial_2 f^2} = \partial_{\bar{2}} f^2 \overline{\partial_3 f^3} - \partial_{\bar{3}} f^2 \overline{\partial_2 f^3}, \tag{9.8}$$
$$\partial_{\bar{3}} f^3 \overline{\partial_1 f^2} - \partial_{\bar{1}} f^3 \overline{\partial_3 f^2} = \partial_{\bar{3}} f^2 \overline{\partial_1 f^3} - \partial_{\bar{1}} f^2 \overline{\partial_3 f^3}, \tag{9.9}$$
$$\partial_{\bar{1}} f^1 \overline{\partial_2 f^3} - \partial_{\bar{2}} f^1 \overline{\partial_1 f^3} = \partial_{\bar{1}} f^3 \overline{\partial_2 f^1} - \partial_{\bar{2}} f^3 \overline{\partial_1 f^1}, \tag{9.10}$$
$$\partial_{\bar{2}} f^1 \overline{\partial_3 f^3} - \partial_{\bar{3}} f^1 \overline{\partial_2 f^3} = \partial_{\bar{2}} f^3 \overline{\partial_3 f^1} - \partial_{\bar{3}} f^3 \overline{\partial_2 f^1}, \tag{9.11}$$
$$\partial_{\bar{3}} f^1 \overline{\partial_1 f^3} - \partial_{\bar{1}} f^1 \overline{\partial_3 f^3} = \partial_{\bar{3}} f^3 \overline{\partial_1 f^1} - \partial_{\bar{1}} f^3 \overline{\partial_3 f^1}. \tag{9.12}$$

Let p be the rank of the 3×3 matrix $(\partial_i f^\alpha)_{1 \leqq i, \alpha \leqq 3}$ and q be the rank of the 3×3 matrix $(\partial_{\bar{i}} f^\alpha)_{1 \leqq i, \alpha \leqq 3}$. We claim that one of p and q is at least 2. Suppose the contrary. Then we can choose $1 \leqq \lambda, \mu \leqq 3$ such that for $1 \leqq \alpha \leqq 3$,

$$\partial f^\alpha = \lambda_\alpha \partial f^\lambda,$$
$$\bar{\partial} f^\alpha = \mu_\alpha \bar{\partial} f^\mu$$

at 0 for some $\lambda_\alpha, \mu_\alpha \in \mathbf{C}$ (which may be zero). It follows that at 0 the 6-form

$$\bigwedge_{\alpha=1}^{3} (df^\alpha \wedge \overline{df^\alpha})$$

is a linear combination of exterior products of $\partial f^\lambda, \bar{\partial} f^\mu, \overline{\partial f^\lambda}, \overline{\bar{\partial} f^\mu}$ and hence must be 0, contradicting the assumption that f is locally diffeomorphic at 0. By applying the transformation $z \mapsto \bar{z}$ if necessary, we can assume without loss of generality that $p \geqq 2$.

Case 1. $p = 3$. By Lemma 3 a), after a linear coordinate change in $w_i, 1 \leqq i \leqq 3$, we can assume without loss of generality that

$$\partial_i f^\alpha = \delta_{i\alpha} \quad \text{for } 1 \leqq i, \alpha \leqq 3.$$

From (9.4)–(9.12) it follows that

$$\begin{aligned} \partial_{\bar{1}} f^1 &= -\partial_{\bar{2}} f^2, & (9.13)\\ \partial_{\bar{3}} f^1 &= 0, \\ 0 &= \partial_{\bar{3}} f^2, \\ \partial_{\bar{1}} f^3 &= 0, \\ -\partial_{\bar{3}} f^3 &= \partial_{\bar{2}} f^2, & (9.14)\\ 0 &= \partial_{\bar{1}} f^2, \\ 0 &= \partial_{\bar{2}} f^3, \\ \partial_{\bar{2}} f^1 &= 0, \\ -\partial_{\bar{1}} f^1 &= \partial_{\bar{3}} f^3. & (9.15) \end{aligned}$$

From (9.13), (9.14), and (9.15) it follows that

$$\partial_{\bar{1}} f^1 = -\partial_{\bar{2}} f^2 = \partial_{\bar{3}} f^3 = -\partial_{\bar{1}} f^1.$$

Hence $\partial_{\bar{i}} f^\alpha = 0$ for $1 \leqq i, \alpha \leqq 3$.

Case 2. $p = 2$. At 0, two of $\partial f^1, \partial f^2, \partial f^3$ are linearly independent. After renumbering f^1, f^2, f^3 if necessary, we can assume without loss of generality that $\partial f^1, \partial f^2$ are linearly independent at 0. After applying a linear coordinate change to $w_i, 1 \leqq i \leqq 3$, we have

$$\begin{aligned} \partial f^1 &= dw_1, \\ \partial f^2 &= dw_2 \end{aligned}$$

at 0; i.e., $\partial_j f^\alpha = \delta_{j\alpha}$ for $1 \leqq j \leqq 3, 1 \leqq \alpha \leqq 2$. Since ∂f^3 is a linear combination of ∂f^1 and ∂f^2 at 0, we have $\partial_3 f^3 = 0$. From (9.5), (9.6), and (9.12) it follows that $\partial_{\bar{3}} f^\alpha = 0$ for $1 \leqq \alpha \leqq 3$. Combining this with $\partial_3 f^\alpha = 0$ for $1 \leqq \alpha \leqq 3$, we conclude that $df^\alpha, \overline{df^\alpha}, 1 \leqq \alpha \leqq 3$, are linear combinations of $dw_1, dw_2, d\bar{w}_1, d\bar{w}_2$ at 0. Hence

$$\bigwedge_{\alpha=1}^{3} (df^\alpha \wedge \overline{df^\alpha})$$

vanishes at 0, contradicting the assumption that f is locally diffeomorphic at 0. Thus Case 2 cannot occur and we have $\partial_{\bar{i}} f^\alpha = 0$ for $1 \leqq i, \alpha \leqq 3$.

10 Strong rigidity and other applications

Theorem 6. $k \geqq 2$. *Suppose g: $N \to M$ is a continuous map of compact Kähler manifolds. If the curvature tensor of M is negative of order k and if the map*

$H_l(N, \mathbf{R}) \to H_l(M, \mathbf{R})$ induced by f is nonzero for some $1 \geqq 2k$, then g is homotopic to a holomorphic or conjugate holomorphic map from N to M.

Proof. There exists a harmonic map $f : N \to M$ which is homotopic to g. Since the map $H_l(N, \mathbf{R}) \to H_l(M, \mathbf{R})$ induced by f is nonzero, it follows that $\operatorname{rank}_{\mathbf{R}} df \geqq 2k$ at some point of N. By Theorem 5, f is either holomorphic or conjugate holomorphic. Q.E.D.

The following theorem follows from Theorem 6 and Lemma 2.

Theorem 7. *Suppose M is a compact Kähler manifold whose curvature tensor is strongly negative. Then for $k \geqq 2$ an element of $H_{2k}(M, \mathbf{Z})$ can be represented by a complex-analytic subvariety of M if it can be represented by the continuous image of a compact Kähler manifold.*

Theorem 8. *Let $g : N \to M$ be a continuous map of compact Kähler manifolds, both of complex dimension $n \geqq 2$. Suppose the curvature tensor of M is adequately negative. Suppose g is of degree 1 and the map $H_{2n-2}(N, \mathbf{R}) \to H_{2n-2}(M, \mathbf{R})$ induced by g is injective. Then g is homotopic to a biholomorphic or conjugate biholomorphic map from N to M.*

Proof. By Theorem 6, g is homotopic to a holomorphic or conjugate holomorphic map $f : N \to M$. Let V be the set of points of N where f is not locally homeomorphic. Since f is of degree 1, $V \neq N$. Suppose V is nonempty. We want to derive a contradiction. V is a complex-analytic subvariety of pure complex codimension 1 in N, because locally V is defined by $\det(\partial z^\alpha / \partial w^i)$ when f is holomorphic and by $\det(\partial \overline{z^\alpha} / \partial w^i)$ when f is conjugate holomorphic, where (z^α) (respectively (w^i)) is a holomorphic local coordinate system of M (respectively N). $f(V)$ is a complex-analytic subvariety in M. Since f is of degree 1, f maps $N - f^{-1}(f(V))$ bijectively onto $M - f(V)$. The complex codimension of $f(V)$ in M is at least two. For otherwise there exists $v \in V$ such that v is an isolated point of $f^{-1}(f(v))$ and, by using a local coordinate chart (w^i) of N at v and by applying the Riemann removable singularity theorem to $w^i \circ f^{-1}$ or $\overline{w^i \circ f^{-1}}$ on $U - f(V)$ for some open neighborhood of $f(v)$ in M, we conclude that f is locally diffeomorphic at v, contradicting $v \in V$. Hence the element in $H_{2n-2}(N, \mathbf{R})$ defined by V is mapped by f to the zero element in $H_{2n-2}(M, \mathbf{R})$, contradicting the injectivity of the map $H_{2n-2}(N, \mathbf{R}) \to H_{2n-2}(M, \mathbf{R})$ induced by f. Q.E.D.

Now Theorems 2 and 4 follow from Theorem 8, Lemma 2, and Propositions 5–8.

STANFORD UNIVERSITY, STANFORD, CALIFORNIA

References

[1] N. ARONSZAJN, A unique continuation theorem for solutions of elliptic partial differential equations or inequalities of second order, J. Math. Pures Appl. **36** (1957), 235–249.

[2] E. BEDFORD and B. A. TAYLOR, Variational properties of the complex Monge-Ampère equation I. Dirichlet principle, Duke Math. J. **45** (1978), 375–403.

[3] E. CALABI and E. VESENTINI, On compact locally symmetric Kähler manifolds, Ann. of Math. **71** (1960), 472–507.

[4] J. EELLS and L. LEMAIRE, A report on harmonic maps, Bull. London Math. Soc. **10** (1978), 1–68.

[5] J. EELLS and J. H. SAMPSON, Harmonic mappings of Riemannian manifolds, Amer. J. Math. **86** (1964), 109–160.

[6] P. HARTMAN, On homotopic harmonic maps, Canad. J. Math. **19** (1967), 673-687.

[7] G. D. MOSTOW, Strong Rigidity of Locally Symmetric Spaces, Ann. of Math. Studies **78** (1973), Princeton University Press.

[8] G. D. MOSTOW and Y -T. SIU, A compact Kähler surface of negative curvature not covered by the ball, Ann. of Math. (to appear in **112** (1980)).

[9] R. SCHOEN and S. T. YAU, On univalent harmonic maps between surfaces, Invent. Math. **44** (1978), 265–278.

[10] Y. T. SIU, Complex-analyticity of harmonic maps and strong rigidity of compact Kähler manifolds (research announcement), Proc. Natl. Acad. Sci. USA **76** (1979), 2107–2108.

Editor's Comments

This is one of the first papers that shows the importance of reformulating a major result on stable vector bundles on Riemann surfaces in terms of differential geometric concepts, i.e., Yang-Mills connections. One reason is that this formulation can be generalized to vector bundles over higher dimensional Kähler manifolds.

Complex Geometry from Riemann to Kähler-Einstein and Calabi-Yau
ALM 38, pp. 411–418

A New Proof of a Theorem of Narasimhan and Seshadri*

Simon Donaldson

1 Introduction

In 1965 Narasimhan and Seshardri proved that the stable holomorphic vector bundles over a compact Riemann surface are precisely those arising from irreducible projective unitary representations of the fundamental group [5], We shall give here a different, more direct, proof of this fact using the differential geometry of connections on holomorphic bundles. This complements, in a small way, the recent paper by Atiyah and Bott [1] in which the result of Narasimhan and Seshadri is used to calculate the cohomology of the moduli spaces of stable bundles, and which we take as a general reference for background and notation.

Let X be a compact Riemann surface with an Hermitian metric normalized to unit volume. If E is a vector bundle over X we write

$$\mu(E) = \text{degree}(E)/\text{rank}(E),$$

where the degree is obtained by evaluating $c_1(E)$ on the fundamental cycle. A holomorphic bundle $\mathscr{E}$ is defined to be *indecomposable* if it cannot be written as a proper direct sum, and to be *stable* if for all proper holomorphic sub-bundles $\mathscr{F} < \mathscr{E}$,

$$\mu(\mathscr{F}) < \mu(\mathscr{E}), \text{ or equivalently } \mu(\mathscr{E}/\mathscr{F}) > \mu(\mathscr{E}).$$

Certainly a stable bundle is indecomposable. The theorem to be proved is:

Theorem. *An indecomposable holomorphic bundle $\mathscr{E}$ over X is stable if and only if there is a unitary connection on $\mathscr{E}$ having constant central curvature $*F = -2\pi i\mu(\mathscr{E})$. Such a connection is unique up to isomorphism.*

* Received October 5, 1982. S. Donaldson, A new proof of a theorem of Narasimhan and Seshadri. J. Differential Geom. 18 (1983), no. 2, 269–277.

Note. If $\deg(\mathscr{E}) = 0$, these connections are flat and so are given by unitary representations of the fundamental group. In the general case it is easy to prove the equivalence of this form of the result with the statement of Narasimhan and Seshadri [1, §6].

2 Definitions and notation, outline of proof

If E is a C^∞ Hermitian vector bundle over X, a unitary connection A on E gives an operator $d_A : \Omega^0(E) \to \Omega^1(E)$ which has a (0,1) component $\bar{\partial}_A : \Omega^0(E) \to \Omega^{0,1}(E)$, and this defines a holomorphic structure $\mathscr{E}_A$ on E (see [1,§5] for a proof that there are sufficiently many local solutions of $\bar{\partial}_A s = 0$). Conversely if $\mathscr{E}$ is a holomorphic structure on E, there is a unique way to define a unitary connection A such that $\mathscr{E} = \mathscr{E}_A$.

A connection on E induces a connection on all associated bundles, in particular, on the bundle of endomorphisms End E. The "gauge group" $\mathscr{G}$ of unitary automorphisms of E acts as a symmetry group on the affine space $\mathscr{Q}$ of all unitary connections on E by

$$u(A) = A - d_A u u^{-1}, \quad u \in \mathscr{G}, \quad A \in \mathscr{Q}.$$

The action extends to the complexification $\mathscr{G}^{\mathbf{C}}$ = group of general linear automorphisms of E:

$$g(A) = A - (\bar{\partial}_A g)g^{-1} + ((\bar{\partial}_a g)g^{-1})^*, \quad g \in \mathscr{G}^{\mathbf{C}}, A \in \mathscr{Q},$$

and connections define isomorphic holomorphic structures precisely when they lie in the same $\mathscr{G}^{\mathbf{C}}$ orbit. Thus the set of orbits parametrize all the holomorphic bundles of the same degree and rank as E (there are no further topological invariants of bundles over X). Given a holomorphic bundle $\mathscr{E}$, we write $0\,(\mathscr{E})$ for the corresponding orbit of connections on the appropriate C^∞ bundle.

Each connection A has a curvature $F(A) \in \Omega^2$ (End E), and $F(A + a) = F(A) + d_A a + a \wedge a$.

The plan of the proof is this: we suppose inductively that the result has been proved for bundles of lower rank (the case of the line bundles being an easy consequence of the Hodge theory), then we choose a minimizing sequence in $0(\mathscr{E})$ for a carefully constructed functional J of the curvature and extract a weakly convergent subsequence. Either the limiting connection is in $0(\mathscr{E})$ and we deduce the result by examining small variations within $0(\mathscr{E})$, or in another orbit $0(\mathscr{F})$ and we deduce that $\mathscr{E}$ is not stable. The main ingredient in this approach is a result of K. Uhlenbeck [6] on the weak compactness of the set of connections with L^2 bounded curvature.

In the intermediate stages of the argument we have to allow generalized connections of class L^2_1 (i.e., which differ from some fixed C^∞ connection by an element of the Hilbert space with norm $\|\alpha\|^2_{L^2_1} = \|\alpha\|^2 + \|\nabla\alpha\|^2$ with curvature in L^2 and gauge transformations in L^2_2. As explained in ([6, §1], [1, §14]) the group actions and properties of curvature which we use extend without essential change (in particular $L^2_2 \hookrightarrow C^0$, so the topology of the bundle is preserved), and it is proved in [1, Lemma 14.8] that each L^2_1 connection defines a holomorphic structure. For simplicity we shall work throughout with these more general objects with only occasional further comment.

Definition of the functional J. The "trace norm" is defined on $n \times n$ Hermitian matrices by

$$\nu(M) = \mathrm{Tr}(M * M)^{1/2} = \sum_{i=1}^{n} |\lambda_i|,$$

where $\{\lambda_i\}$ are the eigenvalues of M, repeated according to multiplicity. We shall need to know that ν defines a norm and that if M is written in blocks: $M = \begin{pmatrix} A & B \\ B^* & D \end{pmatrix}$, then $\nu(M) \geq |\mathrm{Tr}A| + |\mathrm{Tr}D|$. Both properties follow easily from the characterization:

$$\nu(M) = \max_{\{e_i\}} \sum_{i=1}^{n} |\langle Me_i, e_i \rangle|,$$

where $\{e_i\}$ runs over all orthonormal frames for $\mathbf{C}^n$. (There is a complete account of such convex invariant functions in [1, §12].)

Applying ν in each fibre we define, for any smooth self-adjoint section s in Ω^0 (End E),

$$N(s) = \left(\int_X \nu(s)^2 \right)^{1/2}.$$

Then N is a norm equivalent to the usual L^2 norm and so extends to the L^2 cross sections. For an L^2_1 connection A, set

$$J(A) = N\left(\frac{*F}{2\pi i} + \mu.1 \right),$$

where $\mu = \mu(E)$. Thus $J(A) = 0$ if and only if the connection is of the type required by the theorem. For bundles of rank 2 and degree 0, J is essentially the Yang-Mills functional $\|F\|_{L^2}$. For larger ranks the definition of J is chosen to make the inductive step (Lemma 3) carry through easily, although the connections we find in the end are of course Yang-Mills connections. Although J is not smooth, it does have the semi-continuity property: if $A_i \to A$ weakly in L^2_1, so $F(A_i) \to F(A)$ weakly in L^2, then $J(A) \leq \liminf J(A_i)$, because for each $\varepsilon > 0$ we can separate $*F(A)/2\pi i$ from the closed convex set $\{\alpha | N(\alpha + \mu 1) \leq J(A) - \varepsilon\}$ by a hyperplane.

3 Proof of main lemma

This section contains the main part of the argument for which we need to extract the following proposition from [6, Theorem 1.5].

Proposition. *Suppose that* $A_i \in \mathscr{Q}$ *is a sequence of* L^2_1 *connections with* $\|F(A_i)\|_{L^2}$ *bounded. Then there are a subsequence* $\{i'\} \subset \{i\}$ *and* L^2_2 *gauge transformations* $u_{i'}$ *such that* $u_{i'}(A_{i'})$ *converges weakly in* L^2_1.

From this we deduce:

Lemma 1. *Let $\mathscr{E}$ be a holomorphic bundle over X. Then either* $\inf J|_{0(\mathscr{E})}$ *is attained in* $0(\mathscr{E})$ *or there is a holomorphic bundle $\mathscr{F} \not\cong \mathscr{E}$ of the same degree and rank as $\mathscr{E}$ and with* $\inf J|_{0(\mathscr{F})} \leq \inf J_{0(\mathscr{E})}$*;* $\mathrm{Hom}(\mathscr{E}, \mathscr{F}) \neq 0$.

Proof. Pick a minimizing sequence A_i for $J|_{0(\mathscr{E})}$. Since N is equivalent to L^2 norm, we have $\|F(A_i)\|_{L^2}$ bounded and can apply the proposition to deduce that, without loss of generality, $A_i \to B$ weakly in L^2_1 and

$$J(B) \leq \liminf J(A_i) = \inf J|_{0(\mathscr{E})}.$$

Since B defines a holomorphic structure $\mathscr{E}_B$, the lemma will follow if we show $\mathrm{Hom}(\mathscr{E}, \mathscr{E}_B) \neq 0$. (The two alternatives holding as $\mathscr{E} \cong \mathscr{E}_B$ or not.) To see this, define for any $A, A' \in \mathscr{Q}_2$ connection $d_{AA'}$ on the bundle $\mathrm{Hom}(E, E) = E^* \otimes E$ built from the connection A on the left hand factor and A' on the right, with a corresponding

$$\bar{\partial}_{AA'} : \Omega^0(\mathrm{Hom}(E,E)) \to \Omega^{0,1}(\mathrm{Hom}(E,E)).$$

Thus solutions of $\bar{\partial}_{AA'} s = 0$ correspond exactly to elements of $\mathrm{Hom}(\mathscr{E}_A, \mathscr{E}_{A'})$.

If in fact $\mathrm{Hom}(\mathscr{E}, \mathscr{E}_B) = 0$, then $\bar{\partial}_{A_0 B}$ has no kernel, and since it is a first order elliptic operator we have

$$\|\bar{\partial}_{A_0 B} s\|_{L^2} \geq C \|s\|_{L^2_1} \quad \text{for some } C, \text{all } s.$$

By the Sobolev inequality $\|s\|_{L^2_1} \geq \mathrm{Const} \|s\|_{L^4}$ so

$$\|\bar{\partial}_{A_0 B} s\|_{L^2} \geq C_1 \|s\|_{L^4} \text{ say.}$$

On the other hand $L^2_1 \hookrightarrow L^4$ is compact, so $A_i \to B$ in L^4 norm and $\bar{\partial}_{A_0 B} - \bar{\partial}_{A_0 A_i}$ is the algebraic operator $s \to (B - A_i)_{0,1}.s$. Thus

$$\|(\bar{\partial}_{A_0 B} - \bar{\partial}_{A_0 A_i}) s\|_{L^2} \leq C_2 \|A_i - B\|_{L^4} \|s\|_{L^4},$$

so for each i and all s, $\|\bar{\partial}_{A_0 A_i} s\|_{L^2} \geq (C_1 - C_2\|A_i - B\|_{L^4}) \|s\|_{L^4}$.

Since $A_i \to B$ in L^4 norm, this implies that $\mathrm{Hom}(\mathscr{E}, \mathscr{E}_{A_i}) = 0$ for large enough i, contradicting $\mathscr{E}_{A_i} \cong \mathscr{E}$.

4 Curvature and holomorphic extensions

We have to show that if $\mathscr{E}$ is stable, the second alternative of Lemma 1 does not occur. In general if $\alpha : \mathscr{E} \to \mathscr{F}$ is a (nonzero) holomorphic map of bundles over X, there are proper extensions and a factorization (cf. [5, §4]):

$$\begin{array}{ccccccccc}
0 & \longrightarrow & \mathscr{P} & \longrightarrow & \mathscr{E} & \longrightarrow & \mathscr{Q} & \longrightarrow & 0 \\
 & & & & \Big\downarrow \alpha & & \Big\downarrow \beta & & \\
0 & \longleftarrow & \mathscr{N} & \longleftarrow & \mathscr{F} & \longleftarrow & \mathscr{M} & \longleftarrow & 0
\end{array} \qquad (*)$$

With rows exact, rank $\mathscr{Q} = \mathrm{rank}\mathscr{M}$, $\det \beta \not\equiv 0, \deg \mathscr{Q} \leq \mathit{deg}\mathscr{M}$.

The next two lemmas give bounds on the curvature of bundles expressed as extensions; the first exactly as in [1, §8], the second a little stronger, exploiting the special properties of J.

First some generalities: if we have any exact sequence of holomorphic bundles $0 \to \mathscr{S} \to \mathscr{T} \to \mathscr{U} \to 0$, then any unitary connection A on $\mathscr{S}$ has the shape:

$$A = \begin{pmatrix} A_s & \beta \\ -\beta^* & A_{\mathscr{U}} \end{pmatrix} \text{with } A_{\mathscr{S}}, A_u \text{ connections}$$

on $\mathscr{S}, \mathscr{U}$ and β in $\Omega^{0,1}(\mathscr{U}^* \otimes \mathscr{S})$ because $\mathscr{S}$ is a holomorphic subbundle (β is a representative of the extension class in $H^1(\mathscr{U}^* \otimes \mathscr{S})$). In the corresponding curvature matrix

$$F(A) = \begin{pmatrix} F(A_s) - \beta \wedge \beta^* & d\beta \\ -d\beta^* & F(A_u) - \beta^* \wedge \beta \end{pmatrix},$$

where $d : \Omega^1(\mathscr{U}^* \otimes \mathscr{S}) \to \Omega^2(\mathscr{U}^* \otimes \mathscr{S})$ is built from $A_{\mathscr{U}}, A_{\mathscr{S}}$, the quadratic terms have a definite sign (this is the principle that curvature decreases in holomorphic subbundles and increases in quotients). For convenience normalize so that ${}^*\mathrm{Tr}(\beta^* \wedge \beta) = 2\pi i|\beta|^2$.

Conversely, connections on $\mathscr{S}, \mathscr{U}$ and a representative β of the extension class define a unique connection on $\mathscr{F}$, and any nonzero multiple of β also gives a bundle isomorphic to $\mathscr{F}$ (although a different extension class).

Lemma 2. *IF $\mathscr{F}$ is a holomorphic bundle over X which can be expressed as an extension: $0 \to \mathscr{M} \to \mathscr{F} \to \mathscr{N} \to 0$, and if $\mu(\mathscr{M}) \geq \mu(\mathscr{F})$ (so also $\mu(\mathscr{F}) \geq \mu(\mathscr{N})$), then for any unitary connection A on $\mathscr{F}$,*

$$J(A) \geq \mathrm{rk}\, \mathscr{M}(\mu(\mathscr{M}) - \mu(\mathscr{F})) + \mathrm{rk}\, \mathscr{N}(\mu(\mathscr{F}) - \mu(\mathscr{M})), = J_0 \textit{ say},$$

with equality occurring only if the extension splits.

Proof. Using the property of ν on block matrices mentioned in §3 and the notation of the discussion above we have

$$\nu\left(\frac{{}^*F(A)}{2\pi i} + \mu.1\right) \geq \left|\mathrm{Tr}\left(\frac{F(A_m) - \beta \wedge \beta^*}{2\pi i} + \mu.1\right)\right| + \left|\mathrm{Tr}\left(\frac{F(A_n) - \beta^* \wedge \beta}{2\pi i} + \mu.1\right)\right|,$$

where $\mu = \mu(\mathscr{F})$. So

$$\begin{aligned} J(A) &\geq \int_X \nu\left(\frac{{}^*F}{2\pi i} + \mu.1\right) \\ &\geq \left|\int_X \left\{\mathrm{Tr}\left(\frac{{}^*F(A_{\mathscr{M}})}{2\pi i} + \mu.1\right) - |\beta|^2\right\}\right| \\ &\quad + \left|\int_X \left\{\mathrm{Tr}\left(\frac{{}^*F(A_{\mathscr{M}})}{2\pi i} + \mu.1\right) + |\beta|^2\right\}\right|. \end{aligned}$$

But $\int_X \mathrm{Tr}({}^*F(A_M)/2\pi i) = -\deg\mathscr{M} \le \mu \int_X \mathrm{Tr}\, 1_{\mathscr{M}}$ by hypothesis, so the first term on the right is $\mathrm{rk}(\mathscr{M})(\mu(\mathscr{M}) - \mu(\mathscr{F})) + \|\beta\|^2$. Similarly for the second term.

Lemma 3. *Suppose that $\mathscr{E}$ is a stable holomorphic bundle and make the inductive hypothesis that the main theorem has been proved for bundles of lower rank. If $\mathscr{E}$ can be expressed as an extension $0 \to \mathscr{P} \to \mathscr{E} \to \mathscr{Q} \to 0$ (so from the definition of stability $\mu(\mathscr{P}) < \mu(\mathscr{E}) < \mu(\mathscr{Q})$), then there is a connection A on $\mathscr{E}$ with*

$$\begin{aligned} J(A) &< \mathrm{rk}\mathscr{P}(\mu(\mathscr{E}) - \mu(\mathscr{P})) + \mathrm{rk}\mathscr{Q}(\mu(\mathscr{Q}) - \mu(\mathscr{E})) \\ &= J_1 say. \end{aligned}$$

Proof. Observe first that for a general extension $0 \to \mathscr{S} \to \mathscr{T} \to \mathscr{U} \to 0$, the connections in $0(\mathscr{T})$ given by triples $(A_s, A_{\mathscr{U}}, t\beta)$ converge in C^∞ as $t \to 0$ to a connection in $0(\mathscr{S} \oplus \mathscr{U})$. Now it is not hard to prove ([1, §7], [2]) that any holomorphic bundle $\mathscr{P}$ has a canonical "semi-stable filtration"

$$0 = \mathscr{P}_0 < \mathscr{P}_1 < \cdots < \mathscr{P}_k = \mathscr{P}$$

with $\mathscr{P}_i/\mathscr{P}_{i-1}$ semi-stable and $\mu(\mathscr{P}_i/\mathscr{P}_{i-1})$ decreasing with i.

Each quotient $\mathscr{P}_i/\mathscr{P}_{i-1}$ has in turn a filtration [2] with associated quotients $\mathscr{O}_{ij}$ stable and $\mu(\mathscr{O}_{ij}) = \mu(\mathscr{P}_i/\mathscr{P}_{i-1}) < \mu(\mathscr{P}_1)$. If $\mathscr{P} < \mathscr{E}$ are the bundles in the statement of the lemma, we must have $\mu(\mathscr{O}_{ij}) < \mu(\mathscr{E})$ since $\mathscr{E}$ is stable. Also rank $\mathscr{O}_{ij} <$ rank$\mathscr{E}$, so we can suppose inductively that each $\mathscr{O}_{ij}$ has a connection with constant central curvature. Applying the observation to each of the steps in the filtrations we find connections $A^t_{\mathscr{P}} \in 0(\mathscr{P})(t \neq 0)$ converging as $t \to 0$ to $A^0_{\mathscr{P}} \in 0(\otimes_{ij}\mathscr{O}_{ij})$, and ${}^*F(A^0_{\mathscr{P}}) + 2\pi i\Lambda_{\mathscr{P}}$ where $\Lambda_{\mathscr{P}}$ is the constant diagonal matrix with entries the $\mu(\mathscr{O}_{ij}) < \mu(\mathscr{E})$.

Similarly there are $A^t_{\mathscr{Q}} \to A^0_{\mathscr{Q}}, {}^*F(A^0_{\mathscr{Q}}) = 2\pi i\Lambda_{\mathscr{Q}}$, and the entries of $\Lambda_{\mathscr{Q}}$ are greater than $\mu(\mathscr{E})$.

For each t, $A^t_{\mathscr{P}}$ and $A^t_{\mathscr{Q}}$ give an operator d_t on the forms with values in the C^∞ bundle corresponding to $\mathscr{Q}^* \otimes \mathscr{P}$. For $t \neq 0$, choose the harmonic representative β_t of the extension class corresponding to $\mathscr{E}$, i.e., $d_t\beta_t = 0$, scaled to $\|\beta_t\|_{L^2} = 1$.

Since $d_t \to d_0$ as $t \to 0$ and the d_t are elliptic on $\Omega^{0,1}$, there is a uniform bound $\|\beta_t\|_{c^0} <$ Const (because for each t we have the usual elliptic bounds

$$\|\alpha\|_{L^2_{K+1}} \le C_t(\|d_t\alpha\|_{L^2_K} + \|\alpha\|_{L^2_K}),$$

and the C_t can be uniformly bounded since the d_t converge).

By our general discussion the triples $(A^t_{\mathscr{P}}, A^t_{\mathscr{Q}}, s\beta_t)(s, t > 0)$ give connections $A(s,t) \in 0(\mathscr{E})$ with curvature

$$F(s,t) = \begin{pmatrix} F(A^t_{\mathscr{P}}) - s^2\beta_\tau \wedge \beta^*_\tau & 0 \\ 0 & F(A^t_{\mathscr{Q}}) - s^2\beta^*_\tau \wedge \beta_\tau \end{pmatrix}.$$

It is clear that $J(A(s,t)) \to J_1$ as $s, t \to 0$. We have to check that for suitably chosen small s, t, $J(A(s,t)) < J_1$.

Since $\Lambda_{\mathscr{P}} - \mu.1_{\mathscr{P}}$ has all its eigenvalues negative, the same is true for nearby matrices, and for such matrices $\nu() = -\mathrm{Tr}()$. Using the uniform bound on the β_t

and $F(A^t_{\mathscr{P}}) \to \Lambda_{\mathscr{P}}$, together with the corresponding facts for the other component, we have for small s, t,

$$\nu\left(\frac{{}^*F(s,t)}{2\pi i} + \mu.1\right) = J_1 - 2s^2|\beta_t|^2 + \varepsilon(t),$$

where $\varepsilon(t) \to 0$ with t. So

$$J(A(s,t))^2 = \int_X (J_1^2 - 2s^2|\beta_t|^2 + \varepsilon(t))^2.$$

Choosing s so small that $s^4 \int_X |\beta_t|^4$ is much less than $s^2 \int_X |\beta_t|^2 = s^2$ (using the uniform bound again), and then t small enough for the terms in ε to be negligible gives $J(A(s,t)) < J_1$.

5 Proof of the theorem

First, the final clause of Lemma 2 shows that if $\mathscr{E}$ is an indecomposable bundle with a connection of the type required by the theorem (i.e., $J = 0$), then $\mathscr{E}$ must be stable.

Conversely, if $\mathscr{E}$ is stable and the theorem has been proved for bundles of lower rank, then $\inf J|_{0(\mathscr{E})}$ is attained in $0(\mathscr{E})$. For if not, Lemma 1 constructs a bundle $\mathscr{F}$ with $\deg \mathscr{F} = \deg \mathscr{E}$, $\operatorname{rank} \mathscr{F} = \operatorname{rank} \mathscr{E}$, $\operatorname{Hom}(\mathscr{E}, \mathscr{F}) \neq 0$ and $\inf J|_{0(\mathscr{E})} \geq \inf J|_{0(\mathscr{F})}$. In the corresponding diagram $(*)$ we have

$$\mu(\mathscr{M}) \geq \mu(\mathscr{Q}) > \mu(\mathscr{E}) = \mu(\mathscr{F}).$$

So we can apply Lemma 2 to the bottom row of $(*)$ to deduce

$$\inf J|_{0(\mathscr{F})} \geq J_0,$$

and Lemma 3 to the top row to deduce

$$\inf J|_{0(\mathscr{E})} < J_1.$$

But $\operatorname{rk} \mathscr{Q} = \operatorname{rk} \mathscr{M}, \operatorname{rk} \mathscr{P} = \operatorname{rk} \mathscr{N}, \deg \mathscr{Q} > \deg \mathscr{M}, \deg \mathscr{P} < \deg \mathscr{N}$ implies $J_1 \leq J_0$ and we obtain a contradiction.

So suppose $\inf J|_{0(\mathscr{E})}$ is attained at $A \in 0(\mathscr{E})$. The operator $d_A^* d_A$ acting on L_2^2 self-adjoint sections of End E has kernel the constant scalars (because any other element of the kernel would have eigenspaces that would decompose $\mathscr{E}$ holomorphically), and the projection of ${}^*F/(2\pi i)$ onto the scalars is $\mu(\mathscr{E})$. Thus by the Hodge theory (in a version [4] adapted to the case where the coefficients need not be smooth), there is a self-adjoint section $h \in L_2^2$ such that

$$d_A^* d_A h = 2\pi\mu - i^* F(A).$$

For small $t, 1 + th = g_t \in \mathscr{G}^{\mathbf{C}}$. Let $A_t = g_t(A) \in 0(\mathscr{E})$. Then we can compute the curvature

$$\begin{aligned} F(A_t) &= F(A) - \partial_A((\bar\partial_A g)g^{-1}) + \bar\partial_A(g^{-1}(\partial_A g)) \\ &\quad - \bar\partial_A g g^{-2} \partial_A g + g^{-1}\partial_A g \bar\partial_A g g^{-1} \\ &= F(A) - t(\partial_A\bar\partial_A - \bar\partial_A\partial_A)h + q(t,h)\text{say}, \end{aligned}$$

with $\|q(t,h)\|_{L^2} \le C(\|h\|_{L^2_2}).\ t^2$ for small t. Since $d_A^* d_A = i^*(\bar\partial_A \partial_A - \partial_A \bar\partial_A)$ we get

$$N\left(\frac{{}^*F(A_t)}{2\pi i} + \mu\right) = N\left(\frac{{}^*F(A)}{2\pi i} + \mu\right).(1-t) + O(t^2).$$

And if $J(A_t)$ is to be a minimum at $t = 0$ we must have ${}^*F(A)/2\pi i = -\mu$ as required.

Referring again to the paper by Uhlenbeck [6, Corollary 14.] we find that in some gauge the solution is smooth (i.e., there is $u \in \mathscr{G}$ with $u(A)$ smooth). Thus in each stable orbit $0(\mathscr{E})$ there is at least one connection of the required type.

Finally, to see that A is unique up to the action of $\mathscr{G}$ on $0(\mathscr{E})$, recall that any $\tilde g \in \mathscr{G}^{\mathbf{C}}$ can be factored $\tilde g = g.u$ with $g = g^*, u \in \mathscr{G}$, so if A, B are distinct solutions we can suppose $B = g(A)g = g^*$. One computes the formula

$$F(A) = F(B) = \mu.1 \Rightarrow \partial_A \bar\partial_A g^2 = -\{(\bar\partial_A g^2)g^{-1}\}\{(\bar\partial_A g^2)g^{-1}\}^*.$$

Taking the trace $\tau = \mathrm{Tr}(g^2)$ of this gives $\Delta\tau \le 0$ with equality if and only if $\bar\partial_A g^2 = 0$. By the maximum principle the only possibility is $\Delta\tau = 0$ everywhere and $\bar\partial_A g^2 = \partial_A g^2 = 0$. Again since $\mathscr{E}$ is indecomposable, we must have g a constant scalar and $A = B$. This completes the proof of the theorem.

Remark. *One may ask whether a holomorphic vector bundle over a compact Kähler manifold has a distinguished connection. In* [3] *a natural condition is suggested, which reduces in complex dimension one to the condition of the theorem above. In a future article we shall prove that for certain complex surfaces this condition too is precisely related to the algebro-geometric condition of stability.*

I am very grateful to Dr. N. J. Hitchin for explaining this problem to me, and to him and Professor M. F. Atiyah for their guidance.

References

[1] M. F. Atiyah & R. Bott, *The Yang-Mills equations over Riemann surfaces*, Philos. Trans. Roy. Soc. London **A 308** (1982) 523-615.

[2] G. Harder & M. S. Narasimhan, *On the cohomology groups of moduli spaces of vector bundles over curves*, Math. Ann. **212** (1975) 215-248.

[3] S. Kobayashi, *First Chern class and holomorphic tensor fields*, Nagoya Math. J. **77** (1980) 5–11.

[4] M. S. Narasimhan & T. R. Ramadas, *Geometry of SU(2) gauge fields*, Comm. Math. Phys. **67** (1979) 121–136.

[5] M. S. Narasimhan & C. S. Seshadri, *Stable and unitary vector bundles on a compact Riemann surface*, Ann. of Math. **82** (1965) 540-564.

[6] K. K. Uhlenbeck, *Connections with L^p bounds on curvature*, Comm. Math. Phys. **3** (1981) 31–42.

Oxford University

Editor's Comments

Together with the next paper by Uhlenbeck and Yau, this paper of Donaldson established the fundamental result, the Donaldson-Uhlenbeck-Yau theorem, which proves the equivalence between the existence of Hermitian Yang-Mills connections and the stability of vector bundles over compact Kähler manifolds.

Complex Geometry from Riemann to Kähler-Einstein and Calabi-Yau
ALM 38, pp. 421–450

Anti Self-Dual Yang-Mills Connections over Complex Algebraic Surfaces and Stable Vector Bundles*

Simon Donaldson

Introduction

In this paper we trace a correspondence between the algebraic and differential geometry of vector bundles. In algebraic geometry, there is a notion of a *stable* bundle over a projective variety; the equivalence classes of these bundles are parametrized by a variety of moduli, and the bundles that are not stable are, in a sense, exceptional. On the other hand, for a Riemannian 4-manifold there is a notion of a bundle with an anti self-dual connection, and the equivalence classes of these are again parametrized by a moduli space. We show here that, for a projective algebraic surface, these ideas effectively coincide. A moral one could draw, perhaps, is that from this point of view a Riemannian 4-manifold carries as much structure as an algebraic surface.

Results of this kind go back to a theorem of Narasimhan and Seshadri [**18**], extensively discussed by Atiyah and Bott in [**1**]. They identified the stable bundles over an algebraic curve with the irreducible, projectively flat unitary connections. Of course, a connection is flat precisely when its curvature vanishes and the proper generalization of this to higher dimensions was suggested independently by Hitchin and Kobayashi: on any Kähler manifold Z there is an algebraic 'trace' operator Λ (whose definition we recall below) mapping $(1, 1)$ forms to functions, and these writers suggested that the stability of a bundle should be related to the existence

* Received 6 December 1983. S. K. Donaldson, Anti self-dual Yang-Mills connections over complex algebraic surfaces and stable vector bundles. Proc. London Math. Soc. (3) 50 (1985), no. 1, 1–26.

of a compatible unitary connection whose curvature F satisfies

$$\Lambda F = \lambda.1$$

for constant scalar λ. (Here we extend the action of Λ to bundle-valued forms.) Following Kobayashi, we call these *Hermitian-Einstein* connections. When Z has complex dimension 1, these are the connections appearing in Atiyah and Bott's formulation of the theorem of Narasimhan and Seshadri and in complex dimension 2 they are the projectively anti self-dual connections—as was pointed out by Itoh [**10**].

Kobayashi [**13**] and, separately, Lübke [**15**], have shown that any holomorphic bundle which admits an irreducible Hermitian–Einstein connection is stable. (For the purposes of this paper, we say that a $U(r)$-connection is irreducible if the structure group cannot be compressed to $U(k) \times U(r-k)$, for $0 < k < r$.) For completeness, we include a brief proof of this in the case of the problem that we treat, which is as follows. Suppose that a projective algebraic surface X has a given embedding $X \subseteq \mathbb{CP}^N$ and that ω is a Kähler metric on X whose associated cohomology class is dual to the hyperplane section class $[H]$. Then our main result is:

Theorem 1. *A bundle over X is stable, with respect to the projective embedding, if and only if it admits an irreducible Hermitian–Einstein connection with respect to the metric ω. This connection is then unique.*

Our proof falls naturally into three parts which we now describe. Firstly, we shall construct a connection by a limiting process, so we need material on the convergence of a sequence of connections. This appears in §3 and is straightforward, since we are able to call on the general work of Uhlenbeck. Secondly, we need to generate a suitable sequence of connections on a holomorphic bundle, and this is done in §2 by a gradient line method, with some simple arguments mostly turning on the maximum principle for the heat equation, modelled on those of Eells and Sampson. Finally, and perhaps most interestingly, we require a technique for building the geometric condition of stability into our analysis. The Narasimhan–Seshadri theorem can be proved [**4**] by direct variational arguments on functions of Yang–Mills type, and the general setting for this is valid in higher dimensions (as is well known the anti self-dual connections are absolute minima of the Yang–Mills functional), but the technical problems become more daunting. Instead, in §1, we find another variational formulation of the problem which is in many ways better behaved. For any closed $(1,1)$ current T on X we find a way to define a functional M_T on the metrics on a holomorphic bundle, such that if T, T' are homologous, we have an explicit formula for the difference $M_T - M_{T'}$. Taking $T = \omega$, the Kähler metric, we obtain a functional whose critical points are the desired connections, and taking $T' = T_c$, the current corresponding to a hyperplane section C in X, we get a functional that we can understand via the Narasimhan–Seshadri theorem. The key fact which we exploit is a theorem of Mehta and Ramanathan by which we can, in effect, pick the curve C so that the property of (semi) stability is preserved by restriction. This fits the algebrogeometric and analytic theories neatly together and spares us from the details of the definition of stability.

One interesting feature is that we have an example of a variational problem with no simple explicit formula for the Lagrangian. Another is that the methods of our proof and its relation with stability have a very natural abstract general form which we shall explain in §4, and which some readers may prefer to see first. In the rest of this introduction, we assemble for convenience some facts on the theory of algebraic bundles, on Hermitian differential geometry, and on the Yang–Mills functional.

The author acknowledges with thanks the large influence that discussions with M. F. Atiyah and N. J. Hitchin have had on this work. He is grateful also to Professor R. Bott for directing him to Reference [**13**].

Stable vector bundles

Recall first that, by a theorem of Serre, there is a natural equivalence between the *holomorphic* and *algebraic* vector bundles over any projective variety Z. For such bundles we use the definition of stability given by Mumford and Takemoto [**21,19**] which depends upon a choice of embedding $Z \subset \mathbb{CP}^N$. We say that an algebraic vector bundle $\mathscr{E}$ over Z is *semi-stable* if, for each coherent sub-sheaf $\mathscr{S} \subset \mathcal{O}(\mathscr{E})$ with $0 < \operatorname{rank} \mathscr{S} < \operatorname{rank} \mathscr{E}$ the *normalized degrees* $\mu(\mathscr{E}), \mu(\mathscr{S})$ satisfy

$$\mu(\mathscr{S}) \equiv \deg \mathscr{S} / \operatorname{rank} \mathscr{S} \leqslant \deg \mathscr{E} / \operatorname{rank} \mathscr{E} \equiv \mu(\mathscr{E}),$$

where, as usual, we define the *degree* by cutting with the hyperplane section: $\deg \mathscr{E} = c_1(\mathscr{E}).[H]^{n-1} (n = \dim Z)$. If strict inequality holds for every such $\mathscr{S}$ then $\mathscr{E}$ is *stable*. Note that in the case of algebraic curves ($n = 1$) these conditions are independent of the embedding, and in that case we need only consider *sub-bundles* of $\mathscr{E}$. We shall use the following standard properties:

(a) *If $\mathscr{E}, \mathscr{E}'$ are two semi-stable bundles over Z with $\mu(\mathscr{E}) = \mu(\mathscr{E}')$ and if at least one of them is stable then any non-zero holomorphic map $\alpha : \mathscr{E} \to \mathscr{E}'$ is an isomorphism* (*see* [**19**, pp. 170–171] *for the case where* $Z = \mathbb{CP}^N$).

(b) *If* $\dim Z = 1$ *or* 2 *and $\mathscr{E}$ is a bundle over Z which is not semi-stable (respectively not stable) then there is a stable bundle $\mathscr{F}$* with $0 < \operatorname{rank} \mathscr{F} < \operatorname{rank} \mathscr{E}, \mu(\mathscr{F}) > \mu(\mathscr{E})$ (*respectively* $\mu(\mathscr{F}) \geqslant \mu(\mathscr{E})$), *and a non-zero holomorphic map taking $\mathscr{F}$ to $\mathscr{E}$. If* $\dim Z = 1$, *we may suppose that $\mathscr{F}$ is a sub-bundle of $\mathscr{E}$.*

(For, as in [**19**, Chapter II, §§1.1 – 1.2], we may suppose that the destabilizing subsheaf $\mathscr{S}$ has $\mathcal{O}(\mathscr{E})/\mathscr{S}$ torsion free; then for $\dim Z < 3$ this implies that $\mathscr{S} = \mathcal{O}(\mathscr{F})$ is locally free. We may assume that $\mathscr{F}$ is stable by the obvious inductive argument on rank ($\mathscr{F}$).)

Also we need a theorem of Mehta and Ramanathan which says, in part:

Proposition 2 [**16**]. *If* $\dim Z = 2$ *and $\mathscr{E}$ is a semi-stable bundle over Z then for some integer m there is a smooth curve $C \subset Z$ in the linear system $|mH|$ such that the restriction of $\mathscr{E}$ to C is semi-stable.*

Hermitian differential geometry

If Z is a complex Hermitian manifold, the metric form can be written, in normal coordinates at a point, as

$$\omega = \frac{1}{2} i \sum_1^n dz_\alpha \wedge d\bar{z}_\alpha.$$

The natural operator $\Lambda : \Omega_Z^{p,q} \to \Omega_Z^{p-1,q-1}$ acts on $(1,1)$ forms at this point by

$$\Lambda(\Sigma a_{\alpha\beta} dz_\alpha \wedge d\bar{z}_\beta) = -2i\left(\sum_1^n a_{\alpha\alpha}\right)$$

and measures the component of a $(1,1)$ form parallel to ω. Alternatively,

$$\Lambda\alpha.\mathrm{vol} = \frac{1}{(n-1)!}\alpha \wedge \omega^{n-1} \in \Omega_Z^{n,n}.$$

If also the metric is Kähler, the basic Kähler identities are

$$\partial^* = i[\Lambda, \bar\partial], \quad \bar\partial^* = -i[\Lambda, \partial].$$

It follows that the Laplacian on 0-forms is

$$\Delta = d^*d = 2\partial^*\partial = 2\bar\partial^*\bar\partial = 2i \wedge \bar\partial\partial.$$

The positivity of certain terms will be very important in our proof; we often use the fact that, for a $(1,0)$ form $\varphi, i\Lambda\varphi \wedge \bar\varphi = |\varphi|^2 \geqslant 0$.

The Yang-Mills functional in dimension 4 and characteristic classes

In general if $E \to M$ is a $U(r)$ vector bundle over any oriented Riemannian 4-manifold with associated bundle of Lie algebras $\mathfrak{g}_E$ then the Yang-Mills functional, defined on the space $\mathscr{A}$ of connections on E,

$$\|F\|_{L^2}^2 = \int_M |F_A|^2 d\mu \quad \text{for } A \in \mathscr{A} \text{ and } F_A \in \Omega^2(\mathfrak{g}_E),$$

decomposes in two distinct ways.

First, the central component $\mathrm{Tr}\,(F)$ of the curvature is constrained to represent $-2\pi i c_1(E)$ in $H^2(M;\mathbb{R})$; so the norm of this is minimized by the appropriate harmonic 2-form. The norm of the orthogonal component $F^0 = F - (1/r)\mathrm{Tr}\,F.1_E$ (which represents the curvature of the associated projective unitary connection) is related to the characteristic class

$$c_2(E) - ((r-1)/r)c_1(E)^2 \in H^4(M) \cong \mathbb{R}$$

by the identities

$$c_2 - \left(\frac{r-1}{r}\right)c_1^2 = \frac{1}{8\pi^2}\int_M \mathrm{Tr}\,(F^0 \wedge F^0) = \frac{1}{8}\pi^{-2}\{\|F_-^0\|^2 - \|F_+^0\|^2\},$$
$$\|F^0\|^2 = \|F_-^0\|^2 + \|F_+^0\|^2,$$

where F_+^0, F_-^0 are the self-dual and anti self-dual parts (cf. [**2,7**]). The central and trace-free components may be minimized independently and so the absolute minimum of the Yang-Mills functional is achieved when $\mathrm{Tr}\,(F)$ is harmonic and F^0 is self-dual or anti self-dual, depending on the sign of $c_2 - ((r-1)/r)c_1^2$.

In the case when the manifold is a complex Hermitian surface the complex self-dual 2-forms consist of the $(0,2), (2,0)$ forms and the span of the metric ω, and the anti self-dual forms are the '*primitive*' forms $P^{1,1} = \mathrm{Ker}\,\Lambda \subset \Omega^{1,1}$, perpendicular to ω (cf. [**14**] for a more general discussion in arbitrary dimensions).

1.

1.1. Connections on holomorphic bundles, the Yang-Mills equations, and the Yang-Mills flow

It follows from the Newlander-Nirenberg integrability theorem for almost complex manifolds that a holomorphic structure on a $\mathbb{C}^r$ vector bundle $\mathscr{E}$ over some complex manifold Z is equivalent to the existence of a $\bar\partial$-operator,

$$\bar\partial : \Omega^0(\mathscr{E}) \to \Omega^{0,1}(\mathscr{E}), \quad \bar\partial(f.s) = \bar\partial f.s + f.\bar\partial s,$$

whose natural prolongation satisfies $\bar\partial^2 = 0$ (cf. [**2**]). Two $\bar\partial$-operators give isomorphic holomorphic structures if and only if they are conjugate by an automorphism of the bundle.

We say that a connection A is compatible with a holomorphic structure if the $(0,1)$ component of the covariant derivative,

$$d_A = \partial_A + \bar\partial_A : \Omega^0(\mathscr{E}) \to \Omega^1(\mathscr{E}) = \Omega^{1,0}(\mathscr{E}) \oplus \Omega^{0,1}(\mathscr{E}),$$

is the $\bar\partial$-operator above. In particular, if a holomorphic bundle $\mathscr{E}$ is given an Hermitian metric then it is well known that there is a unique connection compatible with the holomorphic and Hermitian structures. If $H = \langle , \rangle_1, K = \langle , \rangle_2$ are two metrics with

$$\langle s,t\rangle_2 = \langle hs,t\rangle_1$$

for an element $h = H^{-1}K \in \operatorname{End}\mathscr{E}$ then the corresponding connections are related by

$$\begin{aligned} \bar\partial_2 &= \bar\partial_1, \\ \partial_2 &= h^{-1} \circ \partial_1 \circ h = \partial_1 + h^{-1}\partial h, \end{aligned}$$

and the associated curvatures $F_1, F_2 \in \Omega^{1,1}(\operatorname{End}\mathscr{E})$ differ by

$$F_2 - F_1 + \bar\partial_1(h^{-1}\partial_1 h).$$

In particular, if $h = 1+\sigma$ is close to 1, the first variation of the curvature is

$$F_2 = F_1 + \bar\partial_1\partial_1\sigma + O(\sigma)^2.$$

Sometimes it is more convenient to take another point of view, corresponding to a parametrization of the metrics on $\mathbb{C}^r$ as $\mathrm{GL}(r,\mathbb{C})/U(r)$. For this we *fix* a metric on a C^∞ bundle E over Z. Then any unitary connection in the subset

$$\mathscr{A}^{(1,1)} = \{A \in \mathscr{A} | F_A^{0,2} = F_A^{2,0} = 0\} \subset \mathscr{A}$$

defines a holomorphic structure on E. The 'complex gauge group' $\mathscr{G}^{\mathbb{C}}$ of general linear automorphisms acts on $\mathscr{A}^{(1,1)}$ by

$$\begin{aligned} \bar\partial_{g(A)} &= g \circ \bar\partial_A \circ g^{-1}, \\ \partial_{g(A)} &= g^{*-1} \circ \partial_A \circ g^*, \quad \text{for} \quad A \in \mathscr{A}^{(1,1)}, g \in \mathscr{G}^{\mathbb{C}}, \end{aligned}$$

extending the action of the unitary gauge group

$$\mathscr{G} = \{g \in \mathscr{G}^{\mathbb{C}} | h(g) = g^* g = 1\},$$

that is, $g^{-1} \circ d_{g(A)} \circ g = \bar{\partial}_A + h^{-1} \circ \partial_A \circ h$. Connections define isomorphic holomorphic structures if and only if they lie in the same $\mathscr{G}^{\mathbb{C}}$ orbit and so this set up is totally equivalent to the other one; in particular, the curvature transforms by

$$\begin{aligned} g^{-1} F_{g(A)} g &= F_A + \bar{\partial}_A (h^{-1} \partial_A h) \\ &= F_A + h^{-1} (\bar{\partial}_A \partial_A h - \bar{\partial}_A h h^{-1} \partial_A h) \end{aligned}$$

(our notation here follows [**1**] and thus differs from [**8**], for example).

Now consider a unitary bundle E, as above, over the algebraic surface X of Theorem 1. For brevity set

$$\hat{F}_A = \Lambda F_A \in \Omega^0(\mathfrak{g}_E)$$

for any unitary connection A on E, and define a number λ by

$$\lambda(E) = \frac{4\pi i}{[H].[H]} \left(\frac{\deg E}{\operatorname{rank} E} \right) = \frac{-4\pi i}{[H].[H]} \mu(E).$$

From obvious considerations, this must be the factor appearing in the definition of an Hermitian-Einstein connection on E.

Then regarding X as a Riemannian 4-manifold, we see that any anti self-dual connection has $F^{0,2} = F^{2,0} = \hat{F} = 0$; in particular, it is an integrable connection defining a holomorphic structure. Slightly more generally, for any integrable connection, the condition that $\operatorname{Tr} F$ be harmonic and F^0 be anti self-dual becomes the Hermitian-Einstein condition $\hat{F} = \lambda.1$. They are absolute minima of the Yang-Mills functional and can plainly only occur if the bundle E satisfies $c_2 - ((r-1)/r)c_1^2 \geqslant 0$. Our theorem will give as a consequence that this inequality (known already in algebraic geometry) is satisfied by the Chern classes of any semi-stable bundle over X. Moreover, we find the interesting fact that the equivalence classes of irreducible unitary representations $\pi_1(X) \to U(r)$ are parametrized, in the form of stable bundles with $c_1 = c_2 = 0$, by an algebraic variety—just as in the Narasimham-Seshadri theory.

Proposition 3. (i) *If $A \in \mathscr{A}^{(1,1)}$ is an integrable connection on a unitary bundle E over X which is a critical point for the Yang-Mills functional then A is a direct sum of Hermitian–Einstein connections.*

(ii) *If a holomorphic bundle $\mathscr{E}$ over X supports an Hermitian-Einstein connection and $\mu(\mathscr{E}) < 0$ then $\mathscr{E}$ has no non-zero holomorphic sections. If $\mu(\mathscr{E}) = 0$ any holomorphic section is covariant constant.*

Proof. (i) The first-order Kähler identities extend to operators on bundle-valued forms:

$$\partial_A^* = i[\Lambda, \bar{\partial}_A], \quad \bar{\partial}_A^* = -i[\Lambda, \partial_\Lambda].$$

The condition that a connection be a critical point for the Yang-Mills functional is $d_A^* F_A = 0$.

For integrable connections F_A is of type $(1,1)$ and the Bianchi identity gives $\bar\partial_A F_A = \partial_A F_A = 0$. Thus

$$d_A^* F_A = -i(\bar\partial_A - \partial_A)\Lambda F_A = 0 \Leftrightarrow d_A \hat{F}_A = 0,$$

and in this case the eigenspaces of the endomorphism $\hat{F} = \Lambda F$ decompose the connection into Hermitian-Einstein components.

(ii) The three Laplacians on sections of $\mathscr{E}$,

$$\Delta_A = d_A^* d_A, \quad \Delta'_A = \partial_A^* \partial_A, \quad \Delta''_A = \bar\partial_A^* \bar\partial_A,$$

are, in general, related in the following way:

$$\Delta_A = i\Delta(\bar\partial_A - \partial_A)(\bar\partial_A + \partial_A) = i\Lambda\bar\partial_A\partial_A - i\Lambda\partial_A\bar\partial_A = \Delta'_A + \Delta''_A,$$
$$\Delta'_A - \Delta''_A - i\Lambda(\bar\partial_A\partial_A + \partial_A\bar\partial_A) = i\hat{F}.$$

So, if A is Hermitian-Einstein, $\hat{F} = \lambda.1$ and

$$\Delta''_A = \frac{1}{2}[\Delta_A - i\lambda.1].$$

If $\mu < 0$, so that $i\lambda < 0$, then Δ'' is a strictly positive operator, and hence there are no non-zero global holomorphic sections. If $\mu = 0$, any holomorphic section is covariant constant.

Corollary 4 (cf. [13, 15]). *Suppose that the main theorem, Theorem* 1, *has been proved for bundles of rank less than* r. *Then any* r*-bundle* $\mathscr{E}$ *which admits an irreducible Hermitian-Einstein connection is a direct sum* $\mathscr{E} = \bigoplus \mathscr{E}_i$ *of stable bundles* $\mathscr{E}_i$ *with* $\mu(\mathscr{E}_i) = \mu(\mathscr{E})$. *In particular,* $\mathscr{E}$ *is semi-stable. If* $\mathscr{E}$ *admits an irreducible Hermitian–Einstein connection, it is stable.*

Proof. If $\mathscr{E}$ is not stable then there is a stable bundle $\mathscr{F}$ of rank less than r with $\mu(\mathscr{F}) \geqslant \mu(\mathscr{E})$ and a non-zero map $\alpha : \mathscr{F} \to \mathscr{E}$.

But

$$\mu(\mathrm{Hom}\,(\mathscr{F}, \mathscr{E})) = \mu(\mathscr{E}) - \mu(\mathscr{F}) \leqslant 0$$

and the Hermitian-Einstein connections on $\mathscr{E}, \mathscr{F}$, which exist by hypothesis, induce an Hermitian-Einstein connection on Hom $(\mathscr{F}, \mathscr{E})$. By Proposition 3 (ii), α is a covariant constant with respect to this connection, and $\mu(\mathscr{E}) = \mu(\mathscr{F})$. This implies that α is the inclusion of a sub-bundle of $\mathscr{E}$ with zero 2nd fundamental form, which contradicts the irreducibility of the connection. The statement of the corollary follows by taking the irreducible constituents of the connection.

All of these facts also hold for bundles over Riemann surfaces [**1, 4**] and, just as in that theory, the degree of instability of a bundle gives a lower bound on the Yang- Mills functional. For example, one has:

Proposition 5. *Let* $\mathscr{E}$ *be an* SL$(2,\mathbb{C})$ *bundle over* X *which is not stable, so there is a line bundle* L *of degree* $d > 0$ *with a holomorphic map* $L \to \mathscr{E}$. *Then, for any connection on* $\mathscr{E}$,

$$\|\hat{F}\|^2 \geqslant 2\pi^2(H.H)d^2,$$

that is,

$$\|F\|^2 \geqslant 8\pi^2 c_2 + 4\pi^2 (H.H) d^2.$$

This statement is arranged for simplicity rather than generality. We shall not use it, so we omit the proof, which can be carried through, for example, by using a birational modification $\tilde{X} \to X$ to reduce to the case when L is a sub-bundle of $\mathscr{E}$.

The Yang-Mills gradient flow

In general, the gradient flow of the Yang-Mills functional $\|F\|^2$ on the space $\mathscr{A}$ of unitary connections corresponds formally to a solution of the evolution equation

$$\frac{\partial A}{\partial t} = -d_A^* F_A.$$

However, if one is interested only in equivalence classes of connections, consider, more generally, solutions of

$$\frac{\partial A}{\partial t} = -d_A^* F_A + d_A(\alpha(t)) \qquad (*)$$

for arbitrary $\alpha(t) \in \Omega^0(\mathfrak{g}_E)$—this corresponding to the infinitesimal action of the automorphism group of the bundle. For example, for any solution of $(*)$,

$$\frac{\partial \|F\|^2}{\partial t} = -2\|d_A^* F_A\|^2,$$

and likewise for the more detailed calculations of §2.

Now in the complex case, if we have an integrable connection $A \in \mathscr{A}^{(1,1)}$, then the gradient vector at A is

$$d_A^* F_A = -i(\bar{\partial}_A - \partial_A)\hat{F}_A$$

(by the Kähler and Bianchi identities) which lies in the tangent space to the orbit of the *complex* group $\mathscr{G}^{\mathbb{C}}$ at A (see §4 for the general context of this). More precisely, if $A_0 \in \mathscr{A}^{(1,1)}$ and $g_t \in \mathscr{G}^{\mathbb{C}}$, for $t \geqslant 0$, satisfy

$$\left(\frac{\partial g}{\partial t}\right) g^{-1} = -i(F_{g(A_0)} - \lambda.1), \quad g_0 = 1$$

(the factor λ being included here for convenience), then $g_t(A_0)$ is a path of connections down the Yang-Mills flow, each representing the same holomorphic structure. Again, we can be more general: if g_t is as above and $h_t = g_t^* g_t$ then

$$\begin{aligned}\frac{\partial h}{\partial t} &= -2ig^*(\hat{F}_t - \lambda)g \\ &= -2ih_t(\hat{F}_0 - \lambda + \Lambda\bar{\partial}_0(h_t^{-1}(\partial_0 h_t)))\end{aligned}$$

(where $\bar{\partial}_0 = \bar{\partial}_{t_0}$ etc.), an equation involving h_t only. Conversely, if we are given a solution h_t to this differential equation, we recover a solution to $(*)$ algebraically by taking $A_t = \tilde{g}_t(A_0)$ for any $\tilde{g}_t \in \mathscr{G}^{\mathbb{C}}$ with $\tilde{g}_t^* \tilde{g}_t = h_t$ (for example, $\tilde{g}_t = h_t^{\frac{1}{2}}$).

The equation for h_t is easier to understand in the equivalent setting of a family of different metrics H_t on a fixed holomorphic bundle $\mathscr{E}$ satisfying

$$\frac{\partial H_t}{\partial t} = -2iH_t(\hat{F}_t - \lambda.1),$$

and of course a stationary point of this Yang-Mills flow is precisely an Hermitian-Einstein connection. The rather trivial discussion above is required merely to reduce us to considering a parabolic equation in §2.

1.2. Another variational formulation

We shall now develop a second variational formulation as described in the introduction and following an initial suggestion of M. F. Atiyah. The functional that we shall find will be defined on the space of metrics, or alternatively, complex gauge transformations $\mathscr{G}^{\mathbb{C}}$, rather than the space of connections.

To motivate this development observe that the linearization of the equation we have to solve to find an Hermitian–Einstein connection is a bundle-valued version of the Poisson equation

$$\Delta u + \rho = 0,$$

whose solutions are the critical points of the classical Dirichlet integral

$$\int |\nabla u|^2 + \rho u d\mu.$$

Thus one is led to ask for a functional in our problem which corresponds roughly to the norm of the first derivative of a metric.

We shall construct a functional of this type by an application of certain 'secondary holomorphic characteristic classes' that were implicitly defined by Bott and Chern in [**3**]. Since these seem not to be well known we work, briefly, in some generality.

Suppose that $\mathscr{E}$ is a holomorphic $\mathbb{C}^r$ vector bundle over some complex manifold Z, and that φ is an invariant p-linear function on $\mathfrak{gl}(r)$ which is, without loss of generality, totally symmetric. Then for any metric H on $\mathscr{E}$ we get a characteristic class representative:

$$\varphi(F_H) \equiv \varphi(F_H, F_H, \ldots, F_H) \in \Omega_Z^{p,p}.$$

Proposition 6. *If H, K are two metrics on $\mathscr{E}$, as above, then for each φ there is an invariant*

$$R(H, K) \in \Omega_Z^{p-1,p-1}/(\operatorname{Im}\partial + \operatorname{Im}\bar{\partial})$$

with the following properties:

(i) *$R(K, K) = 0$ and, for any third metric J,*

$$R(H, K) = R(H, J) + R(J, K);$$

(ii) *if H moves in a smooth 1-parameter family H_t then*

$$\frac{d}{dt}R(H_t, K) = -ip\Phi(H^{-1}\dot{H}, F_H, \ldots, F_H);$$

(iii)
$$i\bar{\partial}\partial R(H,K) = \varphi(F_H) - \varphi(F_K) \in \Omega_Z^{p,p}.$$

Proof. Any two metrics may be joined by a path so, setting $R(K,K) = 0$, we can *define* a representative $(p-1,p-1)$ form for $R(H,K)$ by integrating (ii) along a path. Then we must check that, modulo $\operatorname{Im}\partial + \operatorname{Im}\bar{\partial}$, this definition is independent of the path chosen. By Stokes's theorem this is an infinitesimal problem; we have defined an $\Omega_Z^{p-1,p-1}$-valued 1-form on the space of metrics on $\mathscr{E}$,

$$\theta_H : h \to -ip\varphi(H^{-1}h, F_H, \ldots, F_H),$$

and have to check that its exterior derivative lies in $\operatorname{Im}\partial + \operatorname{Im}\bar{\partial}$. As we have seen the first variation of the curvature is

$$F_{H+k} = F_H + \bar{\partial}\partial(H^{-1}k) + O(k).$$

So, setting $\sigma = H^{-1}h, \tau = H^{-1}k \in \Omega^0(\operatorname{End}\mathscr{E})$, we have

$$\begin{aligned}(\theta_{H+k} - \theta_H)h &= ip\varphi(\tau\sigma, F, F, \ldots, F)\\ &\quad -ip\sum_{j=2}^{p}\varphi(\sigma, F, \ldots, \underset{(j)}{\bar{\partial}\partial\tau}, \ldots, F) + \text{ higher order terms}\end{aligned}$$

(where $F = F_H$ etc.). Skewing this, we see that the exterior derivative of the 1-form evaluated on h, k (tangent vectors in the space of metrics) is

$$\begin{aligned}&ip\varphi([\tau,\sigma], F, \ldots, F) + ip\sum_{j=2}^{p}\varphi(\tau, F, \ldots, \underset{(j)}{\bar{\partial}\partial\sigma}, \ldots, F)\\ &\qquad -ip\sum_{j=2}^{p}\varphi(\sigma, F, \ldots, \underset{(j)}{\bar{\partial}\partial\tau}, \ldots, F).\end{aligned}$$

It is then a simple exercise to show that this form in $\Omega_Z^{p-1,p-1}$ lies in $\operatorname{Im}\partial + \operatorname{Im}\bar{\partial}$, using the Bianchi identity and the following differentiated form of the invariance of φ:

$$\sum_{j=1}^{p}\varphi(A_1, \ldots, [A_j, B], \ldots, A_p) = 0$$

for any $\{A_j\}_{j=1}^{p}, B \in \mathfrak{gl}(p)$.

This procedure serves to define R in such a way that properties (i) and (ii) hold tautologically. To verify (iii) we merely need to check the derivative:

$$\begin{aligned}\frac{d}{dt}\varphi(F_H, \ldots, F_H) &= p\bar{\partial}\partial\{\varphi(H^{-1}H, F_H, \ldots, F_H)\}\\ &= i\bar{\partial}\partial\left\{\frac{d}{dt}R(H_t, K)\right\}.\end{aligned}$$

The invariant R_1 associated to the trace and so to $-2\pi i c_1$ is just

$$R_1(H,K) = \log\det(K^{-1}H).$$

We are really interested in the next case: $R_2(H,K)$ associated to the Killing form $-\mathrm{Tr}\,(AB)$ with

$$i\bar\partial\partial R_2(H,K) = (-\mathrm{Tr}\,(F_H^2)) - (-\mathrm{Tr}\,(F_K^2)),$$

and we ask next how these behave with respect to holomorphic extensions.

If we have a fixed exact sequence of holomorphic bundles

$$0 \longrightarrow \mathscr{E}' \xrightarrow{\ a\ } \mathscr{E} \xrightarrow{\ b\ } \mathscr{E}'' \longrightarrow 0,$$

a metric H on $\mathscr{E}$ induces metrics H', H'' on $\mathscr{E}', \mathscr{E}''$ and a 2nd fundamental form

$$\beta_H \in \Omega^{0,1}(\mathrm{Hom}\,(\mathscr{E}'', \mathscr{E}')).$$

Proposition 7. *For any pair H, K of metrics on $\mathscr{E}$,*

(i) $R_1(H,K) = R_1(H',K') + R_1(H'',K'')$,

(ii) $R_2(H,K) = R_2(H',K') + R_2(H'',K'') - 2i\{\mathrm{Tr}\,(\beta_H \wedge \beta_H^*) - \mathrm{Tr}\,(\beta_K \wedge \beta_K^*)\}$, *where the adjoints $\beta_H^*, \beta_K^* \in \Omega^{1,0}(\mathrm{Hom}\,(\mathscr{E}', \mathscr{E}''))$ are formed with respect to the metrics H, K respectively.*

Proof. Part (i) is easy. For (ii) we merely have to check that the first variations agree. A metric on $\mathscr{E}$ is equivalent to metrics on $\mathscr{E}', \mathscr{E}''$ together with a C^∞ splitting of the exact sequence

$$0 \longrightarrow \mathscr{E}' \underset{a}{\overset{\pi}{\rightleftarrows}} \mathscr{E} \underset{b}{\overset{\lambda}{\rightleftarrows}} \mathscr{E}'' \longrightarrow 0.$$

Any two splittings differ by an element $A \in \mathrm{Hom}\,(\mathscr{E}'', \mathscr{E}')$ (that is, $\pi_2 = \pi_1 + Ab, \lambda_2 = \lambda_1 + aA$), and the change induced in the fundamental form is

$$\beta_2 = \beta_1 + \bar\partial A.$$

The curvature of the connection corresponding to this data is

$$F = \begin{pmatrix} F' - \beta \wedge \beta^* & \partial\beta \\ -\bar\partial\beta^* & F'' - \beta^* \wedge \beta \end{pmatrix}.$$

Suppose that the metrics H', H'' and the splitting (π, λ) move in a 1-parameter family. Then

$$\begin{aligned} \frac{d}{dt}\{R_2' + R_2'' - R_2\} = -2i\{&\mathrm{Tr}\,(\beta \wedge \beta^* H'^{-1}\dot H') + \mathrm{Tr}\,(\beta^* \wedge H''^{-1}\dot H'') \\ &+\mathrm{Tr}\,(\dot A\bar\partial\beta^*) - \mathrm{Tr}\,(\dot A^*\partial\beta)\}. \end{aligned}$$

Now, modulo $\mathrm{Im}\,\partial + \mathrm{Im}\,\bar\partial$,

$$\mathrm{Tr}\,(A\overline{\dot{\bar\partial}\beta}^{\,*}) = -\mathrm{Tr}\,(\bar{\dot\beta}\beta^*), \quad \text{since } \bar{\dot\beta} = \bar{\bar\partial}\dot A,$$

and

$$\mathrm{Tr}\,(\dot A^*\partial\beta) = \mathrm{Tr}\,(\beta(\dot\beta)^*).$$

Moreover,

$$\frac{d}{dt}\beta^* = (\dot{\beta})^* - \beta^* H'^{-1}\dot{H}' + H''^{-1}\dot{H}''\beta^*$$

(since the adjoint $(\ )^*$ depends upon the metrics). So

$$\frac{d}{dt}\{R_2' + R_2'' - R_2\} = 2i\frac{d}{dt}\mathrm{Tr}\,(\beta\beta^*) \mod (\mathrm{Im}\,\partial + \mathrm{Im}\,\bar{\partial}),$$

as required.

Now we can define our functional, loosely analogous to the Dirichlet integral, and we shall do it in the two cases where $\mathscr{E}$ is a holomorphic bundle over the algebraic surface X or a curve C with Kähler metrics (of course all this easily extends to the n-dimensional problem). Set

$$M_\omega(H,K) = \int_X (R_2 - 2\lambda\omega \wedge R_1) \wedge \omega,$$

$$M_C(H,K) = \int_C R_2 - 2\lambda\omega \wedge R_1$$

for metrics H, K on the bundle $\mathscr{E}$ over X, C respectively. Here λ is the normalizing constant as above, while in the curve case, of course, $\lambda = -4\pi i\mu(\mathscr{E})/\mathrm{vol}\,(C)$. These integrals are well defined since the 2-form ω is closed.

Then, in each case, if we fix a metric K as a reference point, we get functionals $M_\omega(-,K), M_C(-,K)$ defined on the space of metrics whose critical points, if they exist, are exactly the Hermitian-Einstein connections; for example, the first variation of $M_\omega(H,K)$ is

$$\delta M_\omega = 2i\int_X \mathrm{Tr}\,(H^{-1}\delta H(F - \lambda\omega)) \wedge \omega,$$

which vanishes for all δH if and only if $(F - \lambda\omega) \wedge \omega = 0$, that is, $\hat{F} = \lambda.1$. Similarly, the time variation for M in our flow of §1.1 satisfies

$$\frac{\partial H}{\partial t} = -2iH_t(\hat{F}_t - \lambda.1),$$

whence

$$\frac{\partial}{\partial t}M(H_t,K) = -i\|\hat{F} - \lambda.1\|^2_{L^2}.$$

Changing the reference point K only changes the functional by a constant.

In the other framework of a fixed metric and variable $\bar{\partial}$-operator our functional goes over to a function

$$m(A_0, g) \in \mathbb{R}, \quad \text{where } A_0 \in \mathscr{A}^{(1,1)}, g \in \mathscr{G}^{\mathbb{C}},$$

invariant, in the obvious way, under the action of $\mathscr{G}$. These functionals are, in a certain sense, convex, as we shall see in the course of the next proof in which we let m be either of m_ω, m_C.

Proposition 8. *If $A_0 \in \mathscr{A}^{(1,1)}$ is an Hermitian-Einstein connection, it gives an absolute minimum of the functional m:*

$$m(A_0, g) \geqslant 0.$$

Moreover, if $\mathscr{E}$ is indecomposable, equality is achieved only in the trivial case when g^*g is a constant scalar.

Proof. By the unitary invariance it suffices to consider the case when

$$g = \exp(s)$$

for some *self-adjoint* section s of End E. Then we shall see that the function

$$m(t) = m(A_0, \exp(st)) \in \mathbb{R}$$

is a *convex* function of $t \in \mathbb{R}$. Since by hypothesis $m'(0) = 0$, this will imply that $m(t) \geqslant m(0)$, as required.

By the construction of m, for definiteness in the surface case,

$$\frac{dm}{dt} = 2i \int_X \operatorname{Tr} \left[\left(F_t - \frac{1}{2}\lambda\omega \right) s \right] \wedge \omega.$$

So

$$\frac{d^2m}{dt^2} = 2i \int_X \operatorname{Tr} \left(\frac{dF_{t.s}}{dt} \right) \wedge \omega,$$

where $F_t = F(A_t) = F(\exp(st)A_0)$. So

$$\frac{dF_t}{dt} = (\bar{\partial}\partial)_t s$$

and

$$\frac{d^2m}{dt^2} = 2i \int_X \operatorname{Tr} ((\bar{\partial}\partial)_t s \wedge s) \wedge \omega.$$

Integrating by parts we see that this is equal to

$$2i \int_X \operatorname{Tr} (\partial_t s \wedge \bar{\partial}_t s) \wedge \omega = 2\|\partial_t s\|^2 = \|d_t s\|^2 \geqslant 0.$$

(Here $\partial_t = \partial_{A_t}$ etc.) So m is convex and, if $\mathscr{E}$ is indecomposable, either s is a constant scalar or $\|d_t s\|^2 > 0$ and the convexity is strict (else the eigenspaces of s decompose $\mathscr{E}$, as in Proposition 3).

Corollary 9. (i) *An Hermitian-Einstein connection on an indecomposable holomorphic bundle is unique up to isomorphism.*

(ii) *If $\mathscr{E}$ is a semi-stable bundle over an algebraic curve C then for each reference point K the functional $M_C(-, K)$ is bounded below.*

Proof. Part (i) follows immediately from Proposition 8, as does (ii) in the case when $\mathscr{E}$ is stable, since we know by the theorem of Narasimhan and Seshadri that an Hermitian-Einstein connection exists. To prove (ii) when $\mathscr{E}$ is semi-stable but not stable use induction on the rank of $\mathscr{E}$—for we may find, as in the introductory remarks, a sub-bundle $\mathscr{E}' \subset \mathscr{E}$ with $\mathscr{E}'$ *stable* and $\mu(\mathscr{E}') = \mu(\mathscr{E})$. Evidently, $\mathscr{E}'' = \mathscr{E}/\mathscr{E}'$ is semi-stable and $\mu(\mathscr{E}'') = \mu(\mathscr{E})$. Now apply Proposition 7 to the sequence

$$0 \to \mathscr{E}' \to \mathscr{E} \to \mathscr{E}'' \to 0$$

to get

$$M_C(H,K) = M_C(H',K') + M_C(H'',K'') + 2\|\beta_H\|^2 - 2\|\beta_K\|^2.$$

By hypothesis $M_C(H',K'), M_C(H'',K'')$ are bounded below and the remaining term on the right exceeds $-2\|\beta_K\|^2$.

One advantage of these functionals is that they allow us to transfer information from a curve to a surface—because under our assumptions the metric is homologically equivalent to the hyperplane section class. More precisely, let $C \subset X$ be any smooth curve in the linear system $|mH|$ on X. Then there is a function $f \in L^1(X)$ such that the equation of currents,

$$\omega = (1/m)T_C + i\bar{\partial}\partial f,$$

is valid (see, for example, [**8**, p. 388]). In fact, we may take

$$f = (1/2\pi m)\log|s|^2,$$

where s is the section of the Hermitian line bundle $H^{\otimes m}$ cutting out C. In particular, $f \to -\infty$ along C and is bounded above.

Then the Kähler metric ω on X induces a metric on C and the normalized degree λ of a bundle $\mathscr{E}$ on X is the same as that of its restriction to C. Thus for any two metrics H, K on $\mathscr{E}$ we have two functionals $M_\omega(H,K), M_C(H,K)$—the latter obtained by restricting to C.

Proposition 10. *We have*

$$M_\omega(H,K) = (1/m)M_C(H,K) - \int_X f(\psi(F_H) - \psi(F_K)),$$

where $\psi(F_H) = \mathrm{Tr}\,(F_H^2) + \frac{1}{2}\lambda \mathrm{Tr}\,(F_H)\wedge\omega \in \Omega_X^{2,2}$.

Proof. In the sense of currents

$$\begin{aligned} M_\omega(H,K) - (1/m)M_C(H,K) &= \langle \omega - (1/m)T_C, R_2 - \frac{1}{2}\lambda R_1\rangle \\ &= \int_X i\bar{\partial}\partial f\left\{R_2 - \frac{1}{2}\lambda\omega R_1\right\}. \end{aligned}$$

Integrating by parts we see that this is equal to

$$\int_X f.i\bar{\partial}\partial\left(R_2 - \frac{1}{2}\lambda\omega R_1\right),$$

and

$$i\bar{\partial}\partial\left(R_2 - \frac{1}{2}\lambda\omega R_1\right) = (-\psi(F_H) - (-\psi(F_K))),$$

by Proposition 6(iii).

2 Properties of the gradient flow

2.1. Short time existence and C^0 bounds

We study the evolution equation for a one-parameter family H_t of metrics on some holomorphic bundle $\mathscr{E}$ over X:

$$\frac{\partial H}{\partial t} = -2iH[\hat{F}_H - \lambda.1]. \tag{A}$$

We suppose that we are given some initial condition H_0 and seek to solve the equation for $t \geqslant 0$. Then, as we explained in §1.1, this is completely equivalent to the equation:

$$\begin{aligned}\frac{\partial h}{\partial t} &= -\{\Delta_0 h + i(\hat{F}_0 h + h\hat{F}_0 - 2\lambda.h) + 2i\Lambda(\bar{\partial}_0 h h^{-1}\partial_0 h)\}, \\ h(0) &= 1\end{aligned} \tag{B}$$

for a positive self-adjoint endomorphism h of a unitary bundle E with integrable connection A_0 (and $\Delta_0 = \Delta_{A_0}$, etc.).

Setting $h = 1 + K$ we see that the linearization of this equation (B), for small K, is

$$\frac{\partial K}{\partial t} = -\Delta_0 K + i(F_0 K + K F_0 - 2\lambda K) + 2i(F_0 - \lambda.1),$$

which is parabolic (of course the sign of our Laplacian Δ is opposite to the classical one).

Proposition 11. *For sufficiently small $\varepsilon > 0$ (depending upon the initial conditions perhaps) the equations* (B), *and so also* (A), *have a smooth solution defined for $0 \leqslant t < \varepsilon$.*

This is a standard fact about non-linear parabolic equations; we may refer to the monograph by Hamilton [**9**, Part 3 and Part 4, §11] for the general theory. To show that the solution can in fact be continued for all positive time we shall exploit the special features of our problem. First we introduce a convenient measure of the 'distance' between two metrics.

Definition 12. For any two Hermitian metrics H, K on a bundle $\mathscr{E}$ set

$$\left.\begin{aligned}\tau(H,K) &= \mathrm{Tr}\,(H^{-1}K), \\ \sigma(H,K) &= \tau(H,K) + \tau(K,H) - 2\mathrm{rank}\ \mathscr{E}\end{aligned}\right\} \in C^\infty(X).$$

It follows immediately from the inequality

$$\lambda + \lambda^{-1} \geqslant 2 \quad \text{for all } \lambda \geqslant 0,$$

that $\sigma(H,K) \geqslant 0$ with equality if and only if $H = K$. The function σ is not quite a metric but it serves almost equally well from our point of view. In particular, a sequence H_i converges to H in the usual C^0 topology if and only if $\operatorname{Sup}_X \sigma(H_i, H) \to 0$. (Compare Corollary 15 below.)

Proposition 13. *If H_t, K_t are two solutions of the evolution equation* (A) *then, writing $\sigma = \sigma(H_t, K_t)$, we have*

$$\frac{\partial \sigma}{\partial t} + \Delta \sigma \leqslant 0.$$

Proof. It suffices to show that $\tau(H_t, K_t)$ has $(\partial\tau/\partial t) + \Delta\tau \leqslant 0$. Now

$$\frac{\partial \tau}{\partial t} = \operatorname{Tr}\left(H^{-1}\frac{\partial K}{\partial t} - H^{-1}\frac{\partial H}{\partial t}H^{-1}K\right).$$

But

$$\frac{\partial H}{\partial t} = -2iH(\hat{F}_H - \lambda)$$

and

$$\frac{\partial K}{\partial t} = -2iK(\hat{F}_K - \lambda),$$

which implies that

$$\frac{\partial \tau}{\partial t} = -2i\operatorname{Tr}\left(h(\hat{F}_k - \hat{F}_H)\right),$$

with $h = H^{-1}K$ as in §1.1. Then, as there,

$$\begin{aligned} F_K &= F_H + \bar{\partial}_H(h^{-1}\partial_H h) \\ &= F_H + h^{-1}(\bar{\partial}_H\partial_H h - (\bar{\partial}_H h)h^{-1}(\partial_H h)). \end{aligned}$$

So

$$\frac{\partial \tau}{\partial t} = -\operatorname{Tr}(\Delta_H h) + 2i\Lambda\operatorname{Tr}\left((\bar{\partial}_H h)h^{-1}(\partial_H h)\right).$$

But $\operatorname{Tr}(\Delta_H h) = \Delta\tau$ and the last term is negative.

Corollary 14. *Any two smooth solutions H_t, K_t to* (A), *which are defined for $0 \leqslant t < \mathscr{E}$, are continuous* (*say*) *at $t = 0$, and have the same initial condition $H_0 = K_0$, agree for all $t \in [0, \varepsilon)$.*

Proof. Apply the maximum principle [**9**, p. 101] to $\sigma(H_t, K_t)$, using Proposition 13.

Corollary 15. *Suppose that a smooth solution H_t to the evolution equation* (A) *is defined for $0 \leqslant t < T$. Then H_t converges in C^0 to some continuous metric H_T as $t \to T$.*

Proof. The space of metrics on the given holomorphic bundle $\mathscr{E}$ is the set of cross sections of a fibre bundle. On each fibre we have a natural metric distance function $d(H, K)$ coming from the description as the homogeneous space,

$\mathrm{GL}(k, \mathbb{C})/U(k)$, and this metric space is *complete*. Moreover the function σ compares uniformly with d, in that $\sigma \leqslant f(d)$ and $d \leqslant F(\sigma)$ for monotone f, F. It suffices then to show that, given $\varepsilon > 0$, we can find $\delta > 0$ such that

$$\operatorname*{Sup}_X \sigma(H_t, H_{t'}) < \varepsilon \quad \text{for all } t, t' > T - \delta,$$

for this implies that the H_t are uniformly Cauchy and converge to a non-degenerate limiting metric H_T. But by continuity at $t = 0$ we can find a δ such that

$$\operatorname*{Sup}_X \sigma(H_t, H_{t'}) < \varepsilon \quad \text{for } 0 < t, t' < \delta.$$

Then the maximum principle and Proposition 13 imply the result for the interval $(T - \delta, T)$, since $\operatorname{Sup}_X \sigma(H_t, H_{t+\tau})$ is a decreasing function of t.

Note. Proposition 13 can also be used to give another proof of the essential uniqueness of an Hermitian-Einstein connection when one exists (cf. [**4**]). Similarly, an easy extension of the arguments below shows that if such a solution exists on an indecomposable holomorphic bundle, corresponding to a metric H, then for any initial condition H_0 the corresponding solution H_t of equation (A) converges in C^∞ to H as $t \to \infty$. For in that case we have a bound of the form

$$\left\| \left(\frac{\partial}{\partial t} + \Delta \right) \sigma(H, H_t) \right\|_{L^1} \leqslant -\text{const}\,.\|\sigma(H, H_t)\|_{L^1},$$

and σ decays exponentially.

2.2. Evolution of the curvature

As we saw in §1.1 our evolution equation (A) is formally gauge-equivalent to the Yang-Mills flow

$$\frac{\partial A}{\partial t} = -d_A^* F.$$

In particular, any equation involving only natural geometric properties of connections which is valid for the latter flow is valid also for the flow (A). Recall then that the curvature F_A of a connection has, on one hand, an abelian component $\operatorname{Tr}(F_A)$ and, on the other hand, an imprimitive component $\hat{F}_A$. Define functions on X by

$$\begin{aligned} e &= |F_A|^2, \\ \hat{e} &= |\hat{F}_A|^2, \\ e_k &= |\nabla_A^k F_A|^2 \quad \text{for } k \geqslant 0, \end{aligned}$$

where ∇^k is the kth iterated covariant derivative. Following the method of Eells and Sampson for the analogous problem in the theory of harmonic maps [**6,9**], we derive differential inequalities for these functions:

Proposition 16. *If $A = A(t)$ is formally gauge-equivalent to a path down the Yang-Mills flow then*

(i) $\left(\frac{\partial}{\partial t}+\Delta\right)\operatorname{Tr}(F_A)=0,$

(ii) $\left(\frac{\partial}{\partial t}+\Delta\right)e\leqslant c(e^{3/2}+3),$

(iii) $\left(\frac{\partial}{\partial t}+\Delta\right)\hat{e}\leqslant 0,$

(iv) $\left(\frac{\partial}{\partial t}+\Delta\right)e_k\leqslant c_k e_k^{\frac{1}{2}}\left(\sum_{i+j=k} e_i^{\frac{1}{2}}(e_j^{\frac{1}{2}}+1)\right)$

for constants c, c_k depending only on the Riemannian metric on X.

Proof. Without loss of generality, we may assume that $\partial A/\partial t=-d_A^*F$. Then

$$\frac{\partial F}{\partial t}=d_A^*d_A^*F=-\Delta_A F,$$

since $d_AF=0$. Taking the bundle trace gives (i). For (ii) it is well known that there is a Weitzenböck formula, for any $\alpha\in\Omega^2(\mathfrak{g}_E)$,

$$\Delta_A\alpha=\nabla_A^*\nabla_A\alpha+\{F_A,\alpha\}+\{R,\alpha\},$$

where R is the curvature of the base manifold and $\{,\}$ denote certain algebraic multilinear expressions.

In general,

$$\Delta|\alpha|^2=2(\nabla_A^*\nabla_A\alpha,\alpha)-2|\nabla_A\alpha|^2.$$

So taking $\alpha=F_A$ we have

$$\begin{aligned}\frac{\partial e}{\partial t}=2\left(\frac{\partial F}{\partial t},F\right)&=-2(\nabla_A^*\nabla_AF,F)+\{R,F,F\}+\{F,F,F\}\\&\leqslant-\Delta e+\{R,F,F\}+\{F,F,F\}\\&\leqslant-\Delta e+c(e^{3/2}+e).\end{aligned}$$

On the other hand, for the component $\hat{F}_A\in\Omega^0(\mathfrak{g}_E)$ in (iii),

$$\frac{\partial\hat{F}}{\partial t}=-\Delta_A\hat{F}=-\nabla_A^*\nabla_A\hat{F},$$

whence

$$\frac{\partial\hat{e}}{\partial t}+\Delta\hat{e}=-2|\nabla_A\hat{F}|^2\leqslant 0.$$

To prove (iv) we consider

$$\begin{aligned}\frac{\partial}{\partial t}\nabla^kF_A&=\nabla^k\left(\frac{\partial F}{\partial t}\right)+\sum_{i+j=k-1}\nabla^i\left\{\frac{\partial A}{\partial t},\nabla^jF\right\}\\&=-\nabla^k(\nabla^*\nabla F+\{F,F\}+\{R,F\})+\sum_{i+j=k-1}\nabla^i\{\nabla F,\nabla^jF\}.\end{aligned}$$

Commuting derivatives we find that

$$\nabla^k(\nabla^*\nabla F) - \nabla^*\nabla(\nabla^k F) = \sum_{i+j=k} \{\nabla^i F, \nabla^j F\}.$$

Hence

$$\left(\frac{\partial}{\partial t} + \nabla^*\nabla\right)(\nabla^k F) = \sum_{i+j=k} \{\nabla^i F, \nabla^j F\} + \sum_{i+j=k} \{\nabla_i R, \nabla_j F\}.$$

So

$$\left(\frac{\partial}{\partial t} + \Delta\right) e_k \leqslant c_k e_k^{\frac{1}{2}} \left(\sum_{i+j=k} e_i^{\frac{1}{2}} (e_j^{\frac{1}{2}} + 1) \right).$$

Corollary 17. *Suppose that a smooth solution to the evolution equation* (A) *is defined for* $0 \leqslant t < T$.

(i) $\operatorname{Sup}_X |\operatorname{Tr} F|, \operatorname{Sup}_X \hat{e}$ *are each bounded for* $t < T$. *This is true even for* $T = \infty$. *In fact,* $\operatorname{Sup}_X \hat{e}$ *is a decreasing function of time and for any* $\varepsilon > 0$,

$$\operatorname*{Sup}_X \hat{e}(t+\varepsilon) \leqslant c(\varepsilon) \int_X \hat{e}(t) d\mu.$$

(ii) *If also the curvature is uniformly bounded for* $t < T, |F_{x,t}| < B$, *then all the covariant derivatives of* F *are uniformly bounded:*

$$|(\nabla^k F)_{x,t}| < B_k, \quad for\ x \in X, t < T.$$

Proof. All of these follow easily from the corresponding parts of Proposition 16. First, since

$$\left(\frac{\partial}{\partial t} + \Delta\right) \operatorname{Tr} F = 0$$

it is well known that $\operatorname{Tr} F$ converges to a harmonic 2-form and so is certainly bounded. Similarly,

$$\left(\frac{\partial}{\partial t} + \Delta\right) \hat{e} \leqslant 0$$

implies, by the maximum principle, that $\operatorname{Sup}_X \hat{e}$ is decreasing; also by the integral formula for the heat flow,

$$\operatorname*{Sup}_X \hat{e}(t+\varepsilon) \leqslant c(\varepsilon) \int_X \hat{e}(t) d\mu,$$

where $c(\varepsilon)$ is the supremum of the heat kernel on X at time ε (cf. [6]).

Likewise for (ii), by hypothesis the statement is true for $k = 1$. Suppose inductively that it is true for $k' < k$. Then

$$\left(\frac{\partial}{\partial t} + \Delta\right) e_k \leqslant M(1 + e_k)$$

say, by Proposition 16(iv). The linear equation

$$\frac{\partial f}{\partial t} + \Delta f = M(1+f), \quad f(0) = e_k(0),$$

has a smooth solution f_t defined for all $t \geqslant 0$; in particular, it is bounded for $t < T$. Then the result follows by applying the maximum principle to $(e_k - f).e^{-Mt}$.

To put ourselves in a position where we can apply Corollary 17 we use two further lemmas, the first of which is a small technicality.

Lemma 18. *Suppose again that H_t satisfies the evolution equation for $0 < t < T$ and that the curvatures $F_t = F_{H_t}$ are bounded in L^q for some $q > 6$. Then in fact the F_t are uniformly bounded for $t < T$.*

Proof. The heat kernel $H_t(x,y)$ for the scalar equation $(\partial/\partial t) + \Delta$ on X is given for small times t and nearby points $x, y \in X$ by

$$\text{Const}\,.t^{-2}e^{-r^2/4t}, \quad r = d(x,y),$$

up to the addition of a smooth function [**6**]. Hence for $t \leqslant T$ and each fixed $x \in X$,

$$\begin{aligned}\|H_t(x,-)\|_{L^p(X)} &\leqslant \frac{\text{const}\,.}{t^2}\left\{\int_0^\infty u^3(t/p)^2 e^{-u^2}du\right\}^{1/p}\\ &\leqslant \text{const}\,.t^{2(1-p)/p}.\end{aligned}$$

So for any $p < 2$,

$$\int_0^T \|H_t(x,-)\|_{L^p(X)}dt \leqslant c_p(T).$$

Now with $e = |F_t|^2$ we know that

$$\frac{\partial e}{\partial t} + e \leqslant \text{const}\,.(e^{3/2} + e).$$

This implies (cf. [**6**]) that

$$e_t \leqslant H_t * e_0 + \text{const}\,.\int_0^t H_{t-\tau} * (e_\tau^{3/2} + e_\tau)d\tau.$$

The first term is everywhere less than $\text{Sup}_X e_0$ and the second is less than

$$\text{const}\,.\int_0^t \|H_{t-r}\|_{L^p}\|e + e^{3/2}\|_{L^{p'}}dt$$

for conjugate indices p, p'. This is certainly bounded for $t < T$ if we can take $p < 2, p' > 2$, that is, if

$$e^{3/2} \text{ is uniformly bounded in } L^{p'} \text{ for } p' > 2,$$

or, what is the same, $|F|$ is bounded in L^q for $q > 6$.

Lemma 19. *Let H_t, for $0 \leqslant t < T$, be any one-parameter family of metrics on a holomorphic bundle $\mathscr{E}$ over X such that*
(i) *H_t converges in C^0 to some continuous metric H_T as $t \to T$,*
(ii) $\operatorname{Sup}_X|\hat{F}_{H_t}|$ *is uniformly bounded for $t < T$.*
Then H_t is bounded in C^1 and F_{H_t} is bounded in L^p for each $p < \infty$, independent of time $t < T$.

Proof. There is no natural C^1 norm on the space of metrics but, in the usual way, any two choices are equivalent. For definiteness choose local holomorphic trivializations of the bundle $\mathscr{E}$ and take the sup of the gradient of the resulting matrix representations of the metrics.

Suppose that on the contrary the H_t are unbounded in C^1 so that for some sequence $t_i \to T, H_i = H_{t_i}$, there are points $x_i \in X$ where $\operatorname{Sup}_X|\nabla H_i| = m_i$ is attained and $m_i \to \infty$. Taking a subsequence we can suppose that the x_i converge in X. Thus we are reduced to a purely local problem with local coordinates $\{z_\alpha\}$ in the polydisc $|z_\alpha| < 1$, say, and matrices $H_i(z_\alpha)$. Translating the coordinates slightly we may suppose that

$$\operatorname*{Sup}_{\underline{z}} |\nabla H_i| = m_i \text{ is attained at } \underline{z} = 0.$$

By assumption $m_i \to \infty$. Rescale to new coordinates $\{w_\alpha\}$ by $w_\alpha = m_i Z_\alpha$; that is, pull back the matrices H_i via the maps

$$\{|w_\alpha| < 1\} \xrightarrow{p_i} \{|w_\alpha| < m_i^{-1}\}$$

to matrices $\tilde{H}(w_\alpha)$ defined for $|w_\alpha| < 1$. With respect to the obvious rescaled metrics,

$$\operatorname*{Sup}_{\underline{w}} |\nabla \tilde{H}_i| = 1 \text{ is attained at } \underline{w} = 0.$$

To see that this contradicts our hypotheses observe that, on the one hand, by condition (i) the total variation of the matrices $\tilde{H}_i$ in $\{|w_\alpha| < 1\}$ tends to zero as $i \to \infty$. On the other hand, by (ii),

$$|\Lambda \hat{F}(\tilde{H}_i)| = |\tilde{H}_i^{-1}(\Delta \tilde{H}_i - i\Lambda \partial \tilde{H}_i \tilde{H}_i^{-1} \bar{\partial} \tilde{H}_i)|$$

is certainly bounded in $\{|w_\alpha| > 1\}$ and so, since $\tilde{H}_i, \nabla \tilde{H}_i$ are bounded, $|\Delta \tilde{H}_i|$ is bounded, independent of i in $|w_\alpha| < 1$. By the properties of the elliptic operator Δ on L^p spaces, the matrices $\tilde{H}_i$ are bounded, independent of i, in L_2^p over a slightly smaller polydisc. Taking $p > 4$, so that $L_2^p \to C^1$ is compact, we deduce that (some subsequence of) the $\tilde{H}_i$ converge strongly in C^1 to $\tilde{H}_\infty$ say. But then the variation of $\tilde{H}_\infty$ is zero, which contradicts

$$|\nabla \tilde{H}_\infty|_{\underline{w}=0} = \lim_{i \to \infty} |\nabla \tilde{H}_i|_{\underline{w}=0} = 1.$$

By this contradiction we deduce that in fact the H_t are bounded in C^1. Then, again by the ellipticity of the Laplacian, $\hat{F}_{H_t}$ uniformly bounded implies H_t bounded in L_2^p and F_{H_t} bounded in L^p.

Proposition 20. *For any holomorphic vector bundle $\mathscr{E}$ over X and any initial metric H_0 on $\mathscr{E}$ the evolution equation* (A),

$$\frac{\partial H}{\partial t} = -2iH(\hat{F}_H - \lambda.1),$$

has a unique smooth solution defined for $0 \leqslant t < \infty$.

Proof. We know that a solution exists for a short time, Proposition 11, and is unique (Corollary 14). Suppose that, on the contrary, it can only be continued for some maximal interval $0 \leqslant t < T$ (as a smooth solution to the differential equation). By Corollaries 14 and 17 (i) the hypotheses of Lemma 19 apply, and so the curvature is bounded in L^p for any $p < \infty$. By Lemma 18 the curvature of the H_t is in fact uniformly bounded, so, by Corollary. 17(ii), every covariant derivative of the curvature is bounded, for $0 \leqslant t < T$. Again by ellipticity this implies that the H_t are bounded in C^k for each k (starting the obvious inductive argument with $k = 1$). Thus the H_t, which we know converge in C^0 as $t \to T$, in fact converge in C^∞ and we can use the short time existence, starting with H_T, to extend the solution to $[0, T + \varepsilon)$.

3 Proof of the theorem

We prove Theorem 1 in two steps. The main step involves *semi-stable* bundles for which we show:

Proposition 21 (cf. [4, Lemma 1]). *Let $\mathscr{E}$ be a semi-stable bundle over the algebraic surface X. There is a holomorphic bundle $\mathscr{E}'$ with the same rank and degree as $\mathscr{E}$ and such that*

(i) *$\mathscr{E}'$ admits an Hermitian-Einstein connection,*

(ii) *there is a non-zero holomorphic map $\alpha : \mathscr{E} \to \mathscr{E}'$.*

Theorem 1 follows immediately from this; for the bundle $\mathscr{E}'$ is semi-stable by Corollary 4 (making the evident hypothesis that the existence theorem has been proved for bundles of lower rank). Then if $\mathscr{E}$ is actually *stable*, as in Theorem 1, the map α is an isomorphism by Property (a) of stable bundles in the introduction, and so $\mathscr{E} \cong \mathscr{E}'$ admits an Hermitian-Einstein connection. The uniqueness was proved in Corollary 9 (see also the note following Corollary 15), and the converse implication then follows inductively (or use the proofs of Kobayashi and Lübke).

The proof of Proposition 21 requires some preparatory work which we can put in the framework of connections over Riemannian 4-manifolds. Suppose that $\tilde{A}$ is a connection on the trivial bundle (so given by a matrix of 1-forms) over the 4-ball $B^4 = \{x \in \mathbb{R}^4 |x| < 1\}$. For simplicity we suppose that $\tilde{A}$ is an SU(r) connection and we work with respect to the standard flat metric on B^4. Then the fundamental theorem of Uhlenbeck on the choice of a 'good gauge' says:

Proposition 22 [24, Theorem 1.3]. *There is an absolute constant $K > 0$ such that if*

$$\|F(\tilde{A})\|_{L^2(B^4)} \leqslant K,$$

then $\tilde{A}$ is gauge-equivalent to a connection with matrix A satisfying:

(i) $d * A = 0$;
(ii) $\|A\|_{L_1^p} \leqslant C_p \|F(A)\|_{L^p} = C_p \|F(\tilde{A})\|_{L^p}$ *for* $p \geqslant 2$.

Corollary 23. *Suppose that* $\tilde{A}_i$ *is a sequence of connections as above and suppose further that the self-dual components* $F_+(\tilde{A}_i)$ *of their curvatures tend to zero in* $C^0(B^4)$. *Then there is another constant* $K_2 \leqslant K$ *such that if*

$$\|F(\tilde{A}_i)\|_{L^2(B^4)} \leqslant K_2,$$

then some subsequence of the corresponding A_i *(as above) converges weakly in* L_1^p, *for all p, on, say, the half-sized ball, to an anti self-dual connection* A_∞.

Proof. Combine the condition $d^* A_i$ with the expression for the curvature,

$$F_+ = (dA + A \wedge A)_+ \in \Omega^2(\mathrm{SU}\,(r)),$$

into the single equation $F_+(A_i) = DA_i + \{A_i, A_i\}$ where $D = d^* + d^+$ is a first-order elliptic operator (cf. [**5,7**]).

We treat the range $2 \leqslant p < 4$; the general case follows by repetition of the argument. Choose a bump function β, equal to 1 in $\frac{1}{2}B^4$ and supported in B^4. Then

$$\begin{aligned}\|D(\beta A_i)\|_{L^p} &\leqslant \mathrm{const}\,.(\|\beta D A_i\|_{L^p} + \|A_i\|_{L^p}) \\ &\leqslant \mathrm{const}\,.(\|\{\beta A_i, A_i\}\| + \|A_i\|_{L^p} + 1)\end{aligned}$$

by the uniform bound on $F_+(A_i)$. We use the same coincidence of Sobolev and Holder inequalities as in [**24**], that is,

$$\|\{\beta A_i, A_i\}\|_{L^p} \leqslant \mathrm{const}\,.\|A_i\|_{L^4} \|\beta A_i\|_{L^q} \quad \text{for } q = 4p/(4-p),$$

and

$$\begin{aligned}\|\beta A_i\|_{L^q} &\leqslant \mathrm{const}\,.\|\beta A_i\|_{L_1^p} \\ &\leqslant \mathrm{const}\,.\|D(\beta A_i)\|_{L^p} \quad \text{by ellipticity.}\end{aligned}$$

So

$$\|\beta A_i\|_{L_1^p} \leqslant C\{\|A_i\|_{L_1^2} \|\beta A_i\|_{L_1^p} + \|A_i\|_{L^p}\}.$$

If K_2 is sufficiently small then so also is $\|A_i\|_{L_1^2}$ and we can rearrange the equation to deduce an L_1^p bound on βA_i from an L^p bound on A_i. Then we can find a subsequence converging weakly in L_1^p to A_∞, say. Then $F_+(A_i) \to 0$ in C^0 implies that A_∞ is anti self-dual and so smooth by elliptic regularity in the chosen gauge: $d^* A_\infty = 0$.

Now we return to our complex case and suppose that $\mathscr{E}$ is a semi-stable bundle over X, as in Proposition 21. By the Theorem of Mehta and Ramanathan cited in the Introduction we can find a smooth curve $C \subseteq X$ in a linear system $|mH|$ such that $\mathscr{E}|_C$ is semi-stable. Then we have, on the one hand, the 'residue formula' from Proposition 10 comparing the functionals M_ω, M_C:

$$M_\omega(H, K) = (1/m) M_C(H, K) - \int_X f(\psi(F_H) - \psi(F_K)).$$

And, on the other hand, if we choose some initial metric H_0 on $\mathscr{E}$, we have the flow H_t of Proposition 20, defined for $0 \leqslant t < \infty$, with

$$\frac{d}{dt} M_\omega(H_t, H_0) = -\|\hat{F} - \lambda.1\|^2_{L^2},$$
$$\frac{d}{dt}\|F\|^2_{L^2} = \frac{d}{dt}\|\hat{F} - \mu.1\|^2_{L^2} = -\|d_H \hat{F}\|^2.$$

The culmination of all this is the next lemma and corollary.

Lemma 24. *The functional $M_\omega(H_t, H_0)$ is bounded below as $t \to \infty$.*

Proof. Since $\mathscr{E}|_C$ is semi-stable the term $M_C(H_t, H_0)$ is bounded below, by Corollary 9(ii). In the integral

$$\int_X f\psi(F_H) = \int_X f\left(\operatorname{Tr} F_H^2 + \frac{1}{2}\lambda \operatorname{Tr} F_H \wedge \omega\right),$$

$\operatorname{Tr} F_H$ is bounded for all time, by Corollary 17(i). Since $f \in L^1(X)$ this term is bounded. Equally

$$\operatorname{Tr} F_H^2 = (|F_H^-|^2 - |\hat{F}_H|^2).\mathrm{Vol}\,.$$

(as in the Introduction and §1.1) and $\hat{F}_H$ is uniformly bounded. Since $f \to -\infty$ along C, $\int f \operatorname{Tr} F_H^2$ is bounded above, whence the term $-\int f\psi(F_H)$ in the residue formula is bounded below. Thus the residue formula implies that $M_\omega(H_t, H_0)$ is bounded below.

Corollary 25. *For the path H_t of metrics on the semi-stable bundle $\mathscr{E}$ over X,*

$$\hat{F}_{H_t} \to \lambda.1 \text{ in } C^0 \text{ as } t \to \infty.$$

Proof. L^2-convergence is immediate from the lemma since $\|\hat{F} - \lambda.1\|^2$ is decreasing and

$$\frac{d}{dt} M_\omega(H_t, H_0) = -8\|\hat{F} - \lambda.1\|^2.$$

Then C^0-convergence follows from Corollary 17 (i).

Now we can complete the proof of Proposition 21. As in §1.1 regard the sequence of connections, defined by the metrics H_t on the semi-stable bundle $\mathscr{E}$ as unitary connection $\tilde{A}_t$, as a fixed C^∞ bundle E. By a combinatorial argument (see [**20; 7; 5**, §3]) we may extend the local convergence result of Proposition 22 to the complement in X of a finite set $\{x_1, x_2, \dots, x_l\}$ via a covering by small balls. This gives a connection A_∞ over $X \backslash \{x_1, \dots, x_l\}$ which is (projectively) anti self-dual,

$$F^0_+(A_\infty) = 0.$$

By Uhlenbeck's theorem on the removability of singularities in Yang-Mills fields ([**25**], again see [**20, 5**] for similar applications) the connection A_∞ extends over all of X to a connection on a, possibly different, bundle E'; and on any compact set $K \subseteq X \backslash \{x_1, \dots, x_l\}$ for some subsequence $t_i \to \infty$, the sequence $\tilde{A}_i = \tilde{A}_{t_i}$ converges, after suitable bundle automorphisms, to A_∞. (The convergence is weak in any $L_1^p(K)$; however, this is good enough to control the continuity of the transition functions and we can transfer all of [**24**, §3] to this case.)

By the integrability theorem, A_∞ defines a holomorphic structure on E', giving a holomorphic bundle $\mathscr{E}'$ say, which admits the Hermitian-Einstein connection A_∞. Furthermore, the 2-forms $\operatorname{Tr}(F_t)$ converge in $C^\infty(X)$ so that $c_1(\mathscr{E}) = c_1(\mathscr{E}')$ in $H^2(X;\mathbb{R})$.

It remains only to show that there is a non-zero holomorphic map from $\mathscr{E}$ to $\mathscr{E}'$. Roughly speaking this is because the existence of holomorphic sections is a closed condition ('semi-continuity of cohomology'). Recall then that for each i we have complex automorphisms $g_i : E \to E$ intertwining $\bar\partial_{A_0}, \bar\partial_{A_i}$ and

$$g_i^{-1} F_{A_i} g_i = F_{A_0} + \bar\partial_{A_0}(h_i^{-1}\partial_{A_0} h_i)$$

for $h_i = g_i^* g_i$. Set $\tau_i = \operatorname{Tr}(h_i) = |g_i|^2$, as in Definition 12. Then by essentially the same calculation as in Proposition 13 we find that

$$\Delta\tau_i \leqslant \text{const}.(|\hat F_i| + |\hat F_0|).\tau_i,$$

that is, $\Delta\tau_i \leqslant \text{const}.\tau_i$. Since all τ_i are positive, a standard argument [**17**, Theorem 5.31] gives

$$\operatorname*{Sup}_X \tau_i \leqslant c.\|\tau_i\|_{L^2}, \quad \text{say.}$$

Now we can multiply the automorphisms g_i by non-zero constant scalars (since the scalars act trivially on $\mathscr{A}^{(1,1)}$) . In particular, we can normalize so that

$$\|g_i\|_{L^4(X)} = \|\tau_i\|_{L^2(X)} = 1.$$

Choose balls B_j containing $x_j (j = 1, \dots, l)$ and so small that

(a) the bundles $\mathscr{E}, \mathscr{E}'$ are holomorphically trivial over each B_j,

(b) $c.\sum_{j=1}^{l} \operatorname{Vol}(B_j) < \frac{1}{2}$, say.

So Condition (b), together with the inequality above, implies that if $K = X \setminus \bigcup_{j=1}^{l} B_j$, then

$$\|g_i\|_{L^4(K)} \geqslant \frac{1}{2}.$$

Also, of course, the g_i are uniformly bounded. Then on K,

$$\bar\partial_{A_0 A_\infty} g_i = (A_i - A_\infty) g_i,$$

where $\bar\partial_{A_0 A_\infty}$ is the operator on $E^* \otimes E'$ defined by A_0 and A_∞ and we have chosen a suitable isomorphism $E|_K \cong E'|_K$. Since the A_i converge to A_∞ weakly in any $L_1^p(K)$, this equation, together with the bound on $|g_i|$, implies that we can find (a subsequence) g_i converging to g_∞, say, weakly in any L_2^p, and $\bar\partial_{A_0 A_\infty} g_\infty = 0$. Since $\|g_i\|_{L^4(K)} \geqslant \frac{1}{2}, g_\infty$ is not identically zero and represents a holomorphic map $\mathscr{E}|_K \to \mathscr{E}'|_K$. Then we see by the theorem of Hartogs that $\alpha = g_\infty$ extends holomorphically over the balls B_j.

4 Metric characterizations of stability

We now explain, briefly and without proofs, how some of the phenomena that we have found in the course of our proof fit into a larger circle of ideas. This follows closely the paper of Atiyah and Bott and is related to results of Kempf and Ness [**11**] and Kirwan [**12**].

Suppose that a Lie group K (in the first instance compact) acts on a complex manifold V by biholomorphic transformations. Then the action extends to the associated complex group $K^{\mathbb{C}}$ and this partitions V into complex orbits $\mathfrak{O}(x) = K^{\mathbb{C}}.x \subseteq V$.

Suppose, further, that V has a K-invariant Kähler metric with associated form $\omega \in \Omega_V^{1,1}$. We regard this as giving V a symplectic structure, so that we can define the Poisson bracket of functions on V and a Lie algebra homomorphism

$$C^\infty(V) \to \text{Hamiltonian vector fields on } V,$$
$$f \to X_f.$$

Under suitable cohomological hypotheses the infinitesimal action of K lifts to a *comoment* map of Lie algebras

$$\hat{\mu} : \mathfrak{k} \to C^\infty(V) \text{ (cf. [26])}$$

with, dually, a *moment map* $\mu : V \to \mathfrak{k}^*$.

In the case when the cohomology class defined by the symplectic form is the reduction of an integral class the co-moment map has a geometric interpretation. In this case there is an Hermitian line bundle $\pi : L \to V$ having a connection with curvature $-2\pi i.\omega$ and the existence of a co-moment map is equivalent to the existence of a lift of the action of $\mathfrak{k}$ to L, preserving the connection and metric. Explicitly, if $w \in \mathfrak{k}$ maps to a vector field $X = X_{\hat{\mu}(w)}$, we let

$$w(\xi) = \{\text{Horizontal lift of } X(\pi\xi)\} + i\hat{\mu}(\pi(\xi)).\xi \quad \text{for } \xi \in L.$$

Again, under suitable general hypotheses, this defines an action of the group K on L.

In our case ω is a form of type $(1,1)$ on V, so L is a *holomorphic* line bundle and we get an action of the complex group $K^{\mathbb{C}}$ covering the action on V. This means that we have a function $m : V \times K^{\mathbb{C}} \to \mathbb{R}$ measuring the distortion of the norm in L caused by $K^{\mathbb{C}}$:

$$|g(\xi)|^2 = e^{m(X,g)}.|\xi|^2 \quad \text{for } g \in K^{\mathbb{C}}, X = \pi(\xi) \in V.$$

Fixing x we see that this defines a K-invariant function $m(x,-)$ on the complex orbit $\mathfrak{O}(x)$ and it follows immediately from the definition of the lifting of the group action that the *critical points of* m *are precisely the zeros of the moment map* μ.

The canonical example of this general set up is when K acts on $\mathbb{CP}^n$ via some representation $K \to \mathrm{SU}(n+1)$, complexifying to $K^{\mathbb{C}} \to \mathrm{SL}(n+1,\mathbb{C})$. We take for L the tautological bundle whose curvature represents the standard Fubini-Study metric on $\mathbb{CP}^n$ (strictly we should take the dual bundle).

This is the basic situation in which stability is defined in algebraic geometry—without regard to the unitary structures. For $x \in \mathbb{CP}^n$ pick a representative $\hat{x} \in \mathbb{C}^{n+1} \backslash \{0\}$ (that is, a point in the fibre of L over x). The action of $K^{\mathbb{C}}$ moves $\hat{x}$ in an orbit $\hat{\mathfrak{O}} \subset \mathbb{C}^{n+1}$ and, by definition,

x is *semi-stable* if and only if the origin $0 \notin \hat{\mathfrak{O}}$,

x is *stable* if and only if the stabilizer of x is finite and the orbit $\hat{\mathfrak{O}}$ is closed in $\mathbb{C}^{n+1}$. These define the stable and semi-stable orbits of $K^{\mathbb{C}}$ in $\mathbb{CP}^n$.

Now there are two simple *metric* characterizations of stability which are in effect given by Kempf and Ness in [**11**] and Kirwan in [**12**].

Proposition. (a) *The following three conditions on an orbit $\mathfrak{O} \subset \mathbb{CP}^n$ are equivalent:*

(i) *$\mathfrak{O}$ is semi-stable;*

(ii) *the distance $d(0, \hat{\mathfrak{O}})$ of the lifted orbit $\hat{\mathfrak{O}}$ from the origin in $\mathbb{C}^{n+1}$ is strictly positive;*

(iii) $\inf |\mu(x)|^2 = 0$ *as x varies over $\mathfrak{O}$.*

(b) *The following three conditions on an orbit $\mathfrak{O}(x) \subset \mathbb{CP}^n$, such that* $\operatorname{Stab}_K x$ *is finite, are equivalent:*

(i) *$\mathfrak{O}$ is stable;*

(ii) *there is a point in $\hat{\mathfrak{O}}$ closest to the origin;*

(iii) *there is a zero of the moment map μ in $\mathfrak{O}$.*

In this case the points of (ii), (iii) *correspond and are unique up to the action of the compact group K.*

Some parts of the proposition—for example, the existence of a minimum in a stable orbit—follow immediately from the definition and an obvious compactness argument. In this case the function m is the logarithm of the ratios of the norms of vectors in $\mathbb{C}^{n+1}$ under the $K^{\mathbb{C}}$ action and so the relation between these 'closest points' and the moment map is part of the general theory. The key to the uniqueness of a critical point of the distance function is a convexity property which again holds in the abstract set up; in that notation we have:

Proposition. *Let $S \in i\mathfrak{k} \subset \mathfrak{k}^{\mathbb{C}}$. Then for any $X \in V$ the function*

$$m(x, \exp(ts)) \in \mathbb{R}$$

is a convex function of $t \in \mathbb{R}$.

Now it is clear how this theory embraces the formal aspects of our proof of the existence of Hermitian-Einstein connections. Following Atiyah and Bott we take the gauge group $\mathscr{G}$ acting on the complex variety $\mathscr{A}^{(1,1)}$ of integrable connections on a unitary bundle E over X (or C). For simplicity suppose that E has degree 0. The orbits of the complexified group $\mathscr{G}^{\mathbb{C}}$ are in one-to-one correspondence with the holomorphic structures on E. We define a symplectic form on $\mathscr{A}$, and hence on $\mathscr{A}^{(1,1)}$, by the skew pairing

$$(a, b) \to \int_X \operatorname{Tr}(a \wedge b) \wedge \omega,$$

$$(a, b) \to \int_C \operatorname{Tr}(a \wedge b), \quad \text{respectively,}$$

(here $a, b \in \Omega^1(\mathfrak{g}_E) = T\mathscr{A}$) and the moment map for the action of $\mathscr{G}$ is, one checks, essentially $A \to \hat{F}_A$ (see [**1**, §9]). Under suitable cohomological assumptions on X we may then lift the action of $\mathscr{G}^{\mathbb{C}}$ to a certain line bundle $\mathscr{L}$ over $\mathscr{A}^{(1,1)}$ with associated distortion factor m just as in §1.2. Then we have established, in the course of our proof, a precise analogue of the finite-dimensional theory above. The connections with connected stabilizers in $\mathscr{G}$ are the *reducible* connections and we have:

Proposition, (a) *The following three conditions on a bundle* $\mathscr{E}$ *are equivalent:*
(i) $\mathscr{E}$ *is semi-stable;*
(ii) m *is bounded below;*
(iii) $\inf \|\hat{F}_A\|^2 = 0$ *as* A *runs over the unitary connections on* $\mathscr{E}$.
(b) *The following three conditions on an indecomposable bundle* $\mathscr{E}$ *are equivalent:*
(i) $\mathscr{E}$ *is stable;*
(ii) *there is a connection on* $\mathscr{E}$ *minimizing* m*;*
(iii) *there is an Hermitian–Einstein connection with* $\hat{F} = 0$ *on* $\mathscr{E}$.
In this case the connections of (ii), (iii) *correspond and are unique up to the action of the gauge group.*

(Strictly, we have only proved that m is bounded below for semi-stable bundles over curves, but it is easy to extend the proof of Corollary 9 (ii) to surfaces, cf. Proposition 5. Similarly, it is easy to see that $\inf m = -\infty$ for bundles that are not semi-stable.)

This precise correspondence is of some interest in view of the fact that the analysis of stability of bundles in algebraic geometry is rather indirect. For example, our theorem gives, as a simple consequence, a proof by transcendental methods of the fundamental fact that the stable bundles are parametrized by a Hausdorff moduli space.

References

[1] M. F. Atiyah and R. Bott, 'The Yang-Mills equations over Riemann surfaces', *Philos. Trans. Roy. Soc. London Ser. A*, 308 (1982), 524–615.

[2] M. F. Atiyah, N. J. Hitchin, and I. M. Singer, 'Self duality in four dimensional Riemannian geometry', *Proc. Roy. Soc. London Ser. A*, 362 (1978), 425–461.

[3] R. Bott and S. S. Chern, 'Hermitian vector bundles and the equidistribution of the zeros of their holomorphic cross-sections', *Acta Math.*, 114 (1968), 71–112.

[4] S. K. Donaldson, 'A new proof of a theorem of Narasimhan and Seshadri', *J. Differential Geom.*, 18 (1983), 269–277.

[5] S. K. Donaldson, 'An application of gauge theory to four dimensional topology', *J. Differential Geom.*, 18 (.1983), 279–315.

[6] J. EELLS and J. H. SAMPSON, 'Harmonic mappings of Riemannian manifolds', *Amer. J. Math.*, 86 (1964), 109–160.

[7] D. FREED, M. H. FREEDMAN, and K. K. UHLENBECK, 'Gauge theories and 4-manifolds', preprint, M.S.R.I. Berkeley, 1983.

[8] P. GRIFFITHS and J. HARRIS, *Principles of algebraic geometry* (Wiley, New York, 1978).

[9] R. S. HAMILTON, *Harmonic maps of manifolds with boundary*, Lecture Notes in Mathematics 471 (Springer, Berlin, 1975).

[10] M. ITOH, 'On the moduli space of anti self dual connections on Kähler surfaces', *Publ. Res. Inst. Math. Sci., Kyoto University*, 19 (1983), 15–32.

[11] G. KEMPF and L. NESS, *On the lengths of vectors in representation spaces*, Lecture Notes in Mathematics 732 (Springer, Berlin, 1982).

[12] F. C. KIRWAN, 'The cohomology of quotients in symplectic and algebraic geometry', D.Phil. Thesis, Oxford, 1983.

[13] S. KOBAYASHI, 'Curvature and stability of vector bundles', *Proc. Japan Acad. Ser. A. Math. Sci.*, 58 (1982), 158–162.

[14] M. LÜBKE, 'Chernklassen von Hermite–Einstein–Vektorbündeln', *Math. Ann.*, 260 (1982), 133–141.

[15] M. LÜBKE, 'Stability of Einstein–Hermitian vector bundles', *Manuscripta Math.*, 42 (1983), 245–257.

[16] V. B. MEHTA and A. RAMANATHAN, 'Semi-stable sheaves on projective varieties and their restriction to curves', *Math. Ann.*, 258 (1982), 213–224.

[17] C. B. MORREY, *Multiple integrals in the calculus of variations* (Springer, Berlin, 1966).

[18] M. S. NARASIMHAN and C. S. SESHADRI, 'Stable and unitary vector bundles on compact Riemann surfaces', *Ann. of Math.*, 82 (1965), 540–567.

[19] C. OKONEK, M. SCHEIDER, and H. SPINDLER, *Vector bundles over complex projective space* (Birkhauser, Boston, 1980).

[20] S. SEDLACEK, 'A direct method for minimising the Yang-Mills functional', *Comm. Math. Phys.*, 86 (1982), 515–528.

[21] F. TAKEMOTO, 'Stable vector bundles on algebraic surfaces', *Nagoya Math. J.*, 47 (1982), 20–48.

[22] C. H. TAUBES, 'Self dual Yang–Mills connections on non-self dual 4-manifolds', *J. Differential Geom.*, 17 (1982), 139–170.

[23] C. H. TAUBES, 'Self dual connections on four-manifolds with indefinite intersection form', *J. Differential Geom.*, to appear.

[24] K. K. UHLENBECK, 'Connections with L^p bounds on curvature', *Comm. Math. Phys.*, 83 (1982), 31–42.

[25] K. K. UHLENBECK, 'Removable singularities in Yang–Mills fields', *Comm. Math. Phys.*, 83 (1982), 11–30.

[26] N. M. J. WOODHOUSE, *Geometric quantization* (Oxford University Press, 1982).

All Souls College
Oxford OX1 4AL

Editor's Comments

This paper is a tour de force by two modern masters, and it is a special combination of a master analyst and a master geometer. It shows the power and courage of the originators in action. Both the results and methods used to prove them are extremely interesting and useful.

As Yau explains in the next paper included in this book, his joint paper with Uhlenbeck is one of the most difficult and important ones he has written, and its methods have influenced the works of several people:

> As was acknowledged by Donaldson, the argument of Uhlenbeck-Yau is most natural and in fact, all the later development for Hermitian Yang-Mills connections for higher dimensional manifolds are based on the procedure of Uhlenbeck-Yau.
>
> Some later paper such as the one by Bando-Siu [5] used the Hermitian Yang-Mills flow to generalize our result, but the essential feature of Uhlenbeck-Yau procedure is still needed in an essential manner....
>
> A few years later, Carlos Simpson [69] generalized the Uhlenbeck-Yau argument to establish similar theorem when the Higgs field was introduced. Hermitian Yang-Mills connections were proposed by me to Edward Witten in 1984 to study heterotic string, which had since became an important subject in mathematical physics. But from the very beginning, we knew the importance of Hermitian Yang-Mills connections, as it provides important Chern number inequalities, and also the conditions for the bundle to be projectively flat.

This paper also partially motivated Yau's conjecture on the equivalence relation between stability and existence of compact Kähler manifolds with the positive first Chern class.

Though almost everyone talks about the importance of the unity of mathematics and its intimate connection with physics, only a few people can actually do this in their own work. This paper sets such an example. Uhlenbeck and Yau started their paper with:

> The Yang-Mills equations were introduced by theoretical physicists and are now accepted as a basic ingredient in particle theory. In the past decade these equations have become important in mathematics in two separate areas. It was observed early on that the twistor formalism

of Penrose and the Atiyah-Singer index theory can be employed to good purpose in order to describe special solutions, called anti-selfdual solutions; see [29], [3]. Using techniques of algebraic geometry, Atiyah-Drinfeld-Hitchin-Manin [2] were able to describe all such solutions over S^4 (or $\mathbb{R}^4$) in terms of holomorphic bundles on $\mathbb{C}P^3$. More recently a simpler description using stable vector bundles was obtained by Donaldson [8]. The techniques of partial differential equations were used by Taubes [25], [26] to construct solutions on arbitrary four-manifolds. These results and compactness theorems of the first author were incorporated into a general theory of Simon Donaldson to obtain beautiful and spectacular results on the differential structures on four-manifolds; see [10], [11]. It is an observation dating essentially back to Yang's formalism of using $\mathbb{C}$ instead of $\mathbb{R}$ that the Taubes solutions give rise to holomorphic vector bundles over Kähler surfaces; see [30].

The theory of holomorphic vector bundles is central to algebraic geometry. The concept of a stable vector bundle was introduced by Mumford in his study of moduli for bundles [18]. This theme has been pursued by several mathematicians, notably Takemoto, Horrocks, Gieseker, Maruyama and Barth. An important contribution was made by Bogomolov [4], ...

Bogomolov's work inspired the work of Miyaoka [17] on the inequality $3c_2 \geq c_1^2$ for algebraic surfaces of general type. Independently, the second author also gave a proof of this inequality at the same time. While the methods of Miyaoka and Bogomolov are essentially based on algebraic geometry, the method of the second author used the techniques of partial differential equations to construct a Kähler-Einstein metric in his solution of the Calabi conjecture in [31]. The Kähler-Einstein method has the advantage of giving insight into the equality case $3c_2 = c_1^2$ as well as generalizing to arbitrary dimension.

The Hermitian Yang-Mills connection plays the same role for vector bundles as the Kähler-Einstein metric for complex manifolds. Historically, Calabi [6] based his conjectures on the same variational formalism used in Yang-Mills theory....

The basic problem is then to demonstrate the existence of such connections in stable bundles.....

In this paper we demonstrate the existence of an Hermitian Yang-Mills connection in stable holomorphc bundles over any compact Kähler manifold....

The above excerpts from the paper most strongly recommend the paper itself to the reader.

Complex Geometry from Riemann to Kähler-Einstein and Calabi-Yau
ALM 38, pp. 453–486

On the Existence of Hermitian Yang-Mills Connections in Stable Vector Bundles*

Karen Uhlenbeck and Shing-Tung Yau

The Yang-Mills equations were introduced by theoretical physicists and are now accepted as a basic ingredient in particle theory. In the past decade these equations have become important in mathematics in two separate areas. It was observed early on that the twistor formalism of Penrose and the Atiyah-Singer index theory can be employed to good purpose in order to describe special solutions, called anti-selfdual solutions; see [29], [3]. Using techniques of algebraic geometry, Atiyah-Drinfeld-Hitchin-Manin [2] were able to describe all such solutions over S^4 (or $\mathbb{R}^4$) in terms of holomorphic bundles on $\mathbb{CP}^3$. More recently a simpler description using stable vector bundles was obtained by Donaldson [8]. The techniques of partial differential equations were used by Taubes [25], [26] to construct solutions on arbitrary four-manifolds. These results and compactness theorems of the first author were incorporated into a general theory of Simon Donaldson to obtain beautiful and spectacular results on the differential structures on four-manifolds; see [10]-[11]. It is an observation dating essentially back to Yang's formalism of using $\mathbb{C}^2$ instead of $\mathbb{R}^4$ that the Taubes solutions give rise to holomorphic vector bundles over Kähler surfaces; see [30].

The theory of holomorphic vector bundles is central to algebraic geometry. The concept of a stable vector bundle was introduced by Mumford in his study of moduli for bundles [18]. This theme has been pursued by several mathematicians, notably Takemoto, Horrocks, Gieseker, Maruyama and Barth. An important contribution was made by Bogomolov [4], who showed that, for a projective variety

* K. Uhlenbeck, S.T. Yau, On the existence of Hermitian Yang-Mills connections in stable vector bundles. Frontiers of the mathematical sciences: 1985 (New York, 1985). Comm. Pure Appl. Math. 39 (1986), no. S, suppl., S257–S293.

M with Pic $(M) = \mathbb{Z}$ and E a stable bundle on M, the inequality

$$\left(c_2 - \frac{k-1}{2k}c_1^2\right) \cup \omega^{n-2} \geqq 0$$

is valid. Here $k = \operatorname{rank} E, c$, and c_2 are the first and second Chern classes, and ω is a polarization for which E is stable.

Bogomolov's work inspired the work of Miyaoka [17] on the inequality $3c_2 \geqq c_1^2$ for algebraic surfaces of general type. Independently, the second author also gave a proof of this inequality at the same time. While the methods of Miyaoka and Bogomolov are essentially based on algebraic geometry, the method of the second author used the techniques of partial differential equations to construct a Kähler-Einstein metric in his solution of the Calabi conjecture in [31]. The Kähler-Einstein method has the advantage of giving insight into the equality case $3c_2 = c_1^2$ as well as generalizing to arbitrary dimension.

The Hermitian Yang-Mills connection plays the same role for vector bundles as the Kähler-Einstein metric for complex manifolds. Historically, Calabi [6] based his conjectures on the same variational formalism used in Yang-Mills theory. He considered the L^2-norm of the curvature of the metric and showed that under the appropriate assumptions the critical metrics must be Kähler-Einstein. The same reasoning leads from Yang-Mills equations to Hermitian Yang-Mills connections. Likewise, the existence of an Hermitian Yang-Mills connection implies the Bogomolov inequality and strengthens the results by giving insight into the case of equality.

The basic problem is then to demonstrate the existence of such connections in stable bundles. This was shown by Donaldson in [9] in the case when M is an algebraic surface. He uses properties of the restriction of the bundle to curves and the theory of secondary characteristic classes. It is difficult to generalize his method directly to higher-dimensional manifolds. These ideas are intertwined with the work of Atiyah-Bott [1] and the convexity of the moment map; see [13]. It is an ultimate goal to extend the entire sphere of ideas to higher dimensions than curves.

In this paper we demonstrate the existence of an Hermitian Yang-Mills connection in stable holomorphic bundles over any compact Kähler manifold. The method is to perturb the equation and to study the limit of the perturbed equation as the perturbation parameter goes to zero. This reduces the problem to the regularity of a blow-up limit when there is no solution. We study this regularity *via* two avenues: estimates for unitary (real) Yang-Mills type equations on the manifold as a real manifold. Alternatively, we reduce the problem to a regularity question for a weakly holomorphic map. Our solution generalized Hartog's theorem in several variables in that we show that a measurable differentiable function of two complex variables which is separately meromorphic if one fixes almost every point in one variable is in fact meromorphic. As an application, one sees that any weakly holomorphic mapping from the polydisc into an algebraic manifold is in fact a rational mapping. It is very likely that it is also true for Kähler manifolds.

Recently, B. Shiffman has also proved the assertion on meromorphic functions by a very different argument. He applied the assertion to study closed positive

(1,1) currents.

1 Preliminary discussion

Let E be a holomorphic vector bundle of rank r over a compact Kähler manifold X. For an Hermitian metric h along the fiber of E, the Hermitian connection $D : \Gamma(E) \to \Gamma(T^*X \otimes E)$ is characterized by the properties

(i) $d(h(s,t)) = h(Ds,t) + h(s,Dt)$,

(ii) $D''s = \bar{\partial}s$, where D'' denotes the (0,1) component of the connection D.

We shall often express D as $d+A$, where $A \in \Gamma(\mathrm{End}\, E \otimes T^*X)$ is the connection matrix. With respect to a local frame $\{e_\alpha\}$, $A = (A_\alpha^\beta)_{1 \leqq \alpha, \beta \leqq n}$ is given by

$$A_\alpha^\beta = (\partial h_{\alpha\bar{\gamma}}) h^{\bar{\gamma}\beta}, \tag{1.1}$$

where $h_{\alpha\bar{\beta}} = h(e_\alpha e_\beta)$ and $(h^{\alpha\bar{\beta}}) = (h_{\alpha\bar{\beta}})^{-1}$.

The curvature $F = dA - A \wedge A$ of the Hermitian connection is a section of End $E \otimes \Lambda^2 T^*X$. For a holomorphic vector bundle it reduces to

$$F = \bar{\partial}A = h^{-1}\bar{\partial}\partial h + h^{-1}\partial h \wedge h^{-1}\bar{\partial}h. \tag{1.2}$$

In particular, the Hermitian curvature forms of a holomorphic vector bundle are of type (1,1).

Conversely, the integrability theorem of Newlander-Nirenberg implies that a complex vector bundle admits a holomorphic structure if there exists a $U(r)$ connection whose curvature is of type (1,1).

Given a Kähler metric g on X, we define an operation tr_g: $\Gamma\,(\mathrm{End}\, E \otimes T^*_{1,1}X) \to \Gamma(\mathrm{End}\, E)$ as follows. For a section $F = (F_\alpha^\beta) \in \Gamma(\mathrm{End}\, E \otimes T^*_{1,1}M)$,

$$\mathrm{tr}\,_g F = (\mathrm{tr}\,_g F_\alpha^\beta)_{1 \leqq \alpha, \beta \leqq n} = \left(\sum g^{j\bar{k}} F^\beta_{\alpha j \bar{k}}\right)_{1 \leqq \alpha, \beta \leqq n}, \tag{1.3}$$

where $F_\alpha^\beta = F^\beta_{\alpha j \bar{k}} dz^j \wedge d\bar{z}^k$. From now on we shall consistently use $1 \leqq \alpha, \beta \leqq r$ to indicate fiber indices and $1 \leqq j, k \leqq n$ to denote base indices. We shall also adopt the following convenient notation:

$$\begin{aligned} F_{j\bar{k}} &= (F^\beta_{\alpha j \bar{k}})_{1 \leqq \alpha, \beta \leqq r}, \\ \mathrm{tr}\,_g F &= \sum_{j,k} g^{j\bar{k}} F_{j\bar{k}}. \end{aligned} \tag{1.4}$$

Definition. A holomorphic vector bundle of rank r over a compact Kähler manifold (X, g) is Hermitian Yang-Mills if there exists an Hermitian metric h for which the Hermitian curvature F satisfies:

$$\mathrm{tr}\,_g F = \mu I, \tag{1.5}$$

where I is the identity endomorphism of E and μ is a constant; in local coordinates:

$$g^{j\bar{k}} F^\beta_{\alpha j \bar{k}} = \mu \delta_\alpha^\beta \quad \text{for all} \quad 1 \leqq \alpha, \beta \leqq r. \tag{1.5a}$$

The Hermitian Yang-Mills condition (1.5) can also be expressed by first lowering the indices using h, i.e.,

$$F_h = (F_{\alpha\bar{\beta}})_{1\leqq\alpha,\beta\leqq n},$$

where

$$F_{\alpha\bar{\beta}} = F_\alpha^\gamma h_{\gamma\bar{\beta}}. \tag{$*$}$$

Now contract by g:

$$\operatorname{tr}_g F_h = \left(\sum_{j,k} g^{j\bar{k}} F_{\alpha\bar{\beta}j\bar{k}}\right) = \sum g^{j\bar{k}} F^\gamma_{\alpha_{j\bar{k}}} h_{\gamma\bar{\beta}}.$$

Then (1.5) is equivalent to $\operatorname{tr}_g F_h = \mu h$, which is

$$\sum h_{\nu\bar{\beta}} g^{j\bar{k}} F^\nu_{\alpha_{j\bar{k}}} = \mu h_{\alpha\bar{\beta}}. \tag{1.6}$$

If we take $E = TX$ to be the holomorphic tangent bundle and $h = g (h_{\nu\bar{\beta}} = g_{\nu\bar{\beta}})$ then (1.6) becomes the familiar Kähler-Einstein condition

$$g_{\nu\bar{\beta}} g^{j\bar{k}} F^\nu_{aj\bar{k}} = \mu g_{\alpha\bar{\beta}} \quad \text{or} \quad \operatorname{Ric} = \mu g. \tag{1.7}$$

The left-hand side is simply the Ricci curvature

$$F_{\alpha\bar{\beta}} = g^{j\bar{k}} F_{\alpha\bar{\beta}j\bar{k}}.$$

The Hermitian-Einstein factor μ can be expressed in terms of invariants of E and X. The first Chern form $C_1(E) = C_1(E, h)$ is given by

$$C_1(E) = \frac{i}{2\pi} \sum_{\alpha,\beta} h^{\alpha\bar{\beta}} F_{\alpha\bar{\beta}} = \frac{i}{2\pi} \sum h^{\alpha\bar{\beta}} F_{\alpha\bar{\beta}j\bar{k}} dz^j \wedge d\bar{z}^k.$$

If ω is the Kähler form given by $(i/2\pi) g_{j\bar{k}} dz^j \wedge d\bar{z}^k$, then

$$\begin{aligned} C_1(E) \wedge *\omega &= \frac{C_1(E) \wedge \omega^{\eta-1}}{(n-1)!} \\ &= \left(\sum h^{\alpha\bar{\beta}} g^{j\bar{k}} F_{\alpha\bar{\beta}j\bar{k}}\right) \frac{\omega^n}{n!}. \end{aligned}$$

Therefore by (1.6) we have

$$C_1(E) \wedge *\omega = \mu \sum h^{\alpha\bar{\beta}} h_{\alpha\bar{\beta}} \frac{\omega^\eta}{\eta!} = \mu (\operatorname{rank} E) \frac{\omega^n}{n!}.$$

Integrating the above, we get

$$\mu = \mu(E) = \frac{1}{\operatorname{vol} X} \frac{\int_X C_1(E) \wedge *\omega}{\operatorname{rank} E} = \frac{1}{(\operatorname{vol} X)} \frac{\deg_\omega E}{\operatorname{rank} E}, \tag{1.8}$$

where

$$\deg_\omega E = \int_X C_1(E) \wedge *\omega. \tag{1.9}$$

Notice that for any Hermitian metrics h and h', we can find a function p such that

$$\int_X [C_1(E,h) - C_1(E,h')] \wedge *\omega = \int_X \partial\bar{\partial} p \wedge *\omega = \int_X d(\bar{\partial} p \wedge *\omega) = 0$$

because $d\omega = 0$ (ω is a Kähler form). Thus $\deg_\omega E$ (hence also μ) is independent of the Hermitian metric h (but does depend on ω).

From now on, we shall call $\mu I - \operatorname{tr}_g F$ the Hermitian Yang-Mills tensor and denote it by H.

For a coherent analytic subsheaf $\mathfrak{F}$ of E, define

$$C_1(\mathfrak{F}) = C_1(\det \mathfrak{F}^{**}), \quad \deg_\omega = \int_X C_1(\mathfrak{F}) \wedge^* \omega \quad \text{and} \quad \mu(\mathfrak{F}) = \deg_\omega \mathfrak{F}/\text{rank } \mathfrak{F}.$$

Definition. A holomorphic vector bundle E over a compact Kähler manifold (X, ω) is (semi)-stable if, for every coherent subsheaf $\mathfrak{F}$ of lower rank, $\mu(\mathfrak{F}) < \mu(E)$ (respectively $\leqq$).

Remark. (i) $\mathfrak{F}^* = \text{Hom }(\mathfrak{F}, \mathcal{O})$, where $\mathcal{O}$ is the structure sheaf of M.

(ii) F is reflexive if and only if $\mathfrak{F}^{**} = (\mathfrak{F}^*)^* = \mathfrak{F}$. A rank one reflexive sheaf is a holomorphic line bundle.

(iii) A reflexive sheaf is locally free (i.e., a holomorphic vector bundle) outside a subvariety of codimension greater than or equal to 2.

(iv) It is enough (in the definition of stability) to consider only reflexive subsheaves.

One has the following:

Proposition (see [19]). *A stable holomorphic vector bundle over a compact projective manifold is simple, i.e.,* $\dim_{\mathbb{C}} \Gamma(M, \operatorname{End} E) = 1$), *namely holomorphic endomorphisms of E are of the form λId_E, with $\lambda \in \mathbb{C}$.*

The proof can be sketched as follows. By tensoring E with H^k for appropriate power of the hyperplane section bundle, $\mu(E \otimes H^k) \in \{-r+1, -r+2, \ldots, -1, 0\}$, where $r = \operatorname{rank} E$. Clearly, $E \otimes H^k$ is stable if and only if E is stable and $E \otimes H^k$ is simple if and only if E is simple. Now if $E \otimes H^k$ is not simple, the structure sheaf $\mathcal{O}$ is a subsheaf and $\mu(\mathcal{O}) = 0 \geqq \mu(E \otimes H^k)$. This will imply that $E \otimes H^k$ is not stable.

It is very likely that the proposition also holds for compact Kähler manifolds. In any case, we shall assume simplicity is part of the definition of stability.

With this definition, the main theorem of this paper is the following:

Theorem. *A stable holomorphic vector bundle over a compact Kähler manifold admits a unique Hermitian Yang-Mills connection.*

As we remarked in the introduction, this generalizes a result of Donaldson's for complex algebraic surfaces. Donaldson uses a heat equation method, secondary characteristic classes, a theorem on the restriction of stable bundles over surfaces to special curves, convergence of connections with L^2-bounded curvatures as unitary connections, and the removable singularities theorem for Yang-Mills equations over real four-manifolds. Our first approach was to try to either extend or simplify his method. It is not hard to see that solutions of the heat equation do exist for all time in any dimension, and one feels that their behavior as $t \to \infty$ is much the same in all dimensions. In fact, our *a priori* estimates in Section 5 do show convergence (as *unitary* connections) off a set of real codimension four. However, the theorem in algebraic geometry on restrictions to curves is missing. Even more serious from the analytic point of view, we know that there is no removable singularities theorem, since we may actually be obtaining sheaf solutions, or solutions which are of necessity singular on a complex submanifold of X of complex codimension two. The analysis of this limit has not been carried out, although we conjecture that there is a "removable singularities" theorem that constructs a sheaf from our limit which is singular off a set of real codimension four.

Our approach is then of necessity different from that of Donaldson. We fix a base metric and its holomorphic structure on E and replace the heat equation by the one-parameter family of nonlinear elliptic equations

$$H = H_0 - g^{\bar{\beta}\alpha}\partial_{\bar{\beta}}(h^{-1}\partial^0_{\alpha}h) = -\varepsilon \log h. \tag{1.10}$$

Here we think of $\varepsilon = e^{-t}$ as replacing the time parameter in the heat equation. Recall that H is the Hermitian Yang-Mills tensor associated with h. The term $\log h$ is only mysterious at first. Recall we have, point-wise over each point of X, two bilinear forms, the original metric and the new metric h. Therefore, for any smooth function $\rho : \mathbb{R}^+ \to \mathbb{R}$ we can talk about $\rho(h)$. As long as h is smooth, so will be $\rho(h)$.

We show that solutions always exist for $0 < \varepsilon < \infty$, and persist to $\varepsilon = 0$ unless a certain very specific disaster happens to the bundle. Analytically, we detect this precisely in the form that, for $\alpha(\varepsilon_j) \to 0, \varepsilon_j \to 0$,

$$\alpha\,(\varepsilon_j) \log h(\varepsilon_j) \underset{j\to 0}{\to} \sum_{j=1}^{J} a_j \pi_j.$$

Here the a_j are positive constants and π_j is the possibly singular orthogonal projection on the regular points of a subsheaf $E_j \subset E$ of rank less than E. Moreover, $E_1 \subset E_2 \subset \cdots \subset E_j$, provides a filtration, rank $E_{j-1} < \operatorname{rank} E_j$. Because of the relationship between the Hermitian Yang-Mills tensor of E and this filtration, as well as the near orthogonal decomposition (which makes the second fundamental form small) we find

$$\frac{C_1(E_j) \cup \omega^{n-1}}{\operatorname{rank} E_j} \geqq \frac{C_1(E_k) \cup \omega^{n-1}}{\operatorname{rank} E_k}$$

for all $j \geqq k \geqq m$. This of course violates the stability condition. This behavior intuitively corresponds to what happens to unstable bundles under one-parameter subgroups as described by Atiyah and Bott [1].

The technical part of the proof is quite straightforward except for one point. We obtain the E_j as "holomorphic" in a very weak sense from a weak convergence argument. Obtaining enough regularity to describe the E_j as sheaves is more difficult. We have two completely different solutions to this difficulty. One uses estimates for connections and curvatures in the holomorphic bundles considered only as unitary bundles which gives enough regularity to insure smooth convergence off a set of real Hausdorff dimension four. The second independent method is a direct proof of a regularity theorem for π_j, which amounts to showing that a L_1^2-weakly holomorphic map into a projective variety is actually meromorphic.

2 Existence of solutions to the perturbed equations

Let E be a holomorphic vector bundle over a compact Kähler manifold X. Fix the background metric h_0 on E; then for any other metric h_1 on E, we define a smooth endomorphic h on E by

$$\langle s,t\rangle_1 = \langle hs,t\rangle_0. \tag{2.1}$$

In terms of local coordinates,

$$h = h_0^{-1}h_1, \tag{2.2}$$

or

$$h_\alpha^\gamma = \sum_\beta h_0^{\gamma\bar\beta}(h_1)_{\alpha\bar\beta}.$$

Clearly, h is selfadjoint and positive. Hence,

$$\langle hs,t\rangle_0 = \langle s,ht\rangle_0, \tag{2.3}$$

and $\langle hs,s\rangle > 0$ if $s \neq 0$.

Conversely, given any selfadjoint positive endomorphism h, equation (2.1) defines an Hermitian metric on E.

From (2.2) we see that the Hermitian connection forms A^1 and A^0 are related by

$$\begin{aligned}A^1 &= h_1^{-1}\partial h_1\\ &= h^{-1}h_0^{-1}\partial(h_0h)\\ &= h^{-1}A^0h + h^{-1}\partial h\\ &= h^{-1}(\partial h + [A^0,h]) + A^0\\ &= A^0 + h^{-1}\partial^0 h.\end{aligned}$$

Hence,

$$F^1 = \bar\partial A^1 = \bar\partial A^0 + \bar\partial(h^{-1}\partial^0 h).$$

The corresponding Hermitian Yang-Mills tensors are related by

$$H - H_0 = -\mathrm{tr}_g\bar\partial(h^{-1}\partial^0 h). \tag{2.4}$$

We simply write (2.4) as

$$H = H_0 - \bar{\partial}(h^{-1}\partial^0 h). \tag{2.5}$$

Thus the existence of the Hermitian Yang-Mills connection is equivalent to the existence of a selfadjoint, positive endomorphism of E satisfying the equation

$$H = H_0 - \bar{\partial}(h^{-1}\partial^0 h) = 0. \tag{2.6}$$

Before we proceed, we introduce some notation. Given any real-valued function ϕ, we define $\phi(h)$ as follows: Diagonalize h with respect to h_0 as $h = \sum_\alpha e^{\lambda_\alpha} e_\alpha \otimes e_\alpha^*$. Then we define $\phi(h)$ to be $\sum_\alpha \phi(e^{\lambda_\alpha}) e_\alpha \otimes e_\alpha^*$. In particular, $\log h = \sum_\alpha \lambda_\alpha e_\alpha \otimes e_\alpha^*$.

To prove the main theorem, we study the following perturbed equation:

$$H + \varepsilon \log h = H_0 - \bar{\partial}(h^{-1}\partial^0 h) + \varepsilon \log h = 0. \tag{2.7}$$

This equation will be studied in these two sections. The problem of letting $\varepsilon \to 0$ will be studied in Section 4.

Before we proceed, we normalize h_0 conformally so that $\operatorname{tr} H_0 = 0$. This can be done because, by (1.8),

$$\int_X \operatorname{tr} H_0 \frac{\omega^n}{n!} = \deg_\omega E - \mu(\operatorname{rank} E)\operatorname{vol} M = 0.$$

(Note that if $h = p h_0$, then $\operatorname{tr} H(h) = \operatorname{tr} H_0 - (\operatorname{rank} E)\Delta p$. The condition for the existence of p such that $\operatorname{tr} H(h) = 0$ is simply given by $\int_X \operatorname{tr} H_0 = 0$.)

To solve (2.7) for large ε, we set up a continuity argument by solving the one-parameter family of equations:

$$L_{\varepsilon,\sigma}(h) = H_0 - \bar{\partial}(h^{-1}\partial^0 h) + \varepsilon \log h - \sigma h^{-1/2} H_0 h^{1/2} = 0, \tag{2.8}$$

with

$$0 \leqq \sigma \leqq 1.$$

Let

$$T_\varepsilon = \{\sigma \in [0,1] | L_{\varepsilon,\sigma}(h) = 0 \text{ has a solution}\}. \tag{2.9}$$

Then clearly T_ε is non-empty since, for $\sigma = 1, h =$ identity solves the equation for $L_{\varepsilon,1}(h) = 0$. As usual, we shall show that T_ε is both open and closed. This implies $T_\varepsilon = [0,1]$ and $0 \in T_\varepsilon$, which means (2.7) has a solution.

Before we go on, we make the following observation:

Proposition 2.1. *Suppose h solves (2.8). Then, under the normalization* $\operatorname{tr} H_0 = 0, \det h = 1$.

Proof. Taking the trace of (2.8), we obtain

$$\operatorname{tr} H_0 - \operatorname{tr} \bar{\partial}(h^{-1}\partial^0 h) + \operatorname{tr}(\varepsilon \log h) - \sigma \operatorname{tr}(h^{-1/2} H_0 h^{1/2}) = 0. \tag{2.10}$$

Since

$$\begin{aligned} \operatorname{tr} \bar{\partial}(h^{-1}\partial^0 h) &= \operatorname{tr}(\bar{\partial}(h^{-1}\partial h) + \partial(h^{-1}(A^0 h - hA^0))) \\ &= \partial\bar{\partial}(\operatorname{tr} \log h), \end{aligned}$$

we have

$$\Delta \text{tr} \log h - \varepsilon \text{tr} \log h = 0.$$

By the maximum principle on compact manifolds, tr $\log h = 0$, and therefore, $\det h = 1$.

In order to see that T_ε is open, one applies the implicit function theorem and hence one wants to study the linearized operator $\delta L_{\varepsilon,\sigma}$. For any C^1 function ϕ, we compute the variation of $\phi(h)$ by

$$\begin{aligned}\delta\phi(h) &= \sum_\alpha \phi'(e^{\lambda_\alpha})e^{\lambda_\alpha}\delta\lambda_\alpha e_\alpha \otimes e_\alpha^* \\ &\quad + \sum_\alpha \phi(e^{\lambda_\alpha})\delta e_\alpha \otimes e_\alpha^* + \sum_\alpha \phi(e^{\lambda_\alpha})e_\alpha \otimes \delta e_\alpha^*.\end{aligned} \tag{2.11}$$

Let $\delta e_\alpha = \sum_\beta \delta a_\alpha^\beta e_\beta$. Then, since $\delta[e_\alpha^*(e_\beta)] = 0, \delta e_\alpha^* = -\sum \delta a_\beta^\alpha e_\beta^*$. Therefore,

$$\delta\phi(h) = \sum_\alpha \phi'(e^{\lambda_\alpha})\delta\lambda_\alpha e_\alpha \otimes e_\alpha^* + \sum_{\alpha,\beta}[\phi(e^{\lambda_\alpha}) - \phi(e^{\lambda_\beta})]\delta a_\alpha^\beta e_\beta \otimes e_\alpha^*. \tag{2.12}$$

From now on we shall define the trace inner product in End E by

$$\langle A, B\rangle = \text{tr } AB^*,$$

where the identification $E = E^*$ is made *via* the base metric h_0. All the constants depend only on the Kähler metric on X and the base metric on E.

Lemma 2.1. *$\langle \delta \log h, \delta h h^{-1}\rangle \geqq (m/(e^m - 1))|\delta h \cdot h^{-1}|^2$, where $m = \max_X |\log h|$.*

Proof. From (2.10) we have

$$\delta \log h = \sum \delta\lambda_\alpha e_\alpha \otimes e_\alpha^* + \sum(\lambda_\alpha - \lambda_\beta)\delta a_\alpha^\beta e_\beta \otimes e_\alpha^*.$$

Furthermore,

$$\delta h \cdot h^{-1} = \sum_\alpha (\delta\lambda_\alpha)e_\alpha \otimes e_\alpha^* + \sum(1 - e^{\lambda_\beta - \lambda_\alpha})\delta a_\alpha^\beta e_\beta \otimes e_\alpha^*.$$

Thus,

$$\langle \delta h h, \delta h \cdot h^{-1}\rangle = \sum_\alpha |\delta\lambda_\alpha|^2 + \sum_{\alpha,\beta}(1 - e^{\lambda_\beta - \lambda_\alpha})(\lambda_\alpha - \lambda_\beta)|\delta a_\alpha^\beta|^2,$$

and

$$|\delta h \cdot h^{-1}|^2 = \sum |\delta\lambda_\alpha|^2 + \sum |\delta a_\alpha^\beta|^2(1 - e^{\lambda_\beta - \lambda_\alpha})^2.$$

It is a simple exercise to check that, for any $\lambda_\alpha, \lambda_\beta$,

$$\frac{\lambda_\alpha - \lambda_\beta}{1 - e^{\lambda_\beta - \lambda_\alpha}} \geqq \frac{m}{e^m - 1},$$

with $m \geqq \max(|\lambda_\alpha|, |\lambda_\beta|)$.

The lemma follows from this fact and the above two inequalities.

Lemma 2.2.

$$\int_X \langle \delta(h^{-1/2} H_0 h^{1/2}), \delta h \cdot h^{-1} \rangle \leqq 2 \|h^{-1/2} H_0 h^{1/2}\|_\infty \int_X |\delta h \cdot h^{-1}|^2.$$

Proof. By computation,

$$\begin{aligned} \delta(h^{-1/2} H_0 h^{1/2}) = &-(h^{-1/2} \delta h^{1/2})[h^{-1/2} H_0 h^{1/2}] \\ &+ [h^{-1/2} H_0 h^{1/2}] h^{-1/2} \delta h^{1/2}. \end{aligned}$$

On the other hard,

$$h^{-1/2} \delta h^{1/2} = \frac{1}{2} \sum \delta\lambda_\alpha e_\alpha \otimes e_\alpha^* + \sum \delta a_\alpha^\beta \left(1 - \exp\left\{\frac{1}{2}(\lambda_\beta - \lambda_\alpha)\right\}\right) e_\beta \otimes e_\alpha^*.$$

Hence,

$$|h^{-1/2} \delta h^{1/2}|^2 = \frac{1}{4} \sum_\alpha |\delta\lambda_\alpha|^2 + \sum |\delta a_\alpha^\beta|^2 \left(1 - \exp\left\{\frac{1}{2}(\lambda_\beta - \lambda_\alpha)\right\}\right)^2 \leqq |\delta h \cdot h^{-1}|^2,$$

and Lemma 2.2 follows.

Lemma 2.3. $\int_X \langle \delta(\bar\partial(h^{-1}\partial^0 h)), (\delta h) \cdot h^{-1} \rangle \geqq e^{-2m} \int_X |\partial^0(\delta h \cdot h^{-1})|^2$.

Proof. By computation,

$$\begin{aligned} \delta(\bar\partial(h^{-1}\partial^0 h)) &= \bar\partial(-h^{-1}\delta h \cdot h^{-1}\delta^0 h + h^{-1}\partial^0(\delta h)) \\ &= \bar\partial(-h^{-1}\delta h \partial^0(h^{-1}) \cdot h + h^{-1}\partial^0(\delta h)) \\ &= \bar\partial(h^{-1}\partial^0(\delta h \cdot h^{-1})h). \end{aligned}$$

Therefore,

$$\begin{aligned} -\int_X \langle \delta(\bar\partial(h^{-1}\partial^0 h)), \delta h \cdot h^{-1} \rangle &= -\int_X \langle \bar\partial(h^{-1}\partial^0(\delta h \cdot h^{-1})h), \delta h \cdot h^{-1} \rangle \\ &= \int_X \langle h^{-1}\partial^0(\delta h \cdot h^{-1})h, \partial^0(\delta h \cdot h^{-1}) \rangle \\ &\geqq e^{-2m} \int_X |\partial^0(\delta h \cdot h^{-1})|^2. \end{aligned}$$

Proposition 2.2. *Let* $\phi = \delta h$ *and* $m = \max|\log h|$. *Then,*

$$\begin{aligned} &\int_X \langle \delta L_{\varepsilon,\delta}(h)(\phi), \phi h^{-1} \rangle \\ &\geqq e^{-2m} \int_X |\partial^0(\phi \cdot h^{-1})|^2 + \frac{\varepsilon m}{e^m - 1} \int_X |\phi \cdot h^{-1}|^2 - 2\sigma e^{2m} \max|H_0| \int_X |\phi \cdot h^{-1}|^2 \end{aligned}$$

Proof. By computation,

$$\begin{aligned}\delta L_{\varepsilon,\sigma}(h)(\phi) &= -\operatorname{tr}\bar{\partial}\delta(h^{-1}\partial^0 h) + \varepsilon\delta(\log h) - \sigma\delta(h^{-1/2}H_0 h^{1/2}) \\ &= \operatorname{tr}\bar{\partial}(h^{-1}\partial^0(\phi\cdot h^{-1})h) + \varepsilon\delta(\log h) - \sigma\delta(h^{-1/2}H_0 h^{1/2}).\end{aligned}$$

The proposition follows from Lemma 2.1, Lemma 2.2 and Lemma 2.3.

Using the normalization tr $H_0 = 0$, we prove in Proposition 2.1 that $\det h = 1$ and tr $\delta h \cdot h^{-1} = 0$. Therefore, $\langle \delta h \cdot h^{-1}, I\rangle = 0$.

Since the kernel of the operator $\bar{\partial}\partial^0$ gives holomorphic sections of End E and E is stable, it consists of a constant multiple of the identity. Therefore, if $K > 0$ is the first non-zero eigenvalue of the operator $-\bar{\partial}^0\partial^0$,

$$\int_X |\partial^0(\phi h^{-1})|^2 \geqq K \int_X |\phi h^{-1}|^2.$$

Hence we have proved the following:

Proposition 2.3. *Let* $\max_X |\log h| \leqq m$ *and* $\phi = \delta h$ *with* $\operatorname{tr}\phi h^{-1} = 0$. *Then,*

$$\int_X \langle \delta L_{\varepsilon,\sigma}(h)\phi, \phi h^{-1}\rangle \geqq C(m,\varepsilon,\sigma) \int_X |\phi\cdot h^{-1}|^2,$$

where

$$C(m,\varepsilon,\sigma) = e^{-2m}K + m(e^m - 1)^{-1}\varepsilon - \sigma e^{2m}\max_X |H_0|.$$

In order to make use of Proposition 2.3, we need an estimate on m.

Lemma 2.4. *Suppose* $H_0 - \bar{\partial}(h^{-1}\partial^0 h) + \varepsilon\log h - h^{-1/2}\tilde{H}_0 h^{1/2} = 0$ *and* $u = \log h$. *Then,*

$$|u||H_0 - \tilde{H}_0| \geqq -\frac{1}{2}\Delta|u|^2 + \varepsilon|u|^2,$$

and

$$m = \max_X |u| \leqq \varepsilon^{-1}\max_X |H_0 - \tilde{H}_0|.$$

Proof. Since $\langle u, h^{-1/2}\tilde{H}_0 h^{1/2}\rangle = \langle u, e^{-u/2}\tilde{H}_0 e^{u/2}\rangle = \langle u, \tilde{H}_0\rangle$, it suffices to prove that

$$\langle u, -\bar{\partial}(h^{-1}\partial h) + \varepsilon\log h\rangle \geqq -\frac{1}{2}\Delta|u|^2 + \varepsilon|u|^2.$$

It is also sufficient to prove this inequality for a dense set of $u \in C^\infty$ (End E). Therefore we can assume u has distinct eigenvalues on an open dense set of full measure. On this open set, we can write

$$u = \sum_\alpha \lambda_\alpha e_\alpha \otimes e_\alpha^*.$$

Further,

$$h = \sum_\alpha e^{\lambda_\alpha} e_\alpha \otimes e_\alpha^*.$$

Let $A_i^{\alpha\beta} = \langle \partial_i^0 e_\alpha, e_\beta \rangle$. Then,

$$\partial_i^0 h = \sum e^{\lambda_\alpha}(\partial_i \lambda_\alpha) e_\alpha \otimes e_\alpha^* + \sum (e^{\lambda_\alpha} - e^{\lambda_\beta}) A_i^{\alpha\beta} e_\beta \otimes e_\alpha^*,$$

and

$$\partial_i^0 u = \sum (\partial_i \lambda_\alpha) e_\alpha \otimes e_\alpha^* + \sum (\lambda_\alpha - \lambda_\beta) A_i^{\alpha\beta} e_\beta \otimes e_\alpha^*.$$

Therefore,

$$\begin{aligned}\langle \partial^0 u, h^{-1}\partial^0 h\rangle &= \sum |\partial \lambda_\alpha|^2 + \sum |A^{\alpha\beta}|^2 (e^{\lambda_\alpha - \lambda_\beta} - 1)(\lambda_\alpha - \lambda_\beta) + \varepsilon |u|^2 \\ &\geqq \sum |\partial \lambda_\alpha|^2.\end{aligned}$$

Further,

$$\begin{aligned}\langle u, -\bar\partial(h^{-1}\partial^0 h) + \varepsilon \log h\rangle &= -\partial\langle u, h^{-1}\partial^0 h\rangle + \langle \partial^0 u, h^{-1}\partial^0 h\rangle + \varepsilon|u|^2 \\ &\geqq -\frac{1}{2}\Delta |u|^2 + \varepsilon |u|^2.\end{aligned}$$

The estimate on $\max_X |u|$ follows from the maximum principle.

By inverting the operator $\left(-\frac{1}{2}\Delta + \varepsilon\right)$, we can also derive the following:

Corollary 2.1. *If $\int_X |u|^2 \leqq m^2$, then*

$$\max_X |u| \leqq C(m + \max |H_0 - \tilde{H}_0|),$$

where C is independent of ε.

Using Lemma 2.4, we see that the constant $C(M, \varepsilon, \sigma)$ in Proposition 2.3 can be estimated in terms of $K, |H_0|, \sigma$ and ε. It is clear that when ε is large (so that m is small) compared with $\max |H_0|, C(m, \varepsilon, \sigma) > 0$. The same conclusion holds when $\sigma = 0$ and $\varepsilon > 0$. This shows that when ε is large or when $\sigma = 0$, the operator $\delta L_{\varepsilon,\sigma}(h)$ is invertible.

Let h_0 be a solution of the equation $L_{\varepsilon,\sigma}(h) = 0$. Then the map $h \to L_{\varepsilon,\sigma}(h)$ can be considered as a nonlinear smooth map from L_k^p-space of selfadjoint operators of E to L_{k-2}^p-space of selfadjoint operators of E. (Note that H is Hermitian with respect to the metric defined by h.) It is easy to check that the openness of T_ε follows if one can prove that the above map maps a small neighborhood of h_0 to a small neighborhood of the zero operator. By the implicit function theorem, it is enough to prove the surjectivity of the linearized operator $\delta L_{\varepsilon,\sigma}(h_0)$. The surjectivity follows by showing that the kernel of its adjoint operator is trivial.

Our second step is to show that T_ε is closed when ε is large. This will require *a priori* estimates. Under the assumption that ε is large, Proposition 2.3 implies that $m = \sup_X |\log h|$ is small. Therefore the closedness of T_ε follows from

Proposition 2.4. *If $H_0 - \bar\partial(h^{-1}\partial^0 h) + \varepsilon \log h - h^{-1/2}\tilde{H}_0 h^{1/2} = 0$ and if $m = \max_X |\log h|$ is small, then*

$$\|h\|_{C^k(X)} \leqq C(m, \|H_0\|_{C^k(X)}, \|\tilde{H}_0\|_{C^k(X)}).$$

Proof. Let $u = \log h$. Then we can rewrite the above equation as

$$-\bar{\partial}(e^{-u}F(u)\partial^0 u) + \varepsilon u = -H_0 + e^{-u/2}\tilde{H}_0 e^{u/2},$$

where $F(u) = \delta e^u/\delta u$ is the matrix of derivatives of exp at u. Since $m = \max_X |\log h|$ is small, $|I - e^{-u}F(u)| = \delta$ is small.

Let $\Delta_0 = \bar{\partial}\partial_0$ and $I - e^{-u}F(u) = G(u)$. Then,

$$-\Delta_0 u + u = (1-\varepsilon)u + \bar{\partial}(G(u)\partial^0 u) - H_0 + e^{-u/2}\tilde{H}_0 e^{u/2}.$$

Since $-\Delta_0 + I$ is an invertible operator from L_1^p to L_{-1}^p, there is a constant $C(p)$ such that

$$\|u\|_{L_1^p} \leqq C(p)[(1-\varepsilon)\|u\|_{L_{-1}^p} + \delta\|u\|_{L_1^p} + \max|H_0| + e^{2m}\max|\tilde{H}_0|].$$

Here we estimated the term

$$\|\bar{\partial}(G(u)\partial^0 u)\|_{L_{-1}^p} \leqq \|G(u)\partial^0 u\|_{L^p} \leqq \delta\|u\|_{L_1^p}.$$

Therefore, if $1 - C(p)\delta \geqq \frac{1}{2}$ or if δ is sufficiently small, we have a uniform L_1^p-estimate on u. Exactly the same argument provides L_k^p-estimates for u and hence the proposition.

Proposition 2.4 shows that T_ε is closed when ε is large. In particular, $\sigma = 0$ is a point in T_ε. In the next section we shall prove the existence of solution for $L_{\varepsilon,0}(h) = 0$ for all $\varepsilon > 0$.

3 Existence of solution for $L_\varepsilon(h) = 0$ for $\varepsilon > 0$

In this section we shall prove existence of a smooth solution for $L_\varepsilon(h) = 0$ where $L_\varepsilon = L_{\varepsilon,0}$. From the previous section we already know the existence for ε large. We also prove that the operator δL_ε is invertible. Hence, the set of $0 < \varepsilon < \infty$ where $L_\varepsilon(h) = 0$ has a solution is a non-empty open set. We shall prove that it is also closed and hence $L_\varepsilon(h) = 0$ always has a solution for $\varepsilon > 0$.

Since the set of ε for which $L_\varepsilon(h) = 0$ is open, we can differentiate the equation with respect to ε and obtain $\delta L_\varepsilon(h)(\delta h) + \log h = 0$. We shall find an estimate for δh.

Proposition 3.1. *Suppose $\delta L_\varepsilon(h)(\delta h) + \log h = 0$. Then, for $\Psi = h^{-1/2}(\delta h)$ $h^{-1/2}$,*

$$-\Delta|\Psi|^2 + \varepsilon|\Psi|^2 \leqq -2\langle \log h, \Psi\rangle. \tag{3.1}$$

Proof. Let

$$\partial^{\bar{A}} = Adh^{1/2} \cdot \bar{\partial} \cdot Adh^{-1/2},$$

and

$$\partial^A = Adh^{-1/2} \cdot \partial^0 \cdot Adh^{1/2}.$$

Then since $\delta H = -\bar{\partial}(h^{-1}\partial^0(\delta h h^{-1})h)$, one verifies that

$$\delta H = -h^{-1/2}(\partial^{\bar{A}}\partial^A(h^{-1/2}\delta h h^{-1/2}))h^{1/2}. \tag{3.2}$$

Hence,

$$\partial^{\bar{A}}\partial^{A}\Psi + \partial^{A}\partial^{\bar{A}}\Psi = -(h^{-1/2}\delta H h^{1/2} + h^{1/2}\delta H h^{-1/2}). \tag{3.3}$$

Since $\delta L_\varepsilon(h)(\delta h) + \log h = 0$, we obtain

$$\begin{aligned}&\partial^{\bar{A}}\partial^{A}\Psi + \partial^{A}\partial^{\bar{A}}\Psi + \varepsilon(h^{-1/2}\delta\log h h^{1/2} + h^{1/2}\delta\log h h^{-1/2})\\ &= -2\log h.\end{aligned} \tag{3.4}$$

Taking the inner product with Ψ, and using the inequalities

$$2\langle\partial^{A}\partial^{\bar{A}}\Psi + \partial^{\bar{A}}\partial^{A}\Psi, \Psi\rangle \leqq \Delta|\Psi|^2, \tag{3.5}$$

and

$$\begin{aligned}&\langle h^{-1/2}(\delta\log h)h^{1/2} + h^{1/2}(\delta\log h)h^{-1/2}, \Psi\rangle\\ &= \langle\delta\log h, \delta h\cdot h^{-1} + h^{-1}\delta h\rangle\\ &\geqq |\Psi|^2,\end{aligned} \tag{3.6}$$

we can derive (3.1) from (3.4).

Corollary 3.1. *If $\delta L_\varepsilon(h)(\delta h) + \log h = 0$ and $m = \max_X \|\log h|$, then $\max_X|\delta h| \leqq C(m)$.*

Proof. By the proposition, we can estimate $\max_X(\Psi)$ in terms of $m+(\int_X|\Psi|^2)^{1/2}$. Therefore, it suffices to estimate

$$\int_X|\Psi|^2 = \int_X|\delta h\cdot h^{-1}|^2.$$

It follows from Proposition 2.3 that we can estimate $\int_X|\delta h\cdot h^{-1}|^2$ in terms of m.

Proposition 3.2. *If $\delta L_\varepsilon(h)(\delta h) + \log h = 0$, then*

$$\|\delta h\|_{L_2^p} \leqq C(m)(\|h\|_{L_2^p} + 1). \tag{3.7}$$

Proof. By direct calculation,

$$\begin{aligned}\delta H = \delta h\cdot h^{-1}(H - H_0) - h(&\bar{\partial}\partial^0(\delta h) - \bar{\partial}(\delta H)h^{-1}\partial^0 h\\ &-\bar{\partial}hh^{-1}\partial^0(\delta h) + \bar{\partial}hh^{-1}(\delta h)h^{-1}\delta^0 h).\end{aligned} \tag{3.8}$$

On the other hand,

$$\delta H = -\varepsilon\delta(\log h) - \log h. \tag{3.9}$$

It follows that we have an equation for $\bar{\partial}\partial^0(\delta h)$. By the preceding corollary, $|\delta h|$ and $|\delta\log h|$ can be estimated by a function of m. Therefore, we have

$$\|\bar{\partial}\partial^0(\delta h)\|_{L^p} \leqq 2e^{-m}\|\delta h\|_{L_1^{2p}}\|h\|_{L_1^{2p}} + C_1(m)\|h\|_{L_1^{2p}} + C_2(m). \tag{3.10}$$

Since the operator $-\bar{\partial}\partial^0 + I$ is invertible, we obtain

$$\|\delta h\|_{L_2^p} \leqq C_3(m)(\|\delta h\|_{L_1^{2p}}\|h\|_{L_1^{2p}} + \|h\|^2_{L_1^{2p}} + 1). \tag{3.11}$$

On the other hand, by interpolation,

$$\|\delta h\|^2_{L_1^{2p}} \leqq \|\delta h\|_{L_2^p}\|\delta h\|_{L^\infty}$$

for p large enough.

Since we have a bound on $|\delta h|$ and $|h|$ in terms of m, we can substitute (3.11) into (3.10) and apply Hölder's inequality to provide a proof of (3.7).

Corollary 3.2. *If there exists a one-parameter solution to* $L_\varepsilon(h_\varepsilon) = 0$ *in an interval* $(\varepsilon_0, \varepsilon_0')$ *such that* $|\log h_\varepsilon| \leqq m$ *uniformly in* ε, *then*

$$\|h_\varepsilon\|_{L_2^p} \leqq C_4(m)(\|h_{\varepsilon_0'}\|_{L_2^p} + 1)$$

for all ε *in* $(\varepsilon_0, \varepsilon_0']$.

Proof. Let $f(\varepsilon) = \|h_\varepsilon\|_{L_2^p}$. Then the proposition gives an inequality:

$$|f'(\varepsilon)| \leqq C(m)(f(\varepsilon) + 1). \tag{3.12}$$

The corollary follows by integrating this inequality.

By Lemma 2.4 and Corollary 2.1, we have then proved the following:

Theorem 3.1. *There exists a solution to the equation* $L_\varepsilon(h) = 0$ *for all* $\varepsilon > 0$. *Moreover, if* $\int_X |\log h_\varepsilon|^2$ *stays bounded for* $\varepsilon \to 0$, *then there exists a solution in an interval in* $(-\varepsilon_0, \infty)$ *for some* $\varepsilon_0 > 0$.

For later purpose, we shall need

Proposition 3.3. *If* h *solves* $L_\varepsilon h = 0$, *and* F_h *is the full curvature tensor for the connection associated with* h, *then*

$$\int_X |F_h|_h^2 \leqq \int_X |F_0|^2,$$

where F_0 *is the full curvature tensor of the base metric.*

Proof. We have

$$\int_X |F_0|^2 - \int_X |F_h|_h^2 = \int_X |H_0|^2 - \int_X |H|_h^2. \tag{3.13}$$

Since $H + \varepsilon \log h = 0$,

$$|H|_h^2 = \operatorname{tr}(hHh^{-1}H^*) = \operatorname{tr}(HH^*) = |H|^2. \tag{3.14}$$

Therefore we need to show

$$\int_X |H_0|^2 \geqq \int_X |H|^2. \tag{3.15}$$

But

$$\begin{aligned} \int_X \langle H_0 - H, H\rangle &= -\varepsilon \int_X \langle \log, h, \bar\partial(h^{-1}\partial^0 h)\rangle \\ &= \varepsilon \int_X \langle \partial^0 \log h, h^{-1}\partial^0 h\rangle \geqq 0, \end{aligned} \tag{3.16}$$

and (3.15) follows by the Schwarz inequality.

4 Proof of the theorem

In order to prove the theorem, we have to prove that if h_ε does not converge, we can produce a subsheaf of E which violates the condition of stability. The subsheaf will be a subbundle outside a closed subvariety of codimension greater than or equal to 2.

The base Hermitian metric allows us to identify a subbundle of E with the Hermitian projection onto that subbundle. Hence it will be a section of End E given by π so that $\pi^2 = \pi = \pi^*$. The holomorphic condition can be identified as

$$(I - \pi)\bar{\partial}\pi = 0. \tag{4.1}$$

This is easily seen to be equivalent to the fact that, for any section f of the subbundle, $\partial_{\bar{\alpha}} f$ is still in the subbundle.

Hence we define a weakly holomorphic subbundle to be a section $\pi \in L_1^2(E, E)$ such that

$$\pi^2 = \pi = \pi^*, \quad (I - \pi)\bar{\partial}\pi = 0. \tag{4.2}$$

We shall prove that π is actually smooth outside a subvariety of complex codimension greater than or equal to 2 and that it defines a coherent holomorphic subsheaf of E. We shall also prove that if h_ε does not converge, we can normalize in a suitable way and obtain a subsequence converging to such a weakly holomorphic subbundle of E.

We shall use the following lemmas:

Lemma 4.1. *For $0 < \sigma \leqq 1$ and $L_\varepsilon(h) = 0$, we have*

$$|h^{-\sigma/2}\partial^0 h^\sigma|^2 + \varepsilon\langle u, h^\sigma\rangle - \frac{1}{\sigma}\Delta|h^\sigma| \leqq -\langle H^0, h^\sigma\rangle, \tag{4.3}$$

where $h^\sigma = e^{\sigma u}$.

Proof. From $\langle L_\varepsilon h, h^\sigma\rangle = 0$ we obtain

$$\langle h^{-1}\partial^0 h, \partial^0 h^\sigma\rangle + \varepsilon\langle u, h^\sigma\rangle - \bar{\partial}\langle h^{-1}\partial^0 h, h^\sigma\rangle = -\langle H_0, h^\sigma\rangle. \tag{4.4}$$

By computation,

$$\bar{\partial}\langle h^{-1}\partial^0 h, h^\sigma\rangle = \sum_j \bar{\partial}(e^{\sigma\lambda_j}\partial\lambda_j) = \frac{1}{\sigma}\Delta\cdot\left(\sum_j e^{\sigma\lambda_j}\right), \tag{4.5}$$

$$\begin{aligned}\langle h^{-1}\partial^0 h, \partial^0 h^\sigma\rangle &= \sigma\sum_j e^{\sigma\lambda_j}|\partial\lambda_j|^2 + \sum_{k\neq j}|A^{jk}|^2(e^{\lambda_k-\lambda_j}-1)(e^{\sigma\lambda_k}-e^{\sigma\lambda_j}) \\ &\geqq \sigma^2\sum_j e^{\sigma\lambda_j}|\partial\lambda_j|^2 + \sum_{k\neq j}|A^{jk}|^2 e^{-\sigma\lambda_k}(e^{\sigma\lambda_j}-e^{\sigma\lambda_k})^2 \qquad (4.6)\\ &= |h^{-\sigma/2}\partial^0 h^\sigma|^2.\end{aligned}$$

Note that we use the following elementary inequality:

$$\begin{aligned}(e^\mu - 1)(e^{\sigma\mu+\sigma\lambda} - e^{\sigma\lambda}) &= e^{\sigma\lambda}(e^\mu - 1)(e^{\sigma\mu} - 1) \\ &\geqq e^{\sigma\lambda}(e^{\sigma\mu}-1)^2 \qquad (4.7)\\ &= e^{-\sigma\lambda}(e^{\sigma\mu+\sigma\lambda} - e^{\sigma\lambda})^2,\end{aligned}$$

where $0 < \sigma \leqq 1, \lambda$ is arbitrary. In the application, we set $\lambda = \lambda_j$, and $\mu = \lambda_j - \lambda_k$.

Proposition 4.1. *Suppose that* $\overline{\lim}_{\varepsilon \to 0} \int_X |\log h_\varepsilon|^2 = \infty$. *Then there exists a subsequence* $\varepsilon_l \to 0$ *and a normalizing constant* $\rho_l \to 0$ *such that, in* L_1^2,

$$\begin{aligned} h_l &= \rho_l h(\varepsilon_l) \leqq I \qquad \text{(as a metric)}, \\ h_l &= \rho_l h(\varepsilon_l) \rightharpoonup h_\infty \neq 0. \end{aligned} \tag{4.8}$$

Finally, $I - \lim_{\sigma \to 0} h_\infty^\sigma$, *represents a weakly holomorphic subbundle.*

Proof. Let $\rho(\varepsilon) = \exp\{-M(\varepsilon)\}$, where $M(\varepsilon)$ is the maximum of the largest eigenvalue of $\log h(\varepsilon)$. Clearly, since tr $\log h(\varepsilon) = 0, M(\varepsilon)$ has the same order as $m(\varepsilon)$.

Then $\rho(\varepsilon)h(\varepsilon) \leqq I$. From Lemma 4.1 and Lemma 2.4 we have

$$\Delta|h(\varepsilon)| \leqq 2 \max_X |H^0||h(\varepsilon)|. \tag{4.9}$$

By applying the maximum principle, we can estimate $\max_X \rho(\varepsilon)|h(\varepsilon)|$ in terms of $(\int_X |\rho(\varepsilon)h(\varepsilon)|^2)^{1/2}$. Therefore, $\int_X |\rho(\varepsilon)h(\varepsilon)|^2$ is bounded from below by a constant independent of ε. This guarantees that any weak L_1^2-limit of any subsequence of $\rho(\varepsilon)h(\varepsilon)$ is nontrivial (as it implies strong convergence in L^2).

In Lemma 4.1, we can replace h by ρh and hence obtain, for $\sigma = 1$,

$$\begin{aligned} \int_X |d(\rho(\varepsilon)h(\varepsilon))|^2 &= 2\int_X |\partial^0(\rho(\varepsilon)h(\varepsilon))|^2 \\ &\leqq 2\int_X |(\rho(\varepsilon)h(\varepsilon))^{-1/2}\partial^0(\rho(\varepsilon)h(\varepsilon))|^2 \\ &\leqq 4\max_X |H^0| \operatorname{vol}(X). \end{aligned} \tag{4.10}$$

Therefore, for some subsequence of $\varepsilon_l \to 0, \rho_l = \rho(\varepsilon_l) \to 0$ and $h_l = \rho_l h_l(\varepsilon_l)$ has a non-zero weak limit h_∞.

The same argument as in (4.10) also shows that $\int_X |dh_l^\sigma|^2$ is bounded by $2\max_X |H^0| \operatorname{vol}(X)$. By taking a diagonal process, we can assume that h_l^σ converges to h_∞^σ almost everywhere and hence h_l^σ converges weakly to h_∞^σ in L_1^2.

The uniform bound on the L_1^2-norm of h_l^σ gives the same bound on h_∞^σ for all σ. It then follows that $I - h_\infty^\sigma$ has a weak limit in L_1^2 for some subsequence $\sigma \to 0$. We call the limit π.

It is easy to show that $\pi = \pi^* = \pi^2$ almost everywhere. It remains to prove the holomorphic condition. Note that $\pi(I - \pi) = 0$ as a distribution; so

$$|(I-\pi)\bar\partial\pi|^2 = |\bar\partial(I-\pi)\pi|^2 = |\pi\partial^0(I-\pi)|^2.$$

In order to show that the later quantity is zero, we notice that, for $0 \leqq \lambda \leqq 1$.

$$\lambda^{-\sigma} \geqq \frac{2s+\sigma}{s}(1-\lambda^s) \geqq 0. \tag{4.11}$$

Hence,

$$h^{-\sigma/2} \geqq \frac{2s+\frac{1}{2}\sigma}{s}(I-h_l^s). \tag{4.12}$$

It follows from Lemma 4.1 and this inequality that

$$\begin{aligned}\int_X |(I-h_l^s)\partial^0 h_l^\sigma| &\leqq \left(\frac{s}{2s+\frac{1}{2}\sigma}\right)^2 \int_X |h_l^{-\sigma/2}\partial^0 h_l^\sigma|^2 \\ &\leqq \left(\frac{s}{2s+\frac{1}{2}\sigma}\right)^2 \max_X |H_0|.\end{aligned} \tag{4.13}$$

Hence for $l \to \infty$, we have

$$\int_X |(I-h_\infty^s)\partial^0 h_\infty^\sigma|^2 \leqq \left(\frac{s}{2s+\sigma}\right)^2 \max |H_0|. \tag{4.14}$$

If we first let $s \to 0$ and then $\sigma \to 0$, we obtain

$$\int_X |\pi\partial^0(I-\pi)|^2 = 0. \tag{4.15}$$

In the next sections we shall show that π represents a holomorphic subsheaf F of E. Here we derive some information about F.

Lemma 4.2. $0 < \text{rank}\ \ F < \text{rank}\ \ E$.

Proof. By Section 3, we can assume $\rho_l \to 0$. Since tr $\log h(\varepsilon_l) = 0$, at least one eigenvalue of $h_l = \rho_l h(\varepsilon_l)$ is smaller than ρ_l. Therefore, h_∞ has at least a one-dimensional kernel and $0 < \text{rank}\ \ h_\infty < \text{rank}\ \ E$.

The lemma follows from the fact that

$$\text{rank}\, h_\infty = \text{rank}\, E - \text{rank}\, F.$$

Proposition 4.2. $[C_1(F)]/\text{rank}\, F \geqq [C_1(E)]/\text{rank}\, E$.

Proof. We compute *via* the Chern-Weil formula. Since the regular points of F are the complements of a complex variety of codimension two, the singular set cannot affect the computation of C_1. In other words, $[C_1(F)] = [C_1(F|\text{regular points})]$. We have the usual formula

$$\begin{aligned}2\pi[C_1(F)] &= \int_X g^{\alpha\bar\beta}(F)\text{tr}\, F^0_{\alpha\bar\beta}(F) \\ &= \int_X g^{\alpha\bar\beta}\text{tr}\,(F^0_{\bar\beta\alpha}(E)\pi) - \int_X |\pi^\perp\partial^0\pi|^2,\end{aligned} \tag{4.16}$$

where $F^0_{\alpha\bar\beta}(F)$ is the curvature of the connection included by the base metric on the regular point of $F \subseteq E, \pi : E \to F$ is the Hermitian projection on the regular points of F in this metric and $\pi^{\perp}\partial^0\pi : F \to F^{\perp}$ is the second fundamental form. Note that $\pi\partial^0\pi = 0$. Therefore, $\pi^{\perp}\partial^0\pi = \partial^0\pi$. Now,

$$\begin{aligned}\int_X g^{\alpha\bar{\beta}}\mathrm{tr}\,(F_{\alpha\bar{\beta}}\pi) &= \int_X \mathrm{tr}\,((H_0 + \mu I)\pi) \\ &= \int_X \mathrm{tr}\,(H_0\pi) + \frac{\mathrm{rank}\ F}{\mathrm{rank}\ E}2\pi[C_1(E)].\end{aligned} \tag{4.17}$$

Proposition 4.2 follows from (4.15), (4.16) and the following inequality:

$$\int_X \mathrm{tr}\,(H_0\pi) \geqq \int_X |\partial^0\pi|^2, \tag{4.18}$$

which can be proved as follows:

Since $\pi = \lim_{\sigma\to 0}\lim_{l\to\infty}(I - h_l^{\sigma})$ weakly in L_1^2 and $\int_X \mathrm{tr}\, H_0 = 0$, we have

$$\int_X \mathrm{tr}\,(H_0\pi) = -\lim_{\sigma\to 0}\lim_{l\to\infty}\int_X \mathrm{tr}\,(H_0 h_l^{\sigma}). \tag{4.19}$$

If H is the Hermitian Yang-Mills tensor associated with the metric $h(\varepsilon_l) = h_l/\rho_l$, then $H_l = -\varepsilon_l \log h(\varepsilon_l)$ and

$$H_l = H_0 - \bar{\partial}(h_l^{-1}\partial^0 h_l). \tag{4.20}$$

Therefore,

$$\begin{aligned}-\int_X \mathrm{tr}\,(H_0 h_l^{\sigma}) &= -\int_X \mathrm{tr}\,(H_l h_l^{\sigma}) - \int_X \mathrm{tr}\,(\bar{\partial}(h_1^{-1}\partial^0 h_l)h_l^{\sigma}) \\ &= \varepsilon_l \int \mathrm{tr}\,(\log h(\varepsilon_l)h_l^{\sigma}) + \int \langle h_l^{-1}\partial^0 h_l, \partial^0 h_l^{\sigma}\rangle.\end{aligned}$$

As $\mathrm{tr}\ \log h(\varepsilon_l) = 0$ and $h_l^{\sigma} = \rho_l^{\sigma} h_l^{\sigma}(\varepsilon_l)), \mathrm{tr}\,(\log h(\varepsilon_l)h_l^{\sigma}) > 0$. Therefore, as in Proposition 4.1,

$$-\int_X \mathrm{tr}\,(H_0 h_l^{\sigma}) \geqq \int_X \langle \partial^0 h_l^{\sigma}, \partial^0 h_l^{\sigma}\rangle. \tag{4.21}$$

The inequality (4.17) now follows from (4.19), (4.10) and the lower semicontinuity

$$\int_X |\partial^0\pi|^2 \leqq \varliminf_{\substack{\sigma\to 0\\ l\to\infty}} \int |\partial^0 h_l^{\sigma}|^2.$$

Theorem 4.1 (Main Theorem). *If E is a stable holomorphic vector bundle over a compact Kähler manifold X, then E admits an Hermitian Yang-Mills connection.*

Proof. By the results of Section 3, there exists a solution $h = h(\varepsilon)$ to the equation $H = \varepsilon \log h$ for $\varepsilon > 0$. Furthermore, if $\int_X |\log h|^2$ has an upper bound independent of ε, then $h(\varepsilon)$ converges to a smooth Hermitian Yang-Mills connection as $\varepsilon \to 0$.

If $\int_X |\log h|^2$ does not remain bounded, then Proposition 4.1 gives a weakly holomorphic subbundle π which represents a subsheaf of rank strictly between zero and rank E. Proposition 4.1 shows that it violates the stability condition.

Appendix. For later purposes it may be desirable to know more precisely the behavior of h_ε when $\varepsilon \to 0$. This can be described in the following theorem for which we give a semi-rigorous proof.

Theorem 4.2. *Assume that $h(\varepsilon)$ solves $H + \varepsilon \log h = 0$ and that $\overline{\lim}_{\varepsilon\to\infty}$ $\int_X |\log h(\varepsilon)|^2 d\mu \to \infty$. Then there exist a subsequence $\varepsilon(l) \to 0$ and a $\delta(l) \to 0$ in $\mathbb{R}^+$ such that in $L_1^2(L(E,E))$ we have the weak convergence:*

$$u_\varepsilon = \delta(l) \log h(\varepsilon(l)) \rightharpoonup u_\infty \neq 0.$$

Moreover, we have u_∞ expressed as a sum:

$$u_\infty = \sum_{j=1}^{J} a_j \pi_j,$$

where $J \geqq 2, a_j > a_{j-1}$, and π_j are weak projections satisfying $\pi_j \pi_k = \delta_{jk}\pi_j$ and $\sum_{j=1}^{J} \pi_j = I$. In addition, $\sum_{j\leqq k} \pi_j$ is weakly holomorphic for $k = 1, 2, \ldots, J$.

Proof. Choose $\delta(\varepsilon)$ to normalize log $h(\varepsilon)$:

$$\int_X |u_\varepsilon|^2 = \delta(\varepsilon)^2 \int_X |\log h(\varepsilon)|^2 = 1.$$

By Corollary 2.2 and Proposition 2.3, if $m(\varepsilon) = \max_X \log |h(\varepsilon)|$ we have both

$$m(\varepsilon) \leqq \varepsilon^{-1} \max_X |H_0|, \tag{4.22}$$

and

$$m(\varepsilon) \leqq C_0(\delta(\varepsilon)^{-1} + \max_X |H_0|).$$

It follows that

$$\begin{aligned}\int_X |u_\varepsilon|^2 \omega^n &\leqq \delta(\varepsilon)^2 m(\varepsilon)^2 \mathrm{vol}\, X \\ &\leqq C_0^2 \mathrm{vol}\ X (1 + \delta(\varepsilon) \max_X |H_0|)^2.\end{aligned} \tag{4.23}$$

By Lemma 2.4 and this inequality we have the additional inequality (assume $m(\varepsilon)$ is large):

$$\begin{aligned}\int_X |\partial^0 u_\varepsilon|^2 \omega^n &= \delta^2(\varepsilon) \int_X |\partial^0 \log h(\varepsilon)|^2 \omega^n \\ &\leqq m(\varepsilon)^2 \delta(\varepsilon)^2 \max_X |H_0| (1 - e^{-m(\varepsilon)})^{-1} \\ &\leqq 2C_0^2 (1 + \delta(\varepsilon) \max_X |H_0|)^2 \max_X |H_0|.\end{aligned}$$

This bounds $u_\varepsilon = \delta(\varepsilon)\log h(\varepsilon)$ in $L_1^2(L(E,E))$. Because $L_1^2(L(E,E))$ is weakly compact, and $\underline{\lim}\delta(\varepsilon) \to 0$, we can pick $\varepsilon_l \in [1,0), \varepsilon_l \to 0$ such that $\delta_l = \delta(\varepsilon_l) \to 0$ and $u_l = u_{\varepsilon_l} = \delta(\varepsilon_l)\log h(\varepsilon_l) \to u_\infty$ in L_1^2.

Suppose for the moment we are on an open set $X_0 \subset X$ in which the eigenspaces of u_l are of constant rank:

$$u_l = \sum_{j=1}^{J} \hat{\lambda}_{l,j}\pi_{j,l} = \sum_{j=1}^{J} \delta(\varepsilon_l)\lambda_{l,j}\pi_{j,l},$$

where $\hat{\lambda}_{l,j} = \delta(\varepsilon)\lambda_{l,j}$ are the normalized eigenvalues. By another application of proof of (2.4),

$$\int_{X_0} |\partial^0\lambda_{j,l}|^2 \leqq m(\varepsilon_l)^2 \max_X |H_0|.$$

Thus, for the normalized eigenvalues,

$$\int_{X_0} |\partial^0\hat{\lambda}_{j,l}|^2\omega^n \leqq \rho(\varepsilon)^2 m(\varepsilon_l)^2 \max_X |H_0| \to 0.$$

It follows that, as $l \to \infty$, the eigenvalues of u_l become constant on X_0. A global version on X states that all the symmetric functions of the eigenvalues of u_l approach constants as $l \to \infty$. Hence the symmetric functions of the eigenvalues of u_∞, and the eigenvalues themselves are constant a.e.

Furthermore, on $X_l \subseteq X$ we have, from (2.2) and (2.3),

$$\int_{X_l} |A_l^{jk}|^2(e^{\lambda_{j,l}-\lambda_{k,l}} - 1)(\hat{\lambda}_{j,l} - \hat{\lambda}_{k,l}) \leqq 2\max_X |H_0|,$$

where $|A_l^{jk}|^2 = |\langle\partial^0 V_{\pi_j} V_{\pi_k}\rangle|^2 = |\langle V_{\pi_j}\bar{\partial}, V_{\pi_k}\rangle|^2$. On the subset $X_l \subset X, \hat{\lambda}_{j,l} > \hat{\lambda}_{k,l}$. So, given that $\delta_l \to 0, \lambda_{j,l} \gg \lambda_{j,l}$. Then $(e^{\lambda_{j,l}-\lambda_{k,l}} - 1) \gg \hat{\lambda}_{j,l} - \hat{\lambda}_{k,l}$.

We conclude that

$$\int_{X_l} |\pi_{k,l}(\bar{\partial}\pi_{j,l})|^2(\hat{\lambda}_{j,l} - \hat{\lambda}_{k,l})^2\omega^n = \int_{X_l} |A^{j,k}|^2(\hat{\lambda}_{j,l} - \hat{\lambda}_{k,l})^2\omega^n \to 0,$$

in case $k > j$. Since $\hat{\lambda}_{j,l} - \hat{\lambda}_{k,l} \to a_j - a_k < 0$ we see that, for $\pi_k = \lim_{l\to\infty}\pi_{k,l}$,

$$\int_X |\pi_k(\bar{\partial}\pi_j)|^2\omega^n = 0$$

for $k > j$. This argument is made completely rigorous by approximating $\sum_{j\leqq k}\pi_k$ by the sequence $\phi_k(u_l)$, where $\phi_k : \mathbb{R} \to \mathbb{R}, \phi_k(\lambda) = 1$ for $\lambda \leqq a_k + \delta, \phi_k(\lambda) = 0$ for $\lambda \geqq a_{k+1} - \delta$. We approximate $\sum_{j>k}\pi_j$ by ${}^*q_k(u)$, where ${}^*q_k(\lambda)\phi_k(\lambda) = 0, {}^*q_k(\lambda) = 0$ for $\lambda \leqq a_k + \delta$, and ${}^*q_k(\lambda) = 1$ for $\lambda \geqq a_{k+1} - \delta$. Then one shows from the same inequalities that ${}^*q_k(u_l) \to {}^*q_k(u_\infty) = \sum_{j>k}\pi_j$ in $L_1^2, \phi_k(u_l) \to \sum_{j\leqq k}\pi_j$ in L_1^2, and $\int_X |({}^*q_k(u_l)\bar{\partial}({}^*q_k(u_l))|^2\omega^n \to 0$.

5 Estimates from Yang-Mills theory

We use the equation $H + \varepsilon \log h = 0$ as an approximation to the equation $H = 0$, a special case of the Yang-Mills equation. A bound on $\max_X |H|$ will lead us to further useful estimates.

Lemma 5.1. *For $\varepsilon > 0$, the Hermitian Yang-Mills tensor $H = H(h(\varepsilon))$ satisfies (in the base norm)*

$$|h^{1/2} H h^{-1/2}| = |H| \leqq \max_X |H_0|.$$

Proof. Since $H = -\varepsilon \log h$, we see luckily that $h^{1/2} H h^{-1/2} = H$, and a max bound on $|\varepsilon \log h|$ is required. From Corollary 2.2,

$$|\log h| \leqq \varepsilon^{-1} \max_X |H_0|$$

as required.

We will be able to see directly that, even as h_ε diverges, the associated connections converge on a dense open set as *unitary connections*, not as holomorphic connections. In one complex dimension, the compactness theorem implies that this convergence is on all of X (see [27]), in two complex dimensions it is known to be valid on the complement of a finite set of points (see [9]). We change our point of view and regard the holomorphic connection $\{\partial^0_{\bar\alpha}, \partial^0_\alpha + h^{-1}\partial^0_\alpha h\}$ as a unitary connection, and extend the results which are known in complex one-and two-dimensional spaces to arbitrary dimension.

The metric for which the new connection is unitary is h. Any $\mathrm{GL}\,(r, \mathbb{C})$ gauge change v for which $h = (vv^*)^{-1}$ will transform h into the base metric. The unitary gauge group freedom lies in the arbitrariness of the choice of v. The possibilities for v are $v = uh^{-1/2}$, where u is an element of the unitary gauge group. If one were to choose a global canonical v, presumably it would be $h^{-1/2}$. Some of the estimates in Section 2 exploit this gauge change in a non-explicit fashion. Let us agree to make this gauge change, so that in the canonical unitary gauge

$$\begin{aligned} D_\alpha &= \partial^0_\alpha + A_\alpha = \partial^0_\alpha + h^{-1/2}\partial^0_\alpha h^{1/2}, \\ D_{\bar\alpha} &= \partial^0_{\bar\alpha} + A_{\bar\alpha} = \partial^0_{\bar\alpha} + h^{1/2}\partial^0_{\bar\alpha} h^{-1/2}, \\ H &= h^{1/2} H h^{-1/2}. \end{aligned}$$

Certainly one reason for the success of our technical method is that the norm of the Hermitian Yang-Mills tensor is not affected by this change. Of course, all our estimates are completely gauge-invariant. It is only necessary to choose the gauges consistently.

The main result we shall use from gauge field theory is an *a priori* estimate for holomorphic connections where a bound on the Hermitian Yang-Mills tensor is known. This is a class of *a priori* estimates, which are proved *via* a scaling inequality and a geometry of X. Balls of radius σ_0 should be nearly Euclidian. The number K depends on the group, in this case on $U(r)$. Recall that the bundle being holomorphic means the $(2,0)$ and hence $(0,2)$ part of the curvature vanishes.

Theorem 5.1 (Uhlenbeck [28]). *Let $E \to X$ be a rank r vector bundle over a compact Kähler manifold X (complex dimension n, real dimension $2n$). Given $n < p < \infty$, there exist $\sigma_0 > 0, K > 0$, and $C_1 < \infty$ such that if ∂^A is a unitary holomorphic connection in E and*

(a) $\max_X |H| \leqq K^2$,

(b) $\sigma^{4-2n} \int_{\delta(x,y) \leqq 4\sigma} |F|^2 \omega^n < K^2$,

(c) $\sigma \leqslant \min(\sigma_0, K)$,

then

$$\int_{\delta(x,y)<2\sigma} |F|^p \omega^n < C_1 \sigma^{(2-p)n} \left(\int_{\delta(x,y) \leqq 4\sigma} |F|^2 \right)^{p/2} + \sigma^{2n} \max_X |H|^p.$$

Here we have set $H = g^{\alpha\bar{\beta}} F_{\alpha\bar{\beta}}$ with no μI factor.

As given, this is not the estimate we need. Part of the proof of the above theorem uses the existence of "good gauges".

Corollary 5.1 (see [27]). *Under the hypothesis of Theorem 5.1, we have the existence of a (unitary) local trivialization in $B_{2\sigma}(y)$ such that $E|B_{2\sigma}(y) = B_{2\sigma}(y) \times \mathbb{C}^N, D = d + A$, and*

$$\int_{\delta(x,y) \leqq 2\sigma} |\text{grad } A|^p + \sigma^{-p} \int_{\delta(x,y) \leqq 2\sigma} |A|^p \leqq C_2 \int_{\delta(x,y) \leqq 2\sigma} |F|^p.$$

How does this lead to an estimate on h? Recall our discussion on the equivalence of holomorphic and unitary approaches.

Proposition 5.1. *Suppose the connection ∂^A arises from the connection ∂^0 via the metric h. Normalize the metric h so that $h \leqq I$. Then under the hypothesis of Theorem 5.1, in $B_{2\sigma}(y)$ we have $h = vv^*$, where A_α satisfies the estimates of Corollary 5.3 and*

$$\partial^0_\alpha v = v A_\alpha.$$

Moreover,

$$\int_{\delta(x,y) \leqq \sigma} |(\partial^0)^2 h|^p \omega^n + \sigma^{-p} \int_{\delta(x,y) \leqq \sigma} |\partial^0 h|^p \omega^n$$
$$\leqq C_3 \left[\sigma^{(2-p)n} \left(\int_{\delta(x,y) \leqq 2\sigma} |F|^2 \right)^{p/2} + \sigma^{2n} \max_X |H|^p + 1 \right].$$

Proof. The terms in σ are scaling constants. We assume the ball is the unit ball and the Kähler metric nearly flat. Moreover, the original connection was C^∞ and smooth. The constant C_3 will depend on ∂^0.

Apply Theorem 5.1 and Corollary 5.1 to the connection $\partial^0_\alpha + h^{-1/2} \partial^0_\alpha h^{1/2}, \partial^0_{\bar{\alpha}} + h^{1/2} \partial^0_{\bar{\alpha}} h^{-1/2}$. Since an arbitrary gauge change for the unitary connection is u, we have $v^{-1}\partial^0_\alpha v = A_\alpha$ for $h^{1/2}u = v$, where A_α is the difference between the original connection ∂^0 and the form for which the estimates in Corollary 5.1 are valid. However, the original connection is uniformly smooth and we have these same estimates for A_α. Elliptic estimates transfer the L^p_1-estimates on A_α in a ball to those on the L^p_2-norm of h in a ball of half the radius.

We state without proof the following convergence theorem. The argument concerning the cover of X follows that of Sedlacek [23] for real dimension four. The global gauge transformations are constructed in Donaldson [10] and Uhlenbeck [27]. The assumption of compactness made there is not necessary, as it can be replaced by a cover which is the finite union of covers made up of disjoint balls. The theorem describes the limiting behavior of our equation $H + \varepsilon \log h = 0$, or of the heat equation.

Theorem 5.2. *Let $E \to X$ be an Hermitian vector bundle over a Kähler manifold X and D_l a sequence of holomorphic connections such that*

(a) $\int_X |F(D_l)|^2 \leqq B$,

(b) $\max_X |H(D_l)| \leqq K$.

Then there exists a subsequence $D_{l'}$, and a closed singular set of finite Hausdorff dimension $2n-4$ such that $D_{l'}|X - \mathscr{S}|$ is gauge equivalent to a sequence of connections which converges (weakly) in $L^p_{1,\mathrm{loc}}(X - \mathscr{S})$ to an $L^p_{1,\mathrm{loc}}$ holomorphic connection on $X - \mathscr{S}$.

Corollary 5.2. *Let E be the Hermitian vector bundle associated to E and the base metric, and B be the holomorphic, unitary connections in E which are $\mathrm{GL}(r, \mathbb{C})$ equivalent to those constructed from the metric solutions to $H + \varepsilon \log h = 0$. Then there exists a subsequence $D_{l'}$ and a singular set of finite Hausdorff dimension $2n-4$ such that $D_{l'}$ are gauge equivalent to a sequence which converges in L^p_1 on $X - \mathscr{S}$. Moreover, $\pi = I - \lim_{s\to 0} \lim_{l\to\infty} h^s$ is meromorphic on $X - \mathscr{S}$.*

Proof. By Lemma 5.1, $\max_X |H|$ is bounded, and by Proposition 3.3, $\int_X |F|^2$ is bounded, so we can apply the theorem. The regularity of A_α is sufficient to show that the cokernel is meromorphic.

Theorem 5.3. *The projection π constructed in Corollary 5.2 represents a holomorphic subsheaf of E.*

Proof. We need only show that the singularities of real codimension $2n-4$ are actually the singularities of a meromorphic function (see [24]) and that a meromorphic function from X into a Grassmannian pulls back the canonical vector bundle to a coherent, reflexive sheaf. This is a much simpler version of the discussion in Section 7 and we refer the reader to that.

As a biproduct of our theory, we have a result concerning the compactness of the space of Hermitian Yang-Mills connections in $E \to X$. Since these parameterize the stable holomorphic bundles $\mathscr{E}$ which are topologically E, this result should relate other methods of compactifying the space of stable bundles. We conjecture that the limit object has a natural structure as a sheaf, although even for complex surfaces the actual construction of the sheaf is unclear.

Theorem 5.4. *Let D_l be a sequence of holomorphic Hermitian Yang-Mills connections in an Hermitian vector bundle E over a compact Kähler manifold X. Then there exists a gauge equivalent subsequence $D_{l'}$, and a closed set $\mathscr{S}$ of (real) Hausdorff dimension $2n-4$ such that $D_{l'}|X-\mathscr{S}| \to D_\infty$. Here D_∞ is a holomorphic Hermitian Yang-Mills connection in $E|X-\mathscr{S}|$.*

For complex surfaces, the removable singularities theorem shows that D_∞ completes to a holomorphic Hermitian Yang-Mills connection in E_∞. However, E_∞ is topologically different from E if $\mathscr{S} \neq \varnothing$ in that at every point in $\mathscr{S}$ at which the convergence fails, of necessity integral multiples of C_2 are absorbed, and the second Chern class of the bundle E_∞ is smaller than that of E.

6 Regularity of weakly holomorphic maps

The purpose of this section is to prove that a weakly holomorphic map into an algebraic manifold is meromorphic. This will be used in the next section to show the regularity of the subsheaf $\mathcal{F}$. To be precise, let us first define a weakly holomorphic map from the ball $B \subset \mathbb{C}^n$ into an algebraic manifold M. We assume that M is isometrically embedded in $\mathbb{CP}^K$ which is embedded in some Euclidian space $\mathbb{R}^N$. The map $F : B \to M \subset \mathbb{R}^N$ is then a vector-valued map and it makes sense to say that it is L_1^2. By definition, this means that there is a sequence of smooth maps F_i from B into $\mathbb{R}^N$ so that both F_i and dF_i converge in $L^2(B)$ and that $\lim_{i\to\infty} F_i = F$. As usual, we define dF to be $\lim_{i\to\infty} dF_i$.

When $n=1$, we define F to be weakly holomorphic if it is L_1^2 and, for almost every point x in B, dF sends the holomorphic tangent space of $\mathbb{C}T_x(B)$ into the holomorphic tangent space at $F(x)$ of $\mathbb{C}T_{F(x)}(M)$. Note that this makes sense because for almost every point x in B, dF sends the tangent space of B to the tangent space of M. (In fact, in this case, F_i can be assumed to have the image in M (see [21])).

When $n > 1$, we call F weakly holomorphic if for any linear holomorphic coordinate system $\{z_1, \ldots, z_n\}$ on B and for almost every point $\{z_2, \ldots, z_n\}$, the restrictions of F to each complex line in the z_1 variable is weakly holomorphic.

Theorem 6.1. *Any weakly holomorphic map into an algebraic manifold is meromorphic.*

Proof. First of all, we prove the theorem when $n=1$.[1] It is clear that we have only to prove the continuity of F.

Let $x \in B$ be any point. Let $B_x(r)$ be the balls of radius r around x and $E_x(r)$ the energy of F over $B_x(r)$. Then we want to prove that, for some $\alpha > 0, E_x(r)r^{-\alpha}$ is bounded as x varies in a compact set of B. The continuity of F will then follow from Morrey's lemma; see [16].

Clearly, we can assume that r is small and $B_x(r) \subset B$. As the total energy is bounded, it suffices to prove the differential inequality

$$\frac{d(r^{-1}E_r)}{dr} = r^{-1}\frac{dE_r}{dr} - r^{-2}E_r \geqq 0.$$

[1] M. Gruter, *Regularity of weak H-surface*, J. Reine Angew. Math., 329, 1981, pp. 1–15.

Therefore, $E_x(r) \leqq rdE/dr$. If the image of the circle of radius r is large, then $\int_0^{2\pi} |dF/d\theta|^2 d\theta \leqq rdE/dr$ is large and the inequality is obvious. Hence we can assume that the length of these curves is uniformly small.

When the length of the image of the circle of radius r is small, it lies in a coordinate chart of M. One can solve the Plateau problem for this curve within this coordinate chart. By the isoperimetric inequality for minimal surfaces, we know that the area of this minimal surface is dominated by the square of the length of the curve. Hence the curve bounds the images of two different disks. In particular, we have a map of the sphere into M. (The map F on $B_x(r)$ is not *a priori* continuous. We simply approximate F in L_1^2-norm by a smooth map into M. This is possible because $n = 1$ (see [22]).) We claim that this map is homotopically trivial.

In fact, by using the uniformization theorem, we may assume that the above map is conformal. In this case, area is the same as energy. Hence the total energy of this map is not greater than $E_x(r)$ plus some multiple of the square of the length of the image circle of radius r. Hence by choosing r small enough, the total energy of the map from the sphere is less than some fixed constant. As was proved in Sacks-Uhlenbeck [20], this constant can be chosen so that the map must be homotopically trivial.

Let Ω be the Kähler form on M. Then, using the Wirtinger inequality and $d\Omega = 0$, we can prove that $\int_{B_x(r)} F^*\Omega$ is not greater than the area of the minimal surface constructed above. In particular, the isoperimetric inequality shows that $\int_{B_x(r)} F^*\Omega$ is not greater than some constant times the square of the length of $F(\partial B_x(r))$.

To prove our differential inequality, it suffices to prove that $\int_{B_x(r)} \tilde{F}^*\Omega$ is $E_x(r)$. Actually, at this moment, $\tilde{F}$ is only a smooth map which approximates the original F in the L_1^2-map. In general, for smooth maps $\tilde{F}$,

$$\int_{B_x(r)} \tilde{F}^*\Omega = \int_{B_x(r)} |\partial\tilde{F}|^2 - \int_{B_x(r)} |\bar{\partial}\tilde{F}|^2.$$

We claim that when $\tilde{F}$ is close enough to our map F in L_1^2-norm, $\int_{B_x(r)} \tilde{F}^*\Omega$ is close to $E_x(r)$.

In fact, given any small $\varepsilon > 0$, we can assume that $\tilde{F}(x)$ is ε-close to $F(x)$ for x outside a set where the energy of $\tilde{F}$ is smaller than ε. For any two points y_1 and y_2 in M which has distance less than ε, we can identify the tangent space of M at y_1 and y_2 by parallel transportation in M along the shortest geodesic joining y_1 and y_2. For each point y in M, let P_y be the projection from the complexified tangent space of $\mathbb{R}^N$ at y to the holomorphic tangent space of M at y. In this terminology, $|(\bar{\partial}\tilde{F})(x)|$ is equal to $|P_{\tilde{F}(x)}\tilde{F}_*(\partial/\partial\bar{z})|$. By assumption, this is close to $|P_{\tilde{F}(x)}F_*(\partial/\partial\bar{z})|$ in L^2-topology.

Outside a set where the energy of F is small, the L^2-norm of $|P_{F(x)}\tilde{F}_*(\partial/\partial\bar{z})|$ is not greater than the L^2-norm $|P_{F(x)}\tilde{F}_*(\partial/\partial\bar{z})|$ plus a small multiple of the L^2-norm of $|F_*(\partial/\partial\bar{z})|$. Therefore, the L^2-norm of $|\bar{\partial}\tilde{F}|$ is not greater than a small constant plus the L^2-norm of $|\bar{\partial}F|$. Since $\bar{\partial}F = 0$ almost everywhere, we conclude that the L^2-norm of $|\bar{\partial}\tilde{F}|$ is small.

A similar reasoning shows that the L^2-norm of $|\partial\tilde{F}|$ is close to the L^2-norm of $|\partial F|$. Hence $\int \tilde{F}^*\Omega$ is close to the energy of F and we have proved our claim.

This claim implies the continuity of F and proves the theorem for $n = 1$. For $n > 1$, we proceed as follows.

Since an algebraic manifold is a submanifold of $\mathbb{CP}^n$, we assume that $M = \mathbb{CP}^n$. By writing $\mathbb{CP}^n$ as $\mathbb{C}^n$ union the plane at infinity, we can write $F = (f_1, \ldots, f_n)$, where we allow the f_i to have values at infinity. Hence Theorem 6.1 can be reduced to the following theorem which is of independent interest. (Clearly the theorem can be generalized to $\mathbb{C}^n$ with $n > 2$.)

Theorem 6.2. *Let f be a measurable function defined on the unit polydisk $D \times D$ in $\mathbb{C}^2$. Suppose that, for some set E of measure zero in $D \times D$, the restriction of f to $D \times D \backslash E$ satisfies the following property. For almost every $z, f(z,w)$ can be extended to be a meromorphic function of w. Similarly, for almost every $w, f(z,w)$ can be extended to be a meromorphic function of z. Then f is equal to a meromorphic function almost everywhere.*

First of all, we can assume that, for almost every $z, f(z,w)$ is a meromorphic function of w.

Lemma 6.1. *Let $E \subset D \times D$ be a subset with measure zero. Then there exists a sequence $\{z_i\} \subset D$ and a subset G of D such that $\mu(G) = \mu(D)$, where μ is the standard measure on D, $(\{z_i\} \times G) \cap E = \varnothing$ and $z_i \to \frac{1}{2} \in D$ as $i \to +\infty$.*

Proof. Since $\mu(E) = 0$, one can find a full measure subset F of D such that, for all $z \subset F, \mu((z \times D) \cap E) = 0$.

We choose a sequence $\{z_i\} \subset F$ such that $z_i \to \frac{1}{2}$ as $i \to +\infty$. If we set $G = D\backslash(\bigcup_{i=1}^{\infty}(z_i \times D) \cap E)$, then $\mu(G) = \mu(D)$ and $(z_i \times G) \cap E = \varnothing$.

By assumption, there is, in case $n = 1$, an $\tilde{f}$ on $D \times D$ such that, for almost all $w \in D, f(\cdot, w)$ is a meromorphic function of z and $E = \{x \in B | f(x) \neq \tilde{f}(x)\}$ has measure zero.

Now we choose z_i as in Lemma 6.1; then, for almost all $w \in D$ and for all $i, \tilde{f}(z_i, w) = f(z_i, w)$.

Taking

$$f_i(z,w) = \sum_{j=1}^{i} f(z_i, w) \prod_{\substack{k=1 \\ k \neq j}} (z - z_k)(z_j - z_k)^{-1},$$

we obtain a sequence of meromorphic functions $f_i(z,w)$, such that $f_i(z_j, w) = f(z_j, w)$ for $1 \leqq j \leqq i$. Now we claim that we can assume that both f_i and $1/f_i$ are in L^p for all $p < 2$, and the L^p-norms are uniformly bounded.

To see this, we notice that for any measurable function f, the integral $\int_{|a|<1} |f - a|^{-p} da\, d\bar{a}$ is bounded independent of f. Therefore, we can apply Fubini's theorem to conclude that the double integral

$$\int_{|a|<1} \left(\int_{|z|<1} \int_{|w|<1} |f_i(z,w) - a|^{-p} dz \wedge d\bar{z} \wedge dw \wedge d\bar{w} \right) da \wedge d\bar{a}$$

is bounded uniformly. Then we can find a_i and b_i so that $|a_i| \leqq \frac{1}{4}, |b_i - \frac{3}{4}| \leqq \frac{1}{4}$, and both $(f_i - a_i)^{-1}$ and $(f_i - b_i)^{-1}$ are L^p integrable for $1 \leqq p < 2$, the L^p-norms are bounded independently of i. Let $g_i = 1 + (a_i - b_i)(f_i - a_i)^{-1}$, then the L^p-norms of both g_i and g_i^{-1} have a uniform upper bound. It is clear that $\{g_i\}$ satisfies all the hypothesis for $\{f_i\}$. Therefore we can assume that both g_i and g_i^{-1} are L^p-integrable for all $p < 2$ and the L^p-norms are uniformly bounded.

Lemma 6.2. *Let ϕ be a meromorphic function defined on the unit disk $D = \{|z| < 1\}$. Then there exists a constant C independent of ϕ, such that*

$$\int_{|z|<3/4} \frac{\partial\phi \wedge \overline{\partial\phi}}{(1+|\phi|^2)^2} \leqq C\left(\int_{|z|<1} [\log^2(1+|\phi|^2) + \log^2(1+|\phi^{-1}|^2)] + 1\right).$$

Proof. Since

$$\frac{\partial^2}{\partial z \partial \bar{z}} \log(1+|\phi|^2) - \frac{\partial\phi \wedge \overline{\partial\phi}}{(1+|\phi|^2)^2}$$

is a linear combination of the δ functions at the poles of ϕ,

$$\int_{|z|<1} p^2 \left[\log^{-1}(1+|\phi|^2)\frac{\partial^2}{\partial z\partial\bar{z}}\log(1+|\phi|^2) - \log^{-1}(1+|\phi|^2)\frac{\partial\phi \wedge \overline{\partial\phi}}{(1+|\phi|^2)^2}\right] = 0,$$

where p is a smooth function with compact support in $|z| < 1$, and equal to 1 on $|z| < \frac{3}{4}$. In the following, we denote by C a constant independent of ϕ.

Integrating by part, we obtain

$$\begin{aligned} -2\textstyle\int_{|z|<1} p\log^{-1}(1+|\phi|^2)\partial p \wedge \bar{\partial}\log(1+|\phi|^2) \\ +\textstyle\int_{|z|<1} p^2\log^{-2}(1+|\phi|^2)\frac{|\phi|^2\partial\phi\wedge\overline{\partial\phi}}{(1+|\phi|^2)^2} \\ -\textstyle\int_{|z|<1} p^2\log^{-1}(1+|\phi|^2)\frac{\partial\phi\wedge\overline{\partial\phi}}{(1+|\phi|^2)^2} = 0. \end{aligned}$$

Clearly the same equations hold if we replace ϕ by ϕ^{-1}. Summing the two equations, we obtain

$$\begin{aligned} &-2\int_{|z|<1} p\log^{-1}(1+|\phi|^2)\frac{\phi\partial p \wedge \overline{\partial\phi}}{1+|\phi|^2} + 2\int_{|z|<1} p\log^{-1}(1+|\phi^{-1}|^2)\frac{\bar{\phi}^{-1}\partial p\wedge\overline{\partial\phi}}{1+|\phi|^2} \\ &+\int_{|z|<1} p^2[|\phi|^2\log^{-2}(1+|\phi|^2) - \log^{-1}(1+|\phi|^2) \\ &+|\phi^{-1}|^2\log^{-2}(1+|\phi^{-1}|^2) - \log^{-1}(1+|\phi^{-1}|^2)]\frac{\partial\phi\wedge\overline{\partial\phi}}{(1+|\phi|^2)^2} = 0. \end{aligned}$$

Let $h(t) = \frac{1}{4}t\log^{-2}(1+t)$ when $t \geqq 1$ and $h(t) = \frac{1}{4}t^{-1}\log^{-2}(1+t^{-1})$ when $t \leqq 1$. Then

$$\begin{aligned} |\phi|^2\log^{-2}(1+|\phi|^2) - \log^{-1}(1+|\phi|^2) + |\phi^{-1}|^2\log^{-2}(1+|\phi^{-1}|^2) \\ -\log^{-1}(1+|\phi^{-1}|^2) \geqq h(|\phi|^2). \end{aligned}$$

Therefore, applying the Schwarz inequality to the previous equation, we find

$$\begin{aligned}&\int_{|z|<1} p^2 h(|\phi|^2)\frac{\partial\phi\wedge\overline{\partial\phi}}{(1+|\phi|^2)^2}\\ &\leqq 4\left(\int_{|z|<1} p^2 h(|\phi|^2)\frac{\partial\phi\wedge\overline{\partial\phi}}{(1+|\phi|^2)^2}\right)^{1/2}\\ &\times\left(\int_{|z|<1}|\nabla p|^2 h(|\phi|^2)^{-1}(|\phi|^2\log^{-2}(1+|\phi|^2)\right.\\ &\left.+|\phi^{-1}|^2\log^{-2}(1+|\phi^{-1}|^2))\right)^{1/2}.\end{aligned}$$

By direct computation we obtain

$$\begin{aligned}&h(|\phi|^2)^{-1}(|\phi|^2\log^{-2}(1+|\phi|^2)+|\phi^{-1}|^2\log^{-2}(1+|\phi^{-1}|^2))\\ &\leqq C(1+\log^2(1+|\phi|^2)+\log^2(1+|\phi^{-1}|^2)).\end{aligned}$$

Since $h(|\phi|^2)\geqq C>0$, we conclude that

$$\int_{|z|\leqq 3/4}\frac{\partial\phi\wedge\overline{\partial\phi}}{(1+|\phi|^2)^2}\leqq C\int_{|z|<1}[1+\log^2(1+|\phi|^2)+\log^2(1+|\phi^{-1}|^2)].$$

Corollary 6.1. *$\iint_{|z|\leqq 3/4,|w|\leqq 3/4}|\nabla f_i|^2/(1+|f_i|^2)^2\leqq C$, where C is independent of f_i.*

Proof. Applying Lemma 6.2 to $f_i(z,\cdot)$ for all $z\in D$, we have

$$\int_{|w|\leqq 3/4}\frac{\partial_w f_i\wedge\overline{\partial_w f_i}}{(1+|f_i|^2)^2}\leqq C\int_{|w|<1}(1+\log^2(1+|f_i|^2)+\log^2(1+|f_i^{-1}|^2)).$$

Therefore,

$$\iint_{\substack{|z|<1\\|w|\leqq 3/4}}\frac{|\nabla_w f_i|^2}{(1+|f_i|^2)^2}\leqq C\iint_{\substack{|z|<1\\|w|<1}}(1+\log^2(1+|f_i|^2)+\log^2(1+|f_i^{-1}|^2)).$$

Similarly, if we replace f by $\tilde{f}$, we get

$$\iint_{\substack{|z|\leqq 3/4\\|w|<1}}\frac{|\nabla_z\tilde{f}_i|^2}{(1+|\tilde{f}_i|^2)^2}\leqq C\iint_{\substack{|z|<1\\|w|<1}}(1+\log^2(1+|\tilde{f}_i|^2)+\log^2(1+|\tilde{f}_i^{-1}|^2))$$

but $f_i=\tilde{f}_i$ a.e., and f_i, f_i^{-1} are uniformly L^p-bounded for $1\leqq p<2$,

$$\iint_{\substack{|z|\leqq 3/4\\|w|\leqq 3/4}}\frac{|\nabla f_i|^2}{(1+|f_i|^2)^2}\leqq C.$$

Now we consider the graphs of f_i as subsets of $D \times D \times \mathbb{CP}^1$; then the above corollary implies that, over each compact subset of $D \times D$, the volumes of the graphs are uniformly bounded, since the volume form of the graph of f_i over $D \times D$ is dominated by the Euclidean volume form of $D \times D$ plus

$$\frac{\partial f_i \wedge \overline{\partial f_i}}{(1+|f_i|^2)^2}(dz \wedge d\bar{z} + dw \wedge \overline{dw}).$$

Therefore, by Bishop's theorem [5], the graphs of $|f_i|$ converge to an analytic subset A of $D \times D \times \mathbb{CP}^1$. Furthermore, for each $(z,w) \in D \times D$, the limit set of $\{f_i(z,w)\}$ is contained in $A \cap ((z,w) \times \mathbb{CP}^1)$. We choose a meromorphic function h on $D \times D \times \mathbb{CP}^1$ so that for each point $x \in (z,w) \times \mathbb{CP}^1$ which is not in the limit set h is holomorphic on $(z,w) \in D \times D$ and $h(A) = 0, h(x) \neq 0$. Define $h_n(y) = h(y) - h(f_n(z',w'))$ for $y \in (z',w') \times \mathbb{CP}^1$; then $h_n = 0$ on the graph of f_n. Now $h_n \to \tilde{h}, \tilde{h}$ vanishes on A, but $\tilde{h}(x) \neq 0$, so $x \notin A$, i.e., the limit set of $\{f_i(z,w)\}$ is just $A \cap ((z,w) \times \mathbb{CP}^1)$. Since the projection of $D \times D \times \mathbb{CP}^1$ onto $D \times D$ has degree one on the graph of each f_i, there exists an analytic subset S of codimension one in $D \times D$ such that $A \setminus (S \times \mathbb{CP}^1)$ is a graph over $D \times D \setminus S$. For each $j, f_i(z_j,w) = f(z_j,w)$ for $i \geqq j$. The limit set of $\{f_i(z_j,w)\}$ consists of only one point $(z_j,w; f(z_j,w))$, i.e., $(z_j,w) \in D^0 \times D^0 \setminus S$.

Hence, we conclude that there exists a meromorphic function g in $D \times D$, such that $g(z_j,w) = f(z_j,w)$ for $j \geqq 1$. Since, for almost all $w \in G \subset D, \tilde{f}(z,w)$ is meromorphic on z, and is equal to $g(z,w)$ at infinite number of points $\{z_i\}, g(z,w) \equiv \tilde{f}(z,w)$ for such $w \in G$, i.e., $g(z,w) = \tilde{f}(z,w)$ a.e.

It follows that f is equal to a meromorphic function almost everywhere. Theorem 6.2 is proved.

Remark. *As was noted by B. Shiffman, a corollary of Theorem* 6.2 *is that every closed locally rectifiable* (1,1) *current is given by a holomorphic chain. This statement was proved by R. Harvey and B. Shiffman in* [12] *with the extra assumption on the support of the current. More recently, Shiffman informed us that he has a different method to demonstrate the statement.*

7 Regularity of weakly holomorphic subbundles

In this section, we describe another approach to prove the regularity of π without using the estimate from Yang-Mills theory. In fact, we shall prove that the weakly holomorphic subbundle defined in Section 4 is given by a holomorphic subsheaf.

Given a holomorphic vector bundle E over a complex manifold M, we can find an open cover $\{O_i\}$ of M so that $E|O_i$ is biholomorphic to $O_i \times \mathbb{C}^r$ and the transition transformations are given by holomorphic maps from $O_i \cap O_j$, to $\mathrm{GL}\,(r,\mathbb{C})$. A holomorphic subbundle of rank k can be given as follows. For each i, it is given by a holomorphic map f_i, from O_i to $G(r,k)$, the Grassmanian of k planes in $\mathbb{C}^r$. The map f_i is equal to f_j up to the automorphism of $G(r,k)$ induced from the transition transformations of the bundle E.

If f_i is only rational, i.e., f_i is defined outside a subvariety of codimension at least two in U_i, then the set $\{f_i\}$ defines a coherent subsheaf of E. It can be seen in the following way. Let $U(k)$ be the universal bundle over $G(r,k), F_i$ be the closure of the graph of f_i. Let π_1^i, π_2^i be the projections to the first and the second factor, respectively. Then $\pi_2^{i*}U(k)$ is a k-bundle on F_i, by the Gauert direct image theorem, $\pi_{i*}^8(\pi_2^{i^*}U(k))$ is a coherent sheaf on O_i, induced by f_i. Obviously, $\pi_{1*}^i(\pi_2^{i^*}U(k))$ and $\pi_{1*}^{\kappa}(\pi_{\psi}^{\kappa^*}U(k))$ are patched up on $O_i \cap O_j$ by the automorphism of $G(r,k)$, since f_i and f_j are.

A weakly holomorphic subbundle is given by a projection map π. The rank of the projection is given by tr π which has to be an integer. Since tr π is L_1^2, it is straightforward to show that the integer is in fact a constant almost everywhere. Let tr $\pi = k$ be this constant. Then, in each coordinate chart O_i, π induces a mapping $\hat{\pi}_i$ from a set of O_i with full measure into $G(r,k)$. We claim that the $\hat{\pi}_i$ can be extended to be a rational mapping from O_i, to $G(r,k)$. Without losing generality, we may assume that $n = 2, O = O_i = D \times D, \hat{\pi} = \hat{\pi}_i$, where D is the unit disk in C^1. As π is L_1^2, for almost every $z \in D$ (respectively $w \in D$), the restriction of π to $\{z\} \times D$ (respectively to $D \times \{w\}$) is L_1^2. Consider π to be a mapping from $D \times D$ to the submanifold V_k of End E consisting of all elements with rank equal k. The well-known approximation theorem (see [22]) tells us that, if $\pi_z = \pi\{z\} \times D$ is L_1^2, π_z can be approximated by L_1^2-bounded smooth mappings F_i from $\{z\} \times D$ into V_k. As discussed in Section 6, we see that $\lim_{i\to\infty} \int_{O\times\Delta} |(I - F_i)\bar{\partial}F_i|^2 = 0$. Define $\hat{F}_i\{z\} \times D \to G(r,k)$ by $\hat{F}_i(z,w) =$ the image $F_i(z,w)(C^r) \in G(r,k)$. We claim that the limit of $\hat{F}_i$ gives a rational extension of $\hat{\pi}_z = \hat{\pi}|_{z\times\Delta}$.

First of all, let us prove that the $\hat{F}_i$ are L_1^2, and their L_1^2-norms are uniformly dominated by those of π_i and so are uniformly bounded. We use the Plücker coordinate for $G(r,k)$. Let $e_1, e_2, \ldots, e_r$, be any unitary frame field which spans $F_i(C^r)$ at any given point $(z,w) \in z \times D$, so that $F_i(e_j) = e_j$ for $j \leqq k$. Since F_i is continuous, $F_i(e_1) \wedge \cdots \wedge F_i(e_k)$ defines the plane in $G(r,k)$ near (z,w) and, at $(z,w), |\nabla[F_i(e_1) \wedge \cdots \wedge F_i(e_k)]|$ is dominated by $|dF_i|$ for any subset $\{i_1, \ldots, i_k\}$ of $\{1, 2, \ldots, n\}$. This implies that the L_1^2-norms of $\hat{F}_i$ are dominated by those of F_i.

Secondly, we prove that $\lim_{i\to+\infty} \int_{O\times\Delta} |\bar{\partial}\hat{F}_i|^2 dw = 0$. This follows from the definition of $\hat{F}_i$ and

$$\lim_{i\to+\infty} \int_{O\times\Delta} |(I - F_i)\bar{\partial}F_i|^2 dw = 0.$$

In fact, let $e_1, \ldots, e_r$ be a holomorphic frame of C^r, so that, at a given point $(z,w), F_i(e_j) = e_j$ for $1 \leqq j \leqq k$, and $e_1, \ldots, e_r$ are orthogonal. One checks that

$$\begin{aligned}
&\bar{\partial}[F_i(e_1) \wedge \cdots \wedge F_i(e_k)] \\
&= \sum_{m=1}^{k} \langle \bar{\partial}F_i(e_j), e_j \rangle F_i(e_1) \wedge \cdots \wedge F_i(e_k) \\
&\quad + \sum_{j=1}^{k} (-1)^{j-1} F_i(e_1) \wedge \cdots \wedge (I - F_i)\bar{\partial}F_i(e_j) \wedge \cdots \wedge F_i(e_k) \\
&= \operatorname{tr}(F_i\bar{\partial}F_i) \cdot F_i(e_1) \wedge \cdots \wedge F_i(e_k) \\
&\quad + \sum_{j=1}^{k} (-1)^{j-1} F_i(e_1) \wedge \cdots \wedge (I - F_i)\bar{\partial}F_i(e_j) \wedge \cdots \wedge F_i(e_k).
\end{aligned}$$

Therefore, $|\bar{\partial}\hat{F}_i|$ is dominated by $|(I - F_i)\bar{\partial}F_i|$ and $|\mathrm{tr}\,(F_i\bar{\partial}F_i)|$, $\mathrm{tr}\,(F_i\bar{\partial}F_i)$ converges to $\mathrm{tr}\,(\pi_z\bar{\partial}\pi_z)$ in L^2 and $\mathrm{tr}\,(\pi_z\bar{\partial}\pi_z) = \mathrm{tr}\,(\bar{\partial}\pi_z)$, $\bar{\partial}\mathrm{tr}\,\pi_z = 0$ since $\mathrm{tr}\,(\pi_z)$ is constant almost everywhere. We conclude that $\int_{Z|\Delta} |\mathrm{tr}\,(F_i\bar{\partial}F_i)|^2 dw \to 0$ as $i \to +\infty$. On the other hand, $\int_{z\times\Delta} |(I - F_i)\bar{\partial}F_i|^2 dw \to 0$ as $i \to +\infty$, hence $\lim_{i\to+\infty} \int_{O\times\Delta} |\bar{\partial}\hat{F}_i|^2 dw = 0$.

Summing the above, we see that the limit of $\hat{F}_i$ gives an extension of $\hat{\pi}_z = \hat{\pi}|_{z\times\Delta}$, which is weakly holomorphic. By the regularity theorem of Section 6, it follows that, for almost every $z \in D, \hat{\pi}|_{O\times\Delta}$ can be extended to be a holomorphic mapping from $\{z\} \times D$ into $G(r,k)$. Similarly, for almost every $w \in D, \hat{\pi}|_{D\times\psi}$ can be extended to be a holomorphic mapping from $D \times \{w\}$ into $G(r,k)$.

Now our claim follows from the regularity theorem of Section 6 for $n = 2$, so does the main theorem.

8 Applications

Our main application is similar to a major application of the solution of the Calabi conjecture: inequalities between Chern numbers of the tangent bundle [31]. The same argument gives the following:

Theorem 8.1. *Let E be a stable holomorphic bundle over a compact n-dimensional Kähler manifold X with respect to the Kähler class ω. Let $r = \mathrm{rank}\ E$. Then $2rC_2(E) \cup \omega^{n-2} \geqq (r+1)C_1^2 \cup \omega^{n-2}$ and the equality holds if and only if the pull-back of E to the universal cover of X is a direct sum of Hermitian holomorphic line bundles where the curvature forms of these line bundles are equal to each other.*

Corollary 8.1. *Let E be a stable holomorphic bundle over a compact Kähler manifold X. Then E is obtained from unitary representation of the fundamental group of X if and only if $C_1^2 \cup \omega^{n-2} = 0$ and $C_2 \cup \omega^{n-2} = 0$.*

Using the connection constructed in this paper, Cho, in his Harvard thesis [7], computed the curvature of the Kähler metric on the moduli space of stable vector bundles. He concluded that the holomorphic sectional curvature must be positive if X is a curve. In particular, if the moduli space is non-singular and compact, it must be simply connected.

References

[1] Atiyah, M. F., and Bott, R., *The Yang-Mills equations over Riemann surfaces*, Phil. Trans. Roy. Soc. London A 308, 1982, pp. 523–615.

[2] Atiyah, M. F., Hitchin, N. J., Drinfeld, V. G., and Manin, Yu J., *Construction of instantons*, Phys. Lett. 65 A 185, 1978.

[3] Atiyah, M. F., Hitchin, N. J., and Singer, I. M., *Self duality in four dimensional Riemannian geometry*, Proc. Roy. Soc. London A 365, 1978, pp. 425–461.

[4] Bogomolov, *Holomorphic tensors and vector bundles*, Izvestia, U.S.S.R., 42, 1978, pp. 1227–1287.

[5] Bishop, E., *Conditions for the analyticity of certain sets*, Michigan-Math. J. 11, 1964, pp. 289-304.

[6] Calabi, E., *On Kähler manifolds with vanishing canonical class*, Algebraic Geometry and Topology Symposium in honour of S. Lefshetz, Princeton University Press, 1955, pp. 78–89.

[7] Cho, K., *Positivity of the curvature of the Weil-Peterson metric on the moduli space of stable vector bundles*, Ph.D. thesis, Harvard Univ., 1985.

[8] Donaldson, S., *Instantons and geometric invariant theory*, Comm. Math. Physics. 93, 1984, pp. 453–460.

[9] Donaldson, S., *Anti-self-dual Yang-Mills connections over complex algebraic surfaces and stable vector bundles*, Proc. London Math. Soc. 50, 1985, pp. 1–26.

[10] Donaldson, S., *An application of gauge theory to four-dimensional topology*, J. Diff. Geom. 18, 1983, pp. 279–315.

[11] Donaldson, S., *Connections, cohomology and the intersection forms of 4-manifolds*, to appear in J. Diff. Geom.

[12] Harvey, R., and Shiffman, B., *A characterization of holomorphic chains*, Ann. Math. 99, 1974, pp. 553–587.

[13] Kirwan, F., *Cohomology of Quotients in Symplectic and Algebraic Geometry*, Math Notes 31, Princeton University Press, 1984.

[14] Kobayashi, S., *Curvature and stability of vector bundles*, Proc. Japan Acd. Ser. A, Math. Sci. 58, 1982, pp. 158–162.

[15] Lubke, M., *Stability of Einstein-Hermitian vector bundles*, Manuscripta Math. 42, 1983, pp. 245–257.

[16] Morrey, C. B., *Multiple Integrals in the Calculus of Variations*, Springer-Verlag, Berlin, 1966.

[17] Miyaoka, Y., *On the Chern numbers of surfaces of general type*, Inventiones Math. 42, 1967, pp. 225–237.

[18] Mumford, D., *Geometric Invariant Theory*, Erg. Math. 34, Springer-Verlag, 1965.

[19] Okonek, C., Schneider, M., and Spindler, H., *Vector Bundles on Complex Projective Space*, Birkhauser, Progress in Math. 3, 1980.

[20] Sacks, J., and Uhlenbeck, K., *The existence of minimal 2-spheres*, Ann. of Math. 113, 1981, pp. 1–24.

[21] Schoen, R., and Uhlenbeck, K., *A regularity theory for harmonic maps*, J. Diff. Geom. 17, 1982, pp. 307–335.

[22] Schoen, R., and Uhlenbeck, K., *Boundary regularity and the Dirichlet problem for harmonic maps*, J. Diff. Geom. 18, 1983, pp. 253–268.

[23] Sedlacek, S., *A direct method for minimizing the Yang-Mills functional on 4-manifolds*, Comm. Math. Phys. 86, 1986, pp. 515-527.

[24] Siu, Y. T., *Extension of meromorphic maps into Kähler manifolds*, Annals of Math. 102, 1975, pp. 421–462.

[25] Taubes, C. H., *Self-dual connections on non-self-dual manifolds*, J. Diff. Geom. 17, 1982, pp. 139.

[26] Taubes, G. H., *Self-dual connections on manifolds with indefinite intersection matrix*, J. Diff. Geom. 19, 1984, pp. 517–560.

[27] Uhlenbeck, K., *Connections with L^p bounds on curvature*, Comm. Math. Phys. 83, 1982, pp. 31–42.

[28] Uhlenbeck, K., *A priori estimates for Yang-Mills fields*, (manuscript in preparation).

[29] Ward, R., *On self-dual gauge-fields*, Physics Letters 61A, 1977, pp. 81–85.

[30] Yang, C. N., *Condition of self-duality for SU(2) gauge fields on Euclidean four-dimensional space*, Phys. Rev. Lett. 38, 1977, pp. 1377–1379.

[31] Yau, S. T., *On the Ricci curvature of a compact Kähler manifold and the complex Monge-Ampere equation*, Comm. Pure and Applied Math. 31, 1978, pp. 339–411.

Editor's Comments

Yau's highly original work in complex geometry and geometric analysis has changed the landscape of Kähler geometry. His work shows the importance of integrating differential geometry, partial differential equations and algebraic geometry. This paper is a historical guided tour by one of its founders. It shows the great unity of mathematics by viewing from Kähler geometry. Since the initial work of Poincaré on rational points on elliptic curves, Weil and Grothendieck had developed the subject of arithmetic algebraic geometry. Calabi-Yau manifolds are higher dimensional analogues of elliptic curves and K3 surfaces and are connected to geometry, topology and analysis, and the arithmetic of Calabi-Yau manifolds is rich and needs to be studied and explored further.

Complex Geometry from Riemann to Kähler-Einstein and Calabi-Yau
ALM 38, pp. 489–522

A Brief History of Kähler Geometry

Shing-Tung Yau*

Contents

* Department of Mathematics, Harvard University, Cambridge, Massachusetts, U.S.A. E-mail: yau@math.harvard.edu. S.T. Yau, A Brief History of Kähler Geometry, ICCM Notices, 3 (2015), no. 2, pp. 1–19.

1 The work of Riemann

Riemann was one of the founders of complex analysis, along with Cauchy. Riemann pioneered several directions in the subject of holomorphic functions:

(1) The idea of using differential equations and variational principle. The major work here is the Cauchy-Riemann equation, and the creation of Dirichlet principle to solve the boundary value problem for harmonic functions. (It took several great mathematicians, such as David Hilbert, to complete this work of Riemann.)

(2) He gave the proof of the Riemann mapping theorem for simply connected domains. This theory of uniformization theorems has been extremely influential. There are methods based on various approaches, including methods of partial differential equations, hypergeometric functions and algebraic geometry. A natural generalization is to understand the moduli space of Riemann surfaces where Riemann made an important contribution by showing that it is a complex variety with dimension $3g - 3$.

(3) The idea of using geometry to understand multivalued holomorphic functions, where he looked at the largest domain that a multivalued holomorphic function can define. He created the concept of Riemann surfaces, where he studied their topology and their moduli space. In fact, he introduced the concept of connectivity of space by cutting Riemann surface into pieces. The concept of Betti number was introduced by him for spaces in arbitrary dimension.

    The idea of understanding analytic problems through topology or geometry has far-reaching consequences. It influenced the later works of Poincaré, Picard, Lefstchetz, Hodge and others.

    Important examples of Riemann's research is to use monodromy groups to study analytic functions. Such study has deep influence on the development of discrete groups in the 20th century. The Riemann-Hilbert problem was inspired by this and up to now, is still an important subject in geometry and analysis. The study of ramified covering and the Riemann-Hurwitz formula gave an efficient technique in algebraic geometry and number theory.

(4) The discovery of Riemann-Roch formula over algebraic curve. The generalizations by Kodaira, Hirzebruch, Grothendieck, Atiyah-Singer have led to tremendous progress in mathematics in the twentieth century.

(5) His study of period integrals related to Abel-Jacobi map and the hypergeometric equations:

$$z(1-z)y'' + [c - (a+b+c)z]y' - aby = 0.$$

(6) The study of Riemann bilinear relations, the Riemann forms and the theta functions. During his study of the periods of Riemann surfaces, he found that the period matrix must satisfy period relations with a suitable invertible skew symmetric integral matrix which is called Riemann matrix later. Riemann realized that the period relations give necessary and sufficient condition for the existence of non-degenerate abelian functions.

(According to Siegel [68], his formulation was incomplete and he did not supply a proof. Later, Weierstrass also failed to establish a complete proof despite his many efforts in this direction. Complete proofs were finally attained by Appell for the case $g = 2$ and by Poincaré for arbitrary g.)

It should be noted that Riemann spent most of his last four years in Italy because he contracted Tuberculosis and needed to avoid the severe winter in Germany. But as a result, he inspired a large group of differential geometers and projective algebraic geometers in Italy. Their works influenced the development of geometry and physics in the 20th century.

First of all, we should say that Riemann was the mathematician that brought us a new concept of space that was not perceived by any mathematician before him. I believe that was the reason that Gauss was so touched by his famous address on the foundations of geometry in 1854. I could not read German and was only able to read this address recently after it was translated into English. I was rather surprised that Riemann had rather liberal view about what geometry is supposed to be.

His guiding principle was nature itself (B. Riemann, *On the Hypotheses Which Lie at the Foundation of Geometry*, 1854.):

> The theorems of geometry cannot be deduced from the general notion of magnitude alone, but only from those properties which distinguished space from other conceivable entities, and these properties can only be found experimentally... . This takes us into the realm of another science-physics.

He thinks a deep understanding of geometry should be based on concepts of physics. And this is indeed the case as we experienced in the past century, especially in the past 50 years development of geometry. Although he was the one who introduced the concept of Riemann surface, which is the largest domain that a multivalued holomorphic function lives in, the precise modern concept was developed much later through the efforts of Klein, Poincaré and others.

While Felix Klein [38] already used atlas to describe Riemann surface, it has to wait until Hermann Weyl [80] who first gave the modern rigorous definition of Riemann surface, in terms of coordinate charts.

It was rather strange that a formal introduction of the concept of complex manifold was quite a bit later. Historically, generalization of one complex variable to several complex variables began by the study of functions on domains in $\mathbb{C}^n$. There were fundamental works of Levi, Oka, and Bergman.

The natural generalization of the concept of two dimensional surfaces to higher dimensional manifolds was done by O. Veblen and J. H. C. Whitehead in 1931–32. H. Whitney (1936) clarified the concept by proving that differentiable manifolds can be embedded into Euclidean space.

However, it was only in 1932 at the International Congress of Mathematicians in Zurich, did Caratheodory study "four dimensional Riemann surface" for its own sake. In 1944, Teichmüller mentioned "komplexe analytische Mannigfaltigkeit" in his work on *"Veränderliche Riemannsce Flächen"*.

Chern was perhaps the first to use the English name "complex manifold" in his work [19].

The general abstract concept of almost complex structure was introduced by Ehresmann and Hopf in the 1940s. In 1948, Hopf [34] proved that the spheres S^4 and S^8 cannot admit almost complex structures.

The concept of Kähler geometry was introduced by Kähler [35] in 1933 where he demanded the Kähler form (which was first constructed by E. Cartan) to have a Kähler potential. Kähler had already observed special properties of such metric. He knew that the Ricci tensor associated to the metric tensor $g_{i\bar{j}}$ can be written rather simply as

$$R_{k\bar{l}} = -\frac{\partial^2}{\partial z_k \partial \bar{z}_l}(\log\det g_{i,\bar{j}}),$$

which gave a globally defined closed form on the manifold.

He knew that it defines a topological invariant for the geometry. It defines a cohomology class independent of the metric. It was found later that, after normalization, it represents the first chern class of the manifold. The simplicity of the Ricci form allows Kähler to define the concept of Kähler-Einstein metric and he wrote down the equation locally in terms of the Kähler potential. He gave examples of the Kähler metric of the ball.

Slightly afterwards, Hodge developed Hodge theory, without knowing the work of Kähler, based on the induced metric from projective space to the algebraic manifolds. He studied the theory of harmonic forms with special attention to algebraic manifolds. The (p, q) decomposition of the differential forms have tremendous influence on the global understanding of Kähler manifolds. A very important observation is that the Hodge Laplacian commutes with the projection operator to the (p, q)-forms and hence the (p, q) decomposition descends to the de Rham cohomology. The theory was soon generalized to cohomology with twisted coefficients.

A very important cohomology with twisted coefficient is cohomology with coefficient in the tangent bundle or cotangent bundle, and their exterior powers. For the first cohomology with coefficient in tangent bundle, Kodaira and Spencer developed the fundamental theory of deformation of geometric structures, which gave far reaching generalization of the works of Riemann, Klein, Teichmüller and others on parametrization of complex structures over Riemann surfaces. They realize that the first cohomology with coefficient on tangent bundle, denoted by $H^1(T)$, parametrize the complex structure infinitesimally and that the second cohomology with coefficient on tangent bundle, denoted by $H^2(T)$, gives rise to obstruction to the deformation. The last statement was made very precisely by Kurinishi using harmonic theory of Hodge-Kodaria. It describes the singular structure of the moduli space locally. Kodaira-Spencer studied how elements in $H^1(T)$ acts on other cohomology, which leads to study of variation of Hodge structures. The Hodge groups can be grouped in an appropriate way to form a natural filtration of the natural de Rham group. The Kodaira-Spencer map plays a very important role in understanding the deformation of such filtrations. Cohomology with coefficient of cotangent bundle or wedge product of cotangent bundle gives to hodge (p, q) forms. The duality of tangent bundle and cotangent bundle gives rise to something called mirror symmetry studied extensively in the last thirty years in relation to the theory of Calabi-Yau manifolds.

A very important tool in complex geometry was the introduction of Chern classes to complex bundles over a manifold and the representation of such classes by curvature of the bundle.

When Chern introduced the concept of Chern classes, he was influenced by the works of Pontryagin classes. In the course of defining Chern classes by de Rham forms given by symmetric polynomial of the curvature form, Chern defined the Chern connection for holomorphic bundles. He also proved that Chern classes of holomorphic bundles are represented by algebraic cycles on algebraic manifolds. This has been the major evidence of the Hodge conjecture: That every (p, p) class can be represented by algebraic cycles.

Chern proved that three different ways to define Chern classes are equivalent. In particular, he proved they are integral classes. Weil explained how they are related to Lie algebra invariant polynomials. Weil remarked that the integrality of Chern classes should play a role in quantum theory. Chern-Weil theory forms a bridge between topology, geometry, and mathematical physics.

The desire to generalize Riemann-Roch formula to higher dimensional algebraic manifolds has been relative slow, until the very powerful methods of sheaf theory was introduced by Leray, and important inputs were given by Weil, Borel and Serre. These basic techniques enabled Hirzebruch to arrive at the important Hirzebruch-Riemann-Roch formula in his 1954 paper [32], which can be stated in the following way:

$$\chi(V, E) = \int_V \mathrm{ch}\,(E)\mathrm{td}\,(V),$$

where E is a holomorphic vector bundle over a projective variety V.

The formulation of this formula by itself is remarkable. Hirzebruch developed the splitting principle and the theory of multiplicative sequences to find important power series of Chern classes. The Todd class is such a power series which is found by Hirzebruch to represent the arithmetic genus of the algebraic manifold, generalizing some old works of Todd in lower dimension. The Chern character $\mathrm{ch}\,(E)$ was invented by him to be a homomorphism from space of holomorphic vector bundles to even dimensional cohomology. The left-hand side of the formula is the Euler characteristic of cohomology with coefficient in E. This beautiful formula was observed by Serre when the algebraic manifold is two dimensional.

In the other direction, Kodaira was the first major mathematician who developed Hodge theory of harmonic forms right after its announcement by Hodge, and he generalized the theory of harmonic forms to manifolds with boundaries, where various boundary conditions have to be imposed.

Perhaps his most important work was his deep understanding that the Bochner argument in Riemannian geometry can be used to prove a vanishing theorem for cohomology classes under curvature condition of the manifold. He realized that the natural place for such vanishing theorem is to deal with cohomology with coefficient on bundle or sheaf. The vanishing theorem of Kodaira says that for positive line bundle L on a compact complex manifold M:

$$H^q(M, K_M \otimes L) = 0$$

for $q > 0$.

Coupled with the following theorem of Serre duality:

$$H^q(M, E) \cong H^{n-q}(M, K \otimes E^*),$$

Kodaira vanishing theorem implies that the Euler characteristic of cohomology with coefficients in a holomorphic vector bundle E with $E \otimes K^*$ positive, is simply the dimension of the group of holomorphic sections of E.

The above mentioned Hirzebruch-Riemann-Roch theorem then gives a formula to compute the dimension of the sections of the holomorphic bundle in terms of Chern numbers defined by Chern classes of the manifold and the bundle. This creates the most basic tool to understand algebraic manifolds.

Kodaira also showed that by blowing up points on the manifold, one can find enough holomorphic sections to separate points of the original manifolds and in fact gives an embedding of the manifold into complex projective space by using holomorphic sections of the bundle.

In particular, he proved that any Kähler manifold, whose Kähler class is defined by the Chern class of a holomorphic line bundle, can be holomorphically embedded into the complex projective space. The theorem of Chow then implies the manifold is in fact defined by an ideal of homogeneous polynomials, and hence an algebraic manifold.

What Kodaira has proved is one of the most spectacular theorems in geometry, and a glorious generalization of the work of Riemann on the condition of a complex torus to be abelian. More importantly the method of proving the Kodaira vanishing theorem has far reaching consequences in complex geometry. It was generalized to noncompact complex manifold, by various mathematicians including C. Morrey, Hörmander, Kohn, Vessintini, and others.

The Kodaira embedding theorem requires a high enough power of the ample line bundle to accomplish the embedding into projective space. An upper bound of this power of the line bundle is not clear from his argument.

Later on, Matsusaka [58, 59] (improved by Kollár-Matsusaka [39]) proved the very-ampleness of mL for an ample line bundle L on an n-dimensional projective variety X, when m is no less than a bound, depending only on the intersection numbers L^n and $K_X \cdot L^{n-1}$ on X.

In 1980s, Kawamata proved his famous basepoint freeness theorem about the pluricanonical systems of minimal models in [36, 37]. This is very important in the study of abundance conjecture. He proved that under the assumption that the numerical Kodaira dimension of a minimal variety X is equal to its Kodaira dimension, the pluricanonical system $|mK_X|$ is basepoint free for large m. This implies the basepoint freeness for minimal models of general type varieties. Later on, in a series paper of Miyaoka and Kawamata, they settled the proof of abundance conjecture for threefolds.

An important unsolved conjecture was proposed by Fujita in 1985, $mL+K_X$ is base-point free for $m \geq n+1$ and is very ample for $m \geq n+2$. Many mathematicians did important work on Fujita's conjecture, including Reider, Ein-Lazarsfeld, Kawamata, and many others. Demailly proved an effective formula for the bound on very ampleness [21]. Angehrn and Siu proved a quadratic bound for basepoint freeness [4].

There are many other contributioins to algebraic geometry made by Japanese algebraic geometers. Mori first introduced the ingenious idea of "bend and break" argument in his proof of Hartshorne conjecture [64]. This leads to his proof of cone theorem in birational geometry and had deep influences in minimal model program. Mukai introduced the Fourier-Mukai transform in 1981 [65]. This became an important tool in the study of derived categories.

2 Calabi conjecture and Kähler-Einstein metrics

The theorems by Kodaira, Matsusaka, Kawamata provide abundance of holomorphic sections for the holomorphic line bundle to embed the manifold into complex projective with higher dimension. An interesting important problem is the zero codimension case where we want to embed X to complex projective space with the same dimension. Hirzebruch and Kodaira [33] conjectured that every algebraic manifold that is homeomorphic to $\mathbb{CP}^n$ is actually biholomorphic to it. They used Hirzebruch-Riemann-Roch formula, but they could only treat the case of odd dimensional manifolds due to the indeterminacy of the sign of the first Chern class. The even dimensional case was finally settled by me [84] in 1976. While the arguments of Kodaria are based on Hilbert space theory, which depends on linear analysis, the argument that I used was nonlinear in nature. It has became an important new tool in complex geometry in the past forty years.

My argument depends on the existence of Kähler- Einstein metrics assuming the first Chern class is either positive, zero or negative. Although the Kähler-Einstein metric was already discussed by Kähler in his 1933 paper [35], where he wrote the equation explicitly, it wasn't until 1954 when Calabi [11] made a formal proposal to prove the existence of Kähler metric with prescribed volume form.

This could be used to proved the existence of Ricci-flat Kähler metric for any polarization if the first Chern class of the manifold is zero. Then Calabi asked the question when the first Chern class of the manifold is either negative or positive. The questions of Calabi were believed to be too good to be true in the old days, as nobody was able to construct an explicit Kähler-Einstein metric on any compact Kähler manifolds with no symmetries.

On 1976, I settled the cases when the first Chern class is either trivial or negative. (Aubin did the work independently for negative first Chern class.) I also considered the case when the manifold can have singularities, as was announced in my talk [87] at 1978 ICM in Helsinki.

2.1 Kähler-Einstein metrics on Fano manifolds

When the first Chern class is positive, it is called a Fano manifold. There are many interesting properties about Fano manifolds. Kollár, Mori and Miyaoka in [40] showed that smooth Fano varieties are rationally connected, in the sense that any two points are connected by a rational curve with (effectively) bounded degree. This implies an effective bound for the degree of the Fano n-fold, with respect to its anti-canonical bundle. Based on the work of Kollár and Matsusaka, it also implies that Fano n-folds form a bounded family.

In this case, there is an obstruction for the existence of Kähler-Einstein metric due to Matsushima [60]: the Lie algebra of the automorphism group of the manifold must be reductive. On [28], Futaki introduced his beautiful invariant defined on this Lie algebra. The Futaki invariant soon became a fundamental tool to study Kähler-Einstein metric on Fano manifolds. On the other hand, It took a long while to find a necessary and sufficient condition for the existence of Kähler-Einstein metric on Fano manifolds. Many people, including Calabi, was misled to believe that the non-existence of nonzero holomorphic vector fields is enough for the existence of Kähler-Einstein metric on Fano manifolds.

Right after I proved the Calabi conjecture on the existence of Kähler metric with prescribed volume form, I tried to work on the problem of the existence of Kähler-Einstein metric on Fano manifolds.

It is clear that based on the (nontrivial) higher order estimates that I had (independently due to Aubin for second order estimate) in the proof of the Calabi conjecture [86], the only missing point is some integral estimate of the Kähler potential. I found it is useful to set up the continuity argument

$$\det\left(g_{i\bar{j}} + \frac{\partial^2 u}{\partial z_i \partial \bar{z}_j}\right) = \exp(h - tu)\det(g_{i\bar{j}}),$$

where $t = 0$ corresponds to a Kähler metric with positive Ricci curvature, as was given by the Calabi conjecture.

A simple calculation shows that the Ricci curvature of all members in the family have positive lower bound. This simplifies the analysis quite a bit as we have experiences with compact manifolds with Ricci curvature bounded from below by positive constant. In 1978, I returned to Stanford from my visit of Berkeley. At that time, I succeeded to convince Stanford mathematics department to hire Y.-T. Siu to come to Stanford from Yale.

We started to think about a proof of the existence of Kähler-Einstein metric by finding some integral estimate of the Kähler potential. Many estimates were found, but they are short of proving the existence of the metric. Some of those estimates can be sharpened if there are symmetries on the manifold, a procedure similar to the way that Moser sharpened the Trudinger inequality on the sphere when there is antipodal symmetry.

In the meanwhile, in 1977, I realized that Bogomolov [8] used the concept of stability of bundles to prove Chern number inequalities for algebraic surfaces which were sharpened by Miyaoka [61] and myself [84] independently. I started to believe there has to be links between the concept of stability with the existence of Hermitian Yang-Mills connections on bundles. When I proved the Calabi conjecture in 1976, I was at UCLA, and had had a fruitful discussion with David Giesecker, who is a great expert on the stability of bundle theory. He re-proved the Chern number inequality of Bogomolov over characteristic p.

The fact that a holomorphic bundle admits an Hermitian Yang-Mills connection if the bundle is polystable was proved by Uhlenbeck-Yau on arbitrary compact Kähler manifolds, and by Simon Donaldson for algebraic surfaces. Simpson observed that the proof of Uhlenbeck-Yau can be used to settle the case when there is a Higg's field. (Up to now, the argument of Uhlenbeck-Yau has been the only argument to prove existence of the Yang-Mills-Higgs equation.)

The Bogomolov inequality is optimal for general stable bundles. But it is not as sharp as the Miyaoka-Yau inequality when applied to the tangent bundle of the manifold. Hence I suspected that existence of Kähler-Einstein metric should be considered as a nonlinear version of the existence of Hermitian Yang- Mills connection, and the stability of bundle should be replaced by manifold stability. Therefore, only in 1980, I realized that the right condition for existence of Kähler-Einstein metric is the stability of the algebraic manifold.

I made the conjecture that the existence of Kähler-Einstein metric is equivalent to stability of manifold. I told all my graduate students about this conjecture, especially to Gang Tian who showed interest in the problem of Kähler-Einstein metric. But it took a long time to convince him of the validity of my conjecture.

There are many ways to define stability of manifolds including the concepts of Chow stability or Hilbert stability. I was not sure which one is correct. But I started to explore it with my students in my seminars. First of all, one had to make sure that algebraic stability, which is defined by embeddings of algebraic manifolds into complex projective space, can be linked to existence of Kähler-Einstein metric.

In fact, in order to link stability condition to algebraic geometry, I [88, p. 139] proposed to prove any Hodge metric on an algebraic manifold can be approximated by normalized Fubini-Study metric induced on the manifold through embedding of the manifold into complex projective space by high powers of an ample line bundle.

I asked Tian to follow this line of argument to finish the first step of my conjecture on the equivalence of stability of Fano manifolds with the existence of Kähler-Einstein metrics.

I suggested Tian to use my method with Siu [70] on the uniformization of Kähler manifolds to produce peak functions to achieve such a goal. (The purpose of that paper with Siu was also embedding of Kähler manifolds.)

The proof was reasonably transparent using technology from my paper with Siu. This became Tian's thesis at Harvard.

The method can be said to be an understanding of the works of Kodaira in the analytic setting. The work was carried out as I expected and it was strengthened by Catlin [13], Zelditch [92] and by Lu [53].

So, we know that we can approximate any Hodge metric by the induced metric of the projective embedding of the manifold into some complex projective spaces. However, there is an ambiguity due to the action of complex projective group. This is of course what geometric invariant theory studies.

It turns out that when I studied first eigenvalue of the Laplacian with Bourguignon and Peter Li [10], we need to find a good position for the embedding upon action by the projective group, which we called the balanced condition. It can be written in the following form:

$$\int_{\sigma(M)} \frac{z_i \bar{z}_j}{|z_0|^2 + \cdots + |z_N|^2} \omega^n = \frac{\mathrm{vol}\,(M)}{N+1} \delta_{ij}$$

for some $\sigma \in SL(N+1, \mathbb{C})$.

With such a condition, we can use the embedding to give a good estimate of the first eigenvalue in terms of the total volume and the degree defined by the Chern form wedge with the Kähler classes.

I suggested this condition as a starting point to my former student Luo to understand the concept of stability required to prove my conjecture on the existence of Kähler-Einstein metric based on stability.

Luo [54] found it effective to change the measure in the above formula defined by the induced measure of the ambient projective space. And it turns out that for a polarized manifold (M, L) if there exists a metric on L such that the Bergman function of L^k is constant for some k, then it is Chow stable.

A theorem of Shouwu Zhang [93] says that the existence of a unique balanced embedding is equivalent to the manifold being Chow-Mumford stable.

My conjecture that the existence of Kähler-Einstein metric is equivalent to stability was announced several times in several conferences and was explicitly written in my article [89] for the proceedings of UCLA conference on differential geometry in 1990.

I also communicated to Tian in detail on how to understand the Futaki invariant in this setting. The final conjecture of mine was solved recently by Chen-Donaldson-Sun [16, 17, 18] based on earlier works of Donaldson including the right algebro-geometric definition of K-stability.

According to Donaldson [23], a Fano manifold is called K-stable if all its non-trivial test configurations (which describe certain degeneration of Kähler manifolds by flat families) have positive Futaki invariants. For a test configuration $\chi \to \mathbb{C}$ with $\mathbb{C}^*$ action, the Futaki invariant F_1 can be found from the total weight w_k of $\mathbb{C}^*$ acting on $H^0(X_0, L^k)$, using

$$\frac{w_k}{kd_k} = F_0 + F_1 k^{-1} + O(k^{-2}),$$

where d_k is the dimension of $H^0(X_0, L^k)$.

But the condition of K-stability is not easy to check, even in the case of surfaces. It would therefore be interesting to prove the existence of balanced condition for high power embeddings of a Fano manifold implies existence of Kähler-Einstein metrics. It is highly desirable to clarify the condition of K-stability so that it can be checked effectively.

2.2 Balanced Metric and Strominger System

Kähler-Einstein metrics are very useful in biregular geometry. We shall discuss it later. However, it cannot answer the important question whether an algebraic manifold is rational or not. The existence of Kähler metric is not a concept that is invariant under birational transformations, while the existence of balanced metric is. The concept of balanced metric was introduced by Michelson [56]. An Hermitian metric is called balanced if its Kähler form satisfied the following equation:

$$d(\omega^{n-1}) = 0$$

and it was proved by Alessandrini and Bassanelli [1] that its existence is invariant under birational transformations. However, there is much more freedom to deform a balanced metric than a Kähler metric. Just demanding that Ricci curvature equal to zero is not enough to determine a unique Balanced metric within the $(n-1, n-1)$ class.

On the other hand, balanced metric comes up naturally in the theory of Heterotic string theory in complex 3-dimension. And (this) balanced condition is related to the concept of supersymmetry. When there is a nonwhere vanishing top dimensional holomorphic 3-form, we look for an Hermitian metric which is balanced, and a stable holomorphic bundle (stable with respect to the balanced metric) whose second Chern Class is equal to the second Chern Class given by the Hermitian metric. Altogether, the following equations of the Strominger system need to be satisfied:

(1) $d(\|\Omega\|_\omega \Omega^2) = 0$.
(2) $F_h^{2,0} = F_h^{0,2} = 0, F_h \wedge \omega^2 = 0$.
(3) $\sqrt{-1}\partial\bar{\partial}\omega = \frac{\alpha'}{4}(\mathrm{tr}\,(R_\omega \wedge R_\omega) - \mathrm{tr}\,(F_h \wedge F_h))$.

It provides a natural generalization of the Calabi- Yau geometry, which couples Hermitian metrics with Hermitian Yang-Mills theory. My belief is that the above system of equations can be solved when the obvious conditions hold. Jun Li and I [45] solved this system on any Calab-Yau manifold by making a deformation from the original Calabi-Yau metric.

For some intrinsically non-Kähler manifold, Fu and I [27] solved the Strominger system based on some ansätz for a 3-dimensional complex manifolds obtained from the Calabi-Eckmann construction. (The construction of the non-Kähler manifolds based on Calabi-Eckmann construction was also observed by Goldstein and Prokushkin [31].) It is a nonsingular complex torus bundle over the K3 surface. The proof of existence of nonsingular solution to the Strominger system given by Fu-Yau [27] is based on nontrivial estimates related to complex Monge-Ampère equations. In order to understand the significance of Strominger system, Tseng and I [76, 77], and later with Tsai [75], developed a new theory of symplectic cohomology which we expect to be dual to this kind of geometry.

Note that the existence of Ricci-flat Kähler metric provides a reduction of holonomic group to a subgroup of $SU(n)$, and according to the work of Candelas-Horowitz-Strominger-Witten [12], provides a supersymmetric model for vacuum solutions for Type II string theory. They called such manifolds to be Calabi-Yau manifolds. The Strominger system was introduced by Strominger to study Heterotic string where the vacuum is a warped product instead of a direct product.

2.3 Questions of Kähler-Einstein Metrics in Algebraic Geometry

There are several interesting consequences of the existence of Kähler-Einstein metric.

2.3.1 Understanding of Kähler-Einstein metrics near singularities

A corollary of the above mentioned theorem of Chen-Donaldson-Sun is that the K-stability of such manifold implies that the tangent bundle is stable with respect to the polarization given by the anticanoncial line bundle. This is an interesting statement that is purely algebraic geometric, for which it would be nice to have a proof based only on algebraic geometry.

Also it implies that a K-stable Fano manifold is biregular to $\mathbb{CP}^n$ if the ratio of its two Chern numbers $c_2c_1^{n-2}$ and c_1^n is the same as $\mathbb{CP}^n$.

Another interesting question is the following: If a smooth algebraic manifold has Kodaira dimension either equal to the dimension of the manifold or $-\infty$, and if it is minimal in the sense in birational geometry and the ratio of two chern numbers $c_2c_1^{n-2}$ and c_1^n is the same as $\mathbb{CP}^n$, then the manifold is either $\mathbb{CP}^n$ or complex ball quotient.

For the case of general type, this is likely to be true. But it will be good to allow singular minimal models and in the case of singular algebraic manifolds, we need to define the Chern numbers suitably. This is related to the question of what is the best Kähler-Einstein metric on an algebraic manifold with singularity.

Let us look at the simplest case when the singularity is isolated. If the Kähler metric is complete at the singularity, it is not hard to prove that the Kähler-Einstein metric is unique. However, when it is not complete, it is not necessarily unique. It depends on the behavior of the volume form near the singularity.

What kind of volume forms are allowed? We need to know that the Ricci form of this volume form is positive definite and that the n-fold product of the Ricci form is asymptotic to this volume form near the singularity. (We may require that the metric defined by the Ricci form should have lower bound on its bisectional curvature.) It would be interesting to classify the asymptotic models of such volume forms. In principe, each of them will give rise to a canonical Kähler-Einstein metric with the given asymptotic behavior of the volume form.

It would be interesting to calculate the contribution of the singularity towards the Chern numbers. An important case is the canonical singularity appearing in the minimal model theory, which we recall below.

Suppose that Y is a normal variety and $f : X \to Y$ be a resolution of the singularities. Then

$$K_X = f^*(K_Y) + \sum_i a_i E_i,$$

where the sum is over the irreducible exceptional divisors and the rational numbers a_i are called the discrepancies.

Then the singularities of Y are called canonical if $a_i \geq 0$ for all i and called terminal if $a_i > 0$ for all i.

A 3-dimensional singularity is terminal of index 1 if and only if it is an isolated composite DuVal (cDV) point in $\mathbb{C}^4$. A 3-dimensional terminal singularity of index $r \geq 2$ is a quotient of an isolated cDV point in $\mathbb{C}^4$.

The important question is to find a good Kähler- Einstein metric in a neighborhood of the cDV singularity which is invariant under the group action. For orbifold singularities, one can use those metrics obtained by pushing down from the nonsingular model before quotient by the group. On the other hand, there may be some other volume form that satisfies the above properties that is distinct from the orbifold construction. The complicated situation is the case that the Ricci form of the volume form may define a metric that is partially going to complete and partially degenerate at the singular point.

It will be important to construct nice model volume form in a neighborhood of the canonical singularities of the manifold whose Ricci form can give rise to a nice metric which is asymptotically Kähler-Einstein.

2.3.2 Kähler-Einstein metrics on quasiprojective varieties and Sasakian-Einstein metrics

In my first paper on the Calabi conjecture, we know that given any Kähler class, we can find a Kähler metric which may degenerate along a divisor whose volume is given by the unique volume defined by the divisor of the pluricanonical sections. How to calculate the second Chern class related to this divisor would be important. The Chern numbers calculated by the degenerate Ricci flat metrics should have residue from the divisor. It would be important to calculate this contribution.

The noncompact version of complete Ricci-flat metric is more complicated, partially because we lack of a good model space to build a good ansatz. At 1978 ICM in Helsinki, I [87] announced the way to build complete noncompact Ricci-flat manifolds.

I conjectured that the manifold can be written as the complement of a divisor D of a compact Kähler manifold M. (It was pointed out by Michael Anderson et al. [3] that we should assume the finiteness of the topology of the manifold, otherwise Taub-NUT manifolds can provide counterexamples.)

My program was to take D to be an anticanonical divisor of M which cannot be contracted to a codimension two subvarity. There will be a holomorphic volume form on M which has poles along D. I expect that this is close enough to provide a necessary and sufficient condition for $M \backslash D$ to admit complete Kähler metric with zero Ricci curvature. When D is nonsingular, I have worked out the program. The details were written up with Tian in two papers [72, 73].

However, when D has normal crossing singularities, the problem is unsolved, largely because we do not have a good model of complete Ricci-flat metric in a neighborhood of D when D has singularity. An important and interesting case is to allow the complete Kähler metric to have certain type of singularities. Besides quotient singularity, we can allow cone singularity.

In the last case, the interesting examples are metric cones over a Sasakian-Einstein manifold. Important progress was made by Gauntlett, Martelli, Sparks and myself starting with [29] on the existence of Sasakian-Einstein metrics. In [29,

55] we gave several obstructions to their existence by studying the Einstein-Hilbert functional restricted to the space of Sasakian-Einstein metrics where it becomes essentially the volume functional. It can further be shown to be a functional of the Reeb vector field associated to the Sasakian structure alone.

We obtain a useful obstruction from the Lichnerowicz bound on the Laplacian [49] which we could identify precisely as the physics criterion of a unitarity bound in the conformal field theory associated to the hypersurface singularity. We also show that the first variation of the volume functional is related to the Futaki invariant on the Kähler orbifold, hence volume minimization (and a-maximization in the physics language) implies vanishing of Futaki invariant. This includes the cases of regular and quasiregular Sasakian structures as classified by Reeb vector orbits. In the irregular case, Collins and Székelyhidi [15] extended the notion of K-semistability to Sasakian structures, showing constant scalar curvature Sasakian metric implies K-semistability and also recovered our results based on the volume functional. The complete classification is still not known, even for complex hypersurfaces with isolated singularity which admits $\mathbb{C}^*$-action.

2.3.3 Compatification of Shimura varieties

Another very important class of Kähler-Einstein metrics on quasi-projective varieties appears on the compactification of Shimura varieties of noncompact type. There is the work of Mumford on giving a toroidal compactification which is nonsingular. In terms of the divisor at infinity, Yi Zhang and I [90] wrote down the behavior of the volume form of the Hermitian symmetric metric in a neighborhood of the divisor.

Here is a summary of my work with Yi Zhang: The positive cone $C(\mathfrak{F}_0)$ of the standard minimal cusp $\mathfrak{F}_0$ of the Siegel space $\mathfrak{H}_g$ can be regarded as the set of all symmetric positive $g \times g$ real matrices. Let $\sum_{\mathfrak{F}_0}$ be any decomposition of $C(\mathfrak{F}_0)$ such that the corresponding Mumford toroidal compactification $\bar{\mathcal{A}}_{g,\Gamma}$ of $\mathcal{A}_{g,\Gamma}$ has normal crossing boundary divisor $D_\infty = \bar{\mathcal{A}}_{g,\Gamma} \backslash \mathcal{A}_{g,\Gamma}$. Let σ be an arbitrary top-dimensional polyhedral cone in $\sum_{\mathfrak{F}_0}$ and Let $D_1, \ldots, D_N (N = \dim_{\mathbb{C}} \mathcal{A}_{g,\Gamma})$ be some different irreducible components of D_∞ corresponding to edges of σ.

Then the volume $\Phi_{g,\Gamma}$ on $\mathcal{A}_{g,\Gamma}$ can be represented by

$$\Phi_{g,\Gamma} = \frac{d\mathcal{V}_g}{(\prod_{j=1}^{N} \|s_i\|_i^2) F_\sigma^{g+1}(\log \|s_1\|_1, \ldots, \log \|s_N\|_N)},$$

where

- $d\mathcal{V}_g$ is a continuous volume form on a partial compactification $\mathcal{U}_{\sigma_{\max}}$ of $\mathcal{A}_{g,\Gamma}$ with $\mathcal{A}_{g,\Gamma} \subset \mathcal{U}_{\sigma_{\max}} \subset \bar{\mathcal{A}}_{g,\Gamma}$,

- the $\| \cdot \|_i$ is a suitable Hermitian metric of the line bundle $[D_i]$ on $\mathcal{A}_{g,n}$ for every integer $i \in [1, N]$,

- the s_i is global section of $\mathcal{O}_{\mathcal{A}_{g,\Gamma}}(D_i)$ such that $D_i = \{s_i = 0\}$,

- the $F_\sigma \in \mathbb{Z}[x_1, \ldots, x_N]$ is a homogenous polynomial of degree g, and the coefficients of F_σ are integers dependent only on Γ and σ together with marking order of edges.

In fact, Yi Zhang and I computed the volume form of the Hermitian symmetric metric as is represented on the coordinate given by the Toroidal compactification. It shows that $K + D$ is nonnegative and positive on $M \backslash D$.

Whether $D_\infty := \bar{\mathcal{A}}_{g,\Gamma} \backslash \mathcal{A}_{g,\Gamma}$ is normal crossing or not, Yi Zhang and I showed that there is always a local model of partial compactification associated to each maximal regular cone σ in the cusp $\mathfrak{F}_0$.

The quotient manifold $\mathfrak{H}_g/(\Gamma \cap U^{\mathfrak{F}_0}(\mathbb{Q}))$ gives an étale map of the Siegel variety. For each maximal cone σ in the cusp $\mathfrak{F}_0$; we have associate exponential maps of the inclusion $\mathfrak{H}_g \subset U^{\mathfrak{F}_0}(\mathbb{C}) \cong \mathbb{C}^n$, so that these maps endow a local model of partial compactification

$$\mathfrak{H}_g/(\Gamma \cap U^{\mathfrak{F}_0}(\mathbb{Q})) \subset (\mathbb{C}^*)^n.$$

The exponential map $\mathbb{C}^n \to (\mathbb{C}^*)^n$ is given by

$$z \mapsto w = (w_1, \ldots, w_n), \text{ where } w_i = \exp(2\pi\sqrt{-1}l_i(z))\forall i,$$

where $\{l_i\}_{i=1}^n$ is the dual base of edges of the cone σ.

The $(w_1, \ldots, w_n)$ gives a local coordinate system of the partial compactification, but it can not be a local coordinate system of $\bar{\mathcal{A}}_{g,\Gamma}$ if the D_∞ is not normal crossing.

The quotient manifold $\mathfrak{H}_g/(\Gamma \cap U^{\mathfrak{F}}(\mathbb{Q}))$ also has an induced Kähler-Einstein metric with volume form

$$\Phi_\sigma = \frac{\left(\frac{\sqrt{-1}}{2}\right)^n 2^{\frac{g(g-1)}{2}} \operatorname{vol}_\Gamma(\sigma)^2 \wedge_{1\leq i\leq n} dw_i \wedge d\overline{w_i}}{(\prod_{1\leq i\leq n} |w_i|^2)(F_\sigma(\log|w_1|, \ldots, \log|w_n|))^{g+1}}.$$

The coefficients of the polynomial F_σ are integers dependent only on Γ and σ, and the function $H := -\log F_\sigma$ must satisfy the following elliptic Monge-Ampére equation

$$\det\left(\frac{\partial^2 H}{\partial x_i \partial x_j}\right)_{i,j} = 2^{\frac{g(g-1)}{2}} \operatorname{vol}_\Gamma(\sigma)^2 \exp((g+1)H)$$

on the domain $\{(x_1, \ldots, x_n) \in \mathbb{R}^n | x_i \leq -C < 0 \forall i\}$.

Note that it was known to me [85] that for a Divisor D in an algebraic manifold M, if $K + D$ is strictly positive, then there is a canonical complete Kähler-Einstein metric on $M \backslash D$ whose volume form behaves like:

$$\Phi \sim \frac{dV}{\prod_{j=1}^k \|s_j\|^2(-\log\|s_j\|)^2}$$

for some integer $k > 0$, where s_i is a section of the line bundle $[D_i]$ if $D = \sum D_i, dV$ is a global smooth volume form on M, and the norm is defined such that its zero set is D and the minus Ricci tensor of M plus the Ricci curvature of the metric on D is positive.

In the above work with Yi Zhang, the volume form is more complicated, because we only know $K + D$ is nonnegative and positive outside D. How to study such quasi-projective manifold? It is important to find the right algebro-geometric and combinatorial conditions on the Chern forms of M and the Chern forms of the various devisors D so that the Ricci curvature of the volume form gives rise to a positive definite Ricci form whose n-fold power is asymptotic to the volume form itself.

An ansatz we propose to construct complete Kähler-Einstein metric on $M \backslash D$ is to construct a volume form described similar to the above Φ_σ, where H satisfies the above Monge-Ampére equation with $x_i = \log |s_i|$. We need to find the condition on the divisors D_i so that Φ_σ gives rise to a positive Ricci form whose n-flod power is asymptotic to Φ_σ. Hopefully we can find a good existence theorem for complete Kähler-Einstein metric with finite volume on $M \backslash D$, when $K + D$ is nonnegative.

The existence of complete Kähler-Einstein metric on the Shimura varieties comes from the Hermitian symmetric domain that covers it. The tangent bundle and various homogeneous bundles over the symmetric domain are invariant under the discrete group acting on the domain. Hence they can be descended to the Shimura varieties. As was explained by Mumford, these bundles and their connections can be extended naturally to the Mumford Toroidal compactification of the Shimura variety. (Mumford proved that the extension satisfies the property of being "good".)

And the Chern forms defined by the connection were identified by Mumford to represent the Chern class of the extended bundles in the sense of distribution. In the case of the cotangent bundle, the cotangent bundle is extended to be $\Omega^1(\log(D))$, where D is the divisor at infinity.

Since the bundles are homogenous, the Chern numbers of these extended bundles are determined by some numerical combination of its curvature tensor at one point times the volume of the Shimura variety. The existence of Kähler-Einstein metric on the Shimura variety shows that the manifold is stable in various senses and the homogeneous bundles are stable with respect to the polarization $K + D$ of the variety. Besides algebraic geometric characterization of Shimura variety, it would be good to characterize algebraic geometrically those holomorphic bundles that are homogenous.

Incidentally, from my observation in 1978 that the positivity of $K + D$ implies the existence of a unique canonical Kähler-Einstein metric on the complement of D. We can find a map from the space of divisors D such that $K + D$ is ample to the space of stable bundles defined by cotangent $\Omega^1(\log(D))$. It will be nice to find conditions on D so that we can weaken the conclusion $K + D > 0$ to $K + D \geq 0$.

For a compact algebraic manifolds M, it can be shown to be a Shimura variety if the canonical line bundle is ample, and if the bundle, formed by symmetric powers of the cotangent (or tangent) bundle tensored by some line bundle so that the determinant bundle is trivial, is irreducible and has nontrivial sections. This is a simple observation (due independently to Kobayashi-Ochiai and myself) [94] because the existence of Kähler-Einstein metric will make this nontrivial section to be a parallel section and hence the holonomic group will be reduced. Algebraic characterization of Shimura varieties allows one to give a simple proof of the theorem of David Kazhdan, that Shimura varieties are invariant under Galois conjugation. Unfortunately our knowledge for noncompact manifold is not good enough to give such a proof in such case.

2.3.4 Explicit construction of Kähler-Einstein metrics and uniformization

For Kähler-Einstein manifolds with negative or zero first Chern class, I proposed [88, p. 139] that the metric can be computed in the following manner: When the canonical line bundle K is ample, we can embed the manifold into the complex projective space by nth power of K.

The embedding can be changed by projective transformation in general. But there was a concept of balanced position (inspired by my work with Bourguignon and Peter Li [10] on first eigenvalue of the Laplacian) that I suggested to my former student Luo [54].

The embedding is unique up to unitary transformation after putting into such balanced position. The induced metric from complex projective space defines a sequence of Kähler metrics on the manifold, which after division by n, will converge to a Kähler-Einstein metric of the manifold.

In the above construction, when the manifold has zero first Chern class (Calabi-Yau manifold), the canonical line bundle should be replaced by any positive line bundle.

The balanced position is achieved by some projective transformation. We expect that the projective transformation depends algebraically on the original embedding of the manifold. The whole procedure should give a reasonable "explicit" form of the Kähler-Einstein metric. Once we obtain an explicit form of the Kähler-Einstein metric, we can compute the uniformization of the manifold.

A simple case is the elliptic curve where we know how to calculate its unique holomorphic 1-form by residue. The absolute value of it gives the Ricci-flat metric on the elliptic curve. We can calculate the uniformization of the elliptic curve, using the period calculation.

Computation of the periods of the holomorphic 1-form is obtained by computing the Picard-Fuchs equation. Once one finds the period, one can obtain a map from the complex line, mod the lattice spanned by the periods, to the elliptic curve. The components of this map is the Weierstrass $\wp$ function and its derivatives. This procedure is classical and went back to Abel, Jacobi and Riemann.

The uniformization of a general algebraic curve: finding a covering holomorphic map from the upper half plane to the curve, is more difficult and is done only for special curves.

Suppose we can calculate the Poincaré metric on the curve, as was explained above, we can calculate this map by studying the periods through the Picard- Fuchs equation. It is of course much more challenging to calculate the uniformization map explicitly in higher dimensions.

Given an algebraic manifold, we know it can be uniformized as a quotient of some classical domain. It is a classical question on how to find such a uniformization.

As mentioned above, we know how to find an algebro-geometric criterion (by using Chern numbers) for an algebraic manifold to be a ball quotient. But we can generalize this criterion to more general manifolds covered by Hermitian symmetric domains.

Once we identify such an algebraic manifold, we need to find suitable multi-valued holomorphic map from the manifold into an Hermitian symmetric domain.

Chenglong Yu, Peng Gao and I proposed the following program for the ball quotient:

(1) Find a reasonably explicit way to construct the Kähler-Einstein metric on the algebraic manifold. This involves having a good understanding of the right projective embedding for the algebraic manifold.

(2) Based on Kähler-Einstein metric, we compute its connection A_h and construct a system of holomorphic linear differential equations

$$ds + \begin{pmatrix} -A_h^T + \frac{1}{n+1}\mathrm{tr}\,(A_h) - a d\hat{z} & \partial a + A_h^T a - a(d\hat{z}\cdot a) \\ d\hat{z} & d\hat{z} a - \frac{1}{n+1}\mathrm{tr}\,(A_h) \end{pmatrix} s = 0,$$

where

$$s = (f^1, f^2, \ldots, f^{n+1})^T, \quad d\hat{z} = (dz^1, dz^2, \ldots, dz^n),$$
$$a = (a_1, \ldots, a_n)^T, \quad a_i = -\Gamma_{ij}^j + \frac{1}{n+1}\sum_k \Gamma_{ki}^k \delta_j^i.$$

Here j doesn't depend on i. The value of a is such that it gives a gauge transformation making the connection matrix holomorphic.

(3) We find a base for the solution of the this system, given by the span of $\{s_1, s_2, \ldots, s_{n+1}\}$. This allows us to define locally a map to the projective space of dimension n. Up to projective transformations we find a map to the complex ball of dimension n as well.

These maps are multivalued functions. The inverse of this map should be given by automorphic forms and one should be able to find information of the discrete group that acts on the ball based on the information of the algebraic variety and the monodromy of the map. An example in the case of elliptic curve is given by the Weierstrass $\wp$ function.

For higher genus curve, if the Kähler-Einstein metric is $e^{2u} dz \wedge d\bar{z}$, then the system above becomes

$$d\begin{pmatrix} f_1 \\ f_2 \end{pmatrix} + \begin{pmatrix} 0 & ((u_z)^2 - u_{zz})dz \\ dz & 0 \end{pmatrix} \begin{pmatrix} f_1 \\ f_2 \end{pmatrix} = 0,$$

or

$$(f_2)'' + ((u_z)^2 - u_{zz}) f_2 = 0.$$

The coefficient in the differential equation gives the Schwarzian derivative for the uniformization map $S(u) = u_{zz} - u_z^2$.

A complete understanding of this program should create many interesting special functions and should be related to the GKZ system and the tautological system introduced by Bong Lian and myself [48] on computing the period of integrals of holomorphic forms.

2.3.5 Relation with birational geometry

The moduli space of algebraic manifolds of general type was studied by Gieseker and Viehweg who proved that they are quasi-projective. The detailed structure of the moduli space is not well understood. However, the canonical Kähler-Einstein metric on the manifold can be useful for such study. It induces a canonical metric on the moduli space which is called Weil-Petersson metric: A Kähler metric on the manifold gives rise to a metric on $H^1(T)$ which is the tangent space of the Kuranishi space of the manifold.

The Weil-Petersson metric can be computed in some cases. But the properties were best understood only for moduli space of curves. There are several metrics defined on the moduli space of curves: Weil-Petersson metric defined by using the Poincaré metric (or the Bergman metric) on the curve, the Teichmüller metric (which was proved by Royden [66] to be equal to the Kobayashi metric), the Carathéodory metric, the Bergman metric and the Kähler-Einstein metric.

The last three metrics can be defined by general method, not just for moduli space of curves. Hence the computation of them is interesting. Although K. Liu, X. Sun and I [50, 51] showed that the last four metrics on the Teichmüller space are all uniformly equivalent to each other, it is likely that the Carathéodory metric is different from the Kobayashi metric. But the precise statement is not known.

However, Liu-Sun-Yau did calculate the asymptotic behavior of the Kähler-Einstein metric on the Teichmüller space. In fact, the minus Ricci tensor of the Weil-Peterson metric defines a complete metric on the Teichmüller space. The Kähler-Einstein metric could be obtained by perturbing from this metric. The Teichmüller metric and the Weil-Petersson metric are computable based on local information of the Riemann surface.

The other metrics are defined by global means. Hence the remarkable theorem of Royden, proving that a metric defined by global means to be equal to locally defined metric, has provided powerful information. It will be very useful to compute all these metrics in the simplest Teichmüller space of genus two curves.

The second class of canonical metrics are those manifolds with zero Kodaira dimension. By taking the absolute value of the pluriholomorphic n-form, and taking roots, we obtain a canonical volume form which may degenerate along a divisor. We can solve the Monge-Ampère equation to obtain some Kähler metric which may be degenerate along the divisor. It will be interesting to know the birational class of these manifolds.

Let us now consider the possibility of using metrics to understand birational geometry. First of all, for two classes of manifolds, there are natural measures that are birational invariant:

For manifolds with Kodaira dimensional equal to zero, we have a canonical volume form by using the absolute value of the pluricanonical form. For manifolds of general type, I introduced 40 years ago [83] an intrinsic measure that is invariant under birational transformations. It was a generalization of the construction given by Kobayashi and Eisenman in 1970s.

In both cases, we can pick any Kähler class and solve the Calabi conjecture with this volume form as prescribed. In the first case, such metric was studied in the second part of my paper on Calabi conjecture. The metric can be proved to be unique and smooth outside the divisor defined by the pluricanoncial form. In the second case, the metric also exists uniquely. But smoothness depends on the measure that I constructed.

In any case, given two birational manifolds of general type M and M', we can find M'' and smooth rational maps from M'' to M and M' respectively. If the pullbacks of polarizations on M and M' are the same on M'', then they are isometric to each other. It is easy to derive from this fact that the group of automorphisms of a manifold of general type is finite.

It is not hard to prove that any algebraic manifold of general type admits a Kähler-Einstein metric with singularity (as was demonstrated by Tsuji and myself thirty years ago). However, in order for such metrics to be useful, one needs to know the singular behavior of the metric. Kähler-Einstein metrics do not respect rational maps. However, the Bergman metric has better behavior under birational transformation.

Let us look at the line bundle K^m where K is the canonical line bundle. For any holomorphic section s of K^m, we can take $2/m$ power of its absolute value, which defines a pseudo-norm on the canonical line bundle. If we normalize its integral to be one and maximize the pseudo-norms among all such s, we obtain a canonical pseudo-norm on the canonical line bundle. It defines a birational invariant volume form. The curvature form of this volume form should define a pseudo-Kähler metric on the manifold. See [20] for a detailed discussion.

We can deform this pseudo-Kähler metric within its class to obtain a pseudo-Kähler metric which is Einstein when it is smooth. When m is large, the Kähler-Einstein metric should be less singular and if we know the singular behavior of the original pseudo-Kähler metric, we should have a way to control the singularity of this pseudo-Kähler-Einstein metric. It should be useful to study birational geometry.

For example, when the Kodaira dimension of the manifold is zero, and the nonzero form s is a section of K^m. The volume form is the absolute value of s to the power of $2/m$. At the nonsingular point of the divisor of s, the local model of the Ricci-flat metric should be the push-forward of the Kähler metric on an m-fold branch cover of the manifold branched along the divisor $s = 0$.

Based on this, one can compute the second Chern form of the Ricci-flat metric degenerate along the divisor $s = 0$. The second Chern form of this degenerate metric wedge with Kähler class to the top dimension is positive unless it is flat. This should give interesting information for manifolds with Kodaira dimension zero.

Many years ago, I conjectured that there are only a finite number of deformation types for compact Kähler manifolds with $c_1 = 0$ at each dimension. The question is still unknown and is getting more and more important in string theory. The minimal model of algebraic manifold with Kodaira dimension zero should play an important role if we want to ask similar questions for such manifolds.

3 Hermitian Yang-Mills connections

Hermitian metric on a complex manifold has a natural generalization to Holomorphic bundles over complex manifolds. Given an Hermitian metric on the bundle, there is a natural connection which preserves the metric and also the $(0,1)$ part of the covariant derivative would be the same as the naturally defined $\bar{\partial}$ operator that depends only on the complex structure of the bundle and the complex manifold. The curvature is a $(1,1)$-form with values in the endomorphism of the bundle.

3.1 Donaldson-Uhlenbeck-Yau correspondence

There is a natural generalization of the Kähler-Einstein condition to this setting by wedging the curvature 2-form with the Kähler form to the top dimension and require it to be a scalar multiple of identity tensor with the volume form. This equation is the natural generalization of anti-self-dual equations for bundles over a Kähler surface.

In fact, around 1977, C. N. Yang [95] was trying to solve the anti-self-dual Yang-Mills equation on $\mathbb{R}^4$, and he showed that it can be reduced to Cauchy-Riemann equations. And therefore he demonstrated that the above equation is part of Yang-Mills equations. It is therefore natural to call such connection to be Hermitian Yang-Mills connection.

The equation became rather well known in the mathematics community after 1977, when people recognized the importance of applications of Kähler-Einstein metric to complex geometry. The proof of the existence of such connections would be clearly different as the Calabi-Yau theorem was based on the complex Monge-Amperè equation which depends only on a scalar. The Hermitian Yang-Mills connection is a vector-valued equation.

In December of 1977, when I was preparing the talk for the ICM in Helsinki, I thought about the possible conditions for existence of Hermitian Yang-Mills connections. I concluded that it had to be related to the slope stability of the holomorphic bundle, as was motivated by the work of Bogomolov and Miyaoka on Chern number inequalities. I was informed much later that this possibility was also believed to be true by Hitchin and Kobayashi.

However, the proof would have to be quite tough as there is no good way to handle such a nonlinear system of elliptic equations. It turns out that Donaldson and Uhlenbeck-Yau were working on this problem independently. I learnt from Hitchin during a trip to England that Donaldson was able to prove the existence for Hermitian connections of any holomorphic vector bundle that can be deformed to the tangent bundle of a K3 surface. (Note that the Ricci-flat metric on a K3 surface provides a natural solution of the Hermitian Yang-Mills connection on the tangent bundle.) This is of course encouraging as it indicates the possibility of the conjecture.

It turns out that Donaldson [22] was concentrated on algebraic surfaces and Uhlenbeck-Yau [79] on arbitrary dimensional Kähler manifolds. While Donaldson used the Bott-Chern form and the Hermitian Yang-Mills flow, Uhlenbeck-Yau constructed a destabilizing sheaf assuming the nonexistence of Hermitian Yang-

Mills connection.

The proof of regularity of the destabilizing subsheaf took nontrivial effort and as a result, our paper appeared later than the work of Donaldson's proof for algebraic surfaces. After we published our work, Donaldson found that some of our formula can be used to re-prove the Uhlenbeck-Yau theorem for algebraic manifolds by restriction of the bundle to hyperplane sections of the algebraic manifold. (It was proved by Maruyama and Mehta-Ramanathan that a stable bundle is stable on a generic hyperplane section.)

This later argument of Donaldson depends intrinsically on the manifold being projective for higher dimensional manifolds. As was acknowledged by Donaldson, the argument of Uhlenbeck-Yau is most natural and in fact, all the later development for Hermitian Yang-Mills connections for higher dimensional manifolds are based on the procedure of Uhlenbeck-Yau.

Some later paper such as the one by Bando-Siu [5] used the Hermitian Yang-Mills flow to generalize our result, but the essential feature of Uhlenbeck-Yau procedure is still needed in an essential manner. It should also be pointed out that the continuity argument used by Uhlenbeck-Yau is just as convenient as the Hermitian Yang-Mills flow.

A few years later, Carlos Simpson [69] generalized the Uhlenbeck-Yau argument to establish similar theorem when the Higgs field was introduced. Hermitian Yang-Mills connections were proposed by me to Edward Witten in 1984 to study heterotic string, which had since became an important subject in mathematical physics. But from the very beginning, we knew the importance of Hermitian Yang-Mills connections, as it provides important Chern number inequalities, and also the conditions for the bundle to be projectively flat.

3.2 Chern number inequalities and characterization of flat bundles

The very first applications was the sharpening of the Chern number inequality of Bogomolov and a very important generalization of the theorem of Seshadri-Narasimhan (1965) that every stable bundle over an algebraic curve is flat if the degree of the bundle is zero. A very remarkable corollary of the existence of Hermitian Yang-Mills connection for stable holomorphic bundle is that such bundle must be projective flat, if the Bogomolov inequality $2rc_2(E) \geq (r-1)c_1(E)^2$ becomes equality.

This can be considered as a generalization of my theorem that the equality of certain Chern numbers can be used to characterize ball quotients. In fact, Carlos Simpson observed that by generalizing this theorem to the Hermitian Yang-Mills-Higgs connection, one can reproduce my previous theorem that an algebraic surface of general type is covered by the ball if the ratio of the two Chern numbers is the same as the projective plane. In fact, by choosing the Higgs field carefully, one can generalize the theorem to characterize quotient of general Hermitian symmetric space assuming suitable stability.

Characterization of flat bundles based on Hermitian Yang-Mills-Higgs connection also allows Simpson to construct variation of Hodge structures. This is remarkable and led me to believe that there is a good connection with the characterization of quotients of more general Hermitian symmetric domains based on the existence of Kähler-Einstein metrics.

It is really remarkable that the construction of stable bundles satisfying certain Chern number equality gives rise to nontrivial projective representation of the fundamental group of the manifold, which we know little about.

In particular, if some natural bundle constructed from the tangent bundle of the manifold is stable with respect to certain polarization and if the numbers defined by wedging the second Chern class of the natural bundle with the polarization $n-2$ times, and the square of the first Chern class of the natural bundle wedged with the polarization $n-2$ times are equal to zero, then the natural bundle admits a flat Hermitian connection, which means that the fundamental group of the manifold has a nontrivial unitary representation, unless the natural bundle is trivial. Note that we do not need to assume existence of Kähler-Einstein metric on the manifold in this setting. Natural bundles are bundles constructed from natural decomposition of tensor product of tangent and cotangent bundles.

It raises an interesting question in this regard: given an algebraic manifold M with a fixed Kähler class, we consider all holomorphic bundles with trivial first Chern class over M which are polystable with respect to this Kähler class. We consider two such bundles equivalent if they become isomorphic to each other after adding trivial bundles. They form a ring consisting of countable number of algebraic subvarieties which are moduli space of the bundles with a fixed Hilbert polynomial.

There is a subring formed by those stable bundles whose second Chern class wedged with the Kähler class to $n-2$ times vanishes. Does the structure of this subring determine the algebraic fundamental group of the manifold? (It is quite likely that we need to consider bundles with Hilbert space fiber in order to obtain information for the full fundamental group.) What is the structure of this subring for Shimura varieties? Can they determine the Shimura variety?

3.3 Generalization to non-Kähler and non-compact manifolds

Since the theory of Uhlenbeck-Yau was generalized by Jun Li and myself to general complex manifolds, we are able to apply it to handle some interesting non-Kähler manifolds. The most notable one was the class VII surfaces of Kodaira. They were studied by Kodaira, Inoue and Bombieri. Kobayashi and Ochiai realized the importance of holomorphic connections for such manifolds. Bogomolov claimed that for such manifolds without curves, they are given by the examples constructed by Inoue. The proof by Bogomolov [6, 9] is not clear.

Jun Li, Fangyang Zheng and I [46] gave a clear proof based on the existence of Hermitian Yang-Mills connections. It should be possible to generalize our argument to handle those class VII surfaces with finite number of curves also. Many years ago, I proposed to study those connections mod the curves long ago. If this proposal is successful, it should complete the Kodaira classification of complex non-Kähler surfaces.

The study of Hermitian Yang-Mills connections over quasi-projective curve was discussed by Simpson. The generalization to the case when the base pair is (M, D) with D nonsingular, is not hard. The case when D is normal crossing divisor is more difficult, and was studied by Takuro Mochizuki [63].

3.4 Analytic criterions for various stability conditions

There is no simple criterion to check whether a bundle is stable or not. In many cases, the existence of Hermitian Yang-Mills connection helps to understand properties of stability of bundles. Slope stability is only one kind of stability that appeared in algebraic geometry. A natural class of stability was introduced by David Gieseker in early 1970s. He compared Hilbert polynomials of the subsheaves.

The analytic analog of Gieseker stability is not well understood, although Conan Leung studied this problem in his PhD thesis [44], under my guidance about 20 years ago. There are a sequence of differential equations which can be considered as a natural generalization of the Hermitian Yang-Mills equations. (Todd classes are part of the equations as Hilbert polynomial need to be expressed.) Assuming the curvature is uniformly bounded, Leung proved that the existence of the equations is equivalent to Gieseker stability of the bundle. This bound of the curvature has not been proved and whether this set of equations is the most natural set of equations is not clear.

As was proposed by me [89], the existence of a Kähler Einstein metric or metrics with constant scalar curvature on an algebraic manifold is related to stability of the algebraic manifold. My former student Luo, followed my suggestion of using the concept of balanced condition to study stability of manifolds. It would be good to relate manifold stability to bundle stability. Now Chen-Donaldson-Sun proved that K-stability of the manifold implies the existence of Kähler-Einstein metric. It implies, in particular, the stability of the tangent bundle of the manifold.

In order to relate two concepts of stability, I propose to define a bundle to be balanced if the sections of the bundle, after twisted by a very ample line bundle, can embed the manifold into a balanced submanifold of the Grassmannian.

Hermitian Yang-Mills bundles are mirror to special Lagrangian submanifolds in the theory of mirror symmetry under the program of Strominger-Yau-Zaslow. Gieseker stability is slightly weaker than slope stability. It may be interesting to know which class of Lagrangian cycles will be their mirror images. By studying stability question for Lagrangian cycles carefully and applying mirror symmetry, Mike Douglas found new concepts of stability of bundles. Based on his work and the works of F. Denef, Douglas- Reinbacher-Yau [24] proposed a conjecture on the existence of stable bundles based on Chern classes of the bundle which can be stated as follows:

Consider an ample class D on a simply connected Calabi-Yau threefold X and an integer $r > 1$ and three classes

$$c_i \in H^{2i}(D, \mathbb{Z}), \quad i = 1, 2, 3,$$

such that

$$\left(2rc_2 - (r-1)c_1^2 - \frac{r^2}{12}c_2(D)\right) = 2r^2 D^2,$$

and

$$(c_1^3 + 3r(rch_3 - ch_2c_1)) < 8\sqrt{2} \cdot r^3D^3.$$

Then there exists a rank r reflexive sheaf V on X stable with respect to some ample class such that

$$c_i(V) = c_i, \quad i = 1, 2, 3.$$

Prior to the SYZ program on mirror symmetry, Kontsevich introduced the concept of homological mirror symmetry, where he introduced the derived category over algebraic manifolds. It was realized later to correspond to branes in string theory. This has been developed into a rich theory. Bridgeland studied the concept of stability of derived category and it is now called Bridgeland stability. It would be interesting to find a suitable analytic counterpart of Bridgeland stability.

4 Mirror symmetry

Supersymmetry provides powerful tools to understand Calabi-Yau manifolds. The intuitions from physics have been powerful. The important concepts introduced by string theorists have deep influence on the geometry of such manifolds. The most important one was the idea of mirror symmetry. It called for the existence of another Calabi-Yau manifold (which we call the mirror manifold) whose Hodge diamond for cohomology is the transpose of the Hodge diamond of the original Calabi-Yau manifold.

4.1 Counting of curves

More importantly the conformal field theory based on one Calabi-Yau manifold is dual to that of its mirror manifold. The Type IIA conformal field theory of Calabi-Yau manifold is isomorphic to the Type IIB theory of the mirror manifold. This is a remarkable theory predicted by Vafa, Dixon and others. But it was Greene-Plesser and Candelas et al. who developed the details of such theory. The most remarkable consequence is that it solved an old problem in enumerative geometry.

The reason is that the Type IIB theory can be computed by deformation theory of Kodaria-Spencer while the type IIA has quantum corrections. The quantum corrections are provided by the rational curves on the Calabi-Yau manifold. Since Type IIA theory of one manifold is isomorphic to the type IIB of the mirror manifold, we can compute the number of rational curves on the Calabi-Yau manifolds by the variation of Hodge structure for its mirror family.

The initial theory was mostly based on physical intuition. But two groups of mathematicians proved such statements rigorously in 1996, by Lian-Liu-Yau [47] and Givental [30] independently. Despite that the proof is rigorous, the intuition from string theory played the most important role. The idea of supersymmetry has become one of the most fundamental philosophies underlying the current modern development of algebraic geometry.

The simplest and most elegant examples of geometric structures showing up in string theory studies are the Calabi-Yau manifolds, where rich structures related to deep string theory and quantum field theory dualities are discovered. This includes the so called Gromov-Witten invariants related to the counting of rational curves mentioned above.

The counting of algebraic curves of higher genus is far more complicated. One approach was initiated by Bershadsky, Cecotti, Ooguri and Vafa (BCOV), who developed a theory called Kodaria-Spencer theory of gravity. The computation gives beautiful predictions based on some master equations. But it suffers from an ambiguity which is called holomorphic ambiguity. Up to now, we still have difficulty to overcome this ambiguity, although some important progress was made by Zinger, Jun Li and their coauthors for genus one or low genus curves.

Modularity of partition functions of the so-called topological strings contains non-trivial arithmetic information of the Gromov-Witten invariants and it is still a mystery to understand the appearance of modular forms completely. Yamaguchi and I were able to demonstrate some polynomial structure on such partition functions [82], which suggested that there is rich algebraic structure behind them. Due to especially its impressive power in enumerative geometry, there was a great desire to understand mirror symmetry mathematically.

4.2 Mathematical approaches to mirror symmetry

As mentioned above, two different approaches were proposed. One is the famous Kontsevich's homological mirror symmetry conjecture [41] which says that the derived category of coherent sheaves of a Calabi-Yau manifold is equivalent to the Fukaya category of its mirror manifold. Fukaya pioneered the research to study the extensive complicated structure of the Florer theory of Lagrangian cycles through A_∞-algebra [26, 25], which is an important ingredient in the conjecture of Konsevich. Another approach was proposed by Strominger-Yau-Zaslow [71] that the Calabi-Yau manifold is fibred by special Lagrangian torus and the mirror manifold is obtained by replacing the torus by its dual torus. Much progress was made by Auroux, Seidel et al. in this direction. It is important that singularities are allowed in the fibration for both topological and more subtle reasons. The SYZ conjecture has much evidence to be true. Gross and Siebert made a lot of progress in the last few years using tropical methods.

Mirror symmetry has inspired many important developments in Kähler geometry. The program of SYZ calls for close relationship between special Lagrangians with bundles. In the paper of Leung-Yau-Zaslow, we explained how, under the SYZ map, equation for special Lagrangian cycle which intersects the SYZ torus at one point can be transformed to an equation for a holomorphic line bundle. The equation turns out to be studied by M. Mariño, R. Minasian, G. Moore and Strominger [57], recently Tristan Collins, Adam Jacob, and myself [14] studied this equation and we can prove the existence for many important cases assuming some form of stability for the $(1,1)$-class. The equation has the form:

$$\mathrm{Im}\,(J+\sqrt{-1}\omega)^n = \tan(\hat{\theta})\mathrm{Re}\,(J+\sqrt{-1}\omega)^n,$$

where $\hat{\theta}$ is a topological constant determined by J and $[\omega]$.

The equation admits supersymmetry and the pair consists of the Kähler class J and the closed $(1,1)$-class ω can be looked as a natural complexfication of the

Kähler class. It defines an open set in the complexified Kähler cone. This may give a good candidate for the mirror of the moduli space of polarized complex structures of its mirror manifold. Note that we like to see the "Kähler moduli" to be isomorphic to the moduli space of the complex structure on its mirror manifold. It is also important to find a suitable discrete group acting on this open set in the Kähler cone. The mirror symmetric version of the special Lagrangian that intersects the SYZ torus for more than a point is supposed to be a higher rank bundle. The equation defined on it is being explored.

One should note that the homological mirror conjecture of Kontsevich has inspired a great deal of study of derived category in geometry. While we may not be used to abstract reasoning of category theory in geometry, we hope that its relationship to SYZ construction may eventually broaden the scope of geometry.

5 Future directions in mathematical physics and arithmetic geometry

In conclusion, we should say that the beautiful subject initiated by Riemann in the nineteenth century on Riemann surfaces had deep influence on the development of complex geometry in the 20th century. While Hodge provided the fundamental structure relating complex analysis with topology via Hodge groups, Kodaira provided fundamental methods to construct holomorphic sections of bundles. With the works of Chern classes and Hirzebruch-Riemann-Roch formula, the works of Hodge and Kodaira have been developed to be most powerful tools in understanding Kähler geometry. The modern development has been emphasizing the use of nonlinear elliptic equations, relating the concept of Kähler-Einstein metrics and Hermitian Yang-Mills equations to various fundamental concepts of stability introduced to study moduli spaces.

The most recent development on Calabi-Yau space due to cooperations between mathematicians and string theorists has been spectacular. Ideas of many fields in mathematics were used. We hope to see some more ideas of number theory in this beautiful subject. For many Calabi-Yau manifolds, the partition functions related to conformal field theory are related to modular forms. For example, it was observed in 1996 by Zaslow and I [91] that the partition function counting rational curves of various degrees in K3 surfaces can be written in terms of η-functions. This was the first time that such modular function appears in counting curves in algebraic geometry. This motivated Göttsche to generalize the Yau-Zaslow formula to general surfaces and for curves of arbitrary genus. This was recently first proved by Y. Tzeng [78] and later by Kool-Shende-Thomas [42]. For Calabi-Yau manifolds with higher dimension, these formulas are much more complicated and their organization is still being explored. In recent work with Zhou and others [2], we were able to show the ring generated by quasi-modular forms associated to $PSL(2,\mathbb{Z})$ or a congruent subgroup therein is isomorphic to the ring of higher genus Gromov-Witten invariants for certain non-compact Calabi-Yau geometries based on the projective plane and also del Pezzo surfaces. This was later extended to the orbifold case [67] and also to include open Gromov-Witten theory in [43].

For more classical arithmetic geometry, we may point out that Serge Lang has noticed long time ago that the importance of Kobayashi hyperbolicity is relevant to the question of Diophanitine problem, such as the Mordell conjecture proved by Faltings for algebraic curves of higher genus. Kobayashi conjectured that for an algebaic manifold of general type, the Kobayashi metric should be non-degenerate in a Zariski open set. In particular, there is a subvariety of the manifold such that all rational curves and elliptic curves are subset of this algebraic subvariety. This is sometimes called the Lang conjecture.

Lang also conjectures that if the manifold is defined over integers, the rational points of the manifold should all be in this subvariety. There was little progress on the Kobayashi-Lang conjecture except in the case of surfaces where Bogomolov [7] and Miyaoka [62] made important contributions. Steven Lu and I studied the differential geometric aspect of it [52].

For algebraic surfaces with positive index, one can find a Finsler metric with strongly negative holomorphic sectional curvature, but the metric may degenerate in some subvarieties. This statement implies the Kobayashi-Lang conjecture. Therefore one would like to make the following conjecture: an algebraic manifold is of general type if and only if it admits a complex Finsler metric which may be degenerate along a subvariety which has strongly negative holomorphic sectional curvature. It is quite possible that Finsler metric may be replaced by Kähler metric. The converse was asked by me, there were some progress due to several people, but only recently Damin Wu and I [81] were able to prove that if an algebraic manifold admits a Kähler metric with strongly negative holomorphic sectional curvature, its canonical line bundle must be ample. (Our original argument assumes manifold to be algebraic, but it was pointed out by Valentino Tosatti and Xiaokui Yang [74] that our argument, which is based on solving Monge-Amperè equation, can work for Kähler manifold also.) It is not hard to generalize the theorem to complete non-compact Kähler manifolds whose holomorphic sectional curvature is bounded by two negative constants. A natural question is that a compact manifold admits a pseudo-Kähler metric (Kähler metric that may degenerate along some subvarities) with strongly negative holomorphic sectional curvature iff the manifold is of general type. Manin has conjectured that Kähler-Einstein metric will play important role in arithmetic geometry. I believe that is the case. There is still much to learn about the relation between complex geometry, algebra and number theory.

References

[1] L. Alessandrini and G. Bassanelli, *Metric properties of manifolds bimeromorphic to compact Kähler spaces*, J. Differential Geom. **37** (1993), 95–121.

[2] M. Alim, E. Scheidegger, S. T. Yau and J. Zhou, *Special polynomial rings, quasi modular forms and duality of topological strings*, Adv. Theor. Math. Phys. **18** (2014), 401.

[3] M. Anderson, P. Kronheimer and C. LeBrun, *Complete Ricci-flat Kähler manifolds of infinite topological type*, Comm. Math. Phys. 125 (1989), 637–642.

[4] U. Angehrn and Y.-T. Siu, *Effective freeness and point separation for adjoint bundles*, Invent. Math. **122** (1995), 291–308.

[5] S. Bando and Y.-T. Siu, *Stable sheaves and Einstein-Hermitian metrics*, Geometry and analysis on complex manifolds, 39–50, World Sci. Publ., River Edge, NJ, 1994.

[6] F. A. Bogomolov, *Classification of surfaces of class VII0 with $b_2 = 0$*, Izv. Akad. Nauk SSSR Ser. Mat. **40** (1976), 273–288.

[7] F. A. Bogomolov, *Families of curves on a surface of general type*, Soviet Math. Dokl. **236** (1977), 1294–1297.

[8] F. A. Bogomolov, *Holomorphic tensors and vector bundles on projective manifolds*, Izv. Akad. Nauk SSSR Ser. Mat. **42** (1978), 1227–1287.

[9] F. A. Bogomolov, *Surfaces of class VII0 and affine geometry*, Izv. Akad. Nauk SSSR Ser. Mat. **46** (1982), 710–761.

[10] J.-P. Bourguignon, P. Li and S.-T. Yau, *Upper bound for the first eigenvalue of algebraic submanifolds*, Comment. Math. Helv. **69** (1994), 199–207.

[11] E. Calabi, *The space of Kähler metrics*, Proc. Int. Congr. Math. Amsterdam 2 (1954), 206–207.

[12] P. Candelas, G. T. Horowitz, A. Strominger and E. Witten, *Vacuum configurations for superstrings*, Nuclear Phys. B **258** (1985), 46–74.

[13] D. Catlin, *The Bergman kernel and a theorem of Tian*, Analysis and geometry in several complex variables (Katata, 1997), 1999, 1–23.

[14] T. C. Collins, A. Jacob and S.-T. Yau, *(1,1) forms with specified Lagrangian phase: A priori estimates and algebraic obstructions*, arXiv:1508.01934.

[15] T. C. Collins and G. Székelyhidi, *K-Semistability for irregular Sasakian manifolds*, arXiv:1204.2230, to appear in J. Differential Geom.

[16] X. Chen, S. Donaldson and S. Sun, *Kähler-Einstein metrics on Fano manifolds. I: Approximation of metrics with cone singularities*, J. Amer. Math. Soc. **28** (2015), 183–197.

[17] X. Chen, S. Donaldson and S. Sun, *Kähler-Einstein metrics on Fano manifolds. II: Limits with cone angle less than 2π*, J. Amer. Math. Soc. **28** (2015), 199–234.

[18] X. Chen, S. Donaldson and S. Sun, *Kähler-Einstein metrics on Fano manifolds. III: Limits as cone angle approaches 2π and completion of the main proof*, J. Amer. Math. Soc. **28** (2015), 235–278.

[19] S.-S. Chern, *Characteristic classes of Hermitian manifolds*, Ann. of Math. **47** (1946), 85–121.

[20] C.-Y. Chi, S.-T. Yau, *A geometric approach to problems in birational geometry*, Proc. Natl. Acad. Sci. USA **105** (2008), 18696–18701.

[21] J.-P. Demailly, *A numerical criterion for very ample line bundles*, J. Differential Geom. **37**, (1993), 323–374.

[22] S. K. Donaldson, *Anti-self-dual Yang-Mills connections over complex algebraic surfaces and stable vector bundles*, Proc. London Math. Soc. (3) **50** (1985), 1–26.

[23] S. K. Donaldson, *Scalar curvature and stability of toric varieties*, J. Differential Geom. **62**(2002), 289–349.

[24] M. Douglas, R. Reinbacher and S.-T. Yau, *Branes, bundles and attactors: Bogomolov and beyond*, arXiv: math/0604597.

[25] K. Fukaya, Y.-G. Oh, H. Ohta and K. Ono. *Lagrangian intersection Floer theory: anomaly and obstruction*, American Mathematical Soc. **46.1** (2010).

[26] K. Fukaya, *Morse homotopy, A_∞-category, and Floer homologies*, Math. Sciences Research Inst. 1993.

[27] J. Fu and S.-T. Yau, *The theory of superstring with flux on non-Kähler manifolds and the complex Monge-Ampère equation*, J. Differential Geom. **78** (2008), 369–428.

[28] A. Futaki, An *obstruction to the existence of Einstein Kähler metrics*, Invent. Math. **73** (1983), 437–443.

[29] J. Gauntlett, D. Martelli, J. Sparks and S.-T. Yau, *Obstructions to the existence of Sasaki-Einstein metrics*, Comm. Math. Phys. 273 (2007), 803–827.

[30] A. Givental, *Equivariant Gromov-Witten invariants.* Internat. Math. Res. Notices (1996), no. 13, 613–663.

[31] E. Goldstein and S. Prokushkin, *Geometric model for complex non-Kähler manifolds with SU(3) structure*, Comm. Math. Phys. **251** (2004), 65–78.

[32] F. Hirzebruch, *Arithmetic genera and the theorem of Riemann-Roch for algebraic varieties*, Proc. Nat. Acad. Sci. U.S.A. **40** (1954), 110–114.

[33] F. Hirzebruch and K. Kodaira, *On the complex projective spaces*, J. Math. Pures Appl. (9) **36** (1957), 201–216.

[34] H. Hopf, *Zur Topologie der komplexen Mannigfaltigkeiten*, Studies and Essays presented to R. Courant, New York (1948), 167–185.

[35] E. Kähler, *Über eine bemerkenswerte Hermitesche Metrik*, Abh. Math. Sem. Hamburg Univ. **9** (1933), 173–186.

[36] Y. Kawamata, *On the finiteness of generators of a pluricanonical ring for a 3-fold of general type*, Amer. J. Math. 106 (1984), no. 6, 1503–1512.

[37] Y. Kawamata, *Pluricanonical systems on minimal algebraic varieties*, Invent. Math. **79** (1985), no. 3, 567–588

[38] F. Klein, *Riemannsche Flächen*, Vorlesungen, gehalten in Göttingen 1891/92. Reprinted by Springer (1986).

[39] J. Kollár and T. Matsusaka, *Riemann-Roch type inequalities*, Amer. J. Math. **105** (1983), 229–252.

[40] J. Kollár, Y. Miyaoka and S. Mori, *Rational connectedness and boundedness of Fano manifolds*, J. Differential Geom. 36 (1992), 765–779.

[41] M. Kontsevich, *Homological algebra of mirror symmetry*, Proceedings of the International Congress of Mathematicians, Vol. **1, 2** (Zürich, 1994), 120–139, Birkhäuser, Basel, 1995.

[42] M. Kool, V. Shende and R. Thomas, *A short proof of the Göttsche conjecture*, Geom. Topol. **15** (2011), 397–406.

[43] S. C. Lau and J. Zhou, *Modularity of Open Gromov-Witten Potentials of Elliptic Orbifolds*, arXiv:1412.1499.

[44] N. C. Leung, *Einstein type metrics and stability on vector bundles*, J. Differential Geom. **45** (1997), 514–546.

[45] J. Li and S.-T. Yau, *The existence of supersymmetric string theory with torsion*, J. Differential Geom. **70** (2005), 143–181.

[46] J. Li, S.-T. Yau and F. Zheng, *A simple proof of Bogomolov's theorem on class* VII_0 *surfaces with* $b_2 = 0$, Illinois J. Math. **34** (1990), 217–220.

[47] B. H. Lian, K. Liu and S.-T. Yau, *Mirror principle. I*, Asian J. Math. **1** (1997), 729–763.

[48] B. H. Lian and S.-T. Yau, *Period integrals of CY and general type complete intersections*, Invent. Math. **191** (2013), 35–89.

[49] A. Lichnerowicz, *Géometrie des groupes de transformations*, Dunod, Paris, 1958.

[50] K. Liu, X. Sun and S.-T. Yau, *Canonical metrics on the moduli space of Riemann surfaces. I*, J. Differential Geom. **68** (2004), 571–637.

[51] K. Liu, X. Sun and S.-T. Yau, *Canonical Metrics on the moduli space of Riemann surfaces. II*, J. Differential Geom. **69** (2005), 163–216.

[52] S-Y. S. Lu and S. T. Yau, *Holomorphic curves in surfaces of general type*, Proc. Nat. Acad. Sci. U.S.A. **87** (1990), no. 1, 80–82.

[53] Z. Lu, *On the lower order terms of the asymptotic expansion of Tian-Yau-Zelditch*, Amer. J. Math. **122** (2000), 235–273.

[54] H. Luo, *Geometric criterion for Gieseker-Mumford stability of polarized manifolds*, J. Differential Geom. **49** (1998), 577–599.

[55] D. Martelli, J. Sparks and S. T. Yau, *Sasaki-Einstein manifolds and volume minimisation*, Commun. Math. Phys. **280** (2008), 611–673.

[56] M. L. Michelsohn, *On the existence of special metrics in complex geometry*, Acta Math. **149** (1982), 261–295.

[57] M. Mariño, R. Minasian, G. Moore and A. Strominger, *Nonlinear instantons from supersymmetric p-branes*, J. High Energy Phys. (2000), no. 1, Paper 5, 32 pp.

[58] T. Matsusaka, *On canonically polarized varieties. II*, Amer. J. Math. **92** (1970), 283–292.

[59] T. Matsusaka, *Polarized varieties with a given Hilbert polynomial, Amer.* J. Math. **94** (1972), 1027–1077.

[60] Y. Matsushima, *Sur la structure du groupe d'homéomorphismes analytiques d'une certaine variété kählérienne*, Nagoya Math. J. **11** (1957), 145–150.

[61] Y. Miyaoka, *On the Chern numbers of surfaces of general type*, Invent. Math. **42** (1977), 225–237.

[62] Y. Miyaoka, *Algebraic surfaces with positive indices*, Classification of algebraic and analytic manifolds (Katata, 1982), 281–301, Progr. Math., 39, Birkhäuser Boston, Boston, MA, 1983.

[63] T. Mochizuki, *Kobayashi-Hitchin correspondence for tame harmonic bundles and an application*, Astérisque No. 309 (2006), viii+117 pp.

[64] S. Mori, *Projective manifolds with ample tangent bundles*, Ann. of Math. (1979), 593–606.

[65] S. Mukai, *Duality between $D(X)$ and $D(\hat{X})$ with its application to Picard sheaves*, Nagoya Mathematical Journal **81** (1981), 153–175.

[66] H. Royden, *Automorphisms and isometries of Teichmüller space*, 1971 Advances in the Theory of Riemann Surfaces (Proc. Conf., Stony Brook, N.Y., 1969) pp. 369–383, Ann. of Math. Studies, No. 66. Princeton Univ. Press, Princeton, N.J.

[67] Y. F. Shen and J. Zhou, *Ramanujan Identities and Quasi-Modularity in Gromov-Witten Theory*, arXiv: 1411.2078.

[68] C. L. Siegel, *Topics in complex function theory*, Interscience Tracts in Pure and Applied Mathematics, No. 25, Vol. I, 1969, ix+186 pp.; Vol. II, 1971, 256 pp.; Vol. Ill, 1973, ix+244 pp.

[69] C. T. Simpson, *Constructing variations of Hodge structure using Yang-Mills theory and applications to uniformization*, J. Amer. Math. Soc. **1** (1988), no. 4, 867–918.

[70] Y. T. Siu and S.-T. Yau, *Complete Kähler manifolds with nonpositive curvature of faster than quadratic decay*, Ann. of Math. (2) **105** (1977), 225–264.

[71] A. Strominger, S.-T. Yau and E. Zaslow, *Mirror symmetry is T-duality*, Nuclear Phys. B **479** (1996), no. 1–2, 243–259.

[72] G. Tian and S.-T. Yau, *Complete Kähler manifolds with zero Ricci curvature. I*, J. Amer. Math. Soc. **3** (1990), 579–609.

[73] G. Tian and S.-T. Yau, *Complete Kähler manifolds with zero Ricci curvature. II*, Invent. Math. **106** (1991), 27–60.

[74] V. Tosatti and X. Yang, *An extension of a theorem of Wu-Yau*, arXiv:1506.01145.

[75] C.-J. Tsai, L.-S. Tseng and S.-T. Yau, *Symplectic cohomologies on phase space*, J. Math. Phys. 53 (2012), no. 9, 095217, 9 pp.

[76] L.-S. Tseng, S.-T. Yau, *Cohomology and Hodge theory on symplectic manifolds: I*, J. Differential Geom. 91 (2012), 383–416.

[77] L.-S. Tseng, S.-T. Yau, *Cohomology and Hodge theory on symplectic manifolds: II*, J. Differential Geom. **91** (2012), 417–443.

[78] Y. Tzeng, *A proof of the Göttsche-Yau-Zaslow formula*, J. Differential Geom. **90** (2012), 439–472.

[79] K. Uhlenbeck and S.-T. Yau, *On the existence of Hermitian Yang-Mills connections in stable vector bundles*, Comm. Pure Appl. Math. **39** (1986), suppl., S257–S293.

[80] H. Weyl, *Die Idee der Riemannschen Fläche*, Teubner, Leipzig (1913).

[81] D. Wu and S.-T. Yau, *Negative holomorphic curvature and positive canonical bundle*, arXiv:1505.05802.

[82] S. Yamaguchi and S. T. Yau, *Topological string partition functions as polynomials*, JHEP **0407** (2004), 047.

[83] S.-T. Yau, *Intrinsic measures of compact complex manifolds*, Math. Ann. **212** (1975), 317–329.

[84] S.-T. Yau, *Calabi's conjecture and some new results in algebraic geometry*, Proc. Nat. Acad. Sci. U.S.A. **74** (1977), 1798–1799.

[85] S.-T. Yau, *Métriques de Kähler-Einstein sur les variétés ouvertes, Astérisque* **58** (1978), 163–167.

[86] S.-T. Yau, *On the Ricci curvature of a compact Kähler manifold and the complex Monge-Ampère equation. I*, Comm. Pure Appl. Math. **31** (1978), 339–411.

[87] S.-T. Yau, *The role of partial differential equations in differential geometry*, Proc. Int. Congr. Math. (Helsinki, 1978), 237–250, Acad. Sci. Fennica, Helsinki, 1980.

[88] S.-T. Yau, *Nonlinear analysis in geometry.* Monographies de L'Enseignement Mathé-matique, **33**. Série des Conférences de l'Union Mathématique Internationale, 8. L'Enseignement Mathématique, Geneva, 1986. 54 pp.

[89] S.-T. Yau, *Open problems in Geometry*, Proc. Symp. Pure Math. **54** (1993) 1–28.

[90] S.-T. Yau and Y. Zhang, *The geometry on smooth toroidal compactifications of Siegel varieties*, Amer. J. Math. **136** (2014), 859–941.

[91] S.-T. Yau and E. Zaslow, *Eric BPS states, string duality, and nodal curves on K3*, Nuclear Phys. B **471** (1996), 503–512.

[92] S. Zelditch, Szegö *kernel and a theorem of Tian*, Internat. Math. Res. Notices (1998), no. 6, 317–331.

[93] S. Zhang, *Heights and reductions of semistable varieties*, Compositio Math. **104** (1996) 77–05.

[94] S. T. Yau, *A splitting theorem and an algebraic geometric characterization of locally Hermitian symmetric spaces*, Comm. Analysis & Geom. **1** (1992), no. 3.

[95] C. N. Yang, *Conditions of Self-Duality for SU(2) gauge fields on euclidean four-dimensional space*, Phys. Rev. Lett. **38** (1977) 1377.

(Note. I intend to give Takagi lectures based on parts of this paper. — Shing-Tung Yau)

Part IV

Selected Commentaries by Shing-Tung Yau

Editor's Comments

As one of the leading mathematicians in the past half century, Yau has worked in many different subjects, in particular, in geometric analysis, and has thought about many issues in mathematics and the mathematics community. His deep mathematical insights will shed light on the origin and history of many mathematical problems, and his personal experiences will enhance the learning of these subjects by the young mathematicians and students and also serve as permanent historical records of some important events in mathematics.

In view of these considerations, it seems not unreasonable to include in this book the many commentaries he wrote for his *Selected Expository Works of Shing-Tung Yau with Commentary.* We hope that the inclusion of these commentaries is also consistent with the rest of the book.

Commentary on Métriques de Kähler-Einstein Sur les Variétés Ouvertes

In 1978 I was on my way from Stanford to attend the International Congress of Mathematicians in Helsinki, where I was to give a plenary talk. I was invited and spent a month at IHES to communicate with our colleagues in France.

I could not speak French at all. A Stanford graduate student, Petric Ames, picked me up after I arrived at the airport and accompanied me to the apartment at IHES.

His parents came from the French part of North Africa. I did not realize that his French carries what the Parisians considered to be an accent. We went to many places in Paris for sightseeing. We met Bernard Saint-Donat on the streets of Paris, when Ames and I decided to watch a movie called "Hitler". B. Saint-Donat wanted to take us to see the great Paris and watch some opera. But P. Ames insisted on watching the movie. It ended up that B. Saint-Donat got angry at the two guys from America who weren't cultured enough to appreciate the great Paris. Well, later we did go to see many great museums in Paris of which I was highly impressed. I never cease to admire French culture every time I visit Paris.

In Paris, I met many mathematicians from America and from other countries. This includes Robert Langlands, Blaine Lawson, Kenneth Ribet, Yoichi Miyaoka, and many mathematicians in France including Jean-Pierre Serre, Pierre Deligne, Jean-Pierre Bourguignon, Marcel Berger, and others.

Nicolaas Kuiper was the director of IHES at that time. He gave me a very warm welcome, inviting me to his home for lunch. Robert Connelly of Cornell University had just found a non-convex polyhedron that was not rigid and flexible in three spaces. N. Kuiper was extremely impressed by this result. He took us to meet an artist in Paris whose artwork included such flexible polyhedrons. Apparently the artist knew how to build such things long ago without knowing that mathematicians were interested in them.

Y. Miyaoka had just finished his proof of the Miyaoka-Yau inequality for algebraic surfaces, based on the ideas of Fedor Bogomolov. I told him what Kähler-Einstein metrics could do, mentioning that the argument also worked in higher dimensions. The log version of the inequality could also be derived. I gave a talk on this at the Polytechnic University, after being invited to do so by my friend J.-P. Bourguignon. He made notes of my talk and recorded my work with S. Y. Cheng on the existence of a Kähler-Einstein metric on the complement of a divisor of normal crossing as long as $K + D$ is positive. (The log Chern number inequality follows from this rather straightforwardly.)

At that time, I had just finished my work with Richard Schoen on the structure of three manifolds with positive scalar curvature, which is related to the positive mass conjecture. I gave a talk on this, and B. Lawson showed great interest in it. R. Schoen and I were also in the process of writing up a proof, which demonstrated that in the category of manifolds with positive scalar curvature, it is possible to perform surgery with codimension not less than three that preserved the property of admitting such metrics. This paper appeared in *Manuscripta Math.* in 1979.

I mentioned our results to B. Lawson before it was published. A year later, B. Lawson wrote a paper that appeared in the *Annals of Mathematics* 1980 with Mikhael Gromov on a different way to perform such a surgery. (It was later discovered that there were some mistakes in the formula they used, although they may have been corrected by now.) R. Schoen and I realized its importance for studying the topology of manifolds with positive scalar curvature, but we did not know the right experts on spin cobordism to transform the surgery result to provide effective classification for simply connected manifolds with positive scalar curvature.

Commentary on the Classical Plateau Problem and the Topology of 3-Manifolds

In the fall of 1977, I was moving from my visiting position at UCLA to another visiting position at Berkeley at the invitation of Chern, who wanted me to try out an offer from Berkeley. I rented a two-bedroom apartment in north Berkeley, where my mother came to stay with me. It was a productive year, working with Richard Schoen and S.-Y. Cheng. I taught a course on the differential geometry of isometric embeddings. (My lecture notes were later given to J. Hong at Fudan University. He wrote a nice book with Han, based partially on my notes.)

Before I started the semester, Chern asked me to go with him to a joint Japan-United States conference on minimal submanifolds and geodesics that he, Osserman, Otsuki, and Obata had arranged. I took the advantage of this trip to visit Hong Kong. I passed Tokyo, flying on United Airlines to Hong Kong, When the airline collected my ticket for the first leg, they mistakenly took my return ticket too. That caused me huge problems when I tried to go from Hong Kong back to Tokyo. But I was still glad that I went, as it was my first time back in Hong Kong since I left in 1969. I was very pleased to see my brother, my sisters, and my friends, although I found out that housing situation in Hong Kong was much worse than in America. This was also my first visit to Japan. The city of Tokyo was as crowded as Hong Kong, so I felt right at home there.

Although I was reasonably famous at that time, as I had just finished the proof of the Calabi conjecture, the Japanese hosts considered age to be important, as do the Chinese. Hence I was not invited to their important banquets nor to the Japanese opera. However, Jim Simons and Blaine Lawson happened to be in Tokyo too, so we went to some bars there, even though I did not drink alcohol. The Vietnam War had not ended long before, and the Japanese were still very much influenced by the American presence in Vietnam. We saw Japanese actors performing cowboy dances and music, which didn't look too different from the kind of things you'd see on American TV. Jim Simons had many interesting comments about them. I met and had nice discussions with a few Japanese mathematicians, including Mrs. Miyaoka (before she married).

I had just finished my work with Bill Meeks on the embedding of the Douglas solution of the Plateau problem. So I talked about that at the conference. But I also announced an interesting conjecture that caught some of my Japanese colleague's interest. I proposed that the first eigenvalue of an embedded closed minimal surface in the unit three sphere must be equal to two. When I told this to E. Calabi in 1979 at the IAS, he was quite excited about the idea. Calabi told my student that this conjecture could offer some good insights into minimal surfaces in a sphere. The conjecture is still not solved. But Choi and Wang took some steps towards a solution ten years later, making a substantial contribution to the theory of minimal surfaces in a sphere.

Commentary on Geometric Bounds on the Low Eigenvalues of a Compact Surface

I gave a presentation with Rick Schoen and Scott Wolpert regarding work we did at Stanford in the 1970s, when both Rick and Scott were graduate students. In 1974, Rick and I were chatting in my office about eigenvalues. Scott wandered into the office with a cup of coffee in his hand, asking what we were doing. So we joked that we could work on any problem on Riemann surfaces, including eigenvalues of the Laplacian for Riemann surfaces. Rick and I then told him how that problem can be related to finding isoperimetric constants for the Poincaré metric. The interesting result is that when the number is greater than $4g-1$, the eigenvalue is always greater than or equal to $1/4$. I'd known that result for quite a while. (It had also been found, independently, by P. Buser.) But I wanted to understand the behavior of eigenvalues lower than $4g-1$. So the three of us looked into that, finding that between $2g-2$ to $4g-1$, the eigenvalues have a positive lower bound depending only on genus. The lower eigenvalues can go to zero when the Riemann surface moves to the boundary of the moduli space of the Riemann surfaces. I thought that was a nice result. But we still did not know how to find the above-mentioned best constant that depends on g only. So I talked about this result at the Hawaii conference, but Scott complained that the writing in this paper was poor, which did not please us. It was, in any case, a result of interest. Yet I wonder what happens in higher dimensional Kähler-Einstein manifolds: Can one replace the length of geodesics by the volume of special Lagrangians? That is a question for future study.

Commentary on Estimates of Eigenvalues of a Compact Riemannian Manifold

Since my graduate student days with S.-Y. Cheng , I have been always interested in estimating the eigenvalues of the Laplacian. When I was at Stanford, I estimated the "Cheeger Constant" for a compact manifold by introducing the concept of visibility of a set with substantial volume. I found an estimate of the first eigenvalue in terms of the lower bound of the Ricci curvature, upper bound of the diameter, and lower bound of the volume. The method was later used by Christ Croke in a skillful way to estimate the Sobolev constant of the manifold. The estimate works for compact manifold with convex boundary. In that case, it amounts to affirming the Poincaré inequality, which says that if we assume the average of a function to be zero, we can dominate the L^2 norm of the function by its energy. In those days, many geometers were not familiar with such basic facts in analysis. I recall that when I gave a talk about my result at Berkeley in 1975, Gromov spent ten minutes complaining that my last statement could not be true until Protter finally explained to him what was going on.

While visiting Berkeley in the fall of 1977, Chern wanted to make sure that I would like Berkeley well enough to accept the job offer they made to me. He asked his student Peter Li, who was from Hong Kong and who had excellent taste, to entertain me. Peter owned a fancy sports car, an Alpha Romeo, when he was a student. Chern asked Peter to drive me around. That turned out to be useful in ways I did not expect. When I had a car accident a year later in Stanford, I needed to buy a new car and I went to negotiate with a car dealership. The salesman did not think much of me until he saw the Alpha Romeo; then his attitude towards me changed drastically.

In 1978, Peter showed me his idea on how to apply my gradient estimate to obtain a good lower estimate of the eigenvalues of the Laplacian for manifolds with nonnegative Ricci curvature. I was impressed by his insight.

He was very excited upon hearing about an American Mathematical Society conference on the spectrum of the Laplacian, which was to be held in Hawaii on March 27–30, 1979—a few days before my 30th birthday. Hawaii was such a beautiful place that most people could not resist the temptation to see it. I'd already been there once, stopping off during a flight from Hong Kong to San Francisco in 1969, and I was eager to go back again. Many geometers attended the 1979 symposium, including Richard Schoen, S.-Y. Cheng, Peter Li, and myself. Peter got a travel agent to line up some rooms and a rental car for all of us. We had a great time driving around Oahu. After the conference, we went to Kauai for sightseeing. It is a beautiful island, and there weren't many tourists then. We enjoyed the tropical scenery very much, while wondering how we might get some coconuts off the tall trees. With sheer power, Rick managed to throw rocks and knock them down, one by one, although we then had to figure out how to open them—a nontrivial task. Our return trip was delayed by a United Airlines strike. All of us stayed in our rental apartment for a few more days. Once, at midnight, some thieves tried to break into the apartment. S.-Y. Cheng said that my snore scared them away.

In any case, one evening, after dinner, I looked carefully at Peter's idea for estimating the first eigenvalue. I found that his argument could be strengthened to prove the more general situation: Assuming a lower bound of the Ricci curvature and the upper bound of the diameter, the gradient estimate can give a lower bound of the first eigenvalue. This was very nice, and combining this idea with the upper bound given by S.-Y. Cheng, it provides a good understanding of the first eigenvalue of the Laplacian for a general manifold. That was a rather unexpected result at that time. We asked two Chinese mathematicians, Zhong and his student Yang, to try to improve our constant based on our method. After proving a conjecture along the lines of our argument, they received a fair amount of recognition in China. I certainly thought that they deserved the recognition. But later the Chinese mathematics community forgot how unexpected the result was before Peter and I did this work. The situation became worse when some Chinese probabilitist claimed to have proved our result without mentioning our prior work and then turned around to attack me.

Similar ideas to ours can be applied to manifolds with different boundary conditions. During my talk at the conference. I mentioned the Pólya conjecture for high eigenvalues of a domain. E. Lieb, among others, was quite interested in this notion. A couple of years later, Peter and I proved the Pólya conjecture in the average. The full conjecture is still not solved, although many people thought that our proof could be extended to prove it by applying the same argument to higher powers of the Laplacian. Peter and I had thought of that approach too but felt the boundary term was too complicated to be handled.

After the conference, I visited MIT, where I stayed at the home of Richard Melrose for a couple of nights. We celebrated his thirtieth birthday in addition to my own. We also knew of two other mathematicians who were born around the same time: Andy Majda and Charles Fefferman.

Commentary on the Total Mass and the Topology of an Asymptotically Flat Spacetime

When I was a graduate student, I audited a course on general relativity. But I did not know the subject well, and did not know how to work on any problem in that area, until I listened to a 1973 talk given by Robert Geroch at a differential geometry conference in Stanford. He was invited by I.M. Singer and S.S. Chern to tell geometers about important problems for them to consider in general relativity. He described a problem of great geometric interest: Can we have a nontrivial metric on three-dimensional linear space, which has nonnegative scalar curvature everywhere and is Euclidean outside a compact set? Everybody was talking about this problem at the Stanford conference and marveled at what we might do with it. It was not apparent at the time that any standard method could handle this problem. Naturally the problem is easily converted to a corollary of a statement that there is no nontrivial metric with nonnegative scalar curvature on the three-dimensional torus. But there is no method to handle this either, although Arthur Fisher and Jerry Marsden used an implicit function theory argument to prove that such a metric with nonnegative scalar curvature cannot appear in a small neighborhood of the standard metric.

I was excited by this simple problem and considered it a challenge for differential geometers to answer this question. I talked with Rick Schoen about the problem often. We made no progress until one day in Berkeley, I invited Rick to my home for dinner, cooked by my mother. While walking on Berkeley's north campus, we suddenly hit upon an idea to handle the second variational formula for stable minimal surfaces in the torus, where we only required the positivity of the scalar curvature of the ambient manifold. Without much effort, we established the conjecture of Geroch that there is no nontrivial metric with non-negative scalar curvature on the three-dimensional torus. Then we pushed further and proved that if a compact three-dimensional manifold admits an incompressible surface with genus greater than zero, it cannot admit a nontrivial metric with nonnegative scalar curvature. After a couple of weeks, we thought we should start to prove the original positive mass conjecture as posed by Geroch in the time-symmetric case. In other words, we wanted to show that for a three-dimensional manifold with nonnegative scalar curvature, the first non-trivial asymptotic coefficient of the metric should have some positivity. This term gave rise to the mass of the three manifold.

We managed to prove the positive mass conjecture in the time-symmetric case using a minimal surface argument and the Gauss-Bonnet theorem for noncompact surfaces. The negativity of the mass provides a barrier to constructing an area-minimizing surface asymptotic to the plane. Once the asymptotic behavior of the surface is understood, we can apply a similar argument for the compact case to get a contradiction. Hence we were able to prove the positivity of the mass for the time-symmetric case.

We were rather pleased by this accomplishment and gave lectures on the theorem we proved at various universities. It drew attention from many physicists. Some of my high school friends, who worked in general relativity, simply could not believe that mathematicians were capable of answering a problem that was of such great interest to physicists. I was told that in the regular international meetings of the general relativists, they had a special section on the topic of the positive mass conjecture each time they met. I was also told that Faddeev, Novikov, and other Russian scholars had also been interested in this problem. The physicists started to become more convinced after Stephen Hawking said that he believed in the validity of our proof.

On the other hand, when Arthur Jaffe asked me to give a talk on the positive mass conjecture at Harvard in 1979, I found several relativists in the audience who were ready to tell me that we had not proved the positive mass conjecture at all. Fortunately, Rick Schoen and I had been working hard. We looked further into the full statement of the positive mass conjecture, considering the non-time-symmetric case as well. We were able to reduce the general statement to the case that we proved in the time-symmetric situation. The reduction turned out to be nontrivial. We made use of an equation first written by a Korean relativist named Jung. Jung did not believe our argument could work because, in general, there is no solution to his equation, which is why he gave up on his approach. However, Rick and I had extensive experience on minimal surface theory, and we knew what could go wrong with the solution. The solution actually exists, although it may blow up along a very interesting surface, which is called the apparent horizon of the three manifold. This kind of singularity did not bother us. Our proof of the positive mass conjecture in the time symmetric case works even when there are apparent horizons. So we were able to move forward to carry out the reduction. The proof demanded a lot of calculations, and it took us the whole summer to finish up the writing. It is still one of the most powerful theorems in classical general relativity. When I told Gromov our general statement of the positive mass conjecture, he did not believe the statement could be valid.

In any case, when S. Deser and L. Smarr learned that we could solve the general case of the positive mass conjecture, they were very friendly to me. I believed the talk that I gave at Harvard was a kind of interview prior to their offering me a professorship. I'm happy to say the "interview" went well. But soon afterward, we found a paper by Jim York called "Why the positive mass conjecture is not proven". What happened was that Jim York did not like the asymptotic condition that we imposed and pointed out that we needed to consider the case when there is a boost on the slice. Rick and I solved this case shortly thereafter. But York's paper had a very nice analysis of the asymptotic behavior of the constrain equations that govern the initial data set for the Einstein equations. It has been quite useful for subsequent work on boosting the slice.

When I visited Hawking in Cambridge in 1979, I learned from him and his postdoctoral fellows about several important problems that are related. One was the positivity of the Bondi mass and the other concerned "positive action", which essentially was the positive mass conjecture for four-dimensional manifolds.

Rick and I solved the positive action conjecture in a short time. But we were not sure about the right definition of the Bondi mass until a year later when I joined the faculty at the IAS in Princeton. I asked Gary Horowitz to work with me as a postdoctoral fellow on general relativity. In 1981, when I was in my office at the IAS, Freeman Dyson knocked on my door saying that he wanted to introduce a young man to me. It was E. Witten, and he started to show me his proof of the positive mass conjecture using Dirac spinors. I was surprised by its elegance. I wrote to Rick, who was visiting Leon Simon in Australia. He managed to reproduce the proof after he read my brief description. Witten's proof, especially the part concerning the proof of the existence of harmonic spinors, needed further justification. This was later carried out by Tom Parker and Cliff Taubes.

Witten's proof followed an approach most physicists can understand, and they were overjoyed that they did not need to understand the theory of minimal surfaces anymore. It immediately inspired a large group of physicists to look into generalization of the positive mass conjecture. The first thing they tried to address was the Bondi mass. Rick and I were rather reluctant to work on the proof of Bondi mass, as we felt that the assumption of the existence of the Bondi coordinate was not clearly valid. In any case, we were under pressure to produce a proof along the lines of our original argument. We succeeded in doing so. Rick was then an assistant professor at the Courant Institute. I decided to work with him on this problem there. We worked hard and did extensive calculations on a Saturday. I forgot that I was invited to a dinner in my honor by a distinguished professor at Rutgers. I suddenly remembered the dinner. But it was already 7 pm and we were still doing the calculation. I was totally embarrassed. But we did prove the positivity of the Bondi mass, and the paper was published in Physical Review Letters. For some reason, very few people quoted our paper. Instead they quoted the approach taken by Gary Horowitz and others using Witten's method. Tom Parker told me that he did not think the proof of Bondi mass based on Witten's approach had been carried out rigorously.

My proof with Rick of the positive mass conjecture had an important consequence. We managed to prove that if the density of matter is large in a certain set, an apparent horizon will form. This justifies popular statements about a black holes forming due to the condensation of matter. I gave a talk about this at Harvard when the university considered hiring me for the second time. We found out much later through Kip Thorne that our argument was related to what he called the "hoop conjecture". But in order to give precise meaning to the term, hoop conjecture, we needed to clarify the concept of quasi-local mass. The definition of quasi-local mass had been an interesting subject–one that I'd heard Roger Penrose talk about on several occasions at Princeton. Both Penrose and Hawking had made serious attempts to define such a quantity. After a long struggle, I have finally found the right definition with Mu-Tao Wang.

I thought about the definition of quasi-local mass for a long time, in relation to the hoop conjecture. After I wrote a paper with E. Witten on the topology of the boundary AdS/CFT geometry in 1990, I had a better idea of how the boundary may influence the geometry of the interior of the manifold. When I lectured on my work with E. Witten at the Chinese University of Hong Kong, L.F. Tam was in the audience. We exchanged some ideas, and I mentioned R. Bartnik's work on quasi-spherical foliation. After a few months, Tam mentioned to me some beautiful result of his on the total mean curvature of the boundary when the boundary is a sphere. Soon he worked with Shi to allow a more general convex boundary. I found the result very interesting. I ran a small seminar on that result at Harvard and concluded that the Shi-Tam result should be interpreted as positivity of certain quasi-local mass. I was informed later that this concept of quasi-local mass was due to Brown-York. But that is only for the time-symmetric case. Melissa Liu and I were able to drop that assumption. But the proof is rather complicated, as we had to treat the boundary terms very carefully. In the last five years, M. T. Wang and later P. N. Chen joined in. We have found a definition that seems to be satisfactory, and that result has given me great satisfaction as well.

Commentary on the Role of Partial Differential Equations in Differential Geometry

This is the first survey paper I ever wrote. I was on the faculty at Stanford at the time. I had just come back from Berkeley after a one-year leave of absence. During my time in Berkeley, I was invited to give a plenary talk at the International Congress of Mathematicians in Helsinki. Shiing-Shen Chern was on the scientific committee, while Armand Borel was the chairman. Apparently they liked my work well enough to recommend that I give a plenary talk.

In the summer of 1978, I decided to return to Stanford after taking a visiting position in Berkeley. I bought a small house in Palo Alto at 2048 Middlefield Avenue. In Berkeley, I was living with my mother in a small apartment. And I need to move all my stuff from Berkeley to Palo Alto.

My friend Shiu-Yuen Cheng and his family were also in Berkeley that year, during his Sloan fellowship. He and his wife went to China while his mother and two children stayed in Berkeley. When I moved, I drove my mother, Cheng's mother, and his two small children with me to my new home at Palo Alto. The next day, on our way back to Berkeley, my small Volkswagen Squareback got out of control on the freeway. A front wheel ran out of the car while we were going about 55 miles per hour. Fortunately, I was able to land the car in a ditch before it went totally out of control. My mother had some injuries, but they were not serious. Cheng's mother made a heroic act to protect the two young kids by embracing them tightly within her two arms. The car was totally ruined, but everybody was fine. The two children were Hsiao-Bing Cheng and Li-Tien Cheng, both of whom are now established mathematicians. Bing was my Ph.D. student. So I did not make any negative contributions to the mathematics community.

Fortunately, I was not thinking too hard on the mathematics related to my talk in Helsinki, and I responded quickly enough to move the car to safety before we had a major crash. In any case, I took off for Europe two days after the accident.

I arrived at IHES and participated in the activities there. I gave quite a few talks on the Calabi conjecture and Kähler-Einstein metrics at the institute. My friend Jean-Pierre Bourguignon was at the Polytechnique, and he took notes for my lectures. In the lectures, I gave talks on applications of Kähler-Einstein metrics to algebraic geometry, and I also gave talks on the second part of my paper on the Calabi conjecture, where I proved the existence and uniqueness of singular Kähler metrics on Kähler manifolds. I have been always interested in singular Kähler metrics because of my previous interest in the Schwarz lemma for holomorphic maps. I also knew that in algebraic geometry, we need to handle manifolds of general type where we can only have Kähler-Einstein metrics that are not strictly positive everywhere. Bourguignon published my lecture (under my name) for the part on quasiprojective manifolds where I constructed complete Kähler metrics with finite volume. (It was published in the French journal, *Asterisque*)

It is not hard to construct singular Kähler-Einstein metrics by taking limits. The problem was in understanding the detailed structures of the singularities, and figuring out how to calculate the contributions of Chern forms due to singularities of the algebraic manifold. I had several discussions with experts in France and in Bonn at that time, including Bernard Tessier, Le Dung Trang, Egbert Brieskorn, and others. But I did not feel that I had a satisfactory answer to these questions. For simple singularities such as orbifolds or conifolds, the problems could be solved, and I gave some interesting applications to prove theorems in algebraic geometry, using singular Kähler-Einstein metrics. Later that year, I suggested some part of the problem of giving deeper meaning of singular Kähler metrics to a student of mine called Richard Klotz for him to work on as his thesis topic. After his Ph.D. thesis on singular Kähler-Einstein metrics was presented at Stanford, he went to work in Wall Street, and I have no idea where he is now.

When I was in Paris, I was invited by the University of Rome to give a lecture. Since I did not have a passport, I regretted that I could not make the trip despite giving a large sum of application fees to the Italian Consulate in Paris to accelerate my visa application. It was too bad that I could not meet my Italian friends at that time.

I spent a month in Bonn, where I met Friedrich Hirzebruch for the second time. (The first time was at Stony Brook in 1972, where he gave lectures on Hilbert modular surfaces. But I had no chance to talk with him at that time.) I was very much impressed by his depth, his originality in mathematics, and by his generosity. When I was a graduate student, I learned about the theory of Chern classes from his book, *Topological Methods in Algebraic Geometry*. I was really pleased to meet him face to face. I was excited to know that he was interested in my result on the Chern number inequality (the so-called Miyaoka-Yau inequality) and the result that the quotient of the ball is characterized by Chern numbers alone among algebraic manifolds. He told me that he would like to nominate me to be a senior Humboldt Professor to visit Bonn for a year. He also told me that John Milnor was nominated by him and did not come to spend his professorship there after it was awarded. He wanted to make sure that I would come. There was no reason that I could not come, so I promised. However, it turned out that when I was offered the position in 1980, my wife was expecting our first baby. So it was simply impossible for me to spend a year in Bonn. I went to Bonn for only one month. I felt very guilty and expressed my deep regret to Hirzebruch many times. Fifteen years ago, he told me that he forgave me. I felt very much relieved. I also met other outstanding mathematicians–Wilhelm Klingenberg, Stefan Hildebrandt, Joel Smoller, Claus Gerhardt, Don Zagier, and others in Bonn.

I went to Helsinki after Bonn. On the long train ride to the airport, I had a lengthy discussion with Tetsuji Shioda about the structure of Chinese characters. He insisted that Chinese characters are absolutely useless. We argued for quite a long time in the train ride. I was glad that when I met him again two years ago, he had changed his mind.

On this journey to Helsinki, I had no idea of what I needed to prepare, I had the impression that one could use the blackboard and lecture like we did in a classroom. But to play it safe, I also prepared a few copies of transparencies, but they were written in a rather informal manner.

When I arrived in Helsinki, I went to an apartment arranged by the Finnish

Mathematics Society for the speakers at the Congress. Nick Katz and Bill Casselman were in my apartment. I had met them at the big AMS-organized algebraic geometry conference in Arcadia in 1974. We had lively conversations on mathematics and were very excited by the great work of Pierre Deligne.

But I started to realize my preparations for my talk were far from adequate, after listening to the first set of lectures by others. The lecture hall was too big for a blackboard talk. I got nervous and decided to cancel my tourist activities, including the concert performed by Mrs. Cheeger and a sightseeing trip to an island.

When I started to prepare my talk, I realized that it would be good to have a picture to go along with my talk. I was grateful to my friend Yum-Tong Siu, who helped me draw a nice picture that I needed. He told me that when he was a graduate student, J. Milnor was writing his famous book on isolated singularities of hypersurfaces. He was the graduate student who provided the pictures for that book. Hence he said it was a trivial thing for him to draw the picture for me. It was done professionally. So finally, and thanks to his help, I had my talk prepared.

Recently I asked B. Casselman about his recollection of the Helsinki Congress. He told me that the memory is deep in his mind. But the most impressive thing for him was the young blond who served us breakfast every day, and she was 6 feet 2 inches tall. For some reason, I have no recollection of that at all, perhaps because I was married not long before and Bill was still single and thus "hungrier". But I remembered that Bill was very nice to loan me his watch for my talk. And right after my first talk, Bill jumped up on the stage. I thought he was excited by my talk and wanted to ask me questions, but he just wanted his watch back.

I gave the talk as described in this paper. It was my first talk before such a large audience. Strangely, I was not disturbed by the environment. And I described the philosophy behind geometric analysis and its development including the Yamabe conjecture (which was not proved at that time), the positive mass conjecture, the Calabi conjecture, and several other important theorems which had just been proved. I outlined how important the role of nonlinear partial differential equations are to geometry, and how we should look at the relationship between them. I discussed how elliptic equations, parabolic equations, and hyperbolic equations are related to different problems in geometry. I also talked about the importance of linear theory to nonlinear theory. When we solve a nonlinear equation, we study its linearized operator. Hence a precise understanding of linear operators is of the utmost importance. The precise estimate and behavior of harmonic functions is also important for studying nonlinear elliptic equation in geometry. I pointed out that the integrability of many nonlinear elliptic equations comes from the solvability of the linearized operators.

I gave a rough classification of methods that were used. Semilinear equations are related to conformal deformations of metrics: the Yamabe problem. Quasilinear elliptic equations related to minimal surface type equations, which were, in turn, related to the positive mass conjecture that R. Schoen and I solved, where we used both conformal deformation and minimal surface theory. Fully nonlinear equations, such as Monge-Ampére equations, are linked to the theory of affine sphere and the Calabi conjectures.

In this talk, I also described the structure theorem for compact Kähler manifolds with nonnegative first Chern class, plus the fact that for Calabi-Yau manifolds, the second Chern class cup product with $n-2$ copies of any Kähler class is nonnegative and is zero if and only if the manifold is covered by the torus. For projective manifolds and a divisor D with normal crossing, the complement of D admits a canonical, complete Kähler-Einstein metric with negative scalar curvature if $K + D > 0$. This statement was due to S.-Y. Cheng and myself. I announced this contribution at the Congress. Many authors repeated the above results a few years later. I also announced the program to construct and classify complete non-compact Kähler manifolds with infinite volume and with zero Ricci curvature. When the divisor at infinity is nonsingular, I proved that the metric exists, according to my proposal. The details of those works were written up first in my paper with S.-Y. Cheng and later in my two papers with Gang Tian.

My Helsinki Congress talk was perhaps the first general talk on the subject of modern geometric analysis delivered to a large audience. A reasonable part of the development of geometric analysis followed this route of thinking. The audience was quite impressed by the program I outlined.

After the Congress, I happened to fly together with Chern and Lipman Bers to London. Bers said that my talk was the second best in the Congress. I guess that was a good compliment. Chern told me that Bers was very skillful in words.

On the other hand, during the Congress, I received two pieces of sad news. H. C. Wang and his wife in Cornell suddenly passed away. They treated me very nicely when I visited them in the spring of 1974, when Cornell was the first department that offered me a tenured job. (Johns Hopkins University offered me a non-tenured associate professor position before the Cornell offer.) I was unmarried at the time; they proposed to help me search for Chinese girlfriends at Cornell. I declined their very nice proposal. The second sad news was the murder of Karel deLeeuw by a deranged graduate student with a hammer at Stanford. I was shocked, especially because deLeeuw was a very nice man and his office was only two doors away from mine at Stanford.

In Helsinki, Frank Adams made a point of telling me that I should go to meet Stephen Hawking, as he had many mathematical questions that F. Adams could not answer. But actually S. Hawking had already invited me to visit him and the purpose of my trip to London was to meet him.

I flew to meet Hawking and spent the whole day talking with him and his students. He could still speak in those days, but I could understand him only through the interpretation help provided by his students. Gary Gibbons, Malcolm Perry, Alan Lapedes, and Don Page were there. It was a very lively discussion. I told him what I knew about the positive mass conjecture and they were very excited about it as R. Schoen and I managed to prove the full conjecture, which was quite a bit beyond the original announcement. They told me they were working on Euclidean gravity and needed the higher dimensional version of the positive mass conjecture. They called it positive action conjecture. And they also told me that it was important to prove the positivity of Bondi mass.

Richard Schoen and I solved the positive action conjecture within a year. Two years later, we proved the Bondi mass conjecture, employing a mixture of techniques from minimal hypersurface and conformal geometry. A key observation was that a stable minimal hypersurface in a manifold with positive scalar curvature can be conformally deformed into one with positive scalar curvature.

Based on our understanding of conformal geometry, R. Schoen and I gave a good construction of surgery on manifolds with positive scalar curvature whose codimension is not less than three. This work appeared in 1979 in the journal, *Manuscripta Mathematica.* In 1980, Mikhail Gromov and Blaine Lawson subsequently arrived at the same result using direct geometric construction. I had told B. Lawson when I met him in IHES in 1978 about my surgery result with Schoen. Apparently some mistakes were found later in their formula. I am not sure what the status of that calculation is at this stage. In any case, they observed that the surgery result immediately reduced the problem of classification of manifolds with positive scalar curvature to problems in spin cobordism. For a simply connected manifold with dimension greater than four, it shows that the existence of metrics with positive scalar curvature depends only on the KO-characteristic numbers. This was carried out successfully by Stephan Stolz later.

R. Schoen and I concluded that manifolds with non-positive curvature cannot admit metrics with positive scalar curvature. At that point, because of a problem of regularity, we could only carry the argument up to dimension eight. Then M. Gromov and B. Lawson took up the argument of Andre Lichnerowicz, using the Dirac operator to generalize it to higher dimensions. I was told that they were so sure that their method was superior to ours that at some seminar talks, they claimed that our methods could not produce accurate results by the example of the connected sum of $\mathbb{C}P^4$ with a torus.

Lichnerowicz's argument of using the Dirac operator on manifolds with positive scalar curvature was rather well known to people who were interested in such manifolds. I learned a great deal about it from Nigel Hitchin in 1971, when I was a visiting member of IAS. In those days, IAS had informal dinners for visiting members. And most unmarried members gathered there, taking dinner frequently. We had many lively discussions on many subjects, including mathematics.

N. Hitchin, Takuro Shintani and many topologists, were around. N. Hitchin was an assistant to Michael Atiyah, who was interested in the topological significance of Dirac spinors. N. Hitchin observed that the vanishing theorem of A. Lichnerowicz had consequences beyond the vanishing of $\hat{A}$ genus. He found that based on the new mod 2 Atiyah-Singer index formula, all the KO-characteristic numbers of manifolds with positive scalar curvature vanish. In particular, he observed that exotic spheres that do not bound spin manifolds cannot admit metrics with positive scalar curvature. I was very impressed by this work.

At the same time, Deane Montgomery, A. Borel, Chung-Tao Yang, and some topologists such as Ted Petrie, Glen Bredon and Wu-Chung Hsiang were interested in the theory of transformations groups on spheres. One day, W.-C. Hsiang declared that he could construct a specific 10-dimensional exotic sphere that admits no circle group action. And he claimed that it was really rare that any homotopic sphere could admit no continuous group of symmetries whatsoever. He then declared to the young mathematicians that one just needed to write a paper of such caliber once a year, and afterwards it would be vacation time. However, within a short time, a paper appeared in the *Proceedings of the AMS* showing that the sphere W.-C. Hsiang constructed actually admits a circle action. Hsiang's construction was based on an abstract method. He had a hard time finding any mistake in his construction. W.-C. Hsiang thought it was beyond his dignity to ask a junior mathematician about his mistake. Instead of asking the authors of the paper about their proof, however, Hsiang wanted me to ask the author who was going to give a talk on the paper at Stony Brook. I ignored his request.

Inspired by that incident, nevertheless, I decided that I could use geometry to study group action. In a short time, I completed two projects along these lines. The first was on the homology of a manifold that admits circle actions. Dennis Sullivan was talking about his theory on de Rham cohomology at IAS. He stressed that de Rham cohomology on the form level carries more information. I thought that one should follow his philosophy and apply it to circle action. I found out that the wedge product of equivariant de Rham forms could give obstruction to manifolds that admit effective circle action. For example, if the circle action has some fixed points, the wedge product of n equivariant closed 1-forms must be trivial as a form. Here n is the dimension of the manifold.

In 1973, I drove across country from Stony Brook to Stanford, where I was supposed to visit for a year. On my way, I passed by Berkeley and spent a day visiting my friend S. Y. Cheng. I told Cheng about my work on partial differential equations related to geometry. Cheng's office was in Evans Hall. I happened to pass through the office of Wu-Yi Hsiang. He asked me to come into his office. Since his work was on group action, I told him about the above observation, and I asked whether he knew a proof based on topology alone. He said that this was a triviality and started to describe his approach on the blackboard. After one hour, he could not produce any proof. Then he said that he needed to go to the restroom. I waited in his office for another half hour. He never showed up, so I left and talked with Cheng again.

Later, I gave a talk in Princeton. Bill Browder liked my result and gave a different proof based on a purely topological argument. I told B. Lawson my simple idea on equivariant forms. He wanted to develop it further. Since that project never came to fruition, I published my work relatively late; it appeared in the journal *Topology* in 1976.

In 1974, I was working with R. Schoen on the uniqueness of harmonic maps. And we realized that the above theorem on group actions could be generalized as the corollaries of uniqueness theorems of harmonic maps mapping into manifolds with nonpositive curvature. In 1975, Chern asked me to give a talk on the result in his geometry seminar. The seminar was packed with about 50 people sitting in the room. When I gave my talk on this description of group action on manifolds using harmonic map, W.-Y. Hsiang stood up, declaring that the subject of group actions belonged to topologists, and there was no need for geometers to interfere. He stormed out of the classroom, packed with people. Instead of being angered by his outburst, I found it amusing, although I was quite young at that time.

My second project on group action was deeper. I wanted to study Lie group actions using the curvature of the manifold, and naturally $SU(2)$ came in. The orbits of the group give rise to a metric on the total space by shrinking the metric along the fibers suitably. If the action has no fixed point, it gives rise to a metric with positive scalar curvature immediately. By using N. Hitchin's observation, I immediately showed that exotic spheres, which do not bound spin manifolds, admit no effective action of $SU(2)$ without a fixed point.

I wanted to remove the condition on the fixed points. I happened to meet B. Lawson in Stony Brook, and I told him about this idea. He was excited by it. We worked on the problem and solved it in a short time. Therefore we proved that exotic spheres, which do not bound spin manifolds, have no effective $SU(2)$ actions.

This is a rather interesting general theorem, which can be applied to any spin manifolds where there is a non-zero KO-characteristic number. I gave a talk at the 1973 Stanford conference about this theorem. My friend S.-Y. Cheng said that it really is a beautiful theorem. Hans Samelson at Stanford liked the theorem too. I believe he liked it enough to propose to Stanford to hire me as a tenured faculty. It is amusing that those topologists who work on this problem try their best not to mention this approach. Incidentally, the theorem shows that the exotic ten sphere that Hsiang was interested in has no continuous non-abelian group action. It would be nice to show that any effective torus action on such an exotic sphere must be of one dimension.

In 1983, I was invited by Chern to run a special year at MSRI. Both R. Schoen and I were there. We gave many new results using minimal surfaces arguments, and we worked quite hard on those results. Sometimes our discussions lasted until 10 pm. Michael Anderson took good notes. There were many points that M. Gromov wanted to clarify from Anderson's notes, which R. Schoen explained to him in detail. Such theorems appeared in Gromov's paper with B. Lawson in a publication of IHES later, without reference to our work. Rick wrote a letter to Gromov to protest; the letter was returned by the post office.

Commentary on the Real Monge-Ampère Equation and Affine Flat Structures

In the summer of 1980, Chern called a big international conference, which was held at the Friendship Hotel of Beijing. Many distinguished mathematicians participated in this event, including Michael Atiyah, Enrico Bombieri, Lars Gårding, Raoul Bott, and others. Most of them came with their wives. S.-Y. Cheng and I also went with our wives. China was so new to all of us! I had just moved from Stanford to join the faculty at the Institute for Advanced Studies in Princeton. My wife was working in San Diego; I visited her every summer. This year she was pregnant with our first baby. There was a big debate in our family as to whether she would come with me to China. The decision was made only about five hours before the trip. A student of mine helped drive us from San Diego to the LA airport. We managed to catch the airplane right before the door was closed. S.-Y. Cheng and his wife were on the same airplane, waiting for us nervously. This was the second time I had visited Beijing and the first time for my wife. It was exciting for both of us: she was able to meet her older sister after twenty years of separation while I was able to watch China gradually recovering from the big damage caused by the "cultural revolution"! The country was still in poverty in most places.

Chern was a great organizer. The guests for the conference really enjoyed the sightseeing trips that he organized. We also helped Enrico Bombieri find places to paint landscape. Bott made many interesting remarks about China. He liked the Chinese name that I made up for him. The Friendship Hotel was owned by the Chinese Academy of Sciences and used to be a residential place for Russian advisors in the old days. My wife and I lived in a big room of the hotel. One day, after lunch, I was approached by a young Chinese mathematician who demanded that I recommend him for a big award given by the government. He asked in such a forceful manner that I got upset and almost fainted. Fortunately, after that incident, the elder Chinese mathematicians made sure that I was left alone. After the conference, my wife and I went to Hangzhou and took a rest near the beautiful West Lake.

During the conference, I lectured on the open problems that I had posed a year earlier at the IAS. These problems attracted much interest in the Chinese mathematics community. I believe that I initiated research in geometric analysis in China. I also talked about my work with Cheng on solving the real Monge-Ampère equation and the concept of affine Kähler geometry. The concept turned out to be important in the later development of the SYZ program on mirror symmetry. But this paper was not popular. Perhaps most people did not recognize papers that appeared in Chinese journals in those days. I reprint it here so that people can access it more easily.

Commentary on Survey on Partial Differential Equations in Differential Geometry

This book "seminar on differential geometry" was a record of what we did during the special year on geometry, 1979–1980, which was organized by me at the Institute for Advanced Study in Princeton.

In April 1978, when I was visiting Berkeley, I got a phone call from A. Borel. At that time, I did not really know him well. I met Borel when I was a temporary member of IAS in 1971. He was a senior faculty and was a towering figure with strict face. So when I received the call, I was somewhat nervous. The first question he asked me was what my salary was. When I answered, he paused for a minute. It turned out that he wanted me to organize a special year on geometry in IAS 1979–1980, and he was willing to help me on several fronts. It was an offer I could not refuse, although the salary he offered was substantially lower than my current salary at that time. From the conversation, I had the impression that I should accept the offer on the spot, but I did not do so over the phone. I first reported to Chern about my conversation with A. Borel. I told him that the salary was too low, and I had no agreement from my wife yet. Chern told me that I had better accept the offer. So I reluctantly accepted the offer. My wife and Stanford University were very generous to let me go, even though it meant I would have to commute between Princeton and San Diego, and that I would have to take a much lower salary.

Prior to the phone call I got from Borel, the only other time I had interacted with him before was in a letter he had sent inviting me to submit my paper on the solution of the Calabi conjecture. Unfortunately, by the time I received his letter, my paper had already been submitted to *Communications in Pure and Applied Mathematics*. So I had to turn down his offer, doing so as politely as I could. To my surprise, he was not unhappy and invited me to submit a future paper on the subject to the *Annals* later.

A. Borel was a great organizer, and he managed to get all the visitors that I suggested, plus some others that he wanted to come. Richard Schoen, Karen Uhlenbeck, Eugenio Calabi, and Leon Simon were the core people that we invited. So were Jean-Pierre Aubin, J.-P. Bourguignon, Peter Li, Robert Bryant, Doris Fischer-Colbrie, and several graduate students of mine including Andrejs Treibergs and others. We also had some shorter-term visitors such as Louis Nirenberg, Y. T. Siu, Enrico Giusti, C. Gerhardt, Jerry Kazdan, Christopher Croke, C. T. Yang, Jeff Cheeger, Isaac Chavel, J. Dodziuk, Robert Finn, Robert Gulliver, Frank Warner, S. Hildebrandt, Malcolm Perry, and Roger Penrose. Of course, there were many postdoctoral fellows in the School of Natural Sciences at IAS. And there were also several young geometers who came from Princeton University. They included S.-Y. Cheng, Chuu-Lian Terng, and Alan Lapedes.

Most of us stayed in the apartments provided by the Institute. (A student of mine shared an apartment with me, since my wife was in California.) Several of us decided to cook dinners together. My poor skill at cooking finally put a stop to that idea. We ended up eating together in restaurants frequently. During the year, we also played ping-pong constantly. I was not a good player and accepted the fact. Enrico Bombieri was much better player than me but did not accept the fact that he was not better than some other players. We also met each Saturday morning to play volleyball. That was fun, and we always had a good time together. We also had parties that were not so formal compared with the other Institute parties. Many young mathematicians were there. Some of the parties were dancing parties, held in my apartment when I was away on travel. I was told by my students that those parties were the most interesting ones. E. Calabi and his wife Juliana, L. Simon and his wife Sandra, Schoen and his wife Doris, Robert Bryant, K. Uhlenbeck, Bob Williams, and many others all had good times, drinking and dancing.

In the spring of 1980, E. Calabi and his wife Juliana invited a few of us to dinner at his home in Philadelphia. The Italian food prepared by Juliana was very nice indeed. We went by several cars. Jean-Pierre Bourguignon told me that he felt very strange that during his long ride with Enrico Bombieri, he was not able to draw Enrico's attention enough to hold any conversation. I told him that I had difficulty understanding Enrico in general, although we went fishing in the pond at IAS. During Enrico's visit to Beijing in 1980, I managed to please him, and he made a few landscape paintings there.

I decided to have three seminars per week: one on minimal surfaces, one on analysis on manifolds, and other one on general topics.

The atmosphere at IAS for this special year was truly inspiring. It was a very fruitful year, as we managed to communicate ideas freely. We digested the ideas that we had and mapped out what we still had to do in the future. We accomplished much research during this special year.

In October of 1979, A. Borel suddenly came to my office, telling me that he understood that Harvard was offering me a position, but I should wait as IAS was thinking of making an offer to me. He communicated this to me in a very serious manner, and I was rather amused by it. In any case, IAS did make me an offer in February of 1980. I turned down the offer from Harvard and kept on thinking about the offer from IAS. In June of 1980, I decided to leave Stanford and accept the offer from IAS. I should say that I left Stanford reluctantly. I am most grateful to the Stanford department, which treated me exceptionally well. Robert Osserman, David Gilbarg, Hans Samuelson, and Ralph Phillips had gone out of their way to help me. They offered me a job with a high salary and no teaching responsibilities. All of them have passed away by now, but their friendship is still memorable to me. As for Harvard's offer, I was much tempted, as two great mathematicians, Heisuke Hironaka and David Mumford, were very warm to me when I came for an interview. The dean had a very nice conversation with me, as well. And they are among the most intelligent people that I know of.

While considering the offers from Harvard, Stanford and IAS, I traveled a lot during this period. My elder brother in Hong Kong was suddenly very ill. I needed to travel back to Hong Kong for a couple of weeks to take care of the situation.

When I decided to accept the job from IAS in 1980, my elder brother's disease became critical. It took me a big effort to move him to America for treatment. Harry Woolf, the director of IAS, was provost of the Johns Hopkins Hospital before he joined IAS. That connection was extremely helpful to me. My brother had a brain tumor deep inside a central, vital area of his brain. The Hong Kong surgeons had no means to treat his tumor. The chief surgeon at Johns Hopkins Hospital, Dr. Long, was a very kind person, and he was willing to treat my brother. But before that, I need to make a special effort to help get my brother into the U.S.

The visa problem delayed the transportation for nine months. The problem came from the HK medical doctor who was treating my brother, saying that he was capable of treating the case. As a result, the America Consulate in HK refused to grant the visa in the beginning. But the visa problem was finally worked out thanks to the great help of Isadore Singer, who was able to gain approval from the U.S. State Department.

I rented an apartment for my mother and my brother in Baltimore. I got tremendous help from my friend, Bun Wong, who was visiting Johns Hopkins University at the time. He asked a graduate student, B. Y. Lin, to help too. Their assistance was invaluable since my mother did not know English. It was a tough time. I commuted between Princeton and Baltimore often in order to take care of my brother.

During this period, A. Borel told me that I should edit and publish the two seminars for the special year in IAS through the book series, Annals of Mathematics Studies.

The two volumes were edited at the same time. I needed to write a survey for the volume. This survey was written while I was visiting Johns Hopkins Hospital seeing my brother and my mother. It was mostly written in the waiting room of the hospital. Both books were finished by the end of 1980. The first book was printed in 1982. It is called *Seminar on Differential Geometry.*

The second book was on minimal surfaces, containing papers by all the speakers who gave talks in the minimal surface seminar. One day, E. Bombieri came to my office and offered to be the sole editor of the book. I gave him the whole box of the manuscripts in their final edited forms. Unfortunately, it took much longer time for that book to come out, and I received many complaints from the authors who had contributed articles. A. Borel was able to step in and help. The book finally went to press. The title of the collection is *Seminar on Minimal Submanifolds.* However, I never received a free copy of the book from Enrico.

Let me now turn to mathematics. In this special year, we were able to direct many young students and postdoctoral fellows to go into fruitful activities in geometric analysis:

Harmonic functions, eigenvalues, eigenfunctions, and heat kernel estimates

S. Y. Cheng, P. Li, R. Schoen, and L. Simon were major contributors. These topics are related to Sobolev inequalities and the isoperimetric inequalities. L. Simon and James Michael had developed a beautiful method to derive Sobolev and Poincaré inequalities on submanifolds with bounded mean curvature. The method and the result had been used extensively by all workers in the field. I also went through my work on proving isoperimetric inequalities in Riemannian manifolds by introducing some angular measure. The constant depends only on the diameter, lower bound of Ricci curvature, and lower bound of volume. This gives the Poincar inequality. The argument was extended by C. Croke to prove the Sobolev inequality on manifolds. An important corollary was that for non-compact manifolds with nonnegative Ricci curvature, Sobolev and Poincaré inequalities hold as long as the volume grows like Euclidean space. P. Li, S.Y. Cheng, and I studied the paper of John Nash on parabolic equations. It is an especially beautiful work, and we started to apply it to our estimate the heat kernel of a manifold. At the end of the special year in 1979, A. Borel told me that he was very much interested in the estimate of heat kernel for some special, locally symmetric space. He asked me how to do that. P. Li, S. Y. Cheng, and I showed him how it works for general manifolds. He was quite impressed by the result. This work appeared later in the American Journal of Mathematics. In estimating the heat kernel, we needed to make a good lower estimate of the injectivity radius, which was of independent interest for later works in Riemannian geometry. This was misquoted by many researchers, who attributed the credit to the later work of Cheeger-Gromov-Taylor, who used the fancier approach of a wave equation. I find it amusing when mathematicians prefer a certain type of proof over another, even though they both give the same statement. In relation to the estimate of injectivity radius, I gave a talk in Berkeley in the 1970s on my result that complete noncompact manifolds with nonnegative Ricci curvature have infinite volume. To achieve the result, I applied geometric analysis to the Bussmann function. M. Gromov did not believe my argument. So I spent the whole day explaining the argument to him in a more geometric fashion. The geometric argument that I offered could be used to provide a way to estimate the injectivity radius for such manifolds in terms of volume of balls when the curvature is bounded. Some people attributed this estimate to Cheeger-Gromov-Taylor, which appeared a few years later. S. Y. Cheng, P. Li, and I also studied heat kernels for minimal submanifolds in spaceform. This is especially interesting for minimal submanifolds in the sphere since it implies trivially, via the Hopf fibration, an upper estimate for heat kernels for complex subvarieties in complex projective space. The estimate for complex subvarieties was claimed by G. Tian ten years later, based on the same method.

Uniformizations of higher dimensional Kähler manifolds, holomorphic sections of line bundles and compactifacations of complete Kähler manifolds

I described several important questions in Kähler geometry during the special year, which set the direction of complex geometry for the next ten years or longer.

I was very much interested in generalizing the classical uniformization theorems to higher dimension. I divided the problems into several categories:

(a) Elliptic case: Y. T. Siu and I proved the Frankel conjecture where holomorphic bisectional curvature is positive. (The counterpart in algebraic geometry, called the Hartshorne conjecture, was solved by Shigefumi Mori.) Various generalizations were proposed. The method of Siu-Yau is based on producing holomorphic curves of low degree by proving the holomorphicity of stable minimal surfaces produced by Sacks-Uhlenbeck. Once the curve is produced, global rigidity of complex projective space followed from classical methods in algebraic geometry. This was the first paper, after the work of Sacks-Uhlenbeck, to study the details of the bubbling of minimal spheres, which was important later for studying of the Gromov-Witten theory. Our argument to prove holomorphicity of stable minimal spheres and our estimate of the index of the minimal spheres was used about five years later by Mario Micallef and Douglas Moore for manifolds with isotropic positive curvature. The later construction of Gromov to use pseudo holomorphic spheres to prove the uniqueness of symplectic structure on the complex projective plane followed the same strategy of the proof of the Frenkel Conjecture. This method should also be credited to the much earlier paper of Hirzebruch-Kodaira, where they proved the rigidity of odd dimensional complex projective spaces.

(b) Parabolic case: I conjectured in 1970 that a non-compact complete complex manifold with positive bisectional curvature is biholomorphic to complex Euclidean space. The motivation for me to pose this conjecture came from the following theorem that I proved using Schwarz lemma, which I developed for non-compact manifolds: On a complete Kähler manifold with non-negative Ricci curvature, there is no non-constant bounded holomorphic function.

I should say that the proof of the Schwarz lemma had been a useful tool for me to work in the field of nonlinear partial differential equations. The proof compares two Kähler metrics by taking the trace of one Kähler metric against the other one, by taking their Laplacian, and by then studying their behavior. This was used in my proof of the second order estimate for the Calabi conjecture in 1975. The proof of the second order estimate was also done independently by J.-P. Aubin in 1976.

Once I proved that there is no nontrivial bounded holomorphic functions on complete Kähler manifolds with nonnegative Ricci curvature, I tried to prove the real analogue of this statement that there is no nontrivial bounded harmonic functions on complete manifolds with nonnegative Ricci curvature. I interpreted bounded harmonic functions as maps into a bounded interval with a Poincaré type metric on the interval. But the argument turned out to be more challenging than I originally thought. I needed to use a trick to squeeze a quadratic term from the equation itself. I carried out this estimate after I had just arrived at Stanford. I still remembered that I tried many different test functions in order to apply the maximal principle. I often missed the beautiful sunset on the Stanford campus and stayed in the office to test those functions, taking late dinners.

When I finally worked out the estimate, I showed it to L. Nirenberg during the

1973 Stanford conference, I was overjoyed to find out that he did not know the estimate. This estimate was developed further and further by me with S. Y. Cheng, especially on the study of the real Monge-Ampére equation for the Minkowski problem, the Bernstein-type problem for affine spheres and the maximal slice in Minkowski space-time. The extension was nontrivial. Some of it was related to the beautiful work of Aleksei Pogorelov on the second order estimate of the real Monge-Ampre equations. (Unfortunately, such estimates are not attributed to the right sources nowadays.) It gave me a good grasp on the fully nonlinear equations, which helped me see how to use the unknown function to construct test functions. This idea was also used by me in the second order estimate for the complex Monge-Ampére equations.

The gradient estimate that I developed had been very important and useful for later developments in geometric analysis. Peter Li found that the gradient estimate can be used to provide an estimate for eigenvalues of the Laplacian. And along with S.Y. Cheng's result, we were able to prove that the first eigenvalue can be controlled completely by the lower bound of Ricci curvature and the upper bound of diameter. I found the result satisfactory.

The gradient estimate that I developed was then generalized by P. Li and me in 1982 to estimate positive solutions of the heat equations. Later in San Diego, when Richard Hamilton was very excited by his Ricci flow, I informed him that a suitable generalization of the Li-Yau inequality would provide a key estimate for his equation. To my surprise, Hamilton succeeded in pushing that estimate much further. It has become the essential tool for treating the Ricci flow. And it provided the key estimate for the treatment of the Poincaré conjecture. The estimate that P. Li and I got was very precise. From the estimate, one can derive the precise behavior of the heat kernel. It can be expressed in terms of the Hamiltonian of the dynamics associated with the Schrödinger operator in spacetime. I believe this was the first time that the WKB approximation method could be turned into a concrete estimate with control. When the potential is replaced by scalar curvature and the operator by the Lapacian of the metric in the Ricci flow, the Li-Yau inequality gives rise to what Grigori Perelman called entropy. The distance function defined by Li-Yau theory was also utilized by G. Perelman extensively in his paper.

Back to my conjecture on the generalization of parabolic uniformization, Y.T. Siu and I worked toward this conjecture by imposing some conditions regarding the decay of curvature. Our work was not too successful as far as the result went. But the method had been useful, and the idea of using singular weight function to produce holomorphic functions or holomorphic sections was important. (Later I taught G. Tian to use such methods to produce holomorphic sections of line bundles and produce peak sections for his thesis concerning the embedding of manifolds.) On the other hand, the hypothesis of my result with Siu was too restrictive, as I discovered during a lecture I gave in the special year. We need to relax the conditions. We found and proposed a gap phenomenon, which I told M. Gromov about. This phenomenon was later attributed to him.

(c) Hyperbolic case: For compact Kähler manifolds with negative sectional curvature, we expect that its universal cover is a bounded domain, and, in particular, it should admit a great deal of bounded holomorphic functions. Unfortunately, up to now, there are still no means of producing such bounded holomorphic functions.

During this special year in IAS, I had many conversations with E. Calabi. I asked him about his work on rigidity theorems that he did on the isometric deformation of hyperbolic space form. After I told him about my work on the rigidity of complex structures on the ball quotient, he was very pleased to tell me about his interpretation of the vanishing theorem of Matsushima for locally symmetric space of higher rank. Calabi said that the key point came from some quadratic terms of curvature that arose in the Bochner formula. I had learned about this vanishing theorem of Matsushima in 1973, when I was in Stony Brook. I talked with H. Garland quite a bit. He managed to find a p-adic analogue of the vanishing theorem.

Commentary on Problem Section

In March 1980, I gave three lectures on open problems in geometry that I considered to be important. My talks aroused much interest in the differential geometry community, and I was rather excited by this group of problems myself.

I still remember that I got the flu during the second lecture. But I insisted on finishing all three lectures, wearing a heavy jacket while giving them. Many people in the audience contributed their opinions. There were very lively conversations indeed. The whole set of problems should be considered a joint effort, although in the final version printed later, I added many more problems to the original set.

In order to encourage Chinese geometers to work on more modern questions, I gave extensive talks on these problems at the 1980 Beijing DD Symposium on differential geometry and differential equations. The conference was organized by Chern in the summer of 1980 and had a rather deep influence on the development of geometry in China.

Many geometers worked on those problems. There were 120 problems in all, and I am pleased that about a quarter of the problems were solved. In 1986, I was asked by Princeton University Press to update this set of problems. My former student Bennett Chow and I spent a month working on this project. But when the manuscript was sent to the Press, it was vetoed by Wu-Chung Hsiang who was then the mathematics editor.

In this commentary, I made some comments myself and also included comments from many friends and colleagues.

Commentary on a Survey on Kähler-Einstein Metric

I settled down at IAS, Princeton in 1980. I brought a few students from Stanford who had not finished their Ph.D. At the same time, I was also taking some new graduate students from Princeton University. Robert Bartnik came first and he helped me edit this survey paper. Then Mark Stern, Leslie Saper, Huai-Dong Cao, B. Chow, and Shigetoshi Bando came. Alex Freire and Wen-Xiang Wang were my last students at Princeton. Jürgen Jost came from Germany to work with me after he got his Ph.D. with S. Hildebrandt in Bonn.

I ran my student seminar at IAS, which made some of my colleagues uncomfortable. They thought that any seminar held there had to be an advanced seminar on new research topics. But my student seminar was composed of new students who did not know much and were asked by me to learn from some of the older literature.

The students also made a lot of noise. I considered it to be a lively atmosphere, while the older gentlemen at IAS felt differently. One student was extremely interested in foosball and tried to buy a table from a shop that was far from Princeton. He took advantage of the free long distance phone time, and I had to stop him. I had a big office on the top floor of Fuld Hall at IAS, and all my students were also on the same floor. So the noise was mainly restricted to that floor. Besides the student seminars, I also ran a member seminar that was more advanced.

During my first year after arriving at IAS, I gave a course on Kähler-Einstein metrics, describing the proof of the existence of Kähler-Einstein metrics in different forms, especially those with singularities. It was simple to extend my existence result for Kähler-Einstein metrics with negative curvature to algebraic manifolds of general type. But one can compute the residue of the Chern numbers only for a special class of manifolds with simple singularities. This holds for orbifolds too. Many mathematicians, including Daniel Burns and Junjiro Noguchi were active in the lectures. J. Noguchi also brought back the lectures to his colleagues in Japan. As a result, quite a few Japanese geometers, such as R. S. Kobayashi, started to contribute to the subject.

In 1982, there was a conference organized by Y. T. Siu in Wisconsin. This survey is based on the talk I gave at that conference. In this survey, I mentioned some of the results discussed in my 1978 Congress talk, which touched on the structure theorem of Calabi-Yau manifolds and the positivity of the second Chern class taken cup product with products of the Kähler form etc. I also discussed the singular cone-like Kähler metrics, those Kähler metrics which degenerate along a divisor, or blow up along a divisor. They gave some new results to algebraic geometry, replacing Chern classes by log Chern classes in the original Chern number inequalities.

A more detailed account of the results was later published with S.Y. Cheng in Contemporary Mathematics, Vol. 49, 1986 by the American Mathematical Society. Unfortunately, some part of that paper was not printed, even though it was submitted to the editor.

After my work on the Calabi conjecture, ideas for constructing singular metrics were further developed by S.Y. Cheng and me in 1978, which appeared in "On the existence of a complete Kähler metric on non-compact complex manifolds and the regularity of Fefferman's equation", Comm. Pure Appl. Math. 33 (1980) 507-544. The ideas behind my two papers with Cheng were reproduced again by G. Tian and me in a paper that appeared in the book, *Mathematical Aspects of String Theory* in 1987.

I also announced the result of K. Uhlenbeck and me on the existence of a solution to the Hermitian Yang-Mills equation on stable holomorphic bundles.

In 1981, Ngaiming Mok graduated from Stanford and arrived at Princeton University, while Jia-Qing Zhong and Hong-Cang Yang came from China, We had lively discussions on several projects that I was thinking about in complex geometry. I introduced them to several important techniques in geometric analysis, especially those regarding estimates for differential equations. Upon my suggestion, they generalized my result with Y. T. Siu on compactifications of complete Kähler manifolds with strongly negative curvature.

In the same year, R. Hamilton succeeded in reaching the first important accomplishment of his work on Ricci flow, proving that the normalized Ricci flow for three-dimensional manifolds with positive Ricci curvature has long time existence and ends up at a manifold with constant positive curvature. I invited him from Cornell to give a series of talks at IAS. It was exciting, and I got my three students, S. Bando, H. D. Cao, and B. Chow, to work on questions related to the Ricci flow. S. Bando observed in his thesis that the flow preserves the positivity of the bisectional curvature for Kähler metrics when the complex dimension is three. I told N. Mok about this result, and he immediately generalized it to higher dimensions by pushing S. Bando's argument further. Following the argument of Bando on using Hamilton's maximal principle, Mok also derived the borderline case, allowing bisectional curvature to be nonnegative instead of strictly positive. It was observed that the Ricci flow will deform such Kähler metric to have strictly positive bisectional curvature unless the metric is Hermitian symmetric. The Bando-Mok-Hamilton result therefore reduces the classification of compact Kähler manifolds with non-negative bisectional curvature to the Frankel conjecture that Siu-Yau proved.

I suggested that H. D. Cao try to reprove the Frankel conjecture by using the Kähler Ricci flow. He could not prove that but managed to prove long time existence. Even though there have been recent claims made by G. Tian and others, I consider this problem unsolved.

B. Chow worked on flows on surfaces driven by roots of curvature.

Peter Li and I also suggested to J. Q. Zhong and H. C. Yang to look at our paper on the estimate of the first eigenvalue of the Laplacian for manifolds with non-negative Ricci curvature. They were able to improve our method to obtain the best constant in the estimate. And they became instantly famous in China, where most people forgot to attribute the original idea to the prior efforts of Li-Yau.

Commentary on Compact Three-Dimensional Kähler Manifolds with Zero Ricci Curvature

I gave this survey during a conference on string theory held at the Argonne National Laboratory in 1984—the same year during which the first so-called "string theory revolution" took place. Physicists working in this area called a major conference to discuss the findings associated with that revolution. The term "Calabi-Yau manifolds" was first coined by four physicists, P. Candelas, G. Horowitz, A. Strominger and E. Witten, as the major model for the string vacuum. They were anxious to know how to construct such manifolds. I was asked by the organizers to give a survey on the theory of Calabi-Yau manifolds in general.

I believe that, besides the toric constructions that I explored with S. Roan a couple of years later, most Calabi-Yau manifolds were constructed from the methods outlined in my 1984 essay. (Most of those methods were well known, although some of them had not been spelled out explicitly in the literature.) I said during the talk that there were at least ten thousand Calabi-Yau manifolds of different topological type. I got this lower estimate out of the blue, answering a question from the audience, without using the computer.

Right before the conference, A. Strominger told me of the importance of constructing Calabi-Yau manifolds with Euler number equal to $+6$ or -6. This number arose because we knew that there were exactly three families of fermions known in nature. And the number of families was known to be half of the absolute value of the Euler number of the Calabi-Yau manifolds. On my fight from San Diego to Chicago, I found out there was a degree $(1,1)$ hypersurface in the product of cubic surfaces and this manifold has Euler number -18. I observed that there is a group of order three acting freely on the manifold. Therefore the quotient manifold has an Euler number of -6.

When I came back to San Diego from the conference, I showed the example to my students J. Li and G. Tian. Tian made a suggestion that my idea can be used to construct a few more examples of Calabi-Yau manifolds with Euler number equal to -6. Later, B. Greene found that all these manifolds are deformations of the original manifold I constructed during my flight to Chicago. Only one manifold of this type was known for a couple of decades until P. Candelas found another one with Euler number equal to -6 a few years ago.

The manifold I found can be considered as a complete intersection of three hypersurfaces in the product of two 3-dimensional complex projective spaces. P. Candelas and his group in UT Austin then generalized the construction using a computer. They looked at various complete intersections within products of many copies of complex projective spaces. Thousands of examples of Calabi-Yau manifolds were found in this way. Interestingly, when they plotted the Hodge number of these examples, they found that there was a primitive pattern of "mirror symmetry" in the class of these manifolds. There were equal numbers of Calabi-Yau manifolds with an Euler number of opposite sign.

That encouraged the researchers to look further into the concept of mirror symmetry, which was discovered through the work of B. Greene and Ronen Plesser [BP], as well as by P. Candelas et al. [CAA]. Mirror symmetry has been a fundamental concept in string theory for the last twenty-five years.

Commentary on Inequality Between Chern Numbers of Singular Kähler Surfaces and Characterization of Orbit Space of Discrete Group of $\mathrm{SU}(2,1)$

This is a description of the work I did with S.-Y. Cheng, which I reported on at a 1984 AMS conference. Cheng and I produced a general construction of singular Kähler-Einstein manifolds with negative scalar curvature. Our construction is modeled on the orbit space of the quotient of the ball by a lattice in $\mathrm{SU}(2,1)$. Such a quotient space may have orbifold points and cusps, which are points at infinity of the quotient manifold. Cheng and I provided an intrinsic characterization of such singular manifolds. The arguments can be generalized to higher dimensional algebraic manifolds. A couple of years later, G. Tian and I carried this effort further, obtaining more details in the work that Cheng and I began.

Commentary on Ricci Flat 3-fold

When I moved from Princeton to San Diego, I decided to train some young Chinese mathematicians. I raised money from my friends in Hong Kong to help fund workshops for faculty members, postdocs and students to spend each summer in San Diego. For various reasons, the workshops happened only during one summer in 1985. The visitors included S. Roan, S. S. Lin and T. P. Liu from Taiwan, postdocs from Peking University and the Academy of Sciences, and friends from different universities in America. We had many lectures given by distinguished professors including Michael Freeman, Richard Hamilton and others. I suggested that Roan work with me, constructing examples of Calabi-Yau manifolds by using the toric methods developed by Mumford et al. This short paper is an outcome of this research. Such a construction was done later in a much more systematical way by Victor Batyrev. He also observed an important method for constructing mirror symmetry for such manifolds. On the other hand, the toric construction of CY manifolds was probably first done here. Later, Roan came to work with me as my postdoc for a couple of years at Harvard, and we wrote a few papers with Brian Greene and Bong Lian. I lobbied strongly for him to be appointed as a senior researcher at the Academic Sinica in Taiwan. He got a nice job there. However, for some reason, he could not believe the validity of the SYZ construction of mirror geometry and went around Taiwan discrediting my students who worked on SYZ.

Commentary on Some Recent Developments in General Relativity

In the summer of 1986, after I had settled down at the University of California, San Diego for two years, I was invited to attend an international conference on general relativity in Stockholm. I went there and met quite a few Chinese scholars who worked on general relativity. I was surprised that there were so many of them who came to this conference. I invited them to a Chinese restaurant for dinner. The Chinese food in Stockholm was not impressive. But the conversation was nice.

After the conference, I flew to northern Sweden to see the midnight sun with Wei-Tou Ni, a physicist who did experiments to test the equivalence principle in general relativity at Tsinghua University of Taiwan.

We spent two nights there on top of a mountain waiting for the sun to come out. But it was too cloudy and we saw nothing. So the two of us spent most of the trip discussing various things. We also visited a museum dedicated to the native cultures of northern Europe.

After Stockholm, I flew to Paris to give a talk at IHES and then drove with Elliot Lieb and his wife to Marseille to attend the International Congress of Mathematical Physicists. The city is by the Mediterranean Sea, and the conference was in Aix-Marseille University in Luminy. I met quite a few distinguished mathematical physicists and enjoyed the scenery tremendously. It was the first time I realized that beaches in Southern Europe had a rather open policy for women in terms of dress code. I met Michael Aizenman who came from Princeton, a Chinese physicist named Han-Ying Guo, and Barry McCoy, who was a professor in the Physics Department at Stony Brook. (His wife had been a classmate of my wife.) There was a conference tour to see an old castle, but Barry declared that he would rather spend time watching modern women on the beach than seeing an old building. So we had a nice dinner at one of the restaurants along the beach. The lodging situation was not optimal, as they put me in a student room. But I enjoyed the company of the people there, as well as the ocean setting.

In my talk at the Stockholm conference, I discussed the positive mass conjecture that R. Schoen and I proved and the work of R. Bartnik, who was my first student at Princeton. He did a fine thesis on the existence of a maximal slice in quite general spacetime, generalizing the work of Claus Gerhardt, using some methods that S.Y. Cheng and I worked out in 1974. I also described my result with R. Schoen on the existence of a trapped surface in a rather general spacetime when the mass density is large enough. This was the first concrete and rigorous mathematical result regarding the existence of black holes, drawing on natural assumptions from physics.

During 1978 and 1979, I gave a course on general relativity at Stanford, lecturing on my work with R. Schoen on the positive mass conjecture and also on the book by George Ellis and Stephen Hawking. Sergiu Klainerman drove quite a few times from Berkeley to seek my advice on what to work on. He wrote his thesis on the solution to some hyperbolic equations based on the Nash-Moser implicit function theorem. I suggested two problems that were related to his thesis: one is to prove the local isometric embedding of surfaces into three-dimensional Euclidean space; the other is to prove that small perturbations of the initial data set for the flat spacetime would lead to a global solution of the Einstein equation, at least when there is no matter.

Apparently, he discussed these problems with L. Nirenberg, and L. Nirenberg asked his student Chang-Shou Lin to take up the first problem. C. S. Lin finished a beautiful thesis on proving the local embedding for surfaces with non-negative curvature.

In order to take on the second problem that I proposed to S. Klainerman, Klainerman first got Demetrios Christodoulou to teach him some general relativity. In the meantime, D. Christodoulou came to IAS to work with me on this problem. I invited him to spend a half year at Princeton. In fact, since my wife was in California and did not stay with me at that period, D. Christodoulou lived in my apartment for some time. He worked with me quite extensively on this problem, even after I moved to the University of California, San Diego, where he came for another half year at my invitation. We mapped out quite a few ideas during that time. I wrote a short paper on the positivity of Hawking energy on surfaces with constant mean curvature. I listed Christodoulou as a co-author because it was meant to be part of a bigger project together. However, since S. Klainerman later intended to be the sole coauthor with D. Christodoulou, despite D. Christodoulou being the major contributor to the work on the dynamic stability of the Minkowski spacetime, I decided to withdraw from the bigger project.

But the short note I wrote on the positivity of Hawking mass led to some interesting consequences. Gerhard Huisken and I used the idea of that paper to prove the existence and uniqueness of foliation with constant mean curvature spheres at spatial infinity where the ADM mass is positive. The idea was also used in the later work of G. Huisken with Tom Ilmanen on what they called the Riemanniann Penrose conjecture.

Commentary on Nonlinear Analysis in Geometry

These were the IMU (International Mathematics Union) lectures I gave on November 20 and 27 and December 1, 1981 at ETH in Zurich at the request of Jürgen Moser. I added a bit more information later when I submitted the final version of the lectures for publication in 1984.

I knew J. Moser well when I visited the Courant Institute in the fall of 1975. I had received a Sloan fellowship in 1974 and was able to spend a year anywhere I wanted. L. Nirenberg invited me to the Courant Institute for a term, using my Sloan fellowship. But it was J. Moser who was my host because L. Nirenberg was on sabbatical. He was pleased that S.Y. Cheng and I had solved the Minkowski problem in higher dimensions when we were at Courant.

J. Moser was in the audience of my Congress lecture in Helsinki in 1978, and he appreciated my work. In 1981, he asked me to give the IMU lecture at ETH, after he moved to Zurich.

In 1980, Jürgen Jost was working with me at IAS as my post-doctoral fellow. He accompanied me to this beautiful city in Switzerland. Since I could not speak German, the companionship with J. Jost was of great help. (Perhaps that explains why I have never been able to understand German.) For those two weeks, I had a good time in Zurich. I gave lectures and interacted with various mathematicians, including Komaravolu S. Chandrasekharan, Eugene Trubowitz, and others. I was very touched by the hospitality of J. Moser, who invited me on several occasions to join him for dinner at his home and at restaurants.

E. Trubowitz was his student. He walked all the way from the institute to the village where J. Moser lived. J. Jost and I took the train instead. Jost and I also went to climb some mountains, and I enjoyed seeing the striking blue sky of Switzerland. There were some Chinese antiques in the stores of Zurich. As a gift to my mother, I bought a set of the Chinese game called Mahjong, which was made of ivory. My mother was very happy with that, as she was fond of playing Mahjong in those days, while she was taking care of my sick brother.

Both K. Chandrasekharan and J. Moser were important persons at the IMU. K. Chandrasekharan was president from 1971–1974, and J. Moser was president from 1982-1986. When they invited me to dinner at a fancy restaurant in Zurich, K. Chandrasekharan insisted that I sit in a particular seat. I did not know why until he told me that it was the seat where several mathematicians sat before they won the Fields Medal. Perhaps he had some inside information that I would be getting the Fields Medal next year.

My three lectures focused on nonlinear equations in geometry. I gave a rather comprehensive survey of the subject of geometric analysis up to 1981. In this survey, I stated that the Kähler-Einstein metric can be approximated in a smooth sense by the induced Fubini metric from some projective embedding of the manifold by a pluri-canonical line bundle or pluri-anticanonical line bundle depending on the sign of the first Chern class. When the first Chern class is zero, a higher power of any ample line bundle will do.

This conjecture paved the way for me to relate the existence of a Kähler-Einstein metric to the geometric stability of the manifold. I remembered that I was rather surprised by the fact that the existence of a Kähler-Einstein metric implies the stability of the tangent bundle for algebraic manifolds. This result is not obvious at all. I speculated that the stability of the tangent bundle should be a condition for the existence of a Kähler-Einstein metric when the first Chern class is positive. When I was at Princeton, I decided that the stability of the tangent bundle is not enough to ensure the existence of a Kähler-Einstein metric when the manifold has positive first Chern class. It also has to be stable as a manifold in the sense of geometric invariant theory. In my student seminar, I asked students to look into this concept.

As a first step towards my conjecture on Kähler-Einstein manifolds, I taught my students, including G. Tian, my conjecture on using projective embedding to approximate a Kähler metric. I spent many hours in my home teaching G. Tian how to do embedding by using the peak function argument. Some of the ideas were in an earlier paper I'd written with Y.-T. Siu. I also invited David Gieseker from UCLA to teach us geometric invariant theory.

At that time, I ran student seminars on Robin Hartshorne's book on algebraic geometry and also on William Fulton's book on intersection theory. D. Gieseker was curious as to what I was going to do with these students. Well, when Jun Li graduated from Harvard, I recommended that he go to UCLA to work with D. Gieseker. That decision worked out well. J. Li learned algebraic geometric methods from D. Giesecker—much more than any of my other former students. He started to look at mathematics from the algebraic geometry point of view, although he still worked with me on geometric analysis once in a while. In 1981, P. Li and I had been very interested in estimating the first eigenvalue of the Laplacian by studying certain conditions related to the concept of the center of gravity defined by the projective embedding. This was explained more transparently by S. Donaldson, who later used the concept of a moment map. Assuming that we chose the measure correctly, we found a concept called the balanced condition. It turned out that this was closely related to geometric stability in the sense of geometric invariant theory. Later upon my suggestion, my former student Huazhang Luo found that the balanced condition inspired by the work of Li-Yau implies Hilbert-Maniford stability. A few years later, S. Donaldson proved that the balanced condition of Li-Yau-Luo can be achieved asymptotically if the manifold admits a Kähler metric with constant scalar curvature.

The study of the existence of a Kähler-Einstein metric with positive scalar curvature dated back to very early days. When I went from a visiting position in Berkeley back to Stanford in 1978, Y.-T. Siu moved from Yale to Stanford at the same time. In that year, we had discussed many topics in mathematics and the Kähler-Einstein metric was one of them. My original proof of the Calabi conjecture was good enough to handle the case with positive scalar curvature, except one point was missing: the zero order estimate of the solution. In the case of zero first Chern class, I developed a method to achieve the zero order estimate, which has since been used by people up to now, except that some authors changed my notations and rewrote the ideas by calling them I or J functionals. Y.-T. Siu and I followed the same basic strategy, and it was clear that everything can be estimated in terms of some integral of the function constructed from the solution. But how to estimate such functionals is the major obstacle. I believed that the question boils down to finding the right definition of algebro-geometric stability in geometric invariant theory so that the integral estimate is equivalent to the condition of geometric stability. This type of argument appeared in my work with K. Uhlenbeck, where we proved that polystable holomorphic bundles admit Hermitian Yang-Mills connections. In my work with K. Uhlenbeck, we found a subsheaf that destabilizes the original bundle. Such a subsheaf violates the condition of Mumford slope stability. We needed to find a similar statement for Kähler-Einstein manifolds. And the reason we had not settled the problem of the existence for Kähler-Einstein manifolds with positive scalar curvature was partially because we had not found the right algebraic geometric definition of stability of the manifold.

After discussing with me during classes and at my home, G. Tian decided to push forward even though there were not enough estimates for the original Monge-Ampère equation. He replaced the missing estimate of the integral by proposing that a certain functional is proper, and he called such a manifold K-stable. The assumption he made was to put some geometric and analytic condition over a certain infinite dimensional space, which was certainly not the goal of my conjecture. I advised him to find conditions on finite dimensional space. I also made some suggestions. Ideas to look for the right geometric stability conditions were developed by several people, starting with the balanced concept introduced by Li-Yau, Shou-Wu Zhang, H. Luo, and Simon Donaldson. The definition provided by S. Donaldson was very nice and it was kind of him to characterize his concept of stability as "K-stable", which was the name used by G. Tian. But this created completely unnecessary confusion in the literature, as Donaldson's condition had not been proven to be the same as Tian's, despite the fact that they were both related to the projective embedding of the Kähler-Einstein metric into complex projective space, as had been proposed by me. But the key importance of Donaldson's definition is that it is expressed purely in terms of concepts from algebraic geometry.

Working in the other direction, Y.-T. Siu and I also took on the rigidity problem that I had proposed for compact Kähler manifolds with negative curvature. I visited Yale several times to work on this problem with Y.-T. Siu. In the beginning, Y.-T. Siu did not believe my conjecture. It was motivated by my previous work on the rigidity of the quotient of the ball, which was a consequence of the Calabi conjecture. I suggested to him that harmonic maps should do the job of proving rigidity, based on my past experience with R. Schoen. Since rigidity can hold only when the dimension of the manifold is greater then one, Y.-T. Siu argued strongly that if my conjecture were true, it should have something to do with the duality of sheaves. After he arrived at Stanford, Siu finally became convinced that my proposal for using harmonic maps should work better. At one point, he met Eric Bedford and learned about an argument that E. Bedford had used in Monge-Ampère equations. The argument allows the Ricci curvature term on the domain manifold in the Bochner formula to be tractable. But the curvature assumption on the image manifold had to be strengthened further.

Siu considered that this was an important achievement and went ahead to write a paper on his own, even though the full conjecture I had made was not completely resolved. (And it was still not resolved at the time of this writing.) I also proposed to him to apply the harmonic map argument to solve some special case of the Hodge conjecture for Shimura varieties. But this was not accomplished.

In 1980, when J. Jost became my postdoctoral fellow at IAS, we looked into the rigidity problem when the Kähler manifold does not have strongly negative curvature. For example, we demonstrated the complex moduli space of products of curves of genus greater than one are the obvious ones obtained by deforming the curves individually. N. Mok also joined in to generalize my work with Jost. Siu, however, advised him not to have further collaboration with us. An interesting observation of my work with Jost was that if there is a continuous map from the Kähler manifold to a curve with genus greater than one, we can change the complex structure of the curve to deform the map to be holomorphic. This statement appeared in the much later talks and survey papers by various authors, which mistakenly gave the credit to Siu.

I was interested in two problems: One was the compactification of complete Kähler manifolds with finite volume whose curvature is pinched between two negative constants. The other was the Frankel conjecture, where we used harmonic maps to create rational curves of low degree. I suggested that Siu work with me on these projects, and we solved them in a short time.

In this survey, I also discussed my work with William H. Meeks III on minimal surfaces and their applications to geometry and topology.

In 1976, I visited UCLA, using my Sloan fellowship. I had just got married and finished the proof of the Calabi conjecture. Then I suddenly met up with W. Meeks whom I had known during his senior year in college and my first year in graduate school in Berkeley. It was a pleasant surprise to see him again. He had long hair and was marching down the hallway of the mathematics department at UCLA, holding a young girl in each arm. He had been interested in minimal surfaces since his Berkeley days. So I suggested that we work together on a problem that I had been thinking about for a while. It was an embedding problem for minimal surfaces. I proposed to use techniques from three-manifold topology. He agreed, and we proved that the classical Douglas solution to the Plateau problem is in fact an embedded solution with no singularity. This is based on the assumption that the boundary curve belongs to a surface with positive mean curvature. This was an important problem since R. Osserman had shown that the Douglas solution is a smooth immersed surface.

I believe this was the first time that some profound technology of three-manifolds was used in a key manner to solve a classical surface problem in differential geometry. Most remarkable (at least to me) was the way we were able to turn it around: using the Douglas solution to solve problems in topology of three-manifolds. Since minimal surfaces provide a canonical way for curves to bound a disk and enjoy nice properties when they intersect, we were able to solve the equivariant Dehn's lemma and equivariant loop theorem.

Soon afterwards, Cameron Gordon was able to put our theorem, along with the work of William Thurston and Hyman Bass, together to solve an old problem of Paul Smith saying that the cyclic group acting on the three-sphere with fixed points must be conjugated to a standard rotation of the three sphere.

In 1978, Robion Kirby called me from Berkeley while I was having breakfast in Palo Alto. He suggested to me that W. Thurston, Meeks and I should announce the solution of the Smith conjecture in the Bulletin of the AMS. I agreed, and he promised to communicate the agreement to Thurston, but there was no further notice afterward. When R. Kirby supported John Morgan to run for president of the American Mathematical Society a few years ago, he said that J. Morgan was the person who "assembled" several works together to prove the Smith conjecture.

In 1980, there was a meeting on the Smith conjecture organized by Samuel Eilenberg, H. Bass, and John Morgan in Columbia. I was invited, but W. Meeks was not. I was not happy about the arrangement, and I requested that W. Meeks be invited as well. When he was finally invited to give the colloquium at Columbia, he was not treated nicely, according to Meeks. During the meeting, W. Thurston described his work on the construction of hyperbolic metrics for sufficiently large manifolds. I was truly excited by the beautiful statement of Thurston's theorem. I asked W. Thurston several questions. Because of this, I was accused by some people at Princeton of being jealous of W. Thurston's work.

When W. Meeks and I wrote up our paper in 1977, I got a letter from Johannes C. C. Nitsche of Minnesota saying that he was excited by our work, which proved the non-singularity of the classical Douglas solution. He asked us to publish our work in his journal, the Archive for Rational Mechanics and Analysis. I was reluctant, but he persisted in asking me. Finally I gave in and sent him the paper in the summer of 1977.

Two years later, I received a letter from the chief editor, James Serrin, saying that the paper was rejected on the grounds that our English was poor. I got help from some American mathematician to improve the English, but the paper was still rejected. I find it ironic that when John Morgan asked to publish this paper in the book that he was editing, I said that it had been submitted long ago to a journal. John Morgan got mad and yelled at me over the phone. W. Meeks and I wrote a few companion papers later, all appearing in good journals. Perhaps J. Serrin was not familiar with the part of our arguments based on the topology of three-manifolds.

We were reasonably new to the PDE aspect of minimal surfaces then and assumed that most of the basic materials in our paper were already known to experts. But we could not find any such references in the literature and, to my surprise, I discovered that most of these materials were in fact not known to the experts. Subsequently, the idea of using a surface of minimal area type for treating three-manifold topology has been employed quite extensively by topologists, though our prior work was rarely cited.

In 1980, when I visited Bonn, I met my old friend L. Simon. I told him what I had done with W. Meeks. He was interested. He had done work with Frederick Almgren that showed how one can minimize area among embedded disks. I saw this approach as being complementary to my work with W. Meeks, so I suggested that we join forces.

W. Meeks, L. Simon, and I made use of ideas from geometric measure theory to solve problems in three-manifold topology. Instead of using an embedded disk, we used an embedded sphere. This approach was different from the approach of Sacks-Uhlenbeck.

We got more precise information about how the resulting minimal surfaces can be joined together to be isotopic to the original surface—a fact that is difficult to obtain through the approach of Sacks-Uhlenbeck.

Also, given a fake disk, which is a contractible subset of the three-manifold that is not a ball, we can enclose it by an embedded minimal sphere with smallest area. This is a remarkable fact, as the existence of such an embedded sphere is independent of the metric. I thought this could be used to resolve the Poincaré conjecture. But I did not succeed.

We did manage to use this nontrivial theorem to prove that if the universal cover of a three-manifold is irreducible, which means that they do not contain a fake disk, the three-manifold itself is still irreducible. We were told that this was a well known problem in three-manifold topology proposed by W. Jaco of Rice University. This theorem was also useful in general relativity as it allowed us to talk about the outermost black hole.

While this work on area-minimizing surfaces was interesting, I found the recent work by Smith and L. Simon particularly exciting. Smith was a Ph.D. student of L. Simon, and they proved a remarkable statement that in any manifold diffeomorphic to three-spheres, there is at least one minimal sphere with index of at most one. The sphere is obtained by a mini-max argument. Later J. Jost carried these arguments further, proving the remarkably precise statement that there are four such minimal spheres.

In 1981, Joachim Hyam Rubinstein from Australia visited IAS. I ran a seminar on applying minimal surfaces theory to the topology of three-manifolds. Michael Freedman also came to visit IAS several times. I was able to get both of them interested in the theory of minimal surfaces. M. Freedman went back to California to work with Joel Hass, a student of R. Kirby, to generalize my theory with W. Meeks to higher genus. Their work is a nice complement to the earlier work of Meeks-Simon-Yau. It is stronger in some directions, but weaker in other directions. For the purposes of topology, the original Meeks-Simon-Yau was actually enough. Since I visited the University of California, San Diego, often those days, I frequently had nice discussions with M. Freedman.

In the meantime, Jon Pitts of Texas A&M proved that any three-manifold admits an embedded minimal surface with index not greater than one. However, he had no control over its genus. For manifolds with positive Ricci curvature I found a method for estimating the genus based on some arguments I had worked out with P. Li and R. Schoen. There is also a rougher estimate for genus if we assume that the scalar curvature is nonnegative. The hope was to find a minimal surface that is a Heegaard surface, and we can estimate its genus. If the estimated genus is small, we can make use of it to prove the Poincaré conjecture when the manifold admits a metric with positive scalar curvature.

Commentary on Three-Dimensional Algebraic Manifolds with $C_1 = 0$ and $\chi = -6$

The construction of Calabi-Yau manifolds with Euler number equal to -6 is important for physics because the number of families of fermions was demonstrated by Witten to be the absolute value of the Euler number of the Calabi-Yau manifold divided by 2. It was known by experiment that the number of families of fermions has to be three. This is the reason I care about Calabi-Yau manifolds with Euler number equal to -6. I constructed the first manifold with this property in 1984 during my trip to the Argonne Lab conference. When I came back from Chicago to San Diego, I suggested that my students use the flop construction to produce more examples of Calabi-Yau manifolds with Euler number equal to -6. Tian and I worked out some details and found manifolds that were still projective after the flop operation. We focused on constructing manifolds with Euler number equal to -6 extending what I had done before. This may be the first published paper on such constructions of Calabi-Yau manifolds. Flop operations become important through the beautiful work of Greene-Morrison on topological change in string theory. On the other hand, my paper with Tian is rarely referred to in the literature, as it does not appear on the web. So I decided to include it here.

Commentary on Uniformization of Geometric Structures

In May 1987, there was a conference in memory of Hermann Weyl at Duke University. I was invited to give a talk on the subject discussed here. On my way to Duke, I went to Washington D. C. for a committee meeting at the NRC (National Research Council), which was organized by the National Academy of Sciences. (That was the last meeting that I attended at the NRC.) I met Phillip Griffiths, who had left Harvard to become provost of Duke a few years before. When I arrived at Duke, P. Griffiths insisted that I stay in his home instead of staying in the hotel arranged by the conference. Later, I found out why he made that proposal. He wanted me to lobby for a certain mathematician to obtain a big offer as a chaired professor at Duke. Although I had high respect for the candidate, I was not comfortable with Griffiths' tactic: He made the dean of Duke call me at his home while he sat in the living room to monitor our conversation.

In 1980, Chuan-Chih Hsiung asked me to reorganize the Journal of Differential Geometry. At first, I was extremely reluctant to do that. After Chern called me several times, I agreed to be one of the editors of the journal. And the other two editors would be B. Lawson and P. Griffiths. We were supposed to help C.-C. Hsiung run the journal. As things evolved over time, most of the work was done by C.-C. Hsiung and me. A couple of years after the Duke University meeting, C.-C. Hsiung decided that since, in his eyes, P. Griffiths was not a professional differential geometer and had not done much work for the journal for a long time, the journal should relieve him of his duty as one of the editors. I was told that P. Griffiths was really upset and called the provost of Lehigh University, who was legally in charge of the journal, to fire C.-C. Hsiung and me. Since C.-C. Hsiung was the founder of the journal, Lehigh University did not see any reason to make such a drastic move. This certainly did not please P. Griffiths, who accused me recently in a major magazine of being a politician. This accusation was ironic because while he has been provost of Duke University, the director of IAS, and involved in possibly every important committee in mathematics, I have not served on any panel in Washington, in the International Mathematical Union, NSF, or MSRI in the past 20 years.

In this talk on the uniformization of geometric structures, I discussed the issue of how to recognize a manifold to be the quotient of a nice manifold by a discrete group, and whether one can give a concrete representation of the uniformization. I mentioned the theorem that R. Schoen and I had just proved: For complete, conformally flat manifolds with positive scalar curvature, any conformal map into the sphere is one to one. Therefore the manifold itself is the quotient of a domain by discrete subgroups of conformal transformations acting on the sphere. We also proved that the complement of the domain in the sphere has small Hausdorff dimension. It is still an interesting question to understand how to characterize such a domain precisely. Incidentally, the injectivity of the developing map for conformally flat manifolds with positive scalar curvature is very remarkable. It is probably the only theorem of this kind that I know of. I am still curious as to whether one can apply the idea to solve the famous Jacobian conjecture.

A similar problem would be to study complex manifolds that can be covered by complex homogeneous manifolds. I proposed to use the Hermitian Yang-Mills solution on a non-Kähler complex manifold based on the work Jun Li and I did, which was a generalization of the work of Uhlenbeck-Yau. I also proposed to study the universal cover of manifolds of general type, which do not contain any rational or elliptic curves. I believed that the universal cover of the manifold should be embeddable canonically into an algebraic manifold, such that the deck transformation group is a subgroup of the group of birational maps of the ambient manifold.

In this survey, I also discussed and generalized the uniformization theorem that I proved before: the characterization of a locally Hermitian symmetric domain by discrete groups.

I used Chern numbers to characterize the quotient of the ball in 1976. I then generalized the characterization to more general locally Hermitian symmetric manifolds (Shimura varieties) by using purely algebraic geometric means. It is remarkable that such an exact characterization can be achieved for compact manifolds. An immediate consequence of this was the famous theorem of David Kazhdan: The Galois conjugation of a Shimura variety is still a Shimura variety. For the non-compact Shimura variety with finite volume, we had Mumford's toroidal compactification, where divisors at infinity satisfy suitable combinatorial algebraic-geometric conditions. James Milne asked me about my algebraic characterization of Shimura varieties when I was at IAS in 1983. However, people working in the Langlands program did not seem to notice this intrinsic algebraic geometric characterization of Shimura varieties.

The simpliest case of the divisor configuration appeared in the toroidal compactification of the Siegel upper space of genus two, after being quotiented by an arithmetic group. My former student, Wen-Xiang Wang, studied this in his thesis at Princeton in connection with the boundary behavior of the Kähler-Einstein metric along the divisor. The general case of higher dimensional Shimura varieties has been investigated by Yi Zhang and myself.

The existence of a canonical Kähler-Einstein metric with singularity on manifolds of general type was already known to me and, independently, to Hajime Tsuji. And I obtained some consequences in algebraic geometry by assuming that singularities of the manifold were reasonably controlled.

Some of these were discussed by me with S. Y. Cheng first and later with G. Tian. I also suggested to G. Tian to consider whether one can apply the metric to study the problem of minimal models in algebraic geometry.

The existence of Hermitian Yang-Mills connections on poly-stable holomorphic bundles would be useful for understanding the geometric structures and uniformization theorems. Some of them were pointed out in Carlos Simpson's thesis at Harvard ("Systems of Hodge Bundles and Uniformization"). This was used to characterize Shimura varieties based on algebraic geometry.

In 1984, K. Uhlenbeck and I finished our work on the Hermitian Yang-Mills equation. That was the year when string theory became popular and physicists were excited by Calabi-Yau manifolds. I told Edward Witten about this work on Hermitian Yang-Mills equations. I felt strongly that they should be useful for string theory, as it is supersymmetric in nature and is also a natural generalization of anti-self-dual instantons on four manifolds. At the time, E. Witten did not see the relevance. But a year later, he changed his mind and wrote an important paper on its significance in heterotic string theory. This idea also appeared in a paper by Andrew Strominger in 1986.

In the fall of 1985, Pit-Mann Wong came to visit me briefly in San Diego and asked me to suggest a significant problem for him to work on. I gave him the handwritten manuscript by Uhlenbeck and me and encouraged him to pursue this line of research further. Wong promised me that he would not show the manuscript to another person without my permission. He was then spending his sabbatical year at Harvard to work with Y. T. Siu, and I soon heard from Siu that he too would like to have a copy the manuscript to read. I refused, as Uhlenbeck and I still needed to make more changes in the manuscript. After three phone calls from Siu, I relented and told him that he could read the manuscript that was already in Wong's possession. A year later, when P. M. Wong called me, I was astonished to find out that, during the year he was at Harvard, he was working with Y. T. Siu to prove the theorem in the manuscript. He told me that he and Siu had discussed my manuscript with K. Uhlenbeck in an open seminar, and had made it known that Uhlenbeck and I had made fatal mistakes in the manuscript. Because I had made it clear to him and to Siu that I was showing the manuscript only to the two of them, I told him that even if there were mistakes, it was only ethical for them to have let me know first. I asked him why he was pursuing the research with Siu when he got the problem from me, but he did not answer my question.

When we sent our final manuscript around, S. Donaldson recognized that some of the formulas in our paper could be helpful to realize another proof for bundles over higher dimensional algebraic manifolds by using induction on dimension of the manifold. (He used the important theorem of V. B. Mehta and A. Ramanathan, which held that the restriction of a stable bundle to a generic hyperplane section is still stable. Hence his argument depended on the manifold being projective algebraic.) But those arguments did not work for Higgs stability and for Hermitian Yang-Mills bundles over non-algebraic manifolds.

A few years later, S. Bando and Y. T. Siu generalized my paper with K. Uhlenbeck to coherent sheaf using Yang-Mills flow, although the key point is still based on the work of Uhlenbeck-Yau. This point was not mentioned in their paper explicitly and many readers were misled to think that the Bando-Siu's paper had a proof independent of ours. There is no clear advantage of Yang-Mills flow over the continuity argument of Uhlenbeck-Yau. Our construction of a destabilization subsheaf is still essential for proving the theorem for dimensions greater than two.

In the fall of 1985, I ran a conference in San Diego called "mathematical aspects of string theory". In the conference, Jun Li and I presented a paper regarding Hermitian Yang-Mills bundles over non-Kähler manifolds. I then went to New Orleans to give a talk. On my way, I gave a talk in Houston, where I met Nicholas Buchdahl. He told me he was trying to push the method of S. Donaldson to work for bundles over non-Kähler surfaces, and he was having some difficulties. I told him about my work with J. Li, which used the method that K. Uhlenbeck and I introduced. Since Buchdahl said that he could not make the S. Donaldson proof work in a non-Kähler setting, and my paper with J. Li had already gone to the publisher, I did not refer to his attempted proof.

I was told later that he was very upset that his work was not mentioned. Of course, in later publications, I mentioned his work after I was told that he was able to push the argument through.

The paper for Hermitian Yang-Mills over non-Kähler surfaces was used by J. Li, Yau and F. Y. Zheng to complete the classification theorem for class VII0 surfaces with no curves. This result was announced by F. Bogomolov first with an extremely complicated proof. (I believe that, to this date, nobody really understood the proof of Bogomolov.) The later claims by Andrei Telemann and others that F. Bogomolov's original argument could be "simplified" were not true, as their argument depended on our arguments in a critical way.

Commentary on Some Remarks on the Quasi-local Mass

In the early 1980s, Christodoulou came to visit me frequently to collaborate on a project aimed at proving the dynamical stability of flat Minkowski spacetime. He stayed in my apartment in Princeton for a few months. During this same period, I had worked with Peter Li in estimating the low eigenvalues of the Laplacian using the concept of the conformal area of a Riemann surface. I observed that the same idea could also be applied to operators with a potential. In fact, if the potential is the Gauss curvature of the surface, and if the estimate only involves the integral of the potential, the inequality is rather clean as it involves only the Euler number of the surface. This is particularly useful for the stability operator associated with the second variational formula of the area of a surface sitting in a three-dimensional manifold with positive scalar curvature – a surface with which Schoen and I had good experience.

The main objective of this paper was to study those surfaces with constant mean curvature that are stable and whose volumes are constrained. The end result is that the Hawking mass defined for these surfaces is positive unless the surface is umbilical and the scalar curvature of the three manifold vanishes along this surface. This is a rather remarkable fact and, as far as I know, it is the only result that establishes the positivity of the Hawking mass. The result was useful in my work with Huisken and Huisken-Ilmanen on the center of mass and the Riemannian Penrose conjecture. Operators with scalar curvature as the potential have become very important in geometry, as was later seen in the work of Perelman where the analog of the Li-Yau inequality was used.

Commentary on a Review of Complex Differential Geometry

This is a review I wrote when I was invited to give a survey talk at the AMS conference on several complex variables held at UC Santa Cruz.

In 1989, my postdoctoral fellow B. Greene and R. Plesser, a Ph.D. student of C. Vafa, along with P. Candelas et al., discovered the intricate concept of "mirror symmetry" for Calabi-Yau manifolds. I was then spending most of my time trying to understand this new concept.

At this conference, I summarized the concept of the geometric uniformization of complex manifolds that I had initiated during my graduate student days and announced at the Williamstown conference on several complex variables in 1975.

After this conference in Santa Cruz, there had been additional work done on this subject by Jun-Muk Hwang, N. Mok, F. Y. Zheng, D.M. Wu, and others for manifolds having properties similar to manifolds with positive curvature. For noncompact manifolds of parabolic type, there has been further work by Albert Chau, Luen-Fai Tam, Bing-Long Chen, Xiping Zhu, Lei Ni and others. They are close to proving my conjecture that every noncompact complete Kähler manifold with positive bisectional curvature is biholomorphic to the complex Euclidean space.

In this survey, I asked the question whether there is only a finite number of topological types of Calabi-Yau manifolds in each dimension. I believe that this was the first time the question was put in print, although I had been curious about it for quite a while.

I was also interested in knowing whether complex manifolds whose universal cover are affine spaces are quotients of these affine spaces by discrete subgroups of affine transformations. This is still unknown. I thought this should also be true if the affine space is replaced by a complex Lie group, related to a theorem of Hsien-Chung Wang.

On the other hand, the progress on complex manifolds of "hyperbolic type" has been much slower. Basically, we need to understand those manifolds covered by a bounded domain.

Such manifolds must have negative first Chern class based on my work with S. Y. Cheng on Kähler-Einstein metrics. I asked in this survey whether the tangent bundle of such a manifold is negative or not—a matter that is still not known. Nor do we know much about the topology of such manifolds, except that it is covered by a Stein manifold and therefore has the homotopy type of a CW complex with dimension not greater than the complex dimension of the manifold. Which raises the question: Is the fundamental group residually finite?

An important related question is whether a compact manifold with strongly negative curvature is covered by a bounded domain.

To answer this question, we need to find enough bounded holomorphic functions on its universal cover to provide an immersion of the universal cover into the Euclidean space. But there has been no progress on this particular problem.

After that, the question is whether the immersion is canonical enough so that we can make it to be an embedding. In fact, I wish that the deck transformations could be realized as birational transformations of the complex projective space.

I also asked in this survey whether one can have higher dimensional generalizations of Kleinian group theory for the subgroup of the projective group that may act properly discontinuously on some domains in the complex projective space.

Commentary on Open Problems in Geometry

There was a large AMS conference at UCLA in 1990, which was organized by S.-Y. Cheng and Robert Greene. It was held in the summer on the UCLA campus. S.-Y. Cheng and R. Greene asked me to come for the whole month to help set up this big conference. A large number of geometers came. It may have been the largest differential geometry conference ever run by the AMS. We decided to dedicate the conference to the celebration of S.-S. Chern's 79th birthday. While I thought this was an obvious proposal, P. Griffiths kept asking about my motives for such a celebration.

I brought my two small children to California with me. We rented a big apartment near the campus. I also rented a huge Cadillac to make sure it was safe to drive in Los Angeles, for I just had a big accident while driving from Harvard to Stony Brook to attend a conference in geometry. Since my mother lived in Monterey Park, which was not far from Westwood, my two sisters also brought their children to stay with us, while my brother Stephen also came to the conference with his son. As a result, we had a big family reunion. Both my mother and the kids were very happy. The kids were excited to play games, making so much noise it was virtually impossible for me to work in the apartment.

Unfortunately, in a short time, when my brother brought my mother for a medical checkup, she was found to have a tumor that seemed serious. On the other hand, it was going to take a long time before she could get a CAT scan. So I called my doctor, Marshall Wolf, in Boston for help. He was very kind and managed to get the chief surgeon at the UCLA hospital to treat my mother. The surgeon decided to perform the operation immediately. But it was found that the cancer had spread badly and could not be surgically removed.

That was very depressing for me. And it was against such a background that I participated in this conference.

I kept on taking care of my mother after the conference. It was very nice and generous of Harvard to allow me to take a leave of absence for one semester (Wilfried Schmid was the chairman who agreed to my request immediately). Friends at the California Institute of Technology (Dinakar Ramakrishnan, Kip S. Thorne, and Thomas Yizhao Hou) proposed that the mathematics department offer me a Fairchild professorship. Caltech did, in fact, offer me a very nice house and a very nice car. But I had already bought a cheap used car. The car was another huge Cadillac, which consisted of large amount of steel. I still had memories of the bad car accident I was involved in the year before. And I was still tied up in a serious court case. I had to fly back to Boston for more than a week to defend myself in court. Fortunately I won the court case and was able to come back to Los Angeles to take care of my mother. I gave up the beautiful mansion Caltech had offered and stayed with my mother in her small apartment. I slept upon a matt on the floor every night, and my mom felt bad about the poor accommodations but was glad that I could spend time with her. She passed away in the next spring.

There were some complications at the conference regarding, for example, the arrangement of the seats for the banquet in honor of Chern. I proposed to establish a medal called the Chern Award, funded by the Journal of Differential Geometry. I contacted Chern ahead of time about the medal named after him. He happily approved the idea. But after I announced the medal, Wu-Yi Hsiang went to complain to Chern, and Chern changed his mind immediately. He got I. Singer to convince me to call the medal off.

During the conference, I gave a couple of talks, despite being busy with my mother and the children. At the request of quite a few geometers, I gave talks on open problems in geometry, and they were well received. This set of open problems may be considered as a continuation of the 120 problems that I had posed during a special year at IAS in 1979. There was a slight overlap with the previous set of problems. In addition, some of the problems came from my own student seminars.

Commentary on the Geometry of Certain Superconformal Field Theory Paradigms

In the fall of 1990, I took a sabbatical leave to visit Tsing Hua University in Taiwan. Both my wife and two young children went with me. My son Isaac was 10 years old and Michael was 8 years old. We had sent them to study Chinese at the Lexington Chinese School, and we taught them Chinese at home too. But it had reached a point when they simply did not want to learn Chinese. They claimed that they were American, and Americans did not need to know Chinese. So my wife and I decided to take them to Taiwan to get a feel for what a Chinese community means.

The plan turned out to be complicated. My wife took an unpaid leave from her laboratory at MIT, and we needed some financial support. I am grateful to President Liu Chao-shiuan of "Tsing Hua University", who worked out a way that I could take the whole family to Hsinchu, where the university is located. However, a year before we planned this trip, my mother suddenly became very ill. She was diagnosed with cancer in the summer of 1989. I got permission from Harvard allowing me to take an unpaid leave to stay with her in Monterey Park, near Los Angeles, so that I could take care of her. My mother's situation improved at first but then worsened, and she finally passed away in June of 1990. It was not until then that our family was able to go to Taiwan.

For the Taiwan visit, I arranged for my student Naichung Conan Leung and postdoctoral fellow Tristan Hübsch to come with me. In the beginning, my two children had to struggle to learn Chinese. My wife worked hard to guide them. Within two months, I was glad that they excelled in their classes. T. Hübsch had been very friendly with our family, and we went out for dinner often. He is a big guy and powerful. I remember that one day a boy became trapped between a bar and a chair in a strange manner. Fortunately, Tristan was around to free the boy.

I gave lectures every week and I talked with Tristan frequently. We spent a lot of time discussing the theory of mirror symmetry. Tristan was writing his book on Calabi-Yau manifolds, and he became very friendly with the secretaries there. We went through the theory of super-spaces. I was invited to go to Korean to give a talk at a conference called algebraic geometry and related topics. I talked about this paper, which was largely prepared by Tristan. I was told that this is still a basic reference on the subject. That was the first time I visited Korea. And my family enjoyed the trip tremendously. We went to Seoul and other places. Professor Jaigyoung Choe, who had been a student of R. Schoen a couple of years before, drove us around. We were able to see a lot of the ancient history of Korea, some of which appeared very close to Chinese culture. It was very sad to see the damage caused by the Japanese invasion of Korea.

As for mathematics, the geometry of super-spaces is still being pursued, recently by E. Witten and R. Donagi in several deep papers.

Commentary on Upper Bound for the First Eigenvalue of Algebraic Submanifolds

In 1974, S.-Y. Cheng told me that M. Berger published a paper asking whether there is an upper bound for the first eigenvalue of the Laplacian of the two-dimensional torus in terms of area alone. This was supposed to be a generalization of the theorem of Hersch on the two-dimensional sphere, which was in turn a generalization of the theorem of Szegö for the two-dimensional disk.

Cheng did not believe the statement and challenged me to prove it. In the spring of 1979, I was lecturing at Stanford on eigenvalues of the Laplacian. I suddenly realized that if we enlarge the space of conformal change and consider instead the space of conformal maps from a surface onto the sphere, the argument of Szegö-Hersch would still apply and yield the desired result. When Paul Yang visited me, I told him the idea, and it took some time for him to write up the paper. A corollary of this theorem says that the degree of any conformal map from the Riemann surface to the sphere can be estimated from above by the product of a lower bound of the first eigenvalue of the Laplacian and its area. There was a theorem by Selberg proving that for Riemann surfaces defined by the quotient of the upper half plane by the congruence subgroup, the first nontrivial eigenvalue of the Laplacian is always greater than $3/16$. Therefore the above estimate gave a natural upper bound of the degree of the Riemann surfaces as defined arithmetically. The argument on the upper bound can be generalized trivially to complete surfaces with finite area, which includes Riemann surfaces with cusps defined by the quotient of the upper half plane by a discrete group. Peter Sarnak told me, twenty years later, that my argument was followed by a Russian whose paper, concerning the noncompact case, came out a couple of years after our paper. In any event, I was rather interested in my argument, as it solved Berger's problem and also gave a bound on the degree of the curve defined by the quotient of the upper half plane by the congruence subgroup.

In 1981 I talked with my friend Peter Li at UC San Diego. Li worked with me to generalize the above result by mapping surfaces into spheres of arbitrary dimension. In doing so, we exploited the large group of conformal automorphisms of the sphere.

We introduced the concept of conformal area to estimate the first eigenvalue of the Laplacian. That was rather nice as it is related to the minimal surfaces in the sphere. This idea of conformal area is quite interesting in itself. It helped the recent proof by André Neves and Fernando C. Marques of the Willmore conjecture. Our theorem was also used by Choi and Schoen to prove compactness of the moduli space of embedded minimal surfaces in the three sphere. The key point here was a theorem by Choi-Wang, which proved that the first eigenvalue of an embedded minimal surface of the three sphere must be greater than one. (They tried to prove my conjecture that the first eigenvalue should be equal to two.) After Peter Li and I finished this work on surfaces in the sphere, we exploited a similar situation where the complex projective space has a large group of automorphisms. Hence we looked into algebraic manifolds that can be mapped to complex projective spaces.

We needed to normalize the mapping suitably. I suggested to my former student, H. Luo, that he look into this paper. He did so, finding a suitable modification of the Li-Yau balanced condition, which can be considered as a stability condition for projective manifolds. When a draft of my paper with Peter Li was written up, Jean-Pierre Bourguignon expressed interest in this work and suggested that he join forces with us on the project. After we waited for a year and received no correspondence from him, we decided to submit the paper, listing all three of us as authors, even though Bourguignon had no input into this work.

Later, I conjectured that for surfaces of higher genus, the k-th highest eigenvalue of the Laplacian also has an upper bound in terms of area and k. During a picnic in Utah in 1989 organized by Peter Li, I challenged N. Korevaar to solve this conjecture. To my surprise, he was able to do it within a year, using a beautiful argument, although the constant was not optimal. Finding the best constant remains an open question.

About eight years ago, Alexander Grigor'yan, Yuri Netrusov, and I found a different way to prove and generalize the work of Korevaar. I think it would be interesting, as well as worthwhile, to find the best constant of higher eigenvalues. For example, which conformal structure can realize the extremal higher eigenvalues? Should minimal surfaces in the sphere be relevant?

Our arguments can actually be generalized to a Laplacian with potential. If the potential happens to be the Gauss curvature, we can use the Gauss-Bonnet formula to get a clean upper estimate. I used this, for example, in the second variational formula for stable surfaces with constant mean curvature in a paper that I published with Demetrios Christodoulou on the positivity of the Hawking mass.

Commentary on Review on Kähler-Einstein Metrics in Algebraic Geometry

I was invited to give a talk in Israel in honor of Prof. F. Hirzebruch's 65th birthday. That was the first time I had visited this old country. While the conference was held in Tel Aviv, I took a day off on my own to visit Jerusalem. There were many places discussed about in the Bible that I saw for the first time. Before this visit, I had a much bigger picture of these sites in my mind. However, I was somewhat disappointed to find out that the actual scale of those famous places was much smaller. On a separate note, it made me somewhat nervous to see all the policemen holding machine guns to guard children near the school. Also, for security reasons, I was questioned rather thoroughly in the airport.

The portion of Israel along the Mediterranean Sea is truly beautiful. I was invited to have lunch at a particularly nice spot on a hill overlooking the blue sea. One thing that struck me during my visit was seeing how the many different cultures had evolved together, and separately, within the same country.

The specific reason for my being there in Tel Aviv was that F. Hirzebruch had started some research institutes in that city. By way of thanks, they held a conference to celebrate his birthday.

In this survey, I discussed the known facts governing the construction of Kähler-Einstein metrics with singularities and their applications to algebraic geometry. This was partially due to F. Hirzebruch's interest in my work on Chern numbers.

Commentary on a Note on the Distribution of Critical Points of Eigenfunctions

I did the work discussed here while visiting Tsinghua university in 1991. The question I wanted to address was finding out how critical points of eigenfunctions are distributed. I was amazed by the fact that practically nothing was known about them. I thought that since eigenfunctions vibrate a lot, the number of critical points of eigenfunctions should go to infinity along with the eigenvalue. During the seminar, I managed to find a way to create a nontrivial critical point of the eigenfunction using the gradient estimate. The existence of the critical point that I found does not come from Morse theory. I reported on that in this paper. I believe that this is the only general theorem on constructing critical points for eigenfunctions, although it works only for two-dimensional surfaces. Based on the argument here, I believe that the number of critical points may grow according to the range of the normalized eigenfunctions. Hence those normalized eigenfunctions that are uniformly bounded may have much fewer critical points. My question, which states that the number of critical points of eigenfunctions should go to infinity, had a counterexample given by Nadirashvili for some perturbed metric on the two-dimensional torus (which appeared in the *Journal of Differential Geometry*), where he found an infinite number of eigenfunctions with only 16 critical points. But my conjecture may still be true for a generic metric. I present this paper here as very few people are aware of this work.

Commentary on Sobolev Inequality for Measure Space

While spending a year at Tsing Hua University in Taiwan in 1991, I lectured on several topics on geometric analysis. I also invited quite a few friends from abroad to lecture on different topics. This includes Robert Bartnik, Gerhard Huisken, Jürgen Jost, Leon Simon, Peter Li and others. They were kind enough to contribute their notes, which along with a few essays written by me and students who attended seminars was published in the book called *"Tsing Hua" Lectures on Geometry and Analysis.* The book was printed by the International Press of Boston. I am somewhat surprised that the book was not more popular, given that most of the essays were beautifully written. I encourage people to read those essays, which are still very useful. In my lectures, I described Michael and Simon's fundamental paper on the Sobolev inequality for minimal subvarieties in Euclidean space. I decided to find out the basic ingredients of that paper by generalizing their work to an abstract metric space with a measure. Nowadays research on measure metric space is popular, but few people know about this essay. Yet I still think it may be useful for people in this area to understand the isoperimetic inequality derived in this paper.

Commentary on Einstein Manifolds with Zero Ricci Curvature

This survey article appeared in the book, *Survey in Differential Geometry*, which was devoted to Einstein manifolds.

There are many examples of Einstein manifolds in the Lorentzian category. However, most of them are non-complete and may admit bad singularities. Finding a vacuum solution to the Einstein equation remains a difficult challenge. The central problem is to construct an Einstein manifold that admits a non-singular spacelike hypersurface which can serve as an initial data set for Einstein's equation, while admitting only black hole type singularities.

Only a very few geodesically-complete examples are known. And they are mostly nondynamical, not emitting any gravitational radiation. There is a procedure called Wick rotation, which can sometimes be used to create nonsingular Euclidean Einstein metrics out of Lorentzian Einstein metrics. One interesting question is whether one can reverse this procedure to create an interesting spacetime out of Euclidean Einstein manifolds.

As a result of the proof of the Calabi conjecture, many examples of Euclidean Einstein manifolds were constructed. It would be great to see some good physics coming out of these metrics. Kaluza-Klein theory allows us to use such manifolds in higher dimensions to study string theory. In this survey, I describe what was known about this class of manifolds after about 15 years of development of string theory.

Commentary on Review of Geometry and Analysis

In 1998, I was asked by the International Mathematics Union to write an article for the celebration of the new millennium. This article was to appear in a book called *Frontier of Mathematics: Frontiers and Perspective.* So I made some comments and gathered information on the subject of geometric analysis.

1 Submanifolds

a. Isometric embeddings: Many of the problems mentioned here are classical and are still unsolved. There is the problem of relating the spectrum of the Laplacian of the surface to the spectrum of the Laplacian of the volume that the surfaces encloses. That is an interesting question to be pursued. But not much has been done so far. In my most recent work with Mu-Tao Wang and PoNing Chen on quasi-local mass, we're still trying to understand canonical embeddings of surfaces into Minkowski spacetime. In that case, it would be nice to study the relationship between the spectra of the surface and the maximal spacelike hypersurface that it bounds. It is also natural to ask whether any two-dimensional surface can be isometrically embedded into Minkowski spacetime.

b. Different geometries: The interplay between different geometries under different groups such as rigid motion, projective transformations, or affine group is certainly very interesting. In particular, it would be nice to answer a simple question: If a closed surface is globally rigid in the sense that any other isometric embedding of it can only differ from the original one by a rigid motion, will the projective image of this surface still be globally rigid? The infinitesimal version of this statement is valid.

c. Minimial surfaces: Tobias H. Colding and William P. Minicozzi II worked toward establishing the limit of sequences of minimal surfaces without bounded area. Their work led to the proof by W. H. Meeks and H. Rosenberg that the only embedded simply connected complete minimal surface in three-dimensional Euclidean space is the helicoid and the plane. Most of the questions asked in this section remain unanswered. But recently Simon Brendle proved the Lawson conjecture that the only embedded minimal torus in the three-sphere is the Clifford torus.

d. Closed extremal surfaces: There are many interesting extremal problems for surfaces in three spaces. It is exciting that the Willmore problem was recently solved by F. Marques and A. Neves. Working in the other direction, G. Huisken and M. Eichmair are studying the extremal problem for the Hawking mass.

e. Motion of surfaces: Perhaps we should take a look at motions of surfaces in Minkowski spacetime, which minimize the energy defined by Mu-Tao Wang, PoNing Chen, and me.

f. Representations of surfaces in Euclidean space: For the Minkowski problem with a boundary, Joel Spruck has looked at a special case of the problem posed here. Bo Guan also studied the problem mentioned in this section.
g. Minimal surfaces in three manifolds: The problems pointed out in this paragraph are interesting, although not yet solved. The question of multiplicities of eigenvalues of minimal submanifolds of the sphere remains an intriguing, unsolved problem.
h. Submanifolds in higher dimensional space: Most problems in this area are unsolved. There are special Lagrangian submanifolds and isoparametric submanifolds that are particularly relevant to recent activity in string theory.

2 Intrinsic geometry

a. Constraints on the full curvature tensor: B. Wilking has done work on constructing metrics with positive sectional curvature. Peter Petersen and Frederick Wilhelm claim to be able to construct metrics with positive sectional curvature on the exotic sphere. But this claim has not been checked thoroughly. S. Brendle and R. Schoen proved the famous quarter pinching problem using Ricci flow, showing that manifolds with quarter-pinched curvature are diffeomorphic to the quotient of the standard sphere.
b. The Ricci tensor: Although much progress was made in the last ten years, most problems mentioned here are still not solved.
c. Scalar curvature: Not much progress has been made.

3 Conclusion

I have mentioned ten open problems here, some of which have been discussed in other places.

Commentary on Geometry and Spacetime

In this note, I explained how modern geometry has advanced our thinking about space-time. I described, for instance, a construction due to Strominger, Zaslow and myself for understanding a very important phenomena: mirror symmetry, which appeared in conformal field theory attached to Calabi-Yau space. This phenomena was discovered by Greene and Plesser and by Candelas et al. in 1989. It has produced spectacular results in both string theory and algebraic geometry. It was only after this discovery that mathematicians started to be impressed by what the string theorists were doing. The calculation of so-called instantons was remarkable, as it resolved some old problems in enumerative geometry. Even though this approach led to successful calculations, it was not initially known how the mirror manifolds themselves led to these results. The discussion up until then had been rather ad hoc.

When Eric Zaslow became my postdoctoral fellow, we started to explore the meaning of mirror symmetry through the newly developed brane theory. Around that time, Andy Strominger came to Harvard to interview for a professorship in the physics department. He had been an old friend, and we started to work on this project with Eric. In the end, it turned out to be a very successful collaboration. We ultimately achieved a good geometric construction of mirror manifolds in a nice, coherent manner. The mirror symmetry possesses a fiber-wise duality along a special Lagrangian torus fibration of the Calabi-Yau manifold over a three dimensional manifold. It is the real version of Fourier-Mukai transform although much still needs to be proven in the future. Nevertheless, the mathematics developed based on our strategy has been powerful.

In 1998, I wrote a paper with Witten on those manifolds that are relevant to the theory of AdS/CFT correspondence. We studied the conformal boundary of complete Einstein manifolds that are asymptotically hyperbolic. We found topological constraints on these manifolds.

Soon afterwards, I realized that some of the arguments could be adapted to my work with Richard Schoen on the formation of black hole due to boundary effect. I wrote a paper that appeared in the journal, ATMP.

I gave a talk on this paper in the mathematics center of the Chinese University of Hong Kong. L. F. Tam showed interest in this work. I advised him to read the paper of Robert Bartnik "Quasi-spherical metrics and prescribed scalar curvature" which appeared in JDG in 1993.

Soon afterwards, Tam showed me what he could do if the boundary is a metric sphere bounding a compact manifolds with nonnegative scalar curvature. He was able to obtain information on its total mean curvature. I was very impressed and encouraged him to generalize his theorem to spheres that are not round. He accomplished it after Shi came to visit him as a postdoctoral fellow. Their paper was accepted by me in JDG in 2002.

In early 2002, I discussed their work in a seminar at Harvard. During the seminar discussion, I concluded that their expression should be related to the concept of quasi-local mass. I had no idea that Brown and York had discussed such a quantity in a 1993 article on quasi-local mass in Physics Review D.

Since the Shi-Tam result only deals with the "time symmetric case" in general relativity, I went ahead with Melissa Liu to extend the work of Shi-Tam to provide a definition of quasi-local mass for a general spacetime that satisfies the local energy condition. The treatment of the general case is rather nontrivial. In fact, I was told that Shi and Tam did not believe that we could do the generalization at the beginning. There was a nontrivial proposition to handle the discontinuity of mean curvature. After I finished the work with Melissa, I gave talks on it and I was informed by the audience of the prior work of Brown-York. I put in the reference to Brown-York after that.

The paper of Liu-Yau was the first paper that can define a quasi-local mass for a general spacetime that is positive. The arguments of Shi-Tam and Liu-Yau were later used extensively by Mu-Tao Wang, Po-Ning Chen and myself where we provided the first good definition of quasi-local mass that is physically reasonable and is positive under natural conditions.

Commentary on Geometry Motivated by Physics

This is an essay written on the occasion of Professor Chen-Ning Yang's 80th birthday celebration was held at Tsinghua University. It was a big occasion. Many distinguished scientists and government officers came for the celebration. Many people, including Professor Tsung-Dao Lee, who did not come, sent flowers and gifts. Lee and Yang were the most celebrated scientists of our generation. They were the first Chinese Nobel Laureates. That meant a great deal to my generation, for we started to feel confident that Chinese could excel in a subject that had been dominated by the Western World for centuries.

Yang visited Hong Kong in 1964 to give a talk. The hall was packed. I was a high school student. Although I was not able to attend the talk, I still managed to feel his influence by reading newspaper accounts. I was finally able to meet him in Stony Brook in 1972 when Chern visited his home and I was the one to drive Chern around in town. They had intensive discussions on the situation of Chinese scientists in America. That was the time when American government was starting to break ice in its relationship with China. I have since been rather friendly with Professor Yang, for after ten years, he started to have close interactions with the Chinese university of Hong Kong, where I attended college. I hope to write about our interactions sometime in the future. I wrote this essay based on my admiration of Yang's work in mathematical physics. I summarize my own work in this brief essay. An important part was the description of the fiber product construction of mirror manifolds based on SYZ construction, which can be used to realize the mirror map from odd-dimensional cohomology of one Calabi-Yau manifold to the even-dimensional cohomology of its mirror partner. I knew this construction a year after the appearance of the SYZ paper. But since there was no concrete calculation, it was not published until this occasion. I still think that verifying this construction would make for an interesting project.

Commentary on Complex Geometry: Its Brief History and Its Future

In 2005, I gave a talk at the University of Science and Technology of China on the occasion of my friend Sheng Gong's 75th birthday. This is the outcome of the survey that collected some of my views on the development of complex geometry.

S. Gong was very happy that I attended his birthday celebration. Prior to that he had visited me many times in San Diego. At one point, when my mother stayed with me briefly in 1985, he came to play the Chinese game Mahjong with her. I am grateful to him for his friendship.

Overall, S. Gong is an excellent complex geometer who is very much underrated by the Chinese mathematics community. I was told that his standing in China was hurt by some personal differences he had with some academicians there.

For my part, I was glad to speak on behalf of S. Gong, as well as to put together this survey, which may be helpful to younger generations.

Commentary on Geometric Aspects of the Moduli Space of Riemann Surfaces

This is a survey paper on the analytic and geometric aspects of the moduli space of Riemann surfaces based on my work with Kefeng Liu and Xiaofeng Sun. Forty years ago, when I was at Stanford, Scott Wolpert was a student who was interested in the Weil-Petersson metric on the moduli space of Riemann surfaces. He told me that his calculations showed that the metric is not complete. At my invitation, Halsey L. Royden came to my seminar to give several talks about the curvature of the metric. I suggested to H. L. Royden at the time that he look into the Weil-Petersson metric defined by Kähler-Einstein metrics. He did interesting calculations on its curvature and, a few years later, he gave some talks in my seminar at the Institute for Advanced Study in Princeton, where a large group of mathematicians working on several complex variables had gathered together. This was around 1982.

But I have also been curious about the structure of the canonical Kähler-Einstein metric on the moduli space itself. Since the Teichmüller space is a domain with a rather non-smooth boundary, the work that I did with S.-Y. Cheng could only prove that the intrinsic Kähler-Einstein metric is almost complete. But I then observed that I could use a result from Clifford Earle on the completeness of the Caratheodory metric for Teichmüller space. The reason is that it is not hard to prove that the Kähler-Einstein metric is dominated from below by the Caratheodory metric. (Later, I managed to prove that the Kähler-Einstein metric is always complete on any bounded domain of holomorphy.) In any case, I was curious about the more detailed structure of the Kähler-Einstein metric on the Teichmüller space. I thought that the negative of the Ricci tensor defined by the Weil-Petersson metric should be a good candidate to model the Kähler-Einstein metric. A great deal of calculation is needed. Only after many years could this dream be realized, thanks to the calculations of two younger mathematicians: K. Liu and X. Sun. In the course of this proof, we also needed to make use of the method that I developed in proving the Calabi conjecture.

Liu, Sun, and I proved that the geometry defined by the Kähler-Einstein metric is bounded. And we also obtained detailed information about the Kähler-Einstein metric near the boundary divisor of the moduli space. This is important for applications of the metric to study the algebro-geometric properties of the moduli space. For example, for the Deligne-Mumford compactification, we proved that the log cotangent bundle is stable. And we managed to obtain some inequalities on the Chern numbers as well.

In the course of this research, we proved that all the known canonical metrics, besides the Weil-Petersson metric, are complete. We also proved that the Weil-Petersson metric is good in the sense of D. Mumford.

Some of this work can be generalized to the higher dimensional moduli space of Kähler-Einstein metrics. Carrying out this analysis in even greater detail could be useful for applications in both algebraic geometry and string theory.

Commentary on Solution of Filtering Problem with Nonlinear Observations

In the nineties, my brother Stephen became interested in problems of nonlinear filtering. His curiosity about such problems was sparked by Roger Brockett of the Department of Applied Sciences at Harvard University. In 1983, Brockett was invited to give a talk at the ICM on how to solve the nonlinear filtering problem by solving an explicit form of a parabolic equation called the Zakai equation. (The probability density can be described by such an equation whose coefficients may involve noise, which by definition is not smooth.) When the coefficients are linear, it produces the well known Kalman filter.

Brockett proposed to use the methods of Lie algebra to solve the equation. My brother made significant progress on this problem. In the end, he proved that in order for Brockett's method to work, the coefficients must assume a certain restrictive form. Stephen asked me for help in treating the general case. First of all, it was not hard to figure out the conditions under which a solution would exist that is positive and integrable. In fact, the solution has to be in such a form because we were looking for the probability density. The analysis is not difficult, and we showed how to estimate the solution based on approximations of solutions of a finite but large ball. Some of the existence theorems with more restriction conditions on coefficients were known. But for control theory, an existence theorem is not the main objective. What is needed for practical purposes is to actually calculate the solutions.

One day at my home in Belmont, we figured out a way to do this using the special structure of the Duncan-Mortensen-Zakai equation to reduce the problem to the calculation of equations with no observation terms. We separated the problem into two parts: one part can be calculated offline and is independent of the information that one gathers. The online part is a simple algebraic calculation. I thought this was a remarkable fact that had not been appreciated by the mathematicians who normally worked on this kind of problem. The actual numerical calculation can be carried out without much trouble when the space dimension is one. Recently, Wen-Wei Lin and his students in Chiao Tung University were able to work with me to carry out the calculation in dimensions up to three. I believe that our method will lead to industrial applications.

Commentary on the Proof of the Poincaré Conjecture

During the summer of 2006, there was a lot of coverage by news reporters regarding Perelman's work on the proof of the Poincaré conjecture. The paper written by Perelman is based on deep ideas he had in geometric analysis, although some of those ideas also drew from the earlier fundamental works of Hamilton and from my previous work with Peter Li and other researchers in fields related to geometric analysis. Apparently some mathematicians, who were not experts on analysis, were not able to appreciate the delicate ideas behind these analytical works. Many reporters were misled by the premature opinions provided to them. As a result, unnecessary controversy was created. I wrote to the *Notices of American Mathematical Society* to clarify my views. This is the letter I submitted.

In the early spring of 2013, Lizhen Ji asked me to write comments about my collected or selected works. I was too busy at the time to take on such a task. At one point, however, I gave in to his request and decided to write comments about my survey articles. Upon tallying them up, I was surprised to see that I had written far more survey articles than I had remembered.

Since I was a child, I have always been interested in history. Hence when I started to write these commentaries, I tried to stick to the facts to the best of my memory. I also consulted friends who participated in these events and looked at letters and emails that I had kept over the past forty years.

This does not mean that there are absolutely no mistakes in the statements. Nevertheless, I believe that these accounts can be interesting — and maybe even important — for students who'd like to know something about how the various papers were written and what my friends and I thought about the approaches we took.

In the course of putting together this collection, I received strong support from Lizhen Ji, Hao Xu, Kefeng Liu, Shiu-Yuen Cheng, and Hung-Hsi Wu. I am also very grateful to the publishers led by Liping Wang, Yushan Deng, and others. My friend Steve Nadis agreed to be the consulting editor for this project. I am extremely thankful for all of their help, without which this project likely would not have materialized.

Shing-Tung Yau
June 30, 2014

Commentary on Perspectives on Geometric Analysis

My teacher, Professor Chern, passed away in December 2004. I gave a talk in his memory on the 70th anniversary of the Chinese Mathematics Society. At the same time, I was invited by Susan Friedlander of the Bulletin of the AMS to write an article for the journal memorializing Chern.

I wrote the essay on geometric analysis and sent it to many people. I did not write the introductory portion at the beginning, and all the people to whom I sent the article showed great interest in my essay. Some people, such as Daniel Stroock, were telling me how much better they understood the philosophy behind geometric analysis after reading my tribute to Chern.

However, after I wrote the introduction to this essay, people started to have different opinions. Apparently, they were misled by some people concerning the statement in my introduction about the difficulties I encountered in China while trying to introduce the subject of geometric analysis in China. The fact is that the accusation against Professor Wang by his student, which I mentioned in the introduction of the essay, was even worse than what I had written. In fact, in 1987, I received a letter from K. C. Chang of Peking University demanding that I write to James Eells, who directed the mathematics division of the theoretical physics center in Trieste at that time, to stop Professor Wang from visiting Trieste on account of Wang's personal behavior that had nothing to do with mathematics. I was astonished by such an accusation, as it would imply that many of my friends in Europe and in America should be condemned as well. Since K. C. Chang was the former teacher of G. Tian, I consulted Tian about Chang's letter. He took Chang's side in this matter, but I decided not to do anything. This upset both K. C. Chang and G. Tian. Other people started to attack me, including D. Stroock. Stroock's statement, which appeared in the New Yorker, showed that he still did not understand the subject of geometric analysis even after reading my article. Nevertheless, the net result of this backlash was that S. Friedlander refused to publish my paper. It has been about eight years since I wrote this long survey, and I am still proud of the effort. I spent a lot of time preparing this article, as it was meant to honor my former teacher, Professor Chern. I also consulted many experts in this subject, who contributed their valuable opinions. I am confident that beginners in geometric analysis, and even others more proficient in the field, could benefit from reading this article.

Commentary on a Survey of Calabi-Yau Manifolds

In this section, I will talk about my proof of the Calabi conjecture–and my journey leading to and from that proof–some of which was described in my book with Steve Nadis, *The Shape of Inner Space.*

When I was a first-year graduate student, I learned from J. Milnor's paper on curvature and the fundamental group. I also read F. Hirzebruch's book, *Topological Methods in Algebraic Geometry*, and Joseph Wolf's book, *On Spaces with Constant Curvature.* I was fascinated by the concept of curvature of Riemannian manifolds. I generalized the paper of Alexandre Preissman, which says that any abelian subgroup in the fundamental group of a compact manifold with negative curvature is cyclic.

When I read this paper, I wondered what would happen if I allowed the manifold to have nonpositive curvature instead of strictly negative curvature. I managed to prove that any solvable subgroup of the fundamental group is represented by a totally geodesic flat torus within the manifold. I did this during the Christmas holiday while reading Milnor's paper in the mathematics department library in Berkeley's Campbell Hall (a building now used by the Astronomy Department). When school started in January 1970, I showed the result to B. Lawson who was teaching an elementary course on differential geometry

He was excited by it and asked me a lot of questions about such manifolds. We started to work on a related problem, a splitting theorem for compact manifolds with nonpositive curvature. When the fundamental group has no center, and if it is the product of two subgroups, we proved that the manifold itself must split as a metric product with each factor carrying the subgroup factor. For some technical reasons, we assumed the manifold is real analytic. We worked quite hard on this problem, and I remembered that we called each other almost every day. We often spent more than an hour in the phone exchanging our ideas. Little did I know that his wife was not happy about all the correspondence. A few years later, after they divorced, I heard from his ex-wife that this had put an additional strain on their relationship.

Both Chern and J. Wolf were senior geometers at Berkeley who happened to be on leave that year. They returned to Berkeley around May. Lawson reported to them what we had done. Chern was the most respected Chinese mathematician, and J. Wolf was his student at Chicago. As mentioned above, I had already gotten to know the name of J. Wolf through his nice book, *Spaces of constant curvature*, which I studied quite thoroughly.

Chern was very happy about our achievement, while J. Wolf was not terribly happy, as he and Detlef Gromoll had been working on the splitting problem independently. But he liked my first result and told me that he had made a conjecture on it in one of his papers. It was difficult to recall exactly what happened. But when Lawson and I submitted the paper to the *Annals of Mathematics*, the first paper was accepted and the second one was rejected. It was accepted later in the *Journal of Differential Geometry*, helped, perhaps, by a recommendation from

Chern.

At that time, Jim Simons was building up the Department of Mathematics at Stony Brook. He was trying to hire B. Lawson, who was taking a sabbatical leave, spending the fall semester of 1970 in Brazil and the spring semester of 1971 at Stony Brook. Simons heard about my story through B. Lawson. He told Lawson that he was ready to hire me when I finished writing the two papers.

During the summer of 1970, I asked Chern to be my advisor. He accepted my request. But within three months, he suggested that I use what I did on manifolds with nonpositive curvature as my thesis.

Reluctantly, I followed his advice, although I felt that I had much more to learn. In the thesis, I wrote about the idea of using geometry to study combinatorial group theory, which was subsequently studied extensively by M. Gromov in the 1980s. (when I was in Hong Kong, I studied the works of Busemann and Alexandroff. I thought that they could be used to handle group theory, similar to what Margulis and Milnor did.) But I wanted to move to other parts of geometry as I thought the world of geometry should be broader, and it should be useful to describe and explain the physical world.

In the academic year of 1969 to 1970, I attended many seminars in Berkeley, including a seminar ran by S. Kobayaski and T. Ochiai. I learned quite a lot of complex geometry from this seminar. During the memorial conference for S. Kobayashi, I found out from T. Ochiai that S. Kobayaski was the chairman of the admission committee for admitting students to Berkeley in 1968. He insisted that Berkeley should admit and offer me the best fellowship. I am most grateful to S. Kobayaski as the fellowship was indeed important for me. At the end of his seminar, S. Kobayaski gave me the full manuscript of his unpublished book on hyperbolic geometry. I was honored to receive the manuscript and read it thoroughly. I was very much impressed by the way he defined his metric. The contents of that manuscript had a significant influence on my later work.

I was starting to learn general relativity and was very much excited by the fact that the Ricci tensor describes matter distribution in the universe. The Einstein equations show that Ricci curvature is exactly the right kind of curvature, giving the correct number of independent variables for the metric.

At the time, I was learning partial differential equations from Charles Morrey. I had already made up my mind that a fruitful way to advance our work in geometry would be through the powerful method of analysis and nonlinear differential equations. Hence I was determined to devote my full energies to the understanding of Ricci curvature.

Because of the Einstein equation, zero Ricci curvature is really a very elegant constraint on the manifold. It describes a vacuum solution to the Einstein equation. The very fundamental question for me was whether we could find a nonsingular compact vacuum whose full curvature is not zero. I felt this problem would provide the key to understanding the Ricci curvature of a manifold.

At the time when I was considering how to construct compact manifolds with non-flat Ricci flat metrics, nobody believed that such manifolds existed. Then I found Calabi's Congress talk in 1954, which described a Kähler geometric version of the problem I had been considering. I was excited to learn that the problem is much easier to handle in Kähler geometry. The Ricci curvature is much more elegant because of the simplicity of the representation of the first Chern class in curvature form. The Einstein equation, which is a tensor equation, is now reduced to a scalar equation. The only problem is that it is a fully nonlinear equation.

Calabi told me that André Weil was interested in the conjecture but decided that the method of nonlinear elliptic equations was not mature enough to tackle it. However, because it provides a key step for understanding Ricci curvature, I concluded that, regardless of whether or not the Calabi conjecture was true, its solution either way would unlock the mysterious structure of Ricci curvature, at least in the Kähler category.

So I felt I needed to solve the Calabi conjecture at all costs. I remember telling B. Lawson, who at that point was not so interested in Kähler geometry, that there was nothing as exciting in mathematics as solving the Calabi conjecture.

After I graduated, I went to the IAS in Princeton. Many young postdocs were there. The most important one for me was Nigel Hitchin, who was an assistant to Michael Atiyah. Most of the young people had dinner together each night at the IAS. We had conversations on many topics, including mathematics. N. Hitchin and I talked about the Calabi conjecture, and we concluded that it was "too good to be true". The only question was how to construct a counterexample.

When I was a graduate student, I learned of the work of Gromoll-Meyer and also the later work of Cheeger-Gromoll, which presented the splitting theorem for manifolds with nonnegative Ricci curvature. I studied the papers of Gromoll-Meyer and Cheeger-Gromoll, hoping that would help me find counterexamples.

If the Calabi conjecture were true, we would get certain metrics with nonnegative Ricci curvature. We wanted to derive some of the consequences of these metrics and see if that might lead to contradictions. The splitting theorem of Cheeger-Gromoll has a natural Kähler analogue, which implies the so-called structure theorem of Calabi-Yau manifolds. This was a fact I recognized, assuming the Calabi conjecture were true.

Hitchin and I also tried to use the curvature representation of the Euler number and index (due to Chern) to derive an inequality between them. This was first done by Chern, J. Milnor, Marcel Berger, and later by Alfred Grey and John A. Thorpe. But N. Hitchin provided a nice refinement. Based on this inequality, Hitchin observed that a strong constraint on K3 surfaces could be obtained. The only metric with nonnegative scalar curvature on K3 is the Kähler Ricci flat metric, which was not known to exist at that time.

These inequalities among characteristic numbers have analogues for general Einstein manifolds. In order to disprove the Calabi conjecture, we needed to use the Kähler versions of these inequalities. The extreme cases of those inequalities are metrics covered by the ball, torus, or complex projective spaces. I tried to use these to produce counterexamples, but I did not succeed.

Ironically, once the Calabi conjecture was proved in 1976, my strategy of finding counterexamples became useful as I could use my failed counterexamples to prove powerful theorems. For example, I showed that for manifolds with ample canonical line bundle, inequalities of the Chern numbers can be derived. The equalities offered a beautiful characterization of the quotient of the ball in terms of Chern numbers. When the first Chern class is zero, the second Chern class is nonnegative when cupped with the Kähler classes. Although I knew such arguments in 1972, I did not consider it worthwhile to publish them unless the Calabi conjecture were true. When I later proved the Calabi conjecture, these statements proved to be both interesting and important.

I finished the proof of Calabi conjecture in 1976, while I was a visiting professor at UCLA. After a couple of weeks, David Gieseker told me that David Mumford was giving two talks and one of them would be at UC, Irvine. At that time, I lived in San Fernando Valley, which is located in northern LA. It took me almost three hours to drive to Irvine to listen to the seminar given by D. Mumford. It was really an important occasion for me because D. Mumford mentioned the work of Fedor Bogomolov, who had established a new Chern number inequality for algebraic surfaces. But the constant was still not optimal. I was excited by it since it looked familiar to me. I went back home that evening and looked up what I had done before in my attempt to disprove the Calabi conjecture. I found that the best constant can be achieved by my method using differential geometry. The method can also be generalized to higher dimensions. That was quite remarkable, as I also found that when the best constant is achieved, the manifold is, in fact, covered by the ball. A corollary of this is that any Kähler manifold homeomorphic to the complex projective space is biholomorphic to the standard complex projective space. In two dimensions, this was a conjecture by Francesco Severi.

I wrote a letter to D. Mumford the next day, explaining the proof. He was delighted. Perhaps that was the reason why Harvard offered me a professorship a couple of years later.

Independently, Yoichi Miyaoka found that the algebraic method of F. Bogomolov can be strengthened to prove the inequality in two dimensions with the best constant. However, algebraic methods, up to now, have not been able to treat the equality case and, therefore, the global rigidity of complex structure on the projective plane has not been proved by purely algebraic geometric means.

But let me go back to 1973. In that summer, when I attended the International Conference on Differential Geometry at Stanford University, I was still trying very hard to come up with a counterexample to the Calabi conjecture. During that conference, I told a few friends that I had found such a counterexample. The news spread quickly, and many people requested that I give an informal talk regarding the counterexample. About thirty people gathered together on the third floor in the Department of Mathematics of Stanford to hear what I had to say. I described how I used the structure theorem of Cheeger-Gromoll to produce a counterexample. I constructed an algebraic surface whose first Chern class is numerically semi-positive. I thought that if the Calabi conjecture were true, I could construct a Kähler metric with nonnegative Ricci curvature. Hence I could apply the structure theorem of Cheeger-Gromoll, and arrive at a contradiction.

Both Chern and E. Calabi thought it was a good counterexample. And Chern said that it was the best achievement of the entire conference. After two months, E. Calabi wrote to me and asked me to produce the details of the example.

I realized immediately that something had gone wrong, essentially because numerical semi-positive is not the same as semi-positive. I worked on fixing that problem and tried very hard to produce many different examples. None of them worked out, and all the arguments broke down in a rather delicate manner. I finally decided that there should be no counterexamples to Calabi's conjecture and turned around to try to prove the conjecture. (The example I used was quite interesting, although I did not publish it.)

In the fall of 1973, I undertook the task of proving the Calabi conjecture. But at that time, techniques for handling elliptic equations were not mature enough to treat fully nonlinear equations. Of course, linear elliptic equations were very well understood. Many estimates related to linear elliptic equations served as the keys for developing tools for solving fully nonlinear equations. Due to the exciting progress of minimal surfaces equations, people were having more success with quasilinear elliptic equations due to the contributions of C. Morrey, John Nash, Ennio De Giorgi, and Jürgen Moser. Working in the other direction, Herbert Federer, Wendell Fleming, Frederick Almgren, and William K. Allard had also developed the powerful method of geometric measure theory. But geometric measure theory is tailor-made for variational problems related to minimal surfaces and is not quite applicable to fully nonlinear equations such as the Monge-Ampère equations.

It so happened that S.Y. Cheng and I, through the lecture notes of Chern, had been following the development of affine geometry, especially the pivotal work of E. Calabi. The target of affine geometry is to study the properties of hypersurfaces in affine space that are invariant under affine groups. Among the important hypersurfaces that were studied extensively, there are affine spheres—convex hypersurfaces whose affine normals converge to a point. When that point is at infinity, meaning that all affine normals are parallel, it is defined by the equation that the determinant of the Hessian of a convex function is constant. This is the real Monge-Ampére equation, which can be considered as the real analogue of the complex Monge-Ampére equation that is needed in the proof of the Calabi conjecture in Kähler geometry. In fact, if we complexified the real geometry, we would obtain exactly the complex Monge-Ampère equation. This fact appeared later in the study of the SYZ (Strominger-Yau-Zaslow) program for the geometric realization of the mirror symmetry of Calabi -Yau manifolds.

In 1980, there was a big international conference held in Beijing organized by Chern called the Differential Equations and Differential Geometry Conference. (Apparently, we had convinced Chern that differential equation should be an important component of geometry.) I gave a talk on open problems in geometry. I thought the subject of the real Monge-Ampère equations defined on affine manifolds offered a good avenue for Chinese mathematicians to begin their research on Kähler-Einstein metrics. S.Y. Cheng and I wrote a joint paper, which was published in the proceedings of this conference, where we discussed affine Kähler geometry. It was our attempt to study affine geometry based on methods of Kähler geometry, a line of inquiry that led to an analogue of the Calabi conjecture. We were mostly focused on nonsingular affine manifolds. Although I thought the topic was interesting, the scope of possible applications was limited as we proved that any compact nonsingular affine Kähler manifolds with zero Ricci curvature must be a torus. When the SYZ program was proposed by Strominger, Zaslow and myself in 1996, I realized that what was needed was an affine Kähler manifold that is singular along a codimension two singular set with prescribed monodromy. The scope of the SYZ geometry is much larger and is still being pursued now. In any case, in this paper in 1980, Cheng and I published the first solution of the Dirichlet problem for real Monge-Ampère equations, the proof of which we knew in 1978. We managed to avoid the estimate of second derivatives up to the boundary for such equations, which was not known at that time. Our method was to blow up the boundary to avoid the boundary estimate. The result is a solution that is smooth in the interior. While the gradient of the solution is Lipschitz up to the boundary, it is not obvious that it would be smooth up to the boundary. Nirenberg announced his solution of this problem with E. Calabi at the Vancouver International Congress of Mathematicians in 1974. But they withdrew it when they found a gap in their second order estimate. The second order boundary estimate was established a few years later by Luis Caffarelli, Joseph Kohn, Louis Nirenberg and Joel Spruck. These are important breakthroughs. It is interesting to compare their results with the earlier work I did with Cheng. We were probably the first to use the interior estimate alone to solve the boundary value problem. On the other hand, they obtained the most impressive boundary regularity, which I tried to find for years. The results are independent as it depends on different regularity assumption. Cheng and I worked on these problems for the sake of geometric applications (interior regularity is good enough for geometric applications). We could have rewritten it without using geometric language, but we didn't, making it hardly accessible to experts of partial differential equations. But many of the estimates that we had achieved are still useful. In fact, I believe that many of the estimates in real Monge-Ampère equations discovered much later can be derived from our geometric estimates.

To us, the great importance of studying affine spheres is that we have a geometric means of understanding the third order estimate in Monge-Ampére equations. This is because the third order derivative of the unknown function defines a rank three tensor in affine geometry, classically called the Pick form. Applying the Bochner method to the Pick form turns out to be fruitful. E. Calabi was the first person to study this. He proved a Bernstein-type theorem for the real Monge-Ampère equation for the parabolic affine sphere of dimension less than five. S.Y. Cheng and I studied this problem extensively. It turned out that the solution to the Bernstein problem for the affine sphere is related to the second order estimate: we needed to estimate the completeness of the affine metric. We succeeded in giving a very nice second order estimate of the real Monge-Ampère equation by following the method I used in the gradient estimate for harmonic functions on manifolds with nonnegative Ricci curvature. We found that it made sense to use the unknown function to construct the cut-off function.

In the spring of 1974, we finished our proof of the Bernstein problem for real Monge-Ampère equations, feeling rather pleased by our achievement. We told Chern and L. Nirenberg, and they liked the result. We also followed our general strategy to solve another Bernstein problem, which E. Calabi also studied: the maximal slice problem. One wants to prove that the only maximal space-like hypersurface in n-dimensional Minkowski spacetime is linear. E. Calabi managed to do this in low dimensions. S.Y. Cheng and I applied our technique and succeeded in proving the completeness of the metric to resolve this Bernstein-type problem. A similar kind of estimate was later used by my student, Robert Bartnik, to generalize this to general spacetime, with Claus Gerhardt earlier taking on the compact space-like version. The results of Gerhardt and Bartnik played an important role in classical general relativity.

After we submitted our paper for publication in *Communications in Pure and Applied Mathematics*, we found that the Russian geometer A. Pogorelov also proved the Bernstein problem for real Monge-Ampère equation using different estimates. At first, we suspected that he had made a mistake in the choice of coordinates in his estimate. But we later realized that his proof was correct, and we really admired the great power of his method. Apparently, he also thought highly of our work and recommended that I give a plenary talk in the Helsinki Congress.

In 1974, S.Y. Cheng graduated from Berkeley and moved to the Courant Institute to be a Courant Instructor, while I was still in Stanford, struggling with the estimates needed for the Calabi conjecture. My friends, Richard Schoen and Leon Simon, had been of great help. Although I started to learn partial differential equations from C. Morrey at Berkeley, I learned most of the classical theory of PDEs from L. Simon through his lecture notes and from the conversations with R. Schoen and him. It was very fortunate that my office was next door to L. Simon's. And Rick came to talk with me often. Rick and I went to dinner together almost every day at a Chinese restaurant called Moon Palace. The three of us wrote an important paper on the curvature estimate of a stable minimal hypersurface. We were very pleased by that paper, but it was rejected by the *Annals of Mathematics*. Leon was especially unhappy about this. The referee admitted ten years later, while he was drunk at a party, that it was his mistake, because he later found out that he and his students needed that estimate badly.

After my experience with the second order estimate of the real Monge-Ampére equation, I started to think about the second order estimate of the complex Monge-Ampére equation in the Calabi conjecture. I compared it with my previous work on the Schwarz lemma that I generalized to prove parabolicity of complete Kähler manifolds with nonnegative Ricci curvature. I needed to prove the completeness of the Kähler metric that was constructed, just the same as what I did with S.Y. Cheng for affine spheres or maximal space-like hypersurfaces.

The Schwarz lemma compares two metrics: one is given and the other one is unknown. But here I could get information only from the Monge-Ampére equation, and I could only keep on differentiating the equation to obtain the information. But it was not hard to see that the kind of terms we obtained were reminiscent of the Schwarz lemma when the linearized operator is used.

The computation for the Laplacian of the trace of metrics did not come up as nicely as in the Schwarz lemma, for the curvature did not help. And that was causing great difficulty for me. But then I remembered the work I did with S.Y. Cheng on affine geometry and maximal space-like hypersurfaces, recognizing that it could be useful to use the unknown potential as an auxiliary cut-off function. So by late 1974 and early 1975, I was able to use the Kähler potential to overcome the bad curvature terms and get an estimate of the metric in terms of the Kähler potential or, more precisely, the unknown function.

In the early spring of 1975, I had obtained all the third order and second order estimates in terms of the zeroth order estimates, but it took more time to prove the zeroth order estimate. Note that although the idea of Calabi on affine geometry can be used effectively to obtain the third order estimate on real Monge-Ampère equation, it was still a nontrivial task to accomplish the similar third order estimate for the Calabi conjecture. For the case of Kähler-Einstein equations with negative scalar curvature, it was a trivial exercise to use the maximal principle to find the zeroth order estimates. Hence I knew the existence of a Kähler-Einstein metric with negative scalar curvature by 1975. A year later, I learned that T. Aubin was also able to solve the existence of a Kähler-Einstein metric with negative scalar curvature independently.

At this time, L. Simon and I had each been awarded a Sloan Fellowship. I was not sure who nominated me. Most likely it was Robert Osserman. He had been very nice to me, and I am very grateful that he recognized me in my early days.

My girlfriend at the time was a postdoctoral fellow at Stanford from 1973 to 1974, but she moved to the Princeton Plasma Physics lab in 1974. I was hoping to visit her by using my Sloan Fellowship. I made the mistake of not writing to the department chair of Princeton directly. Instead, I wrote to W. C. Hsiang at Princeton, who, within a couple of days, told me that there was no office available whatsoever at Princeton, and I should not come. Apparently the office problem was solved a couple of years later when their chairman asked Hsiang to phone me to offer me a professorship.

I was taken aback by Hsiang's reply, but I still needed a host institution. When I met L. Nirenberg in Berkeley in April of 1974, he told me that although he was going to be on sabbatical leave that year, he welcomed me to visit the Courant Institute. Since S.Y. Cheng was then a Courant Instructor, and New York City is not that far from Princeton, I decided to visit Courant Institute. It was a nice surprise for me to find that L. Simon was using his Sloan Fellowship to visit Courant too.

That was a very pleasant visit. In addition to the many prominent professors at the Courant like Jürgen Moser, Joseph Keller, Henry McKean, and S. R. Srinivasa Varadhan, there were also younger Courant Instructors like Eric Bedford.

Courant was then having a special year, which placed the spotlight on hyperbolic equations. They had invited Andrew Majda, James V. Ralston and J. Thomas Beale. The people in hyperbolic equations were naturally installed in the better offices, while L. Simon and I were using a small office. That was actually nicer for us, as we were put in a quieter part of the Institute.

I found that New York people are quite different from those in California. The secretaries were not as friendly as I expected in the beginning, but once they knew us, they became very friendly. I still remember there was one secretary who laughed very loudly. She typed my papers.

For me the first challenge of visiting New York was housing. I could not find any studio with rent less than \$200 per month. And even with that kind of rent, I still could not find any studio because I only planned to visit New York for four months. I was very grateful to J. Moser for letting me use his apartment on Spring

Street, which was under rent control and not far from the Courant Institute. It was an apartment that Moser used when he was too tired to take a train back to his home.

I was really flattered, as J. Moser was a very well known mathematician, and I had the privilege of listening to his Marston Morse Lecture at the IAS in 1972. He helped me move to this furnished one bedroom apartment where the rent was only $50 per month. In any case, J. Moser treated me exceedingly well. He even came to the apartment to fix the plumbing.

I walked every day to the Courant Institute. It was interesting to observe how a car parked on the sidewalk could completely disintegrate in a matter of two days. Chinatown was nearby, and I walked to Chinatown for food often with Cheng's family, and sometimes with E. Bedford. S.Y. Cheng's first son, Bing Cheng, was born in New York. The benefits of the city were excellent during those days. I used to carry Bing during my walks in the neighborhood. It was especially fun that, twenty years later, Bing became my Ph.D. student. I did not cook, and I usually went to Cheng's apartment for dinner and enjoyed the food prepared by his wife almost every night.

It was wonderful that I was able to pursue my research. It was a curious experience for me that many important classes at Courant were held in the evening. I dropped in on some of these classes, including one given by Keller on wave propagation. I continued my work with S.Y. Cheng on real Monge-Ampère equations. And we wrote a few papers in a couple of months, an important one being our solution of the Minkowski equation in higher dimensions.

We were motivated to take on this problem because L. Nirenberg's thesis was on solving this problem in two dimensions, along with the Weyl embedding problem. Louie was traveling during my visit, but his office was very close to mine. Perhaps his spirit inspired us. S.Y. Cheng and I solved the problem in fairly great generality, starting from very weak solutions and then proving the regularity of the weak solutions. After we solved this problem, J. Moser asked us to present the solution in a seminar. He was very pleased. But by the time I was leaving Courant, we learned that A. Pogorelov also solved this problem using different methods. We initially thought there were some regularity issues in his approach, but it turned out to be a very beautiful piece of work. I should mention that I had studied the works of A. Pogorelov and Alexandrov when I was in Hong Kong. I always admired their depth. On the other hand, the great work of A. Pogorelov on the rigidity of closed convex surfaces in three-dimensional space without the regularity assumption still puzzles me.

Besides the Minkowski problem, S.Y. Cheng and I also solved several other problems, including the Dirichlet problem for rather general Monge-Ampère type equations. We also discussed the general theory of partial differential equations on manifolds.

Sometime around Christmas in 1975, I left Courant Institute and went back to Stanford. In the first quarter of 1976, I became quite ill. Since I was still single, I spent a week in L. Simon's apartment to make sure nothing serious happened to me. His wife Sandra was traveling, and his kids were able to take care of themselves. It was great to have a good friend available to help out during a difficult period.

In the meantime, I started to have some ideas about the zeroth order estimate that appears in the Calabi conjecture and worked on it when I felt up to it.

When the second quarter was over, I flew to Princeton to see my girlfriend. Since R. Schoen was graduating that year, I asked my other student, Andrejs Treibergs, to look after my car, which was a Volkswagen Squareback. He seemed pretty happy about taking care of it in my absence. I was also grateful to my secretary, Francis Mak, who was willing to help me store several pieces of luggage in her home.

On this trip, I was successful in winning my girlfriend's heart, and we agreed to get married in Los Angeles since her next job would be at TRW in Los Angeles.

I am lucky to have many good friends who are willing to help on important occasions. I wrote to Barrett O'Neill, whom I got to know at the 1973 Stanford conference, and also to Robert Greene, who was familiar with my work. They were both members of the UCLA mathematics faculty. I told them I would like to visit UCLA for a year, and I could use my Sloan Fellowship for the first quarter. It was settled immediately. My girlfriend was amazed, given that finding jobs in those years was not easy. We decided to get married in early September of 1976. And we needed to move all her stuff from her apartment in the East Coast to the West Coast. On the way, while I was driving, I kept thinking about two problems: the zeroth order estimates of the Calabi conjecture as well as the Poincaré conjecture. For the latter, I was thinking about using minimal surfaces or foliations by surfaces with constant mean curvature. For the portion on minimal surfaces, W. Meeks and I developed many more results after I arrived at UCLA. And the idea to study foliation by constant mean curvature surfaces evolved into a joint effort with Gerhard Huisken. That work was done when he visited UC San Diego to discuss the question of the center of gravity in general relativity.

I was not sure what my fiancée would think if she knew that I was doing research on these problems during our trip. While I was not able to write down my thoughts in any detail, those thoughts finally came together a few months later.

My mother was in Hong Kong and flew all the way to California for the wedding ceremony. And my brother Stephen came from the East Coast in August. We did not invite many people for our wedding. But Chern and his wife insisted on coming from Berkeley. We were very grateful for their kindness. It was not a big, flashy wedding ceremony to show to an important man like Chern.

Not long after I got married, I spent a lot of time to work out the zeroth order estimates. It seemed like a miracle that my ideas converged within a couple of weeks of our wedding. I was able to use the structure of the equation to find integral estimates. The upper estimate was relatively easy, as the potential is close to being plurisubharmonic, and I had a good estimate on the Green's function of the Laplacian. (The estimate of the heat kernel, or Green's function, was rather familiar to me, as I had written a paper on a related subject a couple of year earlier.) The upper estimate then followed from a mean value type inequality.

The difficulty came from the lower estimate. The structure of the equation, which was based on the fact that the Kähler form is closed, enabled me to get the estimate of the unknown. Once I got the estimate for all, it was fairly straightforward to obtain the lower estimate.

UCLA provided a good environment for me. I stayed in a small office right next to R. Greene's office. Besides R. Greene and B. O'Neill, there were many energetic young colleagues: W. Meeks, David Gieseker, Mark Green, Stanley Osher, James Ralston, and Andrew Majda. There were also visitors such as William Huebsch, who gave interesting talks on geometric invariant theory over characteristic p. I had always been interested in what D. Gieseker was doing, especially his work on algebraic surfaces of general type using geometric invariant theory. The question of the stability of the manifold always came back to me, and I told David that I wanted to understand them through Kähler-Einstein metrics. His work with S. Bloch on the positivity of Chern numbers for ample bundles also excited me. Up to that point we still do not have a good way of understanding this last question through differential geometry. I was very impressed by David's work, and I recommended that Stanford make him an offer, although it did not work out.

I taught a research course for the first quarter after I arrived at UCLA, and there were good students. I remember Peter Jones, who was a student of John Garnett at that time. Although he did not attend my class, we had lively discussions that impressed me. I recall that one student in the class was counting how many mistakes I made in my lectures. During the second quarter, I was asked to give a low-level course on calculus for economists. Most of those students were not interested in studying. I forced them to study hard and take the subject seriously. As a result, they gave me very poor teaching evaluations. When the university wanted to offer me a professorship, those teaching evaluations caused a serious problem. UCLA had to call my students at Stanford to testify that I was a fine teacher.

As I mentioned earlier, when I was writing up the proof of the Calabi conjecture, D. Mumford came to give a talk at UC, Irvine. I learned from him about an unsolved problem in algebraic geometry. I used my newly-proved work on the Kähler-Einstein metric to solve several important problems related to this question raised by Mumford. But the most exciting theorem to me was the existence of Kähler Ricci flat metrics on manifolds with zero first Chern class. This beautiful statement, proposed by Calabi more than two decades earlier, is one I had been longing to prove since I was a graduate student.

I was very careful to make sure that all the details of my proof of Calabi's conjecture were accurate. I went through my argument three times, finding everything satisfactory before going public with it in October, 1976 in an announcement that was communicated by Chern to the *Proceedings of the National Academy of Sciences*. Some of the consequences of the Calabi conjecture were so obvious that I did not write up all of them in the announcement, but they appeared in my Helsinki talk in 1978. At that time, my burden was to go around the country to convince experts that my proof was correct. It happened that Isadore Singer invited me to visit MIT for a month. I accepted his offer and went to the East Coast in mid-November. But I wanted to meet E. Calabi first. So I flew to Philadelphia and spent a couple of days at the University of Pennsylvania, where E. Calabi, Jerry Kazdan, and Frank Warner were there attending my lectures throughout. J. Kazdan took detailed notes.

Unfortunately, J. Kazdan shared those notes with Thierry Aubin without letting me know. T. Aubin had found a proof of the existence of the Kähler-Einstein metric for negative scalar curvature independent of mine in 1976. But the notes of J. Kazdan tempted Aubin to go ahead to claim the full Calabi conjecture. It was a noble act for J. Kazdan to acknowledge that he told T. Aubin details of my proof in a note after my published paper. Due to my respect for my mentors at the Courant Institute, I submitted my paper to *Communications of Pure and Applied Mathematics* in late 1976, but it took a long time before it appeared in 1978.

I wanted to talk with L. Nirenberg, whom I always considered to be a mentor. He was not at Courant at that time. So after several arrangements, E. Calabi and I agreed to meet Louis on Christmas day in 1976. We went through the paper carefully, and I was happy that they did not find any mistakes. But we did have great difficulties in finding a place for lunch except in Chinatown.

Before I went to New York, I spent one month at MIT. My mother was staying with my brother Stephen at Harvard, as Stephen was working there as a Benjamin Pierce instructor. They stayed in a small apartment in Somerville.

The time I spent in MIT was somewhat boring, as Singer was extremely busy due to his marital problems and there were not many geometers around. But I did have plenty of time to think in my apartment near MIT. And I actually completed the writing of my paper on the Calabi conjecture in that period. I spent a lot of time working out the case of a Kähler metric with a singularity along a divisor. This took much longer for me to write compared with other parts of that paper. It is funny that a student of mine later made serious attempts to mislead people into thinking that I did not have results on singular Kähler-Einstein metrics, given that roughly half of my first paper was devoted to this topic. Much of the subsequent work on singular Kähler-Einstein metrics is patterned after this paper of mine and my other papers with S.Y. Cheng.

I had a very nice dinner with I. Singer one day in Cambridge during my visit in 1976. He shared with me some exciting work on Yang-Mills instantons. This eventually led to my interest in the subject. Six years later, I completed an important piece of work with Karen Uhlenbeck on Hermitian Yang-Mills connections, which can be considered a generalization of anti-self-dual Yang-Mills connections. I spent much of the rest of my time in Cambridge in Harvard's mathematics department, where I gave lectures and communicated with some algebraic geometers. I met Boris Moishezon and Andrey Todorov, who had left Moscow not long before. We discussed Ricci flat metrics on K3 surfaces. At that time, the proof of the Torelli theorem was a very hot topic, and injectivity of the period map was just shown to be true, even for Kähler manifolds. It was done by Michael Rapoport and Daniel Burns, generalizing the important work of Ilya Piatetski-Shapiro and Igor Shafarevich on the algebraic case. The surjectivity was not quite known in the West yet, but it had been accomplished by Viktor Kulikov in Moscow. In any case, I told A.

Todorov about the metric and that I believed the metric could be helpful in giving an analytic proof of these results. A. Todorov agreed and figured out the best way to proceed. He proposed to make use of a rational pencil of complex structure that can be created based on the Ricci flat metric, as was observed by N. Hitchin. This pencil played an important role in Todorov's proof. Some further details were worked out by Y.-T. Siu. Drawing on ideas from D. Sullivan, Todorov, and me, Siu proved that every K3 surface is a Kähler manifold. The construction of a closed two form with positive (1,1) component was based on the Hahn-Banach theorem. The general idea of using the Hahn-Banach theorem for constructing a geometric object was due to Sullivan. I was fascinated by Sullivan's idea to construct a metric that makes the leaves of a foliation to be minimal surfaces. I suggested to Siu that Sullivan's idea could be used for the construction of the required two-form.

When I was preparing my talk for the ICM in Helsinki, I outlined my program for the generalization of the existence of complete Kähler Ricci flat metrics in noncompact manifolds. This is more difficult than the case of negative scalar curvature.

I explicitly wrote down the program for classifying such manifolds. Demonstrating the existence of such metrics reduces to finding a local ansatz for the metric at infinity. Before I gave the talk in Helsinki, I managed to prove that my ansatz works, assuming the divisor at infinity is nonsingular. I tried and did not succeed in generalizing the ansatz to general divisors. Finally, in 1988, in order to help my former student G. Tian's career, I told him my ideas and asked him to write up the existence theorems for a complete Ricci flat metric on a Zariski open set of a Fano manifold whose complement is a nonsingular divisor.

There were two different groups of researchers who tried to write down the complete Ricci flat metrics explicitly after they knew it existed on K3 surfaces, as was proved by me. One group was led by Gary Gibbons and Stephen Hawking in Cambridge. They were able to use analytical methods to continue the classical solutions to the Einstein equation with a $3+1$ signature, such as the Taub-NUT metrics, Kerr metrics, etc. It was a spectacular achievement. G. Gibbons and S. Hawking used this approach to study Euclidean quantum gravity, enabling them to assign a temperature to a black hole. The asymptotic flat Ricci flat Kähler metrics written by Gibbons and Hawking were later generalized by N. Hitchin using a twistor technique, and Peter Kronheimer gave a more detailed analysis. At around the same time, E. Calabi made a simple ansatz by constructing Kähler Ricci flat metrics on the bundle space over a Fano manifold. E. Calabi's construction included the well-known Eguchi-Hanson metric which, based on its simplicity, is more popular. My ansatz was actually much more general, although the construction is not as explicit. The situation is still not completely resolved when the divisor is singular.

My work in this area was published in the Proceedings of the ICM, Helsinki, 1978. After the Congress in Helsinki, I came back to Stanford. And at my recommendation, Stanford offered Y.-T. Siu a faculty position. After he arrived in the fall of 1978, I suggested that he work on several problems with me. One of the problems was the Frankel conjecture and the other was the rigidity of compact

Kähler manifolds with negative curvature, based on a harmonic maps approach.

But we actually talked a great deal about the existence of Kähler-Einstein metrics with positive scalar curvature. By this time, I had derived many estimates for the integral of the Kähler potential. But they were not powerful enough to give the final answer. When Y.-T. Siu started to work on this problem, he was efficient and understood all these estimates, some by himself and some by me. Y.-T. Siu is, of course, an expert on the Lelong number method and the related approaches. These can be used to some extent in special cases. But the method still fell short of providing the general estimate that we needed. Y.-T. Siu and I did not pursue this topic together after Stanford, although both of us kept the notes of our discussions. We were using the continuity argument, similar to what I did in the Calabi conjecture. It was a simple observation that in the continuity argument, we could keep the Ricci curvature to be strictly positive, an assumption based on the knowledge that the Calabi conjecture is true. So we had a family of Kähler metrics with strictly positive Ricci curvature, which have a uniformly bounded diameter and fixed volume. The problem was to prove the smooth convergence of a subsequence of the metrics. For this family of metrics, one just needed to obtain zeroth order estimates of the unknown Kähler potential. Many of them followed from the estimates I had already proved in the Calabi conjecture.

My estimates in the proof of the Calabi conjecture were later packaged in a different form by T. Aubin and later by G. Tian. They gave these estimates names such as I or J functionals. Since there is a continuous family of such metrics, one can integrate in time t among such functionals, as was discussed already in my paper. But the result was not that different from my original estimate and was not powerful enough to give the desired estimate to solve the equations. The reason was that the obstruction had not been understood yet. Note that the integral estimate sometimes could be improved, based on some arguments due to Lars Hörmander. This is based on the plurisubharmonicity of the unknown function in the coordinate chart after adding the Kähler potential. I learned this kind of argument during my collaboration with Y.-T. Siu, when we worked on embedding manifolds into Euclidean spaces. In 1985, after G. Tian became my student, I told him about the Hörmander type estimate. We obtained an invariant that we called the Alpha invariant. For manifolds with a large group of symmetries, they could be used to prove the existence of the Kähler-Einstein metric. It turned out that, independently, Y.-T. Siu was also using the similar idea of symmetry to improve the zeroth order estimate for the case of cubic surfaces.

All these approaches used the continuity argument. Ironically, in some of his recent talks, G. Tian attributed the continuity argument to T. Aubin, when, in fact, the continuity argument was invented more than 100 years ago. It was certainly known to Juliusz Schauder when he did his famous estimates.

Before 1978, and after the Calabi conjecture was proved, there was a general feeling that the only condition for the existence of Kähler-Einstein metrics with positive scalar curvature is the positivity of the first Chern class plus some conditions on the group of automorphisms due to Yozo Matsushima at Notre Dame. The group is reductive if the metric exists. So it was widely believed by people working on this problem, including S. Bando, T. Mabuchi, A. Futaki, G. Tian, and others, that if the first Chern class is positive, and if there is no holomorphic vector field, then Kähler-Einstein metrics should exist. Mukai found an example and Tian used it to show that the last statement is wrong.

However, I had already realized that the concept of stability was needed for the existence of Kähler-Einstein metrics. The existence of a Kähler-Einstein metric implies the stability of the tangent bundle with respect to the polarization defined by the first Chern class. Hence it is a necessary condition. I was excited by this fact. For it means that the Calabi conjecture implies that for manifolds with either negative or zero first Chern class, the tangent bundle is slope stable in the sense of D. Mumford. This was a fact that had not been proved by algebraic means. I suspected that for manifolds with positive first Chern class, it would not be automatically true and had to be considered as an assumption. I mentioned this in my IMU talk in Zurich in 1981.

The concept of stability has always fascinated me, partly because of the work of F. Bogomolov. After all, he and Y. Miyaoka were able to use this concept to prove the same inequality I proved by Kähler-Einstein metrics. Since the inequality was rather nontrivial, I believed that there was a connection between stability and the existence of Kähler-Einstein metrics.

The first concrete situation involved the generalization of the concept of Kähler-Einstein metrics to the bundle case. In the 1970s, because of I. Singer's great effort to popularize the concept of anti-self-dual bundles on four manifolds, geometers started to look into the subject of constructing them. In 1977, there was the famous work of Atiyah-Hitchin-Singer using the index formula to calculate the dimension of the moduli space. And by 1978, there appeared the work of ADHM (M. Atiyah, Vladimir Drinfeld, N. Hitchin, and Y. Manin) on the explicit construction of them over the four spheres using twistor theory and monoid construction. The similar formulation of anti-self-dual metrics for the case of higher-dimensional manifolds is most natural for Hermitian bundles on Kähler manifolds.

The first good examples were the tangent bundles of Calabi-Yau manifolds. I was interested in generalizing the Kähler-Einstein condition to bundles in 1978 when Y.-T. Siu was at Stanford. But I made little progress until I met K. Uhlenbeck, who had been interested in the Yang-Mills bundle for quite a while. She had proved the most fundamental convergence theorems in this subject from the point of view of partial differential equations. We started to work on the problem related to holomorphic bundles. Of particular interest here was the work of Martin Lubke on the necessary condition that the bundle has to be slope stable in the sense of Mumford if the connection is locally irreducible. So we were building on that assumption that the bundle is slope stable.

In 1981, I was invited to give a plenary talk at the London Mathematical Society's annual meeting in Wales. I had difficulty getting through passport control in London. I did not hold any passport at that time. I was traveling on a white paper issued by the U.S. immigration department. My passport status was stateless, and I had neither a Hong Kong passport nor a US passport. I was delayed in the London airport for an hour. They questioned me as to the reason I came to England. I was using a tourist visa then and had no choice but to tell them which places I wanted to go for sightseeing. It did not help when I misspelled Wales as "Whales". They asked me who my host would be. I mentioned my friend N. Hitchin in Cambridge. So they finally let me go.

In any case, when I met N. Hitchin, he told me that he had this brilliant student, Simon Donaldson, who was able to prove that any stable holomorphic bundle that can be deformed into the tangent bundle admits an anti-self-dual metric (the Ricci flat Calabi-Yau metric). I was quite impressed by that, although I did not manage to meet this brilliant rising star during that trip. But I had been thinking about the problem of the existence of an Hermitian Yang-Mills connection on stable bundles for a while already.

When I came back to Princeton, I talked with K. Uhlenbeck and we worked extensively on this exciting problem. She came to IAS to give the Morse lecture. She gave a beautiful talk about her work on both minimal spheres and gauge theory–efforts that helped lay solid foundations for the future development of geometric analysis. But since the approach was not based on geometric measure theory, F. Almgren was not polite towards her, especially after some drinks. He claimed openly that the work of Sacks-Uhlenbeck was wrong, and he said the same about the work of Morrey. I cannot say that I was pleased with Almgren's behavior during the party in honor of K. Uhlenbeck.

While in Princeton, Karen stayed at my home for a few days, and our work was intense. The regularity estimate for the Hermitian Yang-Mills problem was tough. To this day, the whole procedure to prove the existence of Hermitian Yang-Mills on poly-stable bundles still has to go through the same procedure that we did, except the two-dimensional case due to S. Donaldson, which was based on the Bott-Chern form. While I was working with K. Uhlenbeck on the Hermitian Yang-Mills equation, I was very curious about the concept of Gieseker stability for algebraic surfaces and also the later work by Eckart Viehweg on higher dimensional algebraic manifolds. Both are concepts from geometric invariant theory. What puzzled me was that manifold stability and tangent bundle stability did not seem to be related. I asked D. Gieseker many times about this question, but I did not get a satisfactory answer.

I finally decided that manifold stability is a kind of nonlinear stability, while the tangent bundle is a linearized concept. And as an analogue with the bundle version of Hermitian Yang-Mills, the Kähler-Einstein metric should be linked to manifold stability. Hence at this time, I made the conjecture that manifold stability in the sense of geometric invariant theory should be a proper condition for the existence of a Kähler-Einstein metric when the first Chern class is positive. When I published this conjecture formally in *Open Problems in Geometry* at the 1990 UCLA Conference on Geometry, I added the condition that the tangent bundle should be stable because it is a necessary condition. It is possible that manifold stability implies tangent bundle stability. It could be a nontrivial consequence of the existence of Kähler-Einstein metrics. On the other hand, if it were not true, that might contribute to counterexamples to several of the approaches that were being pursued.

In order to build a program to prove this conjecture, I made a preliminary conjecture at this time: The Kähler-Einstein metric can be approximated by projective embedding by pluri-anti-canonical embeddings. This was needed because projective embedding was how stability began. And we needed to link the two concepts together.

Many people did not believe my "preliminary" conjecture. Note that a consequence of this conjecture is that the induced metric from projective space on a curve of higher genus can have negative curvature. Some experts did not believe it because they were trying to prove the induced metric should have some positivity of curvature. Later I gave the embedding problem to G. Tian as a warm-up problem for the existence of a Kähler-Einstein metric. It took me ten hours to explain why my previous ideas with Y.-T. Siu on asymptotically Euclidean Kähler manifolds could be used to tackle this problem. I explained to him how to use the concept of peak sections, where a similar idea was used by Kodaira. G. Tian wrote up the proof in his thesis, apparently without remembering my discussion with him. The convergence in Tian's thesis was only second order and the higher order convergence was carried out by my other student W.D. Ruan in his thesis.

Once we established that the embeddings were good, the Fubini-Study metric has a natural Kähler potential, which is the log of the sum of the square norm of the sections. This is the Bergman Kernel. When we embedded the algebraic manifolds by the kth power of the anticanonical line bundle, we saw a good expansion of the potential depending on $1/k$. Many authors, especially Lu, Catlin, Donaldson, and Zelditch, studied this problem extensively. My proposal to approximate the Kähler-Einstein metric by projective embedding was later implemented numerically by Donaldson. Also, when I realized the embedding problem was relevant, I thought about the Futaki invariant. Not long after I proved the Calabi conjecture, Akito Futaki generalized the arguments of Y. Matsushima on the automorphism group of Kähler-Einstein metrics with positive scalar curvature. He found a character defined on the Lie algebra of the holomorphic vector fields. The character can be obtained by differentiating the potential defined by the difference between the Kähler form (representing the first Chern class) and the first Chern form defined by the same Kähler metric. Although this was a beautiful result, I thought at the beginning that this might not be a central issue for Kähler-Einstein metrics, as most Kähler manifolds do not admit holomorphic vector fields. However, I soon realized that the conjecture I made about stability was related to the group of projective automorphisms acting on the embedded manifolds. The Futaki invariant became an important concept and should be generalized to study the projective group actions on the manifolds, although the action may not leave the manifolds invariant. I told G. Tian it was the shadow of the action that played the important role, as up to that time, he did not appreciate the importance of Futaki invariants. Later, he took this idea and wrote a note with Weiyue Ding in China, who worked as my postdoctoral fellow at UC San Diego.

When I was in Princeton, I had two students who worked on questions related to Kähler geometry: H.-D. Cao and S. Bando. (There was also J.-Q. Zhong who came from China to study with me.) N. Mok had just graduated from Stanford and was an instructor at Princeton. I got both of them interested in Kähler-Einstein metrics. I told them the importance of the uniqueness of the Kähler-Einstein metric. It was a pleasant surprise to me that a few years later, S. Bando, working with Toshiki Mabuchi, proved the important result that the Kähler-Einstein metric is unique up to an automorphism of the manifold.

In 1982, when I was invited by Jürgen Moser to give the IMU lectures in Zurich, I had already stressed the importance of stability for the existence of Kähler-Einstein metrics. At that time, I was also gaining more familiarity with the embedding of algebraic manifolds into projective spaces and their differential geometric meaning. Peter Li and I were working on estimates of eigenvalues of the Laplacian. For Riemann surfaces, we realized that the optimal conformal embedding gives rise to the best estimate for the first eigenvalue in many cases.

We then generalized the paper to general algebraic manifold embedding into complex projective spaces. We found a condition that was later called the balanced condition. This condition arose from the choice of measure on the submanifold in my argument with P. Li. The balanced condition was later found to be relevant to the stability of the manifold—an idea I had suggested to my former student, H. Luo.

From the period of 1982 to the early 90s, I talked frequently to all my students about the existence of Kähler-Einstein metric based on the stability of the manifold. That is why G. Tian called it a folklore conjecture, as he heard from me on this matter so many times. From the beginning, Tian misled people into calling this his conjecture, until I pointed out to him that I had published this conjecture already. Also I had given a talk for a large audience at the UCLA differential geometry conference in 1990. About twelve years ago, Donaldson circulated a paper, with a title in an important paragraph called "discussion on Tian's conjecture". He mentioned his discussion with Tian on the nice idea of linking the existence of the Kähler-Einstein metric with the stability of Fano manifolds. After the paper was circulated for a while, I was told by one of my student about the preprint. I wrote to Donaldson, with a cc to Tian, about my original conjecture. It took a while to clarify the matter until I showed my published statement in the UCLA conference Proceedings about this conjecture. Finally Tian himself acknowledged the statement I made and suggested to Donaldson to call it the Yau conjecture. But I still received inquires from several mathematicians where I published the statement many years after that discussion. Three years ago, in a talk at Stony Brook, Tian called it a folklore conjecture.

In the fall of 1985, N. Mok and Y.-T. Siu organized a conference at Columbia University on several complex variables. I brought a few students, including G. Tian, J. Li, and others, to attend at the request of N. Mok. Y.-T. Siu gave a talk on the existence of a Kähler-Einstein metric on a special cubic surface. I did not pay much attention to that argument, as Calabi once told me he could construct the metric explicitly on this manifold.

In the fall of 1986, I took all of my graduate students to visit the University of Texas at Austin to try out their offer to me as a chaired professor. Most of my students, including G. Tian's former teacher Kung-Ching Chang, lived in the big house that I rented. One day, J.-Q. Zhong, who was visiting Y.-T. Siu, called me up saying that Y.-T. Siu was extremely mad at G. Tian. He thought G. Tian was copying the result he had presented in a lecture at Columbia. I was taken aback by such an accusation because I had not read what Tian wrote. In any case, I thought those estimates do not reveal the true nature of the problem and were not worth a fight for an outstanding mathematician like Y.-T. Siu.

Actually, G. Tian worked hard when he was a student. He learned a lot from many of the manuscripts I gave him, including those from A. Todorov and from Philip Candelas on the special geometry of Calabi-Yau manifolds. However, many years later, I found out that he should have given more credit to these authors.

Y.-T. Siu wrote me several letters, demanding that I punish G. Tian severely, which I refused to do. My position was that it was hard to verify that G. Tian copied from Y.-T. Siu directly, and the ideas were rather straightforward consequences of the proof of the Calabi conjecture: any control of certain integrals constructed from the unknown function could finish the project. There were ideas, such as Moser-Trudinger estimate, that could do the job, but only for very special manifolds, based on symmetry. Therefore, I did not accept the request of Y.-T. Siu and continued to support G. Tian.

This made Y.-T. Siu very unhappy. And apparently he told many people that I tried to use my student to attack him, which was far from the truth. When G. Tian got an offer from Princeton as an instructor, a senior faculty at Princeton called me, making an attempt to withdraw their offer to G. Tian after listening to the complaint from Y.-T. Siu. In principle, I like to train young students and I like to give them a chance. Due to the heavy attack by Y.-T. Siu, I decided to train G. Tian with all my efforts.

When G. Tian was a student at both UC San Diego and Harvard (when I moved from San Diego, I took four students to Harvard and the rest went to MIT, Brandeis, Northeastern, or stayed in San Diego), I invited G. Tian to my home three days every week where I spent two to three hours teaching him about the major ideas of geometry and especially those about Kähler-Einstein metrics.

Much of the work he published related to Kähler-Einstein metrics grew out of these discussions. However, after he became an established mathematician, some of my friends told me that he accused me of being out of touch with the development of Kähler-Einstein metrics, even claiming as his own some of my results, which had been in print many years earlier.

Tian and I wrote a couple of papers together. One exploited symmetry to increase the Hörmander estimate so that one can achieve the zeroth order estimate. This grew out of the discussions at my home and was useful for many special Fano manifolds.

In the meantime, Alan Nadel did some good work that was helpful for understanding Kähler-Einstein metrics but was still far from the goal. In 1986, I told G. Tian to work on my conjecture using projective embedding to estimate the Kähler-Einstein metric, because this would provide an understanding of the stability of manifolds.

Following my ideas with Y.-T. Siu on constructing sections of holomorphic line bundles and peak sections, Tian was able to establish that particular conjecture of mine. This particular conjecture amounts to understanding the behavior of the Bergman Kernel for the kth power of the ample line bundle when k is large. And I looked at it as a first step toward my conjecture on the connection between the stability of manifolds and Kähler-Einstein metrics.

My effort to support Tian against the wishes of Y.-T. Siu broke a close friendship that I had with Siu. When I entered the Pui Ching middle school in Hong Kong, Siu was graduating and he won practically every award that a student could get. He is naturally one of the greatest heroes of all time at our middle school—a role model for later generations. I was certainly delighted to meet him for the first time in 1975 in Williamstown, Massachusetts during a conference on several complex variables. At that conference, I talked about what a good generalization of the uniformization theorem should be for higher dimensional Kähler manifolds.

Siu showed an immediate interest in these problems. In fact, after eight years, he was still interested in my program by describing it at the Warsaw Congress. We started working together on a holomorphic embedding problem for complete manifolds, which are asymptotically flat. I was visiting the Courant Institute after the Williamstown conference. From there I visited Yale several times to discuss the problem with Siu. We solved the problem, and our paper was published in the *Annals of Mathematics.*

That paper stimulated other research, including studies of the "gap" phenomenon on the decay property. Robert Greene and Hung-Hsi Wu wrote a nice paper on the subject. But after I told M. Gromov about the statement, people started to attribute the theorem on the gap phenomenon to him. I personally do not think my theorem with Y.-T. Siu was particularly good, because the assumption we made was too strong. But the method, itself, was good, and I learned a lot from working with Siu. For one thing, I started to realize how to do holomorphic embeddings for Kähler manifolds. This carried over to my work with Siu on the compactification of a complete Kähler manifold with strongly negative curvature and finite volume.

I taught G. Tian to use similar ideas in his thesis work on the approximation of a Kähler metric by projective embeddings. I also suggested to N. Mok and J. Zhong to look into similar ideas to compactify complete Kähler manifolds with finite volume. Some of the arguments went back to my proof of the non-existence of L^p holomorphic functions on complete Kähler manifolds.

Y.-T. Siu and I had a long-term collaboration until the unfortunate situation of G. Tian taking credit from him. But I felt strongly that young people should be given a chance to make amends and develop, even if they made mistakes early in their careers. Y.-T. Siu clearly did not agree, and the pressure on me had been tremendous. My protection of G. Tian may have, inadvertently, encouraged Tian to become much more aggressive and irresponsible—an unfortunate fact that I did not foresee. I had not realized that the education a young person like G. Tian received during the "cultural revolution" might have been very different from the kind of education typically offered in the western world.

G. Tian was a hard working student. This is demonstrated by the fact that he came to my home to learn ideas from me about three nights per week for three years. But he was eager to show his results, even before his ideas were mature enough to prove those results. After he published the paper with me on the existence of Kähler-Einstein metrics on Fano surfaces with enough symmetries, he claimed that the deformation of such surfaces also admitted such metrics. He put an ambitious title on his paper: *On Calabi's conjecture for complex surfaces with positive first Chern class.* The paper is full of mistakes. Y.-T. Siu pointed out explicitly which sentences had not been proven. I asked G. Tian, and he admitted the mistake but said it was a small mistake. I demanded that he write a correction. After twenty years, I still have not seen that correction. Instead, he claimed to me that he could solve my conjecture on the relation between stability and the existence of Kähler-Einstein metrics. I was impressed and started to promote him vigorously: I recommended him enthusiastically to MIT through I. Singer, who had never heard of him before. I also wrote a very strong letter to the Waterman Prize committee and to the Veblen prize committee. After the decision was made, the chairman of the Veblen Prize committee asked me to help him write up the citation for G. Tian. My letters and the citation were based on the false assumption that G. Tian had solved my conjecture. I owe an apology to the mathematics community for not checking G. Tian's claims thoroughly before I wrote such strong letters of recommendation. He is making similar claims again, unfairly taking credit for the work of S. Donaldson and his colleagues on Kähler-Einstein metrics. An account of this situation was written by X. Chen, S. Donaldson, and S. Sun on September 13, 2013, and posted on Donaldson's website, to rebut some of claims made by Tian "on the grounds of originality, priority, and correctness of the mathematical arguments". [1]

My collaborations with Y.-T. Siu in the seventies certainly had a beneficial influence on both of us: I learned techniques from him regarding several complex variables, while Y.-T. Siu learned geometric ideas from me. Our work on the Frenkel conjecture offered a new method for handling the second variational formula for minimal surfaces in higher codimension, which turned out to be useful for Mario Micallef and John Moore on manifolds with isotopic positive curvature. The way of deforming harmonic maps and the precise analysis of the bubbling phenomena of Sacks-Uhlenbeck were used by Gromov and other scholars who worked on Gromov-Witten invariants.

Y.-T. Siu started to work more extensively on problems related to curvature and rigidity questions in geometry after we discussed my program to prove rigidity using harmonic maps. The study of global rigidity of Kähler manifolds came from my observation that the Mostow rigidity theorem can be applied to the quotient of the ball. This was an observation based on my characterization of such manifolds by the equality of certain Chern numbers, which in turn came from the existence of the Kähler-Einstein metric via the Calabi conjecture. My idea of applying harmonic maps to the strong rigidity problem in geometry dated back to my student days in Berkeley. I recalled that B. Lawson was excited by the idea, and I got a research grant from Stony Brook in 1973 based on this proposal. After my conversation with E. Calabi in 1979 at IAS on the vanishing theorem of Matsushima, I was convinced that the idea could be used to reprove the super rigidity theorem of Margulis. But after Y.-T. Siu published the paper on rigidity by harmonic maps, many people attributed all the related results to him. This includes my work with J. Jost on the statement that a continuous map from a Kähler manifold onto an algebraic curve can be deformed to a holomorphic map

[1] Tian has recently written a response to the Chen-Donaldson-Sun document, which is available at http://www.bicmr.org/ gtian

if we also deform the complex structure of the algebraic curve. Also relevant was my work with J. Jost on carrying out E. Calabi's view of the Matsushima vanishing theorem to prove the super rigidity theorem for locally symmetric space with higher rank. This last work was proved independently by Y.-T. Siu with N. Mok and S. K. Yeung. When the image manifold is replaced by a building, R. Schoen was able to prove in a remarkable manner the regularity result, which is needed for the rigidity argument. The vanishing theorem of Corlette on rank one locally symmetric space, besides real and complex hyperbolic manifolds, can then be treated by the theory of harmonic map. The original theory of G. Margulis on arithmeticity can be extended to cover this class of rank one manifolds, as was observed by Gromov-Schoen.

Commentary on Linear Waves in the Kerr Geometry: A Mathematical Voyage to Black-Hole Physics

I wrote this article with F. Finster, N. Kamran, and J. Smoller. Some seeds of this idea were planted in 1988, when Joel Smoller spent a sabbatical year at Harvard, shortly after I had arrived at the university. Smoller took my advice to shift his research interest into general relativity (GR) that year. In the early stages of his work he listened to me, but soon he was leading the way, and that was certainly a very good thing for the subject of GR. He became an expert, writing outstanding papers with Blake Temple and others. Some years later, around 1995, Felix Finster arrived at Harvard, supported by a German Fellowship. I also convinced him to work in GR, and soon Joel, Felix, and Niky Kamran became my good friends, and the four of us worked together. It pleased me to see how well this collaboration developed. The study of waves in black hole geometries is a difficult subject, having great relevance to black hole stability questions. For non-rotating black holes, linearized stability is relatively easy to establish. But the problem becomes a good deal tougher when the angular velocity is large relative to the mass of the black hole. We developed a highly non-trivial spectral method. We felt that our method had many advantages over the classical energy method, which treats the case of small angular velocity. However, the publication of this paper encountered difficulties from proponents of the alternate method, who insisted that their method was superior, even though they had not—in my opinion—proved a result that was better than ours.

I may mention that the first paper I wrote with Joel was on the proof of the existence of a static spherical solution of coupled Einstein-Yang-Mills solution. It was a joint effort with Mcleod and Wasserman. (The existence of one such non-trivial solution in three plus one dimensions was first found numerically by R. Bartnik and his student J. McKinnon.) We also proved the existence of infinitely many static black hole solutions. I believe that all these solutions would have physical significance, but so far, it has not been found.

Commentary on Recent Development on the Geometry of the Teichmüller and Moduli Spaces of Riemann Surfaces

This survey, written with Kefeng Liu and Xiaofeng Sun, describes our work in understanding the various intrinsic metrics on Teichmüller spaces. This includes the Weil-Petersson metric, the Ricci metric, and the Kähler-Einstein metric, which can all be regarded as metrics on the moduli space of curves. We proved that they behave in accordance with Mumford's theory of good metrics. Hence, we can use them to study the log geometry of the moduli space. We computed their curvature tensor asymptotically. In particular, we proved the dual negativity of the log tangent bundle of the moduli space. The Kähler-Einstein metric allows us to discuss the stability of the log tangent bundle and several related topics that are not accessible by different means.

We also found a canonical way to locally deform holomorphic sections of the pluricanonical bundle over Riemann surfaces.

Commentary on Deformation of Kähler-Einstein Metrics

In 2010, my friend and former student Richard Schoen turned sixty. His former students wanted to put together a book dedicated to him. I thought this was a great idea. Xiaofeng Sun was his student, and we decided to join together to write this article.

Rick and I had worked together extensively until we both left UC San Diego, with him going to Stanford and me to Harvard. We had been close friends and learned a great deal from each other. I still remember that during our Stanford days, we went out to dinner almost every night. We talked about mathematics day and night. We are still very close friends, and I admire his work tremendously.

When we were writing up the paper on the second part of the positive mass conjecture, I spent most of the summer at Stanford working on it. We often worked the whole day and then went to sleep in a house in Los Altos owned by a friend of Doris Fischer-Colbrie. (Doris was Rick's girlfriend at that time and is now his wife.) Rick and I often swam in the swimming pool after 10 pm. At both the Institute for Advanced Study, and at UC San Diego, we gave lectures together on our books, *Lectures on Differential Geometry* and *Lectures on Harmonic Maps*. It was a lot of fun, and those talks sparked interactions with our students and colleagues that helped move the subject of geometric analysis forward.

In this essay, X. Sun and I discuss the deformation of Kähler-Einstein manifolds. In certain Calabi-Yau manifolds we proved that the Kuranishi gauge is equivalent to the divergence gauge. As a result, we can have good canonical deformations of polarization line bundles over Calabi-Yau manifolds. We computed the curvature of the canonical metrics on the direct image sheaves. We also worked on the deformation of Kähler-Einstein metrics with negative scalar curvature.

Commentary on String Theory and the Geometry of the Universe's Hidden Dimensions

Since the rise of string theory in the early eighties, the subject of Calabi-Yau manifolds has become closely connected with a distinct part of theoretical physics and an even bigger part of mathematics. Many books have attempted to describe string theory for the general public. The most popular one was written by Brian Greene, who was my postdoc from 1988 to 1990. But the mathematical part of string theory had not been discussed much in these popular books. I think it is a very beautiful subject and felt that a popular book should be written about it. I asked B. Greene to work on such a book, but he declined saying it would not be an easy job.

In the fall of 2005, Dennis Overbye who wrote for the *New York Times* science section contacted me several times. He proposed to do a profile of me for his section. I agreed, and we had several nice meetings in Cambridge. We often went to a Chinese restaurant called Yenching in Harvard Square, where we had many pleasant discussions on my personal background and the mathematics I did with my colleagues. D. Overbye is a very nice and intelligent person, and I have very fond memories of those encounters. As a result, I was naive enough to think that all reporters are friendly.

Then in 2006, the overheated discussion on Poincaré conjecture showed that the deep theory of geometric analysis had been misinterpreted by many famous mathematicians who shared their (sometimes erroneous) views with the news media. Some reporters exploited such misunderstandings to gather attention and promote their writings. They liked controversy, even if a genuine controversy was not really there. A female reporter shared some false statements with me. Her intentions, apparently, were to induce me to grant an interview with her. When she realized that D. Overbye was going to publish a profile of me in the *New York Times*, she rushed to publish her article. Many elements of her article were drawn directly from Chinese websites, and she quoted statements from people who, for personal reasons, did not get along well with me. To put it mildly, I was not presented in a favorable light.

D. Overbye's article, which appeared in the *New York Times* in December 2006, was much more flattering. In that article, D. Overbye jokingly called me "the emperor of mathematics". A large picture of me appeared along with the full-page article, and I must say it looked rather impressive. I like the picture they selected. On the other hand, it took more than seven hours to get that picture. The *Times* sent two different photographers. The first one spent about three hours taking more than two hundred pictures. The editors at the paper were not impressed with the results and sent another photographer a few days later. That process took another four hours, although the picture they finally selected was excellent.

National Public Radio (NPR) also interviewed me for a short period of time. Hence I had suddenly become famous. This caught the attention of John Brockman, a literary agent who promotes popular science books. He is also the agent for Brian Greene. Brian told me he is a fine person and encouraged me to work with him.

I had some conversations with Sze-Hoi Henry Tye, who was a Cornell professor of physics, and a friend of Brian Greene. Tye is an expert on string cosmology. When I mentioned my intention to write a book for laymen, he recommended Steve Nadis who had worked with him before. Steve lives in Cambridge, not far from my office. Before meeting Steve, I read what he wrote. I found out that he is a great writer although he did not know much mathematics. I thought that this gap in his knowledge could actually be helpful in translating difficult mathematics into layman terms. I immediately decided to work with him. After discussions for several weeks, we focused our topic on the theme that is the title of our book, *The Shape of the Inner Space.* We worked on the book for quite a long time, often late at night or on weekends. Altogether, it took us four years to finish the book.

I was pleased by the final version of our book, although some typos and small mistakes were subsequently found (and corrected before the paperback edition came out). The book was originally supposed to be published by W. W. Norton. But after reading the final version, their editors thought it was too technical for the layman to understand, and we had to move the publication of our book to Basic Books.

We were pleased to get many excellent reviews. In fact, the book won a 2010 Book of the Year Award granted by Foreword Reviews—a Bronze Medal in Philosophy.

I gave many talks about this book all over the world:

1. In Australia, I spoke in cities such as Canberra, Melbourne, and Sydney.

2. In England, I spoke in Cambridge. I was also scheduled to give a talk in London, but a snowstorm forced me to cancel that engagement.

3. I spoke in Singapore, Hong Kong, and Taiwan.

4. Closer to home, I spoke in Cambridge, Mass, at the Smithsonian Institution in Washington, D. C., in College Park, Maryland, and at Caltech, UCLA, Stanford and Berkeley.

I was rather surprised that in all these talks, the audience occupied the full room and asked many interesting questions.

I graduated from Berkeley, and it certainly meant a lot to me to find the whole room packed with faculty members and students when I gave the talk there. The mathematician Steven G. Krantz was there, as well. After the talk, Krantz asked me to write an article for the *Notices of the AMS*, which he edits. I was very happy to do so. The article discussed here (written with Steve Nadis) is based on that talk.

Commentary on Shiing-Shen Chern

When Chern passed away in 2004, I gave several lectures about his work and legacy: The first one was at the big conference ICCM held at the Chinese University of Hong Kong in December 2004, about two weeks after Chern's death. and I later spoke in several other places.

In 2010, Steven Krantz of the *Notices of the AMS* asked me to assemble a collection of articles in memory of Chern. Since 2011 was Chern's 100th anniversary, I was able to convince several friends and students of Chern, including M. Atiyah, I. Singer, F. Hirzebruch, and others, to write about him. The end product was a nice collection of essays on Chern, the man and the mathematician.

In preparing my own article about my mentor, I went through his collected works thoroughly. I learned much more than I had during my previous readings. The history of the development of Chern classes was impressive. It dated back to Hassler Whitney, Eduard Stiefel, and Lev S. Pontryagin on the topological side; to John A. Todd and the Italian geometers on the algebraic geometric side; and to L. Pontryagin on the differential geometric side. (Very few people realize that L. Pontryagin actually defined a topological invariant using the curvature tensor.) It took the great genius of Chern to unify their varied approaches, and because of that the theory of Chern classes became much more powerful in both geometry and topology. Nowadays it is a powerful tool in theoretical physics as well.

Many people did not appreciate the fundamental significance of Chern's paper, "On the characteristic classes of complex sphere bundles and algebraic varieties", which appeared in the *American Journal of Mathematics* in 1953. It offered many important contributions to the theory of Chern classes, a significant one being that Chern classes of holomorphic bundles over algebraic manifolds are represented by holomorphic cycles. Chern's work depended on the splitting principle, which was established by him before the famous and fundamental work of F. Hirzebruch. (In his article in the collection that appeared in the *Notices*, F. Hirzebruch acknowledged that he read this paper of Chern before he finished his thesis.) Chern got valuable advice from A. Weil and Hermann Weyl, which one can probably surmise from reading his papers. The proof of the Gauss-Bonnet theorem was the best example. A. Weil had pointed out to Chern his proof with Carl B. Allendoerfer on the Gauss-Bonnet formula where he made use of a tube around the submanifold. And H. Weyl studied the invariants created from the tube. When I was a student, I was asked by Chern to read this paper of H. Weyl without understanding why Chern was interested in it.

A. Weil's interpretation of Chern's derivation of the Crofton-Santaló formula in integral geometry was very important as it provided the transform that was much later attributed to Israel M. Gelfand, et al. The Chern-Weil homomorphism revealed Weil's deep understanding of Chern classes.

Of course, Chern was a master of Cartan-Kähler theory, which he used to do extensive calculations in geometry. When he returned to China, he pushed this theory vigorously. Unfortunately, few Chinese mathematicians were able to follow his lead and do deep research in this subject.

Many of Chern's papers were influenced by Élie J. Cartan, but I do not think they are as fundamental as the above papers. Chern himself said that he was asked by Weil to look into characteristic classes after his proof of the Gauss-Bonnet theorem. Hence A. Weil, in addition to Blaschke, Cartan and Weyl, had a deep influence on Chern's work.

Commentary on a Survey on Recent Development in Computational Quasi-Conformal Geometry and Its Applications

After studying conformal maps with David Gu using computer graphics, I decided that one needs more flexibility to understand the distortion of geometric figures. Shortly thereafter, Ronald Lui became my postdoctoral fellow, and I suggested that he and Gu look into the construction of quasi-conformal maps. This work led to our development of a new technique called Beltrami holomorphic flow. The computation of quasi-conformal geometry is useful for applications in medical morphometry, as well as in many other areas.

Commentary on Recent Development of Computational Conformal Geometry

In 1998, Xianfeng David Gu studied with me for his Ph.D. program after D. Mumford left Harvard to go to Brown. D. Gu was my first Ph.D. student in computer science.

I suggested that he work on computer graphics through the classical uniformization theorem of conformal geometry. In my whole career, I was always fond of and excited by the uniformization theorem. I was glad that it found applications in computer graphics and, based on that, D. Gu drew some very pretty pictures. During the time when the news media had extensive discussions of the Poincaré conjecture, the graphics produced by Gu, using our technology, appeared as a bunny in the *New York Times* and as a horse in the front page of a book.

We developed many different ways of calculating the uniformization map. We used harmonic maps and developed de Rham calculus on manifolds, which was used to calculate harmonic one-forms. We also used sphere packing methods. The uniformization map can now be computed in real time, thereby becoming quite powerful in practical usage.

In 2001, Gu's friend Yalin Wang came to work with us. In the beginning, most of the three-dimensional models were downloaded from the web. We soon found out they were not realistic enough to test our methods, so we looked for pictures of real three-dimensional objects instead.

Brian mapping served as an important case study for us. The brain has a rather complicated geometry, and yet it can be mapped conformally on the sphere in a reasonable unique manner (modulo the Möbius transformations). This is, of course, very nice, as we could solve the registration problem and use our method to investigate the growth of the brain and associated disease. We were also interested in studying the fiber structure in the interior of the brain. We did volumetric calculations, using harmonic maps and first eigenfunctions of the interior domain.

In late 2001, Tony F. Chan visited me in my office. We had nice discussions, and I told him what we did. He was very interested to learn about our conformal method, and he then proposed to hire Yalin to be his postdoctoral fellow at UCLA. Because we were short on funding at Harvard, Yalin went to join Chan and the UCLA medical imaging group led by Paul Thompson. Through this connection we had access to more interesting brain data, which was provided to us by P. Thompson.

In 2002, I spent a year at Caltech as a Moore visiting professor. D. Gu came to visit me quite a few times. Both he and I gave a few talks on our methods for doing differential form calculations on discrete manifolds, including the treatment of de Rham forms. We spoke of this at seminars held by Thomas Yizhao Hou and Jerrold E. Marsden. We answered many questions asked by the graduate students and postdoctoral fellows in Marsden's group. We were surprised to learn that some of our results later appeared in their papers without any reference to us.

Since I was not a professional computer scientist, it was difficult to apply for funding from the government. But I was told that I could apply for funding from the commerce department. So I did, and after about one-and-a-half years of preparation, and after the judgment of eighteen experts in the field, our group finally got the funding. And we were able to hire Yalin back to Cambridge to work on the computer graphics. Since we needed much more knowledge on human faces, we got Song Zhang to come to work on the 3D camera, which became very important for our group. After that, we were able to get many interesting pictures to test out.

In the meantime, I suggested using more methods from Riemann surface theory to understand computer graphics. An important tool is quasiconformal maps. Working with D. Gu, Ronald Lok Ming Lui, and Y. Wang, we were able to get very good calculations of quasiconformal maps. And the Beltrami coefficient became important for understanding the distortion of geometric figures. Recently D. Gu, R. Lui, and I were able to efficiently calculate the Teichmüller quasiconformal map. These maps are classical, but the new calculations afforded deeper insights.

We are still exploring geometric methods for the computer. It is an interesting area of development whose potential has barely been tapped.

Commentary on Mirror Symmetry and Localizations

In 2003, Harvard hosted an international conference in honor of I. M. Gelfand's 90th birthday. The event was organized by Gelfand's friends and former students, and I was invited to give a talk.

Of course, I had known of his great name since my student days in Hong Kong. I learned about the Gelfand-Naimark theorem in college, considering it one of the most impressive theorems in mathematics—one that changed our view of geometry. In fact, one of the guiding principles of geometric analysis is to study functions over the manifold extensively in order to determine the structure of the space. After I moved to Harvard in 1987, David Kazhdan and Arthur Jaffe arranged for I. M. Gelfand to visit Harvard for a long period of time. He had a young daughter who spent much of her time in my secretary Cindy's office. I frequently attended his seminars at Harvard and MIT. They were certainly quite different from our traditional seminars in America. I think it is good to have many interactions between the audience and the lecturer, which is how Gelfand's seminars typically went.

After I got to know him better, I told him that I was invited by S. Novikov to visit Tbilisi in Georgia. Since I had to pass through Moscow to get to Georgia and had never been to Moscow before, I asked him for help. He wrote a letter for me to the Russian Academy.

My wife and I decided to make an adventure of it. We flew to Moscow in November 1989, leaving our two young kids with my mother-in-law. We arrived in Moscow around 10 pm and had no idea of what to do. After some effort, we found a person who looked like a Georgian who, apparently, was supposed to meet us in the airport lobby. He could not speak English, and we could not understand Russian. He managed to use his body language to get us to walk with him to a car. It took about an hour for us to drive to another airport. The whole time we were nervous—unsure as to whether we had gone with the right person. It was not until we met another American at the second airport, who was waiting to go to Georgia, that we knew we were on the right track.

Our stay in Georgia was nice, and the Georgians showed us warm hospitality: They treated us to big banquets and took us on scenic tours of the Black Sea and the old castles near Tbilisi.

After a few days, We flew back to Moscow on a plane that was so crowded they had to put some extra chairs down in order to seat more people. We arrived at the hotel at around 8 pm, and we were very hungry. There was a big party on the first floor with plenty of food. But they refused to let us buy any food, even though we were ready to pay large amounts of money. And they told us we would not be able to buy any food from the nearby restaurant. So we had no dinner for the whole night. We were put in a room that was so small we could barely move our bodies. It was fortunate that in the morning, we met a Chinese woman scholar who was visiting Moscow from China. She helped us get around. We were finally able to get some food, and she helped us find a good exchange rate for our

American dollars.

The next morning, I went to the Academy to complain about our cramped living conditions at the hotel. They ignored my request at first until I showed them the letter from I. M. Gelfand. After reading that letter, their attitude changed dramatically. We were asked to move to the hotel's penthouse, which was a huge suite that offered plenty of room and beautiful furniture. The only problem was that the bathroom sink was missing. Apart from that, we were extremely happy with our new quarters and dared not complain. We traveled around a bit on our own in the city of Moscow. The subway system was superb. We took a tour of the Red Square, which we admired tremendously. Our Chinese friend got us cheap tickets to watch the Moscow opera. Despite the fact that our seats were far, far away from the stage, we still enjoyed the show. Overall, we really admired the culture in Moscow. But we had great difficulties calling home. We were anxious to check on the safety of our two boys. But we were unable to contact my mother-in-law until we arrived in London.

We had a hard time getting a taxi to the Moscow airport. The best way to get a cab, people had told us, was by showing a pack of cigarettes to the driver. At the airport, we ran into a group of Americans who were sent by McDonalds to set up the first restaurant in Moscow. I was wondering why the Russians would be interested in the Big Mac, though I later learned that this business venture fared quite well.

About 14 years after our trip, we celebrated Gelfand's life and work at the Harvard conference. The organizers held a big dinner party at the Royal East, a restaurant I often went to. It was a lot of fun, and many famous Russian mathematicians were there, including Faddeev, whom I sat next to. I worried that he might not be used to the way the Chinese prepared food, especially upon seeing the Peking duck served with the duck's head on the plate. I was astonished to see him to use his chopsticks skillfully to pick up the duck's head and, within one bite, to finish off the whole head in his mouth. Overall, I think the Russian mathematicians are tremendously original—in mathematics, as well as in other areas.

I was totally amazed that, during the conference, I. M. Gelfand made a big effort, telling various people that he wanted to meet with me. When we finally got together, he told me that I had done the best work on eigenvalues and eigenfunctions of the Laplacian. He also said that I should work with his group. Up to that point, I thought that he might have confused me with other people.

During the conference, I presented work that my colleagues and I had done on mirror symmetry and localizations. The talk covered some of the joint efforts carried out by Kefeng Liu, Bong Lian, Chien-Hao Liu, and me.

I'll briefly discuss the notion of mirror symmetry—an important breakthrough in string theory that has also had an important influence on modern geometry. This idea stems from a physical interpretation of space-time that goes beyond the geometric intuition we had acquired over the preceding hundred years. Supersymmetry is rather counterintuitive, and its consequences for geometry can be spectacular. The most impressive one was the concept of mirror symmetry. The calculations of Candelas et al., which improved upon the previous works of

Greene-Plesser and their own group, have changed modern algebraic geometry significantly. This calculation by itself is full of original ideas about which many mathematicians, including my group of postdoctoral fellows and I, were excited. We made attempts to generalize the scope of the first calculations. These generalizations helped to clarify much of the theoretical research related to mirror symmetry. They are largely based on calculations of the periods of holomorphic n-forms.

A. Klemm, S. Hosono, S. Theisen, and I did the first calculations of mirror symmetry when the moduli space is more than one dimension. (This was independently carried out by Candelas et al.) We realized that the GKZ system, as an extension of the Picard-Fuch's equations, is essential to the calculation. (Here GKZ stands for Gelfand, Kapranov, and Zelevinsky.) We used the Frobenius method to introduce some formal variables to find all periods at the maximum unipotent point. Soon Bong Lian joined us to study the deep consequences of mirror symmetry, where we examined the mirror map itself and found many interesting arithmetic properties. In particular, we discovered that the q expansion of the mirror map gives rise to integers. We proved this in many important cases using the method of Dwork. (The most general case was proved by my former student J. D. Yu when the degree is sufficiently large.) Based on the Schwarzian equation that we derived, B. Lian and I went on to prove some divisibility of the Gromov-Witten invariants, partially answering a famous question of H. Clemens. We studied the linear sigma model introduced by Witten. Then we started to show how the calculations obtained by Candelas et al. can be justified mathematically. This is especially important to demonstrate that the calculations of the Gromov-Witten invariants for the quintic and other known Calabi-Yau manifolds are correct. This is an important problem in classical enumerative geometry. Calculating the number of algebraic curves in a Calabi-Yau manifold dates back to the nineteenth century. The formula of Candelas et al. counts the number of rational curves for each degree. Although the formula appeared to work, it still had to be proved rigorously.

Kefeng Liu is an expert on localization. Bong Lian and I realized that we needed Kefeng's help. So we teamed up with him to use the Atiyah-Bott localization, Witten's idea of the linear sigma model, and Kontsevich's concept of stable mapping spaces in order to construct a rigorous proof of the formula of Candelas et al. The formal variable in the Frobenius method becomes the hyperplane section in the equivariant cohomology. We called this circle of ideas the "Mirror Principle", and it has implications that go far beyond just proving Candelas' formula.

As it turns out, the proof of the Candelas formula was done independently by A. Givental. Although there were misunderstandings on this score, spread by partisans who were largely ignorant of the subject and its history, the essay attached below—written by Bong Lian and Kefeng Liu–explains why the proof of the Candelas formula was done independently by two groups of people: Lian-Liu-Yau and Givental.

Commentary on Non-Kähler Calabi-Yau Manifolds

This is a survey article on non-Kähler Calabi-Yau manifolds that I wrote with Li-Sheng Tseng. I also gave a talk on this subject at the University of Pennsylvania.

In 1986, when I read Andy Strominger's paper on a non-Kähler compactification of heterotic string theory, I was interested to see a nice non-Kähler manifold come up in such a natural setting. Andy has been a good friend for a long time. I was trying to find a nonsingular solution to the system that he wrote down. Andy produced some singular solutions that did not quite satisfy my curiosity. I talked with many of my postdoctoral fellows, including Brian Greene, about constructing nonsingular solutions.

We looked at this for a long time without finding a smooth solution, and Andy told me that it probably does not exist.

In 2004, Jun Li and I constructed a solution for the Strominger system with the base manifold being a standard Calabi-Yau manifold that is Kähler. But the metric that appeared in the Strominger system is not Kähler but balanced. Jun Li and I solved the system by taking small perturbations of the Calabi-Yau metric where the bundle is a perturbation of the direct sum of the tangent bundle plus some trivial line bundles. That was quite nice, although it was not entirely satisfying as it depended on a perturbation of Calabi-Yau metrics. In general, it is still difficult to solve the full system, and we have difficulty finding the moduli space of solutions to the Strominger system.

In 2005, J. X. Fu came from Fudan to Harvard to work with me as my postdoctoral fellow. We started to look at a model of non-Kähler manifolds based on the Calabi-Eckmann construction. It is an elliptic fiber space over a K3 surface that is locally trivial. Fu and I worked out an ansatz to reduce the solution of the Strominger system to a Monge-Ampère type equation on a K3 surface. That was nice though it still left a very nontrivial problem to be solved. We were able to use a few important ideas from my proof of the Calabi conjecture. Some part of that problem, however, is much more complicated. The trick lies in estimating the determinant of the matrix defined by the linearized operator. This was relatively easy in the case of Calabi conjecture. Fu worked hard with me. And through a tough process of estimation, we finally managed to solve the problem. I think it is a solution that we can be proud of. This is the first time that we found a truly non-Kähler solution to the Strominger system, although much more needs to be said about them.

When we finished this solution, the Becker sisters were visiting Harvard, and they got excited by the solution. Li-Sheng Tseng was their postdoc at the time. And the Beckers, Tseng, and I jointly wrote a paper to provide a physical meaning for this solution.

Naturally, Li-Sheng became my postdoctoral fellow, and we had an extremely fruitful exchange of ideas during our collaboration. I really appreciate his devotion and his nice ideas.

J. Fu, Jun Li, and I tried to solve the Strominger system for more examples of non-Kähler manifolds. This remains an interesting, as well as very challenging, problem.

The supersymmetry conditions in the Strominger system requires only the balanced condition, which is weaker than the Kähler condition for the Hermitian form. J. Fu, Jun Li, and I proved that this can be achieved for those manifolds obtained by smoothing out the blowdown of Calabi-Yau manifolds on a suitable configuration of rigid rational curves. The last construction was due to Clemens and Friedman. By blowing down enough rational curves, we can arrive at the connected sum of the products of three spheres. It was conjectured that all Calabi-Yau manifolds should be connected through this procedure.

Thirteen years ago, Richard Thomas, Smith, and I proposed the reverse of this construction by the blowdown of Lagrangian spheres. We expected that the final classes of manifolds are the connected sum of complex projective spaces. And they should be symplectic manifolds that are mirror to those constructed through blowing down of rigid rational curves.

It became an interesting question to find the symplectic mirror of several cohomologies of complex geometry. Tseng and I studied a class of equations that should be mirror to each other. In order to do this, we developed several cohomologies of differential forms for symplectic geometry using some analogs of the del-bar and del-del-bar operators. It turns out that these new symplectic cohomologies are interesting in their own right. C.J. Tsai recently joined us in this project.

Part V
A List of Classic Papers

Complex Geometry from Riemann to Kähler-Einstein and Calabi-Yau
ALM 38, pp. 633–664

Some Classic Papers in Geometric Analysis in the 20th Century

Lizhen Ji and Shing-Tung Yau

Contents

1 Introduction

Complex geometry is an integral part of geometric analysis. In the earlier part of this book, we have reprinted some papers in complex geometry. Naturally these papers could not cover many aspects of the subject of geometric analysis. For the convenience of the reader, we have complied a list of important papers in geometric analysis published in the twentieth century. We hope that this list will complement the previous papers in this book.

Reading classical papers is important. On the other hand, the literature of mathematics is huge and also growing at a rapid speed, and a short list containing papers which have played a fundamental role in geometric analysis will obviously be valuable to many people. We hope that this list can serve as a tour guide to the subject of geometric analysis.

Since identifying and picking important papers is a nontrivial and important problem, we explain briefly the criteria used in the selection. One key aspect of geometric analysis is to combine analysis with the underlying topology and geometry of the spaces under study. Since the algebraic topology and global differential geometry were developed from the early 20th century on, it is reasonable to restrict to papers published after 1900. Since everything needs to be tested by time and fifteen years form probably the minimum period for such a purpose, it is reasonable to restrict to papers published up to the end of the 20th century.

Papers on the list here were selected based on many email correspondences with Prof. Shing-Tung Yau. Besides many papers picked directly by Yau, the editor has also tried to look up various sources and made suggestions. But the final decision was made and the whole list was approved by Yau.

In some cases, major results were first discussed in books, and hence several books are also selected. Otherwise, books are not considered in general. It is probably obvious that it is difficult to make any such list complete[1]. One reason is that there is no precise definition of geometric analysis. In selecting papers for this list, we usually require them to have both components in geometry (differential, complex or algebraic geometry) and analysis (real and complex analysis, differential equations, calculus of variations, and functional analysis). In view of this, papers purely in differential geometry, symplectic geometry, algebraic geometry etc. are not selected in general, though some exceptions were made. Given the consideration about time period above, almost no papers or books published after 1999 are on the list.

[1] We would like to thank Norbert A'Campo and Jürgen Jost for several valuable suggestions.

Since the current list covers many topics and many papers fall into different categories, it is difficult to decide to which section a paper should belong, and different people will generate different divisions even with the same collection of papers. In each section, there are also different ways to order papers. One original general guiding principle of the editor was to put related papers together so that when the reader looks up one paper, they might check nearby ones as people normally do when they browse books in library, and hence the papers are not necessarily arranged according to the years of publication. Later it was suggested to order papers according to time of publication in each section, and to duplicate some papers in multiple sections since they are closely related to several topics.

This is not the first attempt to make a list of great papers in geometric analysis and differential geometry. If a reader is interested in other lists, a google search will turn up several. On the other hand, the current list seems to be more exhaustive in some sense and should reflect the perspective of Prof. Yau on the broad subject of geometric analysis to a certain extent.

2 Spectral theory of Riemannian manifolds

1. H. Weyl, Das asymptotische Verteilungsgesetz linearen partiellen Differentialgleichungen. Math. Ann., 71 (1912) 441–479.
2. S. Minakshisundaram, A. Pleijel, Some properties of the eigenfunctions of the Laplace operator on Riemannian manifolds. Canadian J. Math., 1 (1949) 242–256.
3. G. Polya, G. Szegö, Isoperimetric inequalities in mathematical physics. Annals of Mathematics Studies, no. 27, Princeton University Press, Princeton, N. J., 1951. xvi+279 pp.
4. R. Courant, D. Hilbert, Methods of mathematical physics. vol. I. Interscience Publishers, Inc., New York, N.Y., 1953. xv+561 pp.
5. G. Szego, Inequalities for certain eigenvalues of a membrane of given area. J. Rational Mech. Anal. 3, (1954) 343–356.
6. A. Selberg, Harmonic analysis and discontinuous groups in weakly symmetric Riemannian spaces with applications to Dirichlet series. J. Indian Math. Soc. (N.S.), 20 (1956), 47–87. (*Also appears in section on Lie groups.*)
7. J. Milnor, Eigenvalues of the Laplace operator on certain manifolds. Proc. Natl. Acad. Sci., U.S.A. 51 (1964) 542.
8. A. Selberg, On the estimation of Fourier coefficients of modular forms. 1965 Proc. Sympos. Pure Math., vol. VIII. pp. 1–15 Amer. Math. Soc., Providence, R.I.
9. M. Kac, Can one hear the shape of a drum? Amer. Math. Monthly, 73 (1966) no. 4, part II, 1–23.
10. D. Kazhdan, On the connection of the dual space of a group with the structure of its closed subgroups. Functional analysis and its applications, 1 (1967) 63–65. (*Also appears in section on Lie groups.*)

11. H. McKean, I.M. Singer, Curvature and the eigenvalues of the Laplacian. J. Differential Geometry, 1 (1967) no. 1, 43–69.

12. L. Hörmander, The spectral function of an elliptic operator. Acta Math., 121 (1968), 193–218.

13. J. Cheeger, A lower bound for the smallest eigenvalue of the Laplacian. *Problems in analysis* (Papers dedicated to Salomon Bochner, 1969), pp. 195–199. Princeton Univ. Press, Princeton, N. J., 1970.

14. J. Hersch, Quatre propriétés isopérimétriques de membranes sphériques homogénes. C. R. Acad. Sci. Paris Sér., A-B 270 (1970), A1645–A1648.

15. S.Y. Cheng, Eigenvalue comparison theorems and its geometric applications. Math. Z. 143 (1975), no. 3, 289–297.

16. J. Duistermaat, V. Guillemin, The spectrum of positive elliptic operators and periodic bicharacteristics. Invent. Math., 29 (1975), no. 1, 39–79.

17. S.T. Yau, Isoperimetric constants and the first eigenvalue of a compact Riemannian manifold. Ann. Sci. Ecole Norm. Sup. (4) 8 (1975), no. 4, 487–507.

18. S.Y. Cheng, Eigenfunctions and nodal sets. Comment. Math. Helv. 51 (1976), no. 1, 43–55.

19. R. Langlands, On the functional equations satisfied by Eisenstein series. Lecture Notes in Mathematics, Vol. 544. Springer-Verlag, Berlin-New York, 1976. v+337 pp. (*Also appears in section on Lie groups.*)

20. V. Guillemin, S. Sternberg, Geometric asymptotics. Mathematical Surveys, No. 14. American Mathematical Society, Providence, R.I., 1977. xviii+474 pp.

21. P. van Moerbeke, D. Mumford, The spectrum of difference operators and algebraic curves. Acta Math., 143 (1979), no. 1-2, 93–154.

22. C. Croke, Some isoperimetric inequalities and eigenvalue estimates. Ann. Sci. École Norm. Sup. (4) 13 (1980), no. 4, 419–435.

23. T. Friedrich, Der erste Eigenwert des Dirac-Operators einer kompakten, Riemannschen Mannigfaltigkeit nichtnegativer Skalarkrümmung. Math. Nachr., 97 (1980), 117–146.

24. V. Ivrii, The second term of the spectral asymptotics for a Laplace-Beltrami operator on manifolds with boundary. Functional Anal. Appl., 14 (1980), no. 2, 98–106.

25. P. Li, S.T. Yau, Estimates of eigenvalues of a compact Riemannian manifold. *Geometry of the Laplace operator* (Proc. Sympos. Pure Math., Univ. Hawaii, Honolulu, Hawaii, 1979), pp. 205–239, Proc. Sympos. Pure Math., XXXVI, Amer. Math. Soc., Providence, R.I., 1980.

26. P. Yang, S. T. Yau, Eigenvalues of the Laplacian of compact Riemann surfaces and minimal submanifolds. Ann. Scuola Norm. Sup. Pisa Cl. Sci. (4) 7 (1980), no. 1, 55–63.

27. S.Y. Cheng, P. Li, S.T. Yau, On the upper estimate of the heat kernel of a complete Riemannian manifold. Amer. J. Math., 103 (1981), no. 5, 1021–1063.

28. S. Agmon, Lectures on exponential decay of solutions of second-order elliptic equations: bounds on eigenfunctions of N-body Schrödinger operators. Mathematical Notes, 29. Princeton University Press, Princeton, NJ; University of Tokyo Press, Tokyo, 1982. 118 pp.

29. P. Li, S.T. Yau, A new conformal invariant and its applications to the Willmore conjecture and the first eigenvalue of compact surfaces. Invent. Math., 69 (1982), no. 2, 269–291.

30. H. Choi, A. Wang, A first eigenvalue estimate for minimal hypersurfaces. J. Differential Geom., 18 (1983), no. 3, 559–562.

31. P. Li, S.T. Yau, On the Schrödinger equation and the eigenvalue problem. Comm. Math. Phys., 88 (1983), no. 3, 309–318.

32. B. Simon, Semiclassical analysis of low lying eigenvalues. II. Tunneling. Ann. of Math., (2) 120 (1984), no. 1, 89–118.

33. C. Vafa, E. Witten, Eigenvalue inequalities for fermions in gauge theories. Comm. Math. Phys., 95 (1984), no. 3, 257–276.

34. Y. Colin de Verdière, Ergodicité et fonctions propres du laplacien. Comm. Math. Phys. 102 (1985), no. 3, 497–502.

35. I.M. Singer, B. Wong, S.T. Yau, Stephen S.-T. Yau, An estimate of the gap of the first two eigenvalues in the Schrödinger operator. Ann. Scuola Norm. Sup. Pisa Cl. Sci. (4) 12 (1985), no. 2, 319–333.

36. T. Sunada, Riemannian coverings and isospectral manifolds. Ann. of Math. (2) 121 (1985) 169–186.

37. Y. Colin de Verdiére, Sur la multiplicité de la première valeur propre non nulle du laplacien. Comment. Math. Helv., 61 (1986), no. 2, 254–270.

38. P. Li, S.T. Yau, On the parabolic kernel of the Schrödinger operator. Acta Math., 156 (1986), no. 3-4, 153–201. (*Also appears in section on classification of manifolds.*)

39. J. Sylvester, G. Uhlmann, A global uniqueness theorem for an inverse boundary value problem. Ann. of Math. (2) 125 (1987), no. 1, 153–169.

40. S. Zelditch, Uniform distribution of eigenfunctions on compact hyperbolic surfaces. Duke Math. J., 55 (1987), no. 4, 919–941.

41. H. Donnelly, C. Fefferman, Nodal sets of eigenfunctions on Riemannian manifolds. Invent. Math., 93 (1988), no. 1, 161–183.

42. J. Arthur, Unipotent automorphic representations: conjectures. Orbites unipotentes et représentations, II. Astérisque No. 171-172 (1989), 13–71. (*Also appears in section on spectral theory.*)

43. J. Sjöstrand, M. Zworski, Complex scaling and the distribution of scattering poles. J. Amer. Math. Soc. 4 (1991), no. 4, 729–769.

44. C. Gordon, D. Webb, S. Wolpert, Isospectral plane domains and surfaces via Riemannian orbifolds. Inventiones Mathematicae, 110 (1992) 1–22.

45. N. Korevaar, Upper bounds for eigenvalues of conformal metrics. J. Differential Geom., 37 (1993), no. 1, 73–93.

46. J. Bourguignon, P. Li, S.T. Yau, Upper bound for the first eigenvalue of algebraic submanifolds. Comment. Math. Helv., 69 (1994), no. 2, 199–207.

47. R. Melrose, Geometric scattering theory. Stanford Lectures, Cambridge University Press, Cambridge, 1995. xii+116 pp.

48. A. Grigor'yan, Gaussian upper bounds for the heat kernel on arbitrary manifolds. J. Differential Geom., 45 (1997), no. 1, 33–52.

49. F. Chung, S.T. Yau, Discrete Green's functions. In memory of Gian-Carlo Rota. J. Combin. Theory Ser. A 91 (2000), no. 1-2, 191–214.

50. F. Chung, A. Grigor'yan, S.T. Yau, Higher eigenvalues and isoperimetric inequalities on Riemannian manifolds and graphs. Comm. Anal. Geom., 8 (2000), no. 5, 969–1026.

3 Geodesics, fundamental group, and differential geometry

1. M. Hadamard, Les surfaces à courbures opposées et leur lignes géodésiques. Collected Works, vol. 2, pp. 729–775. J. Math. Pures Appl., 4 (1898), 27–74.

2. E. Cartan, Elie Les systèmes de Pfaff, à cinq variables et les équations aux dérivées partielles du second ordre. Ann. Sci. École Norm. Sup. (3) 27 (1910), 109–192.

3. W. Blaschke, Vorlesungen über Differentialgeometrie. 3 vols., Springer, Grundlehren der mathematischen Wissenschaften 1921-1929 (vol. 1, Elementare Differentialgeometrie; vol. 2, Affine Differentialgeometrie; vol. 3, Differentialgeometrie der Kreise und Kugeln, 1929).

4. A. Preissman, Quelques propriétés globales des espaces de Riemann. Comment. Math. Helv. 15, (1943). 175–216.

5. E. Cartan, Les systèmes différentiels extérieurs et leurs applications géométriques. Actualités Sci. Ind., no. 994. Hermann et Cie., Paris, 1945. 214 pp.

6. J. Milnor, Growth of finitely generated solvable groups. J. Differential Geometry, 2 (1968) 447–449.

7. D. Gromoll, W. Meyer, Periodic geodesics on compact Riemannian manifolds. J. Differential Geometry, 3 (1969) 493–510.

8. D. Gromoll, J. Wolf, Some relations between the metric structure and the algebraic structure of the fundamental group in manifolds of nonpositive curvature. Bull. Amer. Math. Soc., 77 (1971) 545–552.

9. S.T. Yau, On the fundamental group of compact manifolds of non-positive curvature. Ann. of Math. (2) 93 (1971) 579–585.

10. B. Lawson, S.T. Yau, Compact manifolds of nonpositive curvature. J. Differential Geometry, 7 (1972), 211–228.

11. B. Lawson, S.T. Yau, Scalar curvature, non-abelian group actions, and the degree of symmetry of exotic spheres. Comment. Math. Helv., 49 (1974), 232–244.

12. L. Santaló, Integral geometry and geometric probability. With a foreword by Mark Kac. Encyclopedia of Mathematics and its Applications, Vol. 1. Addison-Wesley Publishing Co., Reading, Mass.-London-Amsterdam, 1976. xvii+404 pp.

13. M. Vigué-Poirrier, D. Sullivan, The homology theory of the closed geodesic problem. J. Differential Geometry, 11 (1976), no. 4, 633–644.

14. W. Klingenberg, Lectures on closed geodesics. Grundlehren der Mathematischen Wissenschaften, Vol. 230. Springer-Verlag, Berlin-New York, 1978. x+227 pp.

15. E. Cartan, Geometry of Riemannian spaces. Translated from the French and with a preface by James Glazebrook. With a preface, notes and appendices by Robert Hermann. Lie Groups: History, Frontiers and Applications, Series A, XIII. Math Sci Press, Brookline, MA, 1983. xiv+506 pp.

16. N. Hingston, Equivariant Morse theory and closed geodesics. J. Differential Geom., 19 (1984), no. 1, 85–116.

17. E. Cartan, On manifolds with an affine connection and the theory of general relativity. Translated from the French by Anne Magnon and Abhay Ashtekar. With a foreword by Andrzej Trautman. Monographs and Textbooks in Physical Science, 1. Bibliopolis, Naples, 1986. 199 pp.

18. H. Rademacher, On the average indices of closed geodesics. J. Differential Geom., 29 (1989), no. 1, 65–83.

19. R. Bryant, S.S. Chern, R.B. Gardner, H.L. Goldschmidt, P. Griffiths, Exterior differential systems. Mathematical Sciences Research Institute Publications, 18. Springer-Verlag, New York, 1991. viii+475 pp.

20. J. Franks, Geodesics on S^2 and periodic points of annulus homeomorphisms. Invent. Math., 108 (1992), no. 2, 403–418.

21. V. Bangert, On the existence of closed geodesics on two-spheres. Internat. J. Math., 4 (1993), no. 1, 1–10.

22. E. Cartan, Riemannian geometry in an orthogonal frame. From lectures delivered by Élie Cartan at the Sorbonne in 1926–27. With a preface to the Russian edition by S. P. Finikov. Translated from the 1960 Russian edition by Vladislav V. Goldberg and with a foreword by S. S. Chern. World Scientific Publishing Co., Inc., River Edge, NJ, 2001. xviii+259 pp.

4 Minimal surfaces and submanifolds

1. T. Rado, On Plateau's problem. Ann. of Math. (2) 31 (1930), no. 3, 457–469.

2. J. Douglas, Solution of the problem of Plateau. Trans. Amer. Math. Soc., 33 (1931), no. 1, 263–321.
3. C. Morrey, The problem of Plateau on a Riemannian manifold. Ann. of Math. (2) 49, (1948) 807–851.
4. E. Calabi, Minimal immersions of surfaces in Euclidean spheres. J. Differential Geometry, 1 (1967) 111–125.
5. J. Simons, Minimal varieties in Riemannian manifolds. Ann. of Math. (2) 88 (1968) 62–105.
6. E. Bombieri, E. De Giorgi, E. Giusti, Minimal cones and the Bernstein problem. Invent. Math., 7 (1969) 243–268.
7. S. Hildebrandt, Boundary behavior of minimal surfaces. Arch. Rational Mech. Anal., 35 (1969) 47–82.
8. R. Osserman, Robert A proof of the regularity everywhere of the classical solution to Plateau's problem. Ann. of Math. (2) 91 (1970) 550–569.
9. F. Harvey, B. Lawson, On boundaries of complex analytic varieties. I. Ann. of Math. (2) 102 (1975), no. 2, 223–290.
10. R. Schoen, L. Simon, S.T. Yau, Curvature estimates for minimal hypersurfaces. Acta Math. 134 (1975), no. 3-4, 275–288.
11. R. Hardt, L. Simon, Boundary regularity and embedded solutions for the oriented Plateau problem. Ann. of Math. (2) 110 (1979), no. 3, 439–486.
12. R. Schoen, S.T. Yau, Existence of incompressible minimal surfaces and the topology of three dimensional manifolds with nonnegative scalar curvature. Ann. of Math. (2) 110 (1979), no. 1, 127–142.
13. R. Schoen, S.T. Yau, On the structure of manifolds with positive scalar curvature. Manuscripta Math., 28 (1979), no. 1-3, 159–183. (*Also appears in section on special metrics.*)
14. D. Fischer-Colbrie, R. Schoen, The structure of complete stable minimal surfaces in 3-manifolds of nonnegative scalar curvature. Comm. Pure Appl. Math., 33 (1980), no. 2, 199–211.
15. Y.T. Siu, S.T. Yau, Compact Kähler manifolds of positive bisectional curvature. Invent. Math., 59 (1980), no. 2, 189–204. (*Also appears in section on rigidity of manifolds.*)
16. J. Pitts, Existence and regularity of minimal surfaces on Riemannian manifolds. Mathematical Notes, 27. Princeton University Press, Princeton, N.J.; University of Tokyo Press, Tokyo, 1981. iv+330 pp.
17. J. Sacks, K. Uhlenbeck, The existence of minimal immersions of 2-spheres. Ann. of Math. (2) 113 (1981), no. 1, 1–24.
18. R. Schoen, L. Simon, Regularity of stable minimal hypersurfaces. Comm. Pure Appl. Math., 34 (1981), no. 6, 741–797.
19. R. Harvey, B. Lawson, Calibrated geometries. Acta Math., 148 (1982), 47–157.

20. W. Meeks, L. Simon, S.T. Yau, Embedded minimal surfaces, exotic spheres, and manifolds with positive Ricci curvature. Ann. of Math. (2) 116 (1982), no. 3, 621–659.

21. W. Meeks, S.T. Yau, The classical Plateau problem and the topology of three-dimensional manifolds. The embedding of the solution given by Douglas-Morrey and an analytic proof of Dehn's lemma. Topology, 21 (1982), no. 4, 409–442.

22. L. Simon, Asymptotics for a class of nonlinear evolution equations, with applications to geometric problems. Ann. of Math. (2) 118 (1983), no. 3, 525–571.

23. H. Choi, R. Schoen, The space of minimal embeddings of a surface into a three-dimensional manifold of positive Ricci curvature. Invent. Math., 81 (1985), no. 3, 387–394.

24. M. Gromov, Pseudoholomorphic curves in symplectic manifolds. Invent. Math., 82 (1985), no. 2, 307–347.

25. C.L. Terng, Isoparametric submanifolds and their Coxeter groups. J. Differential Geom., 21 (1985), no. 1, 79–107.

26. M. Gromov, Partial differential relations. Ergebnisse der Mathematik und ihrer Grenzgebiete (3), 9. Springer-Verlag, Berlin, 1986. x+363 pp.

27. H. Wente, Counterexample to a conjecture of H. Hopf. Pacific J. Math. 121 (1986), no. 1, 193–243.

28. W. Meeks, H. Rosenberg, The topology and geometry of embedded surfaces of constant mean curvature. J. Differential Geom., 27 (1988), no. 3, 539–552.

29. M. Micallef, J. Moore, Minimal two-spheres and the topology of manifolds with positive curvature on totally isotropic two-planes. Ann. of Math. (2) 127 (1988), no. 1, 199–227.

30. J. Jost, Embedded minimal surfaces in manifolds diffeomorphic to the three-dimensional ball or sphere. J. Differential Geom., 30 (1989), no. 2, 555–577.

31. G. Lawlor, The angle criterion. Invent. Math., 95 (1989), no. 2, 437–446.

32. D. Hoffman, W. Meeks, Embedded minimal surfaces of finite topology. Ann. of Math. (2) 131 (1990), no. 1, 1–34.

33. G. Thorbergsson, Isoparametric foliations and their buildings. Ann. of Math. (2) 133 (1991), no. 2, 429–446.

34. W. Meeks, H. Rosenberg, The geometry and conformal structure of properly embedded minimal surfaces of finite topology in $\mathbb{R}^3$. Invent. Math., 114 (1993), no. 3, 625–639.

35. W. Meeks, H. Rosenberg, The geometry of periodic minimal surfaces. Comment. Math. Helv., 68 (1993), no. 4, 538–578.

36. L. Simon, Existence of surfaces minimizing the Willmore functional. Comm. Anal. Geom., 1 (1993), no. 2, 281–326.

5 Geometric measure theory

1. H. Federer, W. Fleming, Normal and integral currents. Ann. of Math. (2) 72 (1960) 458–520.
2. H. Federer, Geometric measure theory. Die Grundlehren der mathematischen Wissenschaften, Band 153 Springer-Verlag New York Inc., New York, 1969, xiv+676 pp.
3. W. Allard, On the first variation of a varifold. Ann. of Math. (2) 95 (1972), 417–491.
4. J. Michael, L. Simon, Sobolev and mean-value inequalities on generalized submanifolds of $\mathbb{R}^n$. Comm. Pure Appl. Math., 26 (1973), 361–379.
5. F. Almgren, Q valued functions minimizing Dirichlet's integral and the regularity of area minimizing rectifiable currents up to codimension two. American Mathematical Society, Bulletin. New Series 8 (1983) 327–328.
6. L. Simon, Asymptotics for a class of nonlinear evolution equations, with applications to geometric problems. Ann. of Math. (2) 118 (1983), no. 3, 525–571.
7. L. Simon, Lectures on geometric measure theory. Proceedings of the Centre for Mathematical Analysis, Australian National University, 3. Australian National University, Centre for Mathematical Analysis, Canberra, 1983. vii+272 pp.
8. F. Almgren, Q-valued functions minimizing Dirichlet's integral and the regularity of area-minimizing rectifiable currents up to codimension 2. With a preface by Jean E. Taylor and Vladimir Scheffer. World Scientific Monograph Series in Mathematics, 1. World Scientific Publishing Co., Inc., River Edge, NJ, 2000. xvi+955 pp.

6 Riemann surfaces, moduli spaces, Teichmüller theory and quasi-conformal maps

1. L. Ahlfors, An extension of Schwarz's lemma. Trans. Amer. Math. Soc., 43 (1938), no. 3, 359–364.
2. C. Morrey, On the solutions of quasi-linear elliptic partial differential equations. Trans. Amer. Math. Soc., 43 (1938), no. 1, 126–166. (*also appears in section on nonlinear differential equations.*)
3. O. Teichmüller, Extremale quasikonforme Abbildungen und quadratische Differentiale. Abh. Preuss. Akad. Wiss. Math.-Nat. Kl., 1939, (1940). no. 22, 197 pp.
4. O. Teichmüller, Veränderliche Riemannsche Flächen. Deutsche Math., 7, (1944). 344–359.

5. R. Courant, Dirichlet's principle, conformal mapping, and minimal surfaces. Appendix by M. Schiffer. Interscience Publishers, Inc., New York, N.Y., 1950. xiii+330 pp.

6. A. Weil, Modules des surfaces de Riemann. 1958 Séminaire Bourbaki; 10e année: 1957/1958. Textes des conf Èrences; Exposés 152à 168; 2e éd. corrigée, Exposé 168, 7 pp. Secrétariat mathématique, Paris.

7. L. Ahlfors, L. Bers, Riemann's mapping theorem for variable metrics. Ann. of Math. (2) 72 (1960) 385–404.

8. L. Ahlfors, Some remarks on Teichmüller's space of Riemann surfaces. Ann. of Math. (2) 74 (1961) 171–191.

9. L. Bers, Uniformization by Beltrami equations. Comm. Pure Appl. Math. 14 (1961) 215–228.

10. H. Weyl, The concept of a Riemann surface. Translated from the third German edition by Gerald R. MacLane. ADIWES International Series in Mathematics Addison-Wesley Publishing Co., Inc., Reading, Mass.-London 1964 xi+191 pp.

11. H. Rauch, A transcendental view of the space of algebraic Riemann surfaces. Bull. Amer. Math. Soc., 71 (1965) 1–39.

12. G. Mostow, Quasi-conformal mappings in n-space and the rigidity of hyperbolic space forms. Inst. Hautes Études Sci. Publ. Math. No. 34 (1968) 53–104.

13. H.L. Royden, Automorphisms and isometries of Teichmüller space. 1971 Advances in the Theory of Riemann Surfaces (Proc. Conf., Stony Brook, N.Y., 1969) pp. 369–383 Ann. of Math. Studies, No. 66. Princeton Univ. Press, Princeton, N.J.

14. H. Masur, Extension of the Weil-Petersson metric to the boundary of Teichmüller space. Duke Math. J. 43 (1976), no. 3, 623–635.

15. A. Fathi, F. Laudenbach, V. Poenaru, Travaux de Thurston sur les surfaces. Séminaire Orsay. With an English summary. Astérisque, 66-67. Société Mathématique de France, Paris, 1979. 284 pp.

16. M. Gromov, Hyperbolic manifolds, groups and actions. Riemann surfaces and related topics: Proceedings of the 1978 Stony Brook Conference (State Univ. New York, Stony Brook, N.Y., 1978), pp. 183–213, Ann. of Math. Stud., 97, Princeton Univ. Press, Princeton, N.J., 1981.

17. S. Kerckhoff, The Nielsen realization problem. Ann. of Math. (2) 117 (1983), no. 2, 235–265.

18. R. Mane, P. Sad, D. Sullivan, On the dynamics of rational maps. Ann. Sci. École Norm. Sup. (4) 16 (1983), no. 2, 193–217.

19. S. Wolpert, On the symplectic geometry of deformations of a hyperbolic surface. Ann. of Math. (2) 117 (1983), no. 2, 207–234.

20. D. Sullivan, Quasiconformal homeomorphisms and dynamics. I. Solution of the Fatou-Julia problem on wandering domains. Ann. of Math. (2) 122 (1985), no. 3, 401–418. II. Structural stability implies hyperbolicity for Kleinian groups. Acta Math. 155 (1985), no. 3-4, 243–260.

21. A. Tromba, On a natural algebraic affine connection on the space of almost complex structures and the curvature of Teichmüller space with respect to its Weil-Petersson metric. Manuscripta Math., 56 (1986), no. 4, 475–497.

22. S. Wolpert, Geodesic length functions and the Nielsen problem. J. Differential Geom., 25 (1987), no. 2, 275–296.

23. W. Thurston, On the geometry and dynamics of diffeomorphisms of surfaces. Bull. Amer. Math. Soc. (N.S.) 19 (1988), no. 2, 417–431.

24. P. Pansu, Métriques de Carnot-Carathéodory et quasiisométries des espaces symétriques de rang un. Ann. of Math. (2) 129 (1989), no. 1, 1–60.

25. M. Kontsevich, Intersection theory on the moduli space of curves and the matrix Airy function. Comm. Math. Phys., 147 (1992), no. 1, 1–23.

26. A. Korányi, H. Reimann, Foundations for the theory of quasiconformal mappings on the Heisenberg group. Adv. Math., 111 (1995), no. 1, 1–87.

27. C. McMullen, D. Sullivan, Quasiconformal homeomorphisms and dynamics. III. The Teichmüller space of a holomorphic dynamical system. Adv. Math., 135 (1998), no. 2, 351–395.

28. C. McMullen, Hausdorff dimension and conformal dynamics. II. Geometrically finite rational maps. Comment. Math. Helv., 75 (2000), no. 4, 535–593.

7 Deformation and rigidity of manifolds

1. K. Kodaira, D. Spencer, On deformations of complex analytic structures. I, II. Ann. of Math. (2) 67 (1958) 328–466, On deformations of complex analytic structures. III. Stability theorems for complex structures. Ann. of Math. (2) 71 (1960) 43–76.

2. E. Calabi, E. Vesentini, On compact, locally symmetric Kähler manifolds. Ann. of Math., 71 (1960) 472–507.

3. E. Calabi, On compact, Riemannian manifolds with constant curvature. I, Proc. Sympos. Pure Math., vol. III. pp. 155-180, Amer. Math. Soc., 1961.

4. M. Kuranishi, On the locally complete families of complex analytic structures. Ann. of Math. (2) 75 (1962), 536–577.

5. A. Weil, Remarks on the cohomology of groups. Ann. of Math. (2) 80 (1964) 149–157.

6. J. Cheeger, D. Gromoll, The splitting theorem for manifolds of nonnegative Ricci curvature. J. Differential Geometry, 6 (1971/72), 119–128.

7. J. Cheeger, D. Gromoll, On the structure of complete manifolds of nonnegative curvature. Ann. of Math. (2) 96 (1972), 413–443.

8. G. Mostow, Strong rigidity of locally symmetric spaces. Annals of Mathematics Studies, No. 78. Princeton University Press, Princeton, N.J.; University of Tokyo Press, Tokyo, 1973. v+195 pp.

9. B. Wong, Characterization of the unit ball in $\mathbb{C}^n$ by its automorphism group. Invent. Math., 41 (1977), no. 3, 253–257.

10. Y.T. Siu, The complex-analyticity of harmonic maps and the strong rigidity of compact Kähler manifolds. Ann. of Math. (2) 112 (1980), no. 1, 73–111.

11. Y.T. Siu, S.T. Yau, Compact Kähler manifolds of positive bisectional curvature. Invent. Math., 59 (1980), no. 2, 189–204. (*Also appears in section on minimal surfaces.*)

12. R. Zimmer, Ergodic theory and semisimple groups. Monographs in Mathematics, 81. Birkhäuser Verlag, Basel, 1984. x+209 pp.

13. W. Ballmann, Nonpositively curved manifolds of higher rank. Ann. of Math. (2) 122 (1985), no. 3, 597–609.

14. W. Ballmann, M. Gromov, V. Schroeder, Manifolds of nonpositive curvature. Progress in Mathematics, 61. Birkhäuser Boston, Inc., Boston, MA, 1985. vi+263 pp.

15. K. Burns, R. Spatzier, Manifolds of nonpositive curvature and their buildings. Inst. Hautes Etudes Sci. Publ. Math., No. 65 (1987), 35–59.

16. N. Mok, Uniqueness theorems of Hermitian metrics of seminegative curvature on quotients of bounded symmetric domains. Ann. of Math. (2) 125 (1987), no. 1, 105–152.

17. F.T. Farrell, L. Jones, Negatively curved manifolds with exotic smooth structures. J. Amer. Math. Soc. 2 (1989), no. 4, 899–908.

18. S. Frankel, Complex geometry of convex domains that cover varieties. Acta Math.163 (1989), no. 1-2, 109–149.

19. G. Margulis, Discrete subgroups of semisimple Lie groups. Ergebnisse der Mathematik und ihrer Grenzgebiete, 17. Springer-Verlag, Berlin, 1991. x+388 pp.

20. K. Corlette, Archimedean superrigidity and hyperbolic geometry. Ann. of Math. (2) 135 (1992), no. 1, 165–182.

21. M. Gromov, R. Schoen, Harmonic maps into singular spaces and p-adic superrigidity for lattices in groups of rank one. Inst. Hautes Études Sci. Publ. Math. No. 76 (1992), 165–246.

22. F.T. Farrell, L. Jones, Isomorphism conjectures in algebraic K-theory. J. Amer. Math. Soc., 6 (1993), no. 2, 249–297.

23. J. Jost, S.T. Yau, Harmonic maps and superrigidity. *Differential geometry: partial differential equations on manifolds* (Los Angeles, CA, 1990), 245–280, Proc. Sympos. Pure Math., 54, Part 1, Amer. Math. Soc., Providence, RI, 1993.

24. N. Korevaar, R. Schoen, Sobolev spaces and harmonic maps for metric space targets. Comm. Anal. Geom., 1 (1993), no. 3-4, 561–659.

25. N. Mok, Y.T. Siu, S.K. Yeung, Geometric superrigidity. Invent. Math., 113 (1993), no. 1, 57–83.

26. J. Jost, Equilibrium maps between metric spaces. Calc. Var. Partial Differential Equations, 2 (1994), no. 2, 173–204.

27. G. Besson, G. Courtois, S. Gallot, Entropiés et rigidites des espaces localement symétriques de courbure strictement négative. Geom. Funct. Anal., 5 (1995), no. 5, 731–799.

28. J. Jost, Nonpositive curvature: geometric and analytic aspects. Lectures in Mathematics ETH Zürich. Birkhäuser Verlag, Basel, 1997. viii+108 pp.

8 Classification of manifolds, special metrics and connections

1. M. Berger, Sur les groupes d'holonomie homogène des variétés à connexion affine et des variétés riemanniennes. Bull. Soc. Math. France, 83 (1955), 279–330.

2. E. Calabi, On Kähler manifolds with vanishing canonical class. In: *Algebraic Geometry and Topology. A Symposium in Honor of S. Lefschetz*, Princeton University Press, Princeton, NJ, pp. 78–89. 1957.

3. N. Trudinger, Remarks concerning the conformal deformation of Riemannian structures on compact manifolds. Ann. Scuola Norm. Sup. Pisa (3) 22 (1968) 265–274.

4. D. Sullivan, Cycles for the dynamical study of foliated manifolds and complex manifolds. Invent. Math. 36 (1976), 225–255.

5. S.T. Yau, Calabi's conjecture and some new results in algebraic geometry. Proc. Nat. Acad. Sci. U.S.A. 74 (1977), no. 5, 1798–1799. (*Also appears in section on applications to algebraic geometry.*)

6. S.T. Yau, On the Ricci curvature of a compact Kähler manifold and the complex Monge-Ampère equation. I. Comm. Pure Appl. Math., 31 (1978), no. 3, 339–411. (*Also appears in section on applications to algebraic geometry.*)

7. R. Schoen, S.T. Yau, On the structure of manifolds with positive scalar curvature. Manuscripta Math., 28 (1979), no. 1-3, 159–183. (*Also appears in section on minimal surfaces.*)

8. S.Y. Cheng, S.T. Yau, On the existence of a complete Kähler metric on noncompact complex manifolds and the regularity of Fefferman's equation. Comm. Pure Appl. Math., 33 (1980), no. 4, 507–544.

9. S.T. Yau, The role of partial differential equations in differential geometry. Proceedings of the International Congress of Mathematicians (Helsinki, 1978), pp. 237–250, Acad. Sci. Fennica, Helsinki, 1980.

10. R. Hamilton, Three-manifolds with positive Ricci curvature. J. Differential Geom., 17 (1982), no. 2, 255–306.

11. W. Thurston, Three-dimensional manifolds, Kleinian groups and hyperbolic geometry. Bull. Amer. Math. Soc. (N.S.) 6 (1982), no. 3, 357–381.

12. W. Goldman, The symplectic nature of fundamental groups of surfaces. Adv. in Math. 54 (1984), no. 2, 200–225.

13. G. Huisken, Flow by mean curvature of convex surfaces into spheres. J. Differential Geom. 20 (1984), no. 1, 237–266.

14. R. Schoen, Conformal deformation of a Riemannian metric to constant scalar curvature. J. Differential Geometry 20 (1984) 479–495.

15. J. Wolf, Spaces of constant curvature. Fifth edition. Publish or Perish, Inc., Houston, TX, 1984. xviii+412 pp. (*Also appears in section on Lie groups.*)

16. H.D. Cao, Deformation of Kähler metrics to Kähler-Einstein metrics on compact Kähler manifolds. Invent. Math. 81 (1985), no. 2, 359–372.

17. S.K. Donaldson, Anti self-dual Yang-Mills connections over complex algebraic surfaces and stable vector bundles. Proc. London Math. Soc. (3) 50 (1985), no. 1, 1–26.

18. M. Gage, R. Hamilton, The heat equation shrinking convex plane curves. J. Differential Geom., 23 (1986), no. 1, 69–96.

19. R. Hamilton, Four-manifolds with positive curvature operator. J. Differential Geom., 24 (1986), no. 2, 153–179.

20. P. Li, S.T. Yau, On the parabolic kernel of the Schrödinger operator. Acta Math., 156 (1986), no. 3-4, 153–201. (*Also appears in section on spectral theory.*)

21. W. Thurston, Hyperbolic structures on 3-manifolds. I. Deformation of acylindrical manifolds. Ann. of Math. (2) 124 (1986), no. 2, 203–246.

22. K. Uhlenbeck, S.T. Yau, On the existence of Hermitian Yang-Mills connections in stable vector bundles. Frontiers of the mathematical sciences: 1985 (New York, 1985). Comm. Pure Appl. Math. 39 (1986), no. S, suppl., S257–S293.

23. S.T. Yau, Nonlinear analysis in geometry. Monographies de L'Enseignement Mathématique, 33. Série des Conferences de l'Union Mathematique Internationale, 8. L'Enseignement Mathematique, Geneva, 1986. 54 pp.

24. S.K. Donaldson, Innite determinants, stable bundles and curvature. Duke Math. J., 54 (1987), no. 1, 231–247.

25. N. Hitchin, Stable bundles and integrable systems. Duke Math. J., 54 (1987), no. 1, 91–114.

26. G. Tian, S.T. Yau, Existence of Kähler-Einstein metrics on complete Kähler manifolds and their applications to algebraic geometry. Mathematical aspects of string theory (San Diego, Calif., 1986), 574–628, Adv. Ser. Math. Phys., 1, World Sci. Publishing, Singapore, 1987.

27. B. Osgood, R. Phillips, P. Sarnak, Extremals of determinants of Laplacians. J. Funct. Anal., 80 (1988), no. 1, 148–211.

28. R. Hamilton, The Ricci flow on surfaces. Mathematics and general relativity (Santa Cruz, CA, 1986), 237–262, Contemp. Math., 71, Amer. Math. Soc., Providence, RI, 1988.

29. R. Schoen, S.T. Yau, Conformally flat manifolds, Kleinian groups and scalar curvature. Invent. Math., 92 (1988), no. 1, 47–71.

30. W.X. Shi, Deforming the metric on complete Riemannian manifolds. J. Differential Geom., 30 (1989), no. 1, 223–301.

31. G. Huisken, Asymptotic behavior for singularities of the mean curvature flow. J. Differential Geom., 31 (1990), no. 1, 285–299.

32. G. Tian, On a set of polarized Kähler metrics on algebraic manifolds. J. Differential Geom., 32 (1990), no. 1, 99–130.

33. G. Tian, S.T. Yau, Complete Kähler manifolds with zero Ricci curvature. I. J. Amer. Math. Soc., 3 (1990), no. 3, 579–609.

34. B. Chow, The Ricci flow on the 2-sphere. J. Differential Geom., 33 (1991), no. 2, 325–334.

35. G. Tian, S.T. Yau, Complete Kähler manifolds with zero Ricci curvature. II. Invent. Math., 106 (1991), no. 1, 27–60.

36. C. Simpson, Higgs bundles and local systems. Inst. Hautes Etudes Sci. Publ. Math., No. 75 (1992), 5–95.

37. R. Hamilton, The Harnack estimate for the Ricci flow. J. Differential Geom., 37 (1993), no. 1, 225–243.

38. R. Hamilton, Eternal solutions to the Ricci flow. J. Differential Geom., 38 (1993), no. 1, 1–11.

39. R. Hamilton, A matrix Harnack estimate for the heat equation. Comm. Anal. Geom., 1 (1993), no. 1, 113–126.

40. R. Hamilton, The formation of singularities in the Ricci flow. Surveys in differential geometry, vol. II (Cambridge, MA, 1993), 7–136, Int. Press, Cambridge, MA, 1995.

41. R. Hamilton, Harnack estimate for the mean curvature flow. J. Differential Geom., 41 (1995), no. 1, 215–226.

42. R. Hamilton, A compactness property for solutions of the Ricci flow. Amer. J. Math., 117 (1995), no. 3, 545–572.

43. C. Simpson, Moduli of representations of the fundamental group of a smooth projective variety. I. Inst. Hautes Etudes Sci. Publ. Math., No. 79 (1994), 47–129. II. Inst. Hautes Etudes Sci. Publ. Math. No. 80 (1994), 5–79 (1995)

44. D. Joyce, Compact Riemannian 7-manifolds with holonomy G2. I, II. J. Differential Geom., 43 (1996), no. 2, 291–328, 329–375.

45. N. Leung, Einstein type metrics and stability on vector bundles. J. Differential Geom., 45 (1997), no. 3, 514–546.

46. W. Ruan, Canonical coordinates and Bergmann [Bergman] metrics. Comm. Anal. Geom., 6 (1998), no. 3, 589–631.

47. D. Catlin, The Bergman kernel and a theorem of Tian. Analysis and geometry in several complex variables (Katata, 1997), 1–23, Trends Math., Birkhäuser Boston, Boston, MA, 1999.

48. G. Huisken, C. Sinestrari, Mean curvature flow singularities for mean convex surfaces. Calc. Var. Partial Differential Equations, 8 (1999), no. 1, 1–14.

49. P. Bleher, B. Shiffman, S. Zelditch, Universality and scaling of correlations between zeros on complex manifolds. Invent. Math., 142 (2000), no. 2, 351–395.

50. D. Joyce, Compact manifolds with special holonomy. Oxford Mathematical Monographs. Oxford University Press, Oxford, 2000. xii+436 pp.

51. Z. Lu, On the lower order terms of the asymptotic expansion of Tian-Yau-Zelditch. Amer. J. Math., 122 (2000), no. 2, 235–273.

9 Harmonic forms, harmonic maps, and functional analysis

1. S. Bochner, Harmonic surfaces in Riemann metric. Trans. Amer. Math. Soc., 47, (1940). 146–154.

2. H. Weyl, The method of orthogonal projection in potential theory. Duke Math. J., 7 (1940), 411–444.

3. I. Gelfand, Über absolut konvergente trigonometrische Reihen und Integrale. Rec. Math. [Mat. Sbornik] N. S. 9 (51), (1941), 51–66.

4. W.V. Hodge, The Theory and Applications of Harmonic Integrals. Cambridge University Press, Cambridge, England; Macmillan Company, New York, 1941. ix+281 pp.

5. R. Martin, Minimal positive harmonic functions. Trans. Amer. Math. Soc., 49, (1941). 137–172.

6. S. Bochner, Vector fields and Ricci curvature. Bull. Amer. Math. Soc., 52, (1946). 776–797.

7. A. Milgram, P. Rosenbloom, Harmonic forms and heat conduction. I. Closed Riemannian manifolds. Proc. Nat. Acad. Sci. U. S. A. 37, (1951). 180–184.

8. A. Weil, Sur les theoremes de de Rham. Comment. Math. Helv., 26, (1952). 119–145.

9. K. Kodaira, On a differential-geometric method in the theory of analytic stacks. Proc. Nat. Acad. Sci. U. S. A. 39, (1953). 1268–1273.

10. Y. Yano, S. Bochner, Curvature and Betti numbers. Annals of Mathematics Studies, No. 32. Princeton University Press, Princeton, N. J., 1953. ix+190 pp.

11. P. Dolbeault, Formes différentielles et cohomologie sur une variété analytique complexe. I. Ann. of Math. (2) 64 (1956), 83–130.

12. A. Lichnerowicz, Spineurs harmoniques. C. R. Acad. Sci. Paris, 257 (1963) 7–9.

13. Y. Matsushima, S. Murakami, On vector bundle valued harmonic forms and automorphic forms on symmetric Riemannian manifolds. Ann. of Math., (2) 78 (1963) 365–416.

14. J. Eells, J. Sampson, Harmonic mappings of Riemannian manifolds. Amer. J. Math., 86 (1964) 109–160.

15. A. Grothendieck, On the de Rham cohomology of algebraic varieties. Inst. Hautes Études Sci. Publ. Math., No. 29 (1966) 95–103.

16. P. Griffths, On the periods of certain rational integrals. I, II. Ann. of Math. (2) 90 (1969), 460–495; ibid. (2) 90 (1969) 496–541.

17. P. Deligne, Théorie de Hodge. I. Actes du Congrés International des Mathématiciens (Nice, 1970), Tome 1, pp. 425–430. Gauthier-Villars, Paris, 1971.

18. H. Garland, p-adic curvature and the cohomology of discrete subgroups of p-adic groups. Ann. of Math. (2) 97 (1973), 375–423.

19. W. Schmid, Variation of Hodge structure: the singularities of the period mapping. Invent. Math., 22 (1973), 211–319.

20. P. Deligne, Théorie de Hodge. II. Inst. Hautes Études Sci. Publ. Math. No. 40 (1971), 5–57. Théorie de Hodge. III. Inst. Hautes Études Sci. Publ. Math., No. 44 (1974), 5–77.

21. R. Greene, H. Wu, Integrals of subharmonic functions on manifolds of nonnegative curvature. Invent. Math., 27 (1974), 265–298.

22. N. Hitchin, Harmonic spinors. Advances in Math., 14 (1974) 1–55.

23. S.T. Yau, Harmonic functions on complete Riemannian manifolds. Comm. Pure Appl. Math., 28 (1975), 201–228.

24. M.F. Atiyah, Elliptic operators, discrete groups and von Neumann algebras. Colloque "Analyse et Topologie" en l'Honneur de Henri Cartan (Orsay, 1974), pp. 43–72. Asterisque, No. 32-33, Soc. Math. France, Paris, 1976.

25. S.T. Yau, Some function-theoretic properties of complete Riemannian manifold and their applications to geometry. Indiana Univ. Math. J., 25 (1976), 659–670.

26. H. Karcher, Riemannian center of mass and mollier smoothing. Comm. Pure Appl. Math., 30 (1977), no. 5, 509–541.

27. R. Reilly, Applications of the Hessian operator in a Riemannian manifold. Indiana Univ. Math., 126 (1977) 459–472.

28. R. Schoen, S.T. Yau, On univalent harmonic maps between surfaces. Invent. Math., 44 (1978), no. 3, 265–278.

29. S. Zucker, Hodge theory with degenerating coefficients. L^2 cohomology in the Poincaré metric. Ann. of Math. (2) 109 (1979), no. 3, 415–476.

30. E. Cartan, The theory of spinors. With a foreword by Raymond Streater. A reprint of the 1966 English translation. Dover Books on Advanced Mathematics. Dover Publications, Inc., New York, 1981. v+157 pp.

31. M. Gromov, Curvature, diameter and Betti numbers. Comment. Math. Helv., 56 (1981), no. 2, 179–195.

32. J. Cheeger, M. Goresky, R. MacPherson, L2-cohomology and intersection homology of singular algebraic varieties. Seminar on Differential Geometry, pp. 303–340, Ann. of Math. Stud., 102, Princeton Univ. Press, Princeton, N.J., 1982.

33. M. Giaquinta, E. Giusti, On the regularity of the minima of variational integrals. Acta Math., 148 (1982), 31–46.

34. M. Gromov, Volume and bounded cohomology. Inst. Hautes Études Sci. Publ. Math., No. 56 (1982), 5–99 (1983).

35. R. Schoen, K. Uhlenbeck, A regularity theory for harmonic maps. J. Differential Geom., 17 (1982), no. 2, 307–335.

36. G. de Rham, Differentiable manifolds. Forms, currents, harmonic forms. Translated from the French by F. R. Smith. With an introduction by S. S. Chern. Grundlehren der Mathematischen Wissenschaften, 266. Springer-Verlag, Berlin, 1984. x+167 pp.

37. M. Gromov, B. Lawson, Positive scalar curvature and the Dirac operator on complete Riemannian manifolds. Inst. Hautes Etudes Sci. Publ. Math., No. 58 (1983), 83–196 (1984).

38. M. Giaquinta, E. Giusti, The singular set of the minima of certain quadratic functionals. Ann. Scuola Norm. Sup. Pisa Cl. Sci. (4) 11 (1984), no. 1, 45–55.

39. M. Anderson, R. Schoen, Positive harmonic functions on complete manifolds of negative curvature. Ann. of Math. (2) 121 (1985), no. 3, 429–461.

40. E. Cattani, A. Kaplan, W. Schmid, Degeneration of Hodge structures. Ann. of Math. (2) 123 (1986), no. 3, 457–535.

41. A. Ancona, Negatively curved manifolds, elliptic operators, and the Martin boundary. Ann. of Math. (2) 125 (1987), no. 3, 495–536.

42. E. Looijenga, L^2-cohomology of locally symmetric varieties. Compositio Math., 67 (1988), no. 1, 3–20.

43. R. Penrose, W. Rindler, Spinors and space-time. Vol. 1. Two-spinor calculus and relativistic fields. Cambridge Monographs on Mathematical Physics. Cambridge University Press, Cambridge, 1987. x+458 pp. Vol. 2. Spinor and twistor methods in space-time geometry. Second edition. Cambridge Monographs on Mathematical Physics. Cambridge University Press, Cambridge, 1988. x+501 pp.

44. H. Wu, The Bochner technique in dierential geometry. Math. Rep., 3 (1988), no. 2, i–xii and 289–538.

45. L. Saper, M. Stern, L^2-cohomology of arithmetic varieties. Ann. of Math. (2) 132 (1990), no. 1, 1–69.

46. P. Li, L.F. Tam, Harmonic functions and the structure of complete manifolds. J. Differential Geom., 35 (1992), no. 2, 359–383.
47. J. Jost, S.T. Yau, A nonlinear elliptic system for maps from Hermitian to Riemannian manifolds and rigidity theorems in Hermitian geometry. Acta Math., 170 (1993), no. 2, 221–254. Correction. Acta Math. 173 (1994), no. 2, 307.
48. A. Connes, Noncommutative geometry. Academic Press, Inc., San Diego, CA, 1994. xiv+661 pp.
49. R. Schoen, S.T. Yau, Lectures on differential geometry. Conference Proceedings and Lecture Notes in Geometry and Topology, I. International Press, Cambridge, MA, 1994. v+235 pp.
50. T. Colding, W. Minicozzi, Harmonic functions on manifolds. Ann. of Math. (2) 146 (1997), no. 3, 725–747.
51. R. Schoen, S.T. Yau, Lectures on harmonic maps. Conference Proceedings and Lecture Notes in Geometry and Topology, II. International Press, Cambridge, MA, 1997. vi+394 pp.
52. P. Li, J. Wang, Counting dimensions of L-harmonic functions. Ann. of Math. (2) 152 (2000), no. 2, 645–658.

10 Characteristic classes and index theories

1. S.S. Chern, A simple intrinsic proof of the Gauss-Bonnet formula for closed Riemannian manifolds. Ann. of Math. (2) 45, (1944). 747–752.
2. S.S. Chern, Characteristic classes of Hermitian manifolds. Ann. of Math. (2) 47, (1946). 85–121.
3. A. Borel, F. Hirzebruch, Characteristic classes and homogeneous spaces. I. Amer. J. Math., 80 (1958) 458–538.
4. F. Hirzebruch, Automorphe Formen und der Satz von Riemann-Roch. 1958 Symposium internacional de topologia algebraica International symposium on algebraic topology. pp. 129–144. Universidad Nacional Autónoma de México and UNESCO, Mexico City.
5. M. Atiyah, I.M. Singer, The index of elliptic operators on compact manifolds. Bull. Amer. Math. Soc., 69 (1963) 422–433.
6. F. Hirzebruch, Topological methods in algebraic geometry. Third enlarged edition. New appendix and translation from the second German edition by R. L. E. Schwarzenberger, with an additional section by A. Borel. Die Grundlehren der Mathematischen Wissenschaften, Band 131 Springer-Verlag New York, Inc., New York 1966 x+232 pp.
7. G. Harder, A Gauss-Bonnet formula for discrete arithmetically dened groups. Ann. Sci. École Norm. Sup. (4) 4 (1971), 409–455.
8. V. Patodi, Curvature and the eigenforms of the Laplace operator. J. Differential Geometry, 5 (1971), 233–249.

9. G. Lusztig, Novikov's higher signature and families of elliptic operators. J. Differential Geometry, 7 (1972), 229–256.

10. M. Atiyah, R. Bott, V. Patodi, On the heat equation and the index theorem. Invent. Math., 19 (1973), 279–330.

11. P. Gilkey, Curvature and the eigenvalues of the Laplacian for elliptic complexes. Advances in Math., 10 (1973), 344–382.

12. D. Ray, I. Singer, Analytic torsion for complex manifolds. Ann. of Math. (2) 98 (1973), 154–177.

13. S.S. Chern, J. Simons, Characteristic forms and geometric invariants. Ann. of Math. (2) 99 (1974), 48–69.

14. R. MacPherson, Chern classes for singular algebraic varieties. Ann. of Math. (2) 100 (1974), 423–432.

15. J. Milnor, J. Stashe, Characteristic classes. Annals of Mathematics Studies, No. 76. Princeton University Press, Princeton, N. J.; University of Tokyo Press, Tokyo, 1974. vii+331 pp.

16. M. Atiyah, V. Patodi, I. Singer, Spectral asymmetry and Riemannian geometry. I. Math. Proc. Cambridge Philos. Soc., 77 (1975), 43–69; Spectral asymmetry and Riemannian geometry. II. Math. Proc. Cambridge Philos. Soc. 78 (1975), no. 3, 405–432; Spectral asymmetry and Riemannian geometry. III. Math. Proc. Cambridge Philos. Soc. 79 (1976), no. 1, 71–99.

17. D. Mumford, Hirzebruch's proportionality theorem in the noncompact case. Invent. Math., 42 (1977), 239–272.

18. W. Müller, Analytic torsion and R-torsion of Riemannian manifolds. Adv. in Math., 28 (1978), 233–305.

19. J. Cheeger, Analytic torsion and the heat equation. Ann. of Math. (2) 109 (1979), 259–322.

11 Morse theory and topology of manifolds

1. M. Morse, The calculus of variations in the large. Reprint of the 1932 original. American Mathematical Society Colloquium Publications, 18. American Mathematical Society, Providence, RI, 1996. xii+368 pp.

2. R. Thom, Quelques proprietes globales des varietes differentiables. Comment. Math. Helv., 28 (1954) 17–86.

3. J. Milnor, On manifolds homeomorphic to the 7-sphere. Ann. of Math. (2) 64 (1956), 399–405.

4. K.T. Chen, Integration of paths—a faithful representation of paths by noncommutative formal power series. Trans. Amer. Math. Soc., 89 (1958) 395–407.

5. R. Bott, The stable homotopy of the classical groups, Annals of Mathematics. Second Series, 70 (1959), 313–337.

6. S. Smale, Diffeomorphisms of the 2-sphere. Proc. Amer. Math. Soc., 10 (1959) 621–626.

7. J. Milnor, Morse theory. Based on lecture notes by M. Spivak and R. Wells. Annals of Mathematics Studies, No. 51 Princeton University Press, Princeton, N.J. 1963 vi+153 pp.

8. J. Milnor, Lectures on the h-cobordism theorem. Notes by L. Siebenmann and J. Sondow Princeton University Press, Princeton, N.J. 1965 v+116 pp.

9. S. Novikov, The topology of foliations. Trudy Moskov. Mat. Obshch, 1965, v. 14, p. 248–278

10. E. Brieskorn, Examples of singular normal complex spaces which are topological manifolds. Proceedings of the National Academy of Sciences, 55 (1966) 1395–1397.

11. S. Smale, Differentiable dynamical systems. Bull. Amer. Math. Soc., 73 (1967) 747–817.

12. P. Deligne, P. Griffths, J. Morgan, D. Sullivan, Real homotopy theory of Kähler manifolds. Invent. Math., 29 (1975), no. 3, 245–274.

13. K.T. Chen, Iterated path integrals. Bull. Amer. Math. Soc., 83 (1977), no. 5, 831–879.

14. D. Sullivan, Infinitesimal computations in topology. Inst. Hautes Etudes Sci. Publ. Math., No. 47 (1977), 269–331 (1978).

15. M. Atiyah, N. Hitchin, V. Drinfel'd, Y. Manin, Construction of instantons. Phys. Lett. A 65 (1978), no. 3, 185–187.

16. M. Atiyah, N. Hitchin, I.M. Singer, Self-duality in four-dimensional Riemannian geometry. Proc. Roy. Soc. London Ser. A 362 (1978), no. 1711, 425–461.

17. C. Conley, Isolated invariant sets and the Morse index. CBMS Regional Conference Series in Mathematics, 38. American Mathematical Society, Providence, R.I., 1978. iii+89 pp.

18. M. Gromov, Curvature, diameter and Betti numbers. Comment. Math. Helv., 56 (1981), no. 2, 179–195.

19. M. Atiyah, Convexity and commuting Hamiltonians. Bull. London Math. Soc., 14 (1982), no. 1, 1–15.

20. J. Duistermaat, G. Heckman, On the variation in the cohomology of the symplectic form of the reduced phase space. Invent. Math., 69 (1982), no. 2, 259–268.

21. V. Guillemin, S. Sternberg, Convexity properties of the moment mapping. Invent. Math., 67 (1982), no. 3, 491–513.

22. R. Schoen, S.T. Yau, Complete three-dimensional manifolds with positive Ricci curvature and scalar curvature. *Seminar on Differential Geometry*, pp. 209–228, Ann. of Math. Stud., 102, Princeton Univ. Press, Princeton, N.J., 1982.

23. C. Taubes, Self-dual Yang-Mills connections on non-self-dual 4-manifolds. J. Differential Geom., 17 (1982), no. 1, 139–170.

24. K. Uhlenbeck, Removable singularities in Yang-Mills fields. Comm. Math. Phys., 83 (1982), no. 1, 11–29.

25. E. Witten, Supersymmetry and Morse theory. J. Differential Geom., 17 (1982), no. 4, 661–692

26. M. Atiyah, R. Bott, The Yang-Mills equations over Riemann surfaces. Philos. Trans. Roy. Soc. London Ser. A 308 (1983), no. 1505, 523–615.

27. S. Donaldson, An application of gauge theory to four-dimensional topology. J. Differential Geom., 18 (1983), no. 2, 279–315.

28. A. Hatcher, A proof of the Smale conjecture, $\mathrm{Diff}\,(S^3) \cong O(4)$. Ann. of Math. (2) 117 (1983), no. 3, 553–607.

29. M. Atiyah, R. Bott, The moment map and equivariant cohomology. Topology 23 (1984), no. 1, 1–28.

30. C. Conley, E. Zehnder, Morse-type index theory for flows and periodic solutions for Hamiltonian equations. Comm. Pure Appl. Math., 37 (1984), no. 2, 207–253.

31. J.P. Demailly, Champs magnétiques et inégalités de Morse pour la d''-cohomologie. Ann. Inst. Fourier (Grenoble) 35 (1985), no. 4, 189–229.

32. D. Quillen, Determinants of Cauchy-Riemann operators on Riemann surfaces Functional. Anal. Appl., 19 (1985), no. 1, 31–34.

33. N. Hitchin, The self-duality equations on a Riemann surface. Proc. London Math. Soc. (3) 55 (1987), no. 1, 59–126.

34. A. Floer, Morse theory for Lagrangian intersections. J. Differential Geom., 28 (1988), no. 3, 513–547.

35. M. Goresky, R. MacPherson, Stratied Morse theory. Ergebnisse der Mathematik und ihrer Grenzgebiete (3), 14. Springer-Verlag, Berlin, 1988. xiv+272 pp.

36. A. Floer, Witten's complex and infinite-dimensional Morse theory. J. Differential Geom., 30 (1989), no. 1, 207–221.

37. A. Floer, Symplectic fixed points and holomorphic spheres. Comm. Math. Phys., 120 (1989), no. 4, 575–611.

38. P. Kronheimer, The construction of ALE spaces as hyper-Kähler quotients. J. Differential Geom., 29 (1989), no. 3, 665–683.

39. S. Donaldson, P. Kronheimer, The geometry of four-manifolds. Oxford Mathematical Monographs. Oxford Science Publications. The Clarendon Press, Oxford University Press, New York, 1990. x+440 pp.

40. C. Taubes, Casson's invariant and gauge theory. J. Differential Geom., 31 (1990), no. 2, 547–599.

41. C. LeBrun, Explicit self-dual metrics on $\mathbb{C}P^2\#\cdots\#\mathbb{C}P^2$. J. Differential Geom., 34 (1991), no. 1, 223–253.

42. J. Li, Algebraic geometric interpretation of Donaldson's polynomial invariants. J. Differential Geom., 37 (1993), no. 2, 417–466.
43. J. Smoller, A. Wasserman, S.T. Yau, Existence of black hole solutions for the Einstein-Yang-Mills equations. Comm. Math. Phys., 154 (1993), no. 2, 377–401.
44. P. Kronheimer, T. Mrowka, The genus of embedded surfaces in the projective plane. Math. Res. Lett., 1 (1994), no. 6, 797–808.
45. C. Taubes, The Seiberg-Witten invariants and symplectic forms. Math. Res. Lett., 1 (1994), no. 6, 809–822.
46. E. Witten, Monopoles and four-manifolds. Math. Res. Lett., 1 (1994), no. 6, 769–796.
47. M. Kontsevich, Homological algebra of mirror symmetry. Proceedings of the International Congress of Mathematicians, Vol. 1, 2 (Zürich, 1994), 120–139, Birkhäuser, Basel, 1995.
48. A. Strominger, S.T. Yau, E. Zaslow, Mirror symmetry is T-duality. Nuclear Phys. B 479 (1996), no. 1-2, 243–259.
49. C. Taubes, SW $\Longrightarrow$ Gr: from the Seiberg-Witten equations to pseudo-holomorphic curves. J. Amer. Math. Soc., 9 (1996), no. 3, 845–918.
50. N. Leung, S.T. Yau, E. Zaslow, From special Lagrangian to Hermitian Yang-Mills via Fourier-Mukai transform. Adv. Theor. Math. Phys., 4 (2000), no. 6, 1319–1341.

12 Embedding of manifolds

1. H. Minkowski, Volumen und Oberfläche. Mathematische Annalen, 57 (4) (1903) 447–495.
2. H. Lewy, On differential geometry in the large. I. Minkowski's problem. Trans. Amer. Math. Soc., 43 (1938), no. 2, 258–270.
3. A.D. Alexandrov, On a class of closed surfaces. Recuiel Math., 4 (1938), 69–77.
4. A.V. Pogorelov, On the proof of Weyl's theorem on the existence of a closed analytic convex surface realizing an analytic metric with positive curvature given on the sphere. Uspehi Matem. Nauk (N.S.) 4, (1949). no. 4(32), 183–186.
5. J. Nash, Real algebraic manifolds. Ann. of Math. (2) 56, (1952), 405–421.
6. E. Calabi, Isometric imbedding of complex manifolds. Ann. of Math. (2) 58, (1953). 1–23.
7. L. Nirenberg, TheWeyl and Minkowski problems in differential geometry in the large. Comm. Pure Appl. Math., 6, (1953). 337–394.
8. J. Nash, C^1 isometric imbeddings. Ann. of Math. (2) 60, (1954), 383–396.

9. J. Nash, The imbedding problem for Riemannian manifolds. Ann. of Math. (2) 63 (1956), 20–63.

10. A.V. Pogorelov, A new proof of rigidity of convex polyhedra. Uspehi Mat. Nauk (N.S.) 11 (1956), no. 5(71), 207–208.

11. A.D. Alexandrov, Uniqueness theorems for surfaces in the large I-V, Vestnik Leningrad Univ., 11 (1956), 5–17; 12 (1957), 15–44; 13 (1958), 14–26; 13 (1958), 27–34; 13 (1958), 5–8; English transl. in Am. Math. Soc. Transl., 21 (1962) 341–354, 354–388, 389–403, 403–411, 412–416.

12. E. Calabi, Improper affine hyperspheres of convex type and a generalization of a theorem by K. Jörgens. Michigan Math. J. 5 (1958) 105–126.

13. S. Cohn-Vossen, Some problems of differential geometry in the large. Edited by N. V. Efimov Gosudarstv. Izdat. Fiz.-Mat. Lit., Moscow 1959, 303 pp.

14. L. Nirenberg, Rigidity of a class of closed surfaces. 1963 Nonlinear Problems (Proc. Sympos., Madison, Wis., 1962) pp. 177–193 Univ. of Wisconsin Press, Madison, Wis.

15. N.N. Efimov, Generation of singularites on surfaces of negative curvature. Mat. Sb. (N.S.) 64 (106) (1964) 286–320.

16. A.V. Pogorelov, A regular solution of the n-dimensional Minkowski problem. Dokl. Akad. Nauk SSSR 199 785–788 (Russian); translated as Soviet Math. Dokl., 12 (1971) 1192–1196.

17. E. Calabi, Complete affine hyperspheres. I. Symposia Mathematica, Vol. X (Convegno di Geometria Differenziale, INDAM, Rome, 1971), pp. 19–38. Academic Press, London, 1972.

18. T. Milnor, Efimov's theorem about complete immersed surfaces of negative curvature. Advances in Math., 8 (1972), 474–543.

19. A.V. Pogorelov, Extrinsic geometry of convex surfaces. Translated from the Russian by Israel Program for Scientific Translations. Translations of Mathematical Monographs, Vol. 35. American Mathematical Society, Providence, R.I., 1973. vi+669 pp.

20. S.Y. Cheng, S.T. Yau, On the regularity of the solution of the n-dimensional Minkowski problem. Comm. Pure Appl. Math., 29 (1976), no. 5, 495–516.

21. C.S. Lin, The local isometric embedding in $\mathbb{R}^3$ of 2-dimensional Riemannian manifolds with nonnegative curvature. J. Differential Geom., 21 (1985), no. 2, 213–230.

22. S.Y. Cheng, S.T. Yau, Complete affine hypersurfaces. I. The completeness of affine metrics. Comm. Pure Appl. Math., 39 (1986), no. 6, 839–866.

23. J. Hong, Realization in $\mathbf{R}^3$ of complete Riemannian manifolds with negative curvature. Comm. Anal. Geom., 1 (1993), no. 3-4, 487–514.

24. P. Guan, Y. Li, $C^{1,1}$ estimates for solutions of a problem of Alexandrov. Comm. Pure Appl. Math., 50 (1997), no. 8, 789–811.

13 Nonlinear differential equations

1. H. Lewy, A priori limitations for solutions of Monge-Ampre equations. Trans. Amer. Math. Soc. 37 (1935), no. 3, 417–434.
2. C. Morrey, On the solutions of quasi-linear elliptic partial differential equations. Trans. Amer. Math. Soc., 43 (1938), no. 1, 126–166. (*also appears in section on Riemann surfaces.*)
3. E. De Giorgi, Sulla differenziabilitá e l'analiticita delle estremali degli integrali multipli regolari. Mem. Accad. Sci. Torino. Cl. Sci. Fis. Mat. Nat. (3) 3 (1957) 25–43.
4. H. Lewy, An example of a smooth linear partial differential equation without solution. Annals of Mathematics, 66 (1957) 155–158.
5. J. Nash, Continuity of solutions of parabolic and elliptic equations. Amer. J. Math., 80 (1958) 931–954.
6. J. Moser, A new proof of De Giorgi's theorem concerning the regularity problem for elliptic differential equations. Communications on Pure and Applied Mathematics, 13 (1960) 457–468.
7. J. Moser, On Harnack's theorem for elliptic differential equations. Comm. Pure Appl. Math., 14 (1961) 577–591.
8. J. Moser, On the volume elements on a manifold. Trans. Amer. Math. Soc., 120 (1965) 286–294.
9. J. Moser, On a pointwise estimate for parabolic differential equations. Comm. on Pure and Applied Mathematics, 24 (1971), 727–740.
10. S.Y. Cheng, S.T. Yau, On the regularity of the Monge-Ampere equation $\det(\partial^2 u/\partial x_i \partial x_j) = F(x, u)$. Comm. Pure Appl. Math., 30 (1977), no. 1, 41–68.
11. K. Uhlenbeck, Removable singularities in Yang-Mills fields. Comm. Math. Phys., 83 (1982), no. 1, 11–29.

14 Complex geometry and several complex variables

1. E. Kähler, Über eine bemerkenswerte Hermitesche Metrik. Abh. Math. Sem. Univ., Hamburg, 9 (1933), no. 1, 173–186.
2. W.L. Chow, On compact complex analytic varieties. Amer. J. Math., 71, (1949) 893–914. (*Also appears in section on applications to algebraic geometry.*)
3. S. Bergman, The Kernel Function and Conformal Mapping. Mathematical Surveys, No. 5. American Mathematical Society, New York, N. Y., 1950. vii+161 pp.

4. A. Borel, A. Weil, Representations lineaires et espaces homogénes Kälerians des groupes de Lie compacts. Séminaire Bourbaki, May 1954. (Exposé by J-P. Serre.) (*Also appears in section of Lie groups.*)

5. K. Kodaira, On Kähler varieties of restricted type (an intrinsic characterization of algebraic varieties). Ann. of Math. (2) 60, (1954), 28–48. (*Also appears in section on applications to algebraic geometry.*)

6. H.C. Wang, Closed manifolds with homogeneous complex structure. Amer. J. Math., 76, (1954). 1–32.

7. J.P. Serre, Géométrie algébrique et géométrie analytique. Ann. Inst. Fourier, Grenoble 6 (1955–1956), 1–42.

8. R. Bott, Homogeneous vector bundles. Ann. of Math. (2) 66 (1957), 203–248.

9. A. Newlander, L. Nirenberg, Complex analytic coordinates in almost complex manifolds. Ann. of Math. (2) 65 (1957), 391–404.

10. A. Weil, Introduction á l'étude des variétés kählériennes. Publications de l'Institut de Mathematique de l'Université de Nancago, VI. Actualités Sci. Ind. no. 1267 Hermann, Paris 1958.

11. J. Kohn, Solution of the $\bar{\partial}$-Neumann problem on strongly pseudo-convex manifolds. Proc. Nat. Acad. Sci. U.S.A. 47 (1961) 1198–1202.

12. L. Carleson, Interpolations by bounded analytic functions and the corona problem. Ann. of Math. (2) 76 (1962) 547–559.

13. L.K. Hua, Harmonic analysis of functions of several complex variables in the classical domains. Translated from the Russian by Leo Ebner and Adam Korányi American Mathematical Society, Providence, R.I. 1963, iv+164 pp.

14. J. Kohn, Regularity at the boundary of the $\bar{\partial}$-Neumann problem. Proc. Nat. Acad. Sci. U.S.A. 49 (1963) 206–213.

15. J. Kohn, Harmonic integrals on strongly pseudoconvex manifolds, I, Ann. Math. 78 (1963) 112–148; II, 79 (1964) 450–472.

16. C. Morrey, The $\bar{\partial}$-Neumann problem on strongly pseudo-convex manifolds. 1964 Differential Analysis, Bombay Colloq., 1964 pp. 81–133 Oxford Univ. Press, London.

17. R. Bott, S.S. Chern, Hermitian vector bundles and the equidistribution of the zeroes of their holomorphic sections. Acta Math., 114 (1965) 71–112.

18. L. Hörmander, An introduction to complex analysis in several variables. D. Van Nostrand Co., Inc., Princeton, N.J.-Toronto, Ont.-London 1966 x+208 pp.

19. J. Milnor, Singular points of complex hypersurfaces. Annals of Mathematics Studies, No. 61 Princeton University Press, Princeton, N.J.; University of Tokyo Press, Tokyo, 1968, iii+122 pp.

20. S. Kobayashi, Hyperbolic manifolds and holomorphic mappings. Pure and Applied Mathematics, 2 Marcel Dekker, Inc., New York 1970 ix+148 pp.

21. J. Morrow, K. Kodaira, Complex manifolds. Holt, Rinehart and Winston, Inc., New York-Montreal, Que.-London, 1971. vii+192 pp.

22. S.S. Chern, J.K. Moser, Real hypersurfaces in complex manifolds. Acta Math., 133 (1974), 219–271.

23. C. Fefferman, The Bergman kernel and biholomorphic mappings of pseudoconvex domains. Invent. Math., 26 (1974), 1–65.

24. Y.T. Siu, Analyticity of sets associated to Lelong numbers and the extension of closed positive currents. Invent. Math., 27 (1974), 53–156.

25. E. Bedford, B.A. Taylor, The Dirichlet problem for a complex Monge-Ampére equation. Invent. Math., 37 (1976), no. 1, 1–44.

26. C. Fefferman, Monge-Ampére equations, the Bergman kernel, and geometry of pseudoconvex domains. Ann. of Math. (2) 103 (1976), no. 2, 395–416.

27. S. Kobayashi, Intrinsic distances, measures and geometric function theory. Bull. Amer. Math. Soc., 82 (1976), no. 3, 357–416.

28. D. Phong, E. Stein, Estimates for the Bergman and Szegä projections on strongly pseudoconvex domains. Duke Math. J., 44 (1977), no. 3, 695–704.

29. S.T. Yau, A general Schwarz lemma for Kähler manifolds. Amer. J. Math., 100 (1978), no. 1, 197–203.

30. P. Griffths, J. Harris, Principles of algebraic geometry. Pure and Applied Mathematics. Wiley-Interscience [John Wiley & Sons], New York, 1978. xii+813 pp. (*Also appears in section on applications in algebraic geometry.*)

31. S.S. Chern, Complex manifolds without potential theory. With an appendix on the geometry of characteristic classes. Second edition. Universitext. Springer-Verlag, New York-Heidelberg, 1979. iii+152 pp.

32. J.P. Demailly, Estimations L^2 pour l'operateur $\bar{\partial}$ d'un fibré vectoriel holomorphe semi-positif au-dessus d'une variété kählérienne complète. Ann. Sci. École Norm. Sup. (4) 15 (1982), no. 3, 457–511.

33. C. Fefferman, R. Graham, Conformal invariants. The mathematical heritage of Élie Cartan (Lyon, 1984). Astérisque 1985, Numéro Hors Série, 95–116.

34. D. Catlin, Subelliptic estimates for the $\bar{\partial}$–Neumann problem on pseudoconvex domains. Ann. of Math. (2) 126 (1987), no. 1, 131–191.

15 Applications to algebraic geometry and arithmetic geometry

1. W.L. Chow, On compact complex analytic varieties. Amer. J. Math., 71, (1949) 893–914. (*Also appears in section on complex geometry.*)

2. K. Kodaira, On Kähler varieties of restricted type (an intrinsic characterization of algebraic varieties). Ann. of Math. (2) 60, (1954), 28–48. (*Also appears in section on applications to algebraic geometry.*)

3. K. Kodaira, On compact analytic surfaces. I, II, III. Ann. of Math. (2) 71 (1960) 111–152. (2) 77 (1963), 563–626; 78 (1963) 1963 1–40.

4. F. Hirzebruch, Topological methods in algebraic geometry. Third enlarged edition. New appendix and translation from the second German edition by R. L. E. Schwarzenberger, with an additional section by A. Borel. Die Grundlehren der Mathematischen Wissenschaften, Band 131 Springer-Verlag New York, Inc., New York 1966 x+232 pp.

5. S. Arakelov, An intersection theory for divisors on an arithmetic surface. Izv. Akad. Nauk SSSR Ser. Mat., 38 (1974), 1179–1192.

6. S.T. Yau, Calabi's conjecture and some new results in algebraic geometry. Proc. Natl. Acad. Sci. U.S.A. 74 (1977), no. 5, 1798–1799. (*Also appears in section on special metrics.*)

7. P. Griffths, J. Harris, Principles of algebraic geometry. Pure and Applied Mathematics. Wiley-Interscience [John Wiley & Sons], New York, 1978. xii+813 pp. (*Also appears in section on complex geometry.*)

8. S.T. Yau, On the Ricci curvature of a compact Kähler manifold and the complex Monge-Ampére equation. I. Comm. Pure Appl. Math., 31 (1978), no. 3, 339–411. (*Also appears in section on special metrics.*)

9. A. Todorov, Applications of the Kähler-Einstein-Calabi-Yau metric to moduli of K3 surfaces. Invent. Math., 61 (1980), no. 3, 251–265.

10. F. Bogomolov, Surfaces of class VII_0 and affine geometry. Izv. Akad. Nauk SSSR Ser. Mat., 46 (1982), no. 4, 710–761, 896.

11. G. Faltings, Calculus on arithmetic surfaces. Ann. of Math. (2) 119 (1984), no. 2, 387–424.

12. J.P. Demailly, Mesures de Monge-Ampére et mesures pluriharmoniques. Math. Z., 194 (1987), no. 4, 519–564.

13. J. Bismut, H. Gillet, C. Soulé, Analytic torsion and holomorphic determinant bundles. I. Bott-Chern forms and analytic torsion. Comm. Math. Phys. 115 (1988), no. 1, 49–78. II. Direct images and Bott-Chern forms. Comm. Math. Phys. 115 (1988), no. 1, 79–126. III. Quillen metrics on holomorphic determinants. Comm. Math. Phys. 115 (1988), no. 2, 301–351.

14. H. Gillet, C. Soulé, Arithmetic intersection theory. Inst. Hautes Etudes Sci. Publ. Math., No. 72 (1990), 93–174 (1991).

15. J. Li, S.T. Yau, F. Zheng, A simple proof of Bogomolov's theorem on class VII_0 surfaces with $b_2 = 0$. Illinois J. Math. 34 (1990), no. 2, 217–220.

16. A. Nadel, Multiplier ideal sheaves and Kähler-Einstein metrics of positive scalar curvature. Ann. of Math. (2) 132 (1990), no. 3, 549–596.

17. J.P. Demailly, Regularization of closed positive currents and intersection theory. J. Algebraic Geom., 1 (1992), no. 3, 361–409.

18. J. Bost, H. Gillet, C. Soulé, Heights of projective varieties and positive Green forms. J. Amer. Math. Soc., 7 (1994), no. 4, 903–1027.

19. J.P. Demailly, T. Peternell, M. Schneider, Compact complex manifolds with numerically effective tangent bundles. J. Algebraic Geom., 3 (1994), no. 2, 295–345.

20. S. Zhang, Positive line bundles on arithmetic varieties. J. Amer. Math. Soc., 8 (1995), no. 1, 187–221.

21. Y.T. Siu, Invariance of plurigenera. Invent. Math., 134 (1998), no. 3, 661–673.

16 Lie groups, homogeneous spaces and discrete subgroups

1. A. Borel, A. Weil, Representations linéaires et espaces homogénes Kälerians des groupes de Lie compacts. Séminaire Bourbaki, May 1954. (Exposé by J-P. Serre.) (*Also appears in section on complex geometry.*)

2. H. Huber, Über eine neue Klasse automorpher Funktionen und ein Gitterpunktproblem in der hyperbolischen Ebene. I. Comment. Math. Helv., 30 (1956), 20–62 (1955).

3. A. Selberg, Harmonic analysis and discontinuous groups in weakly symmetric Riemannian spaces with applications to Dirichlet series. J. Indian Math. Soc. (N.S.) 20 (1956), 47–87. (*Also appears in section on spectral theory.*)

4. M. Berger, Les variétés riemanniennes homogénes normales simplement connexes á courbure strictement positive. Ann. Scuola Norm. Sup. Pisa (3) 15 (1961) 179–246.

5. S. Helgason, Dierential geometry and symmetric spaces. Pure and Applied Mathematics, Vol. XII. Academic Press, New York-London, 1962, xiv+486 pp.

6. J. Simons, On the transitivity of holonomy systems. Ann. of Math. (2) 76 (1962) 213–234.

7. W. Baily, A. Borel, Compactication of arithmetic quotients of bounded symmetric domains. Ann. of Math. (2) 84 (1966) 442–528.

8. D. Kazhdan, On the connection of the dual space of a group with the structure of its closed subgroups. Functional analysis and its applications 1 (1967) 63–65. (*Also appears in section on spectral theory.*)

9. A. Borel, Stable real cohomology of arithmetic groups. Ann. Sci. École Norm. Sup. (4) 7 (1974), 235–272 (1975).

10. R. Langlands, On the functional equations satisfied by Eisenstein series. Lecture Notes in Mathematics, Vol. 544. Springer-Verlag, Berlin-New York, 1976. v+337 pp. (*Also appears in section on spectral theory.*)

11. S. Patterson, The limit set of a Fuchsian group. Acta Math., 136 (1976), no. 3-4, 241–273.

12. D. Sullivan, The density at infinity of a discrete group of hyperbolic motions. Inst. Hautes Études Sci. Publ. Math., No. 50 (1979), 171–202.

13. M. Gromov, Groups of polynomial growth and expanding maps. Inst. Hautes Études Sci. Publ. Math., No. 53 (1981), 53–73.

14. Y.T. Siu, S.T. Yau, Compactification of negatively curved complete Kähler manifolds of finite volume. Seminar on Differential Geometry, pp. 363–380, Ann. of Math. Stud., 102, Princeton Univ. Press, Princeton, N.J., 1982.

15. J. Wolf, Spaces of constant curvature. Fifth edition. Publish or Perish, Inc., Houston, TX, 1984. xviii+412 pp. (*Also appears in section on special metrics.*)

16. A. Lubotzky, R. Phillips, P. Sarnak, Ramanujan graphs. Combinatorica 8 (1988), no. 3, 261–277.

17. J. Arthur, Unipotent automorphic representations: conjectures. *Orbites unipotentes et représentations, II. Astérisque*, No. 171-172 (1989), 13–71. (*Also appears in section on spectral theory.*)

18. N. Mok, J. Zhong, Compactifying complete Kähler-Einstein manifolds of finite topological type and bounded curvature. Ann. of Math. (2) 129 (1989), no. 3, 427–470.

17 General relativity and Lorentz geometry

1. H. Lorentz, A. Einstein, H. Minkowski, H. Weyl, The principle of relativity. With notes by A. Sommerfeld. Translated by W. Perrett and G. B. Jeery. A collection of original memoirs on the special and general theory of relativity. Dover Publications, Inc., New York, N. Y., undated. viii+216 pp.

2. H. Weyl, Space, Time, Matter. Dover Publications, 1952, 368 pages.

3. R. Penrose, Techniques of differential topology in relativity. Conference Board of the Mathematical Sciences Regional Conference Series in Applied Mathematics, No. 7. Society for Industrial and Applied Mathematics, Philadelphia, Pa., 1972. viii+72 pp.

4. S.W. Hawking, G.F.R. Ellis, The large scale structure of space-time. Cambridge Monographs on Mathematical Physics, No. 1. Cambridge University Press, London-New York, 1973. xi+391 pp.

5. S.W. Hawking, Particle creation by black holes. Comm. Math. Phys., 43 (1975), no. 3, 199–220.

6. S.Y. Cheng, S.T. Yau, Maximal space-like hypersurfaces in the Lorentz-Minkowski spaces. Ann. of Math. (2) 104 (1976), no. 3, 407–419.

7. R. Schoen, S.T. Yau, On the proof of the positive mass conjecture in general relativity. Comm. Math. Phys. 65 (1979), no. 1, 45–76.

8. R. Schoen, S.T. Yau, Proof of the positive mass theorem. II. Comm. Math. Phys., 79 (1981), no. 2, 231–260.

9. E. Witten, A new proof of the positive energy theorem. Comm. Math. Phys., 80 (1981), no. 3, 381–402.
10. A. Treibergs, Entire spacelike hypersurfaces of constant mean curvature in Minkowski space. Invent. Math., 66 (1982), no. 1, 39–56.
11. S. Chandrasekhar, The mathematical theory of black holes. International Series of Monographs on Physics, 69. Oxford Science Publications. The Clarendon Press, Oxford University Press, New York, 1983. xxi+646 pp.
12. C. Gerhardt, H-surfaces in Lorentzian manifolds. Comm. Math. Phys., 89 (1983), no. 4, 523–553.
13. R. Schoen, S.T. Yau, The existence of a black hole due to condensation of matter. Comm. Math. Phys., 90 (1983), no. 4, 575–579.
14. R. Bartnik, Existence of maximal surfaces in asymptotically flat spacetimes. Comm. Math. Phys., 94 (1984), no. 2, 155–175.
15. D. Christodoulou, S.T. Yau, Some remarks on the quasi-local mass. Mathematics and general relativity (Santa Cruz, CA, 1986), 9–14, Contemp. Math., 71, Amer. Math. Soc., Providence, RI, 1988.
16. R. Bartnik, New definition of quasilocal mass. Phys. Rev. Lett., 62 (1989), no. 20, 2346–2348.
17. D. Christodoulou, S. Klainerman, The global nonlinear stability of the Minkowski space. Princeton Mathematical Series, 41. Princeton University Press, Princeton, NJ, 1993. x+514 pp.
18. B. O'Neill, The geometry of Kerr black holes. A. K. Peters, Ltd., Wellesley, MA, 1995. xviii+381 pp.
19. G. Huisken, T. Ilmanen, The Riemannian Penrose inequality. Internat. Math. Res. Notices, 20 (1997) 1045–1058.
20. C. Gerhardt, Hypersurfaces of prescribed curvature in Lorentzian manifolds. Indiana Univ. Math. J., 49 (2000), no. 3, 1125–1153.

Advanced Lectures in Mathematics (ALM)

Executive editors: Shing-Tung Yau, Kefeng Liu, and Lizhen Ji

www.intlpress.com/alm

ALM 42: Uniformization, Riemann-Hilbert Correspondence, Calabi-Yau Manifolds, and Picard-Fuchs Equations

ALM 41: Handbook of Group Actions, Volume IV

ALM 40: Handbook of Group Actions, Volume III

ALM 39: Hodge Theory and L^2-analysis

ALM 38: Complex Geometry from Riemann to Kähler–Einstein and Calabi–Yau

ALM 37: Proceedings of the Sixth International Congress of Chinese Mathematicians, Volume II

ALM 36: Proceedings of the Sixth International Congress of Chinese Mathematicians, Volume I

ALM 35: The Legacy of Bernhard Riemann After One Hundred and Fifty Years, Volumes I and II

ALM 34: Some Topics in Harmonic Analysis and Applications

ALM 33: Introduction to Modern Mathematics

ALM 32: Handbook of Group Actions, Volume II

ALM 31: Handbook of Group Actions, Volume I

ALM 30: Automorphic Forms and L-functions

ALM 29: Selected Expository Works of Shing-Tung Yau with Commentary, Volume II

ALM 28: Selected Expository Works of Shing-Tung Yau with Commentary, Volume I

ALM 27: Number Theory and Related Areas

ALM 26: Handbook of Moduli, Volume III

ALM 25: Handbook of Moduli, Volume II

ALM 24: Handbook of Moduli, Volume I

ALM 23: Recent Developments in Geometry and Analysis

ALM 22: Differential Geometry: Under the Influence of S.-S. Chern

ALM 21: Advances in Geometric Analysis

ALM 20: Surveys in Geometric Analysis and Relativity

ALM 19: Arithmetic Geometry and Automorphic Forms

ALM 18: Geometry and Analysis, No. 2

ALM 17: Geometry and Analysis, No. 1

ALM 16: Transformation Groups and Moduli Spaces of Curves

ALM 15: An Introduction to Groups and Lattices: Finite Groups and Positive Definite Rational Lattices

ALM 14: Handbook of Geometric Analysis, No. 3

ALM 13: Handbook of Geometric Analysis, No. 2

ALM 12: Cohomology of Groups and Algebraic K-theory

ALM 11: Recent Advances in Geometric Analysis

ALM 10: Trends in Partial Differential Equations

ALM 9: Automorphic Forms and the Langlands Program

ALM 8: Recent Developments in Algebra and Related Areas

ALM 7: Handbook of Geometric Analysis, No. 1

ALM 6: Geometry, Analysis and Topology of Discrete Groups

ALM 4: Variational Principles for Discrete Surfaces

ALM 3: Computational Conformal Geometry

ALM 2: Asymptotic Theory in Probability and Statistics with Applications

ALM 1: Superstring Theory